Lecture Notes in Computer Science 13689

More information about this series at https://link.springer.com/bookseries/558

Shai Avidan · Gabriel Brostow ·
Moustapha Cissé · Giovanni Maria Farinella ·
Tal Hassner (Eds.)

Computer Vision – ECCV 2022

17th European Conference
Tel Aviv, Israel, October 23–27, 2022
Proceedings, Part XXIX

 Springer

Editors
Shai Avidan
Tel Aviv University
Tel Aviv, Israel

Gabriel Brostow ⓘ
University College London
London, UK

Moustapha Cissé
Google AI
Accra, Ghana

Giovanni Maria Farinella ⓘ
University of Catania
Catania, Italy

Tal Hassner ⓘ
Facebook (United States)
Menlo Park, CA, USA

ISSN 0302-9743   ISSN 1611-3349 (electronic)
Lecture Notes in Computer Science
ISBN 978-3-031-19817-5   ISBN 978-3-031-19818-2 (eBook)
https://doi.org/10.1007/978-3-031-19818-2

This Springer imprint is published by the registered company Springer Nature Switzerland AG
The registered company address is: Gewerbestrasse 11, 6330 Cham, Switzerland

Foreword

Organizing the European Conference on Computer Vision (ECCV 2022) in Tel-Aviv during a global pandemic was no easy feat. The uncertainty level was extremely high, and decisions had to be postponed to the last minute. Still, we managed to plan things just in time for ECCV 2022 to be held in person. Participation in physical events is crucial to stimulating collaborations and nurturing the culture of the Computer Vision community.

There were many people who worked hard to ensure attendees enjoyed the best science at the 16th edition of ECCV. We are grateful to the Program Chairs Gabriel Brostow and Tal Hassner, who went above and beyond to ensure the ECCV reviewing process ran smoothly. The scientific program includes dozens of workshops and tutorials in addition to the main conference and we would like to thank Leonid Karlinsky and Tomer Michaeli for their hard work. Finally, special thanks to the web chairs Lorenzo Baraldi and Kosta Derpanis, who put in extra hours to transfer information fast and efficiently to the ECCV community.

We would like to express gratitude to our generous sponsors and the Industry Chairs, Dimosthenis Karatzas and Chen Sagiv, who oversaw industry relations and proposed new ways for academia-industry collaboration and technology transfer. It's great to see so much industrial interest in what we're doing!

Authors' draft versions of the papers appeared online with open access on both the Computer Vision Foundation (CVF) and the European Computer Vision Association (ECVA) websites as with previous ECCVs. Springer, the publisher of the proceedings, has arranged for archival publication. The final version of the papers is hosted by SpringerLink, with active references and supplementary materials. It benefits all potential readers that we offer both a free and citeable version for all researchers, as well as an authoritative, citeable version for SpringerLink readers. Our thanks go to Ronan Nugent from Springer, who helped us negotiate this agreement. Last but not least, we wish to thank Eric Mortensen, our publication chair, whose expertise made the process smooth.

October 2022

Rita Cucchiara
Jiří Matas
Amnon Shashua
Lihi Zelnik-Manor

Preface

Welcome to the proceedings of the European Conference on Computer Vision (ECCV 2022). This was a hybrid edition of ECCV as we made our way out of the COVID-19 pandemic. The conference received 5804 valid paper submissions, compared to 5150 submissions to ECCV 2020 (a 12.7% increase) and 2439 in ECCV 2018. 1645 submissions were accepted for publication (28%) and, of those, 157 (2.7% overall) as orals.

846 of the submissions were desk-rejected for various reasons. Many of them because they revealed author identity, thus violating the double-blind policy. This violation came in many forms: some had author names with the title, others added acknowledgments to specific grants, yet others had links to their github account where their name was visible. Tampering with the LaTeX template was another reason for automatic desk rejection.

ECCV 2022 used the traditional CMT system to manage the entire double-blind reviewing process. Authors did not know the names of the reviewers and vice versa. Each paper received at least 3 reviews (except 6 papers that received only 2 reviews), totalling more than 15,000 reviews.

Handling the review process at this scale was a significant challenge. To ensure that each submission received as fair and high-quality reviews as possible, we recruited more than 4719 reviewers (in the end, 4719 reviewers did at least one review). Similarly we recruited more than 276 area chairs (eventually, only 276 area chairs handled a batch of papers). The area chairs were selected based on their technical expertise and reputation, largely among people who served as area chairs in previous top computer vision and machine learning conferences (ECCV, ICCV, CVPR, NeurIPS, etc.).

Reviewers were similarly invited from previous conferences, and also from the pool of authors. We also encouraged experienced area chairs to suggest additional chairs and reviewers in the initial phase of recruiting. The median reviewer load was five papers per reviewer, while the average load was about four papers, because of the emergency reviewers. The area chair load was 35 papers, on average.

Conflicts of interest between authors, area chairs, and reviewers were handled largely automatically by the CMT platform, with some manual help from the Program Chairs. Reviewers were allowed to describe themselves as senior reviewer (load of 8 papers to review) or junior reviewers (load of 4 papers). Papers were matched to area chairs based on a subject-area affinity score computed in CMT and an affinity score computed by the Toronto Paper Matching System (TPMS). TPMS is based on the paper's full text. An area chair handling each submission would bid for preferred expert reviewers, and we balanced load and prevented conflicts.

The assignment of submissions to area chairs was relatively smooth, as was the assignment of submissions to reviewers. A small percentage of reviewers were not happy with their assignments in terms of subjects and self-reported expertise. This is an area for improvement, although it's interesting that many of these cases were reviewers hand-picked by AC's. We made a later round of reviewer recruiting, targeted at the list of authors of papers submitted to the conference, and had an excellent response which helped provide enough emergency reviewers. In the end, all but six papers received at least 3 reviews.

The challenges of the reviewing process are in line with past experiences at ECCV 2020. As the community grows, and the number of submissions increases, it becomes ever more challenging to recruit enough reviewers and ensure a high enough quality of reviews. Enlisting authors by default as reviewers might be one step to address this challenge.

Authors were given a week to rebut the initial reviews, and address reviewers' concerns. Each rebuttal was limited to a single pdf page with a fixed template.

The Area Chairs then led discussions with the reviewers on the merits of each submission. The goal was to reach consensus, but, ultimately, it was up to the Area Chair to make a decision. The decision was then discussed with a buddy Area Chair to make sure decisions were fair and informative. The entire process was conducted virtually with no in-person meetings taking place.

The Program Chairs were informed in cases where the Area Chairs overturned a decisive consensus reached by the reviewers, and pushed for the meta-reviews to contain details that explained the reasoning for such decisions. Obviously these were the most contentious cases, where reviewer inexperience was the most common reported factor.

Once the list of accepted papers was finalized and released, we went through the laborious process of plagiarism (including self-plagiarism) detection. A total of 4 accepted papers were rejected because of that.

Finally, we would like to thank our Technical Program Chair, Pavel Lifshits, who did tremendous work behind the scenes, and we thank the tireless CMT team.

October 2022

Gabriel Brostow
Giovanni Maria Farinella
Moustapha Cissé
Shai Avidan
Tal Hassner

Organization

General Chairs

Rita Cucchiara   University of Modena and Reggio Emilia, Italy
Jiří Matas   Czech Technical University in Prague, Czech
     Republic
Amnon Shashua   Hebrew University of Jerusalem, Israel
Lihi Zelnik-Manor   Technion – Israel Institute of Technology, Israel

Program Chairs

Shai Avidan   Tel-Aviv University, Israel
Gabriel Brostow   University College London, UK
Moustapha Cissé   Google AI, Ghana
Giovanni Maria Farinella   University of Catania, Italy
Tal Hassner   Facebook AI, USA

Program Technical Chair

Pavel Lifshits   Technion – Israel Institute of Technology, Israel

Workshops Chairs

Leonid Karlinsky   IBM Research, Israel
Tomer Michaeli   Technion – Israel Institute of Technology, Israel
Ko Nishino   Kyoto University, Japan

Tutorial Chairs

Thomas Pock   Graz University of Technology, Austria
Natalia Neverova   Facebook AI Research, UK

Demo Chair

Bohyung Han   Seoul National University, Korea

Social and Student Activities Chairs

Tatiana Tommasi   Italian Institute of Technology, Italy
Sagie Benaim   University of Copenhagen, Denmark

Diversity and Inclusion Chairs

Xi Yin   Facebook AI Research, USA
Bryan Russell   Adobe, USA

Communications Chairs

Lorenzo Baraldi   University of Modena and Reggio Emilia, Italy
Kosta Derpanis   York University & Samsung AI Centre Toronto,
                           Canada

Industrial Liaison Chairs

Dimosthenis Karatzas   Universitat Autònoma de Barcelona, Spain
Chen Sagiv   SagivTech, Israel

Finance Chair

Gerard Medioni   University of Southern California & Amazon,
                           USA

Publication Chair

Eric Mortensen   MiCROTEC, USA

Area Chairs

Lourdes Agapito   University College London, UK
Zeynep Akata   University of Tübingen, Germany
Naveed Akhtar   University of Western Australia, Australia
Karteek Alahari   Inria Grenoble Rhône-Alpes, France
Alexandre Alahi   École polytechnique fédérale de Lausanne,
                           Switzerland
Pablo Arbelaez   Universidad de Los Andes, Columbia
Antonis A. Argyros   University of Crete & Foundation for Research
                           and Technology-Hellas, Crete
Yuki M. Asano   University of Amsterdam, The Netherlands
Kalle Åström   Lund University, Sweden
Hadar Averbuch-Elor   Cornell University, USA

Matthijs Douze	Facebook AI Research, USA
Mohamed Elhoseiny	King Abdullah University of Science and Technology, Saudi Arabia
Sergio Escalera	University of Barcelona, Spain
Yi Fang	New York University, USA
Ryan Farrell	Brigham Young University, USA
Alireza Fathi	Google, USA
Christoph Feichtenhofer	Facebook AI Research, USA
Basura Fernando	Agency for Science, Technology and Research (A*STAR), Singapore
Vittorio Ferrari	Google Research, Switzerland
Andrew W. Fitzgibbon	Graphcore, UK
David J. Fleet	University of Toronto, Canada
David Forsyth	University of Illinois at Urbana-Champaign, USA
David Fouhey	University of Michigan, USA
Katerina Fragkiadaki	Carnegie Mellon University, USA
Friedrich Fraundorfer	Graz University of Technology, Austria
Oren Freifeld	Ben-Gurion University, Israel
Thomas Funkhouser	Google Research & Princeton University, USA
Yasutaka Furukawa	Simon Fraser University, Canada
Fabio Galasso	Sapienza University of Rome, Italy
Jürgen Gall	University of Bonn, Germany
Chuang Gan	Massachusetts Institute of Technology, USA
Zhe Gan	Microsoft, USA
Animesh Garg	University of Toronto, Vector Institute, Nvidia, Canada
Efstratios Gavves	University of Amsterdam, The Netherlands
Peter Gehler	Amazon, Germany
Theo Gevers	University of Amsterdam, The Netherlands
Bernard Ghanem	King Abdullah University of Science and Technology, Saudi Arabia
Ross B. Girshick	Facebook AI Research, USA
Georgia Gkioxari	Facebook AI Research, USA
Albert Gordo	Facebook, USA
Stephen Gould	Australian National University, Australia
Venu Madhav Govindu	Indian Institute of Science, India
Kristen Grauman	Facebook AI Research & UT Austin, USA
Abhinav Gupta	Carnegie Mellon University & Facebook AI Research, USA
Mohit Gupta	University of Wisconsin-Madison, USA
Hu Han	Institute of Computing Technology, Chinese Academy of Sciences, China

Bohyung Han	Seoul National University, Korea
Tian Han	Stevens Institute of Technology, USA
Emily Hand	University of Nevada, Reno, USA
Bharath Hariharan	Cornell University, USA
Ran He	Institute of Automation, Chinese Academy of Sciences, China
Otmar Hilliges	ETH Zurich, Switzerland
Adrian Hilton	University of Surrey, UK
Minh Hoai	Stony Brook University, USA
Yedid Hoshen	Hebrew University of Jerusalem, Israel
Timothy Hospedales	University of Edinburgh, UK
Gang Hua	Wormpex AI Research, USA
Di Huang	Beihang University, China
Jing Huang	Facebook, USA
Jia-Bin Huang	Facebook, USA
Nathan Jacobs	Washington University in St. Louis, USA
C.V. Jawahar	International Institute of Information Technology, Hyderabad, India
Herve Jegou	Facebook AI Research, France
Neel Joshi	Microsoft Research, USA
Armand Joulin	Facebook AI Research, France
Frederic Jurie	University of Caen Normandie, France
Fredrik Kahl	Chalmers University of Technology, Sweden
Yannis Kalantidis	NAVER LABS Europe, France
Evangelos Kalogerakis	University of Massachusetts, Amherst, USA
Sing Bing Kang	Zillow Group, USA
Yosi Keller	Bar Ilan University, Israel
Margret Keuper	University of Mannheim, Germany
Tae-Kyun Kim	Imperial College London, UK
Benjamin Kimia	Brown University, USA
Alexander Kirillov	Facebook AI Research, USA
Kris Kitani	Carnegie Mellon University, USA
Iasonas Kokkinos	Snap Inc. & University College London, UK
Vladlen Koltun	Apple, USA
Nikos Komodakis	University of Crete, Crete
Piotr Koniusz	Australian National University, Australia
Philipp Kraehenbuehl	University of Texas at Austin, USA
Dilip Krishnan	Google, USA
Ajay Kumar	Hong Kong Polytechnic University, Hong Kong, China
Junseok Kwon	Chung-Ang University, Korea
Jean-Francois Lalonde	Université Laval, Canada

Ivan Laptev	Inria Paris, France
Laura Leal-Taixé	Technical University of Munich, Germany
Erik Learned-Miller	University of Massachusetts, Amherst, USA
Gim Hee Lee	National University of Singapore, Singapore
Seungyong Lee	Pohang University of Science and Technology, Korea
Zhen Lei	Institute of Automation, Chinese Academy of Sciences, China
Bastian Leibe	RWTH Aachen University, Germany
Hongdong Li	Australian National University, Australia
Fuxin Li	Oregon State University, USA
Bo Li	University of Illinois at Urbana-Champaign, USA
Yin Li	University of Wisconsin-Madison, USA
Ser-Nam Lim	Meta AI Research, USA
Joseph Lim	University of Southern California, USA
Stephen Lin	Microsoft Research Asia, China
Dahua Lin	The Chinese University of Hong Kong, Hong Kong, China
Si Liu	Beihang University, China
Xiaoming Liu	Michigan State University, USA
Ce Liu	Microsoft, USA
Zicheng Liu	Microsoft, USA
Yanxi Liu	Pennsylvania State University, USA
Feng Liu	Portland State University, USA
Yebin Liu	Tsinghua University, China
Chen Change Loy	Nanyang Technological University, Singapore
Huchuan Lu	Dalian University of Technology, China
Cewu Lu	Shanghai Jiao Tong University, China
Oisin Mac Aodha	University of Edinburgh, UK
Dhruv Mahajan	Facebook, USA
Subhransu Maji	University of Massachusetts, Amherst, USA
Atsuto Maki	KTH Royal Institute of Technology, Sweden
Arun Mallya	NVIDIA, USA
R. Manmatha	Amazon, USA
Iacopo Masi	Sapienza University of Rome, Italy
Dimitris N. Metaxas	Rutgers University, USA
Ajmal Mian	University of Western Australia, Australia
Christian Micheloni	University of Udine, Italy
Krystian Mikolajczyk	Imperial College London, UK
Anurag Mittal	Indian Institute of Technology, Madras, India
Philippos Mordohai	Stevens Institute of Technology, USA
Greg Mori	Simon Fraser University & Borealis AI, Canada

Vittorio Murino   Istituto Italiano di Tecnologia, Italy
P. J. Narayanan   International Institute of Information Technology,
                                 Hyderabad, India
Ram Nevatia   University of Southern California, USA
Natalia Neverova   Facebook AI Research, UK
Richard Newcombe   Facebook, USA
Cuong V. Nguyen   Florida International University, USA
Bingbing Ni   Shanghai Jiao Tong University, China
Juan Carlos Niebles   Salesforce & Stanford University, USA
Ko Nishino   Kyoto University, Japan
Jean-Marc Odobez   Idiap Research Institute, École polytechnique
                                 fédérale de Lausanne, Switzerland
Francesca Odone   University of Genova, Italy
Takayuki Okatani   Tohoku University & RIKEN Center for
                                 Advanced Intelligence Project, Japan
Manohar Paluri   Facebook, USA
Guan Pang   Facebook, USA
Maja Pantic   Imperial College London, UK
Sylvain Paris   Adobe Research, USA
Jaesik Park   Pohang University of Science and Technology,
                                 Korea
Hyun Soo Park   The University of Minnesota, USA
Omkar M. Parkhi   Facebook, USA
Deepak Pathak   Carnegie Mellon University, USA
Georgios Pavlakos   University of California, Berkeley, USA
Marcello Pelillo   University of Venice, Italy
Marc Pollefeys   ETH Zurich & Microsoft, Switzerland
Jean Ponce   Inria, France
Gerard Pons-Moll   University of Tübingen, Germany
Fatih Porikli   Qualcomm, USA
Victor Adrian Prisacariu   University of Oxford, UK
Petia Radeva   University of Barcelona, Spain
Ravi Ramamoorthi   University of California, San Diego, USA
Deva Ramanan   Carnegie Mellon University, USA
Vignesh Ramanathan   Facebook, USA
Nalini Ratha   State University of New York at Buffalo, USA
Tammy Riklin Raviv   Ben-Gurion University, Israel
Tobias Ritschel   University College London, UK
Emanuele Rodola   Sapienza University of Rome, Italy
Amit K. Roy-Chowdhury   University of California, Riverside, USA
Michael Rubinstein   Google, USA
Olga Russakovsky   Princeton University, USA

Mathieu Salzmann	École polytechnique fédérale de Lausanne, Switzerland
Dimitris Samaras	Stony Brook University, USA
Aswin Sankaranarayanan	Carnegie Mellon University, USA
Imari Sato	National Institute of Informatics, Japan
Yoichi Sato	University of Tokyo, Japan
Shin'ichi Satoh	National Institute of Informatics, Japan
Walter Scheirer	University of Notre Dame, USA
Bernt Schiele	Max Planck Institute for Informatics, Germany
Konrad Schindler	ETH Zurich, Switzerland
Cordelia Schmid	Inria & Google, France
Alexander Schwing	University of Illinois at Urbana-Champaign, USA
Nicu Sebe	University of Trento, Italy
Greg Shakhnarovich	Toyota Technological Institute at Chicago, USA
Eli Shechtman	Adobe Research, USA
Humphrey Shi	University of Oregon & University of Illinois at Urbana-Champaign & Picsart AI Research, USA
Jianbo Shi	University of Pennsylvania, USA
Roy Shilkrot	Massachusetts Institute of Technology, USA
Mike Zheng Shou	National University of Singapore, Singapore
Kaleem Siddiqi	McGill University, Canada
Richa Singh	Indian Institute of Technology Jodhpur, India
Greg Slabaugh	Queen Mary University of London, UK
Cees Snoek	University of Amsterdam, The Netherlands
Yale Song	Facebook AI Research, USA
Yi-Zhe Song	University of Surrey, UK
Bjorn Stenger	Rakuten Institute of Technology
Abby Stylianou	Saint Louis University, USA
Akihiro Sugimoto	National Institute of Informatics, Japan
Chen Sun	Brown University, USA
Deqing Sun	Google, USA
Kalyan Sunkavalli	Adobe Research, USA
Ying Tai	Tencent YouTu Lab, China
Ayellet Tal	Technion – Israel Institute of Technology, Israel
Ping Tan	Simon Fraser University, Canada
Siyu Tang	ETH Zurich, Switzerland
Chi-Keung Tang	Hong Kong University of Science and Technology, Hong Kong, China
Radu Timofte	University of Würzburg, Germany & ETH Zurich, Switzerland
Federico Tombari	Google, Switzerland & Technical University of Munich, Germany

James Tompkin	Brown University, USA
Lorenzo Torresani	Dartmouth College, USA
Alexander Toshev	Apple, USA
Du Tran	Facebook AI Research, USA
Anh T. Tran	VinAI, Vietnam
Zhuowen Tu	University of California, San Diego, USA
Georgios Tzimiropoulos	Queen Mary University of London, UK
Jasper Uijlings	Google Research, Switzerland
Jan C. van Gemert	Delft University of Technology, The Netherlands
Gul Varol	Ecole des Ponts ParisTech, France
Nuno Vasconcelos	University of California, San Diego, USA
Mayank Vatsa	Indian Institute of Technology Jodhpur, India
Ashok Veeraraghavan	Rice University, USA
Jakob Verbeek	Facebook AI Research, France
Carl Vondrick	Columbia University, USA
Ruiping Wang	Institute of Computing Technology, Chinese Academy of Sciences, China
Xinchao Wang	National University of Singapore, Singapore
Liwei Wang	The Chinese University of Hong Kong, Hong Kong, China
Chaohui Wang	Université Paris-Est, France
Xiaolong Wang	University of California, San Diego, USA
Christian Wolf	NAVER LABS Europe, France
Tao Xiang	University of Surrey, UK
Saining Xie	Facebook AI Research, USA
Cihang Xie	University of California, Santa Cruz, USA
Zeki Yalniz	Facebook, USA
Ming-Hsuan Yang	University of California, Merced, USA
Angela Yao	National University of Singapore, Singapore
Shaodi You	University of Amsterdam, The Netherlands
Stella X. Yu	University of California, Berkeley, USA
Junsong Yuan	State University of New York at Buffalo, USA
Stefanos Zafeiriou	Imperial College London, UK
Amir Zamir	École polytechnique fédérale de Lausanne, Switzerland
Lei Zhang	Alibaba & Hong Kong Polytechnic University, Hong Kong, China
Lei Zhang	International Digital Economy Academy (IDEA), China
Pengchuan Zhang	Meta AI, USA
Bolei Zhou	University of California, Los Angeles, USA
Yuke Zhu	University of Texas at Austin, USA

Todd Zickler   Harvard University, USA
Wangmeng Zuo   Harbin Institute of Technology, China

Technical Program Committee

Davide Abati	Filippo Aleotti	Sinem Aslan
Soroush Abbasi	Konstantinos P.	Vishal Asnani
Koohpayegani	Alexandridis	Mahmoud Assran
Amos L. Abbott	Motasem Alfarra	Amir Atapour-Abarghouei
Rameen Abdal	Mohsen Ali	Nikos Athanasiou
Rabab Abdelfattah	Thiemo Alldieck	Ali Athar
Sahar Abdelnabi	Hadi Alzayer	ShahRukh Athar
Hassan Abu Alhaija	Liang An	Sara Atito
Abulikemu Abuduweili	Shan An	Souhaib Attaiki
Ron Abutbul	Yi An	Matan Atzmon
Hanno Ackermann	Zhulin An	Mathieu Aubry
Aikaterini Adam	Dongsheng An	Nicolas Audebert
Kamil Adamczewski	Jie An	Tristan T.
Ehsan Adeli	Xiang An	Aumentado-Armstrong
Vida Adeli	Saket Anand	Melinos Averkiou
Donald Adjeroh	Cosmin Ancuti	Yannis Avrithis
Arman Afrasiyabi	Juan Andrade-Cetto	Stephane Ayache
Akshay Agarwal	Alexander Andreopoulos	Mehmet Aygün
Sameer Agarwal	Bjoern Andres	Seyed Mehdi
Abhinav Agarwalla	Jerone T. A. Andrews	Ayyoubzadeh
Vaibhav Aggarwal	Shivangi Aneja	Hossein Azizpour
Sara Aghajanzadeh	Anelia Angelova	George Azzopardi
Susmit Agrawal	Dragomir Anguelov	Mallikarjun B. R.
Antonio Agudo	Rushil Anirudh	Yunhao Ba
Touqeer Ahmad	Oron Anschel	Abhishek Badki
Sk Miraj Ahmed	Rao Muhammad Anwer	Seung-Hwan Bae
Chaitanya Ahuja	Djamila Aouada	Seung-Hwan Baek
Nilesh A. Ahuja	Evlampios Apostolidis	Seungryul Baek
Abhishek Aich	Srikar Appalaraju	Piyush Nitin Bagad
Shubhra Aich	Nikita Araslanov	Shai Bagon
Noam Aigerman	Andre Araujo	Gaetan Bahl
Arash Akbarinia	Eric Arazo	Shikhar Bahl
Peri Akiva	Dawit Mureja Argaw	Sherwin Bahmani
Derya Akkaynak	Anurag Arnab	Haoran Bai
Emre Aksan	Aditya Arora	Lei Bai
Arjun R. Akula	Chetan Arora	Jiawang Bai
Yuval Alaluf	Sunpreet S. Arora	Haoyue Bai
Stephan Alaniz	Alexey Artemov	Jinbin Bai
Paul Albert	Muhammad Asad	Xiang Bai
Cenek Albl	Kumar Ashutosh	Xuyang Bai

Yang Bai
Yuanchao Bai
Ziqian Bai
Sungyong Baik
Kevin Bailly
Max Bain
Federico Baldassarre
Wele Gedara Chaminda
    Bandara
Biplab Banerjee
Pratyay Banerjee
Sandipan Banerjee
Jihwan Bang
Antyanta Bangunharcana
Aayush Bansal
Ankan Bansal
Siddhant Bansal
Wentao Bao
Zhipeng Bao
Amir Bar
Manel Baradad Jurjo
Lorenzo Baraldi
Danny Barash
Daniel Barath
Connelly Barnes
Ioan Andrei Bârsan
Steven Basart
Dina Bashkirova
Chaim Baskin
Peyman Bateni
Anil Batra
Sebastiano Battiato
Ardhendu Behera
Harkirat Behl
Jens Behley
Vasileios Belagiannis
Boulbaba Ben Amor
Emanuel Ben Baruch
Abdessamad Ben Hamza
Gil Ben-Artzi
Assia Benbihi
Fabian Benitez-Quiroz
Guy Ben-Yosef
Philipp Benz
Alexander W. Bergman

Urs Bergmann
Jesus Bermudez-Cameo
Stefano Berretti
Gedas Bertasius
Zachary Bessinger
Petra Bevandić
Matthew Beveridge
Lucas Beyer
Yash Bhalgat
Suvaansh Bhambri
Samarth Bharadwaj
Gaurav Bharaj
Aparna Bharati
Bharat Lal Bhatnagar
Uttaran Bhattacharya
Apratim Bhattacharyya
Brojeshwar Bhowmick
Ankan Kumar Bhunia
Ayan Kumar Bhunia
Qi Bi
Sai Bi
Michael Bi Mi
Gui-Bin Bian
Jia-Wang Bian
Shaojun Bian
Pia Bideau
Mario Bijelic
Hakan Bilen
Guillaume-Alexandre
    Bilodeau
Alexander Binder
Tolga Birdal
Vighnesh N. Birodkar
Sandika Biswas
Andreas Blattmann
Janusz Bobulski
Giuseppe Boccignone
Vishnu Boddeti
Navaneeth Bodla
Moritz Böhle
Aleksei Bokhovkin
Sam Bond-Taylor
Vivek Boominathan
Shubhankar Borse
Mark Boss

Andrea Bottino
Adnane Boukhayma
Fadi Boutros
Nicolas C. Boutry
Richard S. Bowen
Ivaylo Boyadzhiev
Aidan Boyd
Yuri Boykov
Aljaz Bozic
Behzad Bozorgtabar
Eric Brachmann
Samarth Brahmbhatt
Gustav Bredell
Francois Bremond
Joel Brogan
Andrew Brown
Thomas Brox
Marcus A. Brubaker
Robert-Jan Bruintjes
Yuqi Bu
Anders G. Buch
Himanshu Buckchash
Mateusz Buda
Ignas Budvytis
José M. Buenaposada
Marcel C. Bühler
Tu Bui
Adrian Bulat
Hannah Bull
Evgeny Burnaev
Andrei Bursuc
Benjamin Busam
Sergey N. Buzykanov
Wonmin Byeon
Fabian Caba
Martin Cadik
Guanyu Cai
Minjie Cai
Qing Cai
Zhongang Cai
Qi Cai
Yancheng Cai
Shen Cai
Han Cai
Jiarui Cai

Bowen Cai
Mu Cai
Qin Cai
Ruojin Cai
Weidong Cai
Weiwei Cai
Yi Cai
Yujun Cai
Zhiping Cai
Akin Caliskan
Lilian Calvet
Baris Can Cam
Necati Cihan Camgoz
Tommaso Campari
Dylan Campbell
Ziang Cao
Ang Cao
Xu Cao
Zhiwen Cao
Shengcao Cao
Song Cao
Weipeng Cao
Xiangyong Cao
Xiaochun Cao
Yue Cao
Yunhao Cao
Zhangjie Cao
Jiale Cao
Yang Cao
Jiajiong Cao
Jie Cao
Jinkun Cao
Lele Cao
Yulong Cao
Zhiguo Cao
Chen Cao
Razvan Caramalau
Marlène Careil
Gustavo Carneiro
Joao Carreira
Dan Casas
Paola Cascante-Bonilla
Angela Castillo
Francisco M. Castro
Pedro Castro

Luca Cavalli
George J. Cazenavette
Oya Celiktutan
Hakan Cevikalp
Sri Harsha C. H.
Sungmin Cha
Geonho Cha
Menglei Chai
Lucy Chai
Yuning Chai
Zenghao Chai
Anirban Chakraborty
Deep Chakraborty
Rudrasis Chakraborty
Souradeep Chakraborty
Kelvin C. K. Chan
Chee Seng Chan
Paramanand Chandramouli
Arjun Chandrasekaran
Kenneth Chaney
Dongliang Chang
Huiwen Chang
Peng Chang
Xiaojun Chang
Jia-Ren Chang
Hyung Jin Chang
Hyun Sung Chang
Ju Yong Chang
Li-Jen Chang
Qi Chang
Wei-Yi Chang
Yi Chang
Nadine Chang
Hanqing Chao
Pradyumna Chari
Dibyadip Chatterjee
Chiranjoy Chattopadhyay
Siddhartha Chaudhuri
Zhengping Che
Gal Chechik
Lianggangxu Chen
Qi Alfred Chen
Brian Chen
Bor-Chun Chen
Bo-Hao Chen

Bohong Chen
Bin Chen
Ziliang Chen
Cheng Chen
Chen Chen
Chaofeng Chen
Xi Chen
Haoyu Chen
Xuanhong Chen
Wei Chen
Qiang Chen
Shi Chen
Xianyu Chen
Chang Chen
Changhuai Chen
Hao Chen
Jie Chen
Jianbo Chen
Jingjing Chen
Jun Chen
Kejiang Chen
Mingcai Chen
Nenglun Chen
Qifeng Chen
Ruoyu Chen
Shu-Yu Chen
Weidong Chen
Weijie Chen
Weikai Chen
Xiang Chen
Xiuyi Chen
Xingyu Chen
Yaofo Chen
Yueting Chen
Yu Chen
Yunjin Chen
Yuntao Chen
Yun Chen
Zhenfang Chen
Zhuangzhuang Chen
Chu-Song Chen
Xiangyu Chen
Zhuo Chen
Chaoqi Chen
Shizhe Chen

Xiaotong Chen
Xiaozhi Chen
Dian Chen
Defang Chen
Dingfan Chen
Ding-Jie Chen
Ee Heng Chen
Tao Chen
Yixin Chen
Wei-Ting Chen
Lin Chen
Guang Chen
Guangyi Chen
Guanying Chen
Guangyao Chen
Hwann-Tzong Chen
Junwen Chen
Jiacheng Chen
Jianxu Chen
Hui Chen
Kai Chen
Kan Chen
Kevin Chen
Kuan-Wen Chen
Weihua Chen
Zhang Chen
Liang-Chieh Chen
Lele Chen
Liang Chen
Fanglin Chen
Zehui Chen
Minghui Chen
Minghao Chen
Xiaokang Chen
Qian Chen
Jun-Cheng Chen
Qi Chen
Qingcai Chen
Richard J. Chen
Runnan Chen
Rui Chen
Shuo Chen
Sentao Chen
Shaoyu Chen
Shixing Chen

Shuai Chen
Shuya Chen
Sizhe Chen
Simin Chen
Shaoxiang Chen
Zitian Chen
Tianlong Chen
Tianshui Chen
Min-Hung Chen
Xiangning Chen
Xin Chen
Xinghao Chen
Xuejin Chen
Xu Chen
Xuxi Chen
Yunlu Chen
Yanbei Chen
Yuxiao Chen
Yun-Chun Chen
Yi-Ting Chen
Yi-Wen Chen
Yinbo Chen
Yiran Chen
Yuanhong Chen
Yubei Chen
Yuefeng Chen
Yuhua Chen
Yukang Chen
Zerui Chen
Zhaoyu Chen
Zhen Chen
Zhenyu Chen
Zhi Chen
Zhiwei Chen
Zhixiang Chen
Long Chen
Bowen Cheng
Jun Cheng
Yi Cheng
Jingchun Cheng
Lechao Cheng
Xi Cheng
Yuan Cheng
Ho Kei Cheng
Kevin Ho Man Cheng

Jiacheng Cheng
Kelvin B. Cheng
Li Cheng
Mengjun Cheng
Zhen Cheng
Qingrong Cheng
Tianheng Cheng
Harry Cheng
Yihua Cheng
Yu Cheng
Ziheng Cheng
Soon Yau Cheong
Anoop Cherian
Manuela Chessa
Zhixiang Chi
Naoki Chiba
Julian Chibane
Kashyap Chitta
Tai-Yin Chiu
Hsu-kuang Chiu
Wei-Chen Chiu
Sungmin Cho
Donghyeon Cho
Hyeon Cho
Yooshin Cho
Gyusang Cho
Jang Hyun Cho
Seungju Cho
Nam Ik Cho
Sunghyun Cho
Hanbyel Cho
Jaesung Choe
Jooyoung Choi
Chiho Choi
Changwoon Choi
Jongwon Choi
Myungsub Choi
Dooseop Choi
Jonghyun Choi
Jinwoo Choi
Jun Won Choi
Min-Kook Choi
Hongsuk Choi
Janghoon Choi
Yoon-Ho Choi

Yukyung Choi
Jaegul Choo
Ayush Chopra
Siddharth Choudhary
Subhabrata Choudhury
Vasileios Choutas
Ka-Ho Chow
Pinaki Nath Chowdhury
Sammy Christen
Anders Christensen
Grigorios Chrysos
Hang Chu
Wen-Hsuan Chu
Peng Chu
Qi Chu
Ruihang Chu
Wei-Ta Chu
Yung-Yu Chuang
Sanghyuk Chun
Se Young Chun
Antonio Cinà
Ramazan Gokberk Cinbis
Javier Civera
Albert Clapés
Ronald Clark
Brian S. Clipp
Felipe Codevilla
Daniel Coelho de Castro
Niv Cohen
Forrester Cole
Maxwell D. Collins
Robert T. Collins
Marc Comino Trinidad
Runmin Cong
Wenyan Cong
Maxime Cordy
Marcella Cornia
Enric Corona
Huseyin Coskun
Luca Cosmo
Dragos Costea
Davide Cozzolino
Arun C. S. Kumar
Aiyu Cui
Qiongjie Cui

Quan Cui
Shuhao Cui
Yiming Cui
Ying Cui
Zijun Cui
Jiali Cui
Jiequan Cui
Yawen Cui
Zhen Cui
Zhaopeng Cui
Jack Culpepper
Xiaodong Cun
Ross Cutler
Adam Czajka
Ali Dabouei
Konstantinos M. Dafnis
Manuel Dahnert
Tao Dai
Yuchao Dai
Bo Dai
Mengyu Dai
Hang Dai
Haixing Dai
Peng Dai
Pingyang Dai
Qi Dai
Qiyu Dai
Yutong Dai
Naser Damer
Zhiyuan Dang
Mohamed Daoudi
Ayan Das
Abir Das
Debasmit Das
Deepayan Das
Partha Das
Sagnik Das
Soumi Das
Srijan Das
Swagatam Das
Avijit Dasgupta
Jim Davis
Adrian K. Davison
Homa Davoudi
Laura Daza

Matthias De Lange
Shalini De Mello
Marco De Nadai
Christophe De
    Vleeschouwer
Alp Dener
Boyang Deng
Congyue Deng
Bailin Deng
Yong Deng
Ye Deng
Zhuo Deng
Zhijie Deng
Xiaoming Deng
Jiankang Deng
Jinhong Deng
Jingjing Deng
Liang-Jian Deng
Siqi Deng
Xiang Deng
Xueqing Deng
Zhongying Deng
Karan Desai
Jean-Emmanuel Deschaud
Aniket Anand Deshmukh
Neel Dey
Helisa Dhamo
Prithviraj Dhar
Amaya Dharmasiri
Yan Di
Xing Di
Ousmane A. Dia
Haiwen Diao
Xiaolei Diao
Gonçalo José Dias Pais
Abdallah Dib
Anastasios Dimou
Changxing Ding
Henghui Ding
Guodong Ding
Yaqing Ding
Shuangrui Ding
Yuhang Ding
Yikang Ding
Shouhong Ding

Haisong Ding
Hui Ding
Jiahao Ding
Jian Ding
Jian-Jiun Ding
Shuxiao Ding
Tianyu Ding
Wenhao Ding
Yuqi Ding
Yi Ding
Yuzhen Ding
Zhengming Ding
Tan Minh Dinh
Vu Dinh
Christos Diou
Mandar Dixit
Bao Gia Doan
Khoa D. Doan
Dzung Anh Doan
Debi Prosad Dogra
Nehal Doiphode
Chengdong Dong
Bowen Dong
Zhenxing Dong
Hang Dong
Xiaoyi Dong
Haoye Dong
Jiangxin Dong
Shichao Dong
Xuan Dong
Zhen Dong
Shuting Dong
Jing Dong
Li Dong
Ming Dong
Nanqing Dong
Qiulei Dong
Runpei Dong
Siyan Dong
Tian Dong
Wei Dong
Xiaomeng Dong
Xin Dong
Xingbo Dong
Yuan Dong

Samuel Dooley
Gianfranco Doretto
Michael Dorkenwald
Keval Doshi
Zhaopeng Dou
Xiaotian Dou
Hazel Doughty
Ahmad Droby
Iddo Drori
Jie Du
Yong Du
Dawei Du
Dong Du
Ruoyi Du
Yuntao Du
Xuefeng Du
Yilun Du
Yuming Du
Radhika Dua
Haodong Duan
Jiafei Duan
Kaiwen Duan
Peiqi Duan
Ye Duan
Haoran Duan
Jiali Duan
Amanda Duarte
Abhimanyu Dubey
Shiv Ram Dubey
Florian Dubost
Lukasz Dudziak
Shivam Duggal
Justin M. Dulay
Matteo Dunnhofer
Chi Nhan Duong
Thibaut Durand
Mihai Dusmanu
Ujjal Kr Dutta
Debidatta Dwibedi
Isht Dwivedi
Sai Kumar Dwivedi
Takeharu Eda
Mark Edmonds
Alexei A. Efros
Thibaud Ehret

Max Ehrlich
Mahsa Ehsanpour
Iván Eichhardt
Farshad Einabadi
Marvin Eisenberger
Hazim Kemal Ekenel
Mohamed El Banani
Ismail Elezi
Moshe Eliasof
Alaa El-Nouby
Ian Endres
Francis Engelmann
Deniz Engin
Chanho Eom
Dave Epstein
Maria C. Escobar
Victor A. Escorcia
Carlos Esteves
Sungmin Eum
Bernard J. E. Evans
Ivan Evtimov
Fevziye Irem Eyiokur
    Yaman
Matteo Fabbri
Sébastien Fabbro
Gabriele Facciolo
Masud Fahim
Bin Fan
Hehe Fan
Deng-Ping Fan
Aoxiang Fan
Chen-Chen Fan
Qi Fan
Zhaoxin Fan
Haoqi Fan
Heng Fan
Hongyi Fan
Linxi Fan
Baojie Fan
Jiayuan Fan
Lei Fan
Quanfu Fan
Yonghui Fan
Yingruo Fan
Zhiwen Fan

Zicong Fan
Sean Fanello
Jiansheng Fang
Chaowei Fang
Yuming Fang
Jianwu Fang
Jin Fang
Qi Fang
Shancheng Fang
Tian Fang
Xianyong Fang
Gongfan Fang
Zhen Fang
Hui Fang
Jiemin Fang
Le Fang
Pengfei Fang
Xiaolin Fang
Yuxin Fang
Zhaoyuan Fang
Ammarah Farooq
Azade Farshad
Zhengcong Fei
Michael Felsberg
Wei Feng
Chen Feng
Fan Feng
Andrew Feng
Xin Feng
Zheyun Feng
Ruicheng Feng
Mingtao Feng
Qianyu Feng
Shangbin Feng
Chun-Mei Feng
Zunlei Feng
Zhiyong Feng
Martin Fergie
Mustansar Fiaz
Marco Fiorucci
Michael Firman
Hamed Firooz
Volker Fischer
Corneliu O. Florea
Georgios Floros

Wolfgang Foerstner
Gianni Franchi
Jean-Sebastien Franco
Simone Frintrop
Anna Fruehstueck
Changhong Fu
Chaoyou Fu
Cheng-Yang Fu
Chi-Wing Fu
Deqing Fu
Huan Fu
Jun Fu
Kexue Fu
Ying Fu
Jianlong Fu
Jingjing Fu
Qichen Fu
Tsu-Jui Fu
Xueyang Fu
Yang Fu
Yanwei Fu
Yonggan Fu
Wolfgang Fuhl
Yasuhisa Fujii
Kent Fujiwara
Marco Fumero
Takuya Funatomi
Isabel Funke
Dario Fuoli
Antonino Furnari
Matheus A. Gadelha
Akshay Gadi Patil
Adrian Galdran
Guillermo Gallego
Silvano Galliani
Orazio Gallo
Leonardo Galteri
Matteo Gamba
Yiming Gan
Sujoy Ganguly
Harald Ganster
Boyan Gao
Changxin Gao
Daiheng Gao
Difei Gao

Chen Gao
Fei Gao
Lin Gao
Wei Gao
Yiming Gao
Junyu Gao
Guangyu Ryan Gao
Haichang Gao
Hongchang Gao
Jialin Gao
Jin Gao
Jun Gao
Katelyn Gao
Mingchen Gao
Mingfei Gao
Pan Gao
Shangqian Gao
Shanghua Gao
Xitong Gao
Yunhe Gao
Zhanning Gao
Elena Garces
Nuno Cruz Garcia
Noa Garcia
Guillermo
    Garcia-Hernando
Isha Garg
Rahul Garg
Sourav Garg
Quentin Garrido
Stefano Gasperini
Kent Gauen
Chandan Gautam
Shivam Gautam
Paul Gay
Chunjiang Ge
Shiming Ge
Wenhang Ge
Yanhao Ge
Zheng Ge
Songwei Ge
Weifeng Ge
Yixiao Ge
Yuying Ge
Shijie Geng

Zhengyang Geng
Kyle A. Genova
Georgios Georgakis
Markos Georgopoulos
Marcel Geppert
Shabnam Ghadar
Mina Ghadimi Atigh
Deepti Ghadiyaram
Maani Ghaffari Jadidi
Sedigh Ghamari
Zahra Gharaee
Michaël Gharbi
Golnaz Ghiasi
Reza Ghoddoosian
Soumya Suvra Ghosal
Adhiraj Ghosh
Arthita Ghosh
Pallabi Ghosh
Soumyadeep Ghosh
Andrew Gilbert
Igor Gilitschenski
Jhony H. Giraldo
Andreu Girbau Xalabarder
Rohit Girdhar
Sharath Girish
Xavier Giro-i-Nieto
Raja Giryes
Thomas Gittings
Nikolaos Gkanatsios
Ioannis Gkioulekas
Abhiram
   Gnanasambandam
Aurele T. Gnanha
Clement L. J. C. Godard
Arushi Goel
Vidit Goel
Shubham Goel
Zan Gojcic
Aaron K. Gokaslan
Tejas Gokhale
S. Alireza Golestaneh
Thiago L. Gomes
Nuno Goncalves
Boqing Gong
Chen Gong

Yuanhao Gong
Guoqiang Gong
Jingyu Gong
Rui Gong
Yu Gong
Mingming Gong
Neil Zhenqiang Gong
Xun Gong
Yunye Gong
Yihong Gong
Cristina I. González
Nithin Gopalakrishnan
   Nair
Gaurav Goswami
Jianping Gou
Shreyank N. Gowda
Ankit Goyal
Helmut Grabner
Patrick L. Grady
Ben Graham
Eric Granger
Douglas R. Gray
Matej Grcić
David Griffiths
Jinjin Gu
Yun Gu
Shuyang Gu
Jianyang Gu
Fuqiang Gu
Jiatao Gu
Jindong Gu
Jiaqi Gu
Jinwei Gu
Jiaxin Gu
Geonmo Gu
Xiao Gu
Xinqian Gu
Xiuye Gu
Yuming Gu
Zhangxuan Gu
Dayan Guan
Junfeng Guan
Qingji Guan
Tianrui Guan
Shanyan Guan

Denis A. Gudovskiy
Ricardo Guerrero
Pierre-Louis Guhur
Jie Gui
Liangyan Gui
Liangke Gui
Benoit Guillard
Erhan Gundogdu
Manuel Günther
Jingcai Guo
Yuanfang Guo
Junfeng Guo
Chenqi Guo
Dan Guo
Hongji Guo
Jia Guo
Jie Guo
Minghao Guo
Shi Guo
Yanhui Guo
Yangyang Guo
Yuan-Chen Guo
Yilu Guo
Yiluan Guo
Yong Guo
Guangyu Guo
Haiyun Guo
Jinyang Guo
Jianyuan Guo
Pengsheng Guo
Pengfei Guo
Shuxuan Guo
Song Guo
Tianyu Guo
Qing Guo
Qiushan Guo
Wen Guo
Xiefan Guo
Xiaohu Guo
Xiaoqing Guo
Yufei Guo
Yuhui Guo
Yuliang Guo
Yunhui Guo
Yanwen Guo

Akshita Gupta
Ankush Gupta
Kamal Gupta
Kartik Gupta
Ritwik Gupta
Rohit Gupta
Siddharth Gururani
Fredrik K. Gustafsson
Abner Guzman Rivera
Vladimir Guzov
Matthew A. Gwilliam
Jung-Woo Ha
Marc Habermann
Isma Hadji
Christian Haene
Martin Hahner
Levente Hajder
Alexandros Haliassos
Emanuela Haller
Bumsub Ham
Abdullah J. Hamdi
Shreyas Hampali
Dongyoon Han
Chunrui Han
Dong-Jun Han
Dong-Sig Han
Guangxing Han
Zhizhong Han
Ruize Han
Jiaming Han
Jin Han
Ligong Han
Xian-Hua Han
Xiaoguang Han
Yizeng Han
Zhi Han
Zhenjun Han
Zhongyi Han
Jungong Han
Junlin Han
Kai Han
Kun Han
Sungwon Han
Songfang Han
Wei Han

Xiao Han
Xintong Han
Xinzhe Han
Yahong Han
Yan Han
Zongbo Han
Nicolai Hani
Rana Hanocka
Niklas Hanselmann
Nicklas A. Hansen
Hong Hanyu
Fusheng Hao
Yanbin Hao
Shijie Hao
Udith Haputhanthri
Mehrtash Harandi
Josh Harguess
Adam Harley
David M. Hart
Atsushi Hashimoto
Ali Hassani
Mohammed Hassanin
Yana Hasson
Joakim Bruslund Haurum
Bo He
Kun He
Chen He
Xin He
Fazhi He
Gaoqi He
Hao He
Haoyu He
Jiangpeng He
Hongliang He
Qian He
Xiangteng He
Xuming He
Yannan He
Yuhang He
Yang He
Xiangyu He
Nanjun He
Pan He
Sen He
Shengfeng He

Songtao He
Tao He
Tong He
Wei He
Xuehai He
Xiaoxiao He
Ying He
Yisheng He
Ziwen He
Peter Hedman
Felix Heide
Yacov Hel-Or
Paul Henderson
Philipp Henzler
Byeongho Heo
Jae-Pil Heo
Miran Heo
Sachini A. Herath
Stephane Herbin
Pedro Hermosilla Casajus
Monica Hernandez
Charles Herrmann
Roei Herzig
Mauricio Hess-Flores
Carlos Hinojosa
Tobias Hinz
Tsubasa Hirakawa
Chih-Hui Ho
Lam Si Tung Ho
Jennifer Hobbs
Derek Hoiem
Yannick Hold-Geoffroy
Aleksander Holynski
Cheeun Hong
Fa-Ting Hong
Hanbin Hong
Guan Zhe Hong
Danfeng Hong
Lanqing Hong
Xiaopeng Hong
Xin Hong
Jie Hong
Seungbum Hong
Cheng-Yao Hong
Seunghoon Hong

Yi Hong
Yuan Hong
Yuchen Hong
Anthony Hoogs
Maxwell C. Horton
Kazuhiro Hotta
Qibin Hou
Tingbo Hou
Junhui Hou
Ji Hou
Qiqi Hou
Rui Hou
Ruibing Hou
Zhi Hou
Henry Howard-Jenkins
Lukas Hoyer
Wei-Lin Hsiao
Chiou-Ting Hsu
Anthony Hu
Brian Hu
Yusong Hu
Hexiang Hu
Haoji Hu
Di Hu
Hengtong Hu
Haigen Hu
Lianyu Hu
Hanzhe Hu
Jie Hu
Junlin Hu
Shizhe Hu
Jian Hu
Zhiming Hu
Juhua Hu
Peng Hu
Ping Hu
Ronghang Hu
MengShun Hu
Tao Hu
Vincent Tao Hu
Xiaoling Hu
Xinting Hu
Xiaolin Hu
Xuefeng Hu
Xiaowei Hu

Yang Hu
Yueyu Hu
Zeyu Hu
Zhongyun Hu
Binh-Son Hua
Guoliang Hua
Yi Hua
Linzhi Huang
Qiusheng Huang
Bo Huang
Chen Huang
Hsin-Ping Huang
Ye Huang
Shuangping Huang
Zeng Huang
Buzhen Huang
Cong Huang
Heng Huang
Hao Huang
Qidong Huang
Huaibo Huang
Chaoqin Huang
Feihu Huang
Jiahui Huang
Jingjia Huang
Kun Huang
Lei Huang
Sheng Huang
Shuaiyi Huang
Siyu Huang
Xiaoshui Huang
Xiaoyang Huang
Yan Huang
Yihao Huang
Ying Huang
Ziling Huang
Xiaoke Huang
Yifei Huang
Haiyang Huang
Zhewei Huang
Jin Huang
Haibin Huang
Jiaxing Huang
Junjie Huang
Keli Huang

Lang Huang
Lin Huang
Luojie Huang
Mingzhen Huang
Shijia Huang
Shengyu Huang
Siyuan Huang
He Huang
Xiuyu Huang
Lianghua Huang
Yue Huang
Yaping Huang
Yuge Huang
Zehao Huang
Zeyi Huang
Zhiqi Huang
Zhongzhan Huang
Zilong Huang
Ziyuan Huang
Tianrui Hui
Zhuo Hui
Le Hui
Jing Huo
Junhwa Hur
Shehzeen S. Hussain
Chuong Minh Huynh
Seunghyun Hwang
Jaehui Hwang
Jyh-Jing Hwang
Sukjun Hwang
Soonmin Hwang
Wonjun Hwang
Rakib Hyder
Sangeek Hyun
Sarah Ibrahimi
Tomoki Ichikawa
Yerlan Idelbayev
A. S. M. Iftekhar
Masaaki Iiyama
Satoshi Ikehata
Sunghoon Im
Atul N. Ingle
Eldar Insafutdinov
Yani A. Ioannou
Radu Tudor Ionescu

Umar Iqbal
Go Irie
Muhammad Zubair Irshad
Ahmet Iscen
Berivan Isik
Ashraful Islam
Md Amirul Islam
Syed Islam
Mariko Isogawa
Vamsi Krishna K. Ithapu
Boris Ivanovic
Darshan Iyer
Sarah Jabbour
Ayush Jain
Nishant Jain
Samyak Jain
Vidit Jain
Vineet Jain
Priyank Jaini
Tomas Jakab
Mohammad A. A. K.
   Jalwana
Muhammad Abdullah
   Jamal
Hadi Jamali-Rad
Stuart James
Varun Jampani
Young Kyun Jang
YeongJun Jang
Yunseok Jang
Ronnachai Jaroensri
Bhavan Jasani
Krishna Murthy
   Jatavallabhula
Mojan Javaheripi
Syed A. Javed
Guillaume Jeanneret
Pranav Jeevan
Herve Jegou
Rohit Jena
Tomas Jenicek
Porter Jenkins
Simon Jenni
Hae-Gon Jeon
Sangryul Jeon

Boseung Jeong
Yoonwoo Jeong
Seong-Gyun Jeong
Jisoo Jeong
Allan D. Jepson
Ankit Jha
Sumit K. Jha
I-Hong Jhuo
Ge-Peng Ji
Chaonan Ji
Deyi Ji
Jingwei Ji
Wei Ji
Zhong Ji
Jiayi Ji
Pengliang Ji
Hui Ji
Mingi Ji
Xiaopeng Ji
Yuzhu Ji
Baoxiong Jia
Songhao Jia
Dan Jia
Shan Jia
Xiaojun Jia
Xiuyi Jia
Xu Jia
Menglin Jia
Wenqi Jia
Boyuan Jiang
Wenhao Jiang
Huaizu Jiang
Hanwen Jiang
Haiyong Jiang
Hao Jiang
Huajie Jiang
Huiqin Jiang
Haojun Jiang
Haobo Jiang
Junjun Jiang
Xingyu Jiang
Yangbangyan Jiang
Yu Jiang
Jianmin Jiang
Jiaxi Jiang

Jing Jiang
Kui Jiang
Li Jiang
Liming Jiang
Chiyu Jiang
Meirui Jiang
Chen Jiang
Peng Jiang
Tai-Xiang Jiang
Wen Jiang
Xinyang Jiang
Yifan Jiang
Yuming Jiang
Yingying Jiang
Zeren Jiang
ZhengKai Jiang
Zhenyu Jiang
Shuming Jiao
Jianbo Jiao
Licheng Jiao
Dongkwon Jin
Yeying Jin
Cheng Jin
Linyi Jin
Qing Jin
Taisong Jin
Xiao Jin
Xin Jin
Sheng Jin
Kyong Hwan Jin
Ruibing Jin
SouYoung Jin
Yueming Jin
Chenchen Jing
Longlong Jing
Taotao Jing
Yongcheng Jing
Younghyun Jo
Joakim Johnander
Jeff Johnson
Michael J. Jones
R. Kenny Jones
Rico Jonschkowski
Ameya Joshi
Sunghun Joung

Felix Juefei-Xu
Claudio R. Jung
Steffen Jung
Hari Chandana K.
Rahul Vigneswaran K.
Prajwal K. R.
Abhishek Kadian
Jhony Kaesemodel Pontes
Kumara Kahatapitiya
Anmol Kalia
Sinan Kalkan
Tarun Kalluri
Jaewon Kam
Sandesh Kamath
Meina Kan
Menelaos Kanakis
Takuhiro Kaneko
Di Kang
Guoliang Kang
Hao Kang
Jaeyeon Kang
Kyoungkook Kang
Li-Wei Kang
MinGuk Kang
Suk-Ju Kang
Zhao Kang
Yash Mukund Kant
Yueying Kao
Aupendu Kar
Konstantinos Karantzalos
Sezer Karaoglu
Navid Kardan
Sanjay Kariyappa
Leonid Karlinsky
Animesh Karnewar
Shyamgopal Karthik
Hirak J. Kashyap
Marc A. Kastner
Hirokatsu Kataoka
Angelos Katharopoulos
Hiroharu Kato
Kai Katsumata
Manuel Kaufmann
Chaitanya Kaul
Prakhar Kaushik

Yuki Kawana
Lei Ke
Lipeng Ke
Tsung-Wei Ke
Wei Ke
Petr Kellnhofer
Aniruddha Kembhavi
John Kender
Corentin Kervadec
Leonid Keselman
Daniel Keysers
Nima Khademi Kalantari
Taras Khakhulin
Samir Khaki
Muhammad Haris Khan
Qadeer Khan
Salman Khan
Subash Khanal
Vaishnavi M. Khindkar
Rawal Khirodkar
Saeed Khorram
Pirazh Khorramshahi
Kourosh Khoshelham
Ansh Khurana
Benjamin Kiefer
Jae Myung Kim
Junho Kim
Boah Kim
Hyeonseong Kim
Dong-Jin Kim
Dongwan Kim
Donghyun Kim
Doyeon Kim
Yonghyun Kim
Hyung-Il Kim
Hyunwoo Kim
Hyeongwoo Kim
Hyo Jin Kim
Hyunwoo J. Kim
Taehoon Kim
Jaeha Kim
Jiwon Kim
Jung Uk Kim
Kangyeol Kim
Eunji Kim

Daeha Kim
Dongwon Kim
Kunhee Kim
Kyungmin Kim
Junsik Kim
Min H. Kim
Namil Kim
Kookhoi Kim
Sanghyun Kim
Seongyeop Kim
Seungryong Kim
Saehoon Kim
Euyoung Kim
Guisik Kim
Sungyeon Kim
Sunnie S. Y. Kim
Taehun Kim
Tae Oh Kim
Won Hwa Kim
Seungwook Kim
YoungBin Kim
Youngeun Kim
Akisato Kimura
Furkan Osman Kınlı
Zsolt Kira
Hedvig Kjellström
Florian Kleber
Jan P. Klopp
Florian Kluger
Laurent Kneip
Byungsoo Ko
Muhammed Kocabas
A. Sophia Koepke
Kevin Koeser
Nick Kolkin
Nikos Kolotouros
Wai-Kin Adams Kong
Deying Kong
Caihua Kong
Youyong Kong
Shuyu Kong
Shu Kong
Tao Kong
Yajing Kong
Yu Kong

Zishang Kong
Theodora Kontogianni
Anton S. Konushin
Julian F. P. Kooij
Bruno Korbar
Giorgos Kordopatis-Zilos
Jari Korhonen
Adam Kortylewski
Denis Korzhenkov
Divya Kothandaraman
Suraj Kothawade
Iuliia Kotseruba
Satwik Kottur
Shashank Kotyan
Alexandros Kouris
Petros Koutras
Anna Kreshuk
Ranjay Krishna
Dilip Krishnan
Andrey Kuehlkamp
Hilde Kuehne
Jason Kuen
David Kügler
Arjan Kuijper
Anna Kukleva
Sumith Kulal
Viveka Kulharia
Akshay R. Kulkarni
Nilesh Kulkarni
Dominik Kulon
Abhinav Kumar
Akash Kumar
Suryansh Kumar
B. V. K. Vijaya Kumar
Pulkit Kumar
Ratnesh Kumar
Sateesh Kumar
Satish Kumar
Vijay Kumar B. G.
Nupur Kumari
Sudhakar Kumawat
Jogendra Nath Kundu
Hsien-Kai Kuo
Meng-Yu Jennifer Kuo
Vinod Kumar Kurmi

Yusuke Kurose
Keerthy Kusumam
Alina Kuznetsova
Henry Kvinge
Ho Man Kwan
Hyeokjun Kweon
Heeseung Kwon
Gihyun Kwon
Myung-Joon Kwon
Taesung Kwon
YoungJoong Kwon
Christos Kyrkou
Jorma Laaksonen
Yann Labbe
Zorah Laehner
Florent Lafarge
Hamid Laga
Manuel Lagunas
Shenqi Lai
Jian-Huang Lai
Zihang Lai
Mohamed I. Lakhal
Mohit Lamba
Meng Lan
Loic Landrieu
Zhiqiang Lang
Natalie Lang
Dong Lao
Yizhen Lao
Yingjie Lao
Issam Hadj Laradji
Gustav Larsson
Viktor Larsson
Zakaria Laskar
Stéphane Lathuilière
Chun Pong Lau
Rynson W. H. Lau
Hei Law
Justin Lazarow
Verica Lazova
Eric-Tuan Le
Hieu Le
Trung-Nghia Le
Mathias Lechner
Byeong-Uk Lee

Chen-Yu Lee
Che-Rung Lee
Chul Lee
Hong Joo Lee
Dongsoo Lee
Jiyoung Lee
Eugene Eu Tzuan Lee
Daeun Lee
Saehyung Lee
Jewook Lee
Hyungtae Lee
Hyunmin Lee
Jungbeom Lee
Joon-Young Lee
Jong-Seok Lee
Joonseok Lee
Junha Lee
Kibok Lee
Byung-Kwan Lee
Jangwon Lee
Jinho Lee
Jongmin Lee
Seunghyun Lee
Sohyun Lee
Minsik Lee
Dogyoon Lee
Seungmin Lee
Min Jun Lee
Sangho Lee
Sangmin Lee
Seungeun Lee
Seon-Ho Lee
Sungmin Lee
Sungho Lee
Sangyoun Lee
Vincent C. S. S. Lee
Jaeseong Lee
Yong Jae Lee
Chenyang Lei
Chenyi Lei
Jiahui Lei
Xinyu Lei
Yinjie Lei
Jiaxu Leng
Luziwei Leng

Jan E. Lenssen
Vincent Lepetit
Thomas Leung
María Leyva-Vallina
Xin Li
Yikang Li
Baoxin Li
Bin Li
Bing Li
Bowen Li
Changlin Li
Chao Li
Chongyi Li
Guanyue Li
Shuai Li
Jin Li
Dingquan Li
Dongxu Li
Yiting Li
Gang Li
Dian Li
Guohao Li
Haoang Li
Haoliang Li
Haoran Li
Hengduo Li
Huafeng Li
Xiaoming Li
Hanao Li
Hongwei Li
Ziqiang Li
Jisheng Li
Jiacheng Li
Jia Li
Jiachen Li
Jiahao Li
Jianwei Li
Jiazhi Li
Jie Li
Jing Li
Jingjing Li
Jingtao Li
Jun Li
Junxuan Li
Kai Li

Kailin Li
Kenneth Li
Kun Li
Kunpeng Li
Aoxue Li
Chenglong Li
Chenglin Li
Changsheng Li
Zhichao Li
Qiang Li
Yanyu Li
Zuoyue Li
Xiang Li
Xuelong Li
Fangda Li
Ailin Li
Liang Li
Chun-Guang Li
Daiqing Li
Dong Li
Guanbin Li
Guorong Li
Haifeng Li
Jianan Li
Jianing Li
Jiaxin Li
Ke Li
Lei Li
Lincheng Li
Liulei Li
Lujun Li
Linjie Li
Lin Li
Pengyu Li
Ping Li
Qiufu Li
Qingyong Li
Rui Li
Siyuan Li
Wei Li
Wenbin Li
Xiangyang Li
Xinyu Li
Xiujun Li
Xiu Li

Xu Li
Ya-Li Li
Yao Li
Yongjie Li
Yijun Li
Yiming Li
Yuezun Li
Yu Li
Yunheng Li
Yuqi Li
Zhe Li
Zeming Li
Zhen Li
Zhengqin Li
Zhimin Li
Jiefeng Li
Jinpeng Li
Chengze Li
Jianwu Li
Lerenhan Li
Shan Li
Suichan Li
Xiangtai Li
Yanjie Li
Yandong Li
Zhuoling Li
Zhenqiang Li
Manyi Li
Maosen Li
Ji Li
Minjun Li
Mingrui Li
Mengtian Li
Junyi Li
Nianyi Li
Bo Li
Xiao Li
Peihua Li
Peike Li
Peizhao Li
Peiliang Li
Qi Li
Ren Li
Runze Li
Shile Li

Sheng Li
Shigang Li
Shiyu Li
Shuang Li
Shasha Li
Shichao Li
Tianye Li
Yuexiang Li
Wei-Hong Li
Wanhua Li
Weihao Li
Weiming Li
Weixin Li
Wenbo Li
Wenshuo Li
Weijian Li
Yunan Li
Xirong Li
Xianhang Li
Xiaoyu Li
Xueqian Li
Xuanlin Li
Xianzhi Li
Yunqiang Li
Yanjing Li
Yansheng Li
Yawei Li
Yi Li
Yong Li
Yong-Lu Li
Yuhang Li
Yu-Jhe Li
Yuxi Li
Yunsheng Li
Yanwei Li
Zechao Li
Zejian Li
Zeju Li
Zekun Li
Zhaowen Li
Zheng Li
Zhenyu Li
Zhiheng Li
Zhi Li
Zhong Li

Zhuowei Li
Zhuowan Li
Zhuohang Li
Zizhang Li
Chen Li
Yuan-Fang Li
Dongze Lian
Xiaochen Lian
Zhouhui Lian
Long Lian
Qing Lian
Jin Lianbao
Jinxiu S. Liang
Dingkang Liang
Jiahao Liang
Jianming Liang
Jingyun Liang
Kevin J. Liang
Kaizhao Liang
Chen Liang
Jie Liang
Senwei Liang
Ding Liang
Jiajun Liang
Jian Liang
Kongming Liang
Siyuan Liang
Yuanzhi Liang
Zhengfa Liang
Mingfu Liang
Xiaodan Liang
Xuefeng Liang
Yuxuan Liang
Kang Liao
Liang Liao
Hong-Yuan Mark Liao
Wentong Liao
Haofu Liao
Yue Liao
Minghui Liao
Shengcai Liao
Ting-Hsuan Liao
Xin Liao
Yinghong Liao
Teck Yian Lim

Che-Tsung Lin
Chung-Ching Lin
Chen-Hsuan Lin
Cheng Lin
Chuming Lin
Chunyu Lin
Dahua Lin
Wei Lin
Zheng Lin
Huaijia Lin
Jason Lin
Jierui Lin
Jiaying Lin
Jie Lin
Kai-En Lin
Kevin Lin
Guangfeng Lin
Jiehong Lin
Feng Lin
Hang Lin
Kwan-Yee Lin
Ke Lin
Luojun Lin
Qinghong Lin
Xiangbo Lin
Yi Lin
Zudi Lin
Shijie Lin
Yiqun Lin
Tzu-Heng Lin
Ming Lin
Shaohui Lin
SongNan Lin
Ji Lin
Tsung-Yu Lin
Xudong Lin
Yancong Lin
Yen-Chen Lin
Yiming Lin
Yuewei Lin
Zhiqiu Lin
Zinan Lin
Zhe Lin
David B. Lindell
Zhixin Ling

Zhan Ling
Alexander Liniger
Venice Erin B. Liong
Joey Litalien
Or Litany
Roee Litman
Ron Litman
Jim Little
Dor Litvak
Shaoteng Liu
Shuaicheng Liu
Andrew Liu
Xian Liu
Shaohui Liu
Bei Liu
Bo Liu
Yong Liu
Ming Liu
Yanbin Liu
Chenxi Liu
Daqi Liu
Di Liu
Difan Liu
Dong Liu
Dongfang Liu
Daizong Liu
Xiao Liu
Fangyi Liu
Fengbei Liu
Fenglin Liu
Bin Liu
Yuang Liu
Ao Liu
Hong Liu
Hongfu Liu
Huidong Liu
Ziyi Liu
Feng Liu
Hao Liu
Jie Liu
Jialun Liu
Jiang Liu
Jing Liu
Jingya Liu
Jiaming Liu

Jun Liu
Juncheng Liu
Jiawei Liu
Hongyu Liu
Chuanbin Liu
Haotian Liu
Lingqiao Liu
Chang Liu
Han Liu
Liu Liu
Min Liu
Yingqi Liu
Aishan Liu
Bingyu Liu
Benlin Liu
Boxiao Liu
Chenchen Liu
Chuanjian Liu
Daqing Liu
Huan Liu
Haozhe Liu
Jiaheng Liu
Wei Liu
Jingzhou Liu
Jiyuan Liu
Lingbo Liu
Nian Liu
Peiye Liu
Qiankun Liu
Shenglan Liu
Shilong Liu
Wen Liu
Wenyu Liu
Weifeng Liu
Wu Liu
Xiaolong Liu
Yang Liu
Yanwei Liu
Yingcheng Liu
Yongfei Liu
Yihao Liu
Yu Liu
Yunze Liu
Ze Liu
Zhenhua Liu

Zhenguang Liu
Lin Liu
Lihao Liu
Pengju Liu
Xinhai Liu
Yunfei Liu
Meng Liu
Minghua Liu
Mingyuan Liu
Miao Liu
Peirong Liu
Ping Liu
Qingjie Liu
Ruoshi Liu
Risheng Liu
Songtao Liu
Xing Liu
Shikun Liu
Shuming Liu
Sheng Liu
Songhua Liu
Tongliang Liu
Weibo Liu
Weide Liu
Weizhe Liu
Wenxi Liu
Weiyang Liu
Xin Liu
Xiaobin Liu
Xudong Liu
Xiaoyi Liu
Xihui Liu
Xinchen Liu
Xingtong Liu
Xinpeng Liu
Xinyu Liu
Xianpeng Liu
Xu Liu
Xingyu Liu
Yongtuo Liu
Yahui Liu
Yangxin Liu
Yaoyao Liu
Yaojie Liu
Yuliang Liu

Yongcheng Liu

Yuan Liu

Yufan Liu

Yu-Lun Liu

Yun Liu

Yunfan Liu

Yuanzhong Liu

Zhuoran Liu

Zhen Liu

Zheng Liu

Zhijian Liu

Zhisong Liu

Ziquan Liu

Ziyu Liu

Zhihua Liu

Zechun Liu

Zhaoyang Liu

Zhengzhe Liu

Stephan Liwicki

Shao-Yuan Lo

Sylvain Lobry

Suhas Lohit

Vishnu Suresh Lokhande

Vincenzo Lomonaco

Chengjiang Long

Guodong Long

Fuchen Long

Shangbang Long

Yang Long

Zijun Long

Vasco Lopes

Antonio M. Lopez

Roberto Javier
   Lopez-Sastre

Tobias Lorenz

Javier Lorenzo-Navarro

Yujing Lou

Qian Lou

Xiankai Lu

Changsheng Lu

Huimin Lu

Yongxi Lu

Hao Lu

Hong Lu

Jiasen Lu

Juwei Lu

Fan Lu

Guangming Lu

Jiwen Lu

Shun Lu

Tao Lu

Xiaonan Lu

Yang Lu

Yao Lu

Yongchun Lu

Zhiwu Lu

Cheng Lu

Liying Lu

Guo Lu

Xuequan Lu

Yanye Lu

Yantao Lu

Yuhang Lu

Fujun Luan

Jonathon Luiten

Jovita Lukasik

Alan Lukezic

Jonathan Samuel Lumentut

Mayank Lunayach

Ao Luo

Canjie Luo

Chong Luo

Xu Luo

Grace Luo

Jun Luo

Katie Z. Luo

Tao Luo

Cheng Luo

Fangzhou Luo

Gen Luo

Lei Luo

Sihui Luo

Weixin Luo

Yan Luo

Xiaoyan Luo

Yong Luo

Yadan Luo

Hao Luo

Ruotian Luo

Mi Luo

Tiange Luo

Wenjie Luo

Wenhan Luo

Xiao Luo

Zhiming Luo

Zhipeng Luo

Zhengyi Luo

Diogo C. Luvizon

Zhaoyang Lv

Gengyu Lyu

Lingjuan Lyu

Jun Lyu

Yuanyuan Lyu

Youwei Lyu

Yueming Lyu

Bingpeng Ma

Chao Ma

Chongyang Ma

Congbo Ma

Chih-Yao Ma

Fan Ma

Lin Ma

Haoyu Ma

Hengbo Ma

Jianqi Ma

Jiawei Ma

Jiayi Ma

Kede Ma

Kai Ma

Lingni Ma

Lei Ma

Xu Ma

Ning Ma

Benteng Ma

Cheng Ma

Andy J. Ma

Long Ma

Zhanyu Ma

Zhiheng Ma

Qianli Ma

Shiqiang Ma

Sizhuo Ma

Shiqing Ma

Xiaolong Ma

Xinzhu Ma

Gautam B. Machiraju
Spandan Madan
Mathew Magimai-Doss
Luca Magri
Behrooz Mahasseni
Upal Mahbub
Siddharth Mahendran
Paridhi Maheshwari
Rishabh Maheshwary
Mohammed Mahmoud
Shishira R. R. Maiya
Sylwia Majchrowska
Arjun Majumdar
Puspita Majumdar
Orchid Majumder
Sagnik Majumder
Ilya Makarov
Farkhod F.
  Makhmudkhujaev
Yasushi Makihara
Ankur Mali
Mateusz Malinowski
Utkarsh Mall
Srikanth Malla
Clement Mallet
Dimitrios Mallis
Yunze Man
Dipu Manandhar
Massimiliano Mancini
Murari Mandal
Raunak Manekar
Karttikeya Mangalam
Puneet Mangla
Fabian Manhardt
Sivabalan Manivasagam
Fahim Mannan
Chengzhi Mao
Hanzi Mao
Jiayuan Mao
Junhua Mao
Zhiyuan Mao
Jiageng Mao
Yunyao Mao
Zhendong Mao
Alberto Marchisio

Diego Marcos
Riccardo Marin
Aram Markosyan
Renaud Marlet
Ricardo Marques
Miquel Martí i Rabadán
Diego Martin Arroyo
Niki Martinel
Brais Martinez
Julieta Martinez
Marc Masana
Tomohiro Mashita
Timothée Masquelier
Minesh Mathew
Tetsu Matsukawa
Marwan Mattar
Bruce A. Maxwell
Christoph Mayer
Mantas Mazeika
Pratik Mazumder
Scott McCloskey
Steven McDonagh
Ishit Mehta
Jie Mei
Kangfu Mei
Jieru Mei
Xiaoguang Mei
Givi Meishvili
Luke Melas-Kyriazi
Iaroslav Melekhov
Andres Mendez-Vazquez
Heydi Mendez-Vazquez
Matias Mendieta
Ricardo A. Mendoza-León
Chenlin Meng
Depu Meng
Rang Meng
Zibo Meng
Qingjie Meng
Qier Meng
Yanda Meng
Zihang Meng
Thomas Mensink
Fabian Mentzer
Christopher Metzler

Gregory P. Meyer
Vasileios Mezaris
Liang Mi
Lu Mi
Bo Miao
Changtao Miao
Zichen Miao
Qiguang Miao
Xin Miao
Zhongqi Miao
Frank Michel
Simone Milani
Ben Mildenhall
Roy V. Miles
Juhong Min
Kyle Min
Hyun-Seok Min
Weiqing Min
Yuecong Min
Zhixiang Min
Qi Ming
David Minnen
Aymen Mir
Deepak Mishra
Anand Mishra
Shlok K. Mishra
Niluthpol Mithun
Gaurav Mittal
Trisha Mittal
Daisuke Miyazaki
Kaichun Mo
Hong Mo
Zhipeng Mo
Davide Modolo
Abduallah A. Mohamed
Mohamed Afham
  Mohamed Aflal
Ron Mokady
Pavlo Molchanov
Davide Moltisanti
Liliane Momeni
Gianluca Monaci
Pascal Monasse
Ajoy Mondal
Tom Monnier

Aron Monszpart
Gyeongsik Moon
Suhong Moon
Taesup Moon
Sean Moran
Daniel Moreira
Pietro Morerio
Alexandre Morgand
Lia Morra
Ali Mosleh
Inbar Mosseri
Sayed Mohammad
   Mostafavi Isfahani
Saman Motamed
Ramy A. Mounir
Fangzhou Mu
Jiteng Mu
Norman Mu
Yasuhiro Mukaigawa
Ryan Mukherjee
Tanmoy Mukherjee
Yusuke Mukuta
Ravi Teja Mullapudi
Lea Muller
Matthias Müller
Martin Mundt
Nils Murrugarra-Llerena
Damien Muselet
Armin Mustafa
Muhammad Ferjad Naeem
Sauradip Nag
Hajime Nagahara
Pravin Nagar
Rajendra Nagar
Naveen Shankar Nagaraja
Varun Nagaraja
Tushar Nagarajan
Seungjun Nah
Gaku Nakano
Yuta Nakashima
Giljoo Nam
Seonghyeon Nam
Liangliang Nan
Yuesong Nan
Yeshwanth Napolean

Dinesh Reddy
   Narapureddy
Medhini Narasimhan
Supreeth
   Narasimhaswamy
Sriram Narayanan
Erickson R. Nascimento
Varun Nasery
K. L. Navaneet
Pablo Navarrete Michelini
Shant Navasardyan
Shah Nawaz
Nihal Nayak
Farhood Negin
Lukáš Neumann
Alejandro Newell
Evonne Ng
Kam Woh Ng
Tony Ng
Anh Nguyen
Tuan Anh Nguyen
Cuong Cao Nguyen
Ngoc Cuong Nguyen
Thanh Nguyen
Khoi Nguyen
Phi Le Nguyen
Phong Ha Nguyen
Tam Nguyen
Truong Nguyen
Anh Tuan Nguyen
Rang Nguyen
Thao Thi Phuong Nguyen
Van Nguyen Nguyen
Zhen-Liang Ni
Yao Ni
Shijie Nie
Xuecheng Nie
Yongwei Nie
Weizhi Nie
Ying Nie
Yinyu Nie
Kshitij N. Nikhal
Simon Niklaus
Xuefei Ning
Jifeng Ning

Yotam Nitzan
Di Niu
Shuaicheng Niu
Li Niu
Wei Niu
Yulei Niu
Zhenxing Niu
Albert No
Shohei Nobuhara
Nicoletta Noceti
Junhyug Noh
Sotiris Nousias
Slawomir Nowaczyk
Ewa M. Nowara
Valsamis Ntouskos
Gilberto Ochoa-Ruiz
Ferda Ofli
Jihyong Oh
Sangyun Oh
Youngtaek Oh
Hiroki Ohashi
Takahiro Okabe
Kemal Oksuz
Fumio Okura
Daniel Olmeda Reino
Matthew Olson
Carl Olsson
Roy Or-El
Alessandro Ortis
Guillermo Ortiz-Jimenez
Magnus Oskarsson
Ahmed A. A. Osman
Martin R. Oswald
Mayu Otani
Naima Otberdout
Cheng Ouyang
Jiahong Ouyang
Wanli Ouyang
Andrew Owens
Poojan B. Oza
Mete Ozay
A. Cengiz Oztireli
Gautam Pai
Tomas Pajdla
Umapada Pal

Simone Palazzo
Luca Palmieri
Bowen Pan
Hao Pan
Lili Pan
Tai-Yu Pan
Liang Pan
Chengwei Pan
Yingwei Pan
Xuran Pan
Jinshan Pan
Xinyu Pan
Liyuan Pan
Xingang Pan
Xingjia Pan
Zhihong Pan
Zizheng Pan
Priyadarshini Panda
Rameswar Panda
Rohit Pandey
Kaiyue Pang
Bo Pang
Guansong Pang
Jiangmiao Pang
Meng Pang
Tianyu Pang
Ziqi Pang
Omiros Pantazis
Andreas Panteli
Maja Pantic
Marina Paolanti
Joao P. Papa
Samuele Papa
Mike Papadakis
Dim P. Papadopoulos
George Papandreou
Constantin Pape
Toufiq Parag
Chethan Parameshwara
Shaifali Parashar
Alejandro Pardo
Rishubh Parihar
Sarah Parisot
JaeYoo Park
Gyeong-Moon Park

Hyojin Park
Hyoungseob Park
Jongchan Park
Jae Sung Park
Kiru Park
Chunghyun Park
Kwanyong Park
Sunghyun Park
Sungrae Park
Seongsik Park
Sanghyun Park
Sungjune Park
Taesung Park
Gaurav Parmar
Paritosh Parmar
Alvaro Parra
Despoina Paschalidou
Or Patashnik
Shivansh Patel
Pushpak Pati
Prashant W. Patil
Vaishakh Patil
Suvam Patra
Jay Patravali
Badri Narayana Patro
Angshuman Paul
Sudipta Paul
Rémi Pautrat
Nick E. Pears
Adithya Pediredla
Wenjie Pei
Shmuel Peleg
Latha Pemula
Bo Peng
Houwen Peng
Yue Peng
Liangzu Peng
Baoyun Peng
Jun Peng
Pai Peng
Sida Peng
Xi Peng
Yuxin Peng
Songyou Peng
Wei Peng

Weiqi Peng
Wen-Hsiao Peng
Pramuditha Perera
Juan C. Perez
Eduardo Pérez Pellitero
Juan-Manuel Perez-Rua
Federico Pernici
Marco Pesavento
Stavros Petridis
Ilya A. Petrov
Vladan Petrovic
Mathis Petrovich
Suzanne Petryk
Hieu Pham
Quang Pham
Khoi Pham
Tung Pham
Huy Phan
Stephen Phillips
Cheng Perng Phoo
David Picard
Marco Piccirilli
Georg Pichler
A. J. Piergiovanni
Vipin Pillai
Silvia L. Pintea
Giovanni Pintore
Robinson Piramuthu
Fiora Pirri
Theodoros Pissas
Fabio Pizzati
Benjamin Planche
Bryan Plummer
Matteo Poggi
Ashwini Pokle
Georgy E. Ponimatkin
Adrian Popescu
Stefan Popov
Nikola Popović
Ronald Poppe
Angelo Porrello
Michael Potter
Charalambos Poullis
Hadi Pouransari
Omid Poursaeed

Shraman Pramanick
Mantini Pranav
Dilip K. Prasad
Meghshyam Prasad
B. H. Pawan Prasad
Shitala Prasad
Prateek Prasanna
Ekta Prashnani
Derek S. Prijatelj
Luke Y. Prince
Véronique Prinet
Victor Adrian Prisacariu
James Pritts
Thomas Probst
Sergey Prokudin
Rita Pucci
Chi-Man Pun
Matthew Purri
Haozhi Qi
Lu Qi
Lei Qi
Xianbiao Qi
Yonggang Qi
Yuankai Qi
Siyuan Qi
Guocheng Qian
Hangwei Qian
Qi Qian
Deheng Qian
Shengsheng Qian
Wen Qian
Rui Qian
Yiming Qian
Shengju Qian
Shengyi Qian
Xuelin Qian
Zhenxing Qian
Nan Qiao
Xiaotian Qiao
Jing Qin
Can Qin
Siyang Qin
Hongwei Qin
Jie Qin
Minghai Qin

Yipeng Qin
Yongqiang Qin
Wenda Qin
Xuebin Qin
Yuzhe Qin
Yao Qin
Zhenyue Qin
Zhiwu Qing
Heqian Qiu
Jiayan Qiu
Jielin Qiu
Yue Qiu
Jiaxiong Qiu
Zhongxi Qiu
Shi Qiu
Zhaofan Qiu
Zhongnan Qu
Yanyun Qu
Kha Gia Quach
Yuhui Quan
Ruijie Quan
Mike Rabbat
Rahul Shekhar Rade
Filip Radenovic
Gorjan Radevski
Bogdan Raducanu
Francesco Ragusa
Shafin Rahman
Md Mahfuzur Rahman
   Siddiquee
Hossein Rahmani
Kiran Raja
Sivaramakrishnan
   Rajaraman
Jathushan Rajasegaran
Adnan Siraj Rakin
Michaël Ramamonjisoa
Chirag A. Raman
Shanmuganathan Raman
Vignesh Ramanathan
Vasili Ramanishka
Vikram V. Ramaswamy
Merey Ramazanova
Jason Rambach
Sai Saketh Rambhatla

Clément Rambour
Ashwin Ramesh Babu
Adín Ramírez Rivera
Arianna Rampini
Haoxi Ran
Aakanksha Rana
Aayush Jung Bahadur
   Rana
Kanchana N. Ranasinghe
Aneesh Rangnekar
Samrudhdhi B. Rangrej
Harsh Rangwani
Viresh Ranjan
Anyi Rao
Yongming Rao
Carolina Raposo
Michalis Raptis
Amir Rasouli
Vivek Rathod
Adepu Ravi Sankar
Avinash Ravichandran
Bharadwaj Ravichandran
Dripta S. Raychaudhuri
Adria Recasens
Simon Reiß
Davis Rempe
Daxuan Ren
Jiawei Ren
Jimmy Ren
Sucheng Ren
Dayong Ren
Zhile Ren
Dongwei Ren
Qibing Ren
Pengfei Ren
Zhenwen Ren
Xuqian Ren
Yixuan Ren
Zhongzheng Ren
Ambareesh Revanur
Hamed Rezazadegan
   Tavakoli
Rafael S. Rezende
Wonjong Rhee
Alexander Richard

Christian Richardt
Stephan R. Richter
Benjamin Riggan
Dominik Rivoir
Mamshad Nayeem Rizve
Joshua D. Robinson
Joseph Robinson
Chris Rockwell
Ranga Rodrigo
Andres C. Rodriguez
Carlos Rodriguez-Pardo
Marcus Rohrbach
Gemma Roig
Yu Rong
David A. Ross
Mohammad Rostami
Edward Rosten
Karsten Roth
Anirban Roy
Debaditya Roy
Shuvendu Roy
Ahana Roy Choudhury
Aruni Roy Chowdhury
Denys Rozumnyi
Shulan Ruan
Wenjie Ruan
Patrick Ruhkamp
Danila Rukhovich
Anian Ruoss
Chris Russell
Dan Ruta
Dawid Damian Rymarczyk
DongHun Ryu
Hyeonggon Ryu
Kwonyoung Ryu
Balasubramanian S.
Alexandre Sablayrolles
Mohammad Sabokrou
Arka Sadhu
Aniruddha Saha
Oindrila Saha
Pritish Sahu
Aneeshan Sain
Nirat Saini
Saurabh Saini

Takeshi Saitoh
Christos Sakaridis
Fumihiko Sakaue
Dimitrios Sakkos
Ken Sakurada
Parikshit V. Sakurikar
Rohit Saluja
Nermin Samet
Leo Sampaio Ferraz
    Ribeiro
Jorge Sanchez
Enrique Sanchez
Shengtian Sang
Anush Sankaran
Soubhik Sanyal
Nikolaos Sarafianos
Vishwanath Saragadam
István Sárándi
Saquib Sarfraz
Mert Bulent Sariyildiz
Anindya Sarkar
Pritam Sarkar
Paul-Edouard Sarlin
Hiroshi Sasaki
Takami Sato
Torsten Sattler
Ravi Kumar Satzoda
Axel Sauer
Stefano Savian
Artem Savkin
Manolis Savva
Gerald Schaefer
Simone Schaub-Meyer
Yoni Schirris
Samuel Schulter
Katja Schwarz
Jesse Scott
Sinisa Segvic
Constantin Marc Seibold
Lorenzo Seidenari
Matan Sela
Fadime Sener
Paul Hongsuck Seo
Kwanggyoon Seo
Hongje Seong

Dario Serez
Francesco Setti
Bryan Seybold
Mohamad Shahbazi
Shima Shahfar
Xinxin Shan
Caifeng Shan
Dandan Shan
Shawn Shan
Wei Shang
Jinghuan Shang
Jiaxiang Shang
Lei Shang
Sukrit Shankar
Ken Shao
Rui Shao
Jie Shao
Mingwen Shao
Aashish Sharma
Gaurav Sharma
Vivek Sharma
Abhishek Sharma
Yoli Shavit
Shashank Shekhar
Sumit Shekhar
Zhijie Shen
Fengyi Shen
Furao Shen
Jialie Shen
Jingjing Shen
Ziyi Shen
Linlin Shen
Guangyu Shen
Biluo Shen
Falong Shen
Jiajun Shen
Qiu Shen
Qiuhong Shen
Shuai Shen
Wang Shen
Yiqing Shen
Yunhang Shen
Siqi Shen
Bin Shen
Tianwei Shen

Xi Shen
Yilin Shen
Yuming Shen
Yucong Shen
Zhiqiang Shen
Lu Sheng
Yichen Sheng
Shivanand Venkanna
  Sheshappanavar
Shelly Sheynin
Baifeng Shi
Ruoxi Shi
Botian Shi
Hailin Shi
Jia Shi
Jing Shi
Shaoshuai Shi
Baoguang Shi
Boxin Shi
Hengcan Shi
Tianyang Shi
Xiaodan Shi
Yongjie Shi
Zhensheng Shi
Yinghuan Shi
Weiqi Shi
Wu Shi
Xuepeng Shi
Xiaoshuang Shi
Yujiao Shi
Zenglin Shi
Zhenmei Shi
Takashi Shibata
Meng-Li Shih
Yichang Shih
Hyunjung Shim
Dongseok Shim
Soshi Shimada
Inkyu Shin
Jinwoo Shin
Seungjoo Shin
Seungjae Shin
Koichi Shinoda
Suprosanna Shit

Palaiahnakote
  Shivakumara
Eli Shlizerman
Gaurav Shrivastava
Xiao Shu
Xiangbo Shu
Xiujun Shu
Yang Shu
Tianmin Shu
Jun Shu
Zhixin Shu
Bing Shuai
Maria Shugrina
Ivan Shugurov
Satya Narayan Shukla
Pranjay Shyam
Jianlou Si
Yawar Siddiqui
Alberto Signoroni
Pedro Silva
Jae-Young Sim
Oriane Siméoni
Martin Simon
Andrea Simonelli
Abhishek Singh
Ashish Singh
Dinesh Singh
Gurkirt Singh
Krishna Kumar Singh
Mannat Singh
Pravendra Singh
Rajat Vikram Singh
Utkarsh Singhal
Dipika Singhania
Vasu Singla
Harsh Sinha
Sudipta Sinha
Josef Sivic
Elena Sizikova
Geri Skenderi
Ivan Skorokhodov
Dmitriy Smirnov
Cameron Y. Smith
James S. Smith
Patrick Snape

Mattia Soldan
Hyeongseok Son
Sanghyun Son
Chuanbiao Song
Chen Song
Chunfeng Song
Dan Song
Dongjin Song
Hwanjun Song
Guoxian Song
Jiaming Song
Jie Song
Liangchen Song
Ran Song
Luchuan Song
Xibin Song
Li Song
Fenglong Song
Guoli Song
Guanglu Song
Zhenbo Song
Lin Song
Xinhang Song
Yang Song
Yibing Song
Rajiv Soundararajan
Hossein Souri
Cristovao Sousa
Riccardo Spezialetti
Leonidas Spinoulas
Michael W. Spratling
Deepak Sridhar
Srinath Sridhar
Gaurang Sriramanan
Vinkle Kumar Srivastav
Themos Stafylakis
Serban Stan
Anastasis Stathopoulos
Markus Steinberger
Jan Steinbrener
Sinisa Stekovic
Alexandros Stergiou
Gleb Sterkin
Rainer Stiefelhagen
Pierre Stock

Ombretta Strafforello
Julian Straub
Yannick Strümpler
Joerg Stueckler
Hang Su
Weijie Su
Jong-Chyi Su
Bing Su
Haisheng Su
Jinming Su
Yiyang Su
Yukun Su
Yuxin Su
Zhuo Su
Zhaoqi Su
Xiu Su
Yu-Chuan Su
Zhixun Su
Arulkumar Subramaniam
Akshayvarun Subramanya
A. Subramanyam
Swathikiran Sudhakaran
Yusuke Sugano
Masanori Suganuma
Yumin Suh
Yang Sui
Baochen Sun
Cheng Sun
Long Sun
Guolei Sun
Haoliang Sun
Haomiao Sun
He Sun
Hanqing Sun
Hao Sun
Lichao Sun
Jiachen Sun
Jiaming Sun
Jian Sun
Jin Sun
Jennifer J. Sun
Tiancheng Sun
Libo Sun
Peize Sun
Qianru Sun

Shanlin Sun
Yu Sun
Zhun Sun
Che Sun
Lin Sun
Tao Sun
Yiyou Sun
Chunyi Sun
Chong Sun
Weiwei Sun
Weixuan Sun
Xiuyu Sun
Yanan Sun
Zeren Sun
Zhaodong Sun
Zhiqing Sun
Minhyuk Sung
Jinli Suo
Simon Suo
Abhijit Suprem
Anshuman Suri
Saksham Suri
Joshua M. Susskind
Roman Suvorov
Gurumurthy Swaminathan
Robin Swanson
Paul Swoboda
Tabish A. Syed
Richard Szeliski
Fariborz Taherkhani
Yu-Wing Tai
Keita Takahashi
Walter Talbott
Gary Tam
Masato Tamura
Feitong Tan
Fuwen Tan
Shuhan Tan
Andong Tan
Bin Tan
Cheng Tan
Jianchao Tan
Lei Tan
Mingxing Tan
Xin Tan

Zichang Tan
Zhentao Tan
Kenichiro Tanaka
Masayuki Tanaka
Yushun Tang
Hao Tang
Jingqun Tang
Jinhui Tang
Kaihua Tang
Luming Tang
Lv Tang
Sheyang Tang
Shitao Tang
Siliang Tang
Shixiang Tang
Yansong Tang
Keke Tang
Chang Tang
Chenwei Tang
Jie Tang
Junshu Tang
Ming Tang
Peng Tang
Xu Tang
Yao Tang
Chen Tang
Fan Tang
Haoran Tang
Shengeng Tang
Yehui Tang
Zhipeng Tang
Ugo Tanielian
Chaofan Tao
Jiale Tao
Junli Tao
Renshuai Tao
An Tao
Guanhong Tao
Zhiqiang Tao
Makarand Tapaswi
Jean-Philippe G. Tarel
Juan J. Tarrio
Enzo Tartaglione
Keisuke Tateno
Zachary Teed

Ajinkya B. Tejankar
Bugra Tekin
Purva Tendulkar
Damien Teney
Minggui Teng
Chris Tensmeyer
Andrew Beng Jin Teoh
Philipp Terhörst
Kartik Thakral
Nupur Thakur
Kevin Thandiackal
Spyridon Thermos
Diego Thomas
William Thong
Yuesong Tian
Guanzhong Tian
Lin Tian
Shiqi Tian
Kai Tian
Meng Tian
Tai-Peng Tian
Zhuotao Tian
Shangxuan Tian
Tian Tian
Yapeng Tian
Yu Tian
Yuxin Tian
Leslie Ching Ow Tiong
Praveen Tirupattur
Garvita Tiwari
George Toderici
Antoine Toisoul
Aysim Toker
Tatiana Tommasi
Zhan Tong
Alessio Tonioni
Alessandro Torcinovich
Fabio Tosi
Matteo Toso
Hugo Touvron
Quan Hung Tran
Son Tran
Hung Tran
Ngoc-Trung Tran
Vinh Tran

Phong Tran
Giovanni Trappolini
Edith Tretschk
Subarna Tripathi
Shubhendu Trivedi
Eduard Trulls
Prune Truong
Thanh-Dat Truong
Tomasz Trzcinski
Sam Tsai
Yi-Hsuan Tsai
Ethan Tseng
Yu-Chee Tseng
Shahar Tsiper
Stavros Tsogkas
Shikui Tu
Zhigang Tu
Zhengzhong Tu
Richard Tucker
Sergey Tulyakov
Cigdem Turan
Daniyar Turmukhambetov
Victor G. Turrisi da Costa
Bartlomiej Twardowski
Christopher D. Twigg
Radim Tylecek
Mostofa Rafid Uddin
Md. Zasim Uddin
Kohei Uehara
Nicolas Ugrinovic
Youngjung Uh
Norimichi Ukita
Anwaar Ulhaq
Devesh Upadhyay
Paul Upchurch
Yoshitaka Ushiku
Yuzuko Utsumi
Mikaela Angelina Uy
Mohit Vaishnav
Pratik Vaishnavi
Jeya Maria Jose Valanarasu
Matias A. Valdenegro Toro
Diego Valsesia
Wouter Van Gansbeke
Nanne van Noord

Simon Vandenhende
Farshid Varno
Cristina Vasconcelos
Francisco Vasconcelos
Alex Vasilescu
Subeesh Vasu
Arun Balajee Vasudevan
Kanav Vats
Vaibhav S. Vavilala
Sagar Vaze
Javier Vazquez-Corral
Andrea Vedaldi
Olga Veksler
Andreas Velten
Sai H. Vemprala
Raviteja Vemulapalli
Shashanka
   Venkataramanan
Dor Verbin
Luisa Verdoliva
Manisha Verma
Yashaswi Verma
Constantin Vertan
Eli Verwimp
Deepak Vijaykeerthy
Pablo Villanueva
Ruben Villegas
Markus Vincze
Vibhav Vineet
Minh P. Vo
Huy V. Vo
Duc Minh Vo
Tomas Vojir
Igor Vozniak
Nicholas Vretos
Vibashan VS
Tuan-Anh Vu
Thang Vu
Mårten Wadenbäck
Neal Wadhwa
Aaron T. Walsman
Steven Walton
Jin Wan
Alvin Wan
Jia Wan

Jun Wan
Xiaoyue Wan
Fang Wan
Guowei Wan
Renjie Wan
Zhiqiang Wan
Ziyu Wan
Bastian Wandt
Dongdong Wang
Limin Wang
Haiyang Wang
Xiaobing Wang
Angtian Wang
Angelina Wang
Bing Wang
Bo Wang
Boyu Wang
Binghui Wang
Chen Wang
Chien-Yi Wang
Congli Wang
Qi Wang
Chengrui Wang
Rui Wang
Yiqun Wang
Cong Wang
Wenjing Wang
Dongkai Wang
Di Wang
Xiaogang Wang
Kai Wang
Zhizhong Wang
Fangjinhua Wang
Feng Wang
Hang Wang
Gaoang Wang
Guoqing Wang
Guangcong Wang
Guangzhi Wang
Hanqing Wang
Hao Wang
Haohan Wang
Haoran Wang
Hong Wang
Haotao Wang

Hu Wang
Huan Wang
Hua Wang
Hui-Po Wang
Hengli Wang
Hanyu Wang
Hongxing Wang
Jingwen Wang
Jialiang Wang
Jian Wang
Jianyi Wang
Jiashun Wang
Jiahao Wang
Tsun-Hsuan Wang
Xiaoqian Wang
Jinqiao Wang
Jun Wang
Jianzong Wang
Kaihong Wang
Ke Wang
Lei Wang
Lingjing Wang
Linnan Wang
Lin Wang
Liansheng Wang
Mengjiao Wang
Manning Wang
Nannan Wang
Peihao Wang
Jiayun Wang
Pu Wang
Qiang Wang
Qiufeng Wang
Qilong Wang
Qiangchang Wang
Qin Wang
Qing Wang
Ruocheng Wang
Ruibin Wang
Ruisheng Wang
Ruizhe Wang
Runqi Wang
Runzhong Wang
Wenxuan Wang
Sen Wang

Shangfei Wang
Shaofei Wang
Shijie Wang
Shiqi Wang
Zhibo Wang
Song Wang
Xinjiang Wang
Tai Wang
Tao Wang
Teng Wang
Xiang Wang
Tianren Wang
Tiantian Wang
Tianyi Wang
Fengjiao Wang
Wei Wang
Miaohui Wang
Suchen Wang
Siyue Wang
Yaoming Wang
Xiao Wang
Ze Wang
Biao Wang
Chaofei Wang
Dong Wang
Gu Wang
Guangrun Wang
Guangming Wang
Guo-Hua Wang
Haoqing Wang
Hesheng Wang
Huafeng Wang
Jinghua Wang
Jingdong Wang
Jingjing Wang
Jingya Wang
Jingkang Wang
Jiakai Wang
Junke Wang
Kuo Wang
Lichen Wang
Lizhi Wang
Longguang Wang
Mang Wang
Mei Wang

Min Wang
Peng-Shuai Wang
Run Wang
Shaoru Wang
Shuhui Wang
Tan Wang
Tiancai Wang
Tianqi Wang
Wenhai Wang
Wenzhe Wang
Xiaobo Wang
Xiudong Wang
Xu Wang
Yajie Wang
Yan Wang
Yuan-Gen Wang
Yingqian Wang
Yizhi Wang
Yulin Wang
Yu Wang
Yujie Wang
Yunhe Wang
Yuxi Wang
Yaowei Wang
Yiwei Wang
Zezheng Wang
Hongzhi Wang
Zhiqiang Wang
Ziteng Wang
Ziwei Wang
Zheng Wang
Zhenyu Wang
Binglu Wang
Zhongdao Wang
Ce Wang
Weining Wang
Weiyao Wang
Wenbin Wang
Wenguan Wang
Guangting Wang
Haolin Wang
Haiyan Wang
Huiyu Wang
Naiyan Wang
Jingbo Wang

Jinpeng Wang
Jiaqi Wang
Liyuan Wang
Lizhen Wang
Ning Wang
Wenqian Wang
Sheng-Yu Wang
Weimin Wang
Xiaohan Wang
Yifan Wang
Yi Wang
Yongtao Wang
Yizhou Wang
Zhuo Wang
Zhe Wang
Xudong Wang
Xiaofang Wang
Xinggang Wang
Xiaosen Wang
Xiaosong Wang
Xiaoyang Wang
Lijun Wang
Xinlong Wang
Xuan Wang
Xue Wang
Yangang Wang
Yaohui Wang
Yu-Chiang Frank Wang
Yida Wang
Yilin Wang
Yi Ru Wang
Yali Wang
Yinglong Wang
Yufu Wang
Yujiang Wang
Yuwang Wang
Yuting Wang
Yang Wang
Yu-Xiong Wang
Yixu Wang
Ziqi Wang
Zhicheng Wang
Zeyu Wang
Zhaowen Wang
Zhenyi Wang

Zhenzhi Wang
Zhijie Wang
Zhiyong Wang
Zhongling Wang
Zhuowei Wang
Zian Wang
Zifu Wang
Zihao Wang
Zirui Wang
Ziyan Wang
Wenxiao Wang
Zhen Wang
Zhepeng Wang
Zi Wang
Zihao W. Wang
Steven L. Waslander
Olivia Watkins
Daniel Watson
Silvan Weder
Dongyoon Wee
Dongming Wei
Tianyi Wei
Jia Wei
Dong Wei
Fangyun Wei
Longhui Wei
Mingqiang Wei
Xinyue Wei
Chen Wei
Donglai Wei
Pengxu Wei
Xing Wei
Xiu-Shen Wei
Wenqi Wei
Guoqiang Wei
Wei Wei
XingKui Wei
Xian Wei
Xingxing Wei
Yake Wei
Yuxiang Wei
Yi Wei
Luca Weihs
Michael Weinmann
Martin Weinmann

Congcong Wen
Chuan Wen
Jie Wen
Sijia Wen
Song Wen
Chao Wen
Xiang Wen
Zeyi Wen
Xin Wen
Yilin Wen
Yijia Weng
Shuchen Weng
Junwu Weng
Wenming Weng
Renliang Weng
Zhenyu Weng
Xinshuo Weng
Nicholas J. Westlake
Gordon Wetzstein
Lena M. Widin Klasén
Rick Wildes
Bryan M. Williams
Williem Williem
Ole Winther
Scott Wisdom
Alex Wong
Chau-Wai Wong
Kwan-Yee K. Wong
Yongkang Wong
Scott Workman
Marcel Worring
Michael Wray
Safwan Wshah
Xiang Wu
Aming Wu
Chongruo Wu
Cho-Ying Wu
Chunpeng Wu
Chenyan Wu
Ziyi Wu
Fuxiang Wu
Gang Wu
Haiping Wu
Huisi Wu
Jane Wu

Jialian Wu
Jing Wu
Jinjian Wu
Jianlong Wu
Xian Wu
Lifang Wu
Lifan Wu
Minye Wu
Qianyi Wu
Rongliang Wu
Rui Wu
Shiqian Wu
Shuzhe Wu
Shangzhe Wu
Tsung-Han Wu
Tz-Ying Wu
Ting-Wei Wu
Jiannan Wu
Zhiliang Wu
Yu Wu
Chenyun Wu
Dayan Wu
Dongxian Wu
Fei Wu
Hefeng Wu
Jianxin Wu
Weibin Wu
Wenxuan Wu
Wenhao Wu
Xiao Wu
Yicheng Wu
Yuanwei Wu
Yu-Huan Wu
Zhenxin Wu
Zhenyu Wu
Wei Wu
Peng Wu
Xiaohe Wu
Xindi Wu
Xinxing Wu
Xinyi Wu
Xingjiao Wu
Xiongwei Wu
Yangzheng Wu
Yanzhao Wu

Yawen Wu
Yong Wu
Yi Wu
Ying Nian Wu
Zhenyao Wu
Zhonghua Wu
Zongze Wu
Zuxuan Wu
Stefanie Wuhrer
Teng Xi
Jianing Xi
Fei Xia
Haifeng Xia
Menghan Xia
Yuanqing Xia
Zhihua Xia
Xiaobo Xia
Weihao Xia
Shihong Xia
Yan Xia
Yong Xia
Zhaoyang Xia
Zhihao Xia
Chuhua Xian
Yongqin Xian
Wangmeng Xiang
Fanbo Xiang
Tiange Xiang
Tao Xiang
Liuyu Xiang
Xiaoyu Xiang
Zhiyu Xiang
Aoran Xiao
Chunxia Xiao
Fanyi Xiao
Jimin Xiao
Jun Xiao
Taihong Xiao
Anqi Xiao
Junfei Xiao
Jing Xiao
Liang Xiao
Yang Xiao
Yuting Xiao
Yijun Xiao

Yao Xiao
Zeyu Xiao
Zhisheng Xiao
Zihao Xiao
Binhui Xie
Christopher Xie
Haozhe Xie
Jin Xie
Guo-Sen Xie
Hongtao Xie
Ming-Kun Xie
Tingting Xie
Chaohao Xie
Weicheng Xie
Xudong Xie
Jiyang Xie
Xiaohua Xie
Yuan Xie
Zhenyu Xie
Ning Xie
Xianghui Xie
Xiufeng Xie
You Xie
Yutong Xie
Fuyong Xing
Yifan Xing
Zhen Xing
Yuanjun Xiong
Jinhui Xiong
Weihua Xiong
Hongkai Xiong
Zhitong Xiong
Yuanhao Xiong
Yunyang Xiong
Yuwen Xiong
Zhiwei Xiong
Yuliang Xiu
An Xu
Chang Xu
Chenliang Xu
Chengming Xu
Chenshu Xu
Xiang Xu
Huijuan Xu
Zhe Xu

Jie Xu
Jingyi Xu
Jiarui Xu
Yinghao Xu
Kele Xu
Ke Xu
Li Xu
Linchuan Xu
Linning Xu
Mengde Xu
Mengmeng Frost Xu
Min Xu
Mingye Xu
Jun Xu
Ning Xu
Peng Xu
Runsheng Xu
Sheng Xu
Wenqiang Xu
Xiaogang Xu
Renzhe Xu
Kaidi Xu
Yi Xu
Chi Xu
Qiuling Xu
Baobei Xu
Feng Xu
Haohang Xu
Haofei Xu
Lan Xu
Mingze Xu
Songcen Xu
Weipeng Xu
Wenjia Xu
Wenju Xu
Xiangyu Xu
Xin Xu
Yinshuang Xu
Yixing Xu
Yuting Xu
Yanyu Xu
Zhenbo Xu
Zhiliang Xu
Zhiyuan Xu
Xiaohao Xu

Yanwu Xu
Yan Xu
Yiran Xu
Yifan Xu
Yufei Xu
Yong Xu
Zichuan Xu
Zenglin Xu
Zexiang Xu
Zhan Xu
Zheng Xu
Zhiwei Xu
Ziyue Xu
Shiyu Xuan
Hanyu Xuan
Fei Xue
Jianru Xue
Mingfu Xue
Qinghan Xue
Tianfan Xue
Chao Xue
Chuhui Xue
Nan Xue
Zhou Xue
Xiangyang Xue
Yuan Xue
Abhay Yadav
Ravindra Yadav
Kota Yamaguchi
Toshihiko Yamasaki
Kohei Yamashita
Chaochao Yan
Feng Yan
Kun Yan
Qingsen Yan
Qixin Yan
Rui Yan
Siming Yan
Xinchen Yan
Yaping Yan
Bin Yan
Qingan Yan
Shen Yan
Shipeng Yan
Xu Yan

Yan Yan
Yichao Yan
Zhaoyi Yan
Zike Yan
Zhiqiang Yan
Hongliang Yan
Zizheng Yan
Jiewen Yang
Anqi Joyce Yang
Shan Yang
Anqi Yang
Antoine Yang
Bo Yang
Baoyao Yang
Chenhongyi Yang
Dingkang Yang
De-Nian Yang
Dong Yang
David Yang
Fan Yang
Fengyu Yang
Fengting Yang
Fei Yang
Gengshan Yang
Heng Yang
Han Yang
Huan Yang
Yibo Yang
Jiancheng Yang
Jihan Yang
Jiawei Yang
Jiayu Yang
Jie Yang
Jinfa Yang
Jingkang Yang
Jinyu Yang
Cheng-Fu Yang
Ji Yang
Jianyu Yang
Kailun Yang
Tian Yang
Luyu Yang
Liang Yang
Li Yang
Michael Ying Yang

Yang Yang
Muli Yang
Le Yang
Qiushi Yang
Ren Yang
Ruihan Yang
Shuang Yang
Siyuan Yang
Su Yang
Shiqi Yang
Taojiannan Yang
Tianyu Yang
Lei Yang
Wanzhao Yang
Shuai Yang
William Yang
Wei Yang
Xiaofeng Yang
Xiaoshan Yang
Xin Yang
Xuan Yang
Xu Yang
Xingyi Yang
Xitong Yang
Jing Yang
Yanchao Yang
Wenming Yang
Yujiu Yang
Herb Yang
Jianfei Yang
Jinhui Yang
Chuanguang Yang
Guanglei Yang
Haitao Yang
Kewei Yang
Linlin Yang
Lijin Yang
Longrong Yang
Meng Yang
MingKun Yang
Sibei Yang
Shicai Yang
Tong Yang
Wen Yang
Xi Yang

Xiaolong Yang
Xue Yang
Yubin Yang
Ze Yang
Ziyi Yang
Yi Yang
Linjie Yang
Yuzhe Yang
Yiding Yang
Zhenpei Yang
Zhaohui Yang
Zhengyuan Yang
Zhibo Yang
Zongxin Yang
Hantao Yao
Mingde Yao
Rui Yao
Taiping Yao
Ting Yao
Cong Yao
Qingsong Yao
Quanming Yao
Xu Yao
Yuan Yao
Yao Yao
Yazhou Yao
Jiawen Yao
Shunyu Yao
Pew-Thian Yap
Sudhir Yarram
Rajeev Yasarla
Peng Ye
Botao Ye
Mao Ye
Fei Ye
Hanrong Ye
Jingwen Ye
Jinwei Ye
Jiarong Ye
Mang Ye
Meng Ye
Qi Ye
Qian Ye
Qixiang Ye
Junjie Ye

Sheng Ye
Nanyang Ye
Yufei Ye
Xiaoqing Ye
Ruolin Ye
Yousef Yeganeh
Chun-Hsiao Yeh
Raymond A. Yeh
Yu-Ying Yeh
Kai Yi
Chang Yi
Renjiao Yi
Xinping Yi
Peng Yi
Alper Yilmaz
Junho Yim
Hui Yin
Bangjie Yin
Jia-Li Yin
Miao Yin
Wenzhe Yin
Xuwang Yin
Ming Yin
Yu Yin
Aoxiong Yin
Kangxue Yin
Tianwei Yin
Wei Yin
Xianghua Ying
Rio Yokota
Tatsuya Yokota
Naoto Yokoya
Ryo Yonetani
Ki Yoon Yoo
Jinsu Yoo
Sunjae Yoon
Jae Shin Yoon
Jihun Yoon
Sung-Hoon Yoon
Ryota Yoshihashi
Yusuke Yoshiyasu
Chenyu You
Haoran You
Haoxuan You
Yang You

Quanzeng You
Tackgeun You
Kaichao You
Shan You
Xinge You
Yurong You
Baosheng Yu
Bei Yu
Haichao Yu
Hao Yu
Chaohui Yu
Fisher Yu
Jin-Gang Yu
Jiyang Yu
Jason J. Yu
Jiashuo Yu
Hong-Xing Yu
Lei Yu
Mulin Yu
Ning Yu
Peilin Yu
Qi Yu
Qian Yu
Rui Yu
Shuzhi Yu
Gang Yu
Tan Yu
Weijiang Yu
Xin Yu
Bingyao Yu
Ye Yu
Hanchao Yu
Yingchen Yu
Tao Yu
Xiaotian Yu
Qing Yu
Houjian Yu
Changqian Yu
Jing Yu
Jun Yu
Shujian Yu
Xiang Yu
Zhaofei Yu
Zhenbo Yu
Yinfeng Yu

Zhuoran Yu
Zitong Yu
Bo Yuan
Jiangbo Yuan
Liangzhe Yuan
Weihao Yuan
Jianbo Yuan
Xiaoyun Yuan
Ye Yuan
Li Yuan
Geng Yuan
Jialin Yuan
Maoxun Yuan
Peng Yuan
Xin Yuan
Yuan Yuan
Yuhui Yuan
Yixuan Yuan
Zheng Yuan
Mehmet Kerim Yücel
Kaiyu Yue
Haixiao Yue
Heeseung Yun
Sangdoo Yun
Tian Yun
Mahmut Yurt
Ekim Yurtsever
Ahmet Yüzügüler
Edouard Yvinec
Eloi Zablocki
Christopher Zach
Muhammad Zaigham
   Zaheer
Pierluigi Zama Ramirez
Yuhang Zang
Pietro Zanuttigh
Alexey Zaytsev
Bernhard Zeisl
Haitian Zeng
Pengpeng Zeng
Jiabei Zeng
Runhao Zeng
Wei Zeng
Yawen Zeng
Yi Zeng

Yiming Zeng
Tieyong Zeng
Huanqiang Zeng
Dan Zeng
Yu Zeng
Wei Zhai
Yuanhao Zhai
Fangneng Zhan
Kun Zhan
Xiong Zhang
Jingdong Zhang
Jiangning Zhang
Zhilu Zhang
Gengwei Zhang
Dongsu Zhang
Hui Zhang
Binjie Zhang
Bo Zhang
Tianhao Zhang
Cecilia Zhang
Jing Zhang
Chaoning Zhang
Chenxu Zhang
Chi Zhang
Chris Zhang
Yabin Zhang
Zhao Zhang
Rufeng Zhang
Chaoyi Zhang
Zheng Zhang
Da Zhang
Yi Zhang
Edward Zhang
Xin Zhang
Feifei Zhang
Feilong Zhang
Yuqi Zhang
GuiXuan Zhang
Hanlin Zhang
Hanwang Zhang
Hanzhen Zhang
Haotian Zhang
He Zhang
Haokui Zhang
Hongyuan Zhang

Hengrui Zhang
Hongming Zhang
Mingfang Zhang
Jianpeng Zhang
Jiaming Zhang
Jichao Zhang
Jie Zhang
Jingfeng Zhang
Jingyi Zhang
Jinnian Zhang
David Junhao Zhang
Junjie Zhang
Junzhe Zhang
Jiawan Zhang
Jingyang Zhang
Kai Zhang
Lei Zhang
Lihua Zhang
Lu Zhang
Miao Zhang
Minjia Zhang
Mingjin Zhang
Qi Zhang
Qian Zhang
Qilong Zhang
Qiming Zhang
Qiang Zhang
Richard Zhang
Ruimao Zhang
Ruisi Zhang
Ruixin Zhang
Runze Zhang
Qilin Zhang
Shan Zhang
Shanshan Zhang
Xi Sheryl Zhang
Song-Hai Zhang
Chongyang Zhang
Kaihao Zhang
Songyang Zhang
Shu Zhang
Siwei Zhang
Shujian Zhang
Tianyun Zhang
Tong Zhang

Tao Zhang
Wenwei Zhang
Wenqiang Zhang
Wen Zhang
Xiaolin Zhang
Xingchen Zhang
Xingxuan Zhang
Xiuming Zhang
Xiaoshuai Zhang
Xuanmeng Zhang
Xuanyang Zhang
Xucong Zhang
Xingxing Zhang
Xikun Zhang
Xiaohan Zhang
Yahui Zhang
Yunhua Zhang
Yan Zhang
Yanghao Zhang
Yifei Zhang
Yifan Zhang
Yi-Fan Zhang
Yihao Zhang
Yingliang Zhang
Youshan Zhang
Yulun Zhang
Yushu Zhang
Yixiao Zhang
Yide Zhang
Zhongwen Zhang
Bowen Zhang
Chen-Lin Zhang
Zehua Zhang
Zekun Zhang
Zeyu Zhang
Xiaowei Zhang
Yifeng Zhang
Cheng Zhang
Hongguang Zhang
Yuexi Zhang
Fa Zhang
Guofeng Zhang
Hao Zhang
Haofeng Zhang
Hongwen Zhang

Hua Zhang
Jiaxin Zhang
Zhenyu Zhang
Jian Zhang
Jianfeng Zhang
Jiao Zhang
Jiakai Zhang
Lefei Zhang
Le Zhang
Mi Zhang
Min Zhang
Ning Zhang
Pan Zhang
Pu Zhang
Qing Zhang
Renrui Zhang
Shifeng Zhang
Shuo Zhang
Shaoxiong Zhang
Weizhong Zhang
Xi Zhang
Xiaomei Zhang
Xinyu Zhang
Yin Zhang
Zicheng Zhang
Zihao Zhang
Ziqi Zhang
Zhaoxiang Zhang
Zhen Zhang
Zhipeng Zhang
Zhixing Zhang
Zhizheng Zhang
Jiawei Zhang
Zhong Zhang
Pingping Zhang
Yixin Zhang
Kui Zhang
Lingzhi Zhang
Huaiwen Zhang
Quanshi Zhang
Zhoutong Zhang
Yuhang Zhang
Yuting Zhang
Zhang Zhang
Ziming Zhang

Zhizhong Zhang
Qilong Zhangli
Bingyin Zhao
Bin Zhao
Chenglong Zhao
Lei Zhao
Feng Zhao
Gangming Zhao
Haiyan Zhao
Hao Zhao
Handong Zhao
Hengshuang Zhao
Yinan Zhao
Jiaojiao Zhao
Jiaqi Zhao
Jing Zhao
Kaili Zhao
Haojie Zhao
Yucheng Zhao
Longjiao Zhao
Long Zhao
Qingsong Zhao
Qingyu Zhao
Rui Zhao
Rui-Wei Zhao
Sicheng Zhao
Shuang Zhao
Siyan Zhao
Zelin Zhao
Shiyu Zhao
Wang Zhao
Tiesong Zhao
Qian Zhao
Wangbo Zhao
Xi-Le Zhao
Xu Zhao
Yajie Zhao
Yang Zhao
Ying Zhao
Yin Zhao
Yizhou Zhao
Yunhan Zhao
Yuyang Zhao
Yue Zhao
Yuzhi Zhao

Bowen Zhao
Pu Zhao
Bingchen Zhao
Borui Zhao
Fuqiang Zhao
Hanbin Zhao
Jian Zhao
Mingyang Zhao
Na Zhao
Rongchang Zhao
Ruiqi Zhao
Shuai Zhao
Wenda Zhao
Wenliang Zhao
Xiangyun Zhao
Yifan Zhao
Yaping Zhao
Zhou Zhao
He Zhao
Jie Zhao
Xibin Zhao
Xiaoqi Zhao
Zhengyu Zhao
Jin Zhe
Chuanxia Zheng
Huan Zheng
Hao Zheng
Jia Zheng
Jian-Qing Zheng
Shuai Zheng
Meng Zheng
Mingkai Zheng
Qian Zheng
Qi Zheng
Wu Zheng
Yinqiang Zheng
Yufeng Zheng
Yutong Zheng
Yalin Zheng
Yu Zheng
Feng Zheng
Zhaoheng Zheng
Haitian Zheng
Kang Zheng
Bolun Zheng

Haiyong Zheng
Mingwu Zheng
Sipeng Zheng
Tu Zheng
Wenzhao Zheng
Xiawu Zheng
Yinglin Zheng
Zhuo Zheng
Zilong Zheng
Kecheng Zheng
Zerong Zheng
Shuaifeng Zhi
Tiancheng Zhi
Jia-Xing Zhong
Yiwu Zhong
Fangwei Zhong
Zhihang Zhong
Yaoyao Zhong
Yiran Zhong
Zhun Zhong
Zichun Zhong
Bo Zhou
Boyao Zhou
Brady Zhou
Mo Zhou
Chunluan Zhou
Dingfu Zhou
Fan Zhou
Jingkai Zhou
Honglu Zhou
Jiaming Zhou
Jiahuan Zhou
Jun Zhou
Kaiyang Zhou
Keyang Zhou
Kuangqi Zhou
Lei Zhou
Lihua Zhou
Man Zhou
Mingyi Zhou
Mingyuan Zhou
Ning Zhou
Peng Zhou
Penghao Zhou
Qianyi Zhou

Shuigeng Zhou
Shangchen Zhou
Huayi Zhou
Zhize Zhou
Sanping Zhou
Qin Zhou
Tao Zhou
Wenbo Zhou
Xiangdong Zhou
Xiao-Yun Zhou
Xiao Zhou
Yang Zhou
Yipin Zhou
Zhenyu Zhou
Hao Zhou
Chu Zhou
Daquan Zhou
Da-Wei Zhou
Hang Zhou
Kang Zhou
Qianyu Zhou
Sheng Zhou
Wenhui Zhou
Xingyi Zhou
Yan-Jie Zhou
Yiyi Zhou
Yu Zhou
Yuan Zhou
Yuqian Zhou
Yuxuan Zhou
Zixiang Zhou
Wengang Zhou
Shuchang Zhou
Tianfei Zhou
Yichao Zhou
Alex Zhu
Chenchen Zhu
Deyao Zhu
Xiatian Zhu
Guibo Zhu
Haidong Zhu
Hao Zhu
Hongzi Zhu
Rui Zhu
Jing Zhu

Jianke Zhu
Junchen Zhu
Lei Zhu
Lingyu Zhu
Luyang Zhu
Menglong Zhu
Peihao Zhu
Hui Zhu
Xiaofeng Zhu
Tyler (Lixuan) Zhu
Wentao Zhu
Xiangyu Zhu
Xinqi Zhu
Xinxin Zhu
Xinliang Zhu
Yangguang Zhu
Yichen Zhu
Yixin Zhu
Yanjun Zhu
Yousong Zhu
Yuhao Zhu
Ye Zhu
Feng Zhu
Zhen Zhu
Fangrui Zhu
Jinjing Zhu
Linchao Zhu
Pengfei Zhu
Sijie Zhu
Xiaobin Zhu
Xiaoguang Zhu
Zezhou Zhu
Zhenyao Zhu
Kai Zhu
Pengkai Zhu
Bingbing Zhuang
Chengyuan Zhuang
Liansheng Zhuang
Peiye Zhuang
Yixin Zhuang
Yihong Zhuang
Junbao Zhuo
Andrea Ziani
Bartosz Zieliński
Primo Zingaretti

Nikolaos Zioulis
Andrew Zisserman
Yael Ziv
Liu Ziyin
Xingxing Zou
Danping Zou
Qi Zou

Shihao Zou
Xueyan Zou
Yang Zou
Yuliang Zou
Zihang Zou
Chuhang Zou
Dongqing Zou

Xu Zou
Zhiming Zou
Maria A. Zuluaga
Xinxin Zuo
Zhiwen Zuo
Reyer Zwiggelaar

Contents – Part XXIX

Box-Supervised Instance Segmentation with Level Set Evolution

Wentong Li[1], Wenyu Liu[1], Jianke Zhu[1(✉)], Miaomiao Cui[2], Xian-Sheng Hua[2], and Lei Zhang[3]

[1] Zhejiang University, Hangzhou, China
{liwentong,liuwenyu.lwy,jkzhu}@zju.edu.cn
[2] Alibaba Damo Academy, Hangzhou, China
miaomiao.cmm@alibaba-inc.com, xshua@outlook.com
[3] The Hong Kong Polytechnic University, Hong Kong, China
cslzhang@comp.polyu.edu.hk

Abstract. In contrast to the fully supervised methods using pixel-wise mask labels, box-supervised instance segmentation takes advantage of the simple box annotations, which has recently attracted a lot of research attentions. In this paper, we propose a novel single-shot box-supervised instance segmentation approach, which integrates the classical level set model with deep neural network delicately. Specifically, our proposed method iteratively learns a series of level sets through a continuous Chan-Vese energy-based function in an end-to-end fashion. A simple mask supervised SOLOv2 model is adapted to predict the instance-aware mask map as the level set for each instance. Both the input image and its deep features are employed as the input data to evolve the level set curves, where a box projection function is employed to obtain the initial boundary. By minimizing the fully differentiable energy function, the level set for each instance is iteratively optimized within its corresponding bounding box annotation. The experimental results on four challenging benchmarks demonstrate the leading performance of our proposed approach to robust instance segmentation in various scenarios. The code is available at: https://github.com/LiWentomng/boxlevelset.

Keywords: Instance segmentation · Level set · Box supervision

1 Introduction

Instance segmentation aims to obtain the pixel-wise labels of the interested object, which plays an important role in many applications, such as autonomous driving and robotic manipulation. Though having achieved promising performance, most of the existing instance segmentation approaches [9,13,18,43,48] are trained in a supervised manner, which heavily depend on the pixel-wise mask annotations and incur expensive labeling costs.

To deal with this problem, box-supervised instance segmentation takes advantage of the simple box annotation rather than the pixel-wise mask labels, which has recently attracted a lot of research attentions [16,24–26,44,47]. To enable pixel-wise supervision with box annotation, some methods [26,47] focus on generating the pseudo mask labels by an independent network, which needs to employ extra auxiliary salient data [47] or post-processing methods like MCG [39] and CRF [23] to obtain precise pseudo labels. Due to the involved multiple separate steps, the training pipeline becomes complicated with many hyper-parameters. Several recent approaches [16,44] suggest a unified framework using the pairwise affinity modeling, e.g., neighbouring pixel pairs [16] and colour pairs [44], enabling an end-to-end training of the instance segmentation network. The pairwise affinity relationship is defined on the set of partial or all neighbouring pixel pairs, which oversimplifies the assumption that the pixel or colour pairs are encouraged to share the same label. The noisy contexts from the objects and background with similar appearance are inevitably absorbed, leading to inferior instance segmentation performance.

In this paper, we propose a novel single-shot box-supervised instance segmentation approach to address the above limitations. Our approach integrates the classical level set model [7,37] with deep neural network delicately. Unlike the existing box-supervised methods [16,25,26,44], we iteratively learn a series of level set functions for implicit curve evolution within the annotated bounding box in an end-to-end fashion. Different from fully-supervised level set-based methods [15,17,49,54], our proposed approach is able to train the level set functions in a weakly supervised manner using only the bounding box annotations, which are originally used for object detection.

Specifically, we introduce an energy function based on the classical continuous Chan-Vese energy functional [7], and make use of a simple and effective mask supervised method, i.e., SOLOv2 [48], to predict the instance-aware mask map as the level set for each instance. In addition to the input image, the deep structural features with long-range dependencies are introduced to robustly evolve the level set curves towards the object's boundary, which is initialized by a box projection function at each step. By minimizing the fully differentiable energy function, the level set for each instance is iteratively optimized within its corresponding bounding box annotation. Extensive experiments are conducted on four challenging benchmarks for instance segmentation under various scenarios, including general scene, remote sensing and medical images. The leading qualitative and quantitative results demonstrate the effectiveness of our proposed method. Especially, on remote sensing and medical images, our method outperforms the state-of-the-art methods by a large margin.

The highlights of this work are summarized as follows:

1) We propose a novel level set evolution-based approach to instance segmentation. To the best of our knowledge, this is the first deep level set-based method that tackles the problem of box-supervised instance segmentation.
2) We incorporate the deep structural features with the low-level image to achieve robust level set evolution within bounding box region, where a box projection function is employed for level set initialization.
3) Our proposed method achieves new state-of-the-arts of box-supervised instance segmentation on COCO [30] and Pascal VOC [11] datasets, remote sensing dataset iSAID [50] and medical dataset LiTS [3].

2   Related Work

2.1   Box-Supervised Instance Segmentation

The existing instance segmentation methods can be roughly divided into two categories. The first group [9,13,22,55] performs segmentation on the regions extracted from the detection results. Another category [4,5,43,48,51] directly segments each instance in a fully convolutional manner without resorting to the detection results. However, all these methods rely on the expensive pixel-wise mask annotations.

Box-supervised instance segmentation, which only employs the bounding box annotations to obtain pixel-level mask prediction, has recently been receiving increasing attention. Khoreva *et al.* [20] proposed to predict the mask with box annotations under the deep learning framework, which heavily depends on the region proposals generated by the unsupervised segmentation methods like GrabCut [40] and MCG [39]. Based on Mask R-CNN [13], Hsu *et al.* [16] formulated the box-supervised instance segmentation into a multiple instance learning (MIL) problem by making use of the neighbouring pixel-pairwise affinity regularization. BoxInst [44] uses the color-pairwise affinity with box constraint under an efficient RoI-free CondInst framework [43]. Despite the promising performance, the pairwise affinity relationship is built on either partial or all neighbouring pixel pairs with the oversimplified assumption that spatial pixel or color pairs are encouraged to share the same label. This inevitably introduces noises, especially from the nearby background or similar objects. Besides, the recent methods like BBAM [26] and DiscoBox [25] focus on the generation of proxy mask labels, which often require multiple training stages or networks to achieve promising performance. Unlike the above methods, our proposed level set-based approach is learned implicitly in an end-to-end manner, which is able to iteratively align the instance boundaries by optimizing the energy function within the box region.

2.2   Level Set-Based Segmentation

As a classical variational approach, the level set methods [1,37] have been widely used in image segmentation, which can be categorized into two major groups: region-based methods [7,36,45] and edge-based methods [6,33]. The key idea of level set is to represent the implicit curve by an energy function in a higher dimension, which is iteratively optimized by using gradient descent. Some works [15,17,21,49,54] have been proposed to embed the level set into the deep network in an end-to-end manner and achieve promising segmentation results. Wang *et al.* [49] predicted the evolution parameters and evolved the predicted contour by incorporating the user clicks on the boundary points. The energy function is based on the edge-based level set method in [6]. Levelset R-CNN [15] performs the Chan-Vese level set evolution with the deep features based on Mask R-CNN [13], where the original image is not used in the optimization. Yuan *et al.* [54] built a piecewise-constant function to parse each constant sub-region corresponding to a different instance based on the Mumford-Shah

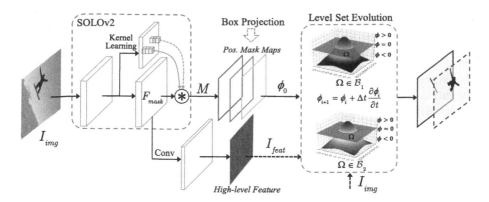

Fig. 1. Overview of our method. Our framework is designed based on SOLOv2 [48]. The positive mask maps M are obtained by level set evolution within the bounding box region. With the iterative energy minimization, the accurate instance segmentation can be obtained with box annotations only. The category branch is not shown here for simpler illustration.

model [36], which achieves instance segmentation by a fully convolutional network. The above methods perform level set evolution between deep features and ground-truth mask in a fully supervised manner, which train the network to predict different sub-regions and get object boundaries. Our proposed approach performs level set evolution only using the box-based annotations without the pixel-wise mask supervision.

Kim *et al.* [21] performed level set evolution in an unsupervised manner, which is mostly related to our proposed approach. To achieve N-class semantic segmentation, it employs the global multi-phase Mumford-Shah function [36] that only evolves on the low-level features of the input image. Our method is based on the Chan-Vese functional [7], which is constrained within the local bounding box with the enriched information from both input image and high-level deep features. Moreover, the initialization of level set is generated automatically for robust curve evolution.

3   Proposed Method

In this section, we present a novel box-supervised instance segmentation method, which incorporates the classical continuous Chan-Vese energy-based level set model [7] into deep neural network. To this end, we introduce an energy function, which enables the neural network to learn a series of level set functions evolving to the instance boundaries implicitly. In specific, we take advantage of an effective mask-supervised SOLOv2 model [48] to dynamically segment objects by locations and predict the instance-aware mask map of full-image size. To facilitate the box-supervised instance segmentation, we treat each mask map as the level set function ϕ for its corresponding object. Furthermore, we make use of both the input image I_{img} and high-level deep features I_{feat} as the input to evolve the level set, where a box projection function is employed to encourage the network to automatically estimate an initial level set ϕ_0 at each step. The level set for each instance is iteratively optimized within its corresponding bounding box annotation. Figure 1 gives the overview of our proposed framework.

3.1 Level Set Model in Image Segmentation

We first give a brief review of the level set methods [7,36,45], which formulate the image segmentation as a consecutive energy minimization problem. In the Mumford-Shah level set model [36], the segmentation of a given image I is obtained by finding a parametric contour C, which partition the image plane $\Omega \subset \mathbb{R}^2$ into N disjoint regions $\Omega_1, \cdots, \Omega_N$. The Mumford-Shah energy functional $\mathcal{F}^{MS}(u, C)$ can be defined as below:

$$\mathcal{F}^{MS}(u_1, \cdots, u_N, \Omega_1, \cdots, \Omega_N) = \sum_{i=1}^{N} (\int_{\Omega_i} (I - u_i)^2 dxdy + \mu \int_{\Omega_i} |\nabla u_i|^2 dxdy + \gamma |C_i|),$$
(1)

where u_i is a piecewise smooth function approximating the input I, ensuring the smoothness inside each region Ω_i. μ and γ are weighted parameters.

Chan and Vese [7] later simplified the Mumford-Shah functional as a variational level set, which has been explored aplenty [31,34,46,52]. Specially, it can be derived as follows,

$$\mathcal{F}^{CV}(\phi, c_1, c_2) = \int_{\Omega} |I(x, y) - c_1|^2 H(\phi(x, y)) dxdy$$
$$+ \int_{\Omega} |I(x, y) - c_2|^2 (1 - H(\phi(x, y))) dxdy + \gamma \int_{\Omega} |\nabla H(\phi(x, y))| dxdy$$
(2)

where H is the Heaviside function and $\phi(x, y)$ is the level set function, whose zero crossing contour $C = \{(x, y) : \phi(x, y) = 0\}$ divides the image space Ω into two disjoint regions, inside contour C: $\Omega_1 = \{(x, y) : \phi(x, y) > 0\}$ and outside contour C: $\Omega_2 = \{(x, y) : \phi(x, y) < 0\}$. In Eq. (2), the first two terms intend to fit the data, and the third term regularizes the zero level contour with a non-negative parameter γ. c_1 and c_2 are the mean values of input $I(x, y)$ inside C and outside C, respectively. The image segmentation is achieved by finding a level set function $\phi(x, y) = 0$ with c_1 and c_2 that minimize the energy $\mathcal{F}^{CV}$.

3.2 Box-Supervised Instance Segmentation

Our proposed method exploits the level set evolution with Chan-Vese energy-based model [7] to achieve high-quality instance segmentation using the box annotations only.

Level Set Evolution within Bounding Box. Given an input image $I(x, y)$, we aim to predict the object boundary curve by evolving a level set implicitly within the region of annotated bounding box $\mathcal{B}$. The mask prediction

$M \in \mathbb{R}^{H \times W \times S^2}$ by SOLOv2 contains $S \times S$ potential instance maps of size $H \times W$. Each potential instance map contains only one instance whose center is at location (i, j). The mask map predicted for the location (i, j) with the category probability $p_{i,j}^* > 0$ is regarded as the positive instance sample. We treat each positive mask map within box $\mathcal{B}$ as the level set $\phi(x, y)$, and its corresponding pixel space of input image $I(x, y)$ is referred as Ω, i.e., $\Omega \in \mathcal{B}$. C is the segmentation boundary with zero level $C = \{(x, y) : \phi(x, y) = 0\}$, which partitions the box region into two disjoint regions, i.e., foreground object and background.

To obtain the accurate boundary for each instance, we learn a series of level sets $\phi(x, y)$ by minimizing the following energy function:

$$\mathcal{F}(\phi, I, c_1, c_2, \mathcal{B}) = \int\limits_{\Omega \in \mathcal{B}} |I^*(x, y) - c_1|^2 \sigma(\phi(x, y)) dx dy$$

$$+ \int\limits_{\Omega \in \mathcal{B}} |I^*(x, y) - c_2|^2 (1 - \sigma(\phi(x, y))) dx dy + \gamma \int\limits_{\Omega \in \mathcal{B}} |\nabla \sigma(\phi(x, y))|\, dx dy, \tag{3}$$

where $I^*(x, y)$ denotes the normalized input image $I(x, y)$, γ is a non-negative weight, and σ denotes the *sigmoid* function that is treated as the characteristic function for level set $\phi(x, y)$. Different from the traditional Heaviside function [7], the *sigmoid* function is much smoother, which can better express the characteristics of the predicted instance and improve the convergence of level set evolution during the training process. The first two items in Eq. (3) force the predicted $\phi(x, y)$ to be uniform both inside region Ω and outside area $\bar{\Omega}$. c_1 and c_2 are the mean values of Ω and $\bar{\Omega}$, which are defined as below:

$$c_1(\phi) = \frac{\int\limits_{\Omega \in \mathcal{B}} I^*(x, y) \sigma(\phi(x, y)) dx dy}{\int\limits_{\Omega \in \mathcal{B}} \sigma(\phi(x, y)) dx dy}, \quad c_2(\phi) = \frac{\int\limits_{\Omega \in \mathcal{B}} I^*(x, y)(1 - \sigma(\phi(x, y))) dx dy}{\int\limits_{\Omega \in \mathcal{B}} (1 - \sigma(\phi(x, y))) dx dy}.$$

$$\tag{4}$$

The energy function $\mathcal{F}$ can be optimized with gradient back-propagation during training. With the time step $t \geqslant 0$, the derivative of energy function $\mathcal{F}$ upon ϕ can be written as follows:

$$\frac{\partial \phi}{\partial t} = -\frac{\partial \mathcal{F}}{\partial \phi} = -\nabla \sigma(\phi)[(I^*(x, y) - c_1)^2 - (I^*(x, y) - c_2)^2 + \gamma div\left(\frac{\nabla \phi}{|\nabla \phi|}\right)],$$

$$\tag{5}$$

where ∇ and div are the spatial derivative and divergence operator, respectively. Therefore, the update of ϕ is computed by

$$\phi_i = \phi_{i-1} + \Delta t \frac{\partial \phi_{i-1}}{\partial t}. \tag{6}$$

The minimization of the above terms can be viewed as an implicit curve evolution along the descent of energy function. The optimal boundary C of the instance is obtained by minimizing the energy $\mathcal{F}$ via iteratively fitting ϕ_i as follows:

$$\inf_{\Omega \in \mathcal{B}} \{\mathcal{F}(\phi)\} \approx 0 \approx \mathcal{F}(\phi_i). \tag{7}$$

Input Data Terms. The energy function in Eq. (3) encourages the curve evolution based on the uniformity of regions inside and outside the object. The input image I_u represents the essential low-level features, including shape, colour, image intensities, etc. However, such low-level features usually vary with illumination variations, different materials and motion blur, making the level set evolution less robust.

In addition to the normalized input image, we take into account the high-level deep features I_f, which embed the image semantic information, to obtain more robust results. To this end, we make full use of the unified and high-resolution mask feature F_{mask} from all FPN levels in SOLOv2, which is further fed into a convolution layer to extract the high-level features I_f. Besides, the features I_f are enhanced by the tree filter [27,41], which employs minimal spanning tree to model long-range dependencies and preserve the object structure. The overall energy function for level set evolution can be formulated as follows:

$$\mathcal{F}(\phi) = \lambda_1 * \mathcal{F}(\phi, I_u, c_{u_1}, c_{u_2}, \mathcal{B}) + \lambda_2 * \mathcal{F}(\phi, I_f, c_{f_1}, c_{f_2}, \mathcal{B}), \qquad (8)$$

where λ_1 and λ_2 are weights to balance the two kinds of features. c_{u_1}, c_{u_2} and c_{f_1}, c_{f_2} are the mean values for input terms I_u and I_f, respectively.

Level Set Initialization. Conventional level set methods are sensitive to the initialization that is usually manually labeled. In this work, we employ a box projection function [44] to encourage the model to automatically generate a rough estimation of the initial level set ϕ_0 at each step.

In particular, we utilize the coordinate projection of ground-truth box to x-axis and y-axis and calculate the projection difference between the predicted mask map and the ground-truth box. Such a simple scheme limits the predicted initialization boundary within the bounding box, providing a good initial state for curve evolution. Let $m^b \in \{0, 1\}^{H \times W}$ denote the binary region by assigning one to the locations in the ground-truth box, and zero otherwise. The mask score predictions $m^p \in (0, 1)^{H \times W}$ for each instance can be regarded as the foreground probabilities. The box projection function $\mathcal{F}(\phi_0)_{box}$ is defined as below:

$$\mathcal{F}(\phi_0)_{box} = \mathcal{P}_{dice}(m_x^p, m_x^b) + \mathcal{P}_{dice}(m_y^p, m_y^b), \qquad (9)$$

where m_x^p, m_x^b and m_y^p, m_y^b denote the x-axis projection and y-axis projection for mask prediction m^p and binary ground-truth region m^b, respectively. $\mathcal{P}_{dice}$ represents the projection operation measured by 1-D dice coefficient [35].

3.3 Training and Inference

Loss Function. The loss function L to train our proposed network consists of two items, including L_{cate} for category classification and L_{inst} for instance segmentation with box annotations:

$$L = L_{cate} + L_{inst}, \qquad (10)$$

Table 1. Performance comparisons on Pascal VOC `val` 2012. "$*$" denotes the results of GrabCut reported from BoxInst [44]. All entries are the results using *box-supervision*.

Methods	Backbone	AP	AP_{25}	AP_{50}	AP_{70}	AP_{75}
GrabCut* [40]	ResNet-101	19.0	-	38.8	-	17.0
SDI [20]	VGG-16	-	-	44.8	-	16.3
Liao *et al.* [28]	ResNet-101	-	-	51.3	-	22.4
Sun *et al.* [42]	ResNet-50	-	-	56.9	-	21.4
BBTP [16]	ResNet-101	23.1	-	54.1	-	17.1
BBTP w/ CRF [16]	ResNet-101	27.5	-	59.1	-	21.9
Arun *et al.* [2]	ResNet-101	-	73.1	57.7	33.5	31.2
BBAM [26]	ResNet-101	-	76.8	63.7	39.5	31.8
BoxInst [44]	ResNet-50	34.3	-	59.1	-	34.2
BoxInst [44]	ResNet-101	36.5	-	61.4	-	37.0
DiscoBox [25]	ResNet-50	-	71.4	59.8	41.7	35.5
DiscoBox [25]	ResNet-101	-	72.8	62.2	45.5	37.5
Ours	ResNet-50	36.3	76.3	64.2	43.9	35.9
Ours	ResNet-101	**38.3**	**77.9**	**66.3**	**46.4**	**38.7**

where L_{cate} is the Focal Loss [29]. For L_{inst}, we employ the presented differentiable level set energy as the optimization objective:

$$L_{inst} = \frac{1}{N_{pos}} \sum_k \mathbb{1}_{\{p_{i,j}^* > 0\}} \{\mathcal{F}(\phi) + \alpha \mathcal{F}(\phi_0)_{box}\}, \qquad (11)$$

where N_{pos} indicates the number of positive samples, and $p_{i,j}^*$ denotes the category probability at target location (i,j). $\mathbb{1}$ represents the indicator function, which ensures only the positive instance mask samples perform the level set evolution. $\mathbb{1}$ is set to one if $p_{i,j}^* > 0$, and zero otherwise. α is the weight parameter, which is set to 3.0 empirically in our implementation.

Inference. It is worth noting that the level set evolution is only employed during training to generate implicit supervisions for network optimization. *The inference process is the same as the original SOLOv2 network.* Given the input image, the mask prediction is directly generated with efficient matrix non-maximum suppression (NMS). Comparing to SOLOv2, our proposed network introduces only one additional convolution layer to generate the high-level features with negligible cost.

4   Experiments

To evaluate our proposed approach, we conduct experiments on four challenging datasets, including Pascal VOC [11] and COCO [30], remote sensing dataset iSAID [50] and medical dataset LiTS [3]. On all datasets, *only box annotations are used during training.*

Table 2. Instance segmentation mask AP (%) on the COCO `test-dev`. "†" denotes the result of BBTP on the COCO `val2017` split. "∗" indicates that the BoxCaseg is trained with box and salient object supervisions.

Method	Backbone	AP	AP_{50}	AP_{75}	AP_S	AP_M	AP_L
Mask-supervised:							
Mask R-CNN [13]	ResNet-101	35.7	58.0	37.8	15.5	38.1	52.4
YOLACT-700 [4]	ResNet-101	31.2	50.6	32.8	12.1	33.3	47.1
PolarMask [51]	ResNet-101	32.1	53.7	33.1	14.7	33.8	45.3
CondInst [43]	ResNet-101	39.1	60.9	42.0	21.5	41.7	50.9
SOLOv2 [48]	ResNet-101	39.7	60.7	42.9	17.3	42.9	57.4
Box-supervised:							
BBTP† [16]	ResNet-101	21.1	45.5	17.2	11.2	22.0	29.8
BBAM [26]	ResNet-101	25.7	50.0	23.3	-	-	-
BoxCaseg∗ [47]	ResNet-101	30.9	54.3	30.8	12.1	32.8	46.3
BoxInst [44]	ResNet-101	33.2	56.5	33.6	16.2	35.3	45.1
BoxInst [44]	ResNet-101-DCN	35.0	**59.3**	35.6	**17.1**	37.2	48.9
Ours	ResNet-101	33.4	56.8	34.1	15.2	36.8	46.8
Ours	ResNet-101-DCN	**35.4**	59.1	**36.7**	16.8	**38.5**	**51.3**

Table 3. Deep variational instance segmentation methods on COCO val. "Sup." denotes the form of supervision, i.e., *Mask* or *Box*. *Our method is only supervised with box annotations yet achieves competitive results.*

Method	Backbone	Sup	AP
DeepSnake [38]	DLA-34 [53]	*Mask*	30.5
Levelset R-CNN [15]	ResNet-50	*Mask*	34.3
DVIS-700 [54]	ResNet-50	*Mask*	32.6
DVIS-700 [54]	ResNet-101	*Mask*	**35.7**
Ours	ResNet-101	*Box*	33.0
Ours	ResNet-101-DCN	*Box*	35.0

4.1 Datasets

Pascal VOC [11]. Pascal VOC consists of 20 categories. As in [16,26,44], the augmented Pascal VOC 2012 [12] dataset is used, which contains 10, 582 images for training and 1, 449 validation images for evaluation.

COCO [30]. COCO has 80 general object classes. Our models are trained on `train2017` (115K images), and evaluated on `val2017` (5K images) and `test-dev` split (20K images).

iSAID [50]. It is a large-scale high-resolution remote sensing dataset for aerial instance segmentation, containing many small objects with complex back-

Table 4. Results of mask AP (%) on iSAID `val`. All models are trained with "1×" schedule (12 epoch) with 600 × 600 input size.

Method	Backbone	Sup	AP	AP$_{50}$	AP$_{75}$
Mask R-CNN [13]	R-50-C4	*Mask*	28.8	51.8	27.7
PolarMask [51]	R-50-FPN	*Mask*	27.2	48.5	27.3
CondInst [43]	R-50-FPN	*Mask*	29.5	54.5	28.3
BoxInst [44]	R-50-FPN	*Box*	17.8	41.4	12.9
Ours	R-50-FPN	*Box*	**20.1**	**41.8**	**16.6**

Table 5. Instance segmentation results on LiTS `val`. All models are trained with "1×" schedule (12 epoch). Our method outperforms BoxInst by 3.8% AP.

Method	Backbone	Sup	AP	AP$_{50}$	AP$_{75}$
Mask R-CNN [13]	R-50-FPN	*Mask*	64.2	81.6	71.0
BoxInst [44]	R-50-FPN	*Box*	40.7	67.8	40.2
Ours	R-50-FPN	*Box*	**44.5**	**78.6**	**45.6**

grounds. The dataset comprises 1,411 images for training and 458 validation images for evaluation with 655,451 instance annotations.

LiTS [3]. The Liver Tumor Segmentation Challenge (LiTS) dataset[1] consists of 130 volume CT scans for training and 70 volume CT scans for testing. We randomly partition all the scans having mask labels into the training and validation dataset with the ratio of 4:1.

4.2   Implementation Details

The models are trained with the AdamW [32] optimizer on 8 NVIDIA V100 GPUs. The training schedules of "1×" and "3×" are the same as `mmdetection` framework [8] with 12 epochs and 36 epochs, respectively. ResNet [14] is employed as the backbone, which is initialized with the ImageNet [10] pretraining weights. For COCO, the initial learning rate is 10^{-4} with 16 images per mini-bath. For Pascal VOC, the initial learning rate is 5×10^{-5} with 8 images per mini-bath. The scale jitter is used where the shorter image side is randomly sampled from 640 to 800 pixels on COCO and Pascal VOC datasets for fair comparison. For iSAID and LiTS, all the models on each dataset are trained with the same settings. COCO-style mask AP (%) is adopted for performance evaluation. Following [25,26], we also report the average precision (AP) at four IoU thresholds (including 0.25, 0.50, 0.70 and 0.75) for the comparison on Pascal VOC dataset. The non-negative weight γ in Eq. 3 is set to 10^{-4} by default.

---

[1] https://competitions.codalab.org/competitions/17094.

4.3   Main Results

We compare our proposed method against the state-of-the-art instance segmentation approaches, including box-supervised and fully mask-supervised methods in different scenarios.

Most box-supervised methods are evaluated on the Pascal VOC dataset. Table 1 reports the comparison results. Our method outperforms BoxInst [44] by 2.0% and 1.8% AP with ResNet-50 and ResNet-101 backbones, respectively, achieving the best performance. For AP_{25} and AP_{50}, our method can obtain 77.9% and 66.3% accuracy, largely outperforming the recent DiscoBox [25] by 5.1% and 4.1%. The high IoU threshold-based AP metrics can reflect the segmentation performance with accurate boundary, which is in line with the practical application. Our approach achieves 38.7% AP_{75} with ResNet-101, which outperforms BoxInst [44] and DiscoBox [25] by 1.7% and 1.2%, respectively.

Table 2 shows the main results on COCO `test-dev` split. Both fully mask-supervised and box-supervised methods are compared in the evaluation. Our method outperforms BBTP [16] by 12.3% AP with the same backbone. In contrast to the recent box-supervised methods, our method outperforms BBAM [26] and BoxCaseg [47] by 7.7% AP and 2.5% AP using ResNet-101. It achieves 33.4% AP and 35.4% AP, which is higher than BoxInst [44] by 0.2% and 0.4% with ResNet-101 and ResNet-101-DCN backbones, respectively. Our approach achieves 16.8% AP_S on small objects, which is slightly lower than BoxInst [44] by 0.3%. This is because small objects lack rich features for level set evolution to distinguish the foreground object and background within the bounding box. However, our method obtains the best results for large objects, largely outperforming BoxInst [44] by 2.4% AP_L using the same ResNet-101-DCN. Our method even performs better than some recent fully mask-supervised methods, such as YOLACT [4] and PolarMask [51]. This shows that our method narrows the performance gap between mask-supervised and box-supervised instance segmentation. Figure 2 visualizes some instance segmentation results on COCO and Pascal VOC datasets.

We then compare our method with other deep variational-based instance segmentation approaches. DeepSnake [38] is based on the classical snake method [19]. Levelset R-CNN [15] and DVIS-700 [54] are also built on level set function. These methods are all fully supervised by the mask annotations. As shown in Table 3, our method achieves comparable results to the fully supervised variational-based methods, and even outperforms DeepSnake [38] and Levelset R-CNN [15].

To further validate the robust performance of our method in more complicated scenarios, we conduct experiments on remote sensing and medical image datasets. In remote sensing, the objects of the same class are densely-distributed. For medical images, the background is highly similar to the foreground. The previous pixel relationship model-based methods are built on the neighbouring pixel pairs. They are easily affected by the noisy context. Our level set-based method drives the curve to fit the object boundary under the guidance of level set minimization, which is more robust. Table 4 and Table 5 show the mask AP results on

Fig. 2. Visualization of instance segmentation results on general scene. The model is trained with only box annotations.

iSAID and LiTS datasets, respectively. It can be clearly seen that our approach outperforms BoxInst [44] by 2.3% AP on iSAID and 3.8% AP on LiTS. Figure 3 and Fig. 4 show several examples of instance segmentation on iSAID and LiTS, respectively. One can see that our method is effective in various scenarios.

4.4 Ablation Experiments

The ablation study is conducted on Pascal VOC dataset to examine the effectiveness of each module in our proposed framework.

Level Set Energy. We firstly investigate the impact of level set energy functional with different settings. Table 6 gives the evaluation results. Our method achieves 19.7% AP only with the box projection function as $\mathcal{F}_{\phi_0}$ to drive the network to initialize the boundary during training. This indicates that the initialization for level set function is effective to generate the initial boundary. When the original image I_u is employed as the input data term in Eq. 8, our method can achieve 22.2% AP. On the other hand, our method achieves better performance with 24.7% AP when the deep high-level features I_f are employed as

Table 6. The impact of **level set energy** with different settings. I_u and I_f denote the input image and high-level feature as the input data terms of energy, respectively. $\mathcal{B}$ and I represent the Ω space of bounding box or the full-image region for level set evolution.

$\mathcal{F}_{\phi 0}$	$\mathcal{F}_\phi(I_u)$	$\mathcal{F}_\phi(I_f)$	$\Omega \in \mathcal{B}$	$\Omega \in I$	AP	AP$_{50}$	AP$_{75}$
✓			✓		19.7	47.4	13.9
✓	✓		✓		22.2	49.5	17.4
✓	✓	✓	✓		**24.7**	**53.3**	**20.8**
✓	✓	✓		✓	21.7	48.4	17.4

Table 7. Different channel number C_{I_f} of high-level features for curve evolution.

C_{I_f}	AP	AP$_{50}$	AP$_{75}$
5	23.3	51.3	18.7
8	24.4	52.1	20.1
9	**24.7**	**53.3**	**20.8**
10	22.0	49.2	17.3
11	21.9	49.4	16.9

Table 8. Training schedules with "1×" single-scale training and "3×" multi-scale training.

Sched.	AP	AP$_{50}$	AP$_{75}$
1×	24.7	53.3	20.8
3×	**34.4**	**62.2.**	**34.6**

Table 9. The effectiveness of tree filter [41] for high-level structural features in level set.

Tree filter	AP	AP$_{50}$	AP$_{75}$
w/o	34.4	62.2	34.6
w.	**36.3**	**64.2.**	**35.9**

the extra input data. This demonstrates that both original image and high-level features can provide useful information for robust level set evolution. Besides, the above results are constrained within the bounding box $\mathcal{B}$ region for curve evolution. When the global region with the full-image size is regarded as the Ω, there is a noticeable performance drop (24.7% vs. 21.7%). This indicates that the bounding box region can make the level set evolution smoother with less noise interference.

Number of Channels for High-Level Feature. Secondly, we investigate the selection of the total number of channels for the output high-level feature I_f. As shown in Table 7, our method obtains better representation with 24.7% AP performance when the number of channels C_{I_f} is set to 9. When $C_{I_f} = 10$, the performance drops (24.7% vs. 22.0%). This indicates that the more channels may introduce uncertain semantic information for level set evolution.

Training Schedule. We evaluate the proposed network using different training schedules. Table 8 shows the results with 12 epochs (1×) and 36 epochs (3×). It can be observed that a longer training schedule benefits the performance of our method. Due to the relatively small size of Pascal VOC compared with COCO (about 1/10), longer training schedule leads to significant improvement (24.7% vs. 34.4%). This implies that level set evolution needs more training time to achieve better convergence for instance segmentation.

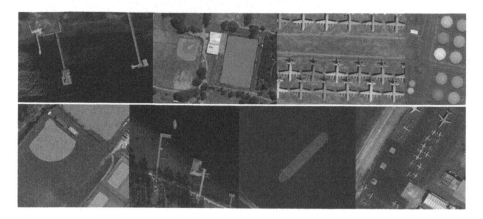

Fig. 3. Visual results of iSAID val. The mask predictions are obtained on the high-resolution remote sensing images only with box supervision.

input image   mask label   high-level feature   mask prediction

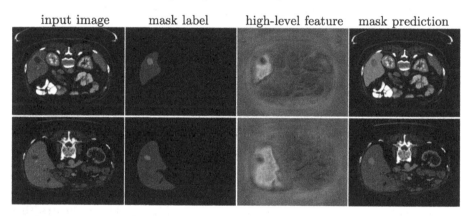

Fig. 4. Visualization examples of LiTS val. The high-level feature represents the input deep feature for level set evolution.

Effectiveness of Deep Structural Feature. We study the impact of tree filter [41], which models long-range dependencies and preserves object structure, on obtaining deep semantic features for level set evolution. Table 9 shows the results. One can see that by applying the tree filter to high-level deep features, +1.9% AP improvement can be achieved.

5   Conclusion

This paper presented a single-shot box-supervised instance segmentation approach that iteratively learns a series of level set functions in an end-to-end fashion. An instance-aware mask map was predicted and used as the level set, and both the original image and deep high-level features were employed as the inputs to evolve the level set curves, where a box projection function was employed to obtain the initial boundary. By minimizing the fully differentiable energy function, the level set for each instance was iteratively optimized within its corresponding bounding box annotation. Extensive experiments were conducted on four challenging benchmarks, and our proposed approach demonstrated leading performance in various scenarios. Our work narrows the performance gap between fully mask-supervised and box-supervised instance segmentation.

Acknowledgments. This work is supported by National Natural Science Foundation of China under Grants (61831015) and Alibaba-Zhejiang University Joint Institute of Frontier Technologies.

References

1. Adalsteinsson, D., Sethian, J.A.: A fast level set method for propagating interfaces. J. Comput. Phys. **118**(2), 269–277 (1995)
2. Arun, A., Jawahar, C.V., Kumar, M.P.: Weakly supervised instance segmentation by learning annotation consistent instances. In: Vedaldi, A., Bischof, H., Brox, T., Frahm, J.-M. (eds.) ECCV 2020. LNCS, vol. 12373, pp. 254–270. Springer, Cham (2020). https://doi.org/10.1007/978-3-030-58604-1_16
3. Bilic, P., et al.: The liver tumor segmentation benchmark (LiTS). arXiv preprint arXiv:1901.04056 (2019)
4. Bolya, D., Zhou, C., Xiao, F., Lee, Y.J.: YOLACT: real-time instance segmentation. In: Proceedings of the IEEE International Conference on Computer Vision, pp. 9157–9166 (2019)
5. Bolya, D., Zhou, C., Xiao, F., Lee, Y.J.: YOLACT++: better real-time instance segmentation. IEEE Trans. Pattern Anal. Mach. Intell. (2020)
6. Caselles, V., Kimmel, R., Sapiro, G.: Geodesic active contours. Int. J. Comput. Vision **22**(1), 61–79 (1997)
7. Chan, T., Vese, L.: Active contours without edges. IEEE Trans. Image Process. **10**(2), 266–277 (2001)
8. Chen, K., et al.: MMDetection: open MMLab detection toolbox and benchmark. arXiv preprint arXiv:1906.07155 (2019)
9. Cheng, T., Wang, X., Huang, L., Liu, W.: Boundary-preserving mask R-CNN. In: Vedaldi, A., Bischof, H., Brox, T., Frahm, J.-M. (eds.) ECCV 2020. LNCS, vol. 12359, pp. 660–676. Springer, Cham (2020). https://doi.org/10.1007/978-3-030-58568-6_39
10. Deng, J., Dong, W., Socher, R., Li, L.J., Li, K., Fei-Fei, L.: Imagenet: a large-scale hierarchical image database. In: Proceedings of IEEE Conference on Computer Vision and Pattern Recognition, pp. 248–255. IEEE (2009)
11. Everingham, M., Van Gool, L., Williams, C.K., Winn, J., Zisserman, A.: The pascal visual object classes (VOC) challenge. Int. J. Comput. Vision **88**(2), 303–338 (2010)
12. Hariharan, B., Arbeláez, P., Bourdev, L., Maji, S., Malik, J.: Semantic contours from inverse detectors. In: Proceedings of IEEE International Conference on Computer Vision, pp. 991–998. IEEE (2011)
13. He, K., Gkioxari, G., Dollar, P., Girshick, R.: Mask R-CNN. In: Proceedings of IEEE International Conference on Computer Vision, pp. 2980–2988 (2017)

14. He, K., Zhang, X., Ren, S., Sun, J.: Deep residual learning for image recognition. In: Proceedings of IEEE Conference on Computer Vision and Pattern Recognition, pp. 770–778 (2016)
15. Homayounfar, N., Xiong, Y., Liang, J., Ma, W.-C., Urtasun, R.: LevelSet R-CNN: a deep variational method for instance segmentation. In: Vedaldi, A., Bischof, H., Brox, T., Frahm, J.-M. (eds.) ECCV 2020. LNCS, vol. 12368, pp. 555–571. Springer, Cham (2020). https://doi.org/10.1007/978-3-030-58592-1_33
16. Hsu, C.C., Hsu, K.J., Tsai, C.C., Lin, Y.Y., Chuang, Y.Y.: Weakly supervised instance segmentation using the bounding box tightness prior. In: Proceedings of Advances in Neural Information Processing Systems, vol. 32, pp. 6582–6593 (2019)
17. Hu, P., Shuai, B., Liu, J., Wang, G.: Deep level sets for salient object detection. In: Proceedings of the IEEE Conference on Computer Vision and Pattern Recognition, pp. 540–549 (2017)
18. Huang, Z., Huang, L., Gong, Y., Huang, C., Wang, X.: Mask scoring R-CNN. In: Proceedings of the IEEE Conference on Computer Vision and Pattern Recognition, pp. 6409–6418 (2019)
19. Kass, M., Witkin, A., Terzopoulos, D.: Snakes: active contour models. Int. J. Comput. Vision **1**(4), 321–331 (1988)
20. Khoreva, A., Benenson, R., Hosang, J., Hein, M., Schiele, B.: Simple does it: Weakly supervised instance and semantic segmentation. In: Proceedings of the IEEE Conference on Computer Vision and Pattern Recognition, pp. 1665–1674 (2017)
21. Kim, B., Ye, J.C.: Mumford-shah loss functional for image segmentation with deep learning. IEEE Trans. Image Process. **29**, 1856–1866 (2019)
22. Kirillov, A., Wu, Y., He, K., Girshick, R.: Pointrend: image segmentation as rendering. In: Proceedings of the IEEE Conference on Computer Vision and Pattern Recognition, pp. 9799–9808 (2020)
23. Krähenbühl, P., Koltun, V.: Efficient inference in fully connected CRFs with gaussian edge potentials. In: Proceedings of Advances in Neural Information Processing Systems, vol. 24 (2011)
24. Kulharia, V., Chandra, S., Agrawal, A., Torr, P., Tyagi, A.: Box2Seg: attention weighted loss and discriminative feature learning for weakly supervised segmentation. In: Vedaldi, A., Bischof, H., Brox, T., Frahm, J.-M. (eds.) ECCV 2020. LNCS, vol. 12372, pp. 290–308. Springer, Cham (2020). https://doi.org/10.1007/978-3-030-58583-9_18
25. Lan, S., et al.: Discobox: weakly supervised instance segmentation and semantic correspondence from box supervision. In: Proceedings of the IEEE International Conference on Computer Vision, pp. 3406–3416 (2021)
26. Lee, J., Yi, J., Shin, C., Yoon, S.: BBAM: bounding box attribution map for weakly supervised semantic and instance segmentation. In: Proceedings of IEEE Conference on Computer Vision and Pattern Recognition, pp. 2643–2652 (2021)
27. Liang, Z., Wang, T., Zhang, X., Sun, J., Shen, J.: Tree energy loss: towards sparsely annotated semantic segmentation. In: Proceedings of IEEE Conference on Computer Vision and Pattern Recognition, pp. 16907–16916 (2022)
28. Liao, S., Sun, Y., Gao, C., KP, P.S., Mu, S., Shimamura, J., Sagata, A.: Weakly supervised instance segmentation using hybrid networks. In: Proceedings of IEEE International Conference on Acoustics, Speech and Signal Processing, pp. 1917–1921. IEEE (2019)
29. Lin, T.Y., Goyal, P., Girshick, R., He, K., Dollár, P.: Focal loss for dense object detection. In: Proceedings of the IEEE International Conference on Computer Vision, pp. 2980–2988 (2017)

30. Lin, T.-Y., et al.: Microsoft COCO: common objects in context. In: Fleet, D., Pajdla, T., Schiele, B., Tuytelaars, T. (eds.) ECCV 2014. LNCS, vol. 8693, pp. 740–755. Springer, Cham (2014). https://doi.org/10.1007/978-3-319-10602-1_48
31. Liu, S., Peng, Y.: A local region-based Chan-Vese model for image segmentation. Pattern Recogn. **45**(7), 2769–2779 (2012)
32. Loshchilov, I., Hutter, F.: Decoupled weight decay regularization. arXiv preprint arXiv:1711.05101 (2017)
33. Malladi, R., Sethian, J.A., Vemuri, B.C.: Shape modeling with front propagation: a level set approach. IEEE Trans. Pattern Anal. Mach. Intell. **17**(2), 158–175 (1995)
34. Maška, M., Daněk, O., Garasa, S., Rouzaut, A., Munoz-Barrutia, A., Ortiz-de Solorzano, C.: Segmentation and shape tracking of whole fluorescent cells based on the Chan-Vese model. IEEE Trans. Med. Imaging **32**(6), 995–1006 (2013)
35. Milletari, F., Navab, N., Ahmadi, S.A.: V-net: fully convolutional neural networks for volumetric medical image segmentation. In: Proceedings of International Conference on 3D Vision (3DV), pp. 565–571 (2016)
36. Mumford, D.B., Shah, J.: Optimal approximations by piecewise smooth functions and associated variational problems. Commun. Pure Appl. Math. (1989)
37. Osher, S., Sethian, J.A.: Fronts propagating with curvature-dependent speed: algorithms based on Hamilton-Jacobi formulations. J. Comput. Phys. **79**(1), 12–49 (1988)
38. Peng, S., Jiang, W., Pi, H., Li, X., Bao, H., Zhou, X.: Deep snake for real-time instance segmentation. In: Proceedings of the IEEE Conference on Computer Vision and Pattern Recognition, pp. 8533–8542 (2020)
39. Pont-Tuset, J., Arbelaez, P., T.Barron, J., Marques, F., Malik, J.: Multiscale combinatorial grouping for image segmentation and object proposal generation. IEEE Trans. Pattern Anal. Mach. Intell. **39**(1), 128–140 (2017)
40. Rother, C., Kolmogorov, V., Blake, A.: Grabcut: interactive foreground extraction using iterated graph cuts. ACM Trans. Graph. (TOG) **23**(3), 309–314 (2004)
41. Song, L., Li, Y., Li, Z., Yu, G., Sun, H., Sun, J., Zheng, N.: Learnable tree filter for structure-preserving feature transform. In: Proceedings of Advances in Neural Information Processing Systems, vol. 32 (2019)
42. Sun, Y., et al.: Weakly supervised instance segmentation based on two-stage transfer learning. IEEE Access **8**, 24135–24144 (2020)
43. Tian, Z., Shen, C., Chen, H.: Conditional convolutions for instance segmentation. In: Vedaldi, A., Bischof, H., Brox, T., Frahm, J.-M. (eds.) ECCV 2020. LNCS, vol. 12346, pp. 282–298. Springer, Cham (2020). https://doi.org/10.1007/978-3-030-58452-8_17
44. Tian, Z., Shen, C., Wang, X., Chen, H.: Boxinst: high-performance instance segmentation with box annotations. In: Proceedings of the IEEE Conference on Computer Vision and Pattern Recognition, pp. 5443–5452 (2021)
45. Vese, L.A., Chan, T.F.: A multiphase level set framework for image segmentation using the mumford and shah model. Int. J. Comput. Vision **50**(3), 271–293 (2002)
46. Wang, X.F., Huang, D.S., Xu, H.: An efficient local Chan-Vese model for image segmentation. Pattern Recogn. **43**(3), 603–618 (2010)
47. Wang, X., Feng, J., Hu, B., Ding, Q., Ran, L., Chen, X., Liu, W.: Weakly-supervised instance segmentation via class-agnostic learning with salient images. In: Proceedings of the IEEE Conference on Computer Vision and Pattern Recognition, pp. 10225–10235 (2021)
48. Wang, X., Zhang, R., Kong, T., Li, L., Shen, C.: Solov2: dynamic and fast instance segmentation. In: Proceedings of Advances in Neural Information Processing Systems, vol. 33, pp. 17721–17732 (2020)

49. Wang, Z., Acuna, D., Ling, H., Kar, A., Fidler, S.: Object instance annotation with deep extreme level set evolution. In: Proceedings of the IEEE Conference on Computer Vision and Pattern Recognition, pp. 7500–7508 (2019)
50. Waqas Zamir, S., et al.: iSAID: a large-scale dataset for instance segmentation in aerial images. In: Proceedings of the IEEE Conference on Computer Vision and Pattern Recognition Workshops, pp. 28–37 (2019)
51. Xie, E., et al.: Polarmask: single shot instance segmentation with polar representation. In: Proceedings of the IEEE Conference on Computer Vision and Pattern Recognition, pp. 12193–12202 (2020)
52. Xu, L., Lu, C., Xu, Y., Jia, J.: Image smoothing via L 0 gradient minimization. In: Proceedings of the SIGGRAPH Asia Conference, pp. 1–12 (2011)
53. Yu, F., Wang, D., Shelhamer, E., Darrell, T.: Deep layer aggregation. In: Proceedings of the IEEE Conference on Computer Vision and Pattern Recognition, pp. 2403–2412 (2018)
54. Yuan, J., Chen, C., Li, F.: Deep variational instance segmentation. In: Proceedings of Advances in Neural Information Processing Systems, vol. 33, pp. 4811–4822 (2020)
55. Zhang, G., et al.: Refinemask: towards high-quality instance segmentation with fine-grained features. In: Proceedings of the IEEE Conference on Computer Vision and Pattern Recognition, pp. 6861–6869 (2021)

Point Primitive Transformer
for Long-Term 4D Point Cloud Video
Understanding

Hao Wen[1], Yunze Liu[1], Jingwei Huang[2], Bo Duan[2], and Li Yi[1,3]([✉])

[1] Tsinghua University, Beijing, China
wenh19@mails.tsinghua.edu.cn
[2] Huawei Technologies, Shenzhen, China
{huangjingwei6,duanbo5}@huawei.com
[3] Shanghai Qi Zhi Institute, Shanghai, China

Abstract. This paper proposes a 4D backbone for long-term point cloud video understanding. A typical way to capture spatial-temporal context is using 4Dconv or transformer without hierarchy. However, those methods are neither effective nor efficient enough due to camera motion, scene changes, sampling patterns, and complexity of 4D data. To address those issues, we leverage the primitive plane as mid-level representation to capture the long-term spatial-temporal context in 4D point cloud videos, and propose a novel hierarchical backbone named Point Primitive Transformer (PPTr), which is mainly composed of intra-primitive point transformers and primitive transformers. Extensive experiments show that PPTr outperforms the previous state of the arts on different tasks.

Keywords: Transformer · Primitive · Long-term point cloud video

1 Introduction

Point cloud videos are ubiquitous in robots and AR systems that act as a window into our dynamically changing 3D world. Being able to record movements in the physical space, point cloud sequences play a key role in comprehending environmental changes and supporting interactions with the world, which can be hardly described by 2D images or static 3D point clouds. Therefore, an intelligent agent must process such a form of data precisely to better model the real world, adapt to environmental changes, and interact with them.

Despite its importance, processing point cloud sequences is a quite challenging task for machines that are largely determined by two aspects: effectiveness and efficiency. Effectiveness refers to the ability to capture long-term spatial-temporal structures. Due to camera motion, scene changes, occlusion changes,

H. Wen and Y. Liu—Equal contribution.

Supplementary Information The online version contains supplementary material available at https://doi.org/10.1007/978-3-031-19818-2_2.

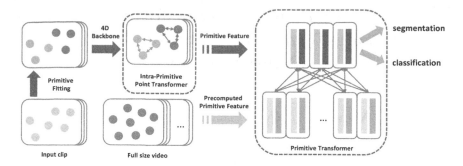

Fig. 1. Architecture of Primitive Point Transformer. On the lower level, PPTr extracts short-term spatial-temporal features through an intra-primitive point transformer for a short video clip around the frame of interest. On the upper level, PPTr extracts long-term spatial-temporal features through a primitive transformer.

and sampling patterns, points between different frames are unstructured and inconsistent, making it difficult to effectively integrate different frames into the underlying spatio-temporal structure. Efficiency refers to how to efficiently process long point cloud videos with limited computing resources. The complexity and dimension of 4D data can easily cause memory and computation explosions. Both challenges grow dramatically as the length of the video increases.

One typical way to tackle the dynamics of point clouds videos is treating the point cloud video as a 4D volume [6], which applies 4D convolution directly after voxelization. It is computationally prohibitive when processing large scenes and long videos. Compared with transformer-based 4D backbones, pure convolution is less effective at capturing long-term spatio-temporal context. However, the existing transformer-based 4D backbone(P4Transformer [10]) also fails to solve the above challenges. The entire point cloud video still needs to be loaded into memory during the training process, which severely limits the length of the point cloud video (for example, a 24GB graphics card can only handle a synthia4D [34] point cloud video of 3 frames). Additionally, even though flat transformers may be able to capture long-term context theoretically, they are difficult to optimize as point numbers increase and usually do not provide much gain in dense prediction tasks, such as 4D semantic segmentation.

Based on the challenges described above, we have several key observations. First, considering the large variety of points, distance point cloud frames should not be extracted at the point level, as this is neither efficient nor effective. Second, a middle-level abstraction representing the underlying geometry spatially and temporally can be better suited for context modeling, which will not only alleviate the need to process raw points for better efficiency but also allow for easier association across frames for a more effective spatial-temporal structure. After revisiting the geometry processing literature, we choose primitive plane as a mid-level representation, which describes the underlying planar structures in a scene and tends to be much more stable across frames.

In this paper, we leverage primitive planes to develop an efficient and effective 4D backbone named Point Primitive Transformer (PPTr). As primitive planes induce a natural scene-primitive-point hierarchy in space, we also design PPTr as a hierarchical transformer operating on two different levels as shown in Fig. 1. On the lower level, PPTr extracts short-term spatial-temporal features through an intra-primitive point transformer for a short video clip around the frame of interest. Primitive planes are used to restrict the spatial support of attention maps in a point-level transformer. Such geometry-aware locality inductive bias is not only beneficial for the optimization of the transformer but also very effective for extracting descriptive and temporally stable geometric features. On the upper level, PPTr extracts long-term spatial-temporal features through a primitive transformer. We allow very efficient consideration of a long sequence by fitting primitives and computing the primitive features in a pre-processing stage. Through the primitive transformer, we could better associate primitives from different frames and effectively integrate long-term context to the frame of interest.

We evaluate our Point Primitive Transformer (PPTr) on several tasks, such as 3D action recognition on MSR-Action [25] and 4D semantic segmentation on Synthia4D [34] and HOI4D [29]. We demonstrate significant improvements over previous method (+1.33% mIoU on synthia4d, +6.28% mIoU on HOI4D and +1.39% accuracy on MSR-Action).

The contributions of this paper are fourfold:

- First, we leverage the primitive plane to capture the long-term spatial-temporal context in 4D point cloud videos and propose a novel backbone named Point Primitive Transformer (PPTr).
- Second, we propose an intra-primitive point transformer for extracting spatially descriptive and temporally stable **short-term** geometric features.
- Third, we propose a primitive transformer to capture **long-term** spatial-temporal features efficiently.
- Fourth, extensive experiments on three datasets show that the proposed Point Primitive Transformer is more effective and efficient than previous state-of-the-art 4D backbones.

2   Related Work

Deep Learning on Point Cloud Video Processing. Different from grid-based RGB video, point cloud video exhibits irregularities and lacks order along the spatial dimension where points emerge inconsistently across time. One approach to deal with that is voxilization. For instance, [6] extends temporal dimension to 3D sparse convolution [15] to extract spatial temporal features on 4D occupancy grids. 3DV [41] proposes a 3D motion representation to encode 3D motion information via temporal rank pooling [12]. Another approach is to perform directly on point sets. MeteorNet [28] adopts PointNet++ [32] to aggregate information from neighbors, while point-track is needed to merge points. PSTNet [11] firstly decomposes spatial and temporal information and proposes a point-based convolution in a hierarchical manner. Following [11,28], P4Transformer [10] proposes 4D Convolution that performs spatial-temporal convolution and captures dynamics of points by self-attention. While like most point-based approaches, they prolong input clip by simply feeding raw points into network, which suffers from limited memory and fails to benefit from long-range temporal dependencies. Based on this, we propose Point Primitive Transformer (PPTr) which enjoys all three properties: point-convolution based, long-term supported and point-track avoided.

Primitive Fitting. Primitive fitting is a long-standing problem of grouping points into specific geometric shapes such as plane, cuboid, cylinder and so on. Such process approximates and abstracts 3D shapes from low-level digitized point data to a succinct high-level parameterized representation. Two mainstream solutions of primitive fitting in geometry community are RANSAC [13,35] and region grow [31,33]. Recently, neural networks have been developed by several works [20,24,38,42,43] to segment primitives. Because primitives extremely simplifies point data while keeps a relatively precise description of 3D geometry, they are widely applied to downstream tasks like instance segmentation [20], reconstruction [5]and animation [37]. For example, [14] utilizes primitive shapes that are rich in underlying structures to reconstruct scanned object and transfer the structural information onto new objects. To directly deal with large-scale scenes, [23] distils organization of point cloud by partitioning heavy points into light shapes, showing the power of such compact yet rich representation. We inject primitives into our network, intending to spatially provide geometric-aware enhancement on local primitive region and temporally leverage long-range information in a memory efficient way.

Transformer Network. Transformer is a powerful deep neural network based on self-attention mechanism [39] and is particularly suitable for modelling long-range dependencies [3,7,36]. It was firstly proposed in [39] for machine translation task and further extended to vision community [1,4,8,9,21,27,40]. Very recently, Swin Transformer [30] proposes a hierarchical design for vision modeling at various scales and yields impressive results. Similar to CNNs [17,22], Swin transformer builds hierarchical feature maps by merging image patches when layers go deeper, and strikes a balance between efficiency and effectiveness by limiting self-attention to local windows while also supporting cross-window connection. In 4D point cloud understanding, prior leading work [10] performs self-attention globally and fails to leverage long-term dependencies effectively. As such, we design a hierarchical Primitive Point Transformer (PPTr) to alleviate ineffectiveness of global-wise attention and introduce intra-primitive point transformer and primitive transformer that perform self-attention at point level and primitive level respectively. Intensive experiments have shown that our network outperforms the state-of-the-art methods for both 4D semantic segmentation and 4D action recognition.

3   Pilot Study: How Does P4Transformer Perform on Long-Term Point Cloud Videos?

4D point cloud video understanding has obtained much attention recently and researchers are actively seeking for backbones to capture descriptive spatial-temporal features. Among them, P4Transformer [10] is the leading one achieving

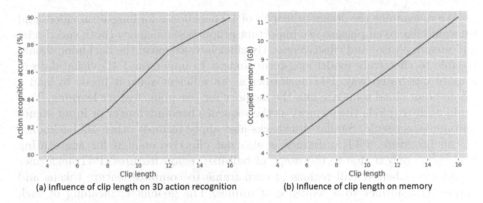

(a) Influence of clip length on 3D action recognition   (b) Influence of clip length on memory

Fig. 2. (a) The performance gain (MSR-Action3D [25]) with the increase of temporal range. (b) The occupied memory with the increase of temporal range. We take the 2080Ti (11 GB) GPU as an example. When the GPU memory cap is reached, the maximum number of frames that can be used is 15, which can only achieve 89% accuracy.

state-of-the-art performance on common tasks including 4D semantic segmentation and 4D action recognition. Briefly speaking, instead of tracking points, P4Transformer uses a point 4D convolution to encode the spatio-temporal local structures in a point cloud video, and utilize the transformer to capture the global appearance and motion information across the entire video. To motivate the necessity of a new backbone, we conduct a pilot study to understand the constraints of P4Transformer for long-term point cloud video understanding.

- We first experiment with the action recognition task on MSR-Action3D [25] dataset. We gradually increase the clip length until our GPU memory cap is reached and examine how well P4transformer performs. The results are shown in Fig. 2.
- We further conduct 4D semantic segmentation experiments on the synthia 4D dataset [34], to verify the effect of Transformer. Specifically, We removed the Transformer in P4Transformer and compared it with the full version.

We can draw mainly two conclusions from the above experiments. First, as shown in Fig. 2, P4Transformer achieves better performance as the clip length increases but is soon restricted by the huge memory cost, and it is hard to apply P4Transformer to very long clips. When the GPU memory cap is reached, the performance still keeps its trend of going up, indicating the huge potential of exploring longer-term videos. Second, in synthia4D [34] semantic segmentation task, we find that P4transformer without Transformer can achieve mIoU of 80.3%, which only drops 2.86% compared with original P4Transformer. This result indicates global spatial-temporal context captured by P4Transformer becomes less useful in 4D dense prediction tasks. This is quite counter-intuitive as the first conclusion indicates the benefit of modeling long-term information. We conjecture that using a flat transformer as in P4Transformer is not effective for long-term spatial-temporal context due to optimization issues.

We re-examine the design principles of 4D backbones for long-term videos and we would like to emphasize two important properties: efficiency (both speed-wise and memory-wise) and effectiveness. By efficiency, we mean the backbone should be able to effectively model long-term context to understand 4D visual data in a more integrated way. P4Transformer is not efficient since it needs to load a whole point cloud sequence into the memory for per-point feature learning. This could easily explode the memory as the sequence becomes longer or input scenes become larger-scale. Similar drawbacks also apply to most other 4D backbones in the literature. P4Transformer is also not effective enough for aggregating long-term context due to the usage of ball-like region features. P4Transformer samples equal-sized ball regions in each frame to compute feature tokens and applies transformer to a sequence of frames. The geometric meaning of such randomly sampled balls could hugely vary in dynamic scenes. This makes it hard to build long-term associations, which is important for long-term context.

4   Method

To develop an efficient and effective backbone for long-term 4D understanding, we draw inspirations from the geometry processing community that primitive planes as some mid-level geometric representations are both compact and stable across time, see Fig. 3. Using primitive planes to model the long-term context not only eases the need to directly deal with the huge number of raw points in a 4D sequence but also facilitates long-term feature association. Furthermore, since primitive planes group points with coherent geometric features, it builds a natural geometry hierarchy (scene-primitive-point) which could be used as a strong inductive bias for powerful yet hard-to-optimize transformer-style architectures. We follow this thought and develop our Point Primitive Transformer.

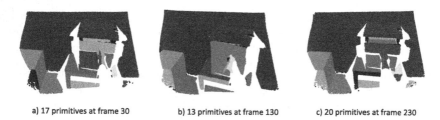

a) 17 primitives at frame 30   b) 13 primitives at frame 130   c) 20 primitives at frame 230

Fig. 3. An illustration of primitive fitting in a HOI4D [29] video. Despite changing view angles and challenging interaction, the primitive fitting remains consistent across time.

Point Primitive Transformer (PPTr) is a two-level hierarchical transformer built upon the geometry hierarchy induced by primitive planes as shown in Fig. 4.

On the lower level, short-term spatial-temporal features are extracted through an intra-primitive point transformer. The intra-primitive point transformer restricts the communication of points within each primitive plane. This design shares a similar flavor with GLOM [19] encouraging aligned features to talk. Also due to the local spatial support, it is more friendly to optimization compared with a global transformer. On the upper level, long-term spatial-temporal features are extracted through a primitive transformer. This is done by jointly analyzing short-term features from the lower level and a memory pool storing pre-computed primitive features from a long video. Pre-computed primitive features allow aggregating long-term spatial-temporal context efficiently and effectively. PPTr is very flexible for both point-wise and sequence-wise inference by simply changing the task head.

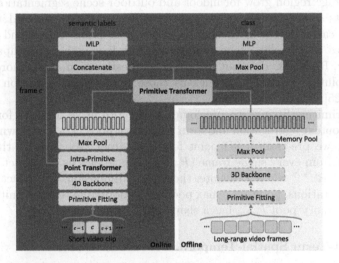

Fig. 4. Pipeline. The backbone consists of two branches: online network and offline pre-computation. **1. Online Branch.** The input to network is a short video clip. After primitive fitting, points are tagged with primitive label, then 4D Backbone is applied and generates per-point features. In the intra-primitive point transformer, points features are enhanced by adaptively adding information from other points inside primitive. Then generate primitive-level representations by maxpooling. In the primitive transformer, clip primitive embeddings (green) perform self-attention with long-term embeddings (yellow) in the memory pool. For semantic segmentation, primitive features are concatenated to corresponding point features then classified into semantic labels. For action recognition, primitive features are merged by maxpooling to a global feature then classified into actions. **2. Offline Branch.** This branch essentially computes primitive level representations of the long-range videos and maintains a memory pool in an offline manner. After primitive fitting, points are fed to a pre-trained 3D backbone. Then maxpool is applied to every primitive region generating primitive-level embeddings in the memory pool. (Color figure online)

In the rest of this section, we will elaborate on the design of PPTr in detail. We start with how we fit primitive planes and how we pre-compute primitive features in Sect. 4.1. Then in Sect. 4.2 and Sect. 4.3, we explain how we extract short-term and long-term spatial-temporal features respectively.

4.1 Primitive Fitting and Feature Pre-computation

We represent a point cloud sequence as $\Psi = \{(P_t, V_t) | t = 1, \ldots, L\}$, where P_t is the point cloud of frame t optionally accompanied with normals V_t. In this phase, we detect planes for each frame (P_t, V_t) and output primitive label $\Xi_t \in \mathbb{R}^{N \times 3}$ and primitive parameters $\Theta_t \in \mathbb{R}^{M \times 4}$, where N is the number of points and M is the number of primitives. We adopt two primitive fitting methods in our study for different datasets: region grow [31] and RANSAC [13].

We leverage region grow for indoor and outdoor scene segmentation. Region grow detects planes based on normal estimation. If not provided with normal V_t, we calculate the normal direction at each point beforehand by linear least squares fitting of a plane over its nearest k neighbors. Compared with region grow, RANSAC does not require normal estimation and is more suitable for low-resolution point clouds such as those for action recognition in MSR-Action3D [25].

After primitive fitting, we pre-compute the primitive features for efficient long-term context aggregation and form a memory pool F_{mem} as shown in Fig. 4. Specifically, we pre-train a 3D point feature learner [10] to solve the task of interest just from every single frame (P_t, V_t). This allows us to extract per-point features $F_t \in \mathbb{R}^{C \times N}$ where C denotes the feature dimension. To extract primitive level representations, point-wise max pooling is adopted for each primitive plane. The final memory pool F_{mem} has a shape of $\mathbb{R}^{C \times M \times L}$.

4.2 Short-Term Spatial-Temporal Feature Extraction

This branch mainly consists of a 4D backbone and an intra-primitive point transformer. The per-point features of each 4D sequence are first extracted using the 4D backbone. Following that, an intra-primitive point transformer is used to extract low-level features. Point features can provide the most fine-grained information, enabling us to better perform dense prediction tasks. The intra-primitive point transformer can not only align point features of similar geometry but also save computational overhead and reduce the optimization difficulty of the transformer.

4D Backbone. Our 4D backbone is built using a UNet structure. Following the state-of-the-art P4Transformer [10], the encoder/decoder is made up of four 4D convolution/deconvolution layers. Given clip Ψ, the convolution layer can be described as:

$$\boldsymbol{f}_t'^{(x,y,z)} = \sum_{\delta_t = -r_t}^{r_t} \sum_{\|(\delta_x, \delta_y, \delta_z)\| \leq r_s} (\boldsymbol{W_d} \cdot (\delta_x, \delta_y, \delta_z, \delta_t)^T) \odot (\boldsymbol{W_f} \cdot \boldsymbol{f}_{t+\delta_t}^{(x+\delta_x, y+\delta_y, z+\delta_z)})$$

(1)

where $(x, y, z) \in P_t$ and $(\delta_x, \delta_y, \delta_z, \delta_t)$ is spatial-temporal offset of kernel and $\cdot$ is matrix multiplication. $f_t^{(x,y,z)} \in \mathbb{R}^{C \times 1}$ is the feature of point at (x, y, z, t), and the temporal aggregation $\sum$ is implemented with sum-pooling and the spatial $\sum$ is max-pooling. r_s and r_t represent temporal and spatial radius. $\boldsymbol{W_d} \cdot (\delta_x, \delta_y, \delta_z, \delta_t)^T$ generates offset weights where $\boldsymbol{W_d} \in \mathbb{R}^{C' \times 4}$ transforms 4D displacements from $\mathbb{R}^{4 \times 1}$ to $\mathbb{R}^{C' \times 1}$, and $\boldsymbol{W_f} \in \mathbb{R}^{C' \times C}$ is a projection matrix. $\odot$ is summation.

Intra-primitive Point Transformer. In this stage, the lower-level feature is extracted by enhancing per-point features obtained from the 4D backbone in a geometry-aware way. Point features are clustered in groups according to their primitive labels given in the primitive fitting phase. Compared with simply grouping by k-NN search in euclidean space [16], primitive-based partition has a more underlying geometric meaning such as normal consistency. As point clouds are sets embedded in a metric space, self-attention is a natural way to build connections among them. By optimizing point embeddings in a geometry-aware manner, our intra-primitive transformer takes advantage of local aggregation rather than global information exchange. It is more friendly to optimization than a global transformer because points within the primitive plane cannot communicate with points outside. After this step, points with similar geometric features are easier to align together, which facilitates subsequent higher-level feature extraction.

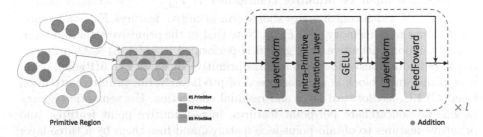

Fig. 5. Left: primitive-based region partition. Points are divided into primitive regions according to primitive labels. Intra-primitive performs self-attention in one primitive region. **Right: Intra-primitive transformer block.** Consisting of intra-primitive attention layer, pre-LayerNorm [2], GELU [18] and residual connection [17].

Specifically, in the layer i, the enhanced point feature F_{out}^i of primitive region i with input embedding set F_{in}^i is computed formally as [39]:

$$Q = W_q \cdot F_{\text{in}}^i, K = W_k \cdot F_{\text{in}}^i, V = W_v \cdot F_{\text{in}}^i$$
$$F_{\text{out}}^i = SA(Q, K, V) = softmax\left(\frac{Q^T K}{\sqrt{C^k}}\right) V \tag{2}$$

where $F_{\text{in}}^i \in \mathbb{R}^{C^i \times N' \times LM}$, C^i, N', L, M represents input dimension, point number per primitive, clip length and primitive number respectively. $W_q, W_k \in \mathbb{R}^{C^k \times C^i}$,

$W_v \in \mathbb{R}^{C^v \times C^i}$ where C^k is the key dimension and C^v is value dimension. Q, K, V are queries, keys and values generated from F_{in}^i. Attention weights $softmax\left(\frac{Q^T K}{\sqrt{C^k}}\right)$ is calculated in the primitive region. The output $F_{out}^i \in \mathbb{R}^{C^v \times N' \times LM}$ is computed as a weighted sum of the values V. As shown in Fig. 5, we build intra-primitive transformer block with layernorm [2], GELU activation [18], one attentive layer and a following feedforward layer [39]. Feedforward is implemented with a two-layer MLP (MultiLayer Perception).

4.3   Long-Term Spatial-Temporal Feature Extraction

After the short-term spatial-temporal feature extraction, primitive transformers are used to jointly analyze short-term features from the lower level and a memory pool containing pre-computed primitive features. This branch can not only reduce the computational cost, but also achieve long-term spatio-temporal information integration.

Primitive Transformer. As demonstrated in Fig. 4, two branches merge here. The output of l layer intra-primitive transformer $F_{out}^l \in \mathbb{R}^{C^l \times N' \times LM}$ is then aggregated by max-pooling operator $MAX\{\cdot\}$ to obtain primitive level feature $F_{out} \in \mathbb{R}^{C^l \times LM}$, where C^l is feature channels, L is the clip length and M is the primitive number. Pre-computed primitive features from memory pool F_{mem} are used to expand the spatio-temporal receptive field of the primitive transformer. Formally, the input of primitive transformer is $F_{in}^{primitive} = [F_{clip} || F_{mem}] \in \mathbb{R}^{C^l \times (L'+L)M}$ which concatenates short-term primitive features F_{clip} and primitive features from memory pool F_{mem}. Note that in the primitive attention layer, spatial-temporal attentive aggregation is performed in $(L' + L) \times M$ primitive regions simultaneously. Identical to intra-primitive shown in Fig. 5(Right), primitive transformer block is also composed of pre-LayerNorm, primitive attention layer, GELU, feedforward layer and residual connection. For semantic segmentation, we concatenate per-point features, intra-primitive point features, and primitive features to obtain point-wise features, and fuse them by a three-layer MLP. For the action recognition task, we use the primitive feature to obtain classification predictions through max-pooling and MLP.

5   Experiments

5.1   4D Semantic Segmentation

Setup. Temporal information can help understand the dynamic objects in the scene, and improve segmentation accuracy and robustness to noise. Due to memory constraints, existing methods only process point cloud videos with a length of 3. Our method can consider a longer temporal range and achieve a more efficient integration of spatio-temporal information. In this task, we fit the scene point cloud into 200 primitives. We use mean IoU (mIoU) % as the evaluation metric.

4D Semantic Segmentation on Synthia 4D Dataset. Setup. Synthia 4D [34] is a synthetic dataset for outdoor autonomous driving. It creates 3D videos with the Synthia dataset, which consists of six videos of driving scenarios in which objects and cameras are moving. We use the same training/validation/test split as previous work, with 19,888/815/1,886 frames, respectively.

Table 1. Evaluation for semantic segmentation on Synthia 4D dataset [34]

Method	Frames	Bldn	Road	Sdwlk	Fence	Vegittn	Pole	Car	T.Sign	Pedstrn	Bicycl	Lane	T.Light	mIoU
3D MinkNet14 [6]	1	89.39	97.68	69.43	86.52	98.11	97.26	93.50	79.45	92.27	0.00	44.61	66.69	76.24
4D MinkNet14 [6]	3	90.13	98.26	73.47	87.19	99.10	97.50	94.01	79.04	**92.62**	0.00	50.01	68.14	77.24
PointNet++ [32]	1	96.88	97.72	86.20	92.75	97.12	97.09	90.85	66.87	78.64	0.00	72.93	75.17	79.35
MeteorNet-m [28]	2	**98.22**	97.79	90.98	93.18	98.31	97.45	94.30	76.35	81.05	0.00	74.09	75.92	81.47
MeteorNet-l [28]	3	98.10	97.72	88.65	94.00	97.98	97.65	93.83	**84.07**	80.90	0.00	71.14	77.60	81.80
P4Transformer [10]	1	96.76	98.23	92.11	95.23	98.62	97.77	95.46	80.75	85.48	0.00	74.28	74.22	82.41
P4Transformer [10]	3	96.73	98.35	94.03	95.23	98.28	98.01	95.60	81.54	85.18	0.00	75.95	**79.07**	83.16
PPTr (ours)	1	97.14	98.42	94.12	97.00	99.59	97.86	98.54	79.68	89.20	0.00	77.26	77.42	83.85
PPTr (ours)	30	98.01	**98.63**	**95.26**	**97.03**	**99.70**	**97.95**	**98.76**	81.99	91.20	0.00	**78.29**	77.09	**84.49**

Result. Table 1 shows our method outperforms the state-of-the-art methods. Our PPTr with 1 frame can achieve 0.69% improvement over the P4Transformer with 3 frames, which demonstrates the effectiveness of the hierarchical structure. When using the memory pool to integrate temporal information from 30 frames, we can achieve 1.33% improvement over previous state-of-the-art methods. It is worth mentioning that our method is the first to integrate point clouds of 30 frames, which is 10 times that of previous methods. And we also demonstrate that longer point cloud sequences are valuable for 4D semantic segmentation.

4D Semantic Segmentation on HOI4D. Setup. In order to further verify the effectiveness of our method, we select the HOI4D dataset for experiments, which is a large-scale 4D egocentric dataset to catalyze the research of category-level human-object interaction. It provides frame-wise annotations for 4D point cloud semantic segmentation. Since the dataset has not been released yet, we sent an email to the author team to request 1000 sequences, which includes 30k frames of the point cloud. The train/test split is the same as HOI4D.

Result. As shown in Table 2, our method outperforms previous methods on this more challenging dataset. Compared with P4transformer, the mIoU goes up from 59.61% to 68.07% and 61.97% to 68.54% in the case of single frame and 3 frames respectively, demonstrating the effectiveness of the hierarchical design again. Due to the limitation of computational resources, P4Transformer can use up to 3 frames, but our method can integrate 30 frames of spatio-temporal information. The improvement from 61.97% to 70.13% further confirms that with our proposed primitive memory pool, we can better leverage the long-term temporal information to boost the 4D segmentation performance.

Table 2. Evaluation for semantic segmentation on HOI4D dataset [29]

Method	Frames	Table	Ground	Metope	Locker	Pliers	Laptop	Safe Deposit	Pillow	Hand and Arm	mIoU
PSTNet [11]	3	57.45	63.38	83.80	44.69	13.71	35.03	51.55	76.30	40.39	51.81
P4Transformer [10]	1	60.84	71.98	86.69	53.89	34.00	65.89	55.87	52.19	55.10	59.61
P4Transformer [10]	3	63.58	66.60	87.17	58.39	32.29	72.03	65.87	57.41	54.36	61.97
PPTr (ours)	1	67.49	74.92	87.92	**62.12**	40.06	69.00	71.39	77.18	62.50	68.07
PPTr (ours)	3	66.78	72.76	88.21	60.83	**41.22**	72.04	73.10	80.64	61.27	68.54
PPTr (ours)	30	**67.76**	**79.55**	**90.67**	59.43	39.43	**72.67**	**73.29**	**84.13**	**64.26**	**70.13**

5.2   3D Action Recognition on MAR-Action3D

Setup. To demonstrate the effect of PPTr, we first conduct experiments on the 3D Action Recognition task. Followed by P4Transformer, we use the MAR-Action3D dataset which consists of 567 human body point cloud videos, including 20 action categories. Our test/train split follows previous work. Each frame is sampled with 2,048 points. As inputs, point cloud videos are split into multiple clips. Video-level labels are used as clip-level labels during training. In order to estimate the video-level probability, we take the mean of all clip-level probability predictions. We fit the human body point cloud into 4 primitives. Due to the small scale of human point cloud videos, we can load the entire point cloud videos at one time, so we can avoid maintaining the long-term memory pool in this case. We use the video classification accuracy as the evaluation metric. We compare our method with the latest 4D backbone for point cloud video including MeteorNet, PSTNet and P4Transformer.

Result. As reported in Table 3, when the number of point cloud frames increases, the classification accuracy can be gradually improved. Our method outperforms all the state-of-the-art methods, demonstrating that our methods can better integrate spacial-temporal information.

5.3   Ablation Study and Discussion

In this section, we first provide an ablation study to verify each component. Then, we provide more analysis to provide an in-depth understanding of our framework.

Efficacy of Intra/Inter-primitive Transformer. We run ablation studies with and without intra/inter-primitive Transformer to quantify its efficacy. We find that PPTr without intra/inter-primitive Transformer results in a 16.73/1.39 accuracy drop on the MSR-Action3D action recognition task. This shows that the intra-primitive transformer is essential in this task. It not only simplifies the optimization difficulty but also aligns similar points, providing good features for the subsequent use of the inter-primitive transformer. Inter-primitive Transformer integrates spatio-temporal information from the entire video, using the complementary information of each frame to further improve classification accuracy.

Table 3. Evaluation for action recognition on MSR-Action3D dataset [25]

Method	Input	Frames	Accuracy
PointNet++ [32]	point	1	61.61
MeteorNet [28]	Point	4	78.11
	Point	8	81.14
	Point	12	86.53
	Point	16	88.21
	Point	24	88.50
PSTNet [11]	Point	4	81.14
	Point	8	83.50
	Point	12	87.88
	Point	16	89.90
	Point	24	91.20
P4Transformer [10]	Point	4	80.11
	Point	8	83.17
	Point	12	87.54
	Point	16	89.56
	Point	24	90.94
PPTr (ours)	Point	4	80.97
	Point	8	84.02
	Point	12	89.89
	Point	16	90.31
	Point	24	**92.33**

Robustness to Primitive-Fitting Hyper-parameters. The performance impacts of different numbers of primitives are provided since primitives are crucial in the framework. On MSR-Action3D, we can achieve 91.5/91.89 accuracy with 2/8 primitives, resulting in a marginal drop. On Synthia 4D, the segmentation mIoUs are 82.98, 84.41, 84.49, 84.28, and 83.56 with a primitive number of 10, 100, 200, 400, and 2000 respectively. Notice when the primitive number varies in a reasonable range from 100 to 400, the segmentation mIoUs vary by no more than 0.21. When the primitive number is 10, the region division is too coarse for fine-grained segmentation. When the primitive number is 2000, the benefit of the spatial hierarchy gets weakened a lot. The network degenerates to a point transformer when further increasing the primitive number to the point number. This shows that different numbers of primitives have a small effect on the results, and all have consistent improvements.

Efficacy of Primitive Representation. Our hierarchical transformer is generic and can be easily applied to mid-level representations other than primitive planes. To confirm the efficacy of primitive planes, we additionally compare primitive planes with two types of mid-level representations, BPSS [26] supervoxels and k-means clusters. Results in the table below show that using BPSS

Table 4. Comparisons between different representations

Method	Synthia 4D [34]	MSRAction [25]
P4Transformer	83.16	90.94
K-means	80.70	91.76
BPSS [26]	83.43	91.98
Ours	**84.49**	**92.33**

supervoxels outperforms P4Transformer but is not as good as using primitive planes while k-means clusters fail to serve as a beneficial mid-level representation on Synthia4D (Table 4).

Offline Branch and Online Branch. The online branch produces fine primitive features with heavy computation while the offline branch produces coarser features efficiently as a surrogate of the online branch so that the network can process long clips with limited computing resources. For the action recognition task where data clips can already be largely fit into the GPU memory, using an online branch only with fine primitive features is preferred. In this case, just using an offline branch or combining the offline and online branches results in marginal performance degradation with accuracy of 92.13 and 92.27 respectively. For the 4D segmentation task, using our online branch independently, the memory could only afford 3 frames and the resulting segmentation mIoU (%) is 84.05. This number goes to 84.49 when assisted by the offline branch covering 30 frames, confirming the value of the offline branch.

6   Conclusions

This paper proposes a 4D backbone for long-term point cloud video understanding. The key idea is to leverage the primitive plane to capture the long-term spatial-temporal context in 4D point cloud videos. Results of experiments showing ablations and state-of-the-art performance on a wide range of 4D tasks including MSR-Action3D action recognition task, 4D semantic segmentation on sythia4D and on HOI4D. This result is very encouraging and suggests future work to explore more possible backbone designs for 4D point cloud understanding.

References

1. Arnab, A., Dehghani, M., Heigold, G., Sun, C., Lučić, M., Schmid, C.: ViViT: a video vision transformer. In: Proceedings of the IEEE/CVF International Conference on Computer Vision, pp. 6836–6846 (2021)
2. Ba, J.L., Kiros, J.R., Hinton, G.E.: Layer normalization. arXiv preprint arXiv:1607.06450 (2016)

3. Brown, T., et al.: Language models are few-shot learners. Adv. Neural. Inf. Process. Syst. **33**, 1877–1901 (2020)
4. Carion, N., Massa, F., Synnaeve, G., Usunier, N., Kirillov, A., Zagoruyko, S.: End-to-end object detection with transformers. In: Vedaldi, A., Bischof, H., Brox, T., Frahm, J.-M. (eds.) ECCV 2020. LNCS, vol. 12346, pp. 213–229. Springer, Cham (2020). https://doi.org/10.1007/978-3-030-58452-8_13
5. Chen, J., Chen, B.: Architectural modeling from sparsely scanned range data. Int. J. Comput. Vision **78**(2), 223–236 (2008)
6. Choy, C.B., Gwak, J., Savarese, S.: 4D spatio-temporal convnets: Minkowski convolutional neural networks. CoRR abs/1904.08755 (2019)
7. Devlin, J., Chang, M.W., Lee, K., Toutanova, K.: Bert: pre-training of deep bidirectional transformers for language understanding. arXiv preprint arXiv:1810.04805 (2018)
8. Dosovitskiy, A., et al.: An image is worth 16x16 words: transformers for image recognition at scale. arXiv preprint arXiv:2010.11929 (2020)
9. Du, H., Yu, X., Zheng, L.: VTNet: visual transformer network for object goal navigation. arXiv preprint arXiv:2105.09447 (2021)
10. Fan, H., Yang, Y., Kankanhalli, M.S.: Point 4D transformer networks for spatio-temporal modeling in point cloud videos. In: IEEE Conference on Computer Vision and Pattern Recognition, CVPR, pp. 14204–14213 (2021)
11. Fan, H., Yu, X., Ding, Y., Yang, Y., Kankanhalli, M.: PSTNet: point spatio-temporal convolution on point cloud sequences. In: International Conference on Learning Representations (2020)
12. Fernando, B., Gavves, E., Oramas, J., Ghodrati, A., Tuytelaars, T.: Rank pooling for action recognition. IEEE Trans. Pattern Anal. Mach. Intell. **39**(4), 773–787 (2016)
13. Fischler, M.A., Bolles, R.C.: Random sample consensus: a paradigm for model fitting with applications to image analysis and automated cartography. Commun. ACM **24**(6), 381–395 (1981)
14. Ganapathi-Subramanian, V., Diamanti, O., Pirk, S., Tang, C., Niessner, M., Guibas, L.: Parsing geometry using structure-aware shape templates. In: 2018 International Conference on 3D Vision (3DV), pp. 672–681. IEEE (2018)
15. Graham, B., van der Maaten, L.: Submanifold sparse convolutional networks. CoRR abs/1706.01307 (2017)
16. Guo, M.H., Cai, J.X., Liu, Z.N., Mu, T.J., Martin, R.R., Hu, S.M.: PCT: point cloud transformer. Comput. Vis. Media **7**(2), 187–199 (2021)
17. He, K., Zhang, X., Ren, S., Sun, J.: Deep residual learning for image recognition. In: Proceedings of the IEEE Conference on Computer Vision and Pattern Recognition, pp. 770–778 (2016)
18. Hendrycks, D., Gimpel, K.: Gaussian error linear units (GELUs). arXiv preprint arXiv:1606.08415 (2016)
19. Hinton, G.: How to represent part-whole hierarchies in a neural network. arXiv preprint arXiv:2102.12627 (2021)
20. Huang, J., Zhang, Y., Sun, M.: Primitivenet: primitive instance segmentation with local primitive embedding under adversarial metric. In: Proceedings of the IEEE/CVF International Conference on Computer Vision, pp. 15343–15353 (2021)
21. Kolesnikov, A., et al.: Big transfer (BiT): general visual representation learning. In: Vedaldi, A., Bischof, H., Brox, T., Frahm, J.-M. (eds.) ECCV 2020. LNCS, vol. 12350, pp. 491–507. Springer, Cham (2020). https://doi.org/10.1007/978-3-030-58558-7_29

22. Krizhevsky, A., Sutskever, I., Hinton, G.E.: Imagenet classification with deep convolutional neural networks. In: Advances in Neural Information Processing Systems, vol. 25 (2012)
23. Landrieu, L., Simonovsky, M.: Large-scale point cloud semantic segmentation with superpoint graphs. In: Proceedings of the IEEE Conference on Computer Vision and Pattern Recognition, pp. 4558–4567 (2018)
24. Li, L., Sung, M., Dubrovina, A., Yi, L., Guibas, L.J.: Supervised fitting of geometric primitives to 3D point clouds. In: Proceedings of the IEEE/CVF Conference on Computer Vision and Pattern Recognition, pp. 2652–2660 (2019)
25. Li, W., Zhang, Z., Liu, Z.: Action recognition based on a bag of 3D points. In: 2010 IEEE computer Society Conference on Computer Vision and Pattern Recognition-Workshops, pp. 9–14. IEEE (2010)
26. Lin, Y., Wang, C., Zhai, D., Li, W., Li, J.: Toward better boundary preserved supervoxel segmentation for 3D point clouds. ISPRS J. Photogramm. Remote. Sens. **143**, 39–47 (2018)
27. Liu, L., Hamilton, W., Long, G., Jiang, J., Larochelle, H.: A universal representation transformer layer for few-shot image classification. arXiv preprint arXiv:2006.11702 (2020)
28. Liu, X., Yan, M., Bohg, J.: Meteornet: deep learning on dynamic 3D point cloud sequences. In: Proceedings of the IEEE/CVF International Conference on Computer Vision, pp. 9246–9255 (2019)
29. Liu, Y., et al.: HOI4D: A 4D Egocentric Dataset for Category-Level Human-Object Interaction. arXiv e-prints, March 2022
30. Liu, Z., et al.: Swin transformer: hierarchical vision transformer using shifted windows. In: Proceedings of the IEEE/CVF International Conference on Computer Vision, pp. 10012–10022 (2021)
31. Marshall, D., Lukacs, G., Martin, R.: Robust segmentation of primitives from range data in the presence of geometric degeneracy. IEEE Trans. Pattern Anal. Mach. Intell. **23**(3), 304–314 (2001)
32. Qi, C.R., Yi, L., Su, H., Guibas, L.J.: Pointnet++: deep hierarchical feature learning on point sets in a metric space. In: Advances in Neural Information Processing Systems, vol. 30 (2017)
33. Rabbani, T., Van Den Heuvel, F., Vosselmann, G.: Segmentation of point clouds using smoothness constraint. Int. Arch. Photogram. Remote Sens. Spatial Inf. Sci. **36**(5), 248–253 (2006)
34. Ros, G., Sellart, L., Materzynska, J., Vazquez, D., Lopez, A.M.: The synthia dataset: a large collection of synthetic images for semantic segmentation of urban scenes. In: Proceedings of the IEEE Conference on Computer Vision and Pattern Recognition, pp. 3234–3243 (2016)
35. Schnabel, R., Wahl, R., Klein, R.: Efficient RANSAC for point-cloud shape detection. In: Computer Graphics Forum, vol. 26, pp. 214–226. Wiley Online Library (2007)
36. Sun, F., et al.: BERT4Rec: sequential recommendation with bidirectional encoder representations from transformer. In: Proceedings of the 28th ACM International Conference on Information and Knowledge Management, pp. 1441–1450 (2019)
37. Thiery, J.M., Guy, É., Boubekeur, T., Eisemann, E.: Animated mesh approximation with sphere-meshes. ACM Trans. Graph. (TOG) **35**(3), 1–13 (2016)
38. Tulsiani, S., Su, H., Guibas, L.J., Efros, A.A., Malik, J.: Learning shape abstractions by assembling volumetric primitives. In: Proceedings of the IEEE Conference on Computer Vision and Pattern Recognition, pp. 2635–2643 (2017)

39. Vaswani, A., et al.: Attention is all you need. In: Advances in Neural Information Processing Systems, vol. 30 (2017)
40. Wang, X., Girshick, R., Gupta, A., He, K.: Non-local neural networks. In: Proceedings of the IEEE Conference on Computer Vision and Pattern Recognition, pp. 7794–7803 (2018)
41. Wang, Y., et al.: 3DV: 3D dynamic voxel for action recognition in depth video. In: Proceedings of the IEEE/CVF Conference on Computer Vision and Pattern Recognition, pp. 511–520 (2020)
42. Yan, S., Yang, Z., Ma, C., Huang, H., Vouga, E., Huang, Q.: HPNet: deep primitive segmentation using hybrid representations. In: Proceedings of the IEEE/CVF International Conference on Computer Vision, pp. 2753–2762 (2021)
43. Zou, C., Yumer, E., Yang, J., Ceylan, D., Hoiem, D.: 3D-PRNN: generating shape primitives with recurrent neural networks. In: Proceedings of the IEEE International Conference on Computer Vision, pp. 900–909 (2017)

Adaptive Agent Transformer for Few-Shot Segmentation

Yuan Wang[1], Rui Sun[1], Zhe Zhang[3,4,5(✉)], and Tianzhu Zhang[1,2,5]

[1] University of Science and Technology of China, Hefei, China
[2] Institute of Artificial Intelligence, Hefei Comprehensive National Science Center, Hefei, China
[3] Beijing Institute of Technology, Beijing, China
cnclepzz@126.com
[4] Lunar Exploration and Space Engineering Center of CNSA, Beijing, China
[5] Deep Space Exploration Laboratory, Beijing, China

Abstract. Few-shot segmentation (FSS) aims to segment objects in a given query image with only a few labelled support images. The limited support information makes it an extremely challenging task. Most previous best-performing methods adopt prototypical learning or affinity learning. Nevertheless, they either neglect to further utilize support pixels for facilitating segmentation and lose spatial information, or are not robust to noisy pixels and computationally expensive. In this work, we propose a novel end-to-end adaptive agent transformer (AAFormer) to integrate prototypical and affinity learning to exploit the complementarity between them via a transformer encoder-decoder architecture, including a representation encoder, an agent learning decoder and an agent matching decoder. The proposed AAFormer enjoys several merits. First, to learn agent tokens well without any explicit supervision, and to make agent tokens capable of dividing different objects into diverse parts in an adaptive manner, we customize the agent learning decoder according to the three characteristics of context awareness, spatial awareness and diversity. Second, the proposed agent matching decoder is responsible for decomposing the direct pixel-level matching matrix into two more computationally-friendly matrices to suppress the noisy pixels. Extensive experimental results on two standard benchmarks demonstrate that our AAFormer performs favorably against state-of-the-art FSS methods.

Keywords: Few-shot segmentation · Semantic segmentation · Transformer

1 Introduction

Semantic segmentation is a fundamental task that has achieved conspicuous achievements attributed to the development in deep neural network, especially

---

Y. Wang and R. Sun—Equal contibution.

Supplementary Information The online version contains supplementary material available at https://doi.org/10.1007/978-3-031-19818-2_3.

S. Avidan et al. (Eds.): ECCV 2022, LNCS 13689, pp. 36–52, 2022.
https://doi.org/10.1007/978-3-031-19818-2_3

fully convolutional network (FCN) [21]. However, it is laborious and time-consuming to gather massive pixel-level annotations as training data. To alleviate the data-hunger issue, considerable works [6,14,16,23] have turned their attention to the semi-supervised setting. However, neither fully supervised models nor semi-supervised models generalize well to novel classes with extremely few exemplars. In contrast, humans can easily identify a new object after only seeing it once. Inspired by this, there has been increasing interest recently on few-shot segmentation (FSS) [26] which can quickly adapt to novel categories.

In this work, we tackle the few-shot segmentation problem, where the goal is to segment objects in a given *query* image I_q while only a few *support* images I_s with corresponding annotations M_s are available. Since there are usually large intra-class variations such as scale, pose or background differences between the support and query images, how to fully exploit limited information from support samples for accurate segmentation is thus extremely challenging.

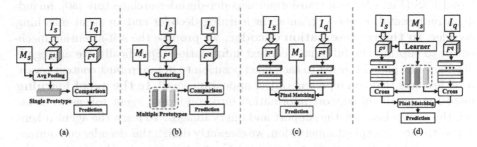

Fig. 1. Different learning formulation for few-shot segmentation. (a) Prototypical learning methods with single prototype. (b) Prototypical learning methods with multiple prototypes. (c) Affinity learning methods. (d) Our proposed AAFormer that absorbs the merits of both prototypical learning and affinity learning methods by modeling adaptive agent tokens for pixel-level matching.

Top-performing FSS methods can be roughly categorized as prototypical learning methods and affinity learning methods. On one hand, prototypical learning methods [10,33,43] adopt masked average pooling to achieve a single prototype in the hope of being robust to noisy pixels, and perform feature comparison between query pixels and a single prototype to segment the desired object, as shown in Fig. 1a. However, these methods inevitably drop the spatial information. Moreover, relying solely on the single prototype focusing on global foreground feature fails to capture the diverse object parts, which are crucial to deal with object occlusion and large variations across images. To alleviate these problems, recent works adopt EM algorithm [38] or clustering [17,20] to generate multiple prototypes for better spatial coverage in foreground regions (see Fig. 1b). However, these methods only conduct matching between obtained prototypes and query features without explicitly exploring the valuable pixel-level support information, which can actually further contribute to the precise segmentation. On the other hand, affinity learning methods [32,40,42] attempts to directly leverage pixel-to-pixel similarity between support features and query features for segmentation (see Fig. 1c). These approaches take advantage of the detailed pixel-level support information and perform well in preserving spatial information. However, direct pixel-level similarity is not only computation prohibitive, but also tends to suffer from the confusion caused by background clutters or noisy pixels because of neglecting contextual information. Overall, The above analysis indicates that prototypical learning methods and affinity learning methods are naturally complementary. The former mines the pixel context information against noisy pixels and is computationally friendly, but fails to further utilize valuable support pixels to facilitate segmentation and loses spatial information, while the latter is just the other way around. Therefore, it is more desirable to integrate these two formulations for exploiting their complementary potential by performing pixel-level matching based on modeling diverse prototypes.

Motivated by the above discussions, we propose an end-to-end Adaptive Agent Transformer (**AAFormer**) to integrate adaptive prototypes as agent into affinity-based FSS (Fig. 1d) via a transformer encoder-decoder architecture [30], including a representation encoder, an agent learning decoder and an agent matching decoder. **In the representation encoder,** we propose the self-attention mechanism to capture the full image context information. Specifically, we aggregate pixel-specific global context to each pixel position to obtain robust context-aware pixel features that can represent object appearance well. **In the agent learning decoder,** we distill support information into condensed agent tokens to establish the bridge between the support and query images. To learn the agent tokens well without any explicit supervision, we elegantly design this decoder customized for the following three characteristics. (a) Context awareness. We introduce the masked cross attention mechanism that only attends agent tokens with support pixels restricted to the foreground region. In this way, agent tokens have the ability to further absorb foreground context from support pixels, and adapt to occlusion and large variations across images. (b) Spatial awareness. To make the part masks activated by the agent tokens more compact rather than dispersive, we model the structural spatial information with the support of distance transformation for agent tokens initialization. In addition, we also introduce the position embedding in the agent learning process to make output agent tokens sensitive to spatial location, and guide the part mask by learning a local activation. (c) Diversity. To avoid the multiple agent tokens focusing on the same object part, we impose the equal partition constraint to expand the discrepancy among part masks. In specific, we allocate foreground pixels evenly over agent tokens benefiting from the initial marginal distribution of the optimal transport algorithm, and attain the optimal transport plan which can be regarded as the refined part masks. In this case, agent tokens can decompose different target objects into diverse and complementary parts in an *adaptive* manner. **In the agent matching decoder,** we decompose the massive pixel-level support-query matching matrix into two more manageable matrices based on obtained agent tokens at a light computational cost, and introduce the alignment matrix for filtering out ambiguous matching caused by noisy pixels. In specific, direct support-query matching is substituted by support-agent matching and agent-query matching. With a limited number of agent tokens, AAFormer efficiently performs pixel-level matching with a drastically reduced complexity compared to previous one. Besides, the alignment matrix guided by context-rich agent tokens can filter out the matching weights between support and query pixels that do not belong to the same object part. In this case, the noisy pixels will be suppressed while the true correspondences enjoy higher weights.

The contributions of our method could be summarized as follows:

- We propose an Adaptive Agent Transformer (AAFormer) for the few-shot segmentation in a unified framework. Specifically, we design the representation encoder to acquire global context-aware pixel features, the agent learning decoder to condense support information into agent tokens for bridging the support and query images, and the agent matching decoder to decompose the direct pixel-level matching matrix into two more computationally-friendly matrices for suppressing the noisy pixels.
- To the best of our knowledge, this is the first work to absorb the merits of both prototypical learning and affinity learning formulation by modeling adaptive agent tokens for pixel-level matching. To learn agent tokens well without any explicit supervision, and to make agent tokens capable of dividing different objects into diverse parts in an adaptive manner, we further customize the agent learning decoder according to the three characteristics of context awareness, spatial awareness and diversity.
- Extensive experimental results with two different backbones on two challenging benchmarks demonstrate that our AAFormer performs favorably against state-of-the-art FSS methods.

2   Related Work

2.1   Semantic Segmentation

Semantic segmentation is a task of assigning each pixel in a given image into a category label, most promising segmentation methods are based on the Fully Convolutional Network (FCN) [21]. Later, many remarkable breakthroughs come from the enlargement of the receptive field. For example, Deeplab [4,5] integrates dilated convolutions combined with pyramid pooling module [44] into the FCN architecture. In addition to CNN based models, some recent works [2,7,8,27,35,45] have applied transformer-based architectures for semantic segmentation [28], and has resulted in comparable performance. For instance, Mask2former [7] treats semantic segmentation as a binary mask prediction task based on set prediction mechanism proposed by DETR [3]. However, these methods usually require massive pixel-level annotations as training data and cannot generalize to novel classes with only a few labelled images. In this paper, we focus on few-shot segmentation to overcome these limitations.

2.2 Few-Shot Segmentation

Few-shot segmentation [26] tackles a challenging task of segmenting novel class query images with only a few labeled support images available. Existing FSS methods can be roughly categorized into two categories: prototypical learning methods and affinity learning methods. For prototypical learning, most methods [10,20,38,39,41,43] adopt masked average pooling to achieve a single prototype and perform feature comparison with query pixels to segment the desired object. For example, PANet [33] performs a prototype alignment regularization that encourages the prototypes to contain more consistent information. However, these methods are prone to inevitably drop the spatial information [17]. To alleviate these problems, recent works [17,20,38] attempt to generate multiple prototypes by EM algorithm or clustering for better spatial coverage in foreground regions. For instance, Zhang *et al.* [39] encode the uncovered support feature for initial prediction as a extra auxiliary prototype to reduce information loss. However, these methods neglect exploring the valuable pixel-level support information, which can actually further contribute to the precise segmentation.

Different from prototypical learning, affinity learning methods [32,39,40,42] attempt to directly leverage pixel-level support-query matching for segmentation. For example, PFENet [29] constructs the class-agnostic prior mask to guide the segmentation by calculating the maximum support-query similarity in high-level features. CyCTR [42] introduces the cycle-consistent attention operation to aggregate beneficial support pixel-level features. However, direct pixel-level similarity is not only computation prohibitive, but also tends to suffer from the confusion caused by background clutters or noisy pixels because of neglecting contextual information. Apart from existing methods, our method absorbs the merits of both prototypical learning and affinity learning methods by modeling adaptive agent tokens for pixel-level matching with a transformer encoder-decoder architecture.

3 The Proposed Approach

3.1 Problem Definition

Widely used episodic meta-training [31] is adopted in few-shot segmentat. Specifically, we denote the training set as $\mathcal{D}_{train}$ and the testing set as $\mathcal{D}_{test}$, the categories of the two sets $\mathbb{C}_{train}$ and $\mathbb{C}_{test}$ are disjoint ($\mathbb{C}_{train} \cap \mathbb{C}_{test} = \emptyset$). To train the model, a set of episodes are sampled from $\mathcal{D}_{train}$, each of which is composed of a support set $\mathcal{S}$ and a query set $\mathcal{Q}$. In the K-shot setting, $\mathcal{S} = \{(I_s^k, M_s^k)\}_{k=1}^K$, where the I_s^k and the M_s^k are the i-th support image and its corresponding ground-truth binary mask. Meanwhile, $\mathcal{Q} = (I_q, M_q)$, where the I_q and M_q are the query image of the same class in $\mathcal{S}$ and its ground-truth, respectively. In each episode, the model makes prediction on the I_q of $\mathcal{Q}$ conditioned on the $\mathcal{S}$, and M_q is provided to supervise the training process.

3.2   Overview

As illustrated in Fig. 2, the proposed AAFormer mainly includes three modules, the representation encoder, the agent learning decoder and the agent matching decoder. Among them, the representation encoder is applied to consider the global context to learn robust features that effectively represent object appearance. The agent learning decoder is responsible for adaptively absorbing contextual information into agent tokens, and makes these regions discovered by learnt agent tokens compact and diverse. The agent matching decoder is used to equip each query pixel with beneficial support information to facilitate the classification. The details are as follows.

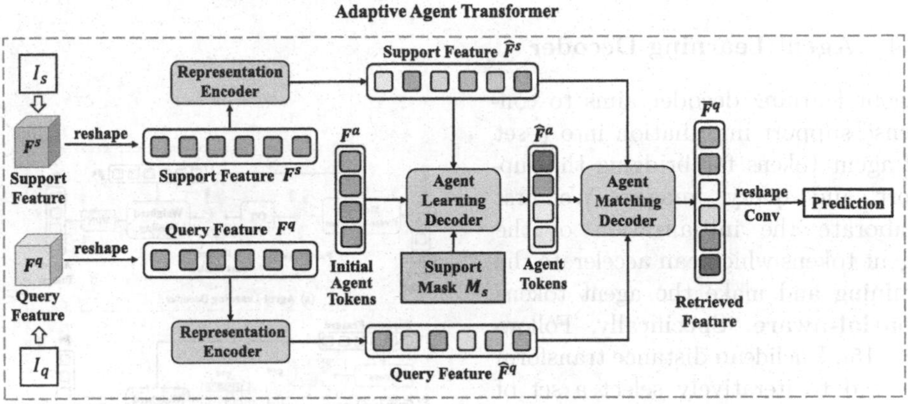

Fig. 2. Framework of our proposed Adaptive Agent Transformer (AAFormer). There are three modules in the AAFormer, i.e., a representation encoder, an agent learning decoder and an agent matching decoder.

3.3   Representation Encoder

We design the representation encoder to acquire robust context-aware pixel features that can represent object appearance well. Given the backbone features $\mathbf{F}^s \in \mathbb{R}^{h \times w \times c}$ and query feature map $\mathbf{F}^q \in \mathbb{R}^{h \times w \times c}$ obtained from the pretrained ResNet [13]. To cope with the difference in the distribution of targets on scale and pose, we adopt self-attention mechanism in the representation encoder to capture the long-range context information. Specifically, we flatten the spatial dimensions of $\mathbf{F}^s$ and $\mathbf{F}^q$ as 1D sequences. Then we obtain the queries, keys and values from $\mathbf{F}^s \in \mathbb{R}^{hw \times c}$ and $\mathbf{F}^q \in \mathbb{R}^{hw \times c}$. Note that we denote the superscript as $*$ and $* \in \{s, q\}$ for brevity. Formally,

$$\mathbf{Q}^* = \mathbf{F}^* \mathbf{W}_*^{\mathcal{Q}}, \quad \mathbf{K}^* = \mathbf{F}^* \mathbf{W}_*^{\mathcal{K}}, \quad \mathbf{V}^* = \mathbf{F}^* \mathbf{W}_*^{\mathcal{V}}, \tag{1}$$

where $\mathbf{W}_*^{\mathcal{Q}} \in \mathbb{R}^{c \times c_k}, \mathbf{W}_*^{\mathcal{K}} \in \mathbb{R}^{c \times c_k}, \mathbf{W}_*^{\mathcal{V}} \in \mathbb{R}^{c \times c_v}$ are linear projections. Then we can calculate the attention weight matrix $\mathbf{S} \in \mathbb{R}^{hw \times hw}$ with the scaled dot-product attention and the output context-aware pixel features are computed through the following equation:

$$\hat{\mathbf{F}}^* = \text{Attention}(\mathbf{Q}^*, \mathbf{K}^*, \mathbf{V}^*) = \text{Softmax}(\frac{\mathbf{Q}^*(\mathbf{K}^*)^\mathsf{T}}{\sqrt{d_k}})\mathbf{V}^*. \tag{2}$$

Among which $\sqrt{d_k}$ is a scaling factor for stabilizing the training and T denotes the transpose operation. Following the standard transformer [30], the Eq. (2) is implemented with the multi-head mechanism and the feed-forward network (FFN) is further applied to obtain the final output $\hat{\mathbf{F}}^s$ and $\hat{\mathbf{F}}^q$. In this way, The obtained pixel features are supported by its global context so that are more robust to background clutters and can better represent object appearance (Fig. 3).

3.4   Agent Learning Decoder

Agent learning decoder aims to condense support information into a set of agent tokens for bridging the support and query images. We first elaborate the initialization of the agent tokens which can accelerate the training and make the agent tokens **spatial-aware**. Specifically, Following [15], Euclidean distance transform is used to iteratively select a set of seed points that far away from each other as well as the boundaries, please refer to the **Supplementary Materials** for specific practices. We then adopt the features at the chosen seed points that distribute uniformly in the masked region as initial agent tokens denoted by $\mathbf{F}^a \in \mathbb{R}^{K \times c}$, where the K is the number of agent tokens.

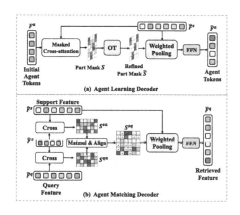

Fig. 3. Illustration of the Agent Learning Decoder (1-st row) and the Agent Matching Decoder (2-nd row).

In order to make agent tokens **context-aware**, we introduce the masked cross-attention between the agent tokens and support features to efficiently aggregate the relevant foreground contextual information into corresponding agent tokens. Concretely, we first calculate a masked attention weight matrix:

$$\mathbf{S} = \text{Softmax}(\frac{\mathbf{Q}^a(\mathbf{K}^s)^\mathsf{T}}{\sqrt{d_k}} + \mathcal{M}), \quad \mathbf{Q}^a = \mathbf{F}^a \mathbf{W}_a^{\mathcal{Q}}, \quad \mathbf{K}^s = \mathbf{F}^s \mathbf{W}_s^{\mathcal{K}}, \tag{3}$$

where the additional attention mask $\mathcal{M}$ at feature location (m, n) is

$$\mathcal{M} = \begin{cases} 0, & \text{if } \mathbf{N}(m, n) = 1 \\ -\infty, & \text{otherwise} \end{cases}, \tag{4}$$

among which the $\mathbf{N} \in \{0,1\}^{K \times hw}$ denote the duplication of $\mathbf{M} \in \mathbb{R}^{1 \times hw}$ from $\mathbb{R}^{1 \times hw}$ to $\mathbb{R}^{K \times hw}$ and the $\mathbf{M}$ is the flattened support mask. The masked cross-attention only attends within the foreground region of the support mask for agent tokens, which not only makes agent tokens rich in foreground context, but also leads to faster convergence [7].

We found that without constraining the agent learning process, multiple agent tokens tend to focus on the same area. For the purpose of learning more **diverse** agent tokens, we further constrain the attention matrix to evenly allocate the foreground pixels to different agent tokens. Concretely, we model the pixels allocating as the Optimal Transport (OT) problem. The goal of the OT problem is to find a transportation plan $\mathbf{T}^*$ at a global minimal transportation cost, which can be solved elegantly using Sinkhorn algorithm with linear pro-

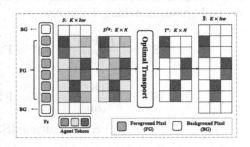

Fig. 4. Process of obtaining the refined part mask via OT algorithm.

gramming [9]. As illustrated in Fig. 4, we are intrested in condensing the foreground support features into different agent tokens. The cost matrix is defined as $(1 - \mathbf{S}^{fg})$, where the $\mathbf{S}^{fg} \in \mathbb{R}^{K \times N}$ is the matrix of similarity between agent tokens and foreground support features, and the N is the amount of the foreground support pixels specified by support mask. The higher similarity in $\mathbf{S}^{fg}$ leads to a lower corresponding transport cost. We denote the transport plan as $\mathbf{T} \in \mathbb{R}^{K \times N}$ and the optimization function is as follows:

$$\max_{\mathbf{T} \in \mathcal{T}} \mathrm{Tr} \left(\mathbf{T}^\mathsf{T} (1 - \mathbf{S}^{fg}) \right) + \epsilon H(\mathbf{T}), \quad H(\mathbf{T}) = -\sum_{ij} \mathbf{T}_{ij} \log \mathbf{T}_{ij}, \qquad (5)$$

where $H(\mathbf{T})$ is the entropy function, and ϵ is the parameter that controls the smoothness of the mapping and is set to be 0.05 in our experiments. We impose the *equal partition* constraints on $\mathbf{T}$:

$$\mathcal{T} = \left\{ \mathbf{T} \in \mathbb{R}_+^{K \times N} \mid \mathbf{T}\mathbf{1} = \frac{1}{K} \cdot \mathbf{1}, \mathbf{T}^\mathsf{T}\mathbf{1} = \frac{1}{N} \cdot \mathbf{1} \right\}, \qquad (6)$$

where $\mathbf{1}$ denotes the vector of all ones in the appropriate dimension. Equation (6) enforces that each agent token is assigned the same number of foreground pixels thus preventing a trivial solution where all pixels are assigned to a single agent token. So that different agent tokens responsible for different areas that are mutual complementary. As shown in Fig. 4, we zero-pad the $\mathbf{T}^*$ to result in the refined part mask $\hat{\mathbf{S}}$. The final output agent tokens are acquired from the weighted sum of $\mathbf{V}^s$:

$$\hat{\mathbf{F}}^a = \mathrm{FFN}(\hat{\mathbf{S}})\mathbf{V}^s \qquad (7)$$

Benefiting from ALD, the output agent tokens decompose different target objects into diverse and complementary parts in an adaptive manner.

3.5   Agent Matching Decoder

Agent matching decoder is designed to equip query pixels with salutary information from support features in a robust and efficient way. Different from previous methods [32,42] that perform dense similarity calculation directly of two branches, we decompose the massive pixel-level support-query matrix into two more mangeable matrices and introduce an extra alignment matrix for filtering out ambiguous matching. Formally:

$$\mathbf{S}^{as} = \frac{\mathbf{Q}^a(\mathbf{K}^s)^{\mathsf{T}}}{\sqrt{d_k}}, \quad \mathbf{Q}^a = \hat{\mathbf{F}}^a \mathbf{W}_a^{\mathcal{Q}}, \quad \mathbf{K}^s = \hat{\mathbf{F}}^s \mathbf{W}_s^{\mathcal{K}}, \tag{8}$$

$$\mathbf{S}^{qa} = \frac{\mathbf{Q}^q(\mathbf{K}^a)^{\mathsf{T}}}{\sqrt{d_k}}, \quad \mathbf{Q}^q = \hat{\mathbf{F}}^q \mathbf{W}_q^{\mathcal{Q}}, \quad \mathbf{K}^a = \hat{\mathbf{F}}^a \mathbf{W}_a^{\mathcal{K}}, \tag{9}$$

$$\mathbf{S}^{qs} = \mathrm{Softmax}(\mathbf{S}^{sa}\mathbf{S}^{aq} + \boldsymbol{\mathcal{A}}), \tag{10}$$

where the $\mathbf{W}_*^{\mathcal{K}}, \mathbf{W}_*^{\mathcal{Q}}$ and $* \in \{s, a, q\}$ denote the linear projection, and the aligning matrix $\boldsymbol{\mathcal{A}} \in \mathbb{R}^{hw \times hw}$ is obtained by

$$\boldsymbol{\mathcal{A}}(i, j) = \begin{cases} 0, & \text{if } \mathrm{argmax}_t \, \mathbf{S}^{as}(t, i) = \mathrm{argmax}_t \, \mathbf{S}^{qa}(j, t) \\ -\infty, & \text{otherwise} \end{cases}, \tag{11}$$

where $(i, j) \in \{1, 2, \dots, hw\}$ and $t \in \{1, 2, \dots, K\}$. In this way, $\boldsymbol{\mathcal{A}}$ filters out these attention weights between the support and query pixels that do not belong to the same agent token. We inplant this support-query correlation into the multi-head attention mechanism within the decoder, and given the support-query attention matrix $\mathbf{S}^{qs}$ we can retrieve the corresponding support features via the weighted sum of $\mathbf{V}^s$ and a FFN:

$$\bar{\mathbf{F}}^q = \mathrm{FFN}((\mathbf{S}^{sq})\mathbf{V}^q), \quad \mathbf{V}^q = \hat{\mathbf{F}}^s \mathbf{W}^{\mathcal{V}}, \tag{12}$$

The obtained $\bar{\mathbf{F}}^q$ is reshaped back to spatial dimensions and processed by a small convolution block to result in the final prediction. The convolution block consists of one 3×3 convolution, one *ReLU* activation and one 1×1 convolution. The proposed alignment matrix injects the contextual information into the pixel-wise matching to filter out the matching weights between support and query pixels that do not belong to the same object part. Besides, the decomposition in Eq. (10) converts the computation complexity from $o(c(hw)^2)$ to $o(K(hw)^2)$, where the c is the hidden dimension of decoder and $K \ll c$, which makes our approach more efficient.

4   Experiments

4.1   Dataset and Evaluation Metric

Dataset. We evaluate our approach on two widely used few-shot segmentation datasets, Pascal-5^i [11] and COCO-20^i [18]. For Pascal-5^i, which consists of the

Table 1. Comparison with other state-of-the-art methods for 1-shot and 5-shot segmentation on Pascal-5^i. The mIoU of each fold and the FB-IoU of four folds are reported. Best results in bold.

Method	Backbone	mIoU(1-shot)					FB-IoU	mIoU(5-shot)					FB-IoU
		5^0	5^1	5^2	5^3	Mean	(1-shot)	5^0	5^1	5^2	5^3	Mean	(5-shot)
PANet [33]	Vgg-16	42.3	58.0	51.1	41.2	48.1	66.5	51.8	64.6	59.8	46.5	55.7	70.7
FWB [24]		47.0	59.6	52.6	48.3	51.9	-	50.9	62.9	56.5	50.1	55.1	-
SG-One [43]		40.2	58.4	48.4	38.4	46.3	63.1	41.9	58.6	48.6	39.4	47.1	65.9
PMM [38]		47.1	65.8	50.6	48.5	53.0	-	50.0	66.5	51.9	47.6	54.0	-
ASR [19]		50.2	66.4	54.3	51.8	55.7	-	53.7	68.5	55.0	54.8	58.0	-
CANet [41]	Res-50	52.5	65.9	51.3	51.9	55.4	66.2	55.5	67.8	51.9	53.2	57.1	69.6
PGNet [40]		56.0	66.9	50.6	50.4	56.0	69.9	57.7	68.7	52.9	54.6	58.5	70.5
PPNet [20]		47.8	58.8	53.8	45.6	51.5	-	58.4	67.8	64.9	56.7	62.0	-
PMM [38]		55.2	66.9	52.6	50.7	56.3	-	56.3	67.3	54.5	51.0	57.3	-
PFENet [29]		61.7	69.5	55.4	56.3	60.8	73.3	63.1	70.7	55.8	57.9	61.9	73.9
SCLNet [39]		63.0	70.0	56.5	57.7	61.8	71.9	64.5	70.9	57.3	58.7	62.9	72.8
ASGNet [17]		58.8	67.9	56.8	53.7	59.3	69.2	63.7	70.6	64.2	57.4	63.9	74.2
MMNet [34]		62.7	70.2	57.3	57.0	61.8	-	62.2	71.5	57.5	**62.4**	63.4	-
RePRI [1]		60.2	67.0	**61.7**	47.5	59.1	-	64.5	70.8	**71.7**	60.3	66.8	-
CWT [22]		56.3	62.0	59.9	47.2	56.4	-	61.3	68.5	68.5	56.6	63.7	-
SAGNN [36]		64.7	69.6	57.0	57.2	62.1	73.2	64.9	70.0	57.0	59.3	62.8	73.3
ASR [19]		55.2	70.4	53.4	53.7	58.2	72.9	59.4	71.9	56.9	55.7	61.0	74.1
CMN [37]		64.3	70.0	57.4	**59.4**	62.8	72.3	65.8	70.4	57.6	60.8	63.7	72.8
CyCTR [42]		67.8	72.8	58.0	58.0	64.2	-	71.1	73.2	60.5	57.5	65.6	-
AAFormer (Ours)	Res-50	**69.1**	**73.3**	59.1	59.2	**65.2**	**73.8**	**72.5**	**74.7**	62.0	61.3	**67.6**	**76.2**
FWB [24]	Res-101	51.3	64.5	56.7	52.2	56.2	-	54.9	67.4	62.2	55.3	59.9	-
DAN [32]		54.7	68.6	57.8	51.6	58.2	62.3	57.9	69.0	60.1	54.9	60.5	63.9
PFENet [29]		60.5	69.4	54.4	55.9	60.1	72.9	62.8	70.4	54.9	57.6	61.4	73.5
ASGNet [17]		59.8	67.4	55.6	54.4	59.3	71.7	64.6	71.3	64.2	57.3	64.4	75.2
RePRI [1]		59.6	68.6	**62.2**	47.2	59.4	-	66.2	71.4	67.0	57.7	65.6	-
CWT [22]		56.9	65.2	61.2	48.8	58.0	-	62.6	70.2	**68.8**	57.2	64.7	-
CyCTR [42]		69.3	72.7	56.5	58.6	64.3	72.9	73.5	74.0	58.6	60.2	66.6	75.0
AAFormer (Ours)	Res-101	**69.9**	**73.6**	57.9	**59.7**	**65.3**	**74.9**	**75.0**	**75.1**	59.0	**63.2**	**68.1**	**77.3**

Pascal VOC 2012 dataset with extra annotations from SBD [12], 20 categories are divided into 4 folds with 5 classes per fold for cross-validation, as done in [26]. For COCO-20^i, we follow the data split protocol in [24] to separate the 80 classes evenly into 4 folds, where each fold contains 60 classes for training and the remaining 20 classes for testing. During inference, 1,000 episodes are randomly sampled from the test split.

Evaluation Metric. Following the previous practices [29,33,41–43], we adopt two evaluation metrics, i.e., mean intersection-over-union (mIoU) and foreground-background IoU (FB-IoU). We mainly focus on the mIoU metric as it reflects the average result over all classes thus alleviating the performance bias of scarce classes.

4.2 Implementation Details

Our models are trained 200 epochs with batch size 4 for Pascal-5^i and 50 epochs with batch size 24 for COCO-20^i. We adopt the ImageNet [25] pretrained ResNet50 and Resnet101 [13] as the backbone to extract features in our experiments for fair comparison. Given the support feature $\mathbf{F}^s \in \mathbb{R}^{h \times w \times c}$ and query feature $\mathbf{F}^q \in \mathbb{R}^{h \times w \times c}$, we set the number of the cross layers in our agent learning decoder to 1, and set 2 in the representation encoder. Please see the **supplementary material** for more implementation details.

Table 2. Comparison with other state-of-the-art methods for 1-shot and 5-shot segmentation on COCO-20^i. The mIoU of each fold and the FB-IoU of four folds are reported. Best results in bold.

Method	Backbone	mIoU(1-shot)					FB-IoU	mIoU(5-shot)					FB-IoU
		2^0	2^1	2^2	2^3	Mean	(1-shot)	2^0	2^1	2^2	2^3	Mean	(5-shot)
PPNet [20]	Res-50	28.1	30.8	29.5	27.7	29.0	-	39.0	40.8	37.1	37.3	38.5	-
PMM [38]		29.5	36.8	28.9	27.0	30.6	-	33.8	42.0	33.0	33.3	35.5	-
MMNet [34]		34.9	41.0	37.2	37.0	37.5	-	37.0	40.3	39.3	36.0	38.2	-
RePRI [1]		31.2	38.1	33.3	33.0	34.0	-	38.5	46.2	40.0	43.6	42.1	-
ASR [19]		30.6	36.7	32.7	35.4	33.9	-	33.1	39.5	34.2	36.2	35.8	-
CMN [37]		37.9	**44.8**	38.7	35.6	39.3	61.7	42.0	**50.5**	41.0	38.9	43.1	63.3
CyCTR [42]		38.9	43.0	39.6	39.8	40.3	-	41.1	48.9	45.2	47.0	45.6	-
FWB [24]	Res-101	17.0	18.0	21.0	28.9	21.2	-	19.1	21.5	23.9	30.1	23.7	-
PFENet [29]		34.3	33.0	32.3	30.1	32.4	58.6	38.5	38.6	38.2	34.3	37.4	61.9
SCLNet [39]		36.4	38.6	37.5	35.4	37.0	-	38.9	40.5	41.5	38.7	39.9	-
CWT [22]		30.3	36.6	30.5	32.2	32.4	-	38.5	46.7	39.4	43.2	42.0	-
SAGNN [36]		36.1	41.0	38.2	33.5	37.2	60.9	40.9	48.3	42.6	38.9	42.7	63.4
AAFormer (Ours)	Res-50	**39.8**	44.6	**40.6**	**41.4**	**41.6**	**67.7**	**42.9**	50.1	**45.5**	**49.2**	**46.9**	**68.2**

4.3   Comparison with State-of-the-Art Methods

Pascal-5^i. In Table 1, we compare our proposed AAFormer with the state-of-the-art few-shot segmentation methods. We consistently observe that our AAFormer outperforms all previous models under both 1-shot and 5-shot settings, which strongly proves the effectiveness of our method. For fair comparison, we report results with the ResNet-50 and Resnet-101 backbones. Specifically, Our approach achieves 65.3% and 68.1% in the 1-shot and 5-shot settings with the ResNet-101 backbone that significantly outperforms the recent prototypical learning methods (e.g., ASGNet), achieving a large margin of 6.0% and 3.7% in mIoU. This is because the prototypic learning methods only leverage the correlation between the prototypes and query features without considering the pixel-level support features, while the agent matching decoder in our method further explores the pixel-wise support information and contributes to accurate segmentation. With the more lightweight ResNet50 backbone, the performance of AAFormer is also in the lead. Compared with the best affinity learning method (CyCTR), our method has a clear lead and obtains 1.0% mIoU gain in the 1-shot setting and 2.0% in the 5-shot setting.

COCO-20^i. In Table 2 we report the comparison on COCO-20^i, which is much more difficult than Pascal-5^i with more complex cases, such as drastic object appearance differences, messy scenes and severe occlusion. In the absence of careful parameter adjustments, our AAFormer also achieves superior results than the existing best performing method (CyCTR), i.e., obtains 1.3% and 1.3% mIoU gain in the 1-shot and 5-shot settings. This demonstrates the stability of our method. We analyze that the performance can also benefit from the proposed representation encoder, which can capture the full image context information for better representing the object appearance to deal with complex cases. While conducting feature processing on the raw backbone features tends to be confused by the background clutters or other interferent.

4.4   Ablation Study and Analysis

To look deeper into our method, we perform a series of ablation studies to analyze each component of our AAFormer, including the representation encoder (REnc), the agent matching decoder (AMD) and the agent learning decoder (ALD). Note that we remove all modules except the

Table 3. Ablation study results. Experiments are conducted on Pascal-5^0 for 1-shot setting with ResNet-50.

REnc	AMD	ALD	mIoU
			62.2
✓			64.9
✓	✓		67.4
✓	✓	✓	**69.1**

(a) Ablation of model components.

Init.	Update	mIoU
DT	K-Means	67.4
DT	ALD(w/o OT)	68.3
DT	ALD(w/ OT)	**69.1**
Learnable	ALD(w/ OT)	68.5

(b) Ablation of agent initialization and update.

encoder with two residual blocks to conduct direct pixel-level matching between the support and query images as our baseline.

Effectiveness of the Representation Encoder. As shown in Table 3a, The introduction of the representation encoder achieves a certain performance lift compared with the baseline, e.g., 2.7% in mIoU. The improvements can be mainly ascribed to the proposed representation encoder that can effectively capture robust context information for representing object appearance well even in complex cases.

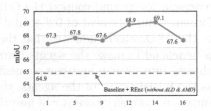

Fig. 5. Comparisons of performance with different number of agent tokens.

Effectiveness of the Agent Matching Decoder. From the comparison between the 2-nd and the 3-rd row of Table 3a, we observe that the agent matching decoder significantly improves the performance, e.g., 2.5% in mIoU. Note that the agent tokens in the 3-rd row are produced from the K-Means clustering. We conclude that this performance gain comes from the alignment matrix in the agent learning decoder, which can filter out the matching weights between support and query pixels that do not belong to the same object part for more accurate segmentation.

Effectiveness of the Agent Learning Decoder. With the utilization of the agent learning decoder, further improvements can be observed, e.g., 2.4% mIoU. This proves that our agent learning decoder can adaptively learn spatial-aware and diverse agent tokens to absorb the local context of support pixels and make the agent matching decoder more effective in reducing background noises.

Analysis of the Agent Learning Process. To explore effectiveness of different ways to learn agents, we evaluate multiple combinations of initialization and update of agent tokens. Naively, we first initialize the agents with the help of distance transformation (DT), and then update them by K-Means algorithm. The way is similar to [17] and the result is displayed in the first row of Table 3b. Then we replace the K-Means algorithm by the agent learning decoder (without

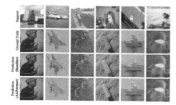

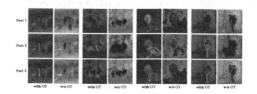

Fig. 6. Qualitative comparison with the baseline. AAFormer can achieve more accurate segmentation.

Fig. 7. Visualization of the learned agent tokens with OT and without OT. As we can see, these agent tokens adaptively decompose the object into different parts benefiting from OT.

OT algorithm) to update the initial agent tokens. We can observe that there is a 0.9% mIoU improvement, showing our proposed ALD can better make the agents represent foreground context than traditional K-Means algorithm. And the performance improvement of 0.8% mIoU can be obtained by adding OT algorithm to the ALD, which indicates that the equal partition constraint brought by OT is beneficial to make the agents learn to adaptively decompose different target objects into diverse and complementary parts. When we use learnable tokens as the initialization of the agents, the mIoU is degraded by 0.6%, which is not surprising as query tokens will lose the structural spatial information compared to DT initialization. Despite, its performance is approximate to DT initialization with ALD (without OT), once again validating that our ALD can produce powerful agent tokens.

Hyperparameter Evaluations. In Fig. 5, we conduct quantitative experiment to analyze how many agent tokens K are better for segmentation. We can observe that the performance continues to grow until $K = 14$, which means that it is sufficient for agent learning decoder by mining fourteen different object parts (Fig. 6).

4.5   Visualizations

Visualization of Learned Agent Tokens. We visualize the part masks activated by agent tokens to qualitatively evaluate the effect of the optimal transport algorithm. As shown in Fig. 7, we can observe that without OT, multiple prototypes tend to focus on the same part containing large background noises. And thanks to the equal partition constraint from OT, the agent tokens successfully divide different target objects into diverse and complementary parts in an adaptive manner. For example, the three part masks in object *dog* (in the fifth column) focus on the head, body, and limbs respectively.

Visualization of Part Correspondence. Under the extreme challenge, we visualize part masks which come from the same class of support-query image pairs and are activated by a specific agent token, as shown in Fig. 8. As we can see, intrinsic semantic correspondence is established between the pair of part masks obtained from the same agent token. For example, in the object *human* (the second column), the part *head* of the query image can accurately match with the corresponding part masks of the support image, even though the objects in the two images have different poses. This proves that our ALD module enables diverse agent tokens to activate on the same latent semantic regions. In this way, agent tokens can adapt to occlusions, different poses and different scales across images.

Visualization of Pixel Correspondence. To vividly present the effect of our AMD module, we visualize differences in pixel correspondences according to whether AMD exits. As shown in Fig. 9, with the utilization of AMD module, the top five pixels corresponding to the query points tend to line in the foreground of support images. While these ones will contain large background noises without the AMD module to perform direct pixel matching. This is in line with the design idea of AMD module, i.e., filtering out the unreasonable matching weights.

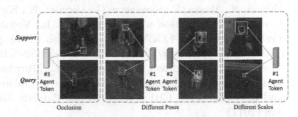

Fig. 8. Visualization of part correspondence. As we can see, the pair of part masks obtained from the same agent token have intrinsic semantic consistency.

Fig. 9. Visualization of pixel correspondence. The green and red arrows point to the top five support pixels that match the query pixel with and without AMD module, respectively. (Color figure online)

5   Conclusion

In this paper, we propose a novel adaptive agent transformer (AAFormer) to integrate prototypical and affinity learning to exploit the complementarity between them via a transformer encoder-decoder architecture. Extensive experimental results on two standard benchmarks demonstrate that AAFormer performs favorably against state-of-the-art FSS methods.

Acknowledgments. This work was partially supported by the National Nature Science Foundation of China (Grant 62022078, 12150007, 62121002), and National Defense Basic Scientific Research Program (JCKY2020903B002).

References

1. Boudiaf, M., Kervadec, H., Masud, Z.I., Piantanida, P., Ben Ayed, I., Dolz, J.: Few-shot segmentation without meta-learning: a good transductive inference is all you need? In: Proceedings of the IEEE/CVF Conference on Computer Vision and Pattern Recognition, pp. 13979–13988 (2021)
2. Bousselham, W., et al.: Efficient self-ensemble framework for semantic segmentation. arXiv preprint arXiv:2111.13280 (2021)
3. Carion, N., Massa, F., Synnaeve, G., Usunier, N., Kirillov, A., Zagoruyko, S.: End-to-end object detection with transformers. In: Vedaldi, A., Bischof, H., Brox, T., Frahm, J.-M. (eds.) ECCV 2020. LNCS, vol. 12346, pp. 213–229. Springer, Cham (2020). https://doi.org/10.1007/978-3-030-58452-8_13
4. Chen, L.C., Papandreou, G., Kokkinos, I., Murphy, K., Yuille, A.L.: DeepLab: semantic image segmentation with deep convolutional nets, atrous convolution, and fully connected CRFs. IEEE Trans. Pattern Anal. Mach. Intell. **40**(4), 834–848 (2017)
5. Chen, L.C., Zhu, Y., Papandreou, G., Schroff, F., Adam, H.: Encoder-decoder with atrous separable convolution for semantic image segmentation. In: Proceedings of the European Conference on Computer Vision (ECCV), pp. 801–818 (2018)
6. Chen, X., Yuan, Y., Zeng, G., Wang, J.: Semi-supervised semantic segmentation with cross pseudo supervision. In: Proceedings of the IEEE/CVF Conference on Computer Vision and Pattern Recognition, pp. 2613–2622 (2021)
7. Cheng, B., Misra, I., Schwing, A.G., Kirillov, A., Girdhar, R.: Masked-attention mask transformer for universal image segmentation. arXiv preprint arXiv:2112.01527 (2021)
8. Cheng, B., Schwing, A., Kirillov, A.: Per-pixel classification is not all you need for semantic segmentation. In: Advances in Neural Information Processing Systems, vol. 34 (2021)
9. Cuturi, M.: Sinkhorn distances: lightspeed computation of optimal transport. In: Advances in Neural Information Processing Systems, vol. 26 (2013)
10. Dong, N., Xing, E.P.: Few-shot semantic segmentation with prototype learning. In: BMVC, vol. 3 (2018)
11. Everingham, M., Van Gool, L., Williams, C.K., Winn, J., Zisserman, A.: The pascal visual object classes (VOC) challenge. Int. J. Comput. Vision **88**(2), 303–338 (2010)
12. Hariharan, B., Arbeláez, P., Girshick, R., Malik, J.: Simultaneous detection and segmentation. In: Fleet, D., Pajdla, T., Schiele, B., Tuytelaars, T. (eds.) ECCV 2014. LNCS, vol. 8695, pp. 297–312. Springer, Cham (2014). https://doi.org/10.1007/978-3-319-10584-0_20
13. He, K., Zhang, X., Ren, S., Sun, J.: Deep residual learning for image recognition. In: Proceedings of the IEEE Conference on Computer Vision and Pattern Recognition, pp. 770–778 (2016)
14. Hu, H., Wei, F., Hu, H., Ye, Q., Cui, J., Wang, L.: Semi-supervised semantic segmentation via adaptive equalization learning. In: Advances in Neural Information Processing Systems, vol. 34 (2021)
15. Irving, B.: maskSLIC: regional superpixel generation with application to local pathology characterisation in medical images. arXiv preprint arXiv:1606.09518 (2016)
16. Koh, J.Y., Nguyen, D.T., Truong, Q.-T., Yeung, S.-K., Binder, A.: SideInfNet: a deep neural network for semi-automatic semantic segmentation with side information. In: Vedaldi, A., Bischof, H., Brox, T., Frahm, J.-M. (eds.) ECCV 2020.

LNCS, vol. 12369, pp. 103–118. Springer, Cham (2020). https://doi.org/10.1007/978-3-030-58586-0_7

17. Li, G., Jampani, V., Sevilla-Lara, L., Sun, D., Kim, J., Kim, J.: Adaptive prototype learning and allocation for few-shot segmentation. In: Proceedings of the IEEE/CVF Conference on Computer Vision and Pattern Recognition, pp. 8334–8343 (2021)

18. Lin, T.-Y., et al.: Microsoft COCO: common objects in context. In: Fleet, D., Pajdla, T., Schiele, B., Tuytelaars, T. (eds.) ECCV 2014. LNCS, vol. 8693, pp. 740–755. Springer, Cham (2014). https://doi.org/10.1007/978-3-319-10602-1_48

19. Liu, B., Ding, Y., Jiao, J., Ji, X., Ye, Q.: Anti-aliasing semantic reconstruction for few-shot semantic segmentation. In: Proceedings of the IEEE/CVF Conference on Computer Vision and Pattern Recognition, pp. 9747–9756 (2021)

20. Liu, Y., Zhang, X., Zhang, S., He, X.: Part-aware prototype network for few-shot semantic segmentation. In: Vedaldi, A., Bischof, H., Brox, T., Frahm, J.-M. (eds.) ECCV 2020. LNCS, vol. 12354, pp. 142–158. Springer, Cham (2020). https://doi.org/10.1007/978-3-030-58545-7_9

21. Long, J., Shelhamer, E., Darrell, T.: Fully convolutional networks for semantic segmentation. In: Proceedings of the IEEE Conference on Computer Vision and Pattern Recognition, pp. 3431–3440 (2015)

22. Lu, Z., He, S., Zhu, X., Zhang, L., Song, Y.Z., Xiang, T.: Simpler is better: few-shot semantic segmentation with classifier weight transformer. In: Proceedings of the IEEE/CVF International Conference on Computer Vision, pp. 8741–8750 (2021)

23. Luo, W., Yang, M.: Semi-supervised semantic segmentation via strong-weak dual-branch network. In: Vedaldi, A., Bischof, H., Brox, T., Frahm, J.-M. (eds.) ECCV 2020. LNCS, vol. 12350, pp. 784–800. Springer, Cham (2020). https://doi.org/10.1007/978-3-030-58558-7_46

24. Nguyen, K., Todorovic, S.: Feature weighting and boosting for few-shot segmentation. In: Proceedings of the IEEE/CVF International Conference on Computer Vision, pp. 622–631 (2019)

25. Russakovsky, O., et al.: Imagenet large scale visual recognition challenge. Int. J. Comput. Vision 115(3), 211–252 (2015)

26. Shaban, A., Bansal, S., Liu, Z., Essa, I., Boots, B.: One-shot learning for semantic segmentation. arXiv preprint arXiv:1709.03410 (2017)

27. Strudel, R., Garcia, R., Laptev, I., Schmid, C.: Segmenter: transformer for semantic segmentation. In: Proceedings of the IEEE/CVF International Conference on Computer Vision, pp. 7262–7272 (2021)

28. Sun, R., Li, Y., Zhang, T., Mao, Z., Wu, F., Zhang, Y.: Lesion-aware transformers for diabetic retinopathy grading. In: Proceedings of the IEEE/CVF Conference on Computer Vision and Pattern Recognition, pp. 10938–10947 (2021)

29. Tian, Z., Zhao, H., Shu, M., Yang, Z., Li, R., Jia, J.: Prior guided feature enrichment network for few-shot segmentation. IEEE Trans. Pattern Anal. Mach. Intell. (2020)

30. Vaswani, A., et al.: Attention is all you need. In: Advances in Neural Information Processing Systems, vol. 30 (2017)

31. Vinyals, O., Blundell, C., Lillicrap, T., Wierstra, D., et al.: Matching networks for one shot learning. In: Advances in Neural Information Processing Systems, vol. 29 (2016)

32. Wang, H., Zhang, X., Hu, Y., Yang, Y., Cao, X., Zhen, X.: Few-shot semantic segmentation with democratic attention networks. In: Vedaldi, A., Bischof, H., Brox, T., Frahm, J.-M. (eds.) ECCV 2020. LNCS, vol. 12358, pp. 730–746. Springer, Cham (2020). https://doi.org/10.1007/978-3-030-58601-0_43

33. Wang, K., Liew, J.H., Zou, Y., Zhou, D., Feng, J.: Panet: few-shot image semantic segmentation with prototype alignment. In: Proceedings of the IEEE/CVF International Conference on Computer Vision, pp. 9197–9206 (2019)
34. Wu, Z., Shi, X., Lin, G., Cai, J.: Learning meta-class memory for few-shot semantic segmentation. In: Proceedings of the IEEE/CVF International Conference on Computer Vision, pp. 517–526 (2021)
35. Xie, E., Wang, W., Yu, Z., Anandkumar, A., Alvarez, J.M., Luo, P.: Segformer: simple and efficient design for semantic segmentation with transformers. In: Advances in Neural Information Processing Systems, vol. 34 (2021)
36. Xie, G.S., Liu, J., Xiong, H., Shao, L.: Scale-aware graph neural network for few-shot semantic segmentation. In: Proceedings of the IEEE/CVF Conference on Computer Vision and Pattern Recognition, pp. 5475–5484 (2021)
37. Xie, G.S., Xiong, H., Liu, J., Yao, Y., Shao, L.: Few-shot semantic segmentation with cyclic memory network. In: Proceedings of the IEEE/CVF International Conference on Computer Vision, pp. 7293–7302 (2021)
38. Yang, B., Liu, C., Li, B., Jiao, J., Ye, Q.: Prototype mixture models for few-shot semantic segmentation. In: Vedaldi, A., Bischof, H., Brox, T., Frahm, J.-M. (eds.) ECCV 2020. LNCS, vol. 12353, pp. 763–778. Springer, Cham (2020). https://doi.org/10.1007/978-3-030-58598-3_45
39. Zhang, B., Xiao, J., Qin, T.: Self-guided and cross-guided learning for few-shot segmentation. In: Proceedings of the IEEE/CVF Conference on Computer Vision and Pattern Recognition, pp. 8312–8321 (2021)
40. Zhang, C., Lin, G., Liu, F., Guo, J., Wu, Q., Yao, R.: Pyramid graph networks with connection attentions for region-based one-shot semantic segmentation. In: Proceedings of the IEEE/CVF International Conference on Computer Vision, pp. 9587–9595 (2019)
41. Zhang, C., Lin, G., Liu, F., Yao, R., Shen, C.: Canet: class-agnostic segmentation networks with iterative refinement and attentive few-shot learning. In: Proceedings of the IEEE/CVF Conference on Computer Vision and Pattern Recognition, pp. 5217–5226 (2019)
42. Zhang, G., Kang, G., Wei, Y., Yang, Y.: Few-shot segmentation via cycle-consistent transformer. arXiv preprint arXiv:2106.02320 (2021)
43. Zhang, X., Wei, Y., Yang, Y., Huang, T.S.: SG-one: similarity guidance network for one-shot semantic segmentation. IEEE Trans. Cybern. **50**(9), 3855–3865 (2020)
44. Zhao, H., Shi, J., Qi, X., Wang, X., Jia, J.: Pyramid scene parsing network. In: Proceedings of the IEEE Conference on Computer Vision and Pattern Recognition, pp. 2881–2890 (2017)
45. Zheng, S., et al.: Rethinking semantic segmentation from a sequence-to-sequence perspective with transformers. In: Proceedings of the IEEE/CVF Conference on Computer Vision and Pattern Recognition, pp. 6881–6890 (2021)

Waymo Open Dataset: Panoramic Video Panoptic Segmentation

Jieru Mei[1], Alex Zihao Zhu[2]([✉]), Xinchen Yan[2], Hang Yan[2], Siyuan Qiao[3],
Liang-Chieh Chen[3], and Henrik Kretzschmar[2]

[1] Johns Hopkins University, Baltimore, USA
[2] Waymo LLC, Mountain View, USA
`alexzzhu@waymo.com`
[3] Google Research, Mountain View, USA

Abstract. Panoptic image segmentation is the computer vision task of finding groups of pixels in an image and assigning semantic classes and object instance identifiers to them. Research in image segmentation has become increasingly popular due to its critical applications in robotics and autonomous driving. The research community thereby relies on publicly available benchmark dataset to advance the state-of-the-art in computer vision. Due to the high costs of densely labeling the images, however, there is a shortage of publicly available ground truth labels that are suitable for panoptic segmentation. The high labeling costs also make it challenging to extend existing datasets to the video domain and to multi-camera setups. We therefore present the Waymo Open Dataset: Panoramic Video Panoptic Segmentation, a large-scale dataset that offers high-quality panoptic segmentation labels for autonomous driving. We generate our dataset using the publicly available Waymo Open Dataset (WOD), leveraging the diverse set of camera images. Our labels are consistent over time for video processing and consistent across multiple cameras mounted on the vehicles for full panoramic scene understanding. Specifically, we offer labels for 28 semantic categories and 2,860 temporal sequences that were captured by five cameras mounted on autonomous vehicles driving in three different geographical locations, leading to a total of 100k labeled camera images. To the best of our knowledge, this makes our dataset an order of magnitude larger than existing datasets that offer video panoptic segmentation labels. We further propose a new benchmark for Panoramic Video Panoptic Segmentation and establish a number of strong baselines based on the DeepLab family of models. We have made the benchmark and the code publicly available, which we hope will facilitate future research on holistic scene understanding. Our dataset can be found at: waymo.com/open.

Keyword: Panoramic video panoptic segmentation

J. Mei—Work done as an intern at Waymo.

Supplementary Information The online version contains supplementary material available at https://doi.org/10.1007/978-3-031-19818-2_4.

1   Introduction

Semantic visual scene understanding has been studied extensively for decades in the field of computer vision [17,37,61,68,77,84]. Researchers have tackled tasks of varying difficulty, ranging from segmenting distinct objects in individual camera images [9,24,26,44] to tracking and segmenting multiple objects in videos [14,69,75]. Robotic applications, such as autonomous driving, have led to new challenges and opportunities for semantic visual scene understanding [12,21].

Modern autonomous vehicles tend to be equipped with multiple cameras and LiDAR scanners. The cameras provide rich semantic information about the scene, whereas the LiDAR scanners capture sparse, but geometrically highly accurate information. Autonomous vehicles need to be able to fuse and interpret the data stream from multiple sensors to build and maintain over time an accurate and consistent estimate of the world. One challenge when tracking and segmenting multiple objects is that objects of interest may leave the field of view of a camera to enter the field of view of another camera across consecutive video frames.

In this paper, we study the new task of video panoptic segmentation [32,35] for autonomous vehicles equipped with multiple cameras. See Fig. 1 for an illus-

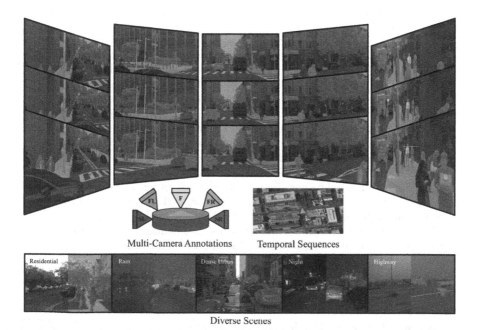

Multi-Camera Annotations   Temporal Sequences

Diverse Scenes

Fig. 1. We provide panoptic segmentation labels for 100k camera images of the Waymo Open Dataset. Our dataset is grouped into 2,860 temporal sequences captured by five cameras, mounted on autonomous vehicles driving in three geographical locations. Instance segmentation labels are consistent both across cameras and over time. Our dataset offers diversity in terms of object classes, locations, weather, and time of day.

tration. Panoptic segmentation enables autonomous vehicles to reason about their surroundings in terms of semantic and geometry properties, such as fine-grained object contours. There are also important offboard applications, including auto-labeling [53,79,87] and camera sensor simulation [10,42,46]. On the one hand, most existing panoptic segmentation datasets [12,49] provide labels for individual camera images. This makes it difficult to train models that fuse information from multiple camera images, either temporally or by leveraging a multi-camera setup. On the other hand, datasets that provide panoptic segmentation labels for video data [32,74] tend to be scarce and much smaller than datasets for object detection and tracking for autonomous driving [21,64].

To bridge this gap, we present a new benchmark dataset for panoptic segmentation based on the popular Waymo Open Dataset (WOD). Specifically, we provide panoptic segmentation labels for video data that are consistent across five cameras mounted on the vehicles. We further present a benchmark that captures the task of multi-camera panoptic segmentation in video data for autonomous driving. Overall, we provide panoptic segmentation labels for 100k camera images, which we group into training (70%), validation (10%) and test (20%) sets. The training set consists of 2,800 sequences, each of which comprises labels for five cameras spanning 1.2 s and five temporal frames. In contrast, our validation and test sets consist of 60 longer sequences, in order to facilitate the evaluation of long-term tracking. Each validation and test sequence consists of 100 temporal frames, spanning the full 20 s of a scene, while also providing labels across all five cameras.

We extend the Segmentation and Tracking Quality (STQ) metric [74] to support our multi-camera setup by computing a weight for pixels depending on the cameras they correspond to. We also extend a state-of-the-art video panoptic segmentation method, ViP-DeepLab [54], to our multi-camera setup by training separate models on each camera view and by training a model on a panorama generated from all views. We present an extensive experimental evaluation on the proposed dataset and metric.

We have open-sourced our full dataset to enhance video panoptic segmentation research, while also opening up the field of panoramic video panoptic segmentation.

2 Related Work

Panoptic Segmentation. The task of panoptic segmentation [35] aims to unify semantic segmentation [26] and instance segmentation [24], requiring assigning a class label and instance ID to all pixels in an image. Modern panoptic segmentation systems could be roughly categorized into top-down (or proposal-based) [34,38,43,52,72,76] and bottom-up (or proposal-free) [11,20,70,71,83] approaches. Our adopted baseline methods belong to the bottom-up category.

Video Panoptic Segmentation. Extending panoptic segmentation to the video domain, Video Panoptic Segmentation (VPS) [32] requires generating the

instance tracking IDs (*i.e.*, temporally consistent instance IDs) along with panoptic segmentation results across video frames. Current VPS datasets are small
scale in terms of semantic classes and sizes. Specifically, Cityscapes-VPS [32]
sparsely annotates (every five frame) Cityscapes [12] video sequences, resulting
in only 3,000 frames with 19 semantic classes for training and testing. Recently,
STEP [74] extends KITTI-MOTS [21,69] and MOTS-Challenge [14,69] for VPS.
However, their annotated datasets are still small-scale (18K annotated frames
with 19 semantic classes for KITTI-STEP, and 2K frames with 8 classes for
MOTChallenge-STEP), and the video sequences are only captured by a single
front-view camera. On the other hand, our annotated dataset presents the first
large-scale VPS annotations and extends to the multi-camera scenario.

Segmentation Benchmarks. There are other popular video segmentation
benchmarks existing in the literature, *e.g.*, VSPW [47] for video semantic segmentation, while MOTS [69] and Youtube-VIS [82] for video instance segmentation. Our benchmark is also related to urban scene understanding, where typical
benchmarks include [2,4–6,12,21,22,28,39–41,49,64,81,85,86,88]. Our work is
most related to WildPASS [81], WildPPS [30] and UP-Drive [50], which also
aim to endow machines with large field-of-view perception. However, building
on top of the large-scale Waymo Open Dataset [64], our benchmark provides
much more high-quality annotated video sequences.

Multi-Camera Multi-Object Tracking. Consistently tracking objects across
multiple cameras, multi-camera multi-object tracking [1,3,13,15,19,27,56,58]
has been a popular research topic in the computer vision community. Typical
benchmarks [7,18,23,36,55,65,78] only track a single class (*e.g.*, people or vehicles) with bounding boxes, while our proposed benchmark demands for pixel-
level tracking and segmentation for multiple classes.

Panoramic Semantic Segmentation. Panoramic semantic segmentation provides surround-view perception [48,62,66,80,81,89], but limited to semantic segmentation without temporal and instance-level understanding. Our work is similar, but additionally tackles video panoptic segmentation. Recently, [51,57]
predict bird's-eye view semantic segmentation using multi-camera inputs.

Table 1. Dataset comparison. Our WOD:PVPS is a new large-scale panoramic video
panoptic segmentation dataset. † indicates panoramas.

Dataset statistics	WOD:PVPS (ours)	WildPASS [81]	WildPPS [30]	Cityscapes-VPS [32]	KITTI-STEP [74]	MOT-STEP [74]
# sequences	2860	-	-	500	50	4
# images	100,000	500†	40†	3,000	19,103	2,075
# tracking classes	8	-	-	8	2	1
# semantic classes	28	8	4	19	19	7
panoramic	✓	✓	✓	✗	✗	✗
panoptic	✓	✗	✓	✓	✓	✓
video panoptic	✓	✗	✗	✓	✓	✓

3 WOD:PVPS Dataset

In this section, we first recap the existing Waymo Open Dataset (WOD) [64], one of the largest and most diverse multi-sensor datasets in the autonomous driving domain. We leverage the existing data that comes with coarse-level annotations (e.g., 2D and 3D bounding boxes) as the foundation, and subsample images for our dataset. We then provide an overview of our WOD:PVPS dataset, including panorama generation, statistics of the semantic classes, and temporal frame sampling. Finally, we explain in details our hybrid scheme to address the challenges in multi-camera and video labeling. We obtain consistent instance IDs across temporal frames and cameras by associating the panoptic labels from each individual image with the existing box-level annotations.

3.1 Dataset Overview

The Waymo Open Dataset contains 1,150 scenes, each consisting of 20 s of data captured 10 Hz (*i.e.*, 10 frames per second, and thus 200 frames per scene). Each data frame in the dataset includes 3D point clouds from the LiDAR devices, images from five cameras (positioned at Front, Front-Left, Front-Right, Side-Left, and Side-Right), and ground truth 3D and 2D bounding boxes annotated by humans in the LiDAR point clouds and camera images, respectively. Each bounding box contains an ID that is unique to that object across the entirety of each scene. For the LiDAR data, this allows for tracking in the whole scene. For the camera data, these IDs are consistent within each camera's images only.

Built on top of the WOD, Our WOD:PVPS dataset consists of 100,000 images with panoptic segmentation labels using a prescribed train, validation, and test set split, subsampled from the existing 1.15 million images. In Table 1, we compare our proposed WOD:PVPS dataset with the public datasets for video panoptic segmentation. Our dataset is the only one that provides panoptic segmentation annotations that are consistent both across multiple cameras and across time. Furthermore, our dataset is much larger both in terms of number of frames and number of semantic classes than existing datasets [32, 74, 81].

Equirectangular Panorama. We reconstruct the equirectangular panorama (220° coverage from five cameras) by stitching each individual camera images as an alternative input format to our dataset. Specifically, we first use the extrinsics and intrinsics from the five cameras provided by WOD to unproject each pixel coordinates to the 3D space. We then set a virtual camera [63] located at the geometric mean of all five camera centers and compute the pixel colors by equirectangular projection from the 3D space with bilinear sampling. For pixels correspond to multiple camera views, we compute the weights based on the distance of each pixel in the panorama to each of the camera views' boundaries. For panoptic labels, we compute labels in each camera view given the camera parameters of five cameras and the virtual camera using the nearest sampling. Then we use the method in Qiao *et al.* [54] to stitch the panorama labels to maintain the view consistency. Finally, we fused the five panorama labels together based on the correspondences and the distances to the camera view's boundaries.

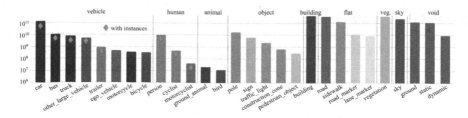

Fig. 2. Histogram of the 28 semantic categories in our dataset in terms of their pixel distributions. The vertical axis denotes the number of pixels for each class in log scale. We provide instance IDs for classes marked with diamonds.

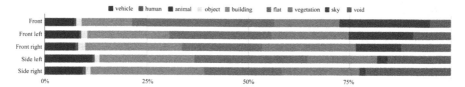

Fig. 3. Super-class distributions for each camera. Each camera sees a different distribution of classes, due to their fixed positions and different field-of-views.

There are more sophisticated methods [60,67] that leverage cross-frame information and the geometry captured from LiDAR sensors to potentially improve panorama generation. We leave this as an open research topic in the future.

Semantic Class Distribution. In total, our dataset contains 28 semantic categories, outlined with their frequency in pixels in Fig. 2. In addition, we provide instance IDs for most of the classes under the `vehicle` and `human` super-classes, as they are major dynamic categories in the autonomous driving space. We also outline the pixel distribution for each camera view in Fig. 3, where we see notable differences in the distributions in each camera. For example, the front camera covers more of `flat` (e.g., road surfaces) and `sky` pixels than the rest of the cameras, while the side left camera covers more `vehicle` pixels due to the ego-vehicle driving on the right hand side of the road. This analysis is important as machine learning models trained on the images captured by a single camera from the existing datasets may not necessarily generalize to the other cameras due to large domain gaps across different cameras. In contrast, our proposed task has an emphasis on the holistic scene understanding, which grants our WOD:PVPS dataset unique value to the research community.

Temporal Frame Sampling for Human Annotations. To maximize the diversity of the images on the training set, we subsample *sparsely* from each scene, labeling chunks of five-frame sequences from all the cameras. We start by randomly selecting 700 out of the 798 scenes. For each scene, which typically has 200 frames, we annotate *four* sets of five-frame sequences, starting at frame indices $\{25, 50, 125, 150\}$ (*i.e.*, we pick 25th, 50th, 125th, and 150th frames as the first frame of each five-frame sequence for annotation). For each set, we further select frames with offsets $\{0, 4, 6, 8, 12\}$ w.r.t. the first frame for annotations. For example, the first set of five-frame sequences will contain frames with indices

$\{25, 29, 31, 33, 37\}$. Our *sparse* sampling strategy facilitates a variety of different sequence lengths, allowing users to train on frame pairs with time difference as small as two frames (0.2 s) and as large as 12 frames (1.2 s). As a result, our training set contains groups of five temporal frames across all five cameras, yielding 2,800 sequences of 25 images (5 temporal frames × 5 cameras), or 70,000 images in total. Finally, we provide the associations between each instance ID and the corresponding 3D LiDAR bounding box, allowing us to compute very long associations (up to 13.7 s between all four sequences), if an object persists across multiple sequences in the same scene.

For the validation and test sets, we aim to enable the testing of long-term consistency across cameras and frames. We therefore *densely* sample frames 5 Hz from chunks of 100 frames across all cameras (*i.e.*, every other two frames are sampled in the 200 frame sequence). We select 20 and 40 scenes for validation and test sets by maintaining diversity in the location, density of object, and time of day distributions of WOD. In contrast to the training set, for each scene selected from the validation and test tests, we *densely* subsample the scenes for these splits by labeling every other frame, resulting in sequences with 100 temporal frames across all five cameras. In the end, our validation set contains annotations for 20 sequences of 500 images (100 temporal frames × 5 cameras), and our test set consists of 40 sequences of 500 images (or totally 10,000 and 20,000 annotations for validation and test sets, respectively). The test set annotations will not be made publicly available, but instead we will prepare a test server to evaluate the held-out test set, once the dataset is released.

3.2   Associating Instance IDs Across Cameras and Frames

In constructing the panoramic video panoptic segmentation dataset, ensuring the annotations have consistent instance IDs across cameras and temporal frames is one of the major challenges. Manual labeling is a straight-forward option, but is time-consuming and expensive at large scales. In addition, it is difficult to develop an effective labeling interface that allows human annotators to iteratively refine instance labels across cameras and temporary frames.

We instead assigned human annotators to label each camera image for panoptic segmentation separately and employed a hybrid scheme that leverages the existing coarse-level annotations in WOD. The coarse-level annotations include (i) 3D bounding boxes with corresponding IDs that are consistent across all frames and cameras; and (ii) 2D bounding boxes with IDs that are consistent across temporal frames, but annotated independently for each camera. Associations were then computed between each instance and its corresponding 3D LiDAR boxes and 2D camera boxes. Instances determined to correspond to the same object are then mapped to the same ID in all frames across cameras. A sample sequence from this process can be found in Fig. 4.

For a given frame with instance labels, 3D point clouds, and 3D bounding boxes, we associate instances with boxes by filtering the LiDAR points within

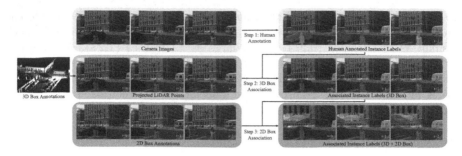

Fig. 4. Labeling and Association Overview. Human annotators first label each camera image for panoptic segmentation separately (step 1). LiDAR points within each ground truth 3D bounding box are then projected to each image, and associate with the single frame instance labels (step 2). For far-range instances without corresponding 3D bounding boxes, we associate the single frame instance labels over time using the ground truth 2D bounding boxes within each camera (step 3). New associations are highlighted in the zoomed-in views at the bottom.

each box, and projecting them onto the image. Association scores are then computed using IoU between the convex hull of the projected LiDAR points and each instance label. Bipartite matching is then applied to match each projected box with an instance label. For 3D driving scenes, points inside the bounding box almost entirely correspond to the instances inside of them, and so these projected LiDAR points have a high overlap with their corresponding instance masks in the image. Our label association step is related to the prior work [29,40], but, our association leverages the ground-truth labeled 3D boxes and only transfers instance IDs rather than fine-grained per-pixel labels.

There are, however, a small number of instances without corresponding LiDAR ground truth boxes due to occlusions, rolling shutter artifacts, and the limited range of the provided LiDAR scans (75 m). We apply an additional matching step by associating the 2D bounding boxes with our instance labels. First, we score matches between 2D boxes and instances by computing the IoU between each 2D box and the tightly-fitting bounding boxes around each instance mask, and then compute associations with bipartite matching.

For boxes with existing 3D associations, we extend these tracks by propagating the existing ID to all other instances that match with the same 2D box. This resolves cases where only a object track misses 3D associations in a few frames. Then, we assign the remaining boxes without any matches to the ID of their corresponding 2D box, if any. Finally, to capture any additional cross-camera associations, we project all of the camera views onto the panorama, and associate instances which overlap in this joint representation.

In order to identify any instances that are still not associated with any ground truth boxes after these steps, we provide an additional mask for these instance pixels indicating that they are not tracked, similar to the *crowd* mask used in single frame instance segmentation labels [12].

4   Benchmark and Evaluation Metrics

In this section, we first describe the task of Panoramic Video Panoptic Segmentation (PVPS). Then we review the evaluation metrics used in the literature, and propose a new metric designed for PVPS with an emphasis on consistent multi-object tracking and segmentation across multiple cameras.

4.1   Problem Definition

We represent a multi-camera video sequence with T frames and M independent camera views as $\{\mathbf{I}_i^{1:T}\}_{i=1}^M$, where $\mathbf{I}_i^t$ is the i-th camera view captured at the t-th time step in the video sequence. Along with the multi-view representation of the full scene, we define the panorama at t-th time step as $\mathbf{I}_{\text{pano}}^t$. In the task of Panoramic Video Panoptic Segmentation (PVPS), we require a mapping f of every pixel (x, y, t, i) in the multi-camera video sequence to a semantic category $c \in \mathbf{C}$ and an instance ID z consistent across camera views and temporal frames. Here, (x, y, t, i) indicates the spatial coordinate (x, y) of the i-th camera view captured at the t-th time step, and $\mathbf{C}$ is the set of semantic categories. Accordingly, we define the mappings f_{id} and f_{sem} for a particular instance ID z and semantic category c in Eq. (1) and Eq. (2), respectively. The mapping functions are the building blocks of our proposed metric introduced in Sect. 4.2.

$$f_{\text{id}}(z) = \{(x, y, i, t) | f(x, y, i, t) = (c, z), c \in \mathbf{C}\}, \tag{1}$$

$$f_{\text{sem}}(c) = \{(x, y, i, t) | f(x, y, i, t) = (c, *), c \in \mathbf{C}\}. \tag{2}$$

Compared to the existing tasks including Video Panoptic Segmentation (VPS) and Panoramic Semantic Segmentation, the proposed task is more challenging in the following aspects. First, each individual camera has its own unique viewpoint and field-of-view such that the semantic class statistics are different across cameras (e.g., see Fig. 3). This leads to a large domain gap between videos captured with different cameras. Second, the instance ID prediction, with the long-term consistency across both time and cameras, requires holistic scene understanding.

4.2   Evaluation Metrics

In this subsection, we overview the existing Video Panoptic Segmentation (VPS) metric: Segmentation and Tracking Quality (STQ) [74], which we extend to evaluate the Panoramic Video Panoptic Segmentation (PVPS) task.

VPS Metric. We use f and g to indicate the prediction and ground-truth mapping, respectively. We define the true positive associations (TPA) [45] of a specific instance as $\text{TPA}(z_f, z_g) = |f_{\text{id}}(z_f) \cap g_{\text{id}}(z_g)|$, where z_f is the predicted instance, $z_g \in \mathbf{G}$ is the ground-truth instance, and $\mathbf{G}$ is the set containing all unique ground-truth instances across cameras and temporal frames. Similarly, false negative associations (FNA) and false positive associations (FPA) can be

Fig. 5. Visualization of the weights tensor for all cameras. Pixels in the blue region have weights 0.5 during evaluation, as they are covered by two cameras. (Color figure online)

defined to compute the Intersection over Union (IoU_{id}) for evaluating tracking quality. Formally, STQ is defined as follows.

$$STQ = (AQ \times SQ)^{\frac{1}{2}}, \tag{3}$$

$$AQ = \frac{1}{|\mathbf{G}|} \sum_{z_g \in \mathbf{G}} \frac{1}{|g_{\text{id}}(z_g)|} \sum_{z_f, |z_f \cap z_g| \neq \emptyset} \text{TPA}(z_f, z_g) \times \text{IoU}_{\text{id}}(z_f, z_g),$$

$$SQ = \frac{1}{|\mathbf{C}|} \sum_{c \in \mathbf{C}} \frac{f_{\text{sem}}(c) \cap g_{\text{sem}}(c)}{f_{\text{sem}}(c) \cup g_{\text{sem}}(c)}.$$

As defined in Eq (3), STQ fairly balances segmentation and tracking performance, and is suitable for evaluating video sequences of arbitrary length. The Association Quality (AQ) measures the association quality for tracking classes, while the Segmentation Quality (SQ) measures the segmentation quality for semantic classes. Specifically, AQ involves the IoU_{id} computation for predicted instance IDs (and further weighted by true positive associations to encourage long-term tracking [74]), while SQ is the typical semantic segmentation metric [16] (*i.e.*, mean IoU_{sem} for predicted semantic classes).

PVPS Metric. We propose to extend the metric STQ [74] for Panoramic Video Panoptic Segmentation (PVPS). However, naïvely adopting STQ for the multi-camera scenario results in a potential issue, where pixels in the overlapping regions covered by multiple cameras will be counted multiple times. Instead, we employ a simple and effective solution by exploiting the pixel-centric property of STQ. In particular, we weight each pixel prediction w.r.t.its coverage by the number of cameras, as determined by the mapping between the camera images and the panorama image. For example, if a pixel is covered by N cameras (in our dataset, $N = 2$), its prediction will contribute $1/N$ when computing AQ, and SQ. We name the resulting metrics as *weighted* STQ (wSTQ), since each pixel prediction takes a different weight depending on its coverage by the number of cameras. In Fig. 5, we visualize the weights for an example of five-camera images.

PS Metric. We also briefly review the metric PQ (panoptic quality) [35] for evaluating image Panoptic Segmentation (PS), since we will build image-level baselines purely trained with image panoptic annotations.

For a particular semantic class c, the sets of true positives (TP_c), false positives (FP_c), and false negatives (FN_c) are formed by matching predictions z_f to the ground-truth masks z_g based on the IoU scores. A minimal threshold of greater than 0.5 IoU is chosen to guarantee unique matching. Formally,

$$PQ_c = \frac{\sum_{(z_f, z_g) \in \text{TP}_c} \text{IoU}(z_f, z_g)}{|\text{TP}_c| + \frac{1}{2}|\text{FP}_c| + \frac{1}{2}|\text{FN}_c|}, \tag{4}$$

where the final PQ is then obtained by averaging PQ_c over semantic classes.

5   Experimental Results

In this section, we introduce our PVPS baselines, which exploit the property of multi-camera images by taking as input either individual camera views or panorama images (generated from all camera views). We then provide extensive experiments on the proposed dataset and metric.

5.1   ViP-DeepLab Extensions as PVPS Baselines

To tackle the new challenging PVPS task, we extend the state-of-art video panoptic segmentation method, ViP-Deeplab [54], to panoramic views.

Baseline Overview. For completeness, we first briefly review ViP-DeepLab [54]. ViP-DeepLab extends the state-of-art image panoptic segmentation model, Panoptic-DeepLab [11], to the video domain. Panoptic-DeepLab employs two separate prediction branches for semantic segmentation [9] and instance segmentation [31], respectively. Both segmentation results are then merged [83] to form the final panoptic segmentation result. To perform video panoptic segmentation, ViP-DeepLab adopts a two-frame image panoptic segmentation framework. Specifically, during training, ViP-DeepLab takes a pair of image frames as input and their panoptic segmentation ground-truths as training target. During inference, ViP-DeepLab performs two-frame image panoptic predictions at each time step, and continues the inference process for every two consecutive frames (*i.e.*, with one overlapping frame at the next time step) in a video sequence. The predictions in the overlapping frames are "stitched" together by propagating instance IDs based on mask IoU between region pairs (*i.e.*, if two masks have high IoU overlap, they will be re-assigned with the same instance ID), and thus temporally consistent IDs are obtained (see Fig. 4 of Qiao *et al.* [54] for an illustration). We refer this post-processing as "panoptic stitching over time".

Baseline Extension for PVPS. We explore several ViP-DeepLab extensions for PVPS, which takes as input individual camera views or panorama images (generated from all camera views). The input types could be different during training and evaluation. Specifically, we define three training schemes: *View*, *Pano*, and *Ensemble-View*. The View scheme refers to the case where ViP-DeepLab is trained with images from all camera views, while Pano means the model is trained with full panorama images. The Ensemble-View scheme refers to the case where we have five camera-specific ViP-DeepLab models, each of which

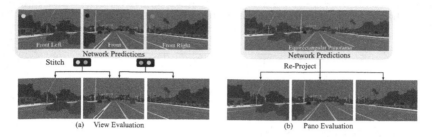

Fig. 6. We experiment with two evaluation schemes: (a) View and (b) Pano. The View evaluation scheme takes individual camera views as input and generates their panoptic predictions, which are then "stitched over cameras" to obtain consistent instance IDs between cameras. The Pano evaluation scheme takes as panorama images as input and generates panoramic panoptic predictions, which are then reprojected back to each camera for evaluation.

is trained and evaluated on their own camera images. We also have two evaluation schemes: *View* and *Pano*. The View scheme refers to the case where the trained model is fed with images from individual camera views and generates the corresponding panoptic predictions for each view. However, the predicted instance IDs are not consistent between cameras, since the predictions are made independently for each view. To generate consistent instance IDs between cameras, we propose a similar method to "panoptic stitching over time": if two masks have high IoU overlap in the overlapping regions between two cameras' field-of-view, we re-assign the same instance ID for them, resulting in the "panoptic stitching over cameras" post-processing method. For the Pano evaluation scheme, the model is fed with panorama images and generates panoramic panoptic predictions. We then re-project panoramic panoptic predictions onto each camera for evaluation. Note that, for the Pano evaluation scheme, the instance IDs are consistent between cameras by nature. We visualize the evaluation schemes in Fig. 6.

Implementation Details. We build our image-based and video-based baselines on top of Panoptic-DeepLab [11] and ViP-DeepLab [54], respectively, using the official code-base [73]. The training strategy follows Panoptic-DeepLab and ViP-DeepLab. Specifically, the models are trained with 32 TPU cores for 60k steps, batch size 32, Adam [33] optimizer and a poly schedule learning rate of 2.5×10^{-4}. We use an ImageNet-1K-pretrained [59] ResNet-50 [25] with stride 16 as the backbone (using atrous convolution [8]). For image-based methods, we use the crop size 1281×1921 during training, while, during inference, we use the whole image (or panorama). We use a similar strategy for the video-based methods, but we use a ResNet-50 backbone with stride 32 and crop size 641×961 due to memory constraints.

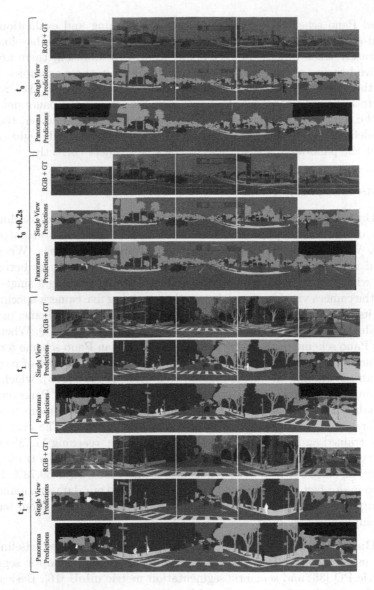

Fig. 7. Comparison of qualitative results from our baseline ViP-DeepLab [54] models over different time intervals. Results show models trained on single images with panoptic stitching over cameras, and trained directly on panorama images. Our baseline models show strong performance for the majority of the scene, although tracking small/distant objects and crowded scenes remains challenging.

5.2 Qualitative Evaluation

In Fig. 7, we provide results from our two ViP-DeepLab baselines, both trained and evaluated on single images and on panorama images (*i.e.*, two models using

View and Pano schemes, respectively, for both training and evaluation), over two (non-adjacent) temporal frames. As shown in the figure, the baseline models accurately track objects in very dense scenes. In addition, there are some qualitative benefits provided by the panorama model in these examples. In particular, the single view model has an inconsistent prediction on the crosswalk in the left and right images for the single view model, but the panorama model attains the full context of the scene and avoids this mistake. Further, the single view models fail to track the car crossing the front right and side right cameras at t_0, but the panorama is again able to track this object correctly.

5.3   Baseline Comparisons

Video-Based Baselines. In Table 2(a), we provide video-based baseline comparisons using ViP-DeepLab [54], evaluated by the proposed weighted STQ (wSTQ). We compare different training and evaluation schemes. When both evaluated with View scheme, training with View scheme performs better than training with Ensemble-View by 0.86% wSTQ. That is, training a single model with all the camera views performs better than training five camera-specific models with its own camera views. Also, when training with View scheme, using the Pano evaluation scheme degrades the performance by 2.91% wSTQ. When training with Pano scheme, using View scheme is better than Pano scheme for evaluation. Due to memory limits, we use a smaller crop size 641×961 of panorama images udring training (from the original resolution 1000×5875), which makes the training and evaluation settings inconsistent. However, the Pano scheme is still valuable, as we observed that the Pano scheme provides more consistent results for large objects that span across multiple cameras. The current best setting is trained and evaluated with the View scheme, reaching 17.78% wSTQ. We observe that our dataset is very challenging in terms of both tracking and segmentation, since our best wAQ is only 8.21% and best wSQ is 39.78%. In summary, two major factors contributed to the low wAQ on our datasets, namely, the scene complexity (e.g., number of objects) and the failure modes of our baselines (see the supplementary doc).

Image-Based Baselines. In Table 2(b), we provide image-based baseline comparisons using Panoptic-DeepLab [11], evaluated by image panoptic segmentation metric PQ [35] and semantic segmentation metric mIoU [16]. Basically, we observe the same trend of image-based baselines and video-based baselines.

5.4   Ablation Studies

Transferability of Models Between Viewpoints. We measure the ability to transfer models between viewpoints. As shown in Table 3, we have the following observations: First, all models, even trained on left side views, perform better on right side views. This phenomenon is due to the ego-vehicle driving on the

Table 2. Quantitative evaluation: the baselines can take different types of inputs: View: individual camera views; Pano: panoramas; Ensemble-View: camera-specific views. Results include (a) video-baseline using ViP-DeepLab, measured by wSTQ; and (b) image-baseline using Panoptic-DeepLab, measured by PQ and mIoU.

<table>
<tr><td colspan="5">(a) Video-baseline Comparison</td><td colspan="4">(b) Image-baseline Comparison</td></tr>
<tr><td>Training Scheme</td><td>Eval Scheme</td><td>wSTQ</td><td>wAQ</td><td>wSQ</td><td>Training Scheme</td><td>Eval Scheme</td><td>PQ</td><td>mIoU</td></tr>
<tr><td>Ensemble-View</td><td>View</td><td>16.92</td><td>7.61</td><td>37.33</td><td>Ensemble-View</td><td>View</td><td>35.70</td><td>48.15</td></tr>
<tr><td>View</td><td>View</td><td>**17.78**</td><td>**8.21**</td><td>38.46</td><td>View</td><td>View</td><td>**40.00**</td><td>**53.64**</td></tr>
<tr><td>View</td><td>Pano</td><td>14.87</td><td>6.13</td><td>36.04</td><td>View</td><td>Pano</td><td>33.65</td><td>50.61</td></tr>
<tr><td>Pano</td><td>View</td><td>17.56</td><td>8.11</td><td>38.04</td><td>Pano</td><td>View</td><td>38.93</td><td>51.65</td></tr>
<tr><td>Pano</td><td>Pano</td><td>15.72</td><td>6.22</td><td>**39.78**</td><td>Pano</td><td>Pano</td><td>36.32</td><td>52.19</td></tr>
</table>

Table 3. View transferability on our video-based baselines, measured by wSTQ. We evaluate models (1st column) trained on a specific view w.r.t.other camera views. The last row, MultiCamera, refers to the model trained with all camera views (*i.e.*, training scheme View), and the last column, All, denotes the evaluation set using all camera views (*i.e.*, evaluation scheme View).

Model \ Eval	Side left	Front left	Front	Front right	Side right	All
Side left	**18.79**	17.41	14.56	19.06	19.40	**16.31**
Front left	16.88	**18.39**	12.84	19.22	18.49	15.36
Front	16.58	18.02	14.54	18.55	18.96	15.98
Front right	16.56	17.36	**14.99**	**19.40**	19.16	16.18
Side right	17.91	16.50	13.16	18.23	**20.47**	15.65
MultiCamera	20.11	19.54	15.63	20.67	21.53	17.78

right side of the road, and providing wider scope, more instances, and smaller objects on the left side (*i.e.*, the left side views are more challenging). Second, the front camera performance is inferior compared to the other cameras. We hypothesize that the front camera captures more diverse and challenging views, *e.g.*, vehicles driving in multiple directions, more dynamic and smaller objects, making tracking more challenging.

6   Conclusion

In this work, we presented a new benchmark called WOD:PVPS. Our benchmark extends video panoptic segmentation to a more challenging multi-camera setting that requires consistent instance IDs both across cameras and over time. Our dataset is an order of magnitude larger than all the existing video panoptic segmentation datasets. We establish several strong baselines evaluated with a new metric, wSTQ, that takes multi-camera, multi-object tracking and segmentation into consideration. We will make our benchmark publicly available, and we hope that it will facilitate future research on panoramic video panoptic segmentation.

References

1. Baqué, P., Fleuret, F., Fua, P.: Deep occlusion reasoning for multi-camera multi-target detection. In: ICCV (2017)
2. Behley, J., et al.: Semantickitti: a dataset for semantic scene understanding of lidar sequences. In: ICCV (2019)
3. Berclaz, J., Fleuret, F., Turetken, E., Fua, P.: Multiple object tracking using k-shortest paths optimization. PAMI **33**(9), 1806–1819 (2011)
4. Brostow, G.J., Fauqueur, J., Cipolla, R.: Semantic object classes in video: a high-definition ground truth database. Pattern Recogn. Lett. **30**(2), 88–97 (2009)
5. Caesar, H., et al.: nuScenes: a multimodal dataset for autonomous driving. In: CVPR (2020)
6. Chang, M.F., et al.: Argoverse: 3D tracking and forecasting with rich maps. In: CVPR (2019)
7. Chavdarova, T., et al.: Wildtrack: a multi-camera HD dataset for dense unscripted pedestrian detection. In: CVPR (2018)
8. Chen, L.C., Papandreou, G., Kokkinos, I., Murphy, K., Yuille, A.L.: Semantic image segmentation with deep convolutional nets and fully connected CRFs. In: ICLR (2015)
9. Chen, L.C., Papandreou, G., Kokkinos, I., Murphy, K., Yuille, A.L.: DeepLab: semantic image segmentation with deep convolutional nets, atrous convolution, and fully connected CRFs. TPAMI **40**(4), 834–848 (2017)
10. Chen, Y., et al.: Geosim: realistic video simulation via geometry-aware composition for self-driving. In: CVPR (2021)
11. Cheng, B., et al.: Panoptic-deeplab: a simple, strong, and fast baseline for bottom-up panoptic segmentation. In: CVPR (2020)
12. Cordts, M., et al.: The cityscapes dataset for semantic urban scene understanding. In: CVPR (2016)
13. Dehghan, A., Modiri Assari, S., Shah, M.: GMMCP tracker: globally optimal generalized maximum multi clique problem for multiple object tracking. In: CVPR (2015)
14. Dendorfer, P., et al.: MOTChallenge: a benchmark for single-camera multiple target tracking. IJCV **129**(4), 845–888 (2020)
15. Eshel, R., Moses, Y.: Homography based multiple camera detection and tracking of people in a dense crowd. In: CVPR (2008)
16. Everingham, M., Van Gool, L., Williams, C.K.I., Winn, J., Zisserman, A.: The pascal visual object classes (VOC) challenge. IJCV **88**(2), 303–338 (2010)
17. Felzenszwalb, P.F., Huttenlocher, D.P.: Efficient graph-based image segmentation. IJCV **59**(2), 167–181 (2004)
18. Ferryman, J., Shahrokni, A.: Pets 2009: dataset and challenge. In: 2009 Twelfth IEEE International Workshop on Performance Evaluation of Tracking and Surveillance, pp. 1–6. IEEE (2009)
19. Fleuret, F., Berclaz, J., Lengagne, R., Fua, P.: Multicamera people tracking with a probabilistic occupancy map. PAMI **30**(2), 267–282 (2007)
20. Gao, N., et al.: SSAP: single-shot instance segmentation with affinity pyramid. In: ICCV (2019)
21. Geiger, A., Lenz, P., Urtasun, R.: Are we ready for autonomous driving? The kitti vision benchmark suite. In: CVPR (2012)
22. Geyer, J., et al.: A2d2: Audi autonomous driving dataset. arXiv preprint arXiv:2004.06320 (2020)

23. Han, X., et al.: MMPTRACK: large-scale densely annotated multi-camera multiple people tracking benchmark (2021)
24. Hariharan, B., Arbeláez, P., Girshick, R., Malik, J.: Simultaneous detection and segmentation. In: Fleet, D., Pajdla, T., Schiele, B., Tuytelaars, T. (eds.) ECCV 2014. LNCS, vol. 8695, pp. 297–312. Springer, Cham (2014). https://doi.org/10.1007/978-3-319-10584-0_20
25. He, K., Zhang, X., Ren, S., Sun, J.: Deep residual learning for image recognition. In: CVPR (2016)
26. He, X., Zemel, R.S., Carreira-Perpiñán, M.Á.: Multiscale conditional random fields for image labeling. In: CVPR (2004)
27. Hofmann, M., Wolf, D., Rigoll, G.: Hypergraphs for joint multi-view reconstruction and multi-object tracking. In: CVPR (2013)
28. Huang, X., Wang, P., Cheng, X., Zhou, D., Geng, Q., Yang, R.: The apolloscape open dataset for autonomous driving and its application. TPAMI **42**(10), 2702–2719 (2019)
29. Huang, X., Wang, P., Cheng, X., Zhou, D., Geng, Q., Yang, R.: The apolloscape open dataset for autonomous driving and its application. PAMI **42**(10), 2702–2719 (2020)
30. Jaus, A., Yang, K., Stiefelhagen, R.: Panoramic panoptic segmentation: towards complete surrounding understanding via unsupervised contrastive learning. In: 2021 IEEE Intelligent Vehicles Symposium (IV), pp. 1421–1427. IEEE (2021)
31. Kendall, A., Gal, Y., Cipolla, R.: Multi-task learning using uncertainty to weigh losses for scene geometry and semantics. In: CVPR (2018)
32. Kim, D., Woo, S., Lee, J.Y., Kweon, I.S.: Video Panoptic Segmentation. In: CVPR (2020)
33. Kingma, D.P., Ba, J.: Adam: a method for stochastic optimization. In: ICLR (2015)
34. Kirillov, A., Girshick, R., He, K., Dollár, P.: Panoptic feature pyramid networks. In: CVPR (2019)
35. Kirillov, A., He, K., Girshick, R., Rother, C., Dollár, P.: Panoptic segmentation. In: CVPR (2019)
36. Kuo, C.-H., Huang, C., Nevatia, R.: Inter-camera association of multi-target tracks by on-line learned appearance affinity models. In: Daniilidis, K., Maragos, P., Paragios, N. (eds.) ECCV 2010. LNCS, vol. 6311, pp. 383–396. Springer, Heidelberg (2010). https://doi.org/10.1007/978-3-642-15549-9_28
37. Ladický, Ľ, Sturgess, P., Alahari, K., Russell, C., Torr, P.H.S.: What, where and how many? Combining object detectors and CRFs. In: Daniilidis, K., Maragos, P., Paragios, N. (eds.) ECCV 2010. LNCS, vol. 6314, pp. 424–437. Springer, Heidelberg (2010). https://doi.org/10.1007/978-3-642-15561-1_31
38. Li, Y., et al.: Attention-guided unified network for panoptic segmentation. In: CVPR (2019)
39. Liang, J., Homayounfar, N., Ma, W.C., Xiong, Y., Hu, R., Urtasun, R.: Polytransform: deep polygon transformer for instance segmentation. In: CVPR (2020)
40. Liao, Y., Xie, J., Geiger, A.: Kitti-360: a novel dataset and benchmarks for urban scene understanding in 2D and 3D. arXiv:2109.13410 (2021)
41. Lin, T.-Y., et al.: Microsoft COCO: common objects in context. In: Fleet, D., Pajdla, T., Schiele, B., Tuytelaars, T. (eds.) ECCV 2014. LNCS, vol. 8693, pp. 740–755. Springer, Cham (2014). https://doi.org/10.1007/978-3-319-10602-1_48
42. Ling, H., Acuna, D., Kreis, K., Kim, S.W., Fidler, S.: Variational amodal object completion. In: NeurIPS (2020)
43. Liu1, H., et al.: An end-to-end network for panoptic segmentation. In: CVPR (2019)

44. Long, J., Shelhamer, E., Darrell, T.: Fully convolutional networks for semantic segmentation. In: CVPR (2015)
45. Luiten, J., et al.: HOTA: a higher order metric for evaluating multi-object tracking. In: IJCV (2020)
46. Mallya, A., Wang, T.-C., Sapra, K., Liu, M.-Y.: World-consistent video-to-video synthesis. In: Vedaldi, A., Bischof, H., Brox, T., Frahm, J.-M. (eds.) ECCV 2020. LNCS, vol. 12353, pp. 359–378. Springer, Cham (2020). https://doi.org/10.1007/978-3-030-58598-3_22
47. Miao, J., Wei, Y., Wu, Y., Liang, C., Li, G., Yang, Y.: VSPW: a large-scale dataset for video scene parsing in the wild. In: CVPR (2021)
48. Narioka, K., Nishimura, H., Itamochi, T., Inomata, T.: Understanding 3D semantic structure around the vehicle with monocular cameras. In: IEEE Intelligent Vehicles Symposium (IV), pp. 132–137. IEEE (2018)
49. Neuhold, G., Ollmann, T., Bulò, S.R., Kontschieder, P.: The mapillary vistas dataset for semantic understanding of street scenes. In: ICCV (2017)
50. Petrovai, A., Nedevschi, S.: Semantic cameras for 360-degree environment perception in automated urban driving. IEEE Trans. Intell. Transp. Syst. (2022)
51. Philion, J., Fidler, S.: Lift, splat, shoot: encoding images from arbitrary camera rigs by implicitly unprojecting to 3D. In: Vedaldi, A., Bischof, H., Brox, T., Frahm, J.-M. (eds.) ECCV 2020. LNCS, vol. 12359, pp. 194–210. Springer, Cham (2020). https://doi.org/10.1007/978-3-030-58568-6_12
52. Porzi, L., Bulò, S.R., Colovic, A., Kontschieder, P.: Seamless scene segmentation. In: CVPR (2019)
53. Qi, C.R., et al.: Offboard 3D object detection from point cloud sequences. In: CVPR (2021)
54. Qiao, S., Zhu, Y., Adam, H., Yuille, A., Chen, L.C.: ViP-DeepLab: learning visual perception with depth-aware video panoptic segmentation. In: CVPR (2021)
55. Ristani, E., Solera, F., Zou, R., Cucchiara, R., Tomasi, C.: Performance measures and a data set for multi-target, multi-camera tracking. In: Hua, G., Jégou, H. (eds.) ECCV 2016. LNCS, vol. 9914, pp. 17–35. Springer, Cham (2016). https://doi.org/10.1007/978-3-319-48881-3_2
56. Ristani, E., Tomasi, C.: Features for multi-target multi-camera tracking and re-identification. In: CVPR (2018)
57. Roddick, T., Cipolla, R.: Predicting semantic map representations from images using pyramid occupancy networks. In: CVPR (2020)
58. Roshan Zamir, A., Dehghan, A., Shah, M.: GMCP-tracker: global multi-object tracking using generalized minimum clique graphs. In: Fitzgibbon, A., Lazebnik, S., Perona, P., Sato, Y., Schmid, C. (eds.) ECCV 2012. LNCS, vol. 7573, pp. 343–356. Springer, Heidelberg (2012). https://doi.org/10.1007/978-3-642-33709-3_25
59. Russakovsky, O., et al.: ImageNet large scale visual recognition challenge. Int. J. Comput. Vision 115(3), 211–252 (2015). https://doi.org/10.1007/s11263-015-0816-y
60. Schönberger, J.L., Zheng, E., Frahm, J.-M., Pollefeys, M.: Pixelwise view selection for unstructured multi-view stereo. In: Leibe, B., Matas, J., Sebe, N., Welling, M. (eds.) ECCV 2016. LNCS, vol. 9907, pp. 501–518. Springer, Cham (2016). https://doi.org/10.1007/978-3-319-46487-9_31
61. Shi, J., Malik, J.: Normalized cuts and image segmentation. PAMI 22(8), 888–905 (2000)
62. Song, S., Zeng, A., Chang, A.X., Savva, M., Savarese, S., Funkhouser, T.: Im2Pano3D: extrapolating 360 structure and semantics beyond the field of view. In: CVPR (2018)

63. Su, Y.C., Grauman, K.: Making 360 video watchable in 2D: learning videography for click free viewing. In: CVPR (2017)
64. Sun, P., et al.: Scalability in perception for autonomous driving: Waymo open dataset. In: CVPR (2020)
65. Tang, Z., et al.: Cityflow: a city-scale benchmark for multi-target multi-camera vehicle tracking and re-identification. In: CVPR (2019)
66. Tateno, K., Navab, N., Tombari, F.: Distortion-aware convolutional filters for dense prediction in panoramic images. In: ECCV (2018)
67. Thrun, S., Montemerlo, M.: The graph slam algorithm with applications to large-scale mapping of urban structures. Int. J. Robot. Res. **25**(5–6), 403–429 (2006)
68. Tu, Z., Chen, X., Yuille, A.L., Zhu, S.C.: Image parsing: unifying segmentation, detection, and recognition. IJCV **63**(2), 113–140 (2005)
69. Voigtlaender, P., et al.: MOTS: multi-object tracking and segmentation. In: CVPR (2019)
70. Wang, H., Luo, R., Maire, M., Shakhnarovich, G.: Pixel consensus voting for panoptic segmentation. In: CVPR (2020)
71. Wang, H., Zhu, Y., Green, B., Adam, H., Yuille, A., Chen, L.-C.: Axial-DeepLab: stand-alone axial-attention for panoptic segmentation. In: Vedaldi, A., Bischof, H., Brox, T., Frahm, J.-M. (eds.) ECCV 2020. LNCS, vol. 12349, pp. 108–126. Springer, Cham (2020). https://doi.org/10.1007/978-3-030-58548-8_7
72. Weber, M., Luiten, J., Leibe, B.: Single-shot panoptic segmentation. In: IROS (2020)
73. Weber, M., et al.: DeepLab2: A TensorFlow Library for Deep Labeling. arXiv: 2106.09748 (2021)
74. Weber, M., et al.: Step: segmenting and tracking every pixel. In: NeurIPS Track on Datasets and Benchmarks (2021)
75. Wu, Y., Lim, J., Yang, M.H.: Online object tracking: a benchmark. In: CVPR (2013)
76. Xiong, Y., et al.: UPSNet: a unified panoptic segmentation network. In: CVPR (2019)
77. Xu, C., Xiong, C., Corso, J.J.: Streaming hierarchical video segmentation. In: Fitzgibbon, A., Lazebnik, S., Perona, P., Sato, Y., Schmid, C. (eds.) ECCV 2012. LNCS, vol. 7577, pp. 626–639. Springer, Heidelberg (2012). https://doi.org/10.1007/978-3-642-33783-3_45
78. Xu, Y., Liu, X., Liu, Y., Zhu, S.C.: Multi-view people tracking via hierarchical trajectory composition. In: CVPR (2016)
79. Yang, B., Bai, M., Liang, M., Zeng, W., Urtasun, R.: Auto4d: learning to label 4D objects from sequential point clouds. arXiv preprint arXiv:2101.06586 (2021)
80. Yang, K., Hu, X., Bergasa, L.M., Romera, E., Wang, K.: Pass: Panoramic annular semantic segmentation. IEEE Trans. Intell. Transp. Syst. **21**(10), 4171–4185 (2019)
81. Yang, K., Zhang, J., Reiß, S., Hu, X., Stiefelhagen, R.: Capturing omni-range context for omnidirectional segmentation. In: CVPR (2021)
82. Yang, L., Fan, Y., Xu, N.: Video instance segmentation. In: ICCV (2019)
83. Yang, T.J., et al.: DeeperLab: Single-Shot Image Parser. arXiv:1902.05093 (2019)
84. Yao, J., Fidler, S., Urtasun, R.: Describing the scene as a whole: joint object detection, scene classification and semantic segmentation. In: CVPR (2012)
85. Yogamani, S., et al.: Woodscape: a multi-task, multi-camera fisheye dataset for autonomous driving. In: ICCV (2019)
86. Yu, F., et al.: BDD100K: a diverse driving dataset for heterogeneous multitask learning. In: CVPR (2020)

87. Zakharov, S., Kehl, W., Bhargava, A., Gaidon, A.: Autolabeling 3D objects with differentiable rendering of SDF shape priors. In: CVPR (2020)
88. Zendel, O., Schörghuber, M., Rainer, B., Murschitz, M., Beleznai, C.: Unifying panoptic segmentation for autonomous driving. In: CVPR (2022)
89. Zhang, C., Liwicki, S., Smith, W., Cipolla, R.: Orientation-aware semantic segmentation on icosahedron spheres. In: ICCV (2019)

TransFGU: A Top-Down Approach to Fine-Grained Unsupervised Semantic Segmentation

Zhaoyuan Yin[1,2], Pichao Wang[2(✉)], Fan Wang[2], Xianzhe Xu[2],
Hanling Zhang[3(✉)], Hao Li[2], and Rong Jin[2]

[1] College of Computer Science and Electronic Engineering, Hunan University,
Changsha, China
zyyin@hnu.edu.cn
[2] Alibaba Group, Hangzhou, China
{pichao.wang,fan.w,xianzhe.xxz,lihao.lh,jinrong.jr}@alibaba-inc.com
[3] School of Design, Hunan University, Changsha, China
jt_hlzhang@hnu.edu.cn

Abstract. Unsupervised semantic segmentation aims to obtain high-level semantic representation on low-level visual features without manual annotations. Most existing methods are bottom-up approaches that try to group pixels into regions based on their visual cues or certain pre-defined rules. As a result, it is difficult for these bottom-up approaches to generate fine-grained semantic segmentation when coming to complicated scenes with multiple objects and some objects sharing similar visual appearance. In contrast, we propose the first top-down unsupervised semantic segmentation framework for fine-grained segmentation in extremely complicated scenarios. Specifically, we first obtain rich high-level structured semantic concept information from large-scale vision data in a self-supervised learning manner, and use such information as a prior to discover potential semantic categories presented in target datasets. Secondly, the discovered high-level semantic categories are mapped to low-level pixel features by calculating the class activate map (CAM) with respect to certain discovered semantic representation. Lastly, the obtained CAMs serve as pseudo labels to train the segmentation module and produce the final semantic segmentation. Experimental results on multiple semantic segmentation benchmarks show that our top-down unsupervised segmentation is robust to both object-centric and scene-centric datasets under different semantic granularity levels, and outperforms all the current state-of-the-art bottom-up methods. Our code is available at https://github.com/damo-cv/TransFGU.

Z. Yin—Work done during an internship at Alibaba Group.
P. Wang—Project lead.

Supplementary Information The online version contains supplementary material available at https://doi.org/10.1007/978-3-031-19818-2_5.

S. Avidan et al. (Eds.): ECCV 2022, LNCS 13689, pp. 73–89, 2022.
https://doi.org/10.1007/978-3-031-19818-2_5

1   Introduction

Given pictures of this world, can a semantic concept be deduced by certain prior rules, or can it be induced from the massive amount of observations? The answer to this question leads to different ways to obtain the semantic concept, and therefore brings different paradigms to the task of unsupervised semantic segmentation, which aims to obtain the pixel-wise classification as the semantic concept without any manual-annotated labels.

One way to tackle unsupervised image segmentation is to group low-level pixels into some semantic groups under the guidance of certain prior knowledge, *i.e.*the bottom-up manner [10, 16, 20, 21, 23, 28, 34], as shown in Fig. 1. Those methods often assume pixels in the same semantic object share a similar representation in the high-level semantic space. However, there is a large gap between low-level pixel and high-level semantic embeddings. Two semantically different objects may be mostly similar in low-level feature space, while the key to distinguishing them lies in some small areas reflecting the uniqueness of a semantic category. *e.g.*, the difference between a horse and a donkey may be found only on ears and legs. Accurate perception of these slight differences is the key to generating fine-grained segmentation, which is very hard to obtain by pixel-level deduction. In contrast, an object that has a large intra-class variance in appearance, *e.g.*the different parts of a person (head, arms, body...), may lead to dissimilar pixel-level features in different parts, which hinder the bottom-up methods from grouping these dissimilar features as an integrated object in high-level semantic space, due to the lack of high-level conceptual understanding of the whole object.

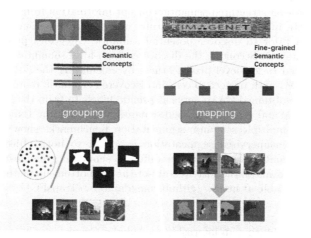

Fig. 1. Bottom-up (left) *vs.*Top-down (right) frameworks. Currently bottom-up manners group feature under the guidance of certain prior knowledge to form the semantic concepts (usually coarse), while our top-down manner maps the fine-grained structural semantic concepts obtained from ImageNet to pixel-level features and generates fine-grained segmentation.

To alleviate these problems, this work proposes a top-down approach to unsupervised semantic segmentation, as shown in Fig. 2. Instead of deducing high-level semantic concepts from low-level visual features, we start from the high-level fine-grained semantic knowledge induced from ImageNet. We benefit from the self-supervised learning method DINO [4] to gain the initial segmentation property of self-attention maps. The semantic representation obtained from DINO is more robust to the object appearance variations. Then we leverage the obtained prior to discover all the potential semantic categories presented in the target dataset, and group them into the desired number of semantic clusters according to their semantic similarity. It makes our method flexible to semantic concepts at different granularity levels. We then project the high-level semantic information to low-level pixel feature space by Grad-CAM [31,32,40], to generate fine-grained active maps for various semantic classes. The obtained active maps serve as pseudo labels for segmentation model training. A bootstrapping mechanism is employed to iteratively refine the quality of pseudo labels and gradually improve the final segmentation performance. The proposed method, named TransFGU, bridges the gap between high-level semantic representation induced by SSL and low-level pixel space, resulting in fine-grained segmentation in both object-centric and scene-centric images without any human annotations.

Through experiments on four public benchmarks, $i.e.$MS-COCO [26], PascalVOC [13], Cityscapes [11], and LIP [14], we show that our method can handle the cases of both foreground-only segmentation and foreground/background segmentation. Moreover, our method can control the semantic granularity level of segmentation by adjusting hyper-parameters, which overcome the limitations of previous methods that they can only produce coarse-level segmentation results. Our method achieves state-of-the-art results on all the benchmarks, surpassing all the current bottom-up methods.

In summary, our contributions are as follows: (i) We propose the first top-down framework for unsupervised semantic segmentation, which directly exploits the semantic information induced from ImageNet to tackle the fine-grained segmentation in an annotation-free way. (ii) We design an effective mechanism that successfully transfers the high-level semantic features obtained from SSL into low-level pixel-wise features to produce high-quality fine-grained segmentation results. (iii) Experiments show that our proposed method achieves state-of-the-art on a variety of semantic segmentation benchmarks including MS-COCO, Pascal-VOC, Cityscapes, and LIP.

2   Related Work

2.1   Self-supervised Representation Learning

Self-supervised representation learning has rapidly gained attention for its promising ability to represent the feature of semantic concepts without additional labels. It is also beneficial to many downstream tasks including detection, segmentation, and tracking. The common paradigm of self-supervised representation learning is to minimize the feature distance between two views of the same object-centric image [3,4,7–9,15,17,24,39]. It allows learning semantic object concepts purely based on images from ImageNet without any annotations. The learned feature can perform well with KNN, which indicates the structural-semantic concepts have been exploited. BYOL [15] shows that representation learning without a single label can be on par or even better than its supervised counterpart. DINO [4], a recent work based on the Transformer, shows that the self-supervised learning method can learn a good representation of structured semantic category information, which has a nice property of focusing on the area of foreground objects. While most of them aim to learn the overall representation from the object-centric image, some others [25,30,35–38] tend to generate pixel-level dense features, which better benefit the tasks that require dense matching like segmentation or detection. However, all of these methods tend to learn the pixel-level correspondence on the category of objects that appear in different views of an image rather than learning the pixel-level semantic concept, so the learned representation embedding feature cannot convert to segmentation mask directly without additional manual labels.

2.2   Unsupervised Segmentation

Most of the current works proposed to tackle unsupervised segmentation start from the observation of pixel-level features and grouping the pixels into different semantic groups with various priors rules that are independent of the training data. [21,23] group the similar pixel extracted from a randomly initialized CNN in both embedding and spatial space while keeping the diversity of embedding features. [16,20,28,29] maximize the mutual information between the pixel-level feature of two views from the same input image to distill the information shared across the image. PiCIE [10] disentangles features between different semantic objects by leveraging two simple rules, i.e.invariance to geometric and equivariance to photometric while transforming two different views of an image and its feature by two asymmetric augmentation processes. These rules often fail to tackle the case with complex scenes, where objects of different categories might share similar appearances and objects of the same category might have a large intra-class appearance variation. Unlike the bottom-up manner, our method is based on the top-down pipeline that obtains fine-grained semantic concepts prior information before performing the segmentation, which is crucial to tackling these two problems.

Besides the bottom-up methods, several works use intermediate features as additional cues to guide the semantic feature grouping, including saliency map [1,2,6,34], super-pixel [21,28], and shape prior [19,22]. MaskContrast [34] leverages two additional unsupervised saliency models to generate pseudo labels of the foreground object in each image and train the final segmentation model with a two-step bootstrapping mechanism, and then cluster all the gathered foreground masks to form the semantic concept by contrast learning. However, these methods assume that objects in an image are salient enough, which is not always true in some complicated scene-based cases and could lead to low-quality segmentation.

3 Methods

Given a set of images $I = [I_1, ..., I_N] \in [0, 1]^{N \times 3 \times H \times W}$ with total N samples, our goal is to generate a group of K segmentation probability maps for each image, denoted as $P = [P_1, ..., P_N] \in [0, 1]^{N \times K \times H \times W}$, where K indicates the predefined number of semantic categories to be segmented in I, and (H, W) are the height and width of the image, respectively. To achieve this goal, a semantic prior model self-supervised trained on ImageNet is introduced to extract the potential top-level semantic information from the whole set of images I, and the extracted top-level semantic features are clustering into K groups based on their semantic similarity, resulting in K semantic concept features. Then, for each $I_i \in I$, a top-down semantic information mapping pipeline is introduced to generate pixel-level semantic pseudo label according to K semantic concept features based on Grad-CAM. An encoder is needed to extract pixel-level features, and a decoder is required to convert the pixel-level features into P_i corresponding to the generated semantic pseudo label. In this section, we will first introduce some basics of the encoder and decoder, then describe the designed top-down pipeline, followed by a bootstrapping process that helps further refine the results.

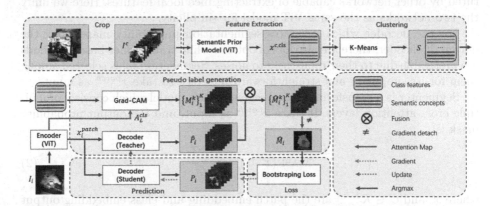

Fig. 2. Our proposed top-down pipeline for unsupervised semantic segmentation. The training samples are first cropped and resized into square patches using sliding windows, and class features for each cropped patch are extracted by ViT pretrained on DINO. The obtained class features are clustered into semantic concept representations, which are mapped to the pixel level to generate the pseudo labels. The quality of pseudo labels is further bootstrapped by the teacher network, based on which a student network is trained to refine the semantic knowledge. The teacher network is updated by the learned student network to produce better pseudo labels for the next round of training.

3.1 Revisiting ViT and Segmenter

Previous works on SSL have shown that ViT [12] models pretrained on ImageNet using methods like DINO [4] can provide surprisingly good properties for segmentation tasks due to its explicit semantic information learned during SSL. Thus, the pre-trained ViT is employed as ENCODER($\cdot$) in our pipeline, for its capability of providing high-level semantic priors through the class token, as well as extracting pixel-level features in image patch tokens. Next, we will introduce some basic notations and terminologies about ViT for easier reference.

ViT is an attention-based model with L layer attention blocks. Each attention block at the l-th layer produces an attention map $A_l \in [0,1]^{(1+hw) \times (1+hw)}$ that can be formulated as:

$$A_l = \text{SOFTMAX}(\frac{Q_l \cdot K_l^{\text{T}}}{\sqrt{d'}}) \tag{1}$$

where $Q_l, K_l \in R^{(1+hw) \times d'}$ are the query and key, respectively, which are mapped and reshaped from the layer input x_l, d' is the dimension of embedding feature in attention block, and (h, w) is the size of the feature map.

As an ENCODER($\cdot$), the final outputs of ViT are denoted as class token x^{cls} and image patch tokens as x^{patch}. It is worth noting that, ViT pretrained with DINO is more irreplaceable for generating high-level semantic priors through class token, however, the model for generating pixel-level features can be substituted by other networks capable of extracting nice local features. Here we unify the two networks as the same ViT model for simplicity.

For the decoder which generates segmentation probability maps using the output of encoder, there are also lots of options while transformer-based methods are preferable due to their robustness to noises. Segmenter [33] is selected among them for its simplicity and straightforward interface to take x^{patch} as input.

Segmenter is a transformer-based segmentation model which consists of multiple cross-attention layers. It takes x^{patch} as input, and the output probability mask is:

$$\text{MASK}(x', C) = \frac{x'C^{\text{T}}}{\sqrt{d}} \tag{2}$$

where x' and $C \in \mathcal{R}^{K \times d}$ are the patch embedding and class embedding output by the final layer in the segmenter, respectively. $\text{MASK}(x', C) \in R^{hw \times K}$ is the downsampled probability maps for K categories, and then it is upsampled to (H, W) to obtain the final mask P. This whole process is denoted as DECODER($\cdot$) in the following sections.

3.2 Top-Down Pipeline

To generate semantic mask P, a ViT model obtained by DINO is first used as a semantic prior model to encode all the potential semantic information in I. In the real-world application, an image might contain abundant and complex semantic concepts instead of a simple object, therefore, applying ViT encoder to the whole image might not be sufficient to attend to all potential semantic priors. Instead, we propose to apply a sliding window cropping operation to each image. For every image I_i, n_c square patches are generated, each with side length $\beta \times \min(H, W)$ where β is a scaling factor taking a series of values. Patches from all images are put together to form a larger "patch image" set $I^c = [I_1^c, \ldots, I_N^c]$ with I_i^c representing the patch set for image I_i. All patches are resized to $(\frac{\min(H,W)}{2}, \frac{\min(H,W)}{2})$ and treated as a whole image before feeding into the ViT encoder.

Class features and patch features are extracted from I^c, denoted as $x^{c,cls} = [x_1^{c,cls}, \ldots, x_N^{c,cls}]$ and $x^{c,patch} = [x_1^{c,patch}, \ldots, x_N^{c,patch}]$, in which:

$$(x_i^{c,cls}, x_i^{c,patch}) = \text{ENCODER}(I_i^c) \tag{3}$$

here the ViT encoder is served as a semantic prior model, $x^{c,cls}$ can be regarded as containing all the potential semantic concepts that have appeared in I, $x^{c,patch}$ is discarded because of its weak semantic prior property. Assuming there are in total K pre-defined semantic categories, we can obtain a set of semantic feature $S = [S_1, \ldots, S_K] \in \mathcal{R}^{K \times d}$ by applying K-means on $x^{c,cls} \in \mathcal{R}^{(N \times n_c) \times d}$ over all the $(N \times n_c)$ features based on Euclidean distance. K can be set as the number of classes, to generate segmentation results at different desired granularity levels.

The extracted semantic features $S_k \in S$ can be seen as a category-wise pseudo semantic label corresponding to a certain granularity level of semantic. Next, the Grad-CAM [31] is adopted to visualize the corresponding category-specific response area to the target pseudo semantic label S. The response heat-map is treated as a coarse semantic area that locates the target semantic object in the pixel-level feature space. Furthermore, these coarse semantic areas are treated as pseudo labels $M = [M_1, \ldots, M_N]$, $M_i \in [0, K]^{H \times W}$, and a decoder will be trained based on these pseudo labels.

To be more specific, inspired by [5], we first obtain the class token x_i^{cls} and patch token x_i^{patch} of the whole image I_i by ViT encoder:

$$(x_i^{cls}, x_i^{patch}) = \text{ENCODER}(I_i), \tag{4}$$

then take attention map from cls token of the last attention block, and take the feature map while discarding the cls token itself. Denote $A_L^{cls} \in [0, 1]^{h \times w}$ as the feature map, gradient is generated on A_L^{cls} w.r.t S_k by maximizing the cosine similarity between x_i^{cls} and each $S_k \in S$:

$$\min(1 - \frac{x_i^{cls} S_k^{\text{T}}}{\sqrt{d}}). \tag{5}$$

We take the gradient value of all the patch locations on A_L^{cls}, and add it back to A_L^{cls} to generate K response maps $[M_i^1, \ldots, M_i^K] \in \mathcal{R}^{K \times h \times w}$, then the pseudo label M_i can be obtained by appling $\arg\max$ operator to each patch location (h', w') across all the K response maps, where $h' \in [0, h], w' \in [0, w]$, and upsample to the target size of (H, W):

$$M_i = \text{UPSAMPLE}(\underset{(h', w')}{\arg\max}(M_i^1, \ldots, M_i^K)), \tag{6}$$

$$M_i^k = \text{GRAD-CAM}(x_i^{cls}, S_k | A_L^{cls}) + A_L^{cls}. \tag{7}$$

If only the foreground objects need to be segmented in an image, we calculate the background probability M_i^{bg} based on all the foreground pseudo labels as follows:

$$M_i^{bg} = \text{RELU}(T^{bg} - \max_{k \in [0,K]} M_i^k) \tag{8}$$

in which T^{bg} is a hyper-parameter that represents the max probability of background, and is set to 0.1 in our experiments. M_i^{bg} is then upsampled and concatenated with M_i. The decoder can be trained by standard cross-entropy loss with its output denoted as segmentation probability maps P:

$$\mathcal{L} = \text{CE}(P, M). \tag{9}$$

3.3 Bootstrapping

Training the decoder directly with the initial pseudo labels may lead to segmentation of low quality because the noise level in pseudo labels can be detrimental. For example, two categories with similar high-level semantic meaning may generate similar response maps on the same image, thus disturbing the ranking of K values on the same location of the pseudo labels and further misleading the learning process. Additionally, the boundary produced by the pseudo labels is not precise enough for certain semantic objects due to the coarse response area in response maps generated by Grad-CAM.

We tackle these problems by a bootstrapping mechanism. First, a teacher-student framework is introduced to refine the pseudo labels progressively. Second, we introduce a set of loss functions to force the model to learn discriminative abilities and prevent the model from over-fitting the noise in the pseudo labels.

Teacher-Student Network. In our design, both teacher and student network have exactly the same architectures as $\text{DECODER}(\cdot)$. As shown in Fig. 2, given an initial pseudo mask M generated by Grad-CAM, its bootstrapped version $\hat{M} = [\hat{M}_1, \ldots, \hat{M}_N]$ can be obtained by aggregating M and the output of the teacher network prediction $\hat{P} = [\hat{P}_1, \ldots, \hat{P}_N] \in [0,1]^{N \times K \times H \times W}$:

$$\hat{P}_i = \text{TEACHER}(x_i^{patch}), \tag{10}$$

$$\hat{M}_i = \text{UPSAMPLE}(\arg\max_{(h',w')}(\hat{M}_i^1, \ldots, \hat{M}_i^K)), \tag{11}$$

$$\hat{M}_i^k = 0.5 \cdot (M_i^k + \hat{P}_i^k), k \in [1, \ldots, K]. \tag{12}$$

where $\hat{P}_i^k \in \mathcal{R}^{h \times w}$ is the k-th probability map in $\hat{P}_i$ before upsampled. With this bootstrapped pseudo labels, the student network is trained with a cross-entropy loss $\text{CE}(P, \hat{M})$ with P as the output segmentation probability of student network, i.e. $P_i = \text{STUDENT}(x_i^{patch})$.

In the next round, the teacher network is updated with the parameter trained from the student network, so it can produce a better prediction of $\hat{P}$, which further improves the bootstrapped pseudo labels $\hat{M}$. The student network in the next round is reset to the initial state and trained with the improved $\hat{M}$. Therefore, the quality of bootstrapped pseudo labels and the output of the student network can be improved progressively as the iteration goes on.

Bootstrapping Loss. A few additional loss functions are further proposed to provide better guidance for training the decoder based on the pseudo label masks.

First, we observe that even though categories with similar semantic meanings are difficult to differentiate thus might confuse the training process, categories with much different semantic meanings are actually easier to identify with high accuracy. Therefore, in addition to the cross-entropy loss which aligns probability masks P and the "correct" pseudo labels $\hat{M}$, we also introduce a set of "wrong" pseudo labels $\hat{M}'$ so that its cross-entropy loss with P should be maximized. Thus a peer loss [27] is defined as below:

$$\mathcal{L}_{peer} = \mathrm{CE}(P, \hat{M}) - \alpha \cdot \mathrm{CE}(P, \hat{M}'). \tag{13}$$

where α is a hyper-parameter, and $\hat{M}'$ is the constructed negative labels by shuffling the K pseudo labels of $\hat{M}$.

Second, it is observed that categories with similar semantic meanings usually generate similar response values at the same location of the pseudo label masks, making it hard to distinguish the correct category from the others and further slow down the training. An uncertainty loss is introduced to diminish such uncertainty. Specifically, this is achieved by maximizing the gap between the largest and the second largest value at each location of the output probability map P:

$$\mathcal{L}_{unc} = 1 - \frac{1}{hw} \sum_{(h', w')} (p' - p''). \tag{14}$$

in which p', p'' are the largest and second largest probability value among the K probabilities at location (h', w').

Third, it is beneficial to keep the representations diversified in the decoder model to learn more meaningful categories information. This is done by a diversity loss which maximizes the summation of all the pairwise distances between the class embeddings of K categories, *i.e.*, minimizing their cosine similarities:

$$\mathcal{L}_{div} = 1 + \frac{1}{K^2} \sum \frac{C \cdot C^T}{\sqrt{d}}. \tag{15}$$

where C is the class embedding in the decoder as in Eq. 2.

The final loss is:

$$\mathcal{L} = \mathcal{L}_{peer} + \omega_1 \cdot \mathcal{L}_{div} + \omega_2 \cdot \mathcal{L}_{unc}. \tag{16}$$

in which ω_1, ω_2 are hyper-parameters to balance between different losses, and set to 1 and 0.3 in our experiments, respectively.

4   Experiments

4.1   Dataset and Evaluation Metrics

We conduct experiments on four semantic segmentation benchmarks including COCO-Stuff, Pascal-VOC, Cityscapes, and LIP, with different definitions of semantic concepts.

COCO-Stuff. COCO-Stuff is a challenging benchmark with scene-centric images, which contains 80 things categories and 91 stuff categories. The total number of training and validation samples is 117,266 and 5,000 images, respectively. Note that the previous works [10,20] only conduct their experiments on the "curated" data (2175 images total) in which the stuff occupied at least 75% pixels of an image. We evaluate our method on three settings: 1) COCO-S-27:all categories in COCO-Stuff merged into 27 superclasses as [10], 2) COCO-S-171: the original 171 things and stuff categories, and 3) COCO-80: only 80 things categories without background stuff.

Cityscapes. Cityscapes contain 5,000 images focusing on street scenes, in which 2,975 and 500 images are used for training and validation. All pixels are categorized into 34 classes. We follow [10] to merge 34 classes into 27 classes after filtering out the 'void' class.

Pascal-VOC. Pascal-VOC contains 1464 images for training and 1,449 for evaluation, with 20 classes as foreground objects to be segmented and the rest pixels as background.

LIP. LIP contains person images cropped from MS-COCO [26]. It has 30,462 training images and 10,000 validation images, with 19 semantic parts defined. The 19 categories are merged into 16 and 5 coarse categories to evaluate the ability of our method to handle the semantic concepts of different granularities. Please refer to supplementary materials for more details.

Evaluation Metric. Following the standard evaluation protocols for unsupervised segmentation used in [10,20,34], we use Hungarian matching algorithm to align indices between our prediction and the ground-truth label over all the images in the validation set. Two metrics are reported for comparison, *i.e.*Intersection over Union (mIoU) and Pixel Accuracy over all the semantic categories.

4.2   Implementation Details

Cropping and Evaluation Protocol. The scaling factor β in sliding window cropping is set to $0.5, 0.4, 0.3, 0.2$, and the step size of the sliding window is set to $\frac{1}{2} \times \beta \times \min(H, W)$.

Foreground prior is introduced in the cropping operation, which is defined as the binarized attention map $\bar{A}_L^{cls}$ obtained from A_L^{cls} on the last attention block by setting the values greater than the mean value of A_L^{cls} as 1 and the rest as 0. Cropped patches are separated into foreground patches and background patches. A patch is a foreground patch when it has more than 50% of pixels belonging to $\bar{A}_L^{cls}$, and a background patch when it contains more than 80% of pixels that belong to $(1 - \bar{A}_L^{cls})$. K-Means is executed on these two groups of patches separately. If a patch is treated as a foreground patch, its probability in those background categories will have a default value of 0, and a background patch is treated in a similar way. The separation of foreground and background patches help generate more accurate semantic clusters and better pseudo labels.

Different cropping protocols are applied to the four datasets, due to their different properties and requirements. On Cityscapes, foreground prior is not involved as there is no definition of foreground/background. On LIP, the images are obtained as bounding boxes around each person without much background area, so we treat all the areas as foreground. On COCO-80, Pascal-VOC and LIP, Eq. (8) is used to generate background probability.

Trainning Setting. We use *ViT-small* 8×8 pre-trained on ImageNet by DINO [4] as encoder and fix the weight during training, and Segmenter [33] with random initial weights as the decoder. We pre-compute and save all the initial pseudo labels and build a pseudo label bank which considerably accelerates the training. More training details are included in the supplemental materials.

Data Augmentation. A set of data augmentations are utilized, such as color jittering, horizontal flipping, Gaussian blur, color dropping, and random resized crop following [4,15]. The cropped image is resized to $(\frac{H}{2}, \frac{W}{2})$ for the following training. We crop and resize the initial pre-computed pseudo label to $(\frac{h}{2}, \frac{w}{2})$ by RoI-Align operator [18].

4.3   Main Results

Baseline. We compare our method with PiCIE [10], IIC [20] and MaskContrast (MC) [34], which can generate segmentation masks without further fine-tuning. Since IIC and PiCIE cannot distinguish foreground objects from the background, they can only be applied on COCO-Stuff and Cityscapes. MaskContrast is only able to deal with the foreground object without background area, so it is only applied to COCO-80 and Pascal-VOC. None of these methods can actually work well on LIP which is essentially a task of fine-grained human parsing. Note that our method can be successfully adapted to all these datasets.

Table 1. Results on four benchmarks. * indicates the results are evaluated on the "curated" samples. † denotes PiCIE trained without auxiliary clustering.

Dataset	Method	mIoU	Acc.	Dataset	Method	mIoU	Acc.
COCO-S-27*	IIC [20]	6.71	21.79	COCO-80	MC [34]	3.73	8.81
	PiCIE† [10]	13.84	48.09		TransFGU	**12.69**	**64.31**
	PiCIE [10]	14.36	49.99	Cityscapes	IIC [20]	6.35	47.88
	TransFGU	**17.47**	**52.66**		PiCIE [10]	12.31	65.50
COCO-S-27	IIC [20]	2.36	21.02		TransFGU	**16.83**	**77.92**
	PiCIE [10]	11.88	37.20	Pascal-VOC	MC [34]	35.00	79.84
	TransFGU	**16.19**	**44.52**		TransFGU	**37.15**	**83.59**
COCO-S-171	IIC [20]	0.64	8.67	LIP-5	TransFGU	25.16	65.76
	PiCIE [10]	4.56	24.66	LIP-16	TransFGU	15.49	60.08
	TransFGU	**11.93**	**34.32**	LIP-19	TransFGU	12.24	42.52

Quantitative Evaluation. The comparison is shown in Table 1. Our method exceeds all the baseline methods on mIoU and pixel accuracy of all datasets. Our method has been adapted to different granularity levels in the same dataset, $e.g.27/171$ categories on COCO-S and 5/16/19 categories in LIP, to make a fair comparison with other methods which mainly work on coarser-level categories. Our method shows a larger margin of performance improvement for the settings with much finer-grained categories. Besides, it can also be adapted to only segment foreground objects, $i.e.$COCO-80, Pascal-VOC, and LIP, demonstrating its all-around flexibility.

Fair Comparison with Prior Arts. The difference in backbone architecture and the pre-training manner to obtain backbone weights influences the model performance. For fair comparison, we conduct the following experiments successively: (1) reproduce IIC with the ResNet-18 backbone fully-supervised trained on ImageNet to align the amount of training data; (2) reproduce IIC and PiCIE with ResNet-50 and ViT-S fully-supervised trained on ImageNet instead of its original ResNet-18 to make backbone parameters comparable with our method; (3) reproduce IIC, PiCIE and MaskContrast with the ResNet-50 backbone trained by DINO to unify the training manner; (4) reproduce the PiCIE and MaskContrast with the ViT backbone trained by DINO instead of ResNet-50 to eliminate the impact of the architecture difference. To obtain the input feature maps of FPN used in PiCIE, we extract pixel features every three attention blocks amount 12 layers and resize them to the target size. The results are shown in Table 2. Our method outperforms all the baseline in each case. Note that the performance of IIC and PiCIE with backbones trained by SSL manner becomes inferior to its original fully-supervised version. It might be due to the feature learned by SSL containing more fine-grained foreground details, which may distract the bottom-up method to find similar features that belong to the same category.

Table 2. Results for the effectiveness of pretrained weight. *, † and ‡ indicates the weights are trained on ImageNet by fully-supervised, DINO and MoCo, respectively.

Dataset	Method	mIoU	Acc.	Dataset	Method	mIoU	Acc.
COCO-S-171	IIC-R18 [20]	0.64	8.67	COCO-S-171	PiCIE-R50† [10]	2.30	13.50
	IIC-R18* [20]	1.22	13.92		PiCIE-ViT† [10]	3.02	18.45
	IIC-R50* [20]	2.15	15.72		TransFGU†	**11.93**	**34.32**
	IIC-R50† [20]	0.98	11.89	Pascal-VOC	MC-R50‡ [34]	35.00	79.84
	PiCIE-R18 [10]	4.56	24.66		MC-R50† [34]	28.72	78.72
	PiCIE-R50 [10]	5.61	29.79		MC-ViT† [34]	31.24	79.18
	PiCIE-ViT* [10]	6.82	31.17		TransFGU†	**37.15**	**83.59**

Qualitative Evaluation. Some qualitative comparisons on different benchmarks are shown in Fig. 3. Our results can show richer semantic information in both scene/object-centric images on all datasets. We obtain much detailed segmentation on foreground objects, especially in complicated scenarios.

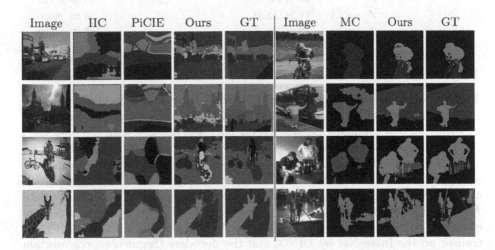

| Image | IIC | PiCIE | Ours | GT | Image | MC | Ours | GT |

Fig. 3. Qualitative comparison on COCO-S-171 (left) and Pascal-VOC (right).

4.4 Ablation Studies

The Bootstrapping Mechanism. Table 3 compares results on MS-COCO of directly using initial pseudo labels ("initial"), training the student network once on the initial pseudo labels without the bootstrapping ("trained"), and the proposed bootstrapping mechanism ("bootstrapped"). Student network trained once on the initial pseudo labels can achieve an improved mIoU over the original pseudo labels, indicating that it can effectively learn meaningful information from the noisy pseudo label. The performance is further improved as more iterations of the teacher-student bootstrapping are introduced.

Table 3. Results for the effectiveness of bootstrap mechanism on MS-COCO with various semantic levels.

Dataset	Initial	Trained	Bootstrapped
COCO-S-27	8.95	13.19	16.19
COCO-S-171	5.05	8.66	11.93
COCO-80	4.23	8.28	12.69

Table 4. Results for the effectiveness of encoder and decoder on COCO-S-171.

Encoder	Decoder	mIoU	Acc.
None	None	5.05	17.33
R50+FPN	Segmenter	9.21	26.17
ViT+FPN	Classifier	7.96	24.17
R50+FPN	Classifier	8.33	25.35

Table 5. Results for different losses on COCO-S-171 and Pascal-VOC.

Dataset	CE	Peer	unc	div	mIoU	Acc.	Dataset	CE	Peer	unc	div	mIoU	Acc.
COCO-S-171	✓				10.49	29.72	Pascal-VOC	✓				34.24	79.85
		✓			10.54	30.46			✓			35.08	80.92
		✓	✓		11.24	33.03			✓	✓		36.46	82.36
		✓		✓	10.96	32.45			✓		✓	35.97	82.03
		✓	✓	✓	11.93	34.32			✓	✓	✓	37.15	83.59

The Effectiveness of Encoder and Decoder. To evaluate the effectiveness of the encoder and decoder used in our top-down pipeline, we conduct three more experiments that gradually replace the encoder and decoder with the ones used in PiCIE [10]. First, we change the encoder from ViT to ResNet-50 followed by an FPN model and keep the decoder as Segmenter. Second, we keep the encoder as ViT and change the decoder from Segmenter to a single layer convolution classifier to map the output dimension of the last attention block from d to K. Last, the encoder and decoder are replaced by ResNet-50 and convolution classifier, respectively. In all the settings, the encoders (ViT, ResNet-50) are trained on the ImageNet by DINO, and the decoders (Segmenter, convolution classifier) are randomly initialized. The results are shown in Table 4. As one can see, in all the settings, the performance can be improved from the initial pseudo label (encoder and decoder are all set to 'None'). Moreover, it's important to use fully-transformer architecture in our pipeline due to its robustness to noises.

The Bootstrapping Loss. We compare different combinations of using the original CE loss in Eq. (9), peer loss in Eq. (13), uncertainty loss in Eq. (14), and diversity loss in Eq. (15) in Table 5. Each of the proposed three bootstrapping losses can effectively improve the performance of the original CE loss, and their combination achieves the best performance.

5   Conclusion

We propose the first top-down framework for unsupervised semantic segmentation, which shows the importance of high-level semantic information in this task. The semantic prior information is learned from large-scale visual data in a self-supervised manner, and then mapped to the pixel-level feature space. By carefully designing the mapping process and the unsupervised training mechanism, we obtain fine-grained segmentation for both foreground and background. Our design also enables the flexible control of granularity levels of the semantic categories, making it possible to generate semantic segmentation results for various datasets with different requirements. The fully unsupervised manner and the flexibility make our method much more practical in real applications.

Acknowledgements. This work was supported by funds for Key R&D Program of Hunan (2022SK2104), Leading plan for scientific and technological innovation of high-tech industries of Hunan (2022GK4010), the National Natural Science Foundation of Changsha (kq2202176), National Key R&D Program of China (2021YFF0900602), the National Natural Science Foundation of China (61672222) and Alibaba Group through Alibaba Research Intern Program.

References

1. Abdal, R., Zhu, P., Mitra, N., Wonka, P.: Labels4Free: unsupervised segmentation using StyleGAN. arXiv preprint arXiv:2103.14968 (2021)
2. Bielski, A., Favaro, P.: Emergence of object segmentation in perturbed generative models. In: Proceedings of the 33rd International Conference on Neural Information Processing Systems, pp. 7256–7266 (2019)
3. Caron, M., Misra, I., Mairal, J., Goyal, P., Bojanowski, P., Joulin, A.: Unsupervised learning of visual features by contrasting cluster assignments. Adv. Neural. Inf. Process. Syst. **33**, 9912–9924 (2020)
4. Caron, M., et al.: Emerging properties in self-supervised vision transformers. In: Proceedings of the International Conference on Computer Vision (ICCV) (2021)
5. Chefer, H., Gur, S., Wolf, L.: Transformer interpretability beyond attention visualization. In: Proceedings of the IEEE/CVF Conference on Computer Vision and Pattern Recognition, pp. 782–791 (2021)
6. Chen, M., Artières, T., Denoyer, L.: Unsupervised object segmentation by redrawing. In: Advances in Neural Information Processing Systems 32 (NIPS 2019), pp. 12705–12716. Curran Associates, Inc. (2019)
7. Chen, T., Kornblith, S., Norouzi, M., Hinton, G.: A simple framework for contrastive learning of visual representations. In: International Conference on Machine Learning, pp. 1597–1607. PMLR (2020)
8. Chen, X., Fan, H., Girshick, R., He, K.: Improved baselines with momentum contrastive learning. arXiv preprint arXiv:2003.04297 (2020)
9. Chen, X., He, K.: Exploring simple siamese representation learning. In: Proceedings of the IEEE/CVF Conference on Computer Vision and Pattern Recognition, pp. 15750–15758 (2021)
10. Cho, J.H., Mall, U., Bala, K., Hariharan, B.: PiCIE: unsupervised semantic segmentation using invariance and equivariance in clustering. In: Proceedings of the IEEE/CVF Conference on Computer Vision and Pattern Recognition, pp. 16794–16804 (2021)
11. Cordts, M., et al.: The cityscapes dataset for semantic urban scene understanding. In: Proceedings of the IEEE Conference on Computer Vision and Pattern Recognition, pp. 3213–3223 (2016)

12. Dosovitskiy, A., et al.: An image is worth 16x16 words: transformers for image recognition at scale. arXiv preprint arXiv:2010.11929 (2020)
13. Everingham, M., Van Gool, L., Williams, C.K., Winn, J., Zisserman, A.: The pascal visual object classes (VOC) challenge. Int. J. Comput. Vision **88**(2), 303–338 (2010)
14. Gong, K., Liang, X., Zhang, D., Shen, X., Lin, L.: Look into person: self-supervised structure-sensitive learning and a new benchmark for human parsing. In: Proceedings of the IEEE Conference on Computer Vision and Pattern Recognition, pp. 932–940 (2017)
15. Grill, J.B., et al.: Bootstrap your own latent: a new approach to self-supervised learning. In: Neural Information Processing Systems (2020)
16. Harb, R., Knöbelreiter, P.: InfoSeg: unsupervised semantic image segmentation with mutual information maximization (2021)
17. He, K., Fan, H., Wu, Y., Xie, S., Girshick, R.: Momentum contrast for unsupervised visual representation learning. In: Proceedings of the IEEE/CVF Conference on Computer Vision and Pattern Recognition, pp. 9729–9738 (2020)
18. He, K., Gkioxari, G., Dollár, P., Girshick, R.: Mask R-CNN. In: Proceedings of the IEEE International Conference on Computer Vision, pp. 2961–2969 (2017)
19. Hwang, J.J., et al.: SegSort: segmentation by discriminative sorting of segments. In: Proceedings of the IEEE/CVF International Conference on Computer Vision, pp. 7334–7344 (2019)
20. Ji, X., Henriques, J.F., Vedaldi, A.: Invariant information clustering for unsupervised image classification and segmentation. In: Proceedings of the IEEE/CVF International Conference on Computer Vision, pp. 9865–9874 (2019)
21. Kanezaki, A.: Unsupervised image segmentation by backpropagation. In: 2018 IEEE International Conference on Acoustics, Speech and Signal Processing (ICASSP), pp. 1543–1547. IEEE (2018)
22. Kim, D., Hong, B.W.: Unsupervised segmentation incorporating shape prior via generative adversarial networks. In: Proceedings of the IEEE/CVF International Conference on Computer Vision, pp. 7324–7334 (2021)
23. Kim, W., Kanezaki, A., Tanaka, M.: Unsupervised learning of image segmentation based on differentiable feature clustering. IEEE Trans. Image Process. **29**, 8055–8068 (2020)
24. Li, C., et al.: Efficient self-supervised vision transformers for representation learning. arXiv preprint arXiv:2106.09785 (2021)
25. Li, X., et al.: Dense semantic contrast for self-supervised visual representation learning. arXiv preprint arXiv:2109.07756 (2021)
26. Lin, T.-Y., et al.: Microsoft COCO: common objects in context. In: Fleet, D., Pajdla, T., Schiele, B., Tuytelaars, T. (eds.) ECCV 2014. LNCS, vol. 8693, pp. 740–755. Springer, Cham (2014). https://doi.org/10.1007/978-3-319-10602-1_48
27. Liu, Y., Guo, H.: Peer loss functions: learning from noisy labels without knowing noise rates. In: International Conference on Machine Learning, pp. 6226–6236. PMLR (2020)
28. Mirsadeghi, S.E., Royat, A., Rezatofighi, H.: Unsupervised image segmentation by mutual information maximization and adversarial regularization. IEEE Robot. Autom. Lett. **6**(4), 6931–6938 (2021)
29. Ouali, Y., Hudelot, C., Tami, M.: Autoregressive unsupervised image segmentation. In: Vedaldi, A., Bischof, H., Brox, T., Frahm, J.-M. (eds.) ECCV 2020. LNCS, vol. 12352, pp. 142–158. Springer, Cham (2020). https://doi.org/10.1007/978-3-030-58571-6_9
30. Pinheiro, P.O., Almahairi, A., Benmalek, R.Y., Golemo, F., Courville, A.C.: Unsupervised learning of dense visual representations. In: NeurIPS (2020)

31. Selvaraju, R.R., Cogswell, M., Das, A., Vedantam, R., Parikh, D., Batra, D.: Grad-cam: visual explanations from deep networks via gradient-based localization. In: Proceedings of the IEEE International Conference on Computer Vision, pp. 618–626 (2017)

32. Shi, X., Khademi, S., Li, Y., van Gemert, J.: Zoom-cam: generating fine-grained pixel annotations from image labels. In: 2020 25th International Conference on Pattern Recognition (ICPR), pp. 10289–10296. IEEE (2021)

33. Strudel, R., Garcia, R., Laptev, I., Schmid, C.: Segmenter: transformer for semantic segmentation. In: Proceedings of the IEEE/CVF International Conference on Computer Vision (2021)

34. Van Gansbeke, W., Vandenhende, S., Georgoulis, S., Van Gool, L.: Unsupervised semantic segmentation by contrasting object mask proposals. In: International Conference on Computer Vision (2021)

35. Wang, X., Zhang, R., Shen, C., Kong, T., Li, L.: Dense contrastive learning for self-supervised visual pre-training. In: Proceedings of the IEEE/CVF Conference on Computer Vision and Pattern Recognition, pp. 3024–3033 (2021)

36. Wang, Z., et al.: Exploring set similarity for dense self-supervised representation learning. arXiv preprint arXiv:2107.08712 (2021)

37. Xie, Z., Lin, Y., Zhang, Z., Cao, Y., Lin, S., Hu, H.: Propagate yourself: exploring pixel-level consistency for unsupervised visual representation learning. In: Proceedings of the IEEE/CVF Conference on Computer Vision and Pattern Recognition, pp. 16684–16693 (2021)

38. Xu, J., Wang, X.: Rethinking self-supervised correspondence learning: a video frame-level similarity perspective. arXiv preprint arXiv:2103.17263 (2021)

39. Yao, Z., Cao, Y., Lin, Y., Liu, Z., Zhang, Z., Hu, H.: Leveraging batch normalization for vision transformers. In: Proceedings of the IEEE/CVF International Conference on Computer Vision, pp. 413–422 (2021)

40. Zou, Y., et al.: PseudoSeg: designing pseudo labels for semantic segmentation. In: International Conference on Learning Representations (ICLR) (2021)

AdaAfford: Learning to Adapt Manipulation Affordance for 3D Articulated Objects via Few-Shot Interactions

Yian Wang[1,2], Ruihai Wu[1,2], Kaichun Mo[3], Jiaqi Ke[1,2], Qingnan Fan[4], Leonidas J. Guibas[3], and Hao Dong[1,2,5(✉)]

[1] CFCS, CS Department, PKU, Beijing, China
{yianwang,wuruihai,kjq001220,hao.dong}@pku.edu.cn
[2] AIIT, PKU, Beijing, China
[3] Stanford University, Stanford, USA
{kaichun,guibas}@cs.stanford.edu
[4] Tencent AI Lab, Bellevue, USA
[5] Peng Cheng Lab, Shenzhen, China
https://hyperplane-lab.github.io/AdaAfford

Abstract. Perceiving and interacting with 3D articulated objects, such as cabinets, doors, and faucets, pose particular challenges for future home-assistant robots performing daily tasks in human environments.

Besides parsing the articulated parts and joint parameters, researchers recently advocate learning manipulation affordance over the input shape geometry which is more task-aware and geometrically fine-grained.

However, taking only passive observations as inputs, these methods ignore many hidden but important kinematic constraints (*e.g.*, joint location and limits) and dynamic factors (*e.g.*, joint friction and restitution), therefore losing significant accuracy for test cases with such uncertainties. In this paper, we propose a novel framework, named AdaAfford, that learns to perform very few test-time interactions for quickly adapting the affordance priors to more accurate instance-specific posteriors. We conduct large-scale experiments using the PartNet-Mobility dataset and prove that our system performs better than baselines. We will release our code and data upon paper acceptance.

1   Introduction

For future home-assistant robots to aid humans in accomplishing diverse everyday tasks, we must equip them with strong capabilities perceiving and interacting with diverse 3D objects in human environments. Articulated objects, such

---

Y. Wang, R. Wu and K. Mo—Equal contribution.

Supplementary Information The online version contains supplementary material available at https://doi.org/10.1007/978-3-031-19818-2_6.

S. Avidan et al. (Eds.): ECCV 2022, LNCS 13689, pp. 90–107, 2022.
https://doi.org/10.1007/978-3-031-19818-2_6

as cabinets, doors, and faucets, are particularly interesting kinds of 3D shapes in our daily lives since agents can interact with them and trigger functionally important state changes of the objects (*e.g.*, push closed the drawer of the cabinet, rotate the handle and pull open the door, turn on/off the water from the faucet by rotating the switch). However, because robots need to understand more semantically complicated part semantics and manipulate articulated parts with higher degree-of-freedoms than rigid objects, it remains a very important yet challenging task to perceive and interact with 3D articulated objects.

Many previous works have investigated the problem of perceiving and interacting with 3D articulated objects. Researchers have been pushing the state-of-the-arts on segmenting articulated parts [32,42], tracking them [30,35], and estimating joint parameters [34,40], enabling robotic systems [2,23,33] to successfully perform sophisticated planning and control over 3D articulated objects.

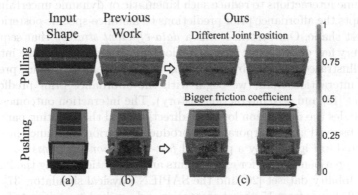

Fig. 1. For robotic manipulation over 3D articulated objects (a), past works [18,36] have demonstrated the usefulness of per-point manipulation affordance (b). However, only observing static visual inputs passively, these systems suffer from intrinsic ambiguities over kinematic constraints. Our *AdaAfford* framework reduces such uncertainties via interactions and quickly adapts instance-specific affordance posteriors (c).

More recently, beyond recognizing the articulated parts and joints, researchers have been proposing learning more task-aware and geometrically fine-grained manipulation affordance over input 3D geometry. Where2Act [18], the most related to our work, learns densely labeled manipulation affordance heatmaps over 3D input partial scans of articulated objects, as illustrated in Fig. 1(b), by performing self-supervised trial-and-error interaction in a physical simulator. There are also many other works leveraging similar dense affordance predictions over 3D scenes [21] and rigid objects [17]. Such densely labeled affordance predictions over 3D data provide more geometrically fine-grained actionable information and can be learned task-specifically given different manipulation actions, showing promises in bridging the perception-interaction gaps for robotic manipulation over large-scale 3D data across different tasks.

However, taking only a single-frame observation of the 3D shape as input (*e.g.*, a single 2D image, a single partial 3D scan), these methods systematically fail to capture many hidden but important kinematic or dynamic factors and therefore predict inaccurate affordance heatmaps, similar to Fig. 1(b), by averaging out such uncertainties. For example, given a fully closed cabinet door with no obvious handle as shown in Fig. 1 (top-row), it is uncertain if the door axis is on the left or right side, which significantly affects the manipulation affordance predictions. Other kinematic uncertainties include joint limits (*e.g.*, push inward or pull outward for a door) and types (*e.g.*, slide or rotate to open a door). Besides, various dynamic or physical parameters (*e.g.*, part mass, joint friction) are also unobservable from single-frame inputs but largely affect manipulation affordance. For example, with increasing friction coefficient for a cabinet drawer (Fig. 1, bottom-row), robots would be able to push the inner board.

In this paper, we propose a novel framework *AdaAfford* learning perform very few test-time interactions to reduce such kinematic or dynamic uncertainties and fastly adapts the affordance prior predictions to instance-specific posteriors given a novel test shape. Our system learns a *data-efficient strategy* that sequentially samples very few uncertain or interesting locations to interact, as the interacting grippers illustrated in Fig. 1(b), according to the current affordance predictions and past interaction trials (we begin with the affordance prior predictions of Where2Act [18] and zero interaction history). The interaction outcomes, each of which includes the interaction location, direction, and the resulting part motion, are then observed and incorporated to produce posterior affordance predictions, as illustrated in Fig. 1(c), by a proposed *fast-adaptation mechanism*.

We set up a benchmark for experiments and evaluations using the large-scale PartNet-Mobility dataset [20] and the SAPIEN physical simulator [37]. We use in total 972 shapes from 15 object categories and conduct experiments for several action types, and randomly sample the kinematic and dynamic parameters for the 3D articulated objects in simulation. Experiments show our method can successfully and efficiently adapt manipulation affordance to novel test shapes with as few as one to four interactions. Quantitative evaluation further proves the effectiveness of our proposed approach.

In summary, our main contributions are the following. 1) we point out and investigate an important limitation of the methods that learn densely labeled visual manipulation affordance – the unawareness of hidden yet important kinematic and dynamic uncertainties; 2) we propose a novel framework *AdaAfford* that learns to perform very few test-time interactions to reduce uncertainties and quickly adapt to predicting an instance-specific affordance posterior; 3) we set up a large-scale benchmark, built upon PartNet-Mobility [20] and SAPIEN [37], for experiments and evaluations, and results demonstrated the effectiveness and efficiency of the proposed approach.

2   Related Work

Visual Affordance on 3D Shapes. Affordance [9] suggests possible ways for agents to interact with objects. Many past works have investigated learning grasp [11,13,15,26,29] and manipulation [17,18,21,27,36,39] affordance for robot-object interaction, while there are also many works studying affordance for hand-object [3,4,12,17,41], object-object [19,31,46], and human-scene [8,16,21,25] interaction scenarios. Among these works, researchers have proposed different representations for visual affordance, including detection locations [15,29], parts [17], keypoints [27], heatmaps [18,21], etc. In this work, we mostly follow the settings in [18] for learning visual affordance heatmaps for manipulating 3D articulated objects. Different from previous works that infer possible agent-object visual affordance heatmaps passively from static visual observations, our framework leverages active interactions to efficiently query uncertain kinematic or dynamic factors for learning more accurate instance-adaptive visual affordance.

Fast Adaption via Few-Shot Interactions. Researchers have explored various approaches [5,7,28,44,45] for fast adaption via few-shot interactions. Many past works have also designed interactive perception methods to figure out object mass [14], dynamic parameters [1,6,10,38], or parameters for known models [43]. Different from these studies proposing general algorithms for policy adaptation or figuring out explicit system parameters for rigid objects, we focus on designing a working solution for our specific task of learning visual affordance heatmaps for manipulating 3D articulated objects with special designs on predicting geometry-grounded interaction proposals and interaction-adaptive affordance predictions.

3   Problem Formulation

Given as input a single-frame 3D partial point cloud observation of an articulated object $O \in \mathbb{R}^{N \times 3}$ (e.g., lifted from a depth scanner with known camera intrinsics), the Where2Act framework [18] directly outputs a per-point manipulation affordance heatmap $A \in [0,1]^N$, where higher scores indicate bigger chances for being interacted with to accomplish a given short-term manipulation task (e.g., pushing, pulling). Additionally, a diverse set of gripper orientations $\{R_1^p, R_2^p, \cdots | R_i^p \in SO(3)\}$ is proposed at each point $p \in O$ suggesting possible ways for robot agents to interact with, each of which also associated with a success likelihood $s_i^p \in [0,1]$. No interaction is allowed at test time in Where2Act and a fixed set of system dynamic parameters is used across all shapes.

We follow most of the Where2Act settings except that we randomly vary the system dynamics and allow test-time interactions over the 3D shape to reduce kinematic or dynamic uncerainties. Our AdaAfford system proposes a few interactions sequentially $\mathcal{I} = \{I_1, I_2, \cdots\}$. Each interaction $I_i = (O_i, p_i, R_i, m_i)$ executes a task-specific hard-coded short-term trajectory defined in Where2Act, parametrized by the interaction point $p_i \in O_i$ and the gripper orientation $R_i \in SO(3)$, and observes a part motion m_i. Starting from the input shape observation $O_1 \leftarrow O$, every interaction I_i where $m_i \neq 0$ changes the part state and thus produces a new shape point cloud input for the next interaction $O_{i+1} \neq O_i$. Leveraging the interaction observations $\mathcal{I}$, our system then adapts the per-point manipulation affordance A predicted by Where2Act to a posterior $A_{\mathcal{I}} \in [0,1]^N$

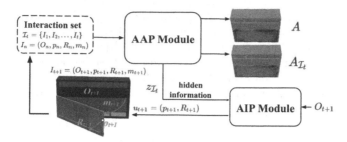

Fig. 2. Method Overview. Starting from the Where2Act [18] predicted affordance prior A, at each timestep $t = 1, 2, \cdots$, we recursively leverage the *Adaptive Interaction Proposal* (AIP) module to propose a next-time interaction action u_{t+1}, observe the interaction outcome m_{t+1}, and feed through the *Adaptive Affordance Prediction* (AAP) module all past few-shot interactions $\mathcal{I}_t$ together with the new one I_{t+1} for adapting to an affordance posterior prediction $A_{\mathcal{I}_{t+1}}$. The procedure iterates until the interaction budget is reached or the AIP module decides to stop.

that reduces uncertainties and provides more accurate instance-specific predictions. For each gripper orientation R_i, we also update the success likelihood score $s_{i,\mathcal{I}}^p \in [0, 1]$ considering the test-time interactions.

4   Method

Our proposed *AdaAfford* framework primarily consists of two modules – an *Adaptive Interaction Proposal* (AIP) module and an *Adaptive Affordance Prediction* (AAP) module. While the AIP module learns a greedy yet effective strategy for sequentially proposing few-shot test-time interactions $\mathcal{I} = \{I_1, I_2, \cdots\}$ revealing hidden information, the AAP module is trained to adapt affordance predictions from Where2Act [18] prior A to a posterior $A_{\mathcal{I}}$ observing the sampled interactions $\mathcal{I}$. We iterate two modules recurrently at test time to produce a sequence of few-shot interactions $\mathcal{I}$ leading to the final affordance posterior prediction $A_{\mathcal{I}}$. During training, we iteratively alternate the training for the two modules until a joint convergence. Below, we first introduce the test-time inference procedure for a brief overview. Next, we describe the input backbone encoders that are shared among all networks in our framework. Then, we describe the detailed architectures and system designs of the two modules. We conclude with the training losses and strategy.

Test-Time Overview. Figure 2 presents an overview of the method. We apply a recurrent structure at test time. Starting from the affordance prediction A, the AIP module proposes the first action for producing the interaction data I_1. Then, at each timestep $t = 1, 2, \cdots$, we feed the current set of interactions $\mathcal{I}_t = \{I_1, \cdots, I_t\}$ as inputs to the AAP module and extract hidden information $z_{\mathcal{I}_t} \in \mathbb{R}^{128}$ that adapts the affordance map prediction to $A_{\mathcal{I}_t}$. The AIP module then takes $z_{\mathcal{I}_t}$ as input and proposes an action $u_{t+1} = (p_{t+1}, R_{t+1})$ composed of the

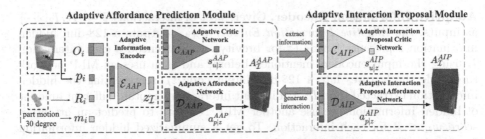

Fig. 3. Network Architecure. Left: the *Adaptive Affordace Prediction* (AAP) module takes as inputs the few-shot interactions $\mathcal{I}$ and predicts the affordance posterior $A_{\mathcal{I}}$. Right: the *Adapative Interaction Proposal* (AIP) module proposes a next-step interaction action $u_{t+1} = (p_{t+1}, R_{t+1})$ (denote the current timestep as t) given the feature $z_{\mathcal{I}}$ extracted from the current interaction observations $\mathcal{I}$.

interaction point p_{t+1} and the gripper orientation R_{t+1} for the next interaction. Performing this action in the environment, we obtain the next-step interaction data $I_{t+1} = (O_{t+1}, p_{t+1}, R_{t+1}, m_{t+1})$ and put it into the interaction set $\mathcal{I}_{t+1} \leftarrow \mathcal{I}_t \cup \{I_{t+1}\}$. We iterate until the interaction budget has been reached or our AIP module decides to stop. When the procedure stops at timestep T, we output the final affordance posterior $A_{\mathcal{I}} = A_{\mathcal{I}_T}$.

Input Encoders. This paragraph details how we encode inputs into features as all the encoder networks in the two modules take the same input entities (*e.g.*, the shape observation O, the interaction action u) and thus share the same architecture. We use the PointNet++ segmentation network [24] to encode the input shape point cloud $O \in \mathbb{R}^{N \times 3}$ into per-point feature maps $f_O \in \mathbb{R}^{N \times 128}$ and denote $f_{p|O} \in \mathbb{R}^{128}$ as the feature at any point $p \in O$. We use Multilayer Perceptron (MLP) networks to encode other vector inputs (*e.g.*, the interaction action u and the part motion m) into $f_a \in R^{128}$. The networks in the following subsections will first encode the inputs into $f_{p|O}$ and f_a, and then concatenate them into $f_I \in R^{256}$. The encoders do not share weights across different modules (Fig. 3).

4.1 Adaptive Affordance Prediction Module

The *Adaptive Affordance Prediction* (AAP) module takes as inputs few-shot interactions $\mathcal{I}$ and predicts the affordance posterior $A_{\mathcal{I}}$. This module is composed of three subnetworks: 1) an *Adaptive Information Encoder* $\mathcal{E}_{AAP}$ that extracts hidden information $z \in \mathbb{R}^{128}$ from a set of interactions $\mathcal{I}$; 2) an *Adaptive Affordance Network* $\mathcal{D}_{AAP}$ that predicts the posterior affordance heatmap $A_{\mathcal{I}}$ conditioned on the hidden information z; and 3) an *Adaptive Critic Network* $\mathcal{C}_{AAP}$ that predicts the AAP action score $s_{u|z}^{AAP} \in [0, 1]$ for an action u conditioned on the hidden information z. Here, an action is represented as $u = (p, R)$ including an interaction point $p \in O$ and a gripper orientation $R \in SO(3)$.

Adaptive Information Encoder. Given a set of interactions $\mathcal{I} = \{I_1, I_2, \cdots\}$ as inputs, the *Adaptive Information Encoder* $\mathcal{E}_{AAP}$ outputs a 128-dim hidden information representation $z_{\mathcal{I}}$ (z for brevity). It first encodes each interaction I_i using the input encoders mentioned before, and then uses an MLP network to encode the features into a 128-dim latent code z_{I_i} representing the hidden information extracted from I_i. As different interactions contain different amount of hidden information, we use another MLP Network to predict an attention score $w_{I_i} \in \mathbb{R}$ for each interaction. To get a summarized hidden information from a set of interactions, we simply computes a weighted average over all z_{I_i}'s according to the weights w_{I_i}'s and use the resulting feature as $z_{\mathcal{I}}$. Formally, we have $z_{\mathcal{I}} \leftarrow \left(\sum_i z_{I_i} \times w_{I_i}\right) / \left(\sum_i w_{I_i}\right)$.

Adaptive Critic Network. Given the object partial point cloud observation O, an arbitrary interaction point $p \in O$, an arbitrary gripper orientation $R \in SO(3)$ and the latent code z, the *Adaptive Critic Network* $\mathcal{C}_{AAP}$ predicts an AAP action score $s_{u|z}^{AAP} \in [0,1]$ indicating the likelihood for the success of the interaction action u given the interaction information z. It first encodes the input $\{O, P, R\}$ using the input encoders as mentioned before and then employs an MLP network to predict AAP action score $s_{u|z}^{AAP}$, taking the concatenated features together with z as inputs. A higher AAP action score $s_{u|z}^{AAP}$ for action u indicates a higher chance for u to succeed in accomplishing the given manipulation task.

Adaptive Affordance Network. Given the input object partial point cloud O, an arbitrary point $p \in O$, and the latent code z, the *Adaptive Affordance Network* $\mathcal{D}_{AAP}$ predicts an actionability score $a_{p|z}^{AAP} \subset [0,1]$ at point p. It first encodes the input $\{O, p\}$ using the aforementioned input encoders and then uses an MLP network that takes the concatenated features together with z as inputs and produces an actionability score $a_{p|z}^{AAP}$ as the output. A higher actionability score $a_{p|z}^{AAP}$ indicates a higher chance to successfully interact on point p.

4.2   Adaptive Interaction Proposal Module

Adaptive Interaction Proposal. (AIP) module proposes an action (denote the current timestep as t) $u_{t+1} = (p_{t+1}, R_{t+1})$ for the next step interaction, given the feature z extracted from the current interaction observations $\mathcal{I}$. This module contains two networks: 1) an *Adaptive Interaction Proposal Affordance Network* $\mathcal{D}_{AIP}$ that predicts an AIP actionability score $a_{p|z}^{AIP} \in \mathbb{R}$ indicating how likely the next-action is worth interacting at point p, and 2) an *Adaptive Interaction Proposal Critic Network* $\mathcal{C}_{AIP}$ predicting an AIP action score $s_{u|z}^{AIP} \in \mathbb{R}$ suggesting the gripper orientation to pick for the next interaction. We leverage the predictions of the two networks to propose the next action $u_{t+1} = (p_{t+1}, R_{t+1})$.

Adaptive Interaction Proposal Critic Network. Given the input object partial point cloud O, an arbitrary interaction point $p \in O$, an arbitrary gripper orientation $R \in SO(3)$, the latent code z, and the AAP action score $s_{u|z}^{AAP}$ produced by $\mathcal{C}_{AAP}$, the *AIP Critic Network* $\mathcal{C}_{AIP}$ predicts the AIP action score

$s_{u|z}^{AIP} \in \mathbb{R}$ of u. It first encodes the inputs $\{O, p, R, s_{u|z}^{AAP}\}$ using the input encoders and then uses an MLP network that takes the concatenated features together with z as inputs and generates an AIP action score $s_{u|z}^{AIP}$ for the action u. A higher AIP action score suggests that the action u may query more unknown yet interesting hidden information and thus is worth exploring next.

Adaptive Interaction Proposal Affordance Network. Given the input partial shape observation O, an arbitrary interaction point $p \in O$, the latent code z, and the AAP actionability score $a_{p|z}^{AAP}$ at point p estimated by $\mathcal{D}_{AAP}$, the *AIP Affordance Network* $\mathcal{D}_{AIP}$ predicts the AIP actionability score $a_{p|z}^{AIP} \in \mathbb{R}$ at point p. It first encodes the inputs $\{O, p, a_{p|z}^{AAP}\}$ using the aforementioned input encoders and then employs an MLP network to predict an AIP actionability score $a_{p|z}^{AIP}$, taking the concatenated features together with z as inputs. A higher AIP actionability score at p indicates more unknown yet helpful hidden information may be obtained by executing an interaction at p.

Next-Step Interaction Action Proposal. In order to propose an action $u_{t+1} = (p_{t+1}, R_{t+1})$ for the next interaction, given the hidden information z and the input shape partial point cloud O, we first obtain the AIP actionability heatmap $A_{p|z}^{AIP}$ for every point $p \in O$ predicted by the *AIP Affordance Network* $\mathcal{D}_{AIP}$ and then select the point $p_{t+1} \leftarrow p_*$ with the highest AIP actionability score $a_{p_*|z}^{AIP}$. Then, we sample 100 random actions $\{u_1, u_2, \cdots, u_{100}\}$ at p using the Where2Act's pre-trained *Action Proposal Network*, use our *AIP critic network* $\mathcal{C}_{AIP}$ to generate the AIP action scores $s_{u_i|z}^{AIP}$ for each action u_i, and then choose the action $u_{t+1} \leftarrow u_*$ with the highest AIP action score $s_{u_*|z}^{AIP}$.

Stopping Criterion for the Few-Shot Interactions. The AIP procedure for generating few-shot interactions stops when a preset budget is reached or the maximal AIP actionability score is below a certain threshold (*e.g.*, 0.05).

4.3 Training and Losses

In brief, for AAP module, we use ground-truth motion m to supervise $\mathcal{E}_{AAP}$ and $\mathcal{C}_{AAP}$, and utilize $\mathcal{C}_{AAP}$ to supervise the training of $\mathcal{D}_{AAP}$. For AIP module, we use AAP module to supervise the training of $\mathcal{C}_{AIP}$ and use it to supervise $\mathcal{D}_{AIP}$. Below, we describe the losses and the training strategy in detail.

AAP Action Scoring Loss. To supervise $\mathcal{C}_{AAP}$, we use a standard binary cross entropy loss, which measures the error between the prediction of $\mathcal{C}_{AAP}$ and target part's ground truth motion m of an interaction I. Specifically, given the hidden information z, a batch of interaction observations $\mathcal{I} = \{I_1, I_2, ..., I_B\}$ where $I_i = \{O_i, u_i, m_i\}$, and the AAP action score prediction $s_{u_i|z}^{AAP}$ for each interaction I_i, the loss is defined as

$$\mathcal{L}_{\mathcal{C}}^{AAP} = -\frac{1}{B} \sum_i r_i \log(s_{u_i|z}^{AAP}) + (1 - r_i) \log(1 - s_{u_i|z}^{AAP})$$

where $r_i = 1$ if $m_i > \tau$ (*e.g.*, $\tau = 0.01$) or $r_i = 0$ rendering a binary discretization for each interaction outcome.

AAP Actionability Scoring Loss. To train $\mathcal{D}_{AAP}$, we apply an $\mathcal{L}_1$ loss to measure the difference from the predicted score $a_{p|z}^{AAP}$ to the ground truth. To estimate the ground truth actionability score for p, we randomly sample 100 actions at p according to pre-trained Where2Act *Action Proposal Network*, predict AAP action scores $s_{u|z}^{AAP}$'s of these actions u's using $\mathcal{C}_{APP}$, and take the average of the top-5 scores as the ground truth actionability score.

AIP Action Scoring Loss. To supervise $\mathcal{C}_{AIP}$, we use an $\mathcal{L}_1$ loss to measure the difference between our predicted AIP action score $s_{u|z}^{AIP}$ and the ground truth AIP action score $gt_{u|z}^{AIP}$. Given a set of interactions $\mathcal{I}_T = \{I_1, I_2, \cdots\}$, to generate $gt_{u_i|z}^{AIP}$ for an interaction action u_i, we respectively encode two interaction subsets $\mathcal{I}_{i-1} = \{I_1, I_2, \cdots, I_{i-1}\}$ and $\mathcal{I}_i = \{I_1, I_2, \cdots, I_i\}$ into latent codes $z_{\mathcal{I}_i}$ and $z_{\mathcal{I}_{i-1}}$. Then feed $z_{\mathcal{I}_i}$ and $z_{\mathcal{I}_{i-1}}$ as the conditional inputs to $\mathcal{C}_{AAP}$ separately and count the difference of $\mathcal{L}_\mathcal{C}^{AAP}$ as the ground truth of AIP action score $gt_{u_i|z_{\mathcal{I}_{i-1}}}^{AIP}$. More concretely, let the AAP action scoring loss conditioned on $z_{\mathcal{I}_i}$ and $z_{\mathcal{I}_{i-1}}$ respectively be $\mathcal{L}_{\mathcal{I}_i}$ and $\mathcal{L}_{\mathcal{I}_{i-1}}$. We define the ground truth AIP action score $gt_{u_i|z_{\mathcal{I}_{i-1}}}^{AIP} \leftarrow \mathcal{L}_{\mathcal{I}_{i-1}} - \mathcal{L}_{\mathcal{I}_i}$. The AIP action score is trained to regress an estimated positive influence of executing u on the AAP action score predictions, where an action giving more influence is preferred as it helps discover more hidden information useful to the task.

AIP Actionability Scoring Loss. To train $\mathcal{D}_{AIP}$, we use another $\mathcal{L}_1$ loss. For each $p \in O$, we sample 100 actions u_i's using the pre-trained Where2Act *Action Proposal Network*, obtain the AIP action scores $s_{u_i|z}^{AIP}$'s of these actions u_i's by $\mathcal{C}_{AIP}$, and use the average of the top-5 scores as the regression target.

Training Strategy. We iteratively train the AAP module and AIP module until a joint convergence since the update of the subnetworks in one module will affect the training of the subnetworks in the other module. More specifically, the update of $\mathcal{C}_{AAP}$ in the AAP module will affect the ground-truth AIP action scores, while the update of $\mathcal{C}_{AIP}$ and $\mathcal{D}_{AIP}$ in the AIP module will change the proposed interactions used to generate z in the AAP module. Therefore, our final solution is to train the AAP and AIP modules iteratively.

5 Experiments

We perform experiments using the large-scale PartNet-Mobility dataset [20] and the SAPIEN simulator [37], and set up several baselines for comparisons. Results demonstrate the effectiveness and superiority of the proposed approach.

5.1 Data and Settings

Data. Following the settings of Where2Act [18], we conduct our experiments in the SAPIEN [37] simulator equipped with NVIDIA PhysX [22] simulation engine and the large scale PartNet-Mobility [20] dataset. We use 972 articulated 3D objects covering 15 object categories, mostly following Where2Act, to carry out the experiments. The dataset is divided into 10 training and 5 testing categories. The shapes in the training categories are further divided into two disjoint sets of training and test shapes. See supplementary for detailed statistics.

Table 1. Quantitative Evaluations. We experiment with three different test-time interaction budgets (*i.e.*, 1, 2, or 4) where numbers are separated by slashes. We use "pushing all" and "pulling all" to denote the experiments over all object categories, while "pulling closed door" and "pushing faucet" refer to the experiments over a single category only. For the experiments over all categories, we report the performance over novel shapes from the training categories (marked with "train cat.") and shapes from novel categories (marked with "test cat.").

		F-score (%)	Sample-Succ (%)
Pushing all (train cat.)	Where2Act	56.44	20.85
	Where2Act-adaptation	64.16/65.42/64.99	20.77/22.72/26.82
	Ours-fps	64.32/69.58/70.99	26.22/27.30/30.65
	Ours-final	**72.78/73.12/75.18**	**33.82/33.23/35.23**
Pushing all (test cat.)	Where2Act	59.95	21.69
	Where2Act-adaptation	51.09/53.28/55.56	19.06/22.27/24.50
	Ours-fps	66.17/67.27/69.08	33.64/35.19/**37.79**
	Ours-final	**77.58/77.63/78.42**	**34.97/36.75**/37.40
Pulling all (train cat.)	Where2Act	31.19	1.92
	Where2Act-adaptation	37.22/38.48/39.13	1.11/2.15/1.62
	Ours-fps	39.88/42.74/43.55	2.78/5.56/4.44
	Ours-final	**42.62/43.87/44.08**	**7.78/9.44/10.55**
Pulling all (test cat.)	Where2Act	36.36	10.00
	Where2Act-adaptation	40.11/45.52/48.80	3.40/6.25/10.17
	Ours-fps	43.67/42.77/48.33	4.35/3.91/4.78
	Ours-final	**49.51/50.00/51.33**	**5.21/7.39/10.45**
Pulling closed door	Where2Act	48.44	4.38
	Where2Act-adaptation	50.21/55.75/56.81	6.60/7.18/6.83
	Ours-fps	**59.79**/63.43/69.13	8.88/11.33/12.10
	Ours-final	57.83/**65.60/79.65**	**10.86/11.57/22.14**
Pushing faucet	Where2Act	64.92	55.46
	Where2Act-adaptation	66.25/62.18/67.15	57.50/52.08/61.70
	Ours-fps	74.19/79.36/77.95	60.44/70.12/77.41
	Ours-final	**77.42/83.06/83.83**	**65.90/81.66/82.14**

Experiment Settings. Following Where2Act [18], we perform experiments over all object categories under different manipulation action types. We train one network for each downstream manipulation task over training shapes from the 10 training object categories and evaluate the performance over test shapes from the training categories and shapes from unseen test categories. Besides, to further demonstrate the effectiveness of our method, we conduct two additional experiments under challenging tasks with clear kinematic ambiguity, each of which is conducted over a single object category: 1) pulling closed doors of cabinets that cannot be easily distinguished which side to pull open; 2) pushing faucets with uncertainties which direction to rotate (clockwise or counter-clockwise). These experiments are particularly interesting yet challenging cases on which previous work Where2Act [18] fail drastically and we hope to test our framework.

Environment Settings. Following Where2Act, we abstract away the robot arm and only use a Franka Panda flying gripper as the robot actuator. The input shape point cloud is assumed to be cleanly segmented out. To generate the input partial point cloud scans, we mount an RGB-D camera with known intrinsic parameters 5-unit-length away pointing to the center of the target object.

To simulate manipulating shapes with uncertain dynamics, we randomly change the following three physical parameters in SAPIEN: 1) the friction of the target part joint, 2) the mass of the target part, and 3) the friction coefficient of the target part surface. For the "pulling closed door" task, we manually select the cabinets whose doors have no clear handle geometry in the PartNet-Mobility dataset [37], and set the poses of those doors to be closed. The gripper cannot tell which side to pull open the door because it is impossible to tell whether the axis position is on the left or right of the door from passive visual observations. For the "pushing faucet" task, we randomly set the rotating direction of the faucet switch to be in one of the following three modes: only clockwise, only counter clockwise, or both ways.

5.2   Baselines and Evaluation Metrics

We set up several baseline and employ two metrics for quantitative comparisons.

Baselines and Ablation Study. We compare our framework with several baselines (see supplementary for more detailed descriptions for the baseline designs):

- **Where2Act:** the original method proposed in [18] where only the pure visual information is used for predicting the visual actionable information and no interaction data is used at all during test time;
- **Where2Act-adaptation:** the Where2Act method augmented with a heuristic based adaptation mechanism to replace the AAP module where given the interaction observations we locally adjust the predictions for similar points;
- **Ours-fps:** a variant of our proposed method that we use FPS to sample over the predicted affordance for interactions instead of the AIP proposals.

We compare to **Where2Act** to show that the few-shot interactions indeed help to remove ambiguities and improve the performance. Furthermore, the

Where2Act-adaptation baseline helps substantiate the effectiveness of our proposed AAP module, while the **Ours-fps** baselines are designed to verify the usefulness of the proposed AIP module.

Besides, we compare to an ablated version of our method to verify the significance of iterative training between the AAP module and the AIP module.

– **Ours w/o iter:** an ablated version that trains the whole framework without the iterative training process.

Evaluation Metrics. Following Where2Act [18], we use the F-score, balancing the precision and recall, to evaluate the predictions of $\mathcal{C}_{AAP}$, and use the sample-successful rate (Sample-Succ) to evaluate the performance of $\mathcal{C}_{AAP}$ and $\mathcal{D}_{AAP}$. To compute the sample successful rate, we apply the learned test-time strategy to fill $\mathcal{I}$ and then use the extracted hidden information z as the conditional input to $\mathcal{C}_{AAP}$ and $\mathcal{D}_{AAP}$. After that, we randomly select a point to interact from the group of points with the top-100 actionability scores $a_{p|z}^{AAP}$, sample 100 actions u_i's at p, obtain $s_{u_i|z}^{AAP}$'s of these actions u_i's predicted by $\mathcal{C}_{AAP}$, and then choose the action u_i with the highest $s_{u_i|z}^{AAP}$ to execute. We perform 10 interaction trials per test shape and report the final sample-succ rate as the percentage of sampling successful interactions in simulation.

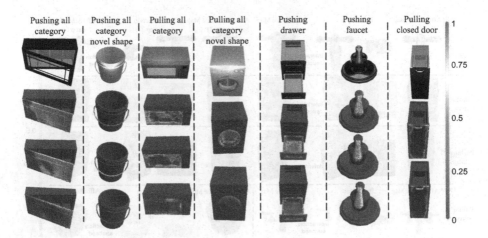

Fig. 4. Example results of adapted affordance predictions given by AAP module under different kinematic and dynamic parameters. The first five columns show the adapted affordance prediction conditioned on increasing joint friction (the first and second columns), part mass (the third column), and friction coefficient on object surface (the fourth and fifth columns). The last two columns respectively show the influence of different rotating directions (*i.e.*, joint limits) and joint axis locations.

5.3   Results and Analysis

Table 1 presents the quantitative comparisons against the baselines showing that our method achieves the best performance in most comparison entries. Specifically, compared to **Where2Act**, we observe that our method can improve the performance evidently with only 1 interaction. Also, the performance increases as the number of interactions increases in most cases. Compared to the **Where2Act-adaptation** baseline, our method with the proposed AAP module shows better performance, revealing that learning an adaptation network works better than using simple heuristics for adaptation. Finally, the superior performance against the **Ours-fps** baseline that use FPS sampled interaction trials further validate that our proposed AIP module is effective in strategically and iteratively picking interaction trials. Our method can generalize well to novel shapes and even shapes from unseen object categories through scores in test-cat.

Figure 4 shows example visualizations for our predicted affordance map posterior given interactions under different hidden kinematic or dynamic information (see the caption for more details explaining the different scenarios). In these figures, it is clear to see that our proposed method successfully adapts the affordance prediction conditioned on different hidden information. The affordance predictions within one shape share the same visual inputs but output different results, showing that our hidden embedding z contains certain information.

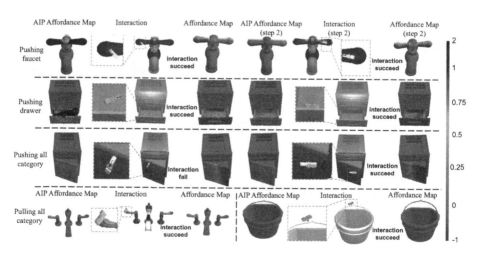

Fig. 5. Example results for the interactions proposed by the AIP module and the corresponding AIP affordance map predictions. In the first three rows, we show the initial and the second AIP affordance maps, the corresponding proposed interactions, and the posterior affordance map predictions. In the last row, we present two more examples that only one interaction is needed. From these results, our AIP module successfully proposes reasonable interactions for querying useful hidden information.

Figure 5 further shows some interaction proposals by our AIP module with its influence on the prediction of AAP affordance map and to the AIP affordance map itself. In the first row, for example, we see that the AIP affordance first proposes to interact at both sides of the faucet since it knows little about the hidden information but at the second timestep proposes the right side as it already learns the left side is actionable. Cases in the first and third rows demonstrate that the past few interactions will influence the selection of future interaction points, justifying the necessity of our recurrent structure for interaction selection. In the last row, we show cases only requiring one step to adapt.

Ablation Study. In Table 2, comparing against **Ours w/o iter** that trains the whole system without the interactive training process, we see that **Ours-final** achieves better results in most cases, which proves the effectiveness of the iterative training scheme. By iteratively alternating the training between the AAP module and the AIP module, the networks would be trained under the distribution of test-time interactions and thus achieve improved performance.

Real-World and Real-Robot Experiments. Finally, we perform real-world and real-robot experiments to show that our method can to some degree work beyond synthetic data. We use a Franka panda robot with a two-finger parallel gripper as the actuator to pull open a cabinet door. Figure 6 presents the results that our system proposes two interaction trials to inquire more information about this real-world cabinet and successfully learns to adapt to the posterior predictions. Please refer to the supplementary materials for a video better illustrating this example, more experiment settings, more example results, and more experiments with additional analysis.

Table 2. Ablation Study. We compare our method to an ablated version, where we remove the iteratively training process. It is clear to see that the iteratively training process helps our framework achieve better results in most cases.

		F-score (%)	Sample-Succ (%)
Pushing all (train cat.)	Ours w/o iter	71.21/72.64/73.16	30.67/31.62/32.56
	Ours-final	**72.78/73.12/75.18**	**33.82/33.23/35.23**
Pushing all (test cat.)	Ours w/o iter	77.24/77.33/77.17	31.03/33.89/**38.83**
	Ours-final	**77.58/77.63/78.42**	**34.97/36.75/37.40**
Pulling all (train cat.)	Ours w/o iter	41.19/42.10/42.81	6.67/7.22/8.33
	Ours-final	**42.62/43.87/44.08**	**7.78/9.44/10.55**
Pulling all (test cat.)	Ours w/o iter	48.31/48.28/50.50	**5.65**/6.52/9.13
	Ours-final	**49.51/50.00/51.33**	5.21/**7.39/10.45**
Pulling closed door	Ours w/o iter	56.74/64.88/**80.64**	9.77/11.50/22.00
	Ours-final	**57.83/65.60**/79.65	**10.86/11.57/22.14**
Pushing faucet	Ours w/o iter	73.81/83.03/**84.32**	61.11/81.60/**84.03**
	Ours-final	**77.42/83.06**/83.83	**65.90/81.66**/82.14

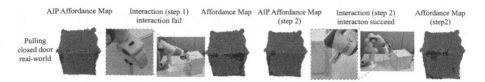

Fig. 6. Real-robot experiment on pulling open a closed door in the real world. We show the AIP affordance map predictions, the AIP proposed interactions, and the AAP posterior predictions, for two interaction trials. The results show that our work could reasonably generalize to real-world scenarios.

6   Conclusion

This work addresses a big limitation of previous works learning visual actionable affordance for manipulating 3D articulated objects – the hidden kinematic or dynamic uncertainties. We propose a novel framework AdaAfford that samples a few test-time interactions for fastly adapting to a more accurate affordance posterior prediction removing such ambiguities. Experimental results validate the effectiveness of our method compared to baseline approaches.

Limitations and Future Works. This work only considers two action types and 3D articulated objects. Future works may study more interaction and data types. Also, we only perform short-term interactions. Future works can investigate how to extend the framework for long-term manipulation trajectories. Future works shall work on considering the robot arm constraints.

Acknowledgements. National Natural Science Foundation of China -Youth Science Fund (No. 62006006). Leonidas and Kaichun were supported by the Toyota Research Institute (TRI) University 2.0 program, NSF grant IIS-1763268, a Vannevar Bush Faculty Fellowship, and a gift from the Amazon Research Awards program. The Toyota Research Institute University 2.0 program (Toyota Research Institute ("TRI") provided funds to assist the authors with their research but this article solely reflects the opinions and conclusions of its authors and not TRI or any other Toyota entity).

References

1. Agrawal, P., Nair, A., Abbeel, P., Malik, J., Levine, S.: Learning to poke by poking: experiential learning of intuitive physics. arXiv preprint arXiv:1606.07419 (2016)
2. Chitta, S., Cohen, B., Likhachev, M.: Planning for autonomous door opening with a mobile manipulator. In: 2010 IEEE International Conference on Robotics and Automation, pp. 1799–1806. IEEE (2010)
3. Corona, E., Pumarola, A., Alenya, G., Moreno-Noguer, F., Rogez, G.: Ganhand: predicting human grasp affordances in multi-object scenes. In: Proceedings of the IEEE/CVF Conference on Computer Vision and Pattern Recognition, pp. 5031–5041 (2020)
4. Fang, K., Wu, T.L., Yang, D., Savarese, S., Lim, J.J.: Demo2vec: reasoning object affordances from online videos. In: Proceedings of the IEEE Conference on Computer Vision and Pattern Recognition, pp. 2139–2147 (2018)

5. Farid, K., Sakr, N.: Few shot system identification for reinforcement learning. arXiv preprint arXiv:2103.08850 (2021)
6. Ferreira, F., Shao, L., Asfour, T., Bohg, J.: Learning visual dynamics models of rigid objects using relational inductive biases. arXiv preprint arXiv:1909.03749 (2019)
7. Finn, C., Abbeel, P., Levine, S.: Model-agnostic meta-learning for fast adaptation of deep networks. In: International Conference on Machine Learning, pp. 1126–1135. PMLR (2017)
8. Fouhey, D.F., Delaitre, V., Gupta, A., Efros, A.A., Laptev, I., Sivic, J.: People watching: human actions as a cue for single view geometry. In: Fitzgibbon, A., Lazebnik, S., Perona, P., Sato, Y., Schmid, C. (eds.) ECCV 2012. LNCS, vol. 7576, pp. 732–745. Springer, Heidelberg (2012). https://doi.org/10.1007/978-3-642-33715-4_53
9. Gibson, J.J.: The theory of affordances. Hilldale USA 1(2), 67–82 (1977)
10. Janner, M., Levine, S., Freeman, W.T., Tenenbaum, J.B., Finn, C., Wu, J.: Reasoning about physical interactions with object-oriented prediction and planning. arXiv preprint arXiv:1812.10972 (2018)
11. Jiang, Z., Zhu, Y., Svetlik, M., Fang, K., Zhu, Y.: Synergies between affordance and geometry: 6-DoF grasp detection via implicit representations. In: Proceedings of Robotics: Science and Systems (RSS) (2021)
12. Kjellström, H., Romero, J., Kragić, D.: Visual object-action recognition: inferring object affordances from human demonstration. Comput. Vis. Image Underst. 115(1), 81–90 (2011)
13. Kokic, M., Kragic, D., Bohg, J.: Learning task-oriented grasping from human activity datasets. IEEE Robot. Autom. Lett. 5(2), 3352–3359 (2020)
14. Kumar, K.N., Essa, I., Ha, S., Liu, C.K.: Estimating mass distribution of articulated objects using non-prehensile manipulation. arXiv preprint arXiv:1907.03964 (2019)
15. Lenz, I., Lee, H., Saxena, A.: Deep learning for detecting robotic grasps. Int. J. Robot. Res. 34(4–5), 705–724 (2015)
16. Li, X., Liu, S., Kim, K., Wang, X., Yang, M.H., Kautz, J.: Putting humans in a scene: learning affordance in 3D indoor environments. In: IEEE Conference on Computer Vision and Pattern Recognition (2019)
17. Mandikal, P., Grauman, K.: Learning dexterous grasping with object-centric visual affordances. In: IEEE International Conference on Robotics and Automation (ICRA) (2021)
18. Mo, K., Guibas, L.J., Mukadam, M., Gupta, A., Tulsiani, S.: Where2act: from pixels to actions for articulated 3D objects. In: Proceedings of the IEEE/CVF International Conference on Computer Vision (ICCV), pp. 6813–6823, October 2021
19. Mo, K., Qin, Y., Xiang, F., Su, H., Guibas, L.: O2O-afford: annotation-free large-scale object-object affordance learning. In: Conference on Robot Learning (CoRL) (2021)
20. Mo, K., et al.: PartNet: a large-scale benchmark for fine-grained and hierarchical part-level 3D object understanding. In: The IEEE Conference on Computer Vision and Pattern Recognition (CVPR), June 2019
21. Nagarajan, T., Grauman, K.: Learning affordance landscapes for interaction exploration in 3D environments. In: NeurIPS (2020)
22. NVIDIA. Nvidia.physx

23. Peterson, L., Austin, D., Kragic, D.: High-level control of a mobile manipulator for door opening. In: Proceedings. 2000 IEEE/RSJ International Conference on Intelligent Robots and Systems (IROS 2000) (Cat. No. 00CH37113), vol. 3, pp. 2333–2338. IEEE (2000)
24. Qi, C.R., Yi, L., Su, H., Guibas, L.J.: Pointnet++: deep hierarchical feature learning on point sets in a metric space. arXiv preprint arXiv:1706.02413 (2017)
25. Qi, W., Mullapudi, R.T., Gupta, S., Ramanan, D.: Learning to move with affordance maps. arXiv preprint arXiv:2001.02364 (2020)
26. Qin, Y., Chen, R., Zhu, H., Song, M., Xu, J., Su, H.: S4G: amodal single-view single-shot se (3) grasp detection in cluttered scenes. In: Conference on Robot Learning, pp. 53–65. PMLR (2020)
27. Qin, Z., Fang, K., Zhu, Y., Fei-Fei, L., Savarese, S.: Keto: learning keypoint representations for tool manipulation. In: 2020 IEEE International Conference on Robotics and Automation (ICRA), pp. 7278–7285. IEEE (2020)
28. Rakelly, K., Zhou, A., Finn, C., Levine, S., Quillen, D.: Efficient off-policy meta-reinforcement learning via probabilistic context variables. In: International Conference on Machine Learning, pp. 5331–5340. PMLR (2019)
29. Redmon, J., Angelova, A.: Real-time grasp detection using convolutional neural networks. In: 2015 IEEE International Conference on Robotics and Automation (ICRA), pp. 1316–1322. IEEE (2015)
30. Schmidt, T., Newcombe, R.A., Fox, D.: Dart: dense articulated real-time tracking. In: Robotics: Science and Systems, Berkeley, CA, vol. 2 (2014)
31. Sun, Yu., Ren, S., Lin, Y.: Object-object interaction affordance learning. Robot. Auton. Syst. **62**(4), 487–496 (2014)
32. Tzionas, D., Gall, J.: Reconstructing articulated rigged models from RGB-D videos. In: Hua, G., Jégou, H. (eds.) ECCV 2016. LNCS, vol. 9915, pp. 620–633. Springer, Cham (2016). https://doi.org/10.1007/978-3-319-49409-8_53
33. Urakami, Y., Hodgkinson, A., Carlin, C., Leu, R., Rigazio, L., Abbeel, P.: Doorgym: a scalable door opening environment and baseline agent. In: Deep RL workshop at NeurIPS 2019 (2019)
34. Wang, X., Zhou, B., Shi, Y., Chen, X., Zhao, Q., Xu, K.: Shape2motion: joint analysis of motion parts and attributes from 3D shapes. In: Proceedings of the IEEE/CVF Conference on Computer Vision and Pattern Recognition, pp. 8876–8884 (2019)
35. Weng, Y., et al.: Captra: category-level pose tracking for rigid and articulated objects from point clouds. In: Proceedings of the IEEE International Conference on Computer Vision (ICCV), pp. 13209–13218, October 2021
36. Wu, R., et al.: VAT-mart: learning visual action trajectory proposals for manipulating 3D ARTiculated objects. In: International Conference on Learning Representations (2022)
37. Xiang, F., et al.: SAPIEN: a simulated part-based interactive environment. In: The IEEE Conference on Computer Vision and Pattern Recognition (CVPR), June 2020
38. Xu, Z., Wu, J., Zeng, A., Tenenbaum, J.B., Song, S.: Densephysnet: learning dense physical object representations via multi-step dynamic interactions. arXiv preprint arXiv:1906.03853 (2019)
39. Xu, Z., He, Z., Song, S.: UMPNet: universal manipulation policy network for articulated objects. IEEE Robot. Autom. Lett. (2022)
40. Yan, Z., et al.: RPM-NET: recurrent prediction of motion and parts from point cloud. ACM Trans. Graph. **38**(6), Article 240 (2019)

41. Yang, L., Zhan, X., Li, K., Xu, W., Li, J., Lu, C.: CPF: learning a contact potential field to model the hand-object interaction. In: Proceedings of the IEEE/CVF International Conference on Computer Vision, pp. 11097–11106 (2021)

42. Yi, L., Huang, H., Liu, D., Kalogerakis, E., Su, H., Guibas, L.: Deep part induction from articulated object pairs. ACM Trans. Graph. **37**(6) (2018)

43. Yu, W., Tan, J., Liu, C.K., Turk, G.: Preparing for the unknown: learning a universal policy with online system identification. arXiv preprint arXiv:1702.02453 (2017)

44. Zhao, T.Z., Nagabandi, A., Rakelly, K., Finn, C., Levine, S.: Meld: meta-reinforcement learning from images via latent state models. arXiv preprint arXiv:2010.13957 (2020)

45. Zhou, W., Pinto, L., Gupta, A.: Environment probing interaction policies. In: 7th International Conference on Learning Representations, ICLR 2019, New Orleans, LA, USA, 6–9 May 2019. OpenReview.net (2019)

46. Zhu, Y., Zhao, Y., Chun Zhu, S.: Understanding tools: task-oriented object modeling, learning and recognition. In: Proceedings of the IEEE Conference on Computer Vision and Pattern Recognition, pp. 2855–2864 (2015)

Cost Aggregation with 4D Convolutional Swin Transformer for Few-Shot Segmentation

Sunghwan Hong[1(✉)], Seokju Cho[1], Jisu Nam[1], Stephen Lin[2], and Seungryong Kim[1]

[1] Korea University, Seoul, Korea
{sung_hwan,seokju_cho,18wltnzzang,seungryong_kim}@korea.ac.kr
[2] Microsoft Research Asia, Beijing, China
stevelin@microsoft.com

Abstract. This paper presents a novel cost aggregation network, called Volumetric Aggregation with Transformers (VAT), for few-shot segmentation. The use of transformers can benefit correlation map aggregation through self-attention over a global receptive field. However, the tokenization of a correlation map for transformer processing can be detrimental, because the discontinuity at token boundaries reduces the local context available near the token edges and decreases inductive bias. To address this problem, we propose a 4D Convolutional Swin Transformer, where a high-dimensional Swin Transformer is preceded by a series of small-kernel convolutions that impart local context to all pixels and introduce convolutional inductive bias. We additionally boost aggregation performance by applying transformers within a pyramidal structure, where aggregation at a coarser level guides aggregation at a finer level. Noise in the transformer output is then filtered in the subsequent decoder with the help of the query's appearance embedding. With this model, a new state-of-the-art is set for all the standard benchmarks in few-shot segmentation. It is shown that VAT attains state-of-the-art performance for semantic correspondence as well, where cost aggregation also plays a central role. Code and trained models are available at https://seokju-cho.github.io/VAT/.

1 Introduction

Semantic segmentation is a fundamental computer vision task that aims to label each pixel in an image with its corresponding class. Substantial progress has been made in this direction with the help of deep neural networks and large-scale datasets containing ground-truth segmentation annotations [3,4,37,46,60]. Manual labeling of pixel-wise segmentation maps, however, requires considerable

---

S. Hong and S. Cho—Equal contribution.

Supplementary Information The online version contains supplementary material available at https://doi.org/10.1007/978-3-031-19818-2_7.

Fig. 1. Our VAT reformulates few-shot segmentation as semantic correspondence. VAT sets a new state-of-the-art in few-shot segmentation, and attains state-of-the-art performance for semantic correspondence as well.

labor, making it difficult to add new classes. Towards reducing reliance on labeled data, attention has increasingly focused on *few*-shot segmentation [48,54], where only a handful of support images and their associated masks are used in predicting the segmentation of a query image.

The key to few-shot segmentation is in making effective use of the few support samples. Many works attempt this by extracting a prototype model from the samples and using it for feature comparison with the query [10,35,57,76]. However, such approaches disregard pixel-level pairwise relationships between support and query features or the spatial structure of features, which may lead to sub-optimal results.

To account for such relationships, we observe that few-shot segmentation can be reformulated as semantic correspondence, which aims to find pixel-level correspondences across semantically similar images which may contain large intra-class appearance and geometric variations [13,14,43]. Recent semantic correspondence models [25,34,41,42,44,49,50,52,64] follow the classical matching pipeline [47,53] of feature extraction, cost aggregation and flow estimation. The cost aggregation stage, where matching scores are refined to produce more reliable correspondence estimates, is of particular importance and has been the focus of much research [6,22,29,34,41,42,51,52]. Recently, CATs [6] proposed to use vision transformers [11] for cost aggregation, but its quadratic complexity to the number of input tokens limits its applicability. It also disregards the spatial structure of matching costs, which may hurt its performance.

In the area of few-shot segmentation, there also exist methods that attempt to leverage pairwise information by refining features through cross-attention [81] or graph attention [67,73,79]. However, they solely rely on raw correlation maps without aggregating the matching scores. As a result, their correspondence may suffer from ambiguities caused by repetitive patterns or background clutters [17,25,27,49,64]. To address this, HSNet [40] aggregates the matching scores with 4D convolutions, but its limited receptive fields prevent long-range context aggregation and it lacks an ability to adapt to the input content due to the use of fixed kernels.

In this paper, we introduce a novel cost aggregation network, called Volumetric Aggregation with Transformers (VAT), that tackles the few-shot segmentation task through a proposed 4D Convolutional Swin Transformer. Specifically, we first extend Swin Transformer [36] and its patch embedding module to handle a high-dimensional correlation map. The patch embedding module is further extended by incorporating 4D convolutions that alleviate issues caused by patch embedding, *i.e.,* limited local context near patch boundaries and low inductive bias. The high-dimensional patch embedding module is designed as a series of overlapping small-kernel convolutions, bringing local contextual information to each pixel and imparting convolutional inductive bias. To further boost performance, we compose our architecture with a pyramidal structure that takes the aggregated correlation maps at a coarser level as additional input at a finer level, providing hierarchical guidance. Our affinity-aware decoder then refines the aggregated matching scores in a manner that exploits the higher-resolution spatial structure given by the query's appearance embedding and finally outputs the segmentation mask prediction.

We demonstrate the effectiveness of our method on several benchmarks [30, 31,54]. Our work attains state-of-the-art performance on all the benchmarks for few-shot segmentation and even for semantic correspondence, highlighting the importance of cost aggregation for both tasks and showing its potential for general matching. We also include ablation studies to justify our design choices.

2   Related Work

Few-Shot Segmentation. Inspired by the few-shot learning paradigm [48,57], which learns to learn a model for a novel task with only a limited number of samples, few-shot segmentation has received considerable attention. Following the success of [54], prototypical networks [57] and numerous other works [10,28, 33,35,45,55,59,68,74,76,77,82] proposed to extract a prototype from support samples, which is used to identify foreground features in the query. In addition, inspired by [80] which observed that simply adding high-level features in feature processing leads to a performance drop, [62] proposed to instead utilize high-level features to compute a prior map that helps to identify targets in the query image. Many variants [59,78] extended this idea of utilizing prior maps to act as additional information for aggregating feature maps.

However, as methods based on prototypes or prior maps have apparent limitations, *e.g.,* disregarding pairwise relationships between support and query features or spatial structure of feature maps, numerous recent works [32,40,67, 73,79] utilize a correlation map to leverage the pairwise relationships between source and query features. Specifically, [67,73,79] use graph attention, HSNet [40] proposes 4D convolutions to exploit multi-level features, and [32] formulates the task as an optimal transport problem. However, these approaches do not provide a means to aggregate the matching scores, solely utilize convolutions for cost aggregation, or use a handcrafted method that is neither learnable nor robust to severe deformations.

Recently, [81] utilized transformers and proposed to use a cycle-consistent attention mechanism to refine the feature maps to become more discriminative, without considering aggregation of matching scores. [59] propose a global and local enhancement module to refine the features using transformers and convolutions, respectively. [39] focuses solely on the transformer-based classifier by freezing the encoder and decoder. Unlike these works, we propose a 4D Convolutional Swin Transformer for an enhanced and efficient cost aggregation.

Semantic Correspondence. The objective of semantic correspondence is to find correspondences between semantically similar images with additional challenges posed by large intra-class appearance and geometric variations [6,34,41]. This is highly similar to the few-shot segmentation setting in that few-shot segmentation also aims to label objects of the same class with large intra-class variation, and thus recent works on both tasks have taken similar approaches. The latest methods [6,22,29,34,41,42,51,52] in semantic correspondence focus on the cost aggregation stage to find reliable correspondences and demonstrated its importance. Among them, [41] proposed to use 4D convolutions for cost aggregation, though exhibiting apparent limitations due to the limited receptive fields of convolutions and lack of adaptability. CATs [6] resolves this issue and sets a new state-of-the-art by leveraging transformers [65] to aggregate the cost volume. However, it disregards the spatial structure of correlation maps and imparts less inductive bias, *i.e.,* translation equivariance, which limits its generalization power [7,8,36]. Moreover, its quadratic complexity may limit applicability when it is used to aggregate correlation maps on its own. In this paper, we propose to resolve the aforementioned issues.

Vision Transformer. Recently, transformer [65], the standard architecture in Natural Language Processing (NLP), has been widely adopted in Computer Vision. Since the pioneering work on ViT [11], numerous works [6,23,36,39,59, 71,81] have adopted transformers to replace CNNs or to be used together with CNNs in a hybrid manner. However, due to quadratic complexity to sequence length, transformers often suffer from large a computational burden. Efficient transformers [24,69,70,75] aim to reduce the computational load via an approximated or simplified self-attention. Swin Transformer [36], a network we extend from, reduces computation by performing self-attention within pre-defined local windows. However, these works inherit the issues caused by patch embedding, which we alleviate by incorporating 4D convolutions.

3 Methodology

3.1 Problem Formulation

The goal of *few*-shot segmentation is to segment objects from unseen classes in a query image given only a few annotated examples [66]. To mitigate the overfitting caused by insufficient training data, we follow the common protocol of *episodic* training [66]. Let us denote the training and test sets as $\mathcal{D}_{\text{train}}$ and $\mathcal{D}_{\text{test}}$, respectively, where the object classes of both sets do not overlap. Under the K-shot setting, multiple *episodes* are formed from both sets, each consisting of a support set $\mathcal{S} = \{(x_s^k, m_s^k)\}_{k=1}^{K}$, where (x_s^k, m_s^k) is k-th support image and its corresponding mask pair, and a query sample (x_q, m_q), where x_q is a query

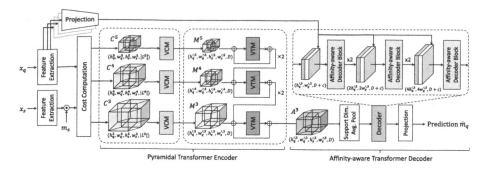

Fig. 2. Overall network architecture. Our network consists of feature extraction and cost computation, a pyramidal transformer encoder, and an affinity-aware transformer decoder.

image and m_q is its paired mask. During training, our model takes a sampled episode from $\mathcal{D}_{\text{train}}$ and learns a mapping from $\mathcal{S}$ and x_q to a prediction m_q. At inference, our model predicts $\hat{m}_q$ given randomly sampled $\mathcal{S}$ and x_q from $\mathcal{D}_{\text{test}}$.

3.2 Motivation and Overview

The key to few-shot segmentation is how to effectively utilize the support samples provided for a query image. While conventional methods [28,59,62,77,81] utilize global- or part-level prototypes extracted from support features, recent methods [32,40,67,73,79,81] instead leverage pairwise matching relationships between query and support. However, exploring such relationships is notoriously challenging due to intra-class variations, background clutters, and repetitive patterns. One of the state-of-the-art methods, HSNet [40], aggregates the matching scores with 4D convolutions. However, solely utilizing convolutions may limit performance due to limited receptive fields or lack of adaptability for convolutional kernels. While there has been no approach to aggregate the matching scores with transformers in few-shot segmentation, CATs [6] proposes cost aggregation with transformers in semantic correspondence, demonstrating the effectiveness of transformers as a cost aggregator. On the other hand, the quadratic complexity of transformers with respect to the number of tokens may limit its utility for segmentation. The absence of operations that impart inductive bias, *i.e.*, translation equivariance, may limit its performance as well. Also, CATs [6] defines the tokens of a correlation map in a way that disregards spatial structure, which is likely to be harmful.

The proposed Volumetric Aggregation with Transformers (VAT) is designed to overcome these problems. In the following, we first describe its feature extraction and cost computation. We then present a general extension of Swin Transformer [36] for cost aggregation. Subsequently, we present 4D Convolutional Swin Transformer for resolving the aforementioned issues. Lastly, we introduce several additional techniques including Guided Pyramidal Processing (GPP) and Affinity-aware Transformer Decoder (ATD) to further boost performance, and combine them to complete the design.

3.3 Feature Extraction and Cost Computation

We extract features from query and support images and compute an initial cost between them following the conventional process [6,17,49,51,52,58,64]. Given query and support images, x_q and x_s, we use a CNN [16,56] to produce a sequence of L feature maps, $\{(F_q^l, F_s^l)\}_{l=1}^L$, where F_q^l and F_s^l denote query and support feature maps at the l-th level. A support mask, m_s, is used to encode segmentation information and filter out the background information as done in [28,40,78]. We obtain a masked support feature as $\hat{F}_s^l = F_s^l \odot \psi^l(m_s)$, where $\odot$ denotes the Hadamard product and $\psi^l(\cdot)$ denotes a function that resizes the given tensor followed by expansion along the channel dimension of the l-th layer.

Given a pair of feature maps, F_q^l and F_s^l, we compute a correlation map using the inner product between l-2 normalized features such that

$$\mathcal{C}^l(i,j) = \text{ReLU}\left(\frac{F_q^l(i) \cdot \hat{F}_s^l(j)}{\|F_q^l(i)\|\|\hat{F}_s^l(j)\|} \right), \tag{1}$$

where i and j denote 2D spatial positions of feature maps. As done in [40], we collect correlation maps computed from all the intermediate features of the same spatial size and stack them to obtain a stacked correlation map $\mathcal{C}^p \in \mathbb{R}^{h_q \times w_q \times h_s \times w_s \times |\mathcal{L}^p|}$, where (h_q, w_q) and (h_s, w_s) are the height and width of the query and support feature maps, respectively, and $\mathcal{L}^p$ is a subset of CNN layer indices $\{1, ..., L\}$ at pyramid layer p, containing correlation maps of identical spatial size.

3.4 Pyramidal Transformer Encoder

In this section, we present 4D Convolutional Swin Transformer for aggregating the correlation maps and then incorporate it into a pyramidal architecture.

Cost Aggregation with Transformers. For a transformer to process a correlation map, a means for token reduction is essential, since it would be infeasible for even an efficient transformer [24,36,69,70,75] to handle a correlation map otherwise. However, when one employs a transformer for cost aggregation, the problem of how to define the tokens for correlation maps, which differ in shape from images, text or features [11,65], is non-trivial. The first attempt to process correlation maps is CATs [6], which reshapes the 4D correlation maps into 2D maps and performs self-attention in 2D. This disregards the spatial structure of correlation maps, i.e., over both support and query, which could limit its performance. To address this, one may treat all the spatial entries, e.g., $h_q \times w_q \times h_s \times w_s$, as tokens and treat $\mathcal{L}^p$ as the feature dimension for tokens. However, this results in a substantial computational burden that increases with larger correlation maps. This prevents the use of standard transformers [11,65] and encourages use of efficient versions as in [24,36,69,70,75]. However, the use of simplified (or approximated) self-attention may be sub-optimal for performance, as will be discussed in Sect. 4.4. Furthermore, as proven in the optical

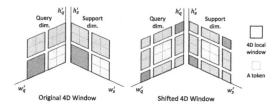

Fig. 3. Illustration of shifted 4D windows in VTM. It computes self-attention within the partitioned windows and considers inter-window interactions by shifting the windows.

flow and semantic correspondence literature [51,58], neighboring pixels tend to have similar correspondences. To preserve the spatial structure of correlation maps, we choose to use Swin Transformer [36] as it not only provides efficient self-attention computation, but also maintains the smoothness property of correlation maps while still providing sufficient long-range self-attention.

To employ Swin Transformer [36] for cost aggregation, we need to extend it to process higher dimensional input, specifically a 4D correlation map. We first follow the conventional patch embedding procedure [11] to embed correlation maps, as they cannot be processed by transformers due to the large number of tokens. However, we extend the patch embedding module to a Volumetric Embedding Module (VEM) which handles higher dimensional inputs, such that $\mathcal{M}^p = \text{VEM}(\mathcal{C}^p)$. Following a procedure similar to patch embedding, we reshape the correlation map to a sequence of flattened 4D windows using a large convolutional kernel, *e.g.*, $16\times16\times16\times16$. Then, we extend the self-attention computations, as shown in Fig. 3, by evenly partitioning the query and support spatial dimensions of $\mathcal{M}^p$ into non-overlapping sub-correlation maps $\mathcal{M}'^{,p} \in \mathbb{R}^{n\times n\times n\times n\times D}$. We compute self-attention within each partitioned sub-correlation map. Subsequently, we shift the windows by a displacement of $\left(\lfloor\frac{n}{2}\rfloor, \lfloor\frac{n}{2}\rfloor, \lfloor\frac{n}{2}\rfloor, \lfloor\frac{n}{2}\rfloor\right)$ pixels from the previously partitioned windows, then perform self-attention within the newly created windows. Then as done in the original Swin Transformer [36], we simply roll the correlation map back to its original form. In computing self-attention, we use relative position bias and take the values from an expanded parameterized bias matrix, following [19,20,36]. We leave the other components of Swin Transformer blocks unchanged, e.g., Layer Normalization (LN) [1] and MLP layers. We call this extension the Volumetric Transformer Module (VTM). To summarize, the overall process is defined as:

$$\mathcal{A}^p = \text{VTM}(\mathcal{M}^p). \tag{2}$$

4D Convolutional Swin Transformer. Although the proposed cost aggregation with transformers can solve the aforementioned issues of using CNNs and the high computational burden of using standard transformers, it may not avoid the issue that other transformers share [11,24,69,70,75]: lack of translation equivariance. This is primarily caused by utilizing non-overlapping operations prior to self-attention computation. Although Swin Transformer alleviates the issue to some extent by using relative positioning bias [36], it provides an

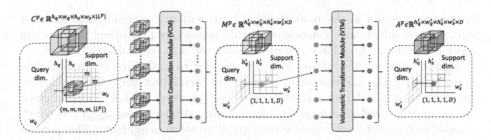

Fig. 4. Overview of 4D Convolutional Swin Transformer. We replace the VEM with VCM and the output undergoes VTM for cost aggregation.

insufficient approximation. We argue that the Volumetric Embedding Module is what needs to be addressed as it leads to several issues. First, the use of large non-overlapping convolution kernels only provides limited inductive bias. Relatively lower translation equivariance is achieved from non-overlapping operations compared to that which are overlapping. This limited inductive bias results in relatively lower generalization power and performance [7,8,36,72]. Furthermore, we argue that for dense prediction tasks, disregarding window boundaries due to non-overlapping kernels hurts overall performance due to discontinuity.

To address the above issues, we replace the Volumetric Embedding Module (VEM) with a module consisting of a series of overlapping convolutions, which we call the Volumetric Convolution Module (VCM). Concretely, we sequentially reduce spatial dimensions of the support and query by applying 4D spatial max-pooling, overlapping 4D convolutions, ReLU, and Group Normalization (GN), where we project the multi-level similarity vector at each 4D position, i.e., projecting a vector size of $|\mathcal{L}^p|$, to an arbitrary fixed dimension denoted as D. Considering receptive fields as a 4D window, i.e., $m \times m \times m \times m$, we obtain a tensor $\mathcal{C}^p \in \mathbb{R}^{h_q'^{,p} \times w_q'^{,p} \times h_s'^{,p} \times w_s'^{,p} \times D}$ from $\mathcal{C}^p$, where $h_s'^{,p}$, $w_s'^{,p}$, $h_q'^{,p}$, and $w_q'^{,p}$ are the processed sizes. Note that a different size of m can be chosen for the support and query spatial dimensions. An overview of VCM is illustrated in Fig. 4. Overall, we define such a process as the following:

$$\mathcal{M}^p = \text{VCM}(\mathcal{C}^p). \tag{3}$$

In this way, our model benefits from additional inductive bias as well as better handling at window boundaries.

Moreover, to stabilize the learning, we propose an additional technique to enforce the networks to estimate residual matching scores as complementary details. We add residual connections in order to expedite the learning process [6, 16,83], accounting for the fact that at the initial phase when the input $\mathcal{M}^p$ is fed, erroneous matching scores are inferred due to randomly-initialized parameters of transformers, which could complicate the learning process as the networks need to learn the complete matching details from random matching scores.

Guided Pyramidal Processing. Following [40,59], we also employ a coarse-to-fine approach through pyramidal processing as illustrated in Fig. 2. Motivated by numerous recent works [6,40,41,81] in both semantic matching and few-shot segmentation which have demonstrated that leveraging multi-level features can boost performance by a large margin, we also use a pyramidal architecture.

In our coarse-to-fine approach, which we refer to as Guided Pyramidal Processing (GPP), the aggregation of a finer-level correlation map $\mathcal{A}^p$ is guided by the aggregated correlation map of the previous (coarser) level $\mathcal{A}^{p+1}$. Concretely, an aggregated correlation map $\mathcal{A}^{p+1}$ is up-sampled into a map $\mathrm{up}(\mathcal{A}^{p+1})$ which is added to the next level's correlation map $\mathcal{A}^p$ to serve as guidance. This process is repeated until the finest-level aggregated map is computed and passed to the decoder. As shown in Table 4, GPP leads to appreciable performance gains.

With GPP, the pyramidal transformer encoder is finally defined as:

$$\mathcal{A}^p = \mathrm{VTM}(\mathrm{VCM}(\mathcal{C}^p) + \mathrm{up}(\mathcal{A}^{p+1})), \tag{4}$$

where $\mathrm{up}(\cdot)$ denotes bilinear upsampling.

3.5   Affinity-Aware Transformer Decoder

Given the aggregated correlation map produced by the pyramidal transformer encoder, a transformer-based decoder generates the final segmentation mask. To improve performance, we propose to conduct further aggregation within the decoder with the aid of the appearance embedding obtained from query feature maps. The query's appearance embedding can help in two ways. First, appearance affinity information is an effective guide for filtering noise in matching scores, as proven in the stereo matching literature, e.g., Cost Volume Filtering (CVF) [18, 58]. In addition, the higher-resolution spatial structure provided by an appearance embedding can be exploited to improve up-sampling quality, resulting in a highly accurate prediction mask $\hat{m}^q$ where fine details are preserved.

For the design of our Affinity-aware Transformer Decoder (ATD), we take the average over the support image dimensions of $\mathcal{A}^p$, concatenate it with the appearance embedding from query feature maps, and then aggregate by transformers [36,65,69,70] with subsequent bilinear interpolation. The process is defined as the following:

$$\hat{m}_q = \mathrm{ATD}([\mathcal{A}'^{,p}, \mathcal{P}(F_q)]), \tag{5}$$

where $\mathcal{A}'^{,p} \in \mathbb{R}^{h_q'^{,p} \times w_q'^{,p} \times D}$ is extracted by average pooling on $\mathcal{A}^p$ over the spatial dimensions of the support image, $\mathcal{P}(\cdot)$ is a linear projection, $\mathcal{P}(F_q) \in \mathbb{R}^{h_q'^{,p} \times w_q'^{,p} \times c}$, and $[\cdot,\cdot]$ denotes concatenation. We sequentially refine the output immediately after bilinear upsampling to recapture fine details and integrate appearance information.

3.6   Extension to K-Shot Setting

Given K pairs of support image and mask $\{(x_s^i, m_s^i)\}_{i=1}^K$ and a query image x_q, our model forward-passes K times to obtain K different query masks $\hat{m}_q^k$. We sum up all the K predictions at each spatial location, and if the sum divided by K exceeds a threshold τ, the location is predicted as foreground, and otherwise it is background.

Table 1. Performance comparison on PASCAL-5^i [54]. Best results in bold, and second best are underlined.

Backbone network	Methods	1-shot							5-shot							# learnable
		5^0	5^1	5^2	5^3	mIoU	FB-IoU	mBA	5^0	5^1	5^2	5^3	mIoU	FB-IoU	mBA	params
ResNet50 [16]	PANet [68]	44.0	57.5	50.8	44.0	49.1	–	–	55.3	67.2	61.3	53.2	59.3	–	–	23.5M
	PFENet [62]	61.7	69.5	55.4	56.3	60.8	73.3	–	63.1	70.7	55.8	57.9	61.9	73.9	–	10.8M
	ASGNet [28]	58.8	67.9	56.8	53.7	59.3	69.2	–	63.4	70.6	64.2	57.4	63.9	74.2	–	10.4M
	CWT [39]	56.3	62.0	59.9	47.2	56.4	–	–	61.3	68.5	68.5	56.6	63.7	–	–	–
	RePRI [2]	59.8	68.3	62.1	48.5	59.7	–	49.0	64.6	71.4	71.1	59.3	66.6	–	43.8	–
	HSNet [40]	64.3	70.7	60.3	60.5	64.0	76.7	53.9	70.3	73.2	67.4	67.1	69.5	80.6	54.5	2.6M
	CyCTR [81]	65.7	71.0	59.5	59.7	64.0	–	–	69.3	73.5	63.8	63.5	67.5	–	–	–
	VAT (ours)	67.6	72.0	62.3	60.1	65.5	77.8	54.4	72.4	73.6	68.6	65.7	70.1	80.9	54.8	3.2M
ResNet101 [16]	FWB [45]	51.3	64.5	56.7	52.2	56.2	–	–	54.8	67.4	62.2	55.3	59.9	–	–	43.0M
	DAN [67]	54.7	68.6	57.8	51.6	58.2	71.9	–	57.9	69.0	60.1	54.9	60.5	72.3	–	–
	PFENet [62]	60.5	69.4	54.4	55.9	60.1	72.9	–	62.8	70.4	54.9	57.6	61.4	73.5	–	10.8M
	ASGNet [28]	59.8	67.4	55.6	54.4	59.3	71.7	–	64.6	71.3	64.2	57.3	64.4	75.2	–	10.4M
	CWT [39]	56.9	65.2	61.2	48.8	58.0	–	–	62.6	70.2	68.8	57.2	64.7	–	–	–
	RePRI [2]	59.6	68.6	62.2	47.2	59.4	–	45.1	66.2	71.4	67.0	57.7	65.6	–	42.0	–
	HSNet [40]	67.3	72.3	62.0	63.1	66.2	77.6	53.9	71.8	74.4	67.0	68.3	70.4	80.6	54.4	2.6M
	CyCTR [81]	67.2	71.1	57.6	59.0	63.7	73.0	–	71.0	75.0	58.5	65.0	67.4	75.4	–	–
	VAT (ours)	70.0	72.5	64.8	64.2	67.9	79.6	54.7	75.0	75.2	68.4	69.5	72.0	83.2	54.8	3.3M

Table 2. Performance comparison on COCO-20^i [31].

Backbone feature	Methods	1-shot							5-shot						
		20^0	20^1	20^2	20^3	Mean	FB-IoU	mBA	20^0	20^1	20^2	20^3	Mean	FB-IoU	mBA
ResNet50 [16]	PMM [76]	29.3	34.8	27.1	27.3	29.6	–	–	33.0	40.6	30.3	33.3	34.3	–	–
	RPMM [76]	29.5	36.8	28.9	27.0	30.6	–	–	33.8	42.0	33.0	33.3	35.5	–	–
	PFENet [62]	36.5	38.6	34.5	33.8	35.8	–	–	36.5	43.3	37.8	38.4	39.0	–	–
	ASGNet [28]	–	–	–	–	34.6	60.4	–	–	–	–	–	42.5	67.0	–
	RePRI [2]	32.0	38.7	32.7	33.1	34.1	–	6.31	39.3	45.4	39.7	41.8	41.6	–	4.21
	HSNet [40]	36.3	43.1	38.7	38.7	39.2	68.2	53.0	43.3	51.3	48.2	45.0	46.9	70.7	53.8
	CyCTR [81]	38.9	43.0	39.6	39.8	40.3	–	–	41.1	48.9	45.2	47.0	45.6	–	–
	VAT (ours)	39.0	43.8	42.6	39.7	41.3	68.8	54.2	44.1	51.1	50.2	46.1	47.9	72.4	54.9

4   Experiments

4.1   Implementation Details

We use ResNet50 and ResNet101 [16] pre-trained on ImageNet [9] and freeze the weights during training, following [40,80]. No data augmentation is used for training, as explained in the supplementary material. We set the input image sizes to 417 or 473, following [2,28]. The window size for Swin Transformer is set to 4. We use AdamW [38] with a learning rate of $5e-4$. Feature maps from conv3_x ($p=3$), conv4_x ($p=4$) and conv5_x ($p=5$) are taken for cost computation. The K-shot threshold τ is set to 0.5 and the embedding dimension D to 128. For appearance affinity, we take the last layers from conv2_x, conv3_x and conv4_x when training on FSS-1000 [30], and conv4_x is excluded when training on PASCAL-5^i [54] and COCO-20^i [31]. We set c to 16, 32, and 64 for conv2_x, conv3_x, and conv4_x.

4.2   Experimental Settings

Datasets. We evaluate our approach on three standard few-shot segmentation datasets, PASCAL-5^i [54], COCO-20^i [31], and FSS-1000 [30]. PASCAL-5^i contains images from PASCAL VOC 2012 [12] with added mask annotations [15].

Table 3. Mean IoU comparison on FSS-1000 [30].

Backbone feature	Methods	mIoU		FB-IoU		mBA	
		1-shot	5-shot	1-shot	5-shot	1-shot	5-shot
ResNet50 [16]	FSOT [32]	82.5	83.8	–	–	–	–
	HSNet [40]	85.5	87.8	91.0	92.5	62.1	63.3
	VAT	**90.1**	**90.7**	**93.8**	**94.2**	**68.3**	**68.4**
ResNet101 [16]	DAN [67]	85.2	88.1	–	–	–	–
	HSNet [40]	86.5	88.5	91.6	92.9	62.4	63.6
	VAT	**90.3**	**90.8**	**94.0**	**94.4**	**68.0**	**68.6**

It consists of 20 object classes, and as done in OSLSM [54], they are evenly divided into 4 folds $i \in \{0, 1, 2, 3\}$ for cross-validation, where each fold contains 5 classes. COCO-20^i contains 80 object classes, and as done for PASCAL-5^i, the dataset is evenly divided into 4 folds of 20 classes each. FSS-1000 is a more diverse dataset consisting of 1000 object classes. Following [30], we divide the 1000 categories into 3 splits for training, validation and testing, which consist of 520, 240 and 240 classes, respectively. For PASCAL-5^i and COCO-20^i, we follow the common evaluation practice [35, 40, 62] and standard cross-validation protocol, where each fold i is used for evaluation with the other folds used for training.

Evaluation Metric. Following common practice [40, 62, 80, 81], we adopt mean intersection over union (mIoU) and foreground-background IoU (FB-IoU) as our evaluation metrics. The mIoU averages over all IoU values for all object classes such that mIoU $= \frac{1}{C} \sum_{c=1}^{C} \text{IoU}_c$, where C is the number of classes in each fold, e.g., $C = 20$ for COCO-20^i. FB-IoU disregards the object classes and instead averages over foreground and background IoU (IoU_F and IoU_B) such that FB $-$ IoU $= \frac{1}{2}(\text{IoU}_F + \text{IoU}_B)$. We additionally adopt Mean Boundary Accuracy (mBA) introduced in [5] to evaluate the model's ability to capture fine details. To measure mBA, we first sample 5 radii in $[3, \frac{w+h}{300}]$ at a uniform interval, where w and h are width and height of input image, and average the segmentation accuracy within each radius from the ground-truth boundary.

4.3 Few-Shot Segmentation Results

Table 1 summarizes quantitative results on PASCAL-5^i [54]. The tests were conducted on two backbone networks, ResNet50 and ResNet101 [16]. The proposed method outperforms the others on almost all the folds in terms of both mIoU and FB-IoU. It surpasses the others, including HSNet [40], in mBA as well, since our ATD helps to improve up-sampling quality by providing higher-level spatial structure for reference. Consistent with this, VAT also attains state-of-the-art performance on COCO-20^i [31], as shown in Table 2. Interestingly, for the most recent dataset specifically created for few-shot segmentation, FSS-1000 [30], VAT outperforms HSNet [40] and FSOT [32] by a large margin, almost a 4.6% increase in mIoU compared to HSNet with ResNet50 as shown in Table 3. VAT sets a new state-of-the-art for all of these benchmarks. We note that our method outperforms HSNet [40] despite having more learnable parameters, which is known to have an inverse relation to generalization power [61], a trend seen in Table 1. With the proposed method, *i.e.*, 4D convolutional Swin Transformer, that is designed to address the issues like lack of inductive bias, VAT can have a larger number of learnable parameters than that of HSNet [40], yet VAT has greater generalization power as well.

4.4 Ablation Study

We conducted ablations on FSS-1000 [30], a large-scale dataset specifically constructed for few-shot segmentation.

Effectiveness of Each Component in VAT. As the baseline model, we take the architecture composed of VEM and the 2D convolution decoder used in HSNet [40]. We then progressively add our components one-by-one as shown in Table 4. Note that we included (**IV**) and (**V**) to show the effectiveness of VCM alone and the performance of a model highly similar to HSNet [40], respectively.

As summarized in Table 4, each component helps to boost performance. Starting from the baseline (**I**), adding Swin Transformer (**II**) brings a large gain, which indicates that Swin Transformer

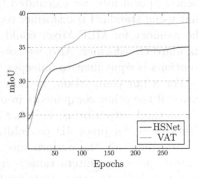

Fig. 5. Convergence comparison. Although VAT starts at a lower mIoU, it quickly exceeds HSNet [40].

effectively performs cost aggregation thanks to its approximated inductive bias and ability to consider spatial structure. When the VEM is replaced by VCM (**III**), we also observe a significant improvement, which confirms that the issues due to non-overlap are alleviated. We note that (**IV**) also highlights the importance of inductive bias. As (**V**) is approximately equivalent to HSNet [40], we first compare it with (**III**), which shows the superiority of the proposed 4D Convolutional Swin Transformer. By including the additional components in (**VI**) and (**VII**), the performance is further boosted. Moreover, we observe a large gain in mBA by adding ATD. This shows that the higher-resolution spatial structure provided by appearance embeddings help to refine the fine details. We additionally provide a visualization of convergence in comparison to HSNet [40] in Fig. 5. Thanks to the early convolutions [72], VAT quickly converges and exceeds HSNet [40] even though it starts at a lower mIOU.

Base Architecture of VTM. As summarized in Table 5, we provide an ablation study to evaluate the effectiveness of different aggregators for VTM. For

Table 4. Ablation study for VAT.

	Components	FSS-1000 [30]			
		mIoU (%)		mBA (%)	
		1-shot	5-shot	1-shot	5-shot
(I)	Baseline	80.0	81.8	56.7	56.9
(II)	+ Swin Trans.	85.4	87.4	58.8	59.5
(III)	+ VCM	87.0	88.6	60.1	61.3
(IV)	Only VCM	86.4	88.0	59.6	60.1
(V)	(IV) + 4D mix	86.4	87.8	59.9	59.6
(VI)	(III) + GPP	87.3	88.8	60.7	61.4
(VII)	+ ATD	90.3	90.8	68.0	68.6

Table 5. Ablation study for VTM. OOM: Out of Memory.

Different aggregators	FSS-1000 [30]		Memory	Run-time
	mIoU (%)	mBA (%)	(GB)	(ms)
Standard transformer [65]	OOM	OOM	84	N/A
MLP-Mixer [63]	OOM	OOM	OOM	N/A
Center-pivot 4D convolutions [40]	88.1	66.5	3.5	52.7
Linear transformer [24]	87.7	66.5	3.5	56.8
Fastformer [70]	87.8	66.4	3.5	122.9
4D Conv. Swin transformer (Ours)	90.3	68.0	3.8	57.3

cost aggregation, there exists a few learnable aggregators, including MLP-, convolution- and transformer-based aggregators, any of which could be used as a base architecture for VTM. It should be noted that the use of standard transformer [65] and MLP-mixer [63] is not feasible due to memory requirements. Specifically, we calculated the memory consumption of each and found that using standard transformer requires approximately 84 GB per batch, while the memory for MLP-Mixer could not be measured as it is much greater than standard transformer. Also, we note that the architecture with center-pivot convolutions is equivalent to a deeper version of the architecture with VCM.

For a fair comparison, we only replace VTM with another aggregator and leave all the other components in our architecture unchanged. We observe that our method outperforms the other aggregators by a large margin. Interestingly, although center-pivot 4D convolution [40] also focuses on locality as in Swin Transformer [36], the performance gap indicates that the ability to adaptively consider pixel-wise interactions is critical. Also, we conjecture that the SW-MSA operation helps to compensate for the lack of global aggregation, which center-pivot convolutions lack. Another interesting point is that Linear Transformer [24] and Fastformer [70], which benefit from the global receptive fields of transformers and approximate the self-attention computation, achieve similar performance.

We additionally provide memory and run-time comparison to other aggregators in Table 5. The results are obtained using a single NVIDIA GeForce RTX 3090 GPU and Intel Core i7-10700 CPU. We observe that VAT is relatively slower and consumes more memory. However, 0.3 GB more memory consumption and 5 ms slower run time is a minor sacrifice for better performance.

Can VAT Also Perform Well on Semantic Correspondence? To tackle the few-shot segmentation task, we reformulated it as finding semantic correspondences under large intra-class variations and geometric deformations. This suggests that the proposed method could be effective for semantic correspondence as well. Here, we compare VAT to other state-of-the-art methods in semantic correspondence.

In order to ensure a fair comparison, we note whether each method leverages multi-level features and fine-tunes the backbone networks. We additionally denote the types of cost aggregation. Note that the only difference we made

Table 6. Quantitative results on SPair-71k [43], PF-PASCAL [14] and PF-WILLOW [13]. *: The results are obtained using pretrained weights provided by authors or taken from papers.

Methods	F.T. Feat.	Data Aug.	Cost aggregation	SPair-71k [43] PCK @ α_{bbox}				PF-PASCAL [14] PCK @ α_{img}				PF-WILLOW [13] PCK @ α_{bbox}		
				0.03	0.05	0.1	0.15	0.03	0.05	0.1	0.15	0.05	0.1	0.15
NC-Net [52]	✓	✗	4D Conv.	–	–	20.1	–	30.9	54.3	78.9	86.0	33.8	67.0	83.7
SCOT [34]	–	✗	OT-RHM	–	–	35.6	–	–	63.1	85.4	92.7	47.8	76.0	87.1
CHM [41]*	✓	✗	4D Conv.	14.9	27.2	46.3	57.5	67.5	80.1	91.6	94.9	52.7	79.4	87.5
MMNet [83]	✓	✗	–	–	–	40.9	–	–	77.6	89.1	94.3	–	–	–
PMNC [26]	✓	✗	4D Conv	–	–	50.4	–	–	71.6	82.4	90.6	–	–	–
DHPF [44]*	✓	✗	RHM	11.0	20.9	37.3	47.5	52.0	75.7	90.7	95.0	49.5	77.6	89.1
	✓	✓	RHM	–	–	39.4	–	–	–	–	–	–	–	–
CATs [6]*	✓	✗	Transformer	10.2	21.6	43.5	55.0	41.6	67.5	89.1	94.9	46.6	75.6	87.5
	✓	✓	Transformer	13.8	27.7	49.9	61.7	49.9	75.4	92.6	96.4	50.3	79.2	90.3
VAT	✓	✗	Transformer	14.9	28.3	48.4	59.1	54.6	72.9	91.1	95.6	46.0	78.8	91.3
	✓	✓	Transformer	19.6	35.0	55.5	65.1	62.7	78.2	92.3	96.2	52.8	81.6	91.4

for this experiment is the objective function for loss computation. Following the common protocol [6,21,41,42,44,83], we use standard benchmarks for this task and our model was trained on the training split of PF-PASCAL [14] when evaluated on the test split of PF-PASCAL [14] and PF-WILLOW [13], and trained on SPair-71k [43] when evaluated on SPair-71k [43]. Experimental setting and implementation details can be found in supplementary material.

As shown in Table 6, VAT either sets a new state-of-the-art [13,43] or attains the second highest PCK [14], indicating the importance of cost aggregation in both few-shot segmentation and semantic correspondence. It also has the potential to benefit general-purpose matching networks as well. Furthermore, when data augmentation is used, we observe a relatively large performance gain compared to DHPF [44], showing that augmentation helps to address the heavy need for data and lack of inductive bias in transformers [6,11]. Although VAT is on par with state-of-the-art on PF-PASCAL [14], we argue that PF-PASCAL [14] is almost saturated, which makes a comparison difficult. Also, it should be noted that for performance on PF-WILLOW [13], VAT outperforms other methods by large margin, which clearly shows superior generalization power of the proposed 4D Convolutional Swin Transformer.

5   Conclusion

In this paper, we presented a novel cost aggregation network for few-shot segmentation. To address issues that arise from tokenization of a correlation map for transformer processing, we proposed a 4D Convolutional Swin Transformer, where a high-dimensional Swin Transformer is preceded by a series of small-kernel convolutions. To boost aggregation performance, we applied transformers within a pyramidal structure, and the output is then filtered and in the subsequent decoder with the help of image's appearance embedding. We have shown that the proposed method attains state-of-the-art performance for all the standard benchmarks for both few-shot segmentation and semantic correspondence, where cost aggregation plays a central role.

Acknowledgements. This research was supported by the MSIT, Korea (IITP-2022-2020-0-01819, ICT Creative Consilience program), and National Research Foundation of Korea (NRF-2021R1C1C1006897).

References

1. Ba, J.L., Kiros, J.R., Hinton, G.E.: Layer normalization. arXiv preprint arXiv:1607.06450 (2016)
2. Boudiaf, M., Kervadec, H., Masud, Z.I., Piantanida, P., Ben Ayed, I., Dolz, J.: Few-shot segmentation without meta-learning: a good transductive inference is all you need? In: Proceedings of the IEEE/CVF Conference on Computer Vision and Pattern Recognition (2021)
3. Chen, L.C., Papandreou, G., Kokkinos, I., Murphy, K., Yuille, A.L.: DeepLab: semantic image segmentation with deep convolutional Nets, Atrous convolution, and fully connected CRFs. IEEE Trans. Patt. Anal. Mach. Intell. (2017)
4. Chen, L.C., Zhu, Y., Papandreou, G., Schroff, F., Adam, H.: Encoder-decoder with atrous separable convolution for semantic image segmentation. In: Proceedings of the European Conference on Computer Vision (ECCV) (2018)
5. Cheng, H.K., Chung, J., Tai, Y.W., Tang, C.K.: CascadePSP: toward class-agnostic and very high-resolution segmentation via global and local refinement. In: CVPR (2020)
6. Cho, S., Hong, S., Jeon, S., Lee, Y., Sohn, K., Kim, S.: CATs: cost aggregation transformers for visual correspondence. In: Thirty-Fifth Conference on Neural Information Processing Systems (2021)
7. Dai, Z., Liu, H., Le, Q., Tan, M.: CoAtNet: marrying convolution and attention for all data sizes. Adv. Neural Inf. Process. Syst. **34**, 3965–3977 (2021)
8. d'Ascoli, S., Touvron, H., Leavitt, M., Morcos, A., Biroli, G., Sagun, L.: ConViT: improving vision transformers with soft convolutional inductive biases. arXiv preprint arXiv:2103.10697 (2021)
9. Deng, J., Dong, W., Socher, R., Li, L.J., Li, K., Fei-Fei, L.: ImageNet: a large-scale hierarchical image database. In: 2009 IEEE Conference on Computer Vision and Pattern Recognition. IEEE (2009)
10. Dong, N., Xing, E.P.: Few-shot semantic segmentation with prototype learning. In: BMVC (2018)
11. Dosovitskiy, A., et al.: An image is worth 16 × 16 words: transformers for image recognition at scale. arXiv preprint arXiv:2010.11929 (2020)
12. Everingham, M., Van Gool, L., Williams, C.K., Winn, J., Zisserman, A.: The pascal visual object classes (VOC) challenge. Int. J. Comput. Vis. (2010). https://doi.org/10.1007/s11263-009-0275-4
13. Ham, B., Cho, M., Schmid, C., Ponce, J.: Proposal flow. In: CVPR (2016)
14. Ham, B., Cho, M., Schmid, C., Ponce, J.: Proposal flow: semantic correspondences from object proposals. IEEE Trans. Patt. Anal. Mach. Intell. (2017)
15. Hariharan, B., Arbeláez, P., Girshick, R., Malik, J.: Simultaneous detection and segmentation. In: Fleet, D., Pajdla, T., Schiele, B., Tuytelaars, T. (eds.) ECCV 2014. LNCS, vol. 8695, pp. 297–312. Springer, Cham (2014). https://doi.org/10.1007/978-3-319-10584-0_20
16. He, K., Zhang, X., Ren, S., Sun, J.: Deep residual learning for image recognition. In: Proceedings of the IEEE Conference on Computer Vision and Pattern Recognition (2016)

17. Hong, S., Kim, S.: Deep matching prior: test-time optimization for dense correspondence. In: Proceedings of the IEEE/CVF International Conference on Computer Vision (ICCV) (2021)
18. Hosni, A., Rhemann, C., Bleyer, M., Rother, C., Gelautz, M.: Fast cost-volume filtering for visual correspondence and beyond. PAMI (2012)
19. Hu, H., Gu, J., Zhang, Z., Dai, J., Wei, Y.: Relation networks for object detection. In: Proceedings of the IEEE Conference on Computer Vision and Pattern Recognition, pp. 3588–3597 (2018)
20. Hu, H., Zhang, Z., Xie, Z., Lin, S.: Local relation networks for image recognition. In: Proceedings of the IEEE/CVF International Conference on Computer Vision, pp. 3464–3473 (2019)
21. Huang, S., Wang, Q., Zhang, S., Yan, S., He, X.: Dynamic context correspondence network for semantic alignment. In: ICCV (2019)
22. Jeon, S., Min, D., Kim, S., Choe, J., Sohn, K.: Guided semantic flow. In: ECCV. Springer (2020). https://doi.org/10.1007/978-3-030-58604-1_38
23. Jiang, W., Trulls, E., Hosang, J., Tagliasacchi, A., Yi, K.M.: COTR: correspondence transformer for matching across images. arXiv preprint arXiv:2103.14167 (2021)
24. Katharopoulos, A., Vyas, A., Pappas, N., Fleuret, F.: Transformers are RNNs: fast autoregressive transformers with linear attention. In: International Conference on Machine Learning, pp. 5156–5165. PMLR (2020)
25. Kim, S., Min, D., Ham, B., Jeon, S., Lin, S., Sohn, K.: FCSS: fully convolutional self-similarity for dense semantic correspondence. In: CVPR (2017)
26. Lee, J.Y., DeGol, J., Fragoso, V., Sinha, S.N.: Patchmatch-based neighborhood consensus for semantic correspondence. In: Proceedings of the IEEE/CVF Conference on Computer Vision and Pattern Recognition (CVPR)
27. Lee, J., Kim, D., Ponce, J., Ham, B.: SFNet: learning object-aware semantic correspondence. In: CVPR (2019)
28. Li, G., Jampani, V., Sevilla-Lara, L., Sun, D., Kim, J., Kim, J.: Adaptive prototype learning and allocation for few-shot segmentation. In: Proceedings of the IEEE/CVF Conference on Computer Vision and Pattern Recognition, pp. 8334–8343 (2021)
29. Li, S., Han, K., Costain, T.W., Howard-Jenkins, H., Prisacariu, V.: Correspondence networks with adaptive neighbourhood consensus. In: CVPR (2020)
30. Li, X., Wei, T., Chen, Y.P., Tai, Y.W., Tang, C.K.: FSS-1000: a 1000-class dataset for few-shot segmentation. In: Proceedings of the IEEE/CVF Conference on Computer Vision and Pattern Recognition (2020)
31. Lin, T.-Y., et al.: Microsoft COCO: common objects in context. In: Fleet, D., Pajdla, T., Schiele, B., Tuytelaars, T. (eds.) ECCV 2014. LNCS, vol. 8693, pp. 740–755. Springer, Cham (2014). https://doi.org/10.1007/978-3-319-10602-1_48
32. Liu, W., Zhang, C., Ding, H., Hung, T.Y., Lin, G.: Few-shot segmentation with optimal transport matching and message flow. arXiv preprint arXiv:2108.08518 (2021)
33. Liu, W., Zhang, C., Lin, G., Liu, F.: CRNet: cross-reference networks for few-shot segmentation. In: Proceedings of the IEEE/CVF Conference on Computer Vision and Pattern Recognition (2020)
34. Liu, Y., Zhu, L., Yamada, M., Yang, Y.: Semantic correspondence as an optimal transport problem. In: Proceedings of the IEEE/CVF Conference on Computer Vision and Pattern Recognition (2020)

35. Liu, Y., Zhang, X., Zhang, S., He, X.: Part-aware prototype network for few-shot semantic segmentation. In: Vedaldi, A., Bischof, H., Brox, T., Frahm, J.-M. (eds.) ECCV 2020. LNCS, vol. 12354, pp. 142–158. Springer, Cham (2020). https://doi.org/10.1007/978-3-030-58545-7_9

36. Liu, Z., et al.: Swin transformer: hierarchical vision transformer using shifted windows. arXiv preprint arXiv:2103.14030 (2021)

37. Long, J., Shelhamer, E., Darrell, T.: Fully convolutional networks for semantic segmentation. In: Proceedings of the IEEE Conference on Computer Vision and Pattern Recognition (2015)

38. Loshchilov, I., Hutter, F.: Decoupled weight decay regularization. arXiv preprint arXiv:1711.05101 (2017)

39. Lu, Z., He, S., Zhu, X., Zhang, L., Song, Y.Z., Xiang, T.: Simpler is better: few-shot semantic segmentation with classifier weight transformer. In: Proceedings of the IEEE/CVF International Conference on Computer Vision (2021)

40. Min, J., Kang, D., Cho, M.: Hypercorrelation squeeze for few-shot segmentation. arXiv preprint arXiv:2104.01538 (2021)

41. Min, J., Kim, S., Cho, M.: Convolutional hough matching networks for robust and efficient visual correspondence. arXiv preprint arXiv:2109.05221 (2021)

42. Min, J., Lee, J., Ponce, J., Cho, M.: Hyperpixel Flow: semantic correspondence with multi-layer neural features. In: Proceedings of the IEEE/CVF International Conference on Computer Vision (2019)

43. Min, J., Lee, J., Ponce, J., Cho, M.: SPair-71k: a large-scale benchmark for semantic correspondence. arXiv preprint arXiv:1908.10543 (2019)

44. Min, J., Lee, J., Ponce, J., Cho, M.: Learning to compose hypercolumns for visual correspondence. In: Vedaldi, A., Bischof, H., Brox, T., Frahm, J.-M. (eds.) ECCV 2020. LNCS, vol. 12360, pp. 346–363. Springer, Cham (2020). https://doi.org/10.1007/978-3-030-58555-6_21

45. Nguyen, K., Todorovic, S.: Feature weighting and boosting for few-shot segmentation. In: Proceedings of the IEEE/CVF International Conference on Computer Vision, pp. 622–631 (2019)

46. Noh, H., Hong, S., Han, B.: Learning deconvolution network for semantic segmentation. In: Proceedings of the IEEE International Conference on Computer Vision (2015)

47. Philbin, J., Chum, O., Isard, M., Sivic, J., Zisserman, A.: Object retrieval with large vocabularies and fast spatial matching. In: CVPR. IEEE (2007)

48. Ravi, S., Larochelle, H.: Optimization as a model for few-shot learning (2016)

49. Rocco, I., Arandjelovic, R., Sivic, J.: Convolutional neural network architecture for geometric matching. In: CVPR (2017)

50. Rocco, I., Arandjelović, R., Sivic, J.: End-to-end weakly-supervised semantic alignment. In: CVPR (2018)

51. Rocco, I., Arandjelović, R., Sivic, J.: Efficient neighbourhood consensus networks via submanifold sparse convolutions. In: Vedaldi, A., Bischof, H., Brox, T., Frahm, J.-M. (eds.) ECCV 2020. LNCS, vol. 12354, pp. 605–621. Springer, Cham (2020). https://doi.org/10.1007/978-3-030-58545-7_35

52. Rocco, I., Cimpoi, M., Arandjelović, R., Torii, A., Pajdla, T., Sivic, J.: Neighbourhood consensus networks. arXiv preprint arXiv:1810.10510 (2018)

53. Scharstein, D., Szeliski, R.: A taxonomy and evaluation of dense two-frame stereo correspondence algorithms. Int. J. Comput. Vis. (2002). https://doi.org/10.1023/A:1014573219977

54. Shaban, A., Bansal, S., Liu, Z., Essa, I., Boots, B.: One-shot learning for semantic segmentation. arXiv preprint arXiv:1709.03410 (2017)

55. Siam, M., Oreshkin, B., Jagersand, M.: Adaptive masked proxies for few-shot segmentation. arXiv preprint arXiv:1902.11123 (2019)
56. Simonyan, K., Zisserman, A.: Very deep convolutional networks for large-scale image recognition. arXiv preprint arXiv:1409.1556 (2014)
57. Snell, J., Swersky, K., Zemel, R.S.: Prototypical networks for few-shot learning. arXiv preprint arXiv:1703.05175 (2017)
58. Sun, D., Yang, X., Liu, M.Y., Kautz, J.: PWC-Net: CNNs for optical flow using pyramid, warping, and cost volume. In: CVPR (2018)
59. Sun, G., Liu, Y., Liang, J., Van Gool, L.: Boosting few-shot semantic segmentation with transformers. arXiv preprint arXiv:2108.02266 (2021)
60. Tao, A., Sapra, K., Catanzaro, B.: Hierarchical multi-scale attention for semantic segmentation. arXiv preprint arXiv:2005.10821 (2020)
61. Tetko, I.V., Livingstone, D.J., Luik, A.I.: Neural network studies, 1. comparison of overfitting and overtraining. J. Chem. Inf. Comput. Sci. **35**, 826–833 (1995)
62. Tian, Z., Zhao, H., Shu, M., Yang, Z., Li, R., Jia, J.: Prior guided feature enrichment network for few-shot segmentation. IEEE Trans. Patt. Anal. Mach. Intell. (2020)
63. Tolstikhin, I.O., et al.: MLP-mixer: an all-MLP architecture for vision. Adv. Neural Inf. Process. Syst. **34**, 24261–24272 (2021)
64. Truong, P., Danelljan, M., Timofte, R.: GLU-Net: global-local universal network for dense flow and correspondences. In: Proceedings of the IEEE/CVF Conference on Computer Vision and Pattern Recognition, pp. 6258–6268 (2020)
65. Vaswani, A., et al.: Attention is all you need. In: Advances in Neural Information Processing Systems (2017)
66. Vinyals, O., Blundell, C., Lillicrap, T., Wierstra, D., et al.: Matching networks for one shot learning. Adv. Neural Inf. Process. Syst. (2016)
67. Wang, H., Zhang, X., Hu, Y., Yang, Y., Cao, X., Zhen, X.: Few-shot semantic segmentation with democratic attention networks. In: Vedaldi, A., Bischof, H., Brox, T., Frahm, J.-M. (eds.) ECCV 2020. LNCS, vol. 12358, pp. 730–746. Springer, Cham (2020). https://doi.org/10.1007/978-3-030-58601-0_43
68. Wang, K., Liew, J.H., Zou, Y., Zhou, D., Feng, J.: PANet: few-shot image semantic segmentation with prototype alignment. In: Proceedings of the IEEE/CVF International Conference on Computer Vision (2019)
69. Wang, S., Li, B.Z., Khabsa, M., Fang, H., Ma, H.: Linformer: Self-attention with linear complexity. arXiv preprint arXiv:2006.04768 (2020)
70. Wu, C., Wu, F., Qi, T., Huang, Y., Xie, X.: Fastformer: additive attention can be all you need. arXiv preprint arXiv:2108.09084 (2021)
71. Wu, S., Wu, T., Lin, F., Tian, S., Guo, G.: Fully transformer networks for semantic image segmentation. arXiv preprint arXiv:2106.04108 (2021)
72. Xiao, T., Singh, M., Mintun, E., Darrell, T., Dollár, P., Girshick, R.: Early convolutions help transformers see better. arXiv preprint arXiv:2106.14881 (2021)
73. Xie, G.S., Liu, J., Xiong, H., Shao, L.: Scale-aware graph neural network for few-shot semantic segmentation. In: Proceedings of the IEEE/CVF Conference on Computer Vision and Pattern Recognition, pp. 5475–5484 (2021)
74. Xie, G.S., Xiong, H., Liu, J., Yao, Y., Shao, L.: Few-shot semantic segmentation with cyclic memory network. In: Proceedings of the IEEE/CVF International Conference on Computer Vision (2021)
75. Xiong, Y., et al.: Nyströmformer: A nyström-based algorithm for approximating self-attention (2021)

76. Yang, B., Liu, C., Li, B., Jiao, J., Ye, Q.: Prototype mixture models for few-shot semantic segmentation. In: Vedaldi, A., Bischof, H., Brox, T., Frahm, J.-M. (eds.) ECCV 2020. LNCS, vol. 12353, pp. 763–778. Springer, Cham (2020). https://doi.org/10.1007/978-3-030-58598-3_45
77. Yang, L., Zhuo, W., Qi, L., Shi, Y., Gao, Y.: Mining latent classes for few-shot segmentation. arXiv preprint arXiv:2103.15402 (2021)
78. Zhang, B., Xiao, J., Qin, T.: Self-guided and cross-guided learning for few-shot segmentation. In: Proceedings of the IEEE/CVF Conference on Computer Vision and Pattern Recognition (2021)
79. Zhang, C., Lin, G., Liu, F., Guo, J., Wu, Q., Yao, R.: Pyramid graph networks with connection attentions for region-based one-shot semantic segmentation. In: Proceedings of the IEEE/CVF International Conference on Computer Vision, pp. 9587–9595 (2019)
80. Zhang, C., Lin, G., Liu, F., Yao, R., Shen, C.: CANet: class-agnostic segmentation networks with iterative refinement and attentive few-shot learning. In: Proceedings of the IEEE/CVF Conference on Computer Vision and Pattern Recognition, pp. 5217–5226 (2019)
81. Zhang, G., Kang, G., Wei, Y., Yang, Y.: Few-shot segmentation via cycle-consistent transformer. arXiv preprint arXiv:2106.02320 (2021)
82. Zhang, H., Ding, H.: Prototypical matching and open set rejection for zero-shot semantic segmentation. In: Proceedings of the IEEE/CVF International Conference on Computer Vision (2021)
83. Zhao, D., Song, Z., Ji, Z., Zhao, G., Ge, W., Yu, Y.: Multi-scale matching networks for semantic correspondence. In: Proceedings of the IEEE/CVF International Conference on Computer Vision, pp. 3354–3364 (2021)

Fine-Grained Egocentric Hand-Object Segmentation: Dataset, Model, and Applications

Lingzhi Zhang[1](✉), Shenghao Zhou[1](✉), Simon Stent[2], and Jianbo Shi[1]

[1] University of Pennsylvania, Philadelphia, USA
zlz@seas.upenn.edu
[2] Toyota Research Institute, Los Altos, USA

Abstract. Egocentric videos offer fine-grained information for high-fidelity modeling of human behaviors. Hands and interacting objects are one crucial aspect of understanding a viewer's behaviors and intentions. We provide a labeled dataset consisting of 11,243 egocentric images with per-pixel segmentation labels of hands and objects being interacted with during a diverse array of daily activities. Our dataset is the first to label detailed hand-object contact boundaries. We introduce a context-aware compositional data augmentation technique to adapt to out-of-distribution YouTube egocentric video. We show that our robust hand-object segmentation model and dataset can serve as a foundational tool to boost or enable several downstream vision applications, including hand state classification, video activity recognition, 3D mesh reconstruction of hand-object interactions, and video inpainting of hand-object foregrounds in egocentric videos. Dataset and code are available at: https://github.com/owenzlz/EgoHOS.

Keywords: Datasets · Egocentric hand-object segmentation · Egocentric activity recognition · Hand-object mesh reconstruction

1 Introduction

Watching someone cooking from a third-person view, we can answer questions such as "what food is the person making?", or "what cooking technique is the person using?" First-person egocentric video, on the other hand, can often show much more detailed information of human behaviors, such as "what finger poses are needed to cut a steak into slices?", "what are the procedures to construct a IKEA table with all the pieces and screws?" Thus, egocentric videos are an essential source of information to study and understand how humans interact with the world at a fine level. In these videos, egocentric viewer's hands and interacting objects are incredibly informative visual cues to understand human behaviors. However, existing tools for extracting these cues are limited, due to lack of robustness in the wild or coarse hand-object representation. Our goal is to

Supplementary Information The online version contains supplementary material available at https://doi.org/10.1007/978-3-031-19818-2_8.

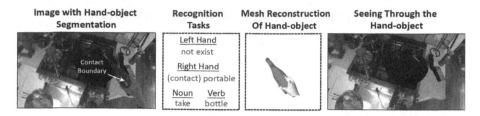

| Image with Hand-object Segmentation | Recognition Tasks | Mesh Reconstruction Of Hand-object | Seeing Through the Hand-object |

Fig. 1. Leftmost image: our proposed dataset enables us to train a robust hand-object segmentation model. We introduce contact boundaries to model the hand-object interaction explicitly. Right: our hand-object segmentation model is helpful for many vision tasks, including recognizing hand state, activities, mesh reconstruction, and seeing-through the hand-object.

create data labels and data argumentation tools for a robust fine-grained egocentric hand-object segmentation system that can generalize in the wild. Utilizing the fine-level interaction segmentation, we show how to construct a high-fidelity model that can serve as a foundation for understanding and modeling human hand-object behaviors.

The first and foremost factor in building a robust egocentric hand-object segmentation model is a good-quality labeled dataset. Previous works [1,33,73] have constructed hand segmentation datasets for egocentric videos. However, the collected data are mostly restricted to in-lab settings or to limited scenes, and lack labels for interacting objects. More recently, 100-DOH [58] made a great effort to label large-scale hand and object interactions in the wild, but the labels for hands and objects are at the bounding box level. To bridge the gap and further advance fine-level understanding of hand-object interactions, we propose a new dataset of 11,243 images with per-pixel segmentation labels. A major characteristic is that our dataset contains very diverse hand-object interaction activities and scenarios, where the frames are sparsely sampled from nearly 1,000 videos in Ego4D [16], EPIC-KITCHEN [8], THU-READ [68], and from our own collected GoPro videos. In addition, we also provide fine-grained labels of whether an object is interacted with by the left hand, right hand, or both hands and whether it is being interacted with directly (in touch) or indirectly.

To serve as an out-of-distribution test set for evaluating in-the-wild performance, we sparsely sampled and labeled 500 additional frames from 30 egocentric videos from YouTube. With our new segmentation dataset, we boost the hand segmentation performance significantly compared to the previous datasets [1,33,73]. Our dataset is the first to label interacting hand-object contact boundaries in egocentric videos. We show this label can improve the detection and segmentation of interaction objects. No matter how diverse our dataset is, we will inevitably encounter new domains with very different illumination, objects, and background clutter. We propose a context-aware data augmentation technique that adaptively composites hand-object pairs into diverse but plausible backgrounds. Our experiments show that our method is effective for out-of-domain adaptation.

We view our hand-object segmentation model as a foundation for boosting or enabling many vision applications, of which we demonstrate three, as shown in Fig. 1. First, we show that recognition tasks can get consistent performance improvement by simply adding reliably segmented hand or object masks as inputs. We experiment with a low-level recognition task to classify the left/right-hand state and a high-level recognition task to understand egocentric video activities by predicting verbs and nouns. Another useful but challenging application is reconstructing hand-object interaction in 3D mesh, which relies on the 2D hand-object masks during optimization. In this application, we integrate our hand-object segmentation model into the mesh reconstruction pipeline [18], and show improvements and generalization for mesh reconstruction of hand-object, compared to its original hand-object segmentation pipeline pretrained on COCO [39]. Finally, we show an interesting application by combining our accurate per-frame hand segmentation and video inpainting [12] to see through hands in egocentric videos, which could help scene understanding models that have not been trained with hands in the foreground. More details of each of these applications are discussed in Sect. 7.

We summarize the contributions of this work as follows: 1) We propose a dataset of 11,243 images with fine-grained per-pixel labels of hand and interacting objects, including interacting object masks, enabling hand segmentation models to generalize much better than previous datasets. 2) We introduce the notion of a dense contact boundary to explicitly model the relationship between hands and interacting objects, which we show helps to improve segmentation performance. 3) We propose a context-aware compositional data augmentation technique, which effectively boosts object segmentation. 4) We demonstrate that our system can serve as a reliable foundational tool to boost or enable many vision applications, such as hand state classification, video activity recognition, 3D reconstruction of hand-object interaction, and seeing through hands in egocentric videos. We will release our dataset, code, and checkpoints to the public for future research.

2 Related Work

2.1 Hand Segmentation

Prior to deep learning, several works have attempted to solve the hand segmentation task. Jedynak et al. [22] used a color statistics-based approach to separate the skin region and the background. Ren and Gu [52] proposed a bottom-up motion-based approach to segment hand and object using the different motion patterns between hands and background in egocentric videos. Following up with [52], Fathi et al. [10] further separates hand and interacting object from the whole foreground by assuming a color histogram prior over hand super-pixels, and uses graph-cut to segment hands and objects. Li and Kitani [31,32] first addressed the hand segmentation problem under various illuminations, and proposed to adaptively select a model that works the best under different illumination scenarios during inference time. Zhu et al. [84] proposed a novel approach by estimating a probability shape mask for a pixel using shape-aware structured forests. Beyond the egocentric viewer, Lee et al. [30] studied the problem of hand disambiguation of multiple people in egocentric videos by modeling the spatial, temporal, and appearance coherency constraints of moving hands.

More recently, many works [1,4,24,34,36,37,50,58,59,61,73] have applied deep networks for hand or object segmentation. Bambach et al. [1] introduced a dataset that contains 48 egocentric video clips for people interacting with others in real environments with over 15,000 labeled hand instances. The authors also proposed CNNs to first detect hand bounding boxes and then use GrabCut [56] to segment the hands. Following up on the same dataset, Urooj and Borji [73] used the RefineNet-ResNet101 [38] to achieve the state-of-the-art hand segmentation performance at the time. To alleviate the generalization issue, Cai et al. [4] proposed the use of a Bayesian CNN to predict the model uncertainty and leveraged the common information of hand shapes to better adapt to an unseen domain. There are also some other dataset efforts regarding hand segmentation. Li et al. [34] proposed the Georgia Tech Egocentric Activity Datasets (GTEA), which includes 625 frames with two-hand labeling and 38 frames with binary labeling. Later, Li et al. [33] extended the dataset (EGTEA) with 1,046 frames with two-hand labels and 12,799 frames with binary masks. Lin et al. [36,37] also explored artificially composited hands with various backgrounds to scale up a large-scale synthetic dataset. Urooj et al. [73] recognized the constrained environment as one big limitation of existing datasets and collected an in-the-wild dataset by sampling frames from YouTube videos (EYTH). Though it is more diverse, it is relatively small with around 2,000 frames sampled from only 3 videos. Since frames are selected by simply sampling at a fixed rate, many frames are similar to each other in appearance. In addition to datasets with per-pixel labels, Shan et al. [58] labeled 100K video frames with bounding box labels for hands and interacting objects. More recently, Shan et al. [59] also proposed to learn hand and hand-held objects segmentation from motion using image and hand location as inputs.

Our work differs from previous works in two main aspects. While previous work mainly focus on egocentric hand segmentation, we take a step further to study not only hand segmentation but also interacting object segmentation. In addition, previous datasets were mainly focused on certain constrained scenes and limited activities. Our proposed dataset includes diverse daily activities with hundreds of human subjects. More detailed comparisons are shown in Sect. 3.

2.2   Hand-Object Interaction

Many works have studied hand-object interaction from different angles other than segmentation. One highly related direction is to model and estimate 3D hand joints [3,5,44,45,53,60,63,66,72,78–80,85] and mesh [26,27,46,51,55,75, 83], object pose [14,28,29,35,43,57,64,71,76,81], or both [6,17–19,69]. A line of works [2,7,21,23,67] have also attempted to generate hand pose conditioned on objects. Mostly related to our work, Hasson et al. [18] and Cao et al. [6] used segmentation masks of hands and interacting objects to compute 2D projection loss in order to optimize the 3D mesh reconstructions of hand-object pairs. However, the instance segmentation model [25] they used to pre-compute hand and object masks are pretrained on COCO [39], which is not tailored to egocentric hand-object segmentation, and thus heavy human intervention is often needed to fix or filter out wrongly predicted masks. Other directions of hand-object inter-action involve using hands as probes for object understanding [15], affordance hotspot reasoning [47,49], or even leveraging visual hand-object understanding for robotic learning [41,42,48]. Overall, we view our work as an orthogonal foun-dational tool for many of these vision tasks.

3 Dataset

Gathering Data from Multiple Sources. A big motivation of this work is that the existing datasets do not support researchers to train a model that generalizes well in the wild. Therefore, we collect data from multiple sources, including 7,458 frames from Ego4d [16], 2,121 frames from EPIC-KITCHENS [8], 806 frames from THU-READ [68], as well as 350 frames of our own collected indoor egocentric videos. This results in a total of 11,243 frames sparsely sampled from nearly 1,000 videos covering a wide range of daily activities in diverse scenarios. We manually select diverse and non-repetitive video frames from the sampled set that contain interesting hand-object interactions to label with per-pixel segments, as shown in Fig. 2. More details on video frame sampling are included in the supplementary materials.

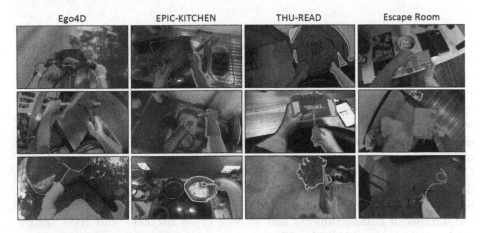

Fig. 2. A selection of images from multiple sources which we label with per-pixel hand and object segments. Color mapping: red → left hand, blue → right hand, green → object interacted by both hands, pink → object interacted by left hand, cyan → object interacted by right hand. (Color figure online)

Annotations. For every image in the dataset, we obtained the following per-pixel mask annotations if applicable: (a) left-hand; (b) right-hand; (c) left-hand object; (d) right-hand object; (e) two-hand object. For each type of interacting object, we also provide two levels of interaction: direct and indirect interaction. We define direct interaction between hand and object if the hand touches the objects, such as the blue, cyan, or pink masks in Fig. 2. Otherwise, we label the object as indirectly interacted with by the hand if the object is being indirectly interacted with, without touching, such as the light cyan masks in the third row of Fig. 2. In this work, we only study directly interacting objects, but we will release the data to support future research into indirect interacting object segmentation. Note that previous works define hand masks in two types: hand only [1,73] and hand with arm [33]. We think both types of labels are useful depending on the application, so we provide both types of hand mask labels for all images in our dataset, where one for hands and another one for the rest of the arms.

Table 1. Egocentric Hand-Object Segmentation Datasets Comparison. Unknown information is denoted with a dash "–". Compared to previous datasets, our proposed datasets cover relatively diverse scenes and activities with fine-grained segmentation labels of both hands and interating objects.

Datasets	Label	#Frames	#Hands	#Objects	Objects	Interaction	L/R Hand	#Subjects	#Activities
100-DOH [33]	Box.	100K	189.6K	110.1K	Yes	Yes	Yes	–	–
EGTEA [33]	Seg.	13,847	–	–	Yes	No	No	32	1
EgoHand [1]	Seg.	4,800	15,053	–	Yes	–	Yes	4	4
EYTH [73]	Seg.	1,290	2,600	–	No	No	No	–	–
Ours	Seg.	11,243	20,701	17,568	Yes	Yes	Yes	100+	300+

Comparison with Existing Datasets. In Table 1, we compare our proposed dataset with existing labeled datasets. 100-DOH [58] also provide a large volume of labelled images and objects, but its labels are at the bounding box level and not tailored towards egocentric images only. Although 100-DOH [58] has made a great effort to improve the generalization of hand-object bounding box detection, we think that having the segmentation prediction is particularly useful or necessary for many downstream vision applications, such as mesh reconstruction of hand-object interaction and seeing through the hands, as shown in Sect. 7. Compared to other segmentation datasets, one important characteristic of our dataset is that our images cover diverse activities and many human subjects. Since we do not have frame-level semantic labels, our conservative estimation of the number of human subjects and activity types are 300+ and 100+ respectively, according to the video IDs/names in the datasets [8,16,68]. Both the number of subjects and activities are orders of magnitude larger than previous segmentation datasets. In addition, unlike previous segmentation datasets, we are also the first to provide per-pixel mask labels for the interacting objects.

4   Hand-Object Contact Boundary

A key challenge of hand-object segmentation is the explicit understanding and modeling of the relationship between the hand and the interacting object. Segmenting the object purely based on appearance, as in traditional segmentation tasks, would not properly solve our problem. The reason is that the same object requires segmentation in certain frames but not in the others, depending on whether the hand is in contact with the object. To this end, we propose to explicitly model the interaction relationship between hand and object by introducing the notion of a dense contact boundary.

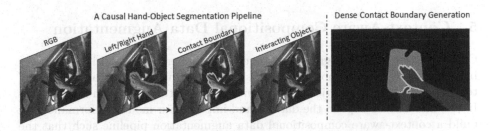

Fig. 3. Left: an overview of our causal hand-object segmentation pipeline. **Right**: a demo to show how dense contact boundary is defined.

Conceptually, the dense contact boundary is defined to be the contact region between the hand and the interacting object. In our implementation, we first dilate both the labeled hand and the object masks in an image, then find the overlapped region between the two dilated masks, and finally binarize the overlapped region as our pseudo-ground truth for contact boundary, as shown in the yellow region in Fig. 3. With such a pipeline, we automatically generate supervision on the contact boundary for all images, where we could train the network to make prediction for it with standard binary cross entropy loss.

The advantages of explicitly predicting a dense contact boundary for interacting object segmentation are: 1) the contact boundary could provide a cue as to whether there is an interacting object for a given hand mask; 2) it also provides a clearer hand-object separation cue to improve segmentation accuracy. Our experiments show that the contact boundary helps the segmentation model to achieve a higher averaged object mask mIoU, and more ablation studies are shown in Sect. 6.2. Other advantages of the contact boundary besides boosting segmentation performance include: 1) the contact boundary segmentation contains crucial information for many downstream tasks, such as activity recognition and 3D mesh modeling of hand and object; 2) it could also provide potential metrics for evaluating segmentation, specifically for object-hand segmentation during an interaction.

We experiment with one hand-object segmentation pipeline that uses dense interaction boundary as an intermediate stage output. We sequentially predict first the left/right hand, then the contact boundary, and finally the interacting object in three stages, as shown in the left of Fig. 3. In each stage, we concatenate the outputs from previous stages as additional inputs. For example, the left/right hand masks are concatenated with the RGB image as inputs to predict the contact boundary; and in the last stage, the RGB image, hand masks, and contact boundary masks are concatenated as inputs to predict the interacting object masks. Our model is built by sequentially stacking networks, which we tried both a convolutional architecture (ResNet-18 backbone [20] and HRNet head [74]) and a transformer architecture (Swin-L backbone [40] and UperNet head [77]). Note that we do not focus on the architecture and loss design in this work, and more training details are described in the supplementary materials.

5   Context-Aware Compositional Data Augmentation

Copying-and-pasting foreground instances at different locations into different background scenes has shown to be a simple and effective data augmentation technique for object detection and instance segmentation, as shown in [9,13,82]. In order to further expand the dataset and improve our model performance, we build a context-aware compositional data augmentation pipeline such that the new composite image has semantically consistent foreground (hand-object) and background context.

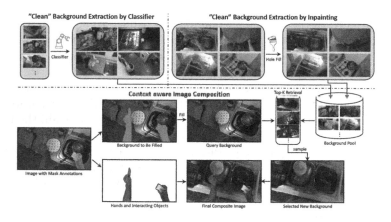

Fig. 4. An overview of our context-aware compositional data augmentation pipeline.

Our overall pipeline design is shown in Fig. 4. In the first step, we need to find the "clean" background scenes that do not contain any hands or interacting objects. The reason is that the image should only contain one egocentric viewer's hands and interacting objects after the composition. To this end, we propose two ways to generate "clean" background. The first is to build a simple binary classifier that finds the frames with no hands from a large pool of video frames, as shown in the top left of Fig. 4. The second way is to remove the existing hand-object using an image inpainting model [65] and the labeled segmentation masks, as shown in the top right of Fig. 4. Both approaches enable us generate a large pool of "clean" background candidates. On the other hand, when given an image with hand-object segmentation masks, we first inpaint the hand-object regions using the inpainting model [65] to generate the "clean" query background, and then use it to retrieve the top-K similar background scenes from the "clean" background candidate pool based on deep features extracted by [62]. Finally, multiple background scenes are sampled from the top-K retrieved background images, as shown in the bottom of Fig. 4. Overall, our designed context-aware image composition pipeline allows us to generate semantically consistent hand-object and context as much as needed. In the experiments, we show the effectiveness of our proposed data augmentation technique.

6   Experiments on Hand-Object Segmentation

In this section, we first make comparison studies with the existing datasets on hand segmentation, and then we discuss the benchmark performance of the hand-object segmentation with an ablation study. In order to evaluate the in-the-wild segmentation performance, we sparsely sampled 500 frames from 30 collected Youtube egocentric videos to label as our out-of-distribution test set. In the following segmentation experiments, all results are evaluated on this test set unless otherwise specified. All of our models are trained and evaluated using the MMSegmentation codebase[1].

6.1   Two-Hand Segmentation

Previous hand segmentation datasets have different definitions of hand labels, such as left/right hand [1], binary hand [73] or binary hand + arm [33]. Since our datasets provide all these types of hand labels, we compare individually with the previous datasets in their settings. For a fair comparison, we train the same ResNet-18 backbone [20] and HRNet head [74] on each dataset, select the best checkpoints based on the validation set, and finally compute the results on the same held-out test set.

Table 2. Left/Right hand segmentation.

Datasets	mIoU	mPrec	mRec	mF1
EgoHand [1]	10.68/33.28	43.61/43.20	12.39/59.16	19.30/49.93
100-DOH [58] + BoxInst [70]	36.30/37.51	50.06/61.63	56.91/48.94	53.27/54.55
Ours	76.29/77.00	83.39/87.06	89.97/86.95	86.55/87.00
+ CCDA	79.73/82.17	84.26/90.38	93.68/90.04	88.72/90.21

In the first type of labeling, EgoHand [1] labeled the left and right hands of people in egocentric activities. Similarly, 100-DOH [58] also labeled large-scale left/right hands but with only bounding box annotations. We compare with

---

[1] MMSegmentation github: https://github.com/open-mmlab/mmsegmentation.

Table 3. Binary hand segmentation.

Datasets	mIoU	mPrec	mRec	mF1
EgoHand [1]	56.51	76.33	68.52	72.22
100-DOH [58]+ BoxInst [70]	69.50	84.80	79.67	82.00
EYTH [73]	75.94	85.17	87.51	86.32
Ours	83.18	89.34	92.34	90.82
+ CCDA	85.45	90.11	94.3	92.15

Table 4. Binary hand + Arm segmentation.

Datasets	mIoU	mPrec	mRec	mF1
EGTEA [33]	33.26	38.24	71.87	49.92
Ours	92.46	96.67	95.50	96.08
+ CCDA	95.20	97.68	97.40	97.54

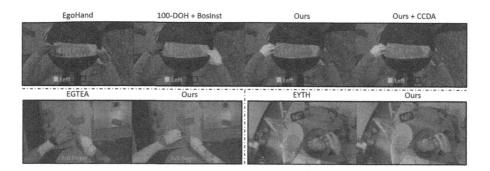

Fig. 5. A qualitative comparison between segmentation models trained on previous datasets and our proposed dataset. The **top row** shows the comparison with EgoHand and 100-DOH + BoxInst in Left/Right Hand segmentation, where red and cyan indicate left hand and right hand respectively. The **bottom left** shows the comparison with EGTEA on binary Hand + Arm segmentation. The **bottom right** shows the comparison with EYTH on binary Hand segmentation.

100-DOH by training a weakly supervised segmentation model, BoxInst [70], which learns to segment objects given bounding box annotation. To make the hand segmentation performance as good as possible for 100-DOH, we pre-trained BoxInst on 2000 frames sampled from EPIC-KITCHEN. As shown in Table 2, the model trained on our dataset significantly outperforms the model trained on EgoHand and 100-DOH with BoxInst. From the visual results, as shown in the first row of Fig. 5, we observe that models trained on EgoHand and 100-DOH often generate wrong mask categorical labels, which causes significantly lower mIoU (mean Intersection over Union) for left/right hand segmentation. When we binarize the predicted left/right hand masks of EgoHand and 100-DOH and evaluate them on the binary hand segmentation task, the performance gap bridges closer to us, as shown in Table 3. This again shows that mis-classification of left/right hand is indeed a major issue that causes low mIoU in Table 2 for the models trained on [1,58].

The other two datasets EYTH [73] and EGTEA [33] provide only binary mask labels for both hands without differentiating between left or right. EYTH [73] labeled only the hand region, and EGTEA [33] labeled both hand and arm regions. In Table 3 and Table 4, the quantitative results show that the model trained on our datasets also outperforms previous datasets by an obvious margin in both "hand" and "hand + arm" settings, and visual comparisons are shown in the bottom of Fig. 5. In all these hand segmentation settings, we observe that our context-aware compositional data augmentation (CCDA) consistently improves the hand segmentation performance quantitatively. More qualitative comparisons are included in the supplementary materials.

6.2 Hand and Interacting Object Segmentation

Since our dataset is the first to provide mask labels for interacting objects, we discuss the benchmark performance of hand-object segmentation with an ablation study in this section. In this task, we assign hands to left and right categories, and objects to three categories based on the interacting hand: left-hand object, right-hand object, and two-hand object. A naive solution is to train a segmentation network that decodes five channels of outputs in parallel, as shown in the 1st row of Table 5. However, this might not be ideal, since parallel decoding of outputs does not leverage any explicit understanding of the hand-object relationship, as discussed in Sect. 4. Thus, we propose to try to sequentially decode the hand first, and then use predicted left/right hand mask information to explicitly guide the interacting object segmentation, as shown in the 3rd row of Table 5. We also studied adding contact boundary (CB) as intermediate guide information, and found that it effectively boosts the object segmentation performance, as shown in the comparison between 3rd and 5th rows. More details about contact boundary are discussed in Sect. 4. Finally, we evaluated the effectiveness of our context-aware compositional data augmentation (CCDA) by integrating it on top of both parallel and sequential models. As shown in the comparison between rows $1^{st}, 3^{rd}, 5^{th}$ and rows $2^{nd}, 4^{th}, 6^{th}$, CCDA slightly improves the left/right hand segmentation and significantly boosts the object segmentation performance. We think the reasons are that compositional augmentation enables the network to learn the pixel grouping of objects more easily when placing them into many different background. More details are on how we choose the quantity of augmented images are are discussed in the supplementary materials. The qualitative results for dense contact boundary prediction and hand-object segmentation in diverse activities are shown in the supplemental materials.

Table 5. A quantitative ablation study on the hand-object segmentation.

Models	Left hand	Right hand	Left-hand object	Right-hand object	Two-hand object
Para. Decode	69.08	73.50	48.67	36.21	37.46
Para. Decode + CCDA	77.57	81.06	54.83	38.48	39.14
Seq. Decode	73.17	80.56	54.83	38.48	39.14
Seq. Decode + CCDA	87.70	88.79	58.32	40.18	46.24
Seq. Decode + CB	77.25	81.17	59.05	40.85	49.94
Seq. Decode + CB + CCDA	87.70	88.79	62.20	44.40	52.77

7   Applications

7.1   Boosting Hand State Classification and Activity Recognition

Understanding the hand state and recognizing types of activities in egocentric videos are important for human behavior analysis. Similarly to 100-DOH [58], we define hand states for both left and right hands as the following: (contact with) portable, (contact with) stationary, no-contact, self-contact, and not-exist. The goal of this task is to classify a correct state for each of the two hands of the egocentric viewer, where we use two classification heads to handle this. To this end, we labeled the hand states of 3,531 frames from EPIC-KITCHEN [8] dataset with diverse hand-object interaction. During training, we adopt 8:1:1 ratio to split train, val, and test sets. As shown in Table 6, by adding hand mask and hand-object masks into the input channel, a classifier with the same backbone [62] could effectively improve its classification performance compared to the baseline that uses RGB images only. For the video activity recognition, we used a subset data of EPIC-KITCHEN Action Recognition benchmark [8] as well as its evaluation protocol. With the SlowFast network [11], we show that by adding hand masks into training, the top-1 classification accuracy of "verbs/nouns" boosts from 23.95%/36.77% to 25.98%/37.04%. A visual illustration of these two tasks is shown in the supplemental.

Table 6. Quantitative results for left/right hand state classification.

Models	Accuracy	Precision	Recall	F1 score
Baseline	78.53%/74.29%	43.02%/37.70%	36.86%/30.60%	37.53%/32.02%
+ Hand mask	84.18%/83.33%	64.33%/66.16%	57.31%/56.34%	59.42%/59.00%
+ Hand & object mask	86.72%/83.33%	68.18%/69.04%	57.12%/56.08%	60.35%/59.64%

7.2   Improved 3D Mesh Reconstruction of Hand-Object Interaction

Mesh reconstruction of hand-object interaction is a useful but very challenging task. One mainstream approach [6,18] to solve this task is to jointly optimize the 3D scale, translation and rotation of a given 3D object model, as well as the MANO parameters [54] for hand. Such optimization process often relies on the estimated hand and object segmentation masks to compute the 3D mesh to 2D projection error. In previous works, researchers [6,18] leverage the 100-DOH's [58] detector to localize the hand and interacting object at bounding box level, and then use PointRend [25] pre-trained on COCO [39] to segment the hand and object masks. The interacting object mask is assigned by a heuristic that the object mask with highest confidence score is the one in interaction.

In this work, we integrate our robust hand-object segmentation model into the previous mesh reconstruction pipeline [18]. Since our hand-object segmentation could generalize better than the previous segmentation component, we enable the hand-object reconstruction generalize in more diverse scenarios with

Image with Hand-object Seg.	Mesh (Original View)	Mesh (Rotated View)	Image with Hand-object Seg.	Mesh (Original View)	Mesh (Rotated View)

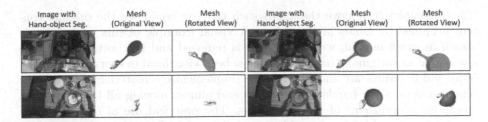

Fig. 6. Visual comparison between 3D mesh reconstruction of hand-object interaction. The **left** results are from the original code of [18], where they use 100-DOH [58] detector with PointRend [25] to compute hand-object masks. The **right** results are computed by integrating our hand-object segmentation into [18] for mesh optimization.

→ Time

Fig. 7. A qualitative demo to show the application of seeing through the hand in egocentric videos. This application is enabled by our robust per-frame hand segmentation together with the video inpainting model [12]. The top row are the frames with predicted hand segmentation masks, and the bottom row shows the "see through" frames at the corresponding timestamp. More video results are shown in the supplemental.

higher visual fidelity. As shown in the first row of Fig. 6, the previous segmentation pipeline oftentimes fails to segment the complete object, and thus the object was optimized into a wrong 3D pose, while our accurate hand-object segmentation enables the object mesh reconstruction to be more accurate. In the second row of Fig. 6, we observe that the previous segmentation pipeline sometimes completely misses the interacting object at the bounding box detection stage, and thus no segmentation and 3D mesh could be generated. In contrast, our pipeline provides higher recall on the object detection, and thus is able to recover the object mesh, as shown in the bottom right of Fig. 6.

7.3   Seeing Through the Hand in Egocentric Videos

Finally, in this work, we propose a new interesting application, where the goal is to see through the hand in egocentric videos. With our robust per-frame segmentation of hand masks, we use the recent flow-guided video inpainting algorithm

[12] to completely remove the hands such that we could see the original content occluded by hands in the videos. A visual example of this application is shown in supplemental, where the hand is removed and the bottles and fridge layers that are originally occluded can now been visualized in every video frame. More video results are included in the supplementary materials. In the egocentric videos, since hands are prevalent and almost moving all the time, they create large occlusions of visual contents. The practical use of our "hand see through" system is that we could potentially enable the vision system analyze more previously occluded information, for example, in the future AR system.

8   Conclusion

We created a fine-grained egocentric hand-object segmentation dataset and synthetic data augmentation method to 1) enable robustness against out-of-distribution domain change and 2) support downstream tasks. Our labeled dataset of 11,243 images contains both per-pixel segmentation labels of hand and interacting objects and dense contact boundaries. Our context-aware compositional data augmentation technique significantly improves segmentation performance, especially for interacting objects. We show that our robust hand-object segmentation model can serve as a foundational tool for several vision applications, including hand state classification, activity recognition, 3D mesh reconstruction of hand-object interaction, and seeing through the hand in egocentric videos.

Acknowledgment. This research is based on work supported by Toyota Research Institute and Adobe Gift Fund. The views and conclusions contained herein are those of the authors and should not be interpreted as representing the official policies, either expressed or implied, of the sponsors.

References

1. Bambach, S., Lee, S., Crandall, D.J., Yu, C.: Lending a hand: detecting hands and recognizing activities in complex egocentric interactions. In: Proceedings of the IEEE International Conference on Computer Vision, pp. 1949–1957 (2015)
2. Brahmbhatt, S., Handa, A., Hays, J., Fox, D.: ContactGrasp: functional multi-finger grasp synthesis from contact. In: 2019 IEEE/RSJ International Conference on Intelligent Robots and Systems (IROS), pp. 2386–2393. IEEE (2019)
3. Brahmbhatt, S., Tang, C., Twigg, C.D., Kemp, C.C., Hays, J.: ContactPose: a dataset of grasps with object contact and hand pose. In: Vedaldi, A., Bischof, H., Brox, T., Frahm, J.-M. (eds.) ECCV 2020. LNCS, vol. 12358, pp. 361–378. Springer, Cham (2020). https://doi.org/10.1007/978-3-030-58601-0_22
4. Cai, M., Lu, F., Sato, Y.: Generalizing hand segmentation in egocentric videos with uncertainty-guided model adaptation. In: Proceedings of the IEEE/CVF Conference on Computer Vision and Pattern Recognition, pp. 14392–14401 (2020)
5. Cai, Y., Ge, L., Cai, J., Yuan, J.: Weakly-supervised 3D hand pose estimation from monocular RGB images. In: Proceedings of the European Conference on Computer Vision (ECCV), pp. 666–682 (2018)

6. Cao, Z., Radosavovic, I., Kanazawa, A., Malik, J.: Reconstructing hand-object interactions in the wild. In: Proceedings of the IEEE/CVF International Conference on Computer Vision, pp. 12417–12426 (2021)
7. Corona, E., Pumarola, A., Alenya, G., Moreno-Noguer, F., Rogez, G.: GanHand: predicting human grasp affordances in multi-object scenes. In: Proceedings of the IEEE/CVF Conference on Computer Vision and Pattern Recognition, pp. 5031–5041 (2020)
8. Damen, D., et al.: Scaling egocentric vision: the epic-kitchens dataset. In: Proceedings of the European Conference on Computer Vision (ECCV), pp. 720–736 (2018)
9. Fang, H.S., Sun, J., Wang, R., Gou, M., Li, Y.L., Lu, C.: InstaBoost: boosting instance segmentation via probability map guided copy-pasting. In: Proceedings of the IEEE/CVF International Conference on Computer Vision, pp. 682–691 (2019)
10. Fathi, A., Ren, X., Rehg, J.M.: Learning to recognize objects in egocentric activities. In: CVPR 2011, pp. 3281–3288. IEEE (2011)
11. Feichtenhofer, C., Fan, H., Malik, J., He, K.: SlowFast networks for video recognition. In: Proceedings of the IEEE/CVF International Conference on Computer Vision, pp. 6202–6211 (2019)
12. Gao, C., Saraf, A., Huang, J.-B., Kopf, J.: Flow-edge guided video completion. In: Vedaldi, A., Bischof, H., Brox, T., Frahm, J.-M. (eds.) ECCV 2020. LNCS, vol. 12357, pp. 713–729. Springer, Cham (2020). https://doi.org/10.1007/978-3-030-58610-2_42
13. Ghiasi, G., et al.: Simple copy-paste is a strong data augmentation method for instance segmentation. In: Proceedings of the IEEE/CVF Conference on Computer Vision and Pattern Recognition, pp. 2918–2928 (2021)
14. Gkioxari, G., Malik, J., Johnson, J.: Mesh R-CNN. In: Proceedings of the IEEE/CVF International Conference on Computer Vision, pp. 9785–9795 (2019)
15. Goyal, M., Modi, S., Goyal, R., Gupta, S.: Human hands as probes for interactive object understanding. arXiv preprint arXiv:2112.09120 (2021)
16. Grauman, K., et al.: Ego4D: around the world in 3,000 hours of egocentric video. arXiv preprint arXiv:2110.07058 (2021)
17. Hasson, Y., Tekin, B., Bogo, F., Laptev, I., Pollefeys, M., Schmid, C.: Leveraging photometric consistency over time for sparsely supervised hand-object reconstruction. In: Proceedings of the IEEE/CVF Conference on Computer Vision and Pattern Recognition, pp. 571–580 (2020)
18. Hasson, Y., Varol, G., Schmid, C., Laptev, I.: Towards unconstrained joint hand-object reconstruction from RGB videos. In: 2021 International Conference on 3D Vision (3DV), pp. 659–668. IEEE (2021)
19. Hasson, Y., et al.: Learning joint reconstruction of hands and manipulated objects. In: Proceedings of the IEEE/CVF Conference on Computer Vision and Pattern Recognition, pp. 11807–11816 (2019)
20. He, K., Zhang, X., Ren, S., Sun, J.: Deep residual learning for image recognition. In: Proceedings of the IEEE Conference on Computer Vision and Pattern Recognition, pp. 770–778 (2016)
21. Jiang, H., Liu, S., Wang, J., Wang, X.: Hand-object contact consistency reasoning for human grasps generation. arXiv preprint arXiv:2104.03304 (2021)
22. Jones, M.J., Rehg, J.M.: Statistical color models with application to skin detection. Int. J. Comput. Vis. **46**(1), 81–96 (2002). https://doi.org/10.1023/A:1013200319198

23. Karunratanakul, K., Yang, J., Zhang, Y., Black, M.J., Muandet, K., Tang, S.: Grasping field: learning implicit representations for human grasps. In: 2020 International Conference on 3D Vision (3DV), pp. 333–344. IEEE (2020)
24. Kim, S., Chi, H.G.: First-person view hand segmentation of multi-modal hand activity video dataset. BMVC 2020 (2020)
25. Kirillov, A., Wu, Y., He, K., Girshick, R.: Pointrend: Image segmentation as rendering. In: Proceedings of the IEEE/CVF Conference on Computer Vision and Pattern Recognition, pp. 9799–9808 (2020)
26. Kulon, D., Guler, R.A., Kokkinos, I., Bronstein, M.M., Zafeiriou, S.: Weakly-supervised mesh-convolutional hand reconstruction in the wild. In: Proceedings of the IEEE/CVF Conference on Computer Vision and Pattern Recognition, pp. 4990–5000 (2020)
27. Kulon, D., Wang, H., Güler, R.A., Bronstein, M., Zafeiriou, S.: Single image 3D hand reconstruction with mesh convolutions. arXiv preprint arXiv:1905.01326 (2019)
28. Kundu, A., Li, Y., Rehg, J.M.: 3D-RCNN: instance-level 3D object reconstruction via render-and-compare. In: Proceedings of the IEEE Conference on Computer Vision and Pattern Recognition, pp. 3559–3568 (2018)
29. Kuo, W., Angelova, A., Lin, T.-Y., Dai, A.: Mask2CAD: 3D shape prediction by learning to segment and retrieve. In: Vedaldi, A., Bischof, H., Brox, T., Frahm, J.-M. (eds.) ECCV 2020. LNCS, vol. 12348, pp. 260–277. Springer, Cham (2020). https://doi.org/10.1007/978-3-030-58580-8_16
30. Lee, S., Bambach, S., Crandall, D.J., Franchak, J.M., Yu, C.: This hand is my hand: a probabilistic approach to hand disambiguation in egocentric video. In: Proceedings of the IEEE Conference on Computer Vision and Pattern Recognition Workshops, pp. 543–550 (2014)
31. Li, C., Kitani, K.M.: Model recommendation with virtual probes for egocentric hand detection. In: Proceedings of the IEEE International Conference on Computer Vision, pp. 2624–2631 (2013)
32. Li, C., Kitani, K.M.: Pixel-level hand detection in ego-centric videos. In: Proceedings of the IEEE Conference on Computer Vision and Pattern Recognition, pp. 3570–3577 (2013)
33. Li, Y., Liu, M., Rehg, J.M.: In the eye of beholder: joint learning of gaze and actions in first person video. In: Proceedings of the European Conference on Computer Vision (ECCV), pp. 619–635 (2018)
34. Li, Y., Ye, Z., Rehg, J.M.: Delving into egocentric actions. In: Proceedings of the IEEE Conference on Computer Vision and Pattern Recognition, pp. 287–295 (2015)
35. Lim, J.J., Pirsiavash, H., Torralba, A.: Parsing IKEA objects: fine pose estimation. In: Proceedings of the IEEE International Conference on Computer Vision, pp. 2992–2999 (2013)
36. Lin, F., Martinez, T.: Ego2Hands: a dataset for egocentric two-hand segmentation and detection. arXiv preprint arXiv:2011.07252 (2020)
37. Lin, F., Wilhelm, C., Martinez, T.: Two-hand global 3D pose estimation using monocular RGB. In: Proceedings of the IEEE/CVF Winter Conference on Applications of Computer Vision, pp. 2373–2381 (2021)
38. Lin, G., Milan, A., Shen, C., Reid, I.: RefineNet: multi-path refinement networks for high-resolution semantic segmentation. In: Proceedings of the IEEE Conference on Computer Vision and Pattern Recognition, pp. 1925–1934 (2017)

39. Lin, T.-Y., et al.: Microsoft COCO: common objects in context. In: Fleet, D., Pajdla, T., Schiele, B., Tuytelaars, T. (eds.) ECCV 2014. LNCS, vol. 8693, pp. 740–755. Springer, Cham (2014). https://doi.org/10.1007/978-3-319-10602-1_48

40. Liu, Z., et al.: Swin transformer: hierarchical vision transformer using shifted windows. In: Proceedings of the IEEE/CVF International Conference on Computer Vision, pp. 10012–10022 (2021)

41. Mandikal, P., Grauman, K.: DexVIP: learning dexterous grasping with human hand pose priors from video. In: 5th Annual Conference on Robot Learning (2021)

42. Mandikal, P., Grauman, K.: Learning dexterous grasping with object-centric visual affordances. In: IEEE International Conference on Robotics and Automation, ICRA 2021, 30 May–5 June 2021, Xi'an, China, pp. 6169–6176. IEEE (2021)

43. Michel, F., et al.: Global hypothesis generation for 6D object pose estimation. In: Proceedings of the IEEE Conference on Computer Vision and Pattern Recognition, pp. 462–471 (2017)

44. Moon, G., Chang, J.Y., Lee, K.M.: V2V-PoseNet: voxel-to-voxel prediction network for accurate 3D hand and human pose estimation from a single depth map. In: Proceedings of the IEEE Conference on Computer Vision and Pattern Recognition, pp. 5079–5088 (2018)

45. Mueller, F., et al.: GANerated hands for real-time 3D hand tracking from monocular RGB. In: Proceedings of the IEEE Conference on Computer Vision and Pattern Recognition, pp. 49–59 (2018)

46. Muller, L., Osman, A.A., Tang, S., Huang, C.H.P., Black, M.J.: On self-contact and human pose. In: Proceedings of the IEEE/CVF Conference on Computer Vision and Pattern Recognition, pp. 9990–9999 (2021)

47. Nagarajan, T., Feichtenhofer, C., Grauman, K.: Grounded human-object interaction hotspots from video. In: Proceedings of the IEEE/CVF International Conference on Computer Vision, pp. 8688–8697 (2019)

48. Nagarajan, T., Grauman, K.: Shaping embodied agent behavior with activity-context priors from egocentric video. Adv. Neural Inf. Process. Syst. 34, 29794–29805 (2021)

49. Nagarajan, T., Li, Y., Feichtenhofer, C., Grauman, K.: Ego-topo: environment affordances from egocentric video. In: Proceedings of the IEEE/CVF Conference on Computer Vision and Pattern Recognition, pp. 163–172 (2020)

50. Narasimhaswamy, S., Nguyen, T., Nguyen, M.H.: Detecting hands and recognizing physical contact in the wild. Adv. Neural Inf. Process. Syst. 33, 7841–7851 (2020)

51. Pavlakos, G., et al.: Expressive body capture: 3D hands, face, and body from a single image. In: Proceedings of the IEEE/CVF Conference on Computer Vision and Pattern Recognition, pp. 10975–10985 (2019)

52. Ren, X., Gu, C.: Figure-ground segmentation improves handled object recognition in egocentric video. In: 2010 IEEE Computer Society Conference on Computer Vision and Pattern Recognition, pp. 3137–3144. IEEE (2010)

53. Romero, J., Kjellström, H., Kragic, D.: Hands in action: real-time 3D reconstruction of hands in interaction with objects. In: 2010 IEEE International Conference on Robotics and Automation, pp. 458–463. IEEE (2010)

54. Romero, J., Tzionas, D., Black, M.J.: Embodied hands: modeling and capturing hands and bodies together. ACM Trans. Graph. (Proc. SIGGRAPH Asia) 36(6) (2017)

55. Rong, Y., Shiratori, T., Joo, H.: FrankMocap: fast monocular 3D hand and body motion capture by regression and integration. arXiv preprint arXiv:2008.08324 (2020)

56. Rother, C., Kolmogorov, V., Blake, A.: "GrabCut" interactive foreground extraction using iterated graph cuts. ACM Trans. Graph. (TOG) **23**(3), 309–314 (2004)
57. Sahasrabudhe, M., Shu, Z., Bartrum, E., Alp Guler, R., Samaras, D., Kokkinos, I.: Lifting autoencoders: unsupervised learning of a fully-disentangled 3D morphable model using deep non-rigid structure from motion. In: Proceedings of the IEEE/CVF International Conference on Computer Vision Workshops (2019)
58. Shan, D., Geng, J., Shu, M., Fouhey, D.F.: Understanding human hands in contact at internet scale. In: Proceedings of the IEEE/CVF Conference on Computer Vision and Pattern Recognition, pp. 9869–9878 (2020)
59. Shan, D., Higgins, R., Fouhey, D.: COHESIV: contrastive object and hand embedding segmentation in video. Adv. Neural Inf. Process. Syst. **34**, 5898–5909 (2021)
60. Sharp, T., et al.: Accurate, robust, and flexible real-time hand tracking. In: Proceedings of the 33rd Annual ACM Conference on Human Factors in Computing Systems, pp. 3633–3642 (2015)
61. Shilkrot, R., Narasimhaswamy, S., Vazir, S., Hoai, M.: WorkingHands: a hand-tool assembly dataset for image segmentation and activity mining. In: BMVC, p. 258 (2019)
62. Simonyan, K., Zisserman, A.: Very deep convolutional networks for large-scale image recognition. arXiv preprint arXiv:1409.1556 (2014)
63. Sridhar, S., Oulasvirta, A., Theobalt, C.: Interactive markerless articulated hand motion tracking using RGB and depth data. In: Proceedings of the IEEE International Conference on Computer Vision, pp. 2456–2463 (2013)
64. Sun, X., et al.: Pix3D: dataset and methods for single-image 3D shape modeling. In: Proceedings of the IEEE Conference on Computer Vision and Pattern Recognition, pp. 2974–2983 (2018)
65. Suvorov, R., et al.: Resolution-robust large mask inpainting with Fourier convolutions. In: Proceedings of the IEEE/CVF Winter Conference on Applications of Computer Vision, pp. 2149–2159 (2022)
66. Tagliasacchi, A., Schröder, M., Tkach, A., Bouaziz, S., Botsch, M., Pauly, M.: Robust articulated-ICP for real-time hand tracking. In: Computer Graphics Forum, vol. 34, pp. 101–114. Wiley Online Library (2015)
67. Taheri, O., Ghorbani, N., Black, M.J., Tzionas, D.: GRAB: a dataset of whole-body human grasping of objects. In: Vedaldi, A., Bischof, H., Brox, T., Frahm, J.-M. (eds.) ECCV 2020. LNCS, vol. 12349, pp. 581–600. Springer, Cham (2020). https://doi.org/10.1007/978-3-030-58548-8_34
68. Tang, Y., Tian, Y., Lu, J., Feng, J., Zhou, J.: Action recognition in RGB-D egocentric videos. In: 2017 IEEE International Conference on Image Processing (ICIP), pp. 3410–3414. IEEE (2017)
69. Tekin, B., Bogo, F., Pollefeys, M.: H+O: unified egocentric recognition of 3D hand-object poses and interactions. In: Proceedings of the IEEE/CVF Conference on Computer Vision and Pattern Recognition, pp. 4511–4520 (2019)
70. Tian, Z., Shen, C., Wang, X., Chen, H.: BoxInst: high-performance instance segmentation with box annotations. In: Proceedings of the IEEE/CVF Conference on Computer Vision and Pattern Recognition, pp. 5443–5452 (2021)
71. Tulsiani, S., Gupta, S., Fouhey, D.F., Efros, A.A., Malik, J.: Factoring shape, pose, and layout from the 2D image of a 3D scene. In: Proceedings of the IEEE Conference on Computer Vision and Pattern Recognition, pp. 302–310 (2018)
72. Tzionas, D., Ballan, L., Srikantha, A., Aponte, P., Pollefeys, M., Gall, J.: Capturing hands in action using discriminative salient points and physics simulation. Int. J. Comput. Vis. **118**(2), 172–193 (2016). https://doi.org/10.1007/s11263-016-0895-4

73. Urooj, A., Borji, A.: Analysis of hand segmentation in the wild. In: Proceedings of the IEEE Conference on Computer Vision and Pattern Recognition, pp. 4710–4719 (2018)

74. Wang, J., et al.: Deep high-resolution representation learning for visual recognition. IEEE Trans. Pattern Anal. Mach. Intell. **43**(10), 3349–3364 (2020)

75. Xiang, D., Joo, H., Sheikh, Y.: Monocular total capture: posing face, body, and hands in the wild. In: Proceedings of the IEEE/CVF Conference on Computer Vision and Pattern Recognition, pp. 10965–10974 (2019)

76. Xiang, Y., Schmidt, T., Narayanan, V., Fox, D.: PoseCNN: a convolutional neural network for 6D object pose estimation in cluttered scenes. arXiv preprint arXiv:1711.00199 (2017)

77. Xiao, T., Liu, Y., Zhou, B., Jiang, Y., Sun, J.: Unified perceptual parsing for scene understanding. In: Proceedings of the European Conference on Computer Vision (ECCV), pp. 418–434 (2018)

78. Yang, L., Yao, A.: Disentangling latent hands for image synthesis and pose estimation. In: Proceedings of the IEEE/CVF Conference on Computer Vision and Pattern Recognition, pp. 9877–9886 (2019)

79. Ye, Q., Yuan, S., Kim, T.-K.: Spatial attention deep net with partial PSO for hierarchical hybrid hand pose estimation. In: Leibe, B., Matas, J., Sebe, N., Welling, M. (eds.) ECCV 2016. LNCS, vol. 9912, pp. 346–361. Springer, Cham (2016). https://doi.org/10.1007/978-3-319-46484-8_21

80. Yuan, S., et al.: Depth-based 3d hand pose estimation: from current achievements to future goals. In: Proceedings of the IEEE Conference on Computer Vision and Pattern Recognition, pp. 2636–2645 (2018)

81. Zhang, J.Y., Pepose, S., Joo, H., Ramanan, D., Malik, J., Kanazawa, A.: Perceiving 3D human-object spatial arrangements from a single image in the wild. In: Vedaldi, A., Bischof, H., Brox, T., Frahm, J.-M. (eds.) ECCV 2020. LNCS, vol. 12357, pp. 34–51. Springer, Cham (2020). https://doi.org/10.1007/978-3-030-58610-2_3

82. Zhang, L., Wen, T., Min, J., Wang, J., Han, D., Shi, J.: Learning object placement by inpainting for compositional data augmentation. In: Vedaldi, A., Bischof, H., Brox, T., Frahm, J.-M. (eds.) ECCV 2020. LNCS, vol. 12358, pp. 566–581. Springer, Cham (2020). https://doi.org/10.1007/978-3-030-58601-0_34

83. Zhou, Y., Habermann, M., Xu, W., Habibie, I., Theobalt, C., Xu, F.: Monocular real-time hand shape and motion capture using multi-modal data. In: Proceedings of the IEEE/CVF Conference on Computer Vision and Pattern Recognition, pp. 5346–5355 (2020)

84. Zhu, X., Jia, X., Wong, K.-Y.K.: Pixel-level hand detection with shape-aware structured forests. In: Cremers, D., Reid, I., Saito, H., Yang, M.-H. (eds.) ACCV 2014. LNCS, vol. 9006, pp. 64–78. Springer, Cham (2015). https://doi.org/10.1007/978-3-319-16817-3_5

85. Zimmermann, C., Brox, T.: Learning to estimate 3D hand pose from single RGB images. In: Proceedings of the IEEE International Conference on Computer Vision, pp. 4903–4911 (2017)

Perceptual Artifacts Localization
for Inpainting

Lingzhi Zhang[1]([✉]), Yuqian Zhou[2], Connelly Barnes[2], Sohrab Amirghodsi[2], Zhe Lin[2], Eli Shechtman[2], and Jianbo Shi[1]

[1] University of Pennsylvania, Philadelphia, USA
zlz@seas.upenn.edu
[2] Adobe Research, Cambridge, USA

Abstract. Image inpainting is an essential task for multiple practical applications like object removal and image editing. Deep GAN-based models greatly improve the inpainting performance in structures and textures within the hole, but might also generate unexpected artifacts like broken structures or color blobs. Users perceive these artifacts to judge the effectiveness of inpainting models, and retouch these imperfect areas to inpaint again in a typical retouching workflow. Inspired by this workflow, we propose a new learning task of automatic segmentation of inpainting perceptual artifacts, and apply the model for inpainting model evaluation and iterative refinement. Specifically, we first construct a new inpainting artifacts dataset by manually annotating perceptual artifacts in the results of state-of-the-art inpainting models. Then we train advanced segmentation networks on this dataset to reliably localize inpainting artifacts within inpainted images. Second, we propose a new interpretable evaluation metric called Perceptual Artifact Ratio (PAR), which is the ratio of objectionable inpainted regions to the entire inpainted area. PAR demonstrates a strong correlation with real user preference. Finally, we further apply the generated masks for iterative image inpainting by combining our approach with multiple recent inpainting methods. Extensive experiments demonstrate the consistent decrease of artifact regions and inpainting quality improvement across the different methods. Dataset and code are available at: https://github.com/owenzlz/PAL4Inpaint

1 Introduction

Deep GAN-based image synthesis methods have been continuously improving image inpainting performance [4,30,36,41,64,66,68,73,74] for practical applications like object removal and image editing. Due to the ill-posed nature of image inpainting tasks, when encountering large holes or complex structures [41] within the hole, image inpainting becomes extremely challenging. Along

---

Supplementary Information The online version contains supplementary material available at https://doi.org/10.1007/978-3-031-19818-2_9.

with almost all state-of-the-art algorithms, inpainting artifacts tend to appear in the generated images. Those artifacts mostly include broken structures or color bleeding in the traditional patch synthesis methods [2], imperfect structures like disconnected or distorted lines, GAN-based generation artifacts or color blobs. In typical retouching workflows, users tend to judge the inpainting performance by those artifacts, and fix them by drawing masks on those regions and re-runing the automatic inpainting tools. Therefore, localizing and segmenting those artifacts is intuitively and naturally beneficial for inpainting algorithm evaluation and performance improvement.

Intuitively, finding more or larger artifacts within the hole area indicates a worse inpainting performance. Traditionally, image inpainting is regarded as an image reconstruction and restoration problem, and commonly-used metrics like PSNR, MSE and LPIPS [69] are utilized to compare the inpainted result to the original image in terms of content or pixel similarity. However, in many cases, image inpainting is used for foreground object removal [9,10]. Users prefer a visually plausible background generation rather than a faithful foreground reconstruction. Other quantitative metrics like Frechet Incept Distance (FID) [17,35] and Paired/Unpaired Inception Discriminative Score (P/U-IDS) [74] are computed on the entire images over large evaluation datasets. We are lacking in an intuitive metric which is more interpretable, operates on localized hole regions, and supports single result evaluation. Therefore, an automatic and reliable artifacts segmentation network may fill the gap.

In practical inpainting applications, users may choose to manually fix those artifacts by re-masking perceptually bad regions and re-running the models. Intuitively, after a couple of iterations, inpainting results are expected to be largely improved compared with the initial ones. Iterative hole filling has been studied in deep learning pipelines [13,34,66], and is shown to outperform one-pass inpainting. But the masks used in each iteration are either unreliable ones [66] learned with image reconstruction loss or predefined eroded masks [13,34]. Hence, an automatic artifacts segmentation network can effectively detect the perceptual artifacts in each iteration, and make the iterative filling run in a more efficient and effective way.

Although these inpainting artifacts are easily identifiable by humans, very few studies [66] have developed models to automatically detect and localize these artifacts in inpainting results. Researchers have studied identifying manipulated or synthesized images [5,8,32,48,50,50,71], edited image regions [5], or the entire inpainted image regions [22,23,53]. However, automatic localization of those artifacts within the inpainted holes was seldom discussed. This is mainly because a representative and well-organized dataset consisting of image inpainting results and artifact annotations is not yet available. Using the knowledge and expertise of professional photographers, deep networks can learn to efficiently detect and segment these artifacts.

In this paper, inspired by a typical user workflow when using inpainting tools, we assume that automatic perceptual artifacts segmentation for image inpainting will potentially benefit algorithm evaluation and boost inpainting performance. To verify our hypothesis, we collect inpainting results generated by multiple state-of-the-art deep inpainting models and annotate pixel-wise artifacts with a team of human professionals, and benchmark the dataset using advanced segmentation networks. Our proposed artifacts localization network outputs a binary mask highlighting the artifacts region. This mask can be used to: (1) compute the occupation ratio over the hole mask to evaluate and compare different inpainting algorithms on single test image without ground truth, and (2) achieve iterative filling to progressively improve inpainting performance. In summary, our contributions are in three folds (Fig. 1):

- We study the importance of a novel task, inpainting artifacts segmentation. Given its strengths in inpainting evaluation and result refinement, we construct a dataset consisting of 4,795 inpainting results with per-pixel perceptual artifacts annotations. We further benchmark the dataset using multiple segmentation network structures and analyze the human subjective factors in detail. Extensive experiments demonstrate its robustness on state-of-the-art inpainting models.
- We present the Perceptual Artifact Ratio (PAR) calculated from the artifact area detected inside the hole. PAR is an interpretable, intuitive, simple yet effective evaluation metric for comparing inpainting algorithms on a single image without ground truth. Our metric makes it possible to automatically evaluate object removal performance. Our user study also shows that PAR correlates more strongly with real user preferences than other metrics.
- We applied the artifacts segmentation network to iterative filling pipeline. After each iteration, we visualize that the detected artifact regions are consistently shrinking for all the tested inpainting models, and the results are refined with better structures and colors. Another user study suggests that iterative filling using our proposed artifacts masks will likely not degrade the inpainting performance and in many cases improve it.

2   Related Work

2.1   Image Inpainting

Classical image inpainting methods include diffusion-based methods [1,3] that propagate information from the boundary inwards to fill the hole, and patch-based methods [2] that search for the reference region to fill the hole. On the rise of deep learning, researchers proposed deep models to improve the inpainting performance from diverse angles, such as attention mechanism [19,28,29,41,45, 55,62,63], loss function and discriminator design [19,58,59,62,65], progressive [13,24,24,25,66,67] or multiscale [52,57,60,65] architectures, use of intermediate guide respresentation [12,26,33,38,39,47,51,56,62], and multimodal plausible outputs [4,14,30,36,64,68,73,75]. Among these works, ProFill [66], CoMod-GAN [74], and LaMa [41] are the most recent leading models. ProFill [66] proposed to implicitly learn a confidence map that guides the generator to iteratively

Image with Hole | Original Fill with Artifacts Localization | Iterative Fill | Image with Hole | Original Fill with Artifacts Localization | Iterative Fill

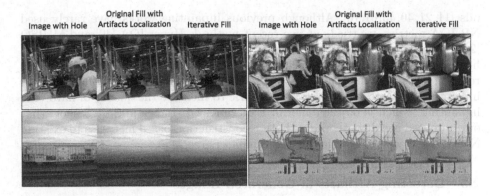

Fig. 1. Visual examples to show that our segmentation network can reliably localize the perceptual artifacts region, as indicated by the pink boundary in the second columns. Given the artifacts localization, we enable the inpainting model [40] iteratively fill on the artifacts region to obtain better inpainting quality, as shown in the third columns.

fill the hole, as well as a attention-guided refinement module to upsample the output. CoMod-GAN [74] leveraged the StyleGAN architecture [20] to conditionally synthesize filled region, where their filled content could be creative and not necessarily existed in the context. Finally, LaMa [40] integrated the fast Fourier convolution [6] to effectively capture global contextual information, and set the new state-of-the-arts. The goal of our work is to detect and localize the perceptual artifacts in the filled images independent of the inpainting models, and thus is in an orthogonal direction to these previous inpainting works.

2.2   Image Inpainting Quality Assessment

There are two types of commonly used metrics for image inpainting. The first quantifies the performance for a whole dataset of generated images. These metrics include Frechet Incept Distance (FID) [17,35] and Paired/Unpaired Inception Discriminative Score (P/U-IDS) [74], which measure the distance between the distribution of generated and real images using the deep Inception features [42]. For single image quality assessment, previous works often treat inpainting as a reconstruction task and thus compare the filled image with the original image using the reconstruction metrics, such as MSE, SSIM, PSNR or LPIPS [69]. This is reasonable only when holes are sampled on the background region. When holes largely overlap with or totally cover a foreground object, the current models would mostly fill the hole using background pixels, which is totally irrelevant to the original content. In these cases, the filled region of object removal could look natural and realistic, but could be totally different from the original object. Thus, reconstruction metric would no longer be a proper metric. Other potential assessments for measuring single image inpainting quality of object removal are No-Reference Image Quality Assessment (NR-IQA) methods [11,21,40,43,61,70]. Although previous inpainting works have rarely used NR-IQA metrics, we tried out two recent methods Hyper-IQA [40] and MUISQ [21], and found that MUISQ [21] has a relatively reasonable correlation with human perception compared to the other existing metrics for measureing object removal inpainting quality. In this work, we aim to use the area size of localized inpainting artifacts as a no-reference metric to measure the quality of hard case inpainting, which is object removal. Experimental studies in Sect. 5 show that our proposed metric outperforms both reconstruction-based metrics and existing NR-IQA in terms of correlation with human perception.

2.3 Detecting Artifacts in Generated Images

Other related works include detecting the generated/fake images and localizing the manipulated region in the image. One line of works [5,8,32,48,50,71] have studied training a binary classifier to classify the generated images, and Wang et al. [50] has shown surprising generalization on diverse and unseen model outputs by detecting the common artifacts in CNNs/GANs. Chai et al. [5] proposed patch-based classifier to localize the region that causes the fake image detectable. Another line of works [7,18,37,49] have proposed techniques to detect general image manipulations, such JPEG and resampling, other than GANs. In the image inpainting domain, Li et al. [23] first proposed to use the high-pass filter CNNs to detect the inpainting region given the filled image. Later, Wu et al. [53] and Li et al. [22] further improves the generalization of the mask detection to diverse inpainting models by proposing novel architecture or explicitly process high-frequency noise residual. Although all these works are related to us, a fundamental difference is that we aim to detect the perceptual artifacts that are judged by humans rather than simply detecting high-frequency noise/artifacts in the generated images. More specifically, in the inpainting context, our system detects the perceptual artifact region rather than the whole mask region, where perceptual artifact region is often a small subset of the mask. Thus, our work is essentially a different task from [22,23,53].

3   Dataset Labeling and Statistics

In order to train a system that can detect the perceptual artifacts in the inpainted images, we build a dataset that consists of 4,795 images with per-pixel perceptual artifacts labels from humans. We use three leading inpainting models ProFill [66], CoMod-GAN [74], and LaMa [41] to generate images to label. A labeling interface and a few examples of the labeled images are shown in the left and right of Fig. 5, respectively. During labeling, we provide the users a filled image without showing the original image, and ask users to label regions with perceptual artifacts on their tablets. We intentionally do not include the original images in the interface, since otherwise users might have bias to compare everything with the original content in the hole. As we have discussed in Sect. 2.1, the filled image could look natural and realistic, even though it's very different from the original image. We

Fig. 2. The left is an illustration of interface that users label on the inpainted image, where the blue bounding box indicates the region the users should inspect. On the right, we show a few examples of bad fill labels from users, where the blue and pink boundaries indicate the holes and user labels, respectively. (Color figure online)

also put dilated bounding box around the hole region to help users more easily find the labeling region and focus on it. We intentionally do not indicate the hole mask in the image, so that the workers do not have any bias labeling around the hole boundary, and thus can purely make judgement based on the perceptual quality. In this case, since workers do not know where the hole is, their labeling might go over the hole boundary. However, this is not an issue, as we simply intersect the hole masks with the human labels as a post-process. In addition, we provide a duplicate of the filled image in the interface, so that when workers can see the unbrushed filled image as reference during brushing (Fig. 2).

From the labeling perspective, one fundamental challenge is that this task is highly subjective compared to traditional segmentation, since different people may have different opinions or standards to judge perceptual artifacts. Therefore, in order to standardize the labeling as much as possible, we recruit and train a professional team for this task. We use two rounds of checks to avoid "missed labeled" or "overly labeled" regions. In the first round, the professional workers cross check the results with each other. In the second round, a single human expert checks through all the labels. On average, approximately 10% of labels have been rectified during the checking process. In addition, in order to generate high-quality labels, we recruited five human experts with photography or design background in our team to label these images. Since these workers are heavy users of image editing tools, i.e. Photoshop, their labeling criterion could better reflect the common unsatisfactory/retouch regions in the hole filling process.

Among these 4,795 images, there are 832 images that have nearly perfect fills, and thus workers did not label anything on these images. Although these images do not have segmentation labels, adding them into training could effectively help the network avoid predicting false positives. In terms of size of the labeling region, we found that the averaged ratio of "perceptual artifacts region/hole mask region" is 29.67%. This number once again shows that detecting perceptual

Table 1. An ablation study of the segmentation model, and human performance.

Models	IoU	Precision	Recall	Fscore
ResNet-50 backbone [16] + HRNet head [46]	41.35	58.45	58.56	58.51
Swin-B backbone [31] + Uper head [54]	44.20	63.01	59.69	61.30
ResNet-50 backbone [16] + PSPNet head [72]	46.04	59.78	66.71	63.05
- Perfect Filled Images	43.83	64.92	57.43	60.94
- Pretrained Weights	44.93	66.22	58.29	62.00
+ Hole Mask	45.96	66.07	60.16	62.98
+ Pseudo Pretraining	46.44	62.01	64.91	63.43
+ Pseudo Pretraining & Real Images	**46.77**	59.59	**68.49**	**63.73**
Human Subject A	45.60	**75.07**	53.73	62.64
Human Subject B	42.21	60.40	58.36	59.36
Human Subject C	36.85	61.47	47.93	53.86

artifacts is fundamentally different from detecting the hole mask in [22,23,53]. We plan to release our dataset to the community for future research.

4   Perceptual Artifacts Segmentation

In this section, we discuss the details of our segmentation model along with extensive ablation studies. During training, we used "8:1:1" ratio to randomly split the train/val/test set. In total, we have 3,836 training images, 480 validation images, and 479 test images. In each model training, we use the validation set to select the best checkpoint, and evaluate the performance on the test set. All of our models are trained and evaluated using the MMSegmentation codebase[1].

4.1   Ablation Studies

In the ablation study, we first tried out a few advance segmentation backbones/heads, such as HRNet [46] head, PSPNet [72] head, ResNet-50 backbone [16], and Swin Transformer [31] backbone, as shown in the top 3 rows (excluding the header) of Table 1. However, we do not observe obvious improvement when using the more complex backbones or heads for our task, after several trials of training comparison. We think a major potential reason is that our segmentation performance of the simpler backbone [16] and head [72] is nearly saturated given the highly subjective labels, and thus simply adding capacity or complexity of backbone does not improves much. This is discussed more in details when we compare with human performance in Sect. 4.2. Thus, we chose ResNet-50 backbone [16] + PSPNet head [72] as our base network, due to its simplicity and efficiency. The rest of ablation studies all shared the same base network for fair comparison, and thus the results should be compared with 3rd row (Fig. 3).

---

[1] MMSegmentation github: https://github.com/open-mmlab/mmsegmentation.

Besides the network backbones, we also studied other aspects that might potentially affect the segmentation performance. As we mentioned in Sect. 3, 832 images in the labeled dataset have almost perfect fill and thus have no mask labeling, and thus we wonder whether having these images in the training would be helpful. In the 4th row, we can see that the model trained without using these images indeed has worse performance, which concludes that adding perfect fills to training is important. All of our models starts training based on the checkpoints pretrained on ADE20K [76], and we show that the performance also decreases obviously without pretrained weights, as shown in the 5th row. Another intuitive thing is to concatenate the hole mask in the input, as it could theoretically help the network quickly localize the potential artifacts region. However, as shown in the 6th row, our experiments show that adding the mask into input channel does not actually boost the segmentation performance, and thus we decide not to use it for the simplicity purpose.

We also studied the possibility of generating pseudo labels on large scale unlabeled images for the pretraining purpose. Inspired by BoxInst [44], which used bounding box masks as weak supervision to train instance segmentation, we aim to find some similar "enlarged" masks covering the artifacts region as our pseudo labels. Initially, we tried using the hole mask as weak supervision, but realize that the network quickly overfits on the high-frequency artifacts on the hole boundary, which is not useful for our purpose. To this end, we used a pretrained artifacts segmentation network to generate artifacts mask regions on 100K unlabeled images, and then enlarged the segmented masks by some random dilation iterations to cover the perceptual artifacts region. The results in the 7th row show that such pretraining strategy slight improves the performance. Finally,

Fig. 3. Qualitative results of the predicted bad fill segmentation on six different inpainting model outputs. The pink and blue boundaries indicate the predicted bad fill region and the hole region, respectively. Please feel free to zoom in to see the details. (Color figure online)

we also tried adding the same quantity of real images into training, where the masks are empty for these real images. The 8th row shows that this is also useful to further boost the performance.

4.2   Analysis on Human Perceptual Judgement

As mentioned in Sect. 3, the labeled dataset from our labeling team has been carefully checked and thus have relatively high quality. In order to better understand how subjective human judgements are and the human performance bound, we ask three more human subjects to label on the 479 test images. Then, we compare the labels from these three human subjects with the previous labels from the labeling team, which are shown in the last three rows of Table 1. Regarding the three workers' background, human subject A has worked on this task before but not these images, and human subjects B & C have never worked on this task before but are taught by the labeling team with a bunch of labeled examples. Thus, human subject A should theoretically have better understanding of the task as well as the labeling criterion of the labeling team, compared to the other two subjects. All of them have photography or design background.

Interestingly, the results show that our segmentation model reaches and even surpasses the best human subject on all metrics except for precision. This infers that our model actually learns a better understanding of averaged judgement criterion of the labeling team, compared to each individual human. On the other hand, these results also indicate that humans have very subjective opinions on the labeling the artifacts regions, as the quantitative scores deviate obviously from each other. A visual illustration of different people's labels on the same filled image is shown in Fig. 4, and we include more examples like this in the supplemental. Since our segmentation performance surpasses the human performance, this indicates that our segmentation model reaches to a near saturation point for this highly subjective segmentation task. This might also explain why more complex backbone [31] or other tweaks of data or training do not provide significant performance improvement, as we observed in the ablation study.

Fig. 4. A visual comparison between labels from multiple human subjects on the same filled image. Our segmentation result is shown in the last column.

5 Evaluating Inpainting Quality for Object Removal

5.1 Motivation

It has been widely discussed [62,63] that image inpainting lacks good evaluation metrics, especially for single image quality assessment. Previous works mostly treat image inpainting as a kind of restoration task, so the reconstruction metrics such as MSE, SSIM, PSNR, and LPIPS [69], are often used to quantify the similarity between the filled image and the original image. Thinking carefully, we realize that reconstruction metrics might reasonably measure inpainting performance only when the holes are not very large and on the background region. When the holes largely overlap with or cover the foreground objects, most inpainitng algorithms would fill the hole regions by using the background context, where the object is oftentimes completely removed from the image. In these scenarios, reconstruction metrics are no longer proper metrics to gauge the inpainting quality, since the filled region could be totally irrelevant to the original pixels inside the hole. As shown in Fig. 2, when removing the person from the image, output A is visually more plausible than output B, but somehow all the existing reconstruction metrics make opposite judgement. Embarrassingly, object removal is arguably the most frequently used applicable scenario for inpainting algorithms. Thus, it means we really lack good metric for assessing inpainting quality in this scenario. This motivates us to think if the perceptual artifacts localization could be used as a no-reference metric to evaluate inpainting quality in the object removal scenario.

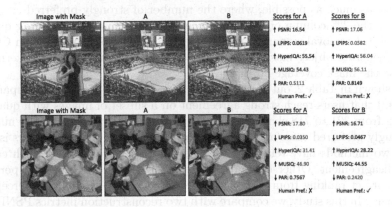

Fig. 5. A visual illustration of filled outputs by two inpainting models [33,66], with the corresponding metric scores. The red scores indicate the preferred choice according to each metric.

5.2 Metric Definition

Since our segmentation model could generalize reasonably well to diverse and unseen inpainting methods, we start to wonder whether the size of the detected artifacts region can be used as a metric to assess the inpainting quality. Basically, we assume that an image with good inpainting quality should have relatively smaller the perceptual artifacts region, and vice versa. We name this metric as Perceptual Artifacts Ratio (PAR), which is the ratio of "size of the perceptual artifact region/size of the input hole". The metric computation procedure is that we first run the segmentation model on the filled image and then compute PAR for any filled images, without the need of using the original image. During the quality comparison between different inpainting models, we simply evaluate which inpainting ouptuts have smaller artifacts region among the comparisons.

5.3   Correlation with Human Perception

In order to evaluate how well our PAR metric correlates with human perception, we collected user preferences on the filled images between four pairs of inpainting methods. Among these user comparisons, two pairs of comparisons happen between two strong inpainting models, as shown in the first two rows of Table 2, and another two pairs are between one strong and one relatively weak model, as shown in the bottom two rows of Table 2. In each pair of methods comparison, we show users two filled images with randomized order, and ask users to pick the preferred image out of the two options. The user studies were conducted on Amazon Mechanical Turk (AMT), where we asked five users to vote on each image. Finally, we consider that one filled image is strongly preferred than the other, only if 4 out of 5 users reach an agreement. In this study, we only used the strongly preferred image pairs as human preference ground truth to reduce the noise as much as possible, where the number of strongly preferred cases are shown in the 2nd column of Table 2. Since we are evaluating inpainting quality in the object removal scenarios, we use Mask R-CNN [15] pretrained on COCO [27] to generate object masks, and dilate three iterations with 5×5 kernel to increase the mask coverage on the object.

As shown in Table 2, out of 1,000 images for each pair of method comparison, we found that users reach strong agreement on a subset of images with quantity ranging from 321 to 718 shown in the 2nd column. The reason why the number of strongly preferred cases of "LaMa vs. ProFill" are less than the others is that these two methods have relatively closer inpainting performance, which causes more disagreement. Other columns in Table 2 basically indicates the percentage of correct ranking from each metric, with respect to the human perceptual judgement. In this study, we compare with two reconstruction metrics PSNR and LPIPS [69], as well as two NR-IQA metrics Hyper-IQA [40] and MUSIQ [21]. Overall, the quantitative results show that our PAR metric outperforms all these existing metrics for assessing inpainting quality in object removal scenarios.

5.4   PAR Analysis with Hole Size and Scene Types

We claimed that inpainting artifacts mostly appear in larger holes and complex scene structures. Using our pretrained artifacts segmentation model, we also studied how PAR would change with respect to the hole size in two scenarios:

Table 2. Quantitative results for measuring the correlation between different metrics and human perceptual judgement.

Comparisons	No. Pairs	PSNR	LPIPS [69]	HyperIQA [40]	MUSIQ [21]	PAR (Ours)
LaMa vs. ProFill	321	56.70 %	62.31 %	39.97%	65.11%	**65.42 %**
LaMa vs. CoMod-GAN	367	48.77 %	48.77 %	51.50%	55.31%	**69.21 %**
ProFill vs. EdgeConnect	560	23.92 %	11.96 %	56.39%	49.62%	**79.82 %**
LaMa vs. EdgeConnect	718	44.71 %	43.45 %	35.71%	71.72%	**72.70 %**
Overall	1966	41.50%	38.55 %	45.24%	61.28%	**72.89 %**

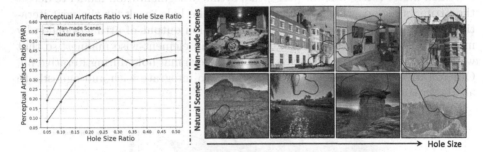

Fig. 6. Left: relationship between PAR and hole size for both man-made scenes and natural scenes. **Right**: some visual examples of segmented perceptual artifacts region (pink boundary) with varying hole (blue boundary) size. Inpainting models like LaMa produce more artifacts when the hole is larger and scenes are more complex. (Color figure online)

man-made scenes and natural scenes. In the places2 testing dataset, we sampled man-made scenes from the categories, such as building, room, shop, stadium, studio, factory and so on. On the other hand, natural scenes are sampled from categories, such as sky, land, mountain, forest, garden, pasture, beach, desert, and so on. We sampled 2,000 test images for both scenarios and randomly placed stroke holes of specific sizes on them. Then we run LaMa to fill the hole. The relationship between PAR and hole size for natural or man-made scenes is shown in Fig. 6. Our conclusions from the figure are: (1) As the hole size increases, LaMa has a higher possibility of generating inpainting artifacts. (2) Inpainting models like Lama struggles more to complete man-made structures than natural scenes. We believed that this rule applies to other inpainting algorithms as well.

6 Making Inpainting Models Iterative

Modern inpainting algorithms have shown consistent performance improvement over the last few years. However, when inpainting large holes, we often still observe that the inpainting models could often perfectly fill a partial region of the hole while generating obvious artifacts on the other regions. Given this observation, an intuitive idea is that: if the perceptual artifacts region can be reliably segmented out, can we enable the inpainting refill on the artifacts region? In this section, we discuss how we make the inpainting models iteratively fill on the artifacts region, and its effectiveness to improve the inpainting quality.

6.1 Iterative Fill Pipeline

In Fig. 7, we show an overview of our iterative fill pipeline. The input image with hole is first fed into an inpainting model to generate a filled image. Then, the filled image is fed into our perceptual artifacts segmentation model to detect the artifacts region, which are converted into the hole mask for the next iteration inpainting. We post-process the segmentation output of artifacts region by multiplying it with the original hole mask in an element-wise manner, so that we ensure not to change any pixels outside the original hole during iterative fill. Our iterative fill pipeline is extremely simple to integrate with and agnostic to all the inpainting models.

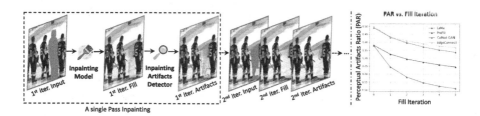

Fig. 7. Left: an overview pipeline of our iterative fill. **Right**: curves that show predicted perceptual artifacts ratio consistently decreases over the fill iteration for all inpainting models.

6.2 Performance Improvement by Iterative Fill

We evaluate the performance of the iterative fill from two aspects. First, we compute the size of the detected artifacts region or PAR over the fill iterations, as shown in the right of Fig. 7. We observe that the detected artifacts region consistently decreases as more iterative fill happens. This indicates that our iterative fill indeed improves the filled image quality, such that less perceptual artifacts are detected. Here, we show up to 5th iterative fill in the main paper, and put analysis on more iterations of refill in the supplemental. We also conducted a user study that shows whether users think the 5th iteratively filled image are better, same, or worse than the original fill for four inpainting models. As shown in Table 3, our iterative fill pipeline improves approximately 30% of images compared to the original fill, and rarely make the filled images worse, especially for the best model LaMa [40]. This implies that our system can be safely integrated into these inpainting models to boost inpainting quality. All of our user studies are conducted on AMT. In each inpainting method, we uniformly sampled 500 images from the testset, which result in 2,000 images in total. We asked 20 turkers to carefully check on each image and averaged the preference. We do not use the traditional metrics to quantify the performance between original fill and iterative fill, since we found that these metric scores between them are too close and sometimes random, which does not reflect much information. We have more discussion on this in the supplemental.

Table 3. A user study to show the comparison between original fill and the 5th refill.

Models	Preferred original fill	Same	Preferred iterative fill
EdgeConnect	53 (10.6%)	258 (51.6%)	189 (37.8%)
CoMod-GAN	45 (9.0%)	334 (66.8%)	121(24.2%)
ProFill	14 (2.8%)	337 (67.4%)	149 (29.8%)
LaMa	9 (1.8%)	341 (68.2%)	150 (30.0%)

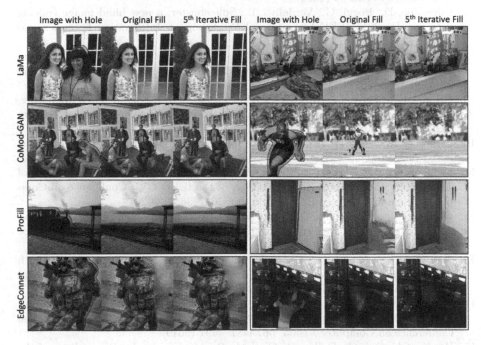

Fig. 8. Qualitative comparison between the original fill and the 5th iterative fill.

We show visual comparison between original fill and 5th iterative fill in Fig. 8. We observe that iterative fill could oftentimes help the inpainting models refine both structure and texture in many cases. However, due to the limitation of the inpainting algorithms themselves, the predicted perceptual artifact regions would not always reach to zero and thus would still leave some artifacts in the image.

References

1. Ballester, C., Bertalmio, M., Caselles, V., Sapiro, G., Verdera, J.: Filling-in by joint interpolation of vector fields and gray levels. IEEE TIP (2001)
2. Barnes, C., Shechtman, E., Finkelstein, A., Goldman, D.B.: Patchmatch: a randomized correspondence algorithm for structural image editing. ACM Trans. Graph. **28**(3), 24 (2009)
3. Bertalmio, M., Sapiro, G., Caselles, V., Ballester, C.: Image inpainting. In: SIGGRAPH (2000)
4. Cai, W., Wei, Z.: Piigan: generative adversarial networks for pluralistic image inpainting. IEEE Access **8**, 48451–48463 (2020)
5. Chai, L., Bau, D., Lim, S.-N., Isola, P.: What makes fake images detectable? understanding properties that generalize. In: Vedaldi, A., Bischof, H., Brox, T., Frahm, J.-M. (eds.) ECCV 2020. LNCS, vol. 12371, pp. 103–120. Springer, Cham (2020). https://doi.org/10.1007/978-3-030-58574-7_7
6. Chi, L., Jiang, B., Mu, Y.: Fast fourier convolution. Advances in Neural Information Processing Systems 33 (2020)
7. Cozzolino, D., Poggi, G., Verdoliva, L.: Splicebuster: a new blind image splicing detector. In: 2015 IEEE International Workshop on Information Forensics and Security (WIFS), pp. 1–6. IEEE (2015)
8. Cozzolino, D., Thies, J., Rössler, A., Riess, C., Nießner, M., Verdoliva, L.: Forensictransfer: weakly-supervised domain adaptation for forgery detection. arXiv preprint arXiv:1812.02510 (2018)
9. Criminisi, A., Perez, P., Toyama, K.: Object removal by exemplar-based inpainting. In: 2003 IEEE Computer Society Conference on Computer Vision and Pattern Recognition, 2003. Proceedings, vol. 2, p. II. IEEE (2003)
10. Criminisi, A., Pérez, P., Toyama, K.: Region filling and object removal by exemplar-based image inpainting. IEEE Trans. Image Process. **13**(9), 1200–1212 (2004)
11. Fang, Y., Zhu, H., Zeng, Y., Ma, K., Wang, Z.: Perceptual quality assessment of smartphone photography. In: Proceedings of the IEEE/CVF Conference on Computer Vision and Pattern Recognition, pp. 3677–3686 (2020)
12. Guo, X., Yang, H., Huang, D.: Image inpainting via conditional texture and structure dual generation. In: Proceedings of the IEEE/CVF International Conference on Computer Vision, pp. 14134–14143 (2021)
13. Guo, Z., Chen, Z., Yu, T., Chen, J., Liu, S.: Progressive image inpainting with full-resolution residual network. In: Proceedings of the 27th ACM International Conference on Multimedia, pp. 2496–2504 (2019)
14. Han, X., Wu, Z., Huang, W., Scott, M.R., Davis, L.S.: Finet: compatible and diverse fashion image inpainting. In: Proceedings of the IEEE/CVF International Conference on Computer Vision, pp. 4481–4491 (2019)
15. He, K., Gkioxari, G., Dollár, P., Girshick, R.: Mask r-cnn. In: Proceedings of the IEEE International Conference on Computer Vision, pp. 2961–2969 (2017)

16. He, K., Zhang, X., Ren, S., Sun, J.: Deep residual learning for image recognition. In: Proceedings of the IEEE Conference on Computer Vision and Pattern Recognition, pp. 770–778 (2016)
17. Heusel, M., Ramsauer, H., Unterthiner, T., Nessler, B., Hochreiter, S.: Gans trained by a two time-scale update rule converge to a local nash equilibrium. In: Advances in Neural Information Processing Systems 30 (2017)
18. Huh, M., Liu, A., Owens, A., Efros, A.A.: Fighting fake news: image splice detection via learned self-consistency. In: Proceedings of the European Conference on Computer Vision (ECCV), pp. 101–117 (2018)
19. Iizuka, S., Simo-Serra, E., Ishikawa, H.: Globally and locally consistent image completion. ACM Trans. Graph. (ToG) **36**(4), 1–14 (2017)
20. Karras, T., Laine, S., Aittala, M., Hellsten, J., Lehtinen, J., Aila, T.: Analyzing and improving the image quality of stylegan. In: Proceedings of the IEEE/CVF Conference on Computer Vision and Pattern Recognition, pp. 8110–8119 (2020)
21. Ke, J., Wang, Q., Wang, Y., Milanfar, P., Yang, F.: Musiq: Multi-scale image quality transformer. In: Proceedings of the IEEE/CVF International Conference on Computer Vision, pp. 5148–5157 (2021)
22. Li, A., et al.: Noise doesn't lie: towards universal detection of deep inpainting. arXiv preprint arXiv:2106.01532 (2021)
23. Li, H., Huang, J.: Localization of deep inpainting using high-pass fully convolutional network. In: Proceedings of the IEEE/CVF International Conference on Computer Vision, pp. 8301–8310 (2019)
24. Li, J., He, F., Zhang, L., Du, B., Tao, D.: Progressive reconstruction of visual structure for image inpainting. In: Proceedings of the IEEE/CVF International Conference on Computer Vision, pp. 5962–5971 (2019)
25. Li, J., Wang, N., Zhang, L., Du, B., Tao, D.: Recurrent feature reasoning for image inpainting. In: Proceedings of the IEEE/CVF Conference on Computer Vision and Pattern Recognition, pp. 7760–7768 (2020)
26. Liao, L., Xiao, J., Wang, Z., Lin, C.-W., Satoh, S.: Guidance and evaluation: semantic-aware image inpainting for mixed scenes. In: Vedaldi, A., Bischof, H., Brox, T., Frahm, J.-M. (eds.) ECCV 2020. LNCS, vol. 12372, pp. 683–700. Springer, Cham (2020). https://doi.org/10.1007/978-3-030-58583-9_41
27. Lin, T.-Y., Maire, M., Belongie, S., Hays, J., Perona, P., Ramanan, D., Dollár, P., Zitnick, C.L.: Microsoft COCO: common objects in context. In: Fleet, D., Pajdla, T., Schiele, B., Tuytelaars, T. (eds.) ECCV 2014. LNCS, vol. 8693, pp. 740–755. Springer, Cham (2014). https://doi.org/10.1007/978-3-319-10602-1_48
28. Liu, G., Reda, F.A., Shih, K.J., Wang, T.-C., Tao, A., Catanzaro, B.: Image inpainting for irregular holes using partial convolutions. In: Ferrari, V., Hebert, M., Sminchisescu, C., Weiss, Y. (eds.) ECCV 2018. LNCS, vol. 11215, pp. 89–105. Springer, Cham (2018). https://doi.org/10.1007/978-3-030-01252-6_6
29. Liu, H., Jiang, B., Xiao, Y., Yang, C.: Coherent semantic attention for image inpainting. In: Proceedings of the IEEE/CVF International Conference on Computer Vision, pp. 4170–4179 (2019)
30. Liu, H., Wan, Z., Huang, W., Song, Y., Han, X., Liao, J.: Pd-gan: probabilistic diverse gan for image inpainting. In: Proceedings of the IEEE/CVF Conference on Computer Vision and Pattern Recognition, pp. 9371–9381 (2021)
31. Liu, Z., Lin, Y., Cao, Y., Hu, H., Wei, Y., Zhang, Z., Lin, S., Guo, B.: Swin transformer: hierarchical vision transformer using shifted windows. In: Proceedings of the IEEE/CVF International Conference on Computer Vision, pp. 10012–10022 (2021)

32. Marra, F., Gragnaniello, D., Cozzolino, D., Verdoliva, L.: Detection of gan-generated fake images over social networks. In: 2018 IEEE Conference on Multimedia Information Processing and Retrieval (MIPR), pp. 384–389. IEEE (2018)
33. Nazeri, K., Ng, E., Joseph, T., Qureshi, F.Z., Ebrahimi, M.: Edgeconnect: Generative image inpainting with adversarial edge learning. arXiv preprint arXiv:1901.00212 (2019)
34. Oh, S.W., Lee, S., Lee, J.Y., Kim, S.J.: Onion-peel networks for deep video completion. In: Proceedings of the IEEE/CVF International Conference on Computer Vision, pp. 4403–4412 (2019)
35. Parmar, G., Zhang, R., Zhu, J.Y.: On buggy resizing libraries and surprising subtleties in fid calculation. arXiv preprint arXiv:2104.11222 (2021)
36. Peng, J., Liu, D., Xu, S., Li, H.: Generating diverse structure for image inpainting with hierarchical vq-vae. In: Proceedings of the IEEE/CVF Conference on Computer Vision and Pattern Recognition, pp. 10775–10784 (2021)
37. Rao, Y., Ni, J.: A deep learning approach to detection of splicing and copy-move forgeries in images. In: 2016 IEEE International Workshop on Information Forensics and Security (WIFS), pp. 1–6. IEEE (2016)
38. Ren, Y., Yu, X., Zhang, R., Li, T.H., Liu, S., Li, G.: Structureflow: Image inpainting via structure-aware appearance flow. In: Proceedings of the IEEE/CVF International Conference on Computer Vision, pp. 181–190 (2019)
39. Song, Y., Yang, C., Shen, Y., Wang, P., Huang, Q., Kuo, C.C.J.: Spg-net: Segmentation prediction and guidance network for image inpainting. arXiv preprint arXiv:1805.03356 (2018)
40. Su, S., et al.: Blindly assess image quality in the wild guided by a self-adaptive hyper network. In: Proceedings of the IEEE/CVF Conference on Computer Vision and Pattern Recognition, pp. 3667–3676 (2020)
41. Suvorov, R., et al.: Resolution-robust large mask inpainting with fourier convolutions. arXiv preprint arXiv:2109.07161 (2021)
42. Szegedy, C., Vanhoucke, V., Ioffe, S., Shlens, J., Wojna, Z.: Rethinking the inception architecture for computer vision. In: Proceedings of the IEEE Conference on Computer Vision and Pattern Recognition, pp. 2818–2826 (2016)
43. Talebi, H., Milanfar, P.: Nima: neural image assessment. IEEE Trans. Image Process. **27**(8), 3998–4011 (2018)
44. Tian, Z., Shen, C., Wang, X., Chen, H.: Boxinst: High-performance instance segmentation with box annotations. In: Proceedings of the IEEE/CVF Conference on Computer Vision and Pattern Recognition, pp. 5443–5452 (2021)
45. Wan, Z., Zhang, J., Chen, D., Liao, J.: High-fidelity pluralistic image completion with transformers. arXiv preprint arXiv:2103.14031 (2021)
46. Wang, J., Sun, K., Cheng, T., Jiang, B., Deng, C., Zhao, Y., Liu, D., Mu, Y., Tan, M., Wang, X., et al.: Deep high-resolution representation learning for visual recognition. IEEE Trans. Pattern Anal. Mach. Intell. **43**(10), 3349–3364 (2020)
47. Wang, N., Ma, S., Li, J., Zhang, Y., Zhang, L.: Multistage attention network for image inpainting. Pattern Recogn. **106**, 107448 (2020)
48. Wang, R., et al.: Fakespotter: a simple yet robust baseline for spotting ai-synthesized fake faces. arXiv preprint arXiv:1909.06122 (2019)
49. Wang, S.Y., Wang, O., Owens, A., Zhang, R., Efros, A.A.: Detecting photoshopped faces by scripting photoshop. In: Proceedings of the IEEE/CVF International Conference on Computer Vision, pp. 10072–10081 (2019)
50. Wang, S.Y., Wang, O., Zhang, R., Owens, A., Efros, A.A.: Cnn-generated images are surprisingly easy to spot... for now. In: Proceedings of the IEEE/CVF Conference on Computer Vision and Pattern Recognition, pp. 8695–8704 (2020)

51. Wang, T., Ouyang, H., Chen, Q.: Image inpainting with external-internal learning and monochromic bottleneck. In: Proceedings of the IEEE/CVF Conference on Computer Vision and Pattern Recognition, pp. 5120–5129 (2021)
52. Wang, Y., Tao, X., Qi, X., Shen, X., Jia, J.: Image inpainting via generative multi-column convolutional neural networks. arXiv preprint arXiv:1810.08771 (2018)
53. Wu, H., Zhou, J.: Giid-net: generalizable image inpainting detection network. In: 2021 IEEE International Conference on Image Processing (ICIP), pp. 3867–3871. IEEE (2021)
54. Xiao, T., Liu, Y., Zhou, B., Jiang, Y., Sun, J.: Unified perceptual parsing for scene understanding. In: Ferrari, V., Hebert, M., Sminchisescu, C., Weiss, Y. (eds.) ECCV 2018. LNCS, vol. 11209, pp. 432–448. Springer, Cham (2018). https://doi.org/10.1007/978-3-030-01228-1_26
55. Xie, C., et al.: Image inpainting with learnable bidirectional attention maps. In: Proceedings of the IEEE/CVF International Conference on Computer Vision, pp. 8858–8867 (2019)
56. Xiong, W., et al.: Foreground-aware image inpainting. In: Proceedings of the IEEE/CVF Conference on Computer Vision and Pattern Recognition, pp. 5840–5848 (2019)
57. Yang, C., Lu, X., Lin, Z., Shechtman, E., Wang, O., Li, H.: High-resolution image inpainting using multi-scale neural patch synthesis. In: Proceedings of the IEEE Conference on Computer Vision and Pattern Recognition, pp. 6721–6729 (2017)
58. Yang, J., Qi, Z., Shi, Y.: Learning to incorporate structure knowledge for image inpainting. In: Proceedings of the AAAI Conference on Artificial Intelligence, vol. 34, pp. 12605–12612 (2020)
59. Yeh, R.A., Chen, C., Yian Lim, T., Schwing, A.G., Hasegawa-Johnson, M., Do, M.N.: Semantic image inpainting with deep generative models. In: Proceedings of the IEEE Conference on Computer Vision and Pattern Recognition, pp. 5485–5493 (2017)
60. Yi, Z., Tang, Q., Azizi, S., Jang, D., Xu, Z.: Contextual residual aggregation for ultra high-resolution image inpainting. In: Proceedings of the IEEE/CVF Conference on Computer Vision and Pattern Recognition, pp. 7508–7517 (2020)
61. Ying, Z., Niu, H., Gupta, P., Mahajan, D., Ghadiyaram, D., Bovik, A.: From patches to pictures (paq-2-piq): mapping the perceptual space of picture quality. In: Proceedings of the IEEE/CVF Conference on Computer Vision and Pattern Recognition, pp. 3575–3585 (2020)
62. Yu, J., Lin, Z., Yang, J., Shen, X., Lu, X., Huang, T.S.: Generative image inpainting with contextual attention. In: Proceedings of the IEEE Conference on Computer Vision and Pattern Recognition, pp. 5505–5514 (2018)
63. Yu, J., Lin, Z., Yang, J., Shen, X., Lu, X., Huang, T.S.: Free-form image inpainting with gated convolution. In: Proceedings of the IEEE/CVF International Conference on Computer Vision, pp. 4471–4480 (2019)
64. Yu, Y., et al.: Diverse image inpainting with bidirectional and autoregressive transformers. arXiv preprint arXiv:2104.12335 (2021)
65. Zeng, Y., Fu, J., Chao, H., Guo, B.: Learning pyramid-context encoder network for high-quality image inpainting. In: Proceedings of the IEEE/CVF Conference on Computer Vision and Pattern Recognition, pp. 1486–1494 (2019)
66. Zeng, Yu., Lin, Z., Yang, J., Zhang, J., Shechtman, E., Lu, H.: High-resolution image inpainting with iterative confidence feedback and guided upsampling. In: Vedaldi, A., Bischof, H., Brox, T., Frahm, J.-M. (eds.) ECCV 2020. LNCS, vol. 12364, pp. 1–17. Springer, Cham (2020). https://doi.org/10.1007/978-3-030-58529-7_1

67. Zhang, H., Hu, Z., Luo, C., Zuo, W., Wang, M.: Semantic image inpainting with progressive generative networks. In: Proceedings of the 26th ACM International Conference on Multimedia, pp. 1939–1947 (2018)
68. Zhang, L., Wang, J., Shi, J.: Multimodal image outpainting with regularized normalized diversification. In: Proceedings of the IEEE/CVF Winter Conference on Applications of Computer Vision, pp. 3433–3442 (2020)
69. Zhang, R., Isola, P., Efros, A.A., Shechtman, E., Wang, O.: The unreasonable effectiveness of deep features as a perceptual metric. In: Proceedings of the IEEE Conference on Computer Vision and Pattern Recognition, pp. 586–595 (2018)
70. Zhang, W., Ma, K., Yan, J., Deng, D., Wang, Z.: Blind image quality assessment using a deep bilinear convolutional neural network. IEEE Trans. Circuits Syst. Video Technol. **30**(1), 36–47 (2018)
71. Zhang, X., Karaman, S., Chang, S.F.: Detecting and simulating artifacts in gan fake images. In: 2019 IEEE International Workshop on Information Forensics and Security (WIFS), pp. 1–6. IEEE (2019)
72. Zhao, H., Shi, J., Qi, X., Wang, X., Jia, J.: Pyramid scene parsing network. In: Proceedings of the IEEE Conference on Computer Vision and Pattern Recognition, pp. 2881–2890 (2017)
73. Zhao, L., et al.: Uctgan: diverse image inpainting based on unsupervised cross-space translation. In: Proceedings of the IEEE/CVF Conference on Computer Vision and Pattern Recognition, pp. 5741–5750 (2020)
74. Zhao, S., et al.: Large scale image completion via co-modulated generative adversarial networks. arXiv preprint arXiv:2103.10428 (2021)
75. Zheng, C., Cham, T.J., Cai, J.: Pluralistic image completion. In: Proceedings of the IEEE/CVF Conference on Computer Vision and Pattern Recognition, pp. 1438–1447 (2019)
76. Zhou, B., Zhao, H., Puig, X., Fidler, S., Barriuso, A., Torralba, A.: Scene parsing through ade20k dataset. In: Proceedings of the IEEE Conference on Computer Vision and Pattern Recognition, pp. 633–641 (2017)

2D Amodal Instance Segmentation Guided by 3D Shape Prior

Zhixuan Li[1,2], Weining Ye[2], Tingting Jiang[1,2(✉)], and Tiejun Huang[2]

[1] Advanced Institute of Information Technology, Peking University, Hangzhou, China
{zhixuanli,ttjiang}@pku.edu.cn
[2] National Engineering Research Center of Visual Technology, School of Computer
Science, Peking University, Beijing, China
{ywning,tjhuang}@pku.edu.cn

Abstract. Amodal instance segmentation aims to predict the complete mask of the occluded instance, including both visible and invisible regions. Existing 2D AIS methods learn and predict the complete silhouettes of target instances in 2D space. However, masks in 2D space are only some observations and samples from the 3D model in different viewpoints and thus can not represent the real complete physical shape of the instances. With the 2D masks learned, 2D amodal methods are hard to generalize to new viewpoints not included in the training dataset. To tackle these problems, we are motivated by observations that (1) a 2D amodal mask is the projection of a 3D complete model, and (2) the 3D complete model can be recovered and reconstructed from the occluded 2D object instances. This paper builds a bridge to link the 2D occluded instances with the 3D complete models by 3D reconstruction and utilizes 3D shape prior for 2D AIS. To deal with the diversity of 3D shapes, our method is pretrained on large 3D reconstruction datasets for high-quality results. And we adopt the unsupervised 3D reconstruction method to avoid relying on 3D annotations. In this approach, our method can reconstruct 3D models from occluded 2D object instances and generalize to new unseen 2D viewpoints of the 3D object. Experiments demonstrate that our method outperforms all existing 2D AIS methods.

Keywords: Amodal · Occlusion · Instance segmentation

1 Introduction

Different from *visible* instance segmentation (VIS) [1,9,14,31] which only predicts the visible region of each instance, *amodal* instance segmentation (AIS) [19,30,37] task poses a harder challenge that demands to predict both the visible and occluded parts. AIS has many potential applications, including auto-driving [27], automatic checkout in the market [7] and image editing [36].

The concept of AIS was proposed in 2016 [19], and several datasets [5,7,13, 27,38] have been provided. Most of the existing *amodal* methods [5,7,13,17,19,

© The Author(s), under exclusive license to Springer Nature Switzerland AG 2022
S. Avidan et al. (Eds.): ECCV 2022, LNCS 13689, pp. 165–181, 2022.
https://doi.org/10.1007/978-3-031-19818-2_10

27, 34, 38] are developed based on *visible* instance segmentation methods [9, 18] that directly minimize the discrepancy between amodal prediction and ground-truth masks. Recently, some methods consider the characteristics of the amodal problem itself and propose new solutions. For example, relative depth order of instances is used to help comprehend the scene [36, 37]. Weakly supervised methods are proposed [23, 25, 36] without needing ground-truth amodal mask while taking ground-truth visible mask as input and supervision.

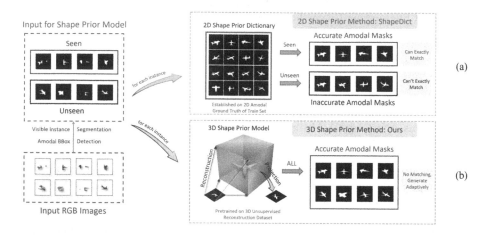

Fig. 1. Overview and comparison of the 2D shape prior dictionary (SPD) based method ShapeDict [30] and our proposed 3D shape prior generation-based method. For the left input RGB images, the first stage of both methods conduct instance segmentation to obtain *amodal* bounding boxes and *visible* masks for each instance. (a) ShapeDict regards each visible mask as *query* for SPD and *retrieves* the matched amodal shape prior masks. Due to the limited diversity of prestored shape prior, the retrieved shape prior is more appropriate for samples that *have been seen* (green box) in the dictionary rather than the *unseen ones* (red box). (b) Our proposed method adaptively generates 3D shape prior models from visible masks and performs projection for 2D amodal masks without needing for prestoring.

Besides, a natural solution is to use the shape-prior knowledge for handling the occluded instance, which lacks the shape and appearance information of the invisible region. In 2020, ShapeDict [30] proposes to utilize 2D shape prior knowledge to deal with the amodal segmentation problem. As shown in Fig. 1(a), ShapeDict first establishes a 2D shape prior dictionary (SPD) by applying the K-means algorithm on the ground-truth 2D amodal masks in the training set, and takes the cluster centers as shape priors. During inference, the closet shape prior to the predicted visible mask is retrieved from the SPD and used for amodal segmentation. However, this nearest neighbor search approach can only work for amodal masks having been seen during training (as shown in the *green* box of Fig. 1(a)). Otherwise, it will fetch inappropriate shape prior and lead to wrong amodal segmentation. For example, an occluded airplane photographed from a new viewpoint, whose amodal mask is not stored in the SPD, cannot be correctly matched. Therefore, ShapeDict is limited to the number and variety of shape prior masks stored in the dictionary, making it hard to generalize on unseen viewpoints. In this paper, we consider *if it is possible to adaptively generate the shape prior masks rather than prestoring the shape prior masks in a dictionary, and tackle the challenges of viewpoint changes?*

With these problems in mind, we hope to learn the shape prior in the 3D space, which is a unified representation of 2D masks from all viewpoints and can generalize to new perspectives. Meanwhile, to avoid the shortcoming of the SPD method in ShapeDict, we hope to generate the 3D shape prior with learned shape knowledge adaptively and need no requirement for a prestored shape dictionary. To achieve these two purposes, we propose to reconstruct the complete 3D shape prior from the 2D occluded instance, as shown in Fig. 1(b). To accomplish 3D reconstruction, either multi-view images as input or 3D models as supervision signals are usually needed. Unfortunately, both are *not available* in any existing 2D AIS datasets. However, the good news is that, in recent years, single-view unsupervised 3D reconstruction methods [12,20,24] can avoid the requirements of multiple views and 3D model for training, which makes 2D amodal datasets usable for 3D reconstruction. For single-view unsupervised 3D reconstruction methods, the 3D model is first reconstructed from the single-view input RGB image and then projected along the estimated viewpoints to 2D masks. During reconstruction, only the 2D visible mask is provided as the supervision signal and the 3D reconstruction model is supervised *indirectly by the consistency between 2D projection of the 3D model and 2D visible mask*. In our method, the 2D amodal mask is used as the supervision signal for the single-view unsupervised 3D reconstruction.

In this work, we propose Amodal 3D Network (A3D), a novel coarse-to-fine architecture that combines category-specific 3D shape prior with 2D AIS. As shown in Fig. 2, for an input RGB image, we first apply the visible instance segmentation method to obtain visible masks of each instance. Next, we use a two-branch structure, in which the upper branch utilizes an *Encoder Decoder Network* for Category-specific 3D shape prior reconstruction, and the lower branch predicts camera viewpoint parameters by the *Viewpoint Estimator*. Then the *Differentiable Render* projects the 3D shape prior model according to the predicted viewpoint to the 2D coarse amodal mask. Finally, the *Region-specific Edge Refine module* refines the edges with the guidance of the visible mask and predicts the final amodal mask. With this coarse-to-fine pipeline, A3D can benefit from the power of 3D shape prior modelling and 2D edge refinement at the same time.

It is worth noting that our 3D shape prior reconstruction method *only requires 2D amodal masks as ground truth, without supervision signals like 3D models*, which are expensive to obtain. Because the 3D reconstruction module plays a crucial role in our method, we need to ensure that the reconstruction module can generate high-quality 3D shape prior models when facing 2D occluded instances. We design a pretrain-and-finetune pipeline that the single-view 3D reconstruction module is first pretrained on a large 3D reconstruction dataset like ShapeNet [2] for common shape *in unsupervised approach without using 3D annotations.* Then we conduct finetuning on the 2D AIS dataset for specific shape knowledge learning.

The effectiveness of our proposed method A3D is evaluated on several challenging datasets, including D2SA for market goods, KINS for person and vehicle, and COCOA-cls for life scene. We achieve state-of-the-art on all datasets.

We summarize our final contributions as follows:

1. A new method A3D is proposed for AIS, which utilizes the single-view unsupervised 3D reconstruction for 3D shape prior learning, to tackle the problem that 2D amodal segmentation methods are hard to generalize on new viewpoints. To our best knowledge, it is the first time the 3D shape prior is used for 2D AIS.
2. A coarse-to-fine pipeline is designed, which learns the 3D coarse shape prior and then refine edges with region-specific loss. It is end-to-end trainable and profits from both 3D and 2D information.

2   Related Work

2.1   2D Instance Segmentation

Amodal Instance Segmentation. Comparing to visible instance segmentation, due to the shape of both visible and occluded regions needing to be predicted, the Amodal Instance Segmentation (AIS) task has fewer clues to infer the complete silhouette of instance and more ambiguity because of the occlusion. Existing methods mainly solve the task in 4 ways, including (1) directly minimizing between the prediction [5,19,27,38] and the target (2) using relative depth order to comprehend the relationship between different objects [36,37] (3) mutual helping from visible and amodal masks [7,17] and (4) using pre-stored shape prior knowledge [23,30]. In the meanwhile, several datasets have been proposed including realistic ones [7,27,38] and synthetic ones [5,13]. In addition, some papers [23,36] are working on a relevant task *amodal completion,* which aims to predict the complete amodal shape based on the given visible mask, while there are no visible masks given in the AIS task.

All of the existing AIS algorithms are working in 2D space, which lacks comprehension of the real shape in 3D space. In this paper, our method learns the shape knowledge in 3D space to overcome the drawbacks of 2D AIS methods.

2.2   Unsupervised Single View 3D Reconstruction

Based on deep learning, supervised 3D reconstruction methods [4,32,33] are relying on high-quality 3D models for supervision, which are expensive to build. And because 2D supervision signals like segmentation masks are easier to obtain, unsupervised 3D model reconstruction methods are more popular. The pipeline of unsupervised 3D reconstruction consists of two steps, including 3D model reconstruction from the 2D image and rendering the 3D model into the 2D space, which is called *rasterization*. Whether the *rasterization* is differentiable decides whether the second rendering step can be included in the deep learning model with end-to-end training.

Before 2018, most methods [15,35] take *rasterization by discrete assignment* and cannot be trained end-to-end for the whole network. To *make the rasterization differentiable*, in 2018, NMR [16] proposes an approximate gradient approach to make the backward gradient progress in rasterization differentiable. SoftRas [24] and DIB-R [3] methods make both of the forward and backward steps in rasterization differentiable and can be trained in an end-to-end manner. Based on the differentiable rasterization technique, UMR [20] utilizes the semantic parts consistency between 2D and 3D spaces as supervision. SMR [12] proposes landmark and interpolation consistency for self-supervision.

Because our method aims to represent and learn the complete shape in 3D space, we utilize the unsupervised single-view algorithm to gain a deeper understanding of the real shape in 3D space and help with 2D AIS with the reconstructed complete 3D shape.

3   Amodal 3D Network (A3D)

In this section, we develop a novel coarse-to-fine structure by combining the strength of 3D shape prior reconstruction and 2D edge refinement. We will first show the overall architecture of A3D and introduce each stage of A3D in detail.

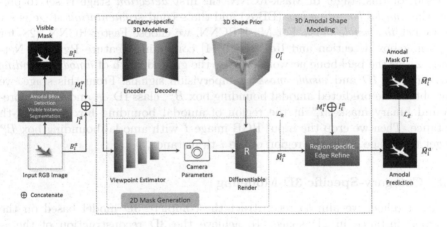

Fig. 2. The pipeline of our proposed Amodal 3D network (A3D).

3.1   Overall Architecture

Given an input image $I \in R^{H \times W \times 3}$ containing N instances, for the i-th instance, 2D AIS algorithms aim to predict the class ID $c_i \in \{1, 2, ..., K\}$ and 2D amodal masks M_i^a.

The overall architecture is illustrated in Fig. 2. We take Amodal BBox Detection and Visible Instance Segmentation as the first stage to predict amodal bounding box B_i^a and the visible mask M_i^v. Then the RGB image I_i^a is cropped by using B_i^a. For each instance, we concatenate M_i^v and I_i^a as input to the second proposed *3D Amodal Shape Modeling* (3D-ASM) stage for predicting the final amodal segmentation mask. 3D-ASM contains three important modules including *Category-specific 3D Modeling, 2D Mask Generation* and *Region-specific Edge Refine*. (1) The *Category-specific 3D Modeling* module reconstructs the 3D complete model as shape prior. (2) In the *2D Mask Generation* module, the *Viewpoint Estimator* first predicts the camera parameters and apply the transformation to the reconstructed 3D shape prior model to the appropriate observation viewpoint. Then a *Differentiable Render* projects the 3D shape prior along the predicted viewpoint to obtain a coarse amodal segmentation mask. (3) Finally, the *Region-specific Edge Refine* module utilizes the visible mask, whose edge is accurate because the appearance information of the visible region is available to modify edges of the amodal mask. The whole network is end-to-end trainable.

We will introduce the details of our network in the following sections.

3.2   Amodal BBox Detection and Visible Instance Segmentation

In this stage, we aim to predict the amodal bounding boxes and the visible masks for each instance. Following [7,30,37] we choose the popular Mask-RCNN [9] method for this stage. In Mask-RCNN, the first *detection* stage is set to predict the *amodal bounding box* (BBox) and the second *segmentation* stage is set to predict *the visible mask*. For Mask-RCNN, we choose Faster-RCNN [28] for bounding box detection and ResNet-50 [11] combining Feature Pyramid Network [21] as the backbone network. We use the ground truth of *amodal bounding box, category ID* and *visible mask* as supervision signals. From this stage, we can obtain the predicted amodal bounding box B_i^a, class ID c_i, and visible foreground binary mask M_i^v in the region of amodal bounding box for the i-th instance. Then we crop the input RGB image I with amodal bounding box B_i^a to get the image I_i^a in the region of the i-th instance.

3.3   Category-Specific 3D Modeling

In this module, we aim to reconstruct the complete 3D model based on the occluded instance in 2D space. To achieve the 3D reconstruction of the i-th occluded instance, traditional 3D reconstruction methods [4,32,33] requires either *multiple-view inputs* or *3D supervision signals*, which are not available in any existing 2D amodal segmentation datasets [7,27,38]. Therefore we choose to use the *single-view unsupervised* 3D reconstruction methods [3,12,20,24] considering the dataset limitation.

In the single-view unsupervised 3D reconstruction framework, the 3D model is first reconstructed from 2D inputs (2D → 3D) and then projected to 2D space (3D → 2D) for 2D amodal mask predictions. The 3D reconstruction network is *indirectly* supervised by the ground-truth of the 2D amodal masks. There are not any 3D models used as supervision signals.

For the i-th instance, the input for 3D reconstruction is $A_i = [I_i^a, M_i^v]$, which is the concatenation of the image region I_i^a and visible mask M_i^v. We use the simple and classic *Encoder Decoder* structure following [16,24] for 3D shape modeling, leaving room for improvement by using more complex models. The *Encoder* contains five *conv-bn-relu* blocks for visible feature extraction, and three *fully connected (fc)* layers for linearly feature mapping. There is also a *classification branch* taking the feature from Encoder output and predicts the category of each instance, making the 3D shape reconstruction in a class-specific manner. It is worthy to notice that we use one model to handle all categories.

We take a sphere as the initial object model $O_i^0(V_i^0)$, in which V_i^0 is the initial vertices. The *Decoder* consists of two *fc-relu* blocks to predict the offset $\triangle V_i$ between reconstructed 3D model and initial 3D model. Finally we can obtain the vertices $V_i^r = V_i^0 + \triangle V_i$ and the reconstructed 3D object $O_i^r(V_i^r)$. The detailed network architecture is described in the supplementary.

3.4   2D Mask Generation

For the i-th reconstructed 3D shape prior model O_i^r, if we want to obtain the 2D amodal mask, it is necessary to transform the 3D model with the correct parameters of the camera and project the 3D model to the camera plane. In this module, we take a *Viewpoint Estimator* to predict the camera parameters and utilize a *Differentiable Render* for projection.

Following SMR [12], we use an Encoder Network to construct the *Viewpoint Estimator*, which consists of five *conv-bn-relu* blocks and three *fully-connected* layers. The *Viewpoint Estimator* predicts the camera parameters $[e_i, d_i, (a_i^x, a_i^y)]$ representing elevation e_i, distance d_i and azimuth (a_i^x, a_i^y) in Cartesian coordinates, in which azimuth $a_i = arctan2(a_i^x, a_i^y)$. With the predicted viewpoint $[e_i, d_i, (a_i^x, a_i^y)]$, the 3D model is transformed appropriately. The *Viewpoint Estimator* is supervised *indirectly* by the ground-truth of 2D amodal masks, and no ground truth of viewpoints is used for supervision.

Finally to project the transformed 3D shape prior model O_i^r for the coarse 2D amodal mask $\widetilde{M}_i^a$, we utilize the *Differentiable Render* SoftRas [24], which can maintain the gradient flow for end-to-end training.

3.5   Region-Specific Edge Refine

In previous modules, we have obtained the 3D reconstructed shape prior model O_i^r and the projected 2D coarse amodal mask $\widetilde{M}_i^a$. However, the quality of the existing 3D reconstruction method is affected by the number of vertices in the initial sphere O_i^0 and topological changes like holes which are hard to be learned in the mesh format. Therefore we design the *Region-specific Edge Refine* module to use the 2D amodal image region I_i^a and visible mask M_i^v to improve the edge of the amodal mask $\widetilde{M}_i^a$, because the appearance textures are only available in the visible region.

We take 3 repeated *conv-bn-relu* layers as the module architecture with kernel size=3. This module takes the concatenation of coarse 2D amodal mask $\widetilde{M}_i^a$, the 2D amodal image region I_i^a and visible mask M_i^v as input, and uses visible mask M_i^v to help with loss function, which is designed to punish more on the visible edge and less on the occluded edge. Visualization example of Edge Refine are shown in Fig. 3.

Fig. 3. Visualization of region-specific edge refine. $\oplus$ means concatenation.

3.6   Loss Functions

In this section, we will introduce loss functions for our 3D modelling and the edge refinement modules.

3D Modeling. To get rid of dependence on the expensive 3D model annotation, we choose to use the unsupervised 3D reconstruction method without needing 3D models as supervision signals. We train both the Category-specific 3D Modeling module and Region-specific Edge Refine module simultaneously because only the ground-truth 2D amodal masks $\overline{M}_i^a$ are available for supervision. Therefore the loss function is designed to encourage the predicted coarse amodal mask $\widetilde{M}_i^a$ being close to $\overline{M}_i^a$, which indirectly supervises the quality of reconstructed 3D shape prior model O_i^r. The loss function for unsupervised 3D reconstruction is:

$$\mathcal{L}_\mathcal{R} = \frac{1}{N} \sum_{i=1}^{N} (1 - IoU(\widetilde{M}_i^a, \overline{M}_i^a)) \tag{1}$$

where IoU computes the *intersection over union* between the predicted coarse amodal mask $\widetilde{M}_i^a$ and the ground-truth amodal mask $\overline{M}_i^a$.

Edge Refine. In the *Region-specific Edge Refine* module, the loss function of the *Region-specific Amodal Edge Refine* module is designed as following:

$$\mathcal{L}_E = \frac{1}{N} \sum_{i=1}^{N} (\sum_{p \in S_i} \mathcal{L}_B(\widehat{M}_{i,p}^a, \overline{M}_{i,p}^a) + \lambda \sum_{p \notin S_i} \mathcal{L}_B(\widehat{M}_{i,p}^a, \overline{M}_{i,p}^a)) \tag{2}$$

where N is the instance number in the input image. For the i-th instance, S_i represents the visible region indicated by the ground-truth visible mask, and p denotes the pixel p. $\mathcal{L}_B$ is the *Binary Cross Entropy* loss function, computing the difference between the predicted values of pixel p from refined prediction $\widehat{M}_{i,p}^a$ and ground-truth mask $\overline{M}_{i,p}^a$. We set the loss weight $\lambda = 0.5$ to rely more on the visible region for edge refinement.

3.7   Pretrain and Finetune

In previous subsections, the whole framework of A3D and loss functions are introduced. In this subsection, a carefully designed pretrain and finetune strategy is introduced to improve the performance of 3D shape reconstruction for better 2D amodal segmentation results. It is worth noting that the ground truth of 3D models and viewpoints of the pretrain and finetune datasets are never used, to make our method applicable in real applications.

We first pretrain our A3D network on the 3D reconstruction dataset by unsupervised approach (as described in Sect. 3.3), and then finetune on the train set of 2D AIS dataset, finally conduct inference on its test set.

To handle the problem that the categories of pretrain and finetune datasets are different, including overlapped and non-overlapped categories, we deal with them in different approaches. For overlapped categories, we reuse the network weights after pretraining as the initialization for training on the finetune dataset. With the weights being reused, the category-specific knowledge can be transfered for the overlapped categories between pretraining and finetune datasets. For non-overlapped categories, the weight parameters are randomly initialized [10]. The performance of overlapped and non-overlapped categories are shown in Table 3.

Besides, to improve the performance of 3D reconstruction supervised by single view, in the pretrain process, we use the cross-view technique [24], which requires the 3D model reconstructed from two viewpoints for the same object to be similar. The cross-view technique only additionally uses the correspondence information that two images from different viewpoints corresponds to the same object, and *never* takes the ground truth of viewpoints and 3D models as supervisions. The details are shown in the supplementary. The cross-view technique is *optional*.

4   Experiments

In order to evaluate our proposed method, extensive experiments have been conducted on three public amodal segmentation datasets, including COCOA-cls, KINS and D2SA, as well as a 3D dataset ShapeNet. Our method is compared with several SOTA amodal segmentation methods, and results show the advantage of our approach.

4.1   Datasets

2D AIS Datasets. For 2D AIS task, we conduct experiments on a large-scale synthetic dataset ShapeNet [2] and three real 2D AIS datasets including COCOA-cls [7], KINS [27] and D2SA [7] for scenes of *outdoor, street and indoor supermarkets*. ShapeNet dataset contains 735,432 instances for training and 210,288 instances for testing. There are 13 categories, including various objects, and each object is rendered in 24 different viewpoints. We randomly occlude the RGB image and mask to simulate the occluded inputs. We do not use any 3D shape and viewpoint supervision signals in all experiments even they are available. COCOA-cls dataset, which annotates a subpart of COCO [22] dataset with amodal masks, has 3,501 images and 10,592 instances in 80 categories. KINS dataset is the biggest street scene amodal dataset, which can be applied on tasks like auto driving, built on KITTI dataset [8] with re-annotated amodal masks. KINS has two super-classes, including person and vehicle, and seven sub-classes with 7,474 and 7,517 images for training and testing. D2SA dataset is built upon D2S [6] dataset with amodal mask re-annotated, including plenty kinds of goods placed in different postures and occlusion approaches on a rotatable supermarket platform with varying light conditions. D2SA contains 5,600 images totally and 28,720 instances in 60 classes.

Pretraining Datasets. For using the pretrain and finetune strategy claimed in Sect. 3.7, both ShapeNet and PASCAL3D+ [29] are used. PASCAL3D+ dataset contains 55,867 3D models in 12 categories, which is more than 39,405 3D models in ShapeNet dataset, providing richer shape knowledge for 3D reconstruction.

4.2   Implementation Details and Evaluation Metric

Our method is implemented based on the Pytorch [26] framework. For all of the 2D amodal segmentation methods used in our experiments, we use the same configuration following ShapeDict [30]. For our proposed A3D Network, the learning rate is set to 0.0001, and we use the Adam algorithm for gradient descent with 64 batches. All experiments are conducted on a single 2080Ti GPU card. All categories including *rigid and non-rigid* are used in all experiments. Faster-RCNN with ResNet-50 is used for all methods for object detection.

We choose the *mean Intersection over Union* (mIoU) and *mean Average Precision* (mAP) as metrics for performance evaluation. It is worthy to notice that the commonly chosen metric *mean Average Precision* (mAP) measures the performance of two sub-tasks in *Amodal Instance Segmentation* simultaneously, including *Object Detection* and *Semantic Segmentation*. Therefore for ShapeDict dataset we only report mIoU because there are only one object in each image and mAP which measures the performance of object detection is not reported.

4.3   2D Amodal Instance Segmentation

This section evaluates 2D AIS methods on the challenging ShapeNet dataset and three amodal datasets, including COCOA-cls, D2SA and KINS for different scenes. Following state-of-the-art methods are used for comparison.

(1) Mask-RCNN [9] is trained using ground-truth *amodal* bounding boxes and masks to show the transferability of the visible instance segmentation method on the amodal problem. (2) ORCNN [7] is a two-branch approach that predicts and supervises the visible, amodal and occluded region at the same time. (3) BCNet [17] decouples the occluding and occluded instances combining graph

convolution. (4) ShapeDict [30] establishes a 2D shape prior dictionary by clustering the ground-truth amodal masks and uses the *query-and-retrieve* approach to provide prior knowledge. Besides, we also take Deocclusion [36], a weakly-supervised amodal completion method, for comparision to show the performance gap between weakly and fully supervised AIS methods.

Table 1. Results (mIoU) on the ShapeNet dataset. For each category, bold performance is the best, and the second-best is underlined. The subscript numbers are the subtraction results between ours and the second-best methods. SU means the Supervision signal type. W and F mean weakly and fully supervised.

Methods	SU	Airplane	Bench	Dresser	Car	Chair	Display	Lamp	Speaker	Rifle	Sofa	Table	Phone	Vessel	mIoU
Deocclusion [36]CVPR'20	W	24.9	67.4	45.3	58.8	83.7	78.4	77.9	15.2	_48.7_	48.1	39.5	23.8	71.9	52.2
Mask-RCNN [9]ICCV'17	F	73.4	66.0	92.4	_93.5_	**89.3**	**90.0**	77.4	88.5	30.0	86.0	73.1	89.8	80.5	79.2
ORCNN [7]WACV'19	F	71.5	61.1	92.0	92.7	_88.8_	88.8	_79.5_	88.7	32.8	85.6	72.5	89.0	80.0	78.7
BCNet [17]CVPR'21	F	73.0	_75.1_	_93.8_	89.4	86.6	88.7	**81.6**	_90.2_	32.8	83.4	_77.5_	88.7	74.8	78.2
ShapeDict [30]AAAI'21	F	_75.2_	68.5	93.7	**93.6**	88.4	_89.3_	78.1	88.6	34.4	_87.3_	74.8	_90.7_	_80.9_	_80.3_
Ours (no pretrain)	F	**77.9**	**80.8**	**94.2**	92.8	79.7	87.5	67.8	**90.5**	**69.9**	**90.3**	**86.2**	**92.1**	**81.3**	**83.9**

Table 2. Results (mIoU and mAP) on the 2D AIS datasets. SU means the Supervision signal type. W and F mean weakly and fully supervised. FLOPs and Params measure the computational efficiency and model size.

Method	SU	mIoU ↑			mAP ↑			FLOPs(G) ↓	Params(M) ↓
		D2SA	KINS	COCOA-cls	D2SA	KINS	COCOA-cls		
Deocclusion [36]CVPR'20	W	73.8	59.2	39.2	61.7	27.5	19.9	160.4	44.1
Mask-RCNN [9]ICCV'17	F	74.6	60.1	63.8	63.6	30.0	33.7	160.4	44.1
ORCNN [7]WACV'19	F	74.1	55.1	57.6	64.2	30.6	28.0	229.4	46.8
BCNet [17]CVPR'21	F	74.9	44.0	15.1	50.9	22.1	16.2	263.5	63.2
ShapeDict [30]AAAI'21	F	_75.0_	_63.7_	_64.5_	_70.3_	_32.1_	_35.4_	271.3	48.0
Ours (w/o pretrain)	F	74.7	61.4	64.2	68.5	31.4	34.9	229.8	57.2
Ours (w/ pretrain)	F	**78.4**	**65.5**	**67.4**	**73.5**	**36.2**	**40.6**	229.8	57.2

The comparison results of AIS methods are shown in Table 1 for ShapeNet dataset and Table 2 for the three 2D AIS datasets, including D2SA, KINS and COCOA-cls. In Table 1, all methods are trained on the train set of ShapeNet and there are no extra data used for pretraining in our method. Our method achieves the best performance on nine categories. Compared with the methods of the second-best performance, A3D gains significant IoU improvement on *Bench, Rifle and Table* with 5.7%, 21.6% and 8.7% respectively, which shows the effectiveness of our proposed A3D method. Figure 4 shows the qualitative results of our A3D method on the ShapeNet dataset.

In Table 2, all methods except our methods are trained on the train split of respective datasets, and our method additionally uses ShapeNet and PASCAL3D+ for pretraining to make our method applicable in real applications. Our A3D network outperforms all methods with certain advantages for mIoU

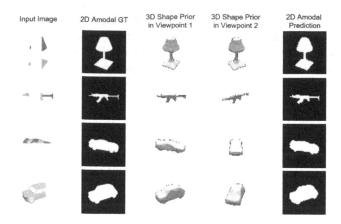

Fig. 4. Visualization result of our method on the ShapeNet dataset. The reconstructed 3D shape prior models are shown from two viewpoints.

and mAP. In terms of the number of FLOPs and Parmeters, our method is comparable to previous work. More visualizations for D2SA, KINS and COCOA-cls datasets are given in the supplementary.

4.4   Effectiveness of Pretraining

Table 3. Ablation study results (mAP) of pretraining N, S, P and S+P means no pretraining, pretraining with ShapeNet, with PASCAL3D+, and with both ShapeNet & PASCAL3D+. #Overlapped means the number of categories overlapped between pretrain and finetune datasets. SU means supervision signal (W and F for weakly and fully supervised). Category number is noted in the brackets after each dataset name.

Index	Methods	SU	D2SA (60)				KINS (7)				COCOA-cls (80)			
			N	S	P	S+P	N	S	P	S+P	N	S	P	S+P
#Overlapped			–	2	2	4	–	1	4	4	–	13	12	19
1	Deocclusion	W	61.7	61.9	62.1	62.3	27.5	27.9	28.2	28.8	19.9	20.4	20.9	21.3
2	Mask-RCNN	F	63.6	63.9	64.2	64.8	30.0	30.5	30.8	31.1	33.7	33.9	34.2	34.6
3	ORCNN	F	64.2	64.8	65.1	65.5	30.6	30.9	31.2	31.8	28.0	28.3	28.7	29.1
4	BCNet	F	50.9	51.2	51.5	51.9	22.1	22.6	22.8	23.0	16.2	16.8	17.1	17.4
5	ShapeDict	F	**70.3**	70.5	70.9	71.2	**32.1**	32.2	32.4	32.5	**35.4**	35.7	36.1	36.4
6	Ours	F	68.5	**71.4**	**72.6**	**73.5**	31.4	**33.7**	**35.2**	**36.2**	34.9	**37.4**	**39.2**	**40.6**

In Table 2, all the previous methods do not use the pretrain dataset while our method does. To make a fair comparison, we design an experiment such that each previous method can also take advantage of the pretrain dataset. Specifically, in pretrain process, each previous method can take all images in the pretrain dataset, each of which contains one non-occluded object with white background and the corresponding 2D amodal mask, as training data. With this pretrain strategy, all the previous methods can also make use of the pretrain dataset and thus the comparison between previous methods and ours is fair. We compare the performance of all methods with & without pretraining on three 2D AIS datasets. Results are shown in Table 3.

Comparing methods in different lines, we can conclude that by directly adding the 3D representation without pretraining, our method outperforms the baseline method Mask-RCNN but cannot beat ShapeDict. This is because the 2D AIS datasets provide not enough supervision for training 3D reconstruction in our method. In the last line of our method, for each finetune dataset, the performance increases with more pretrain data used. Meanwhile the performance of 2D AIS methods (Table 3, Line #1 to #5) do not increase much with pretraining. This is because for both pretrain datasets, each input RGB image contains only one non-occluded object with white background, which is easy to be segmented and not very helpful for amodal segmentation. However the pretrain datasets are very helpful for 3D reconstruction, making our method gains much improvements. With limited number of overlapped categories between pretrain and finetune datasets, our method can still achieve good performances.

4.5 Ablation Study

In this section, we conduct ablation experiments to validate the effectiveness of our proposed modules and pretraining for 3D reconstruction.

Effectiveness of 3D Modeling and Edge Refine. In our proposed A3D network, we design a Category-specific 3D Modeling module for 3D shape prior generation and a Region-specific Edge Refine module for 2D edge refinement. In this section, we validate the effectiveness of the two proposed modules on the ShapeNet dataset. (1) The Mask-RCNN method, which directly predicts the 2D amodal mask from the input image, is the baseline method. (2) If the Category-specific 3D Modeling module is added, 4% mIoU improvement will be obtained, which shows the effectiveness of 3D modeling. (3) Then after further combining the Region-specific Edge Refine module, the performance can boost 0.7% mIoU, which improves the quality of some details of the edge and not drastically modifies the predicted mask. Visualizations of the effectiveness are shown in Fig. 3, and Edge Refine can improve the quality of boundary to some extent.

Effectiveness of Cross-View Technique. Cross-view technique supervises the reconstructed 3D model from two viewpoints to be consistent. It does not use any additional supervision signals like viewpoint information or 3D model, only use the correspondence of two images from the same object of different viewpoints. We evaluate the performance without and with cross-view technique on ShapeNet dataset, and the mIoU results are 82.5% and 83.9% respectively. Cross-view technique brings 1.4% improvement with the reconstruction consistency between different viewpoints.

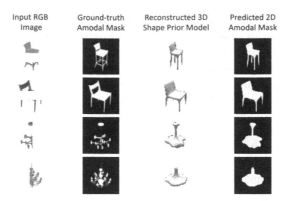

Fig. 5. Examples of chairs and lamps. In each four-tuple, images from left to right are input RGB images, ground-truth amodal masks, reconstructed 3D shape prior models, and predicted amodal masks.

4.6   Methodology Limitation

As shown in Table 1, for the categories of *Chair* and *Lamp*, ours A3D method fails to perform well with large margin, dropping for 9.6% and 11.7% IoU compared with the best performance. As illustrated in Fig. 5, there are plenty of holes in the *Chair* category and complicated structures in the *Lamp* category. However, in our A3D method, the 3D shape prior model is reconstructed by predicting vertices offset from the initial sphere, remaining the topology unchanged. Complicated structures in both *Chair* and *Lamp* categories require the topology changes, where our A3D network is incapable at present. We leave this problem to future work.

5   Conclusion

In this paper, we propose a novel coarse-to-fine Amodal 3D (A3D) network. A3D is a brand new framework which for the first time tackles the 2D AIS problem by reconstructing the 3D complete shape prior model. With the benefits of 3D modelling, A3D can alleviate the shortcoming that 2D AIS methods are difficult to generalize on untrained new viewpoints of the occluded 3D object. Our A3D achieves state-of-the-art performance on multiple AIS datasets.

Acknowledgments. This work was partially supported by the Natural Science Foundation of China under contracts 62088102. This work was also partially supported by Qualcomm. We also acknowledge High-Performance Computing Platform of Peking University for providing computational resources.

References

1. Bolya, D., Zhou, C., Xiao, F., Lee, Y.J.: Yolact: real-time instance segmentation. In: Proceedings of the IEEE/CVF International Conference on Computer Vision, pp. 9157–9166 (2019)
2. Chang, A.X., et al.: ShapeNet: an information-rich 3D model repository. arXiv preprint arXiv:1512.03012 (2015)
3. Chen, W., et al.: Learning to predict 3D objects with an interpolation-based differentiable renderer. Adv. Neural Inf. Process. Syst. **32**, 9609–9619 (2019)
4. Choy, C.B., Xu, D., Gwak, J.Y., Chen, K., Savarese, S.: 3D-R2N2: a unified approach for single and multi-view 3D object reconstruction. In: Leibe, B., Matas, J., Sebe, N., Welling, M. (eds.) ECCV 2016. LNCS, vol. 9912, pp. 628–644. Springer, Cham (2016). https://doi.org/10.1007/978-3-319-46484-8_38
5. Ehsani, K., Mottaghi, R., Farhadi, A.: Segan: segmenting and generating the invisible. In: Proceedings of the IEEE Conference on Computer Vision and Pattern Recognition, pp. 6144–6153 (2018)
6. Follmann, P., Bottger, T., Hartinger, P., Konig, R., Ulrich, M.: MVTec D2S: densely segmented supermarket dataset. In: Proceedings of the European Conference on Computer Vision (ECCV), pp. 569–585 (2018)
7. Follmann, P., König, R., Härtinger, P., Klostermann, M., Böttger, T.: Learning to see the invisible: end-to-end trainable Amodal instance segmentation. In: 2019 IEEE Winter Conference on Applications of Computer Vision (WACV), pp. 1328–1336. IEEE (2019)
8. Geiger, A., Lenz, P., Urtasun, R.: Are we ready for autonomous driving? the kitti vision benchmark suite. In: 2012 IEEE Conference on Computer Vision and Pattern Recognition, pp. 3354–3361. IEEE (2012)
9. He, K., Gkioxari, G., Dollár, P., Girshick, R.: Mask R-CNN. In: Proceedings of the IEEE International Conference on Computer Vision, pp. 2961–2969 (2017)
10. He, K., Zhang, X., Ren, S., Sun, J.: Delving deep into rectifiers: surpassing human-level performance on imageNet classification. In: 2015 IEEE International Conference on Computer Vision (ICCV), pp. 1026–1034 (2015)
11. He, K., Zhang, X., Ren, S., Sun, J.: Deep residual learning for image recognition. In: Proceedings of the IEEE Conference on Computer Vision and Pattern Recognition, pp. 770–778 (2016)
12. Hu, T., Wang, L., Xu, X., Liu, S., Jia, J.: Self-supervised 3D mesh reconstruction from single images. In: Proceedings of the IEEE/CVF Conference on Computer Vision and Pattern Recognition, pp. 6002–6011 (2021)
13. Hu, Y.T., Chen, H.S., Hui, K., Huang, J.B., Schwing, A.G.: SAIL-VOS: Semantic Amodal instance level video object segmentation-a synthetic dataset and baselines. In: Proceedings of the IEEE/CVF Conference on Computer Vision and Pattern Recognition, pp. 3105–3115 (2019)
14. Huang, Z., Huang, L., Gong, Y., Huang, C., Wang, X.: Mask scoring R-CNN. In: Proceedings of the IEEE/CVF Conference on Computer Vision and Pattern Recognition, pp. 6409–6418 (2019)
15. Kanazawa, A., Tulsiani, S., Efros, A.A., Malik, J.: Learning category-specific mesh reconstruction from image collections. In: Proceedings of the European Conference on Computer Vision (ECCV), pp. 371–386 (2018)
16. Kato, H., Ushiku, Y., Harada, T.: Neural 3D mesh renderer. In: Proceedings of the IEEE Conference on Computer Vision and Pattern Recognition, pp. 3907–3916 (2018)

17. Ke, L., Tai, Y.W., Tang, C.K.: Deep occlusion-aware instance segmentation with overlapping bilayers. In: Proceedings of the IEEE/CVF Conference on Computer Vision and Pattern Recognition, pp. 4019–4028 (2021)
18. Li, K., Hariharan, B., Malik, J.: Iterative instance segmentation. In: Proceedings of the IEEE Conference on Computer Vision and Pattern Recognition, pp. 3659–3667 (2016)
19. Li, K., Malik, J.: Amodal instance segmentation. In: Leibe, B., Matas, J., Sebe, N., Welling, M. (eds.) ECCV 2016. LNCS, vol. 9906, pp. 677–693. Springer, Cham (2016). https://doi.org/10.1007/978-3-319-46475-6_42
20. Li, X., et al.: Self-supervised single-view 3D reconstruction via semantic consistency. In: Vedaldi, A., Bischof, H., Brox, T., Frahm, J.-M. (eds.) ECCV 2020. LNCS, vol. 12359, pp. 677–693. Springer, Cham (2020). https://doi.org/10.1007/978-3-030-58568-6_40
21. Lin, T.Y., Dollár, P., Girshick, R., He, K., Hariharan, B., Belongie, S.: Feature pyramid networks for object detection. In: Proceedings of the IEEE Conference on Computer Vision and Pattern Recognition, pp. 2117–2125 (2017)
22. Lin, T.-Y., et al.: Microsoft COCO: common objects in context. In: Fleet, D., Pajdla, T., Schiele, B., Tuytelaars, T. (eds.) ECCV 2014. LNCS, vol. 8693, pp. 740–755. Springer, Cham (2014). https://doi.org/10.1007/978-3-319-10602-1_48
23. Ling, H., Acuna, D., Kreis, K., Kim, S.W., Fidler, S.: Variational Amodal object completion. In: NeurIPS (2020)
24. Liu, S., Li, T., Chen, W., Li, H.: Soft rasterizer: a differentiable renderer for image-based 3d reasoning. In: Proceedings of the IEEE/CVF International Conference on Computer Vision, pp. 7708–7717 (2019)
25. Nguyen, K., Todorovic, S.: A weakly supervised Amodal segmenter with boundary uncertainty estimation. In: Proceedings of the IEEE/CVF International Conference on Computer Vision, pp. 7396–7405 (2021)
26. Paszke, A., et al.: Pytorch: an imperative style, high-performance deep learning library. In: Wallach, H., et al. (eds.) Advances in Neural Information Processing Systems, vol. 32, pp. 8024–8035. Curran Associates, Inc. (2019)
27. Qi, L., Jiang, L., Liu, S., Shen, X., Jia, J.: Amodal instance segmentation with kins dataset. In: Proceedings of the IEEE/CVF Conference on Computer Vision and Pattern Recognition, pp. 3014–3023 (2019)
28. Ren, S., He, K., Girshick, R., Sun, J.: Faster R-CNN: towards real-time object detection with region proposal networks. Adv. Neural Inf. Process. Syst. **28**, 91–99 (2015)
29. Xiang, Y., Mottaghi, R., Savarese, S.: Beyond pascal: a benchmark for 3D object detection in the wild. In: IEEE Winter Conference on Applications of Computer Vision, pp. 75–82. IEEE (2014)
30. Xiao, Y., Xu, Y., Zhong, Z., Luo, W., Li, J., Gao, S.: Amodal segmentation based on visible region segmentation and shape prior. arXiv preprint arXiv:2012.05598 (2020)
31. Xie, E., et al.: Polarmask: single shot instance segmentation with polar representation. In: Proceedings of the IEEE/CVF Conference on Computer Vision and Pattern Recognition, pp. 12193–12202 (2020)
32. Xie, H., Yao, H., Sun, X., Zhou, S., Zhang, S.: Pix2vox: context-aware 3D reconstruction from single and multi-view images. In: Proceedings of the IEEE/CVF International Conference on Computer Vision, pp. 2690–2698 (2019)
33. Xie, H., Yao, H., Zhang, S., Zhou, S., Sun, W.: Pix2vox++: multi-scale context-aware 3D object reconstruction from single and multiple images. Int. J. Comput. Vis. **128**(12), 2919–2935 (2020)

34. Yan, X., Wang, F., Liu, W., Yu, Y., He, S., Pan, J.: Visualizing the invisible: occluded vehicle segmentation and recovery. In: Proceedings of the IEEE/CVF International Conference on Computer Vision, pp. 7618–7627 (2019)
35. Yan, X., Yang, J., Yumer, E., Guo, Y., Lee, H.: Perspective transformer nets: learning single-view 3D object reconstruction without 3D supervision. arXiv preprint arXiv:1612.00814 (2016)
36. Zhan, X., Pan, X., Dai, B., Liu, Z., Lin, D., Loy, C.C.: Self-supervised scene de-occlusion. In: Proceedings of the IEEE/CVF Conference on Computer Vision and Pattern Recognition, pp. 3784–3792 (2020)
37. Zhang, Z., Chen, A., Xie, L., Yu, J., Gao, S.: Learning semantics-aware distance map with semantics layering network for Amodal instance segmentation. In: Proceedings of the 27th ACM International Conference on Multimedia, pp. 2124–2132 (2019)
38. Zhu, Y., Tian, Y., Metaxas, D., Dollár, P.: Semantic Amodal segmentation. In: Proceedings of the IEEE Conference on Computer Vision and Pattern Recognition, pp. 1464–1472 (2017)

Data Efficient 3D Learner via Knowledge Transferred from 2D Model

Ping-Chung Yu$^{(\boxtimes)}$, Cheng Sun, and Min Sun

National Tsing Hua University, Hsinchu, Taiwan
{pingchungyu,chengsun}@gapp.nthu.edu.tw, sunmin@ee.nthu.edu.tw

Abstract. Collecting and labeling the registered 3D point cloud is costly. As a result, 3D resources for training are typically limited in quantity compared to the 2D images counterpart. In this work, we deal with the data scarcity challenge of 3D tasks by transferring knowledge from strong 2D models via RGB-D images. Specifically, we utilize a strong and well-trained semantic segmentation model for 2D images to augment RGB-D images with pseudo-label. The augmented dataset can then be used to pre-train 3D models. Finally, by simply fine-tuning on a few labeled 3D instances, our method already outperforms existing state-of-the-art that is tailored for 3D label efficiency. We also show that the results of mean-teacher and entropy minimization can be improved by our pre-training, suggesting that the transferred knowledge is helpful in semi-supervised setting. We verify the effectiveness of our approach on two popular 3D models and three different tasks. On ScanNet official evaluation, we establish new state-of-the-art semantic segmentation results on the data-efficient track.

Keywords: Knowledge transfer · 3D semantic segmentation · Point cloud recognition · 3D pre-training · Label efficiency

1 Introduction

Nowadays, 3D sensors are in demand by applications like AR/VR, 3D reconstruction, and autonomous driving. To have a high-level scene understanding (*e.g.*, recognition, semantic segmentation) on the captured 3D data, deep-learning-based models are typically employed for their outstanding performance. As 3D sensors become easier accessible, the architecture of deep 3D models [8,14,29,30,36,39,42] also progresses steadily for better result qualities. In this work, we investigate an orthogonal direction to model architecture design—we present a novel 3D models pre-training strategy to improve performance in a model agnostic manner.

In 2D vision tasks, model pre-training on ImageNet [10] has become a commonly applied strategy for achieving better performances across different downstream tasks. However, there is no standard large-scale dataset like ImageNet to

Supplementary Information The online version contains supplementary material available at https://doi.org/10.1007/978-3-031-19818-2_11.

pre-train 3D models due to the considerable effort to acquire and label a diverse set of point cloud data compared to the 2D counterpart. As a result, 3D models are typically trained from scratch, which hinders the performance, especially under the data scarcity scenario of the registered point cloud.

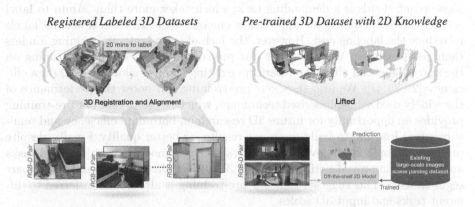

Fig. 1. Left-panel: collecting labeled 3D data is challenging due to the need for (a) robust 3D registration and alignment processes, and (b) a time-consuming human labeling process. Right-panel: a large amount of RGB-D data serve as the bridge between 2D and 3D knowledge. First, pseudo semantic labels are generated by applying an off-the-shelf 2D model on RBG images. Next, the pseudo-labeled 2D data are lifted to 3D using the associated depth map. Finally, we pre-train a 3D model with the large amount of pseudo-labeled 3D data.

To avoid the burden of data labeling, self-supervised learning has emerged as an alternative for pre-training 2D models without labels [3,6,7,16,18,19]. To reproduce this trend in 3D data, PointContrast [41] uses contrastive loss to learn the correspondence between two point clouds with visual overlap, improving results on the downstream 3D tasks. However, the diversity and scale of 3D datasets, even releasing the dependency on labeling, are still not comparable to the 2D datasets. For instance, ScanNet [9], which is used in PointContrast's pre-training, has only about a thousand indoor scenes, while ImageNet has more than a million images covering a thousand different classes. As a result, the accuracy improvement by self-supervised learning on point clouds is still limited for 3D tasks.

To address the issue of limited resources of point clouds, we present a novel 3D model pre-training approach via transferring the learned knowledge of a 2D model via RGB-D datasets (see Fig. 1). The single view depth sensor is much cheaper than ever and could be widely popular as a built-in function of phones to capture various scenes. Using RGB-D data as the bridge to transfer knowledge from strong 2D models to 3D models is thus a valuable direction to explore. Specifically, we employ a 2D semantic segmentation model, which is trained on a large and diverse scene parsing dataset, to augment the RGB-D images with pseudo-labels. We then train a 3D model to take the 3D point cloud lifted from

RGB-D as input and reproduce the pseudo-labels. By doing so, the 3D models can learn from the strong 2D teacher model and also see a large variety of scenes captured by the RGB-D data.

We demonstrate the effectiveness of our pre-training on semantic segmentation for scene point cloud of the popular ScanNet dataset. Annotating 3D scene point clouds is a demanding task, which takes more than 20 min to label a single scene. Some recent approaches emerges to learn from fewer 3D labels to reduce the labeling cost. However, the lack of large-scale pre-training hinders their performances. We show that our pre-training with simple fine-tuning on the scarce label can already outperform existing results tailored for 3D data efficiency [21,26,41]. We also show our pre-training can boost the performance of the widely used semi-supervised techniques, which suggests that our pre-training provides an opportunity for future 3D research on both data efficiency and semi-supervised learning to build upon our results for better quality. Finally, despite pre-training on a scene level, we also evaluate our models on object-level tasks and observe improved performances on 3D object classification and shape part segmentation. This suggests that our pre-training is also well transferred to different tasks and input 3D scales.

We summarize our contributions as follows:

- We introduce a pre-training strategy to transfer knowledge from a strong 2D scene parsing model via RGB-D images to 3D models.
- We demonstrate the effectiveness of our pre-training under limited data scenario across two models (*i.e.*, O-CNN [37], SparseConv [14]) and three different tasks (*i.e.*, 3D object classification, point-cloud part segmentation, and indoor point-cloud segmentation).
- By simply finetuning our pre-trained model on a few labels, we establish new state-of-the-art results on ScanNet [9] official evaluation on data-efficient setups, verifying that our pre-training results in data-efficient 3D learners.

2   Related Work

Deep 3D Models for Point Cloud Understanding. As the 3D analysis studies flourish, several deep models are proposed to extract point cloud features for a high-level understanding. Existing approaches can be classified into point-based and voxel-based methods. PointNet [29] and PointNet++ [30] are the pioneering point-based methods to apply multi-layer perceptron layers directly on point clouds. After that, several convolution-based models [1,24,36,39] are proposed, achieving better quality. Recently, attention-based models [42] have become a new effective way for point cloud processing. On the other hand, the voxel-based method is attractive for its computational-friendly 3D data representations. A discretized step is typically applied on the point cloud before employing the voxel-based models [28,40].

As the cubic memory complexity limits the resolution of dense voxel grids, 3D sparse CNNs [8,13,14] emerges to achieve a feasible space-time complexity for scene-level point clouds, where the CNNs are working on occupied voxels only.

OctNet [32] and O-CNN [37] further use the octree data structure to achieve a higher grid resolution efficiently.

As the deep architectures for 3D point clouds progress steadily, all of them are still trained from scratch due to the lack of a large-scale point cloud dataset for pre-training. To sidestep the issue of 3D resources, this work presents a model agnostic pre-training strategy using 2D resources to improve performances.

Data-Efficient 3D. Data-efficient learning restricts the amount and the variety of labeled data of the target task, which is helpful for tasks on scene-level point clouds. Registered scene point clouds are hard to acquire and time-consuming to label (*e.g.*, 20 more minutes for a ScanNet scene). Thus, data-efficient solutions are always welcome. Existing works have explored self-supervised pre-training and semi-supervised learning using unlabeled scene point clouds to improve performance. PointContrast [41] sub-samples partial scans and use contrastive learning to pre-train 3D models to identify point correspondents. CSC [21] further improves the pre-training by incorporating spatial scene contexts into the objective. As negative pairs sampling in contrastive learning could be ambiguous, ViewPointBN [27] proposed to use correlation as the objective instead. The aforementioned self-supervised pre-training achieves good label efficiency in a target task agnostic manner. In OTOC [26], pseudo-label-based semi-supervised learning is employed to simultaneously learn from both labeled and unlabeled data, achieving superior label efficiency. We use the abundant 2D resources to pre-train instead of 3D, achieving state-of-the-art label efficiency on scene point clouds.

Knowledge Transferred from 2D. Most recently, DepthContrast [45] and Contrastive Pixel-to-point [25] are also proposed to pre-train 3D models using 2D RGB-D datasets. DepthContrast trains 3D models to discriminate 3D point clouds projected from different RGB-D images. Contrastive Pixel-to-point treat each 3D point as instance and apply contrastive loss to learn from pixel features extracted by a trained 2D model.

One challenge of these contrastive-based methods is the ambiguity of the selected negative pairs. Negative pairs sampling is a crucial problem for contrastive pre-training to avoid collapsing solutions [7,11]. The sampling strategy requires careful designs [6,19] in the contrastive pre-training, even for images. As the diversity of the employed indoor pre-training data is still limited, the model could have difficulty differentiate two room with similar setup or two 3D points belonging to the same stuff (*e.g.*, walls). Conversely, we directly train the 3D models to reproduce the pseudo labels generated by a strong and well-trained image scene parser, which is more straightforward but non-trivial to learn as we have clear and informative supervision for each 3D point.

2D3DNet [12] trains 3D models with only 2D supervision, which requires the 2D teacher to be a specialist for the downstream task and the 3D model has to co-work with the 2D teacher to achieve good results. Our pre-trained 3D model does not work directly in the downstream 3D task, but it is generally benefit to different downstream task and our 3D model works independently after pre-training.

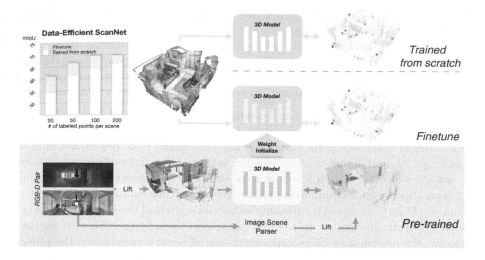

Fig. 2. Overview of our approach. **(Below)** We use a strong and well-trained image scene parser to augment single view RGB-D datasets with pseudo-labels, which is used to pre-train a 3D model in an architecture agnostic manner. **(Top)** Our pre-training improves the results of the limited annotation training.

3   Approach

Our goal is to pre-train 3D models in an architecture agnostic manner. An overview of the proposed approach is illustrated in Fig. 2. Below, we first give a general introduction to 3D models in Sect. 3.1. In Sect. 3.2, we detail the proposed pre-training approach. We also apply our pre-trained model on some semi-supervised techniques in Sect. 3.3.

3.1   3D Model

Our pre-training approach is model agnostic and does not require specific 3D architecture designs. Here, we describe a general 3D encoder-decoder to extract deep features from point clouds. A 3D point cloud $x^{(\mathrm{pts})}$ represents a 3D scene or object by a set of 3D coordinates (with additional color or normal features). For voxel-based models, a pre-process input layer before the first model layer is needed to discretize the coordinates into a regular 3D grid. The encoder E, consisting of a sequence of 3D convolution layers, batch normalizations, and nonlinear activation (*e.g.*, ReLU), is a bottom-up way to map the point clouds into a down-sampled latent features $z^{(\mathrm{e})} = E(x^{(\mathrm{pts})})$, which has the high-level understanding of the input scene or object. The decoder D then upsamples and incorporates the low-level feature into $z^{(\mathrm{e})}$ to have a holistic point-wise deep feature $z^{(\mathrm{d})} = D(z^{(\mathrm{e})})$.

Classification. To classify a whole input point cloud into a pre-defined set of classes, we discard the decoder D as we do not need point-level features. The high-level features $z^{(e)}$ are first aggregated into a single latent vector via global max pooling, which is then followed by a classification head ClsHead to map the latent vector into a categorical distribution: $y^{(cls)} =$ ClsHead(GlobalMaxPool($z^{(e)}$)), where the ClsHead consists of a linear layer and a Softmax layer.

Point-Level Semantic Segmentation. The semantic segmentation head simply maps the point features into per-point class distribution by a linear and a Softmax layer: $y^{(ss)} = \text{SegHead}(z^{(d)})$.

3.2   Knowledge Transferred from 2D

Our approach aims to pre-train a 3D model by transferring the knowledge of a strong 2D model learned from a large-scale 2D dataset. However, the 2D model $F^{(2D)}$ takes images $x^{(img)}$ as input while the 3D model takes point clouds $x^{(pts)}$. To learn from 2D models, we use RGB-D dataset $\{(x^{(img)}, d^{(img)})\}$ as the bridge, where $d^{(img)}$ is the depth map of a image. Our idea is to use a well-trained 2D model to generate pseudo-label for the images in the RGB-D dataset

$$t = F^{(2D)}\left(x^{(img)}\right),\qquad(1)$$

and then we lift the image to point cloud using their depth maps

$$x^{(img2pts)} = \text{Lift}\left(x^{(img)}, d^{(img)}\right)\qquad(2)$$

so that we can train the 3D model using the augmented RGB-D dataset $\{(x^{(img2pts)}, t)\}$. Below, we introduce the explored two different sources of RGB-D data and detail our pre-training strategy.

Lifting Perspective Images. With the provided depth maps and the camera intrinsic K, we can lift a 2D image coordinate $[u, v]$ to the 3D camera coordinate $[x, y, z]$ by:

$$\begin{bmatrix} x \\ y \\ z \end{bmatrix} = d \cdot K^{-1} \begin{bmatrix} u \\ v \\ 1 \end{bmatrix},\qquad(3)$$

where d is the depth value of a pixel and $[u, v, 1]^{\top}$ is the homogeneous coordinate.

Lifting Panoramic Images. Panoramic images cover more information in one shot attributed to the omnidirectional field of view compared to perspective images. Unlike perspective depth maps where z-values are recorded, panoramic depth maps directly record the distance between the observed 3D points to camera. We can lift a panoramic image to 3D by:

$$\begin{cases} x = d \cdot \cos(v) \cdot \cos(u) \; ; \\ y = d \cdot \cos(v) \cdot \sin(u) \; ; \\ z = -d \cdot \sin(v) \; , \end{cases}\qquad(4)$$

where d is the recorded distance of a pixel in the panoramic depth maps, and $u \in [-\pi, \pi]$, $v \in [-\frac{\pi}{2}, \frac{\pi}{2}]$ are the panoramic image coordinate in UV space.

Learning from 2D Scene Parser via Soft Pseudo-Label. Image scene parsing aims to classify each pixel of images, which provides a thorough and detailed understanding of the captured scenes. Besides, existing training corpora for image scene parsing [2,48] is abundant, with more than 10k images covering a large variety of scenes and classes. Scene parsing models [5,22,31,44,46,49] have also progressed steadily and achieved strong performances. In this work, we adopt DPT [31] for its outstanding performance, and we find it generalized well on both perspective and panoramic images.

We first attach a point semantic segmentation head SegHead$^{(\text{pre})}$ on the 3D encoder decoder to classify C classes of the trained DPT

$$\boldsymbol{y}^{(\text{img2pts})} = \text{SegHead}^{(\text{pre})} \left(D \left(E \left(\boldsymbol{x}^{(\text{img2pts})} \right) \right) \right). \tag{5}$$

Instead of the one-hot decision, soft labels are learned as the target during pre-training. Soft-label is first introduced in a knowledge distillation method [20], where each pixel or point is labeled by the categorical distribution over classes. We use cross-entropy loss as the training objective for the 3D model to learn from the soft-label:

$$\mathcal{L}^{(\text{pretrain})} = \frac{1}{N} \sum_{i=1}^{N} \sum_{c=1}^{C} \left(-t_i[c] \log \left(\boldsymbol{y}_i^{(\text{img2pts})}[c] \right) \right), \tag{6}$$

where i is the index to the N 3D points, and t_i is the soft-label probability of the pixel corresponding to the i-th 3D point.

Downstream Tasks. After pre-training, we discard the pre-training head SegHead$^{(\text{pre})}$ and directly used the trained encoder E and decoder D in the downstream tasks.

3.3   Semi-supervised

Semi-supervised learning is helpful to learn from scarce labels, where the improvement by our pre-training can be additive to it to achieve an even better result. We use two simple and commonly used semi-supervised techniques in this work. First, entropy minimization [15] encourages concentration of the predicted probability distribution on the unlabeled data:

$$\mathcal{L}^{(\text{mini-entropy})} = \frac{1}{N} \sum_{i=1}^{N} \sum_{c=1}^{C} \left(-\boldsymbol{y}_i[c] \log \left(\boldsymbol{y}_i[c] \right) \right). \tag{7}$$

Second, we use mean-teacher [35] to guide our model, where the weights of teacher model θ_t' is obtained by exponential moving averaging (EMA) of the weight of student model θ_t across training step t:

$$\theta_t' = \alpha \theta_{t-1}' + (1 - \alpha)\theta_t, \tag{8}$$

where α is a smoothing hyperparameter. We use mean squared error to encourage consistency between our predictions y and teacher predictions y'

$$\mathcal{L}^{(\text{consistency})} = \frac{1}{N} \sum_{i=1}^{N} \|y_i - y'_i\|^2 . \tag{9}$$

Different from the 2d teacher model we mention in Sect. 1, the mean-teacher here is a typical semi-supervised technique.

4   Experiments

In this section, we evaluate our pre-training strategy under the data scarcity scenario. The implementation details are provided in Sect. 4.1. To demonstrate the benefits of our approach, we conduct a series of experiments on scene understanding task in Sect. 4.2 and the shape analysis tasks in Sect. 4.4. We show the results of our pre-training on different 3D models in Sect. 4.3 and present the ablation experiments in Sect. 4.5.

4.1   Implement Details

Pre-training Images Dataset. For perspective pre-training, we utilize SUNRGB-D [34], which consists of more than 10,000 RGB-D indoor images and the corresponding pixel-wise semantic labels from 37 categories. In addition to the depth maps, SUNRGB-D also provides the intrinsic which allow us to lift the 3D scenes. We split the dataset into 5,285 training sets and 5,050 validation sets in the pre-training procedure. We only used the provided ground-truth semantic maps in ablation experiments and used the generated pseudo-label by our approach in all other experiments. For panoramic pre-training, Matterport3D [4] provides various indoor scenes captured by panoramic images, each of which covers a omnidirectional field of view. Matterport3D contains 10,800 panoramic views with corresponding depth maps from 90 building-scale scenes, including 61 scenes for training, 11 for validation, and 18 for testing following official data splits.

Image Scene Parser. We employ DPT [31] as a teacher model in our pre-training. The DPT is first trained on the ADE20K [48] dataset, which has 20k images with ground-truth semantic segmentation maps covering 150 different classes. In pre-training, the DPT is fixed and used to generate soft pseudo-labels to transfer its knowledge to the 3D models.

3D Models. SparseConv [14] is trained by Adam optimizer [23] for scene semantic segmentation. The models are trained on NVIDIA GTX 1080Ti GPUs for 200 epochs with batch size 4 and learning rate 0.001. O-CNN [37] is implemented for shape analysis and indoor scene semantic segmentation. During lifted scenes pre-training, the projected-points are formed as 512^3 resolution of leaf octants as the same as indoor scene semantic segmentation. For part segmentation, we train networks for 600 epochs with batch size 32, leaning rate 0.025 and weight decay 0.0001. For scene semantic segmentation, we train models for 500 epochs with batch size 4 and leaning rate 0.05. Both tasks adopt the SGD optimizer with a momentum 0.9 and use polynomial schedulers powered by 0.9. The models are trained on several NVIDIA Tesla V100-SXM2 GPUs. As the setup for part segmentation, the classification model is trained for 300 epochs with batch size 32 and learning rate 0.05. The O-CNNs are adopted with 32^3, 64^3 and 512^3 resolution of leaf octants for object classification, shape part segmentation and scene semantic segmentation, respectively. The point-wise predictions are interpolated by linear for part segmentation and nearest for scene semantic segmentation.

Data Augmentation. We apply random rotation, random scaling, random elastic distortion, random color contrast, and random color jittering to the input point cloud as data augmentation.

4.2   Data Efficient Scene Semantic Segmentation

We validate the effectiveness of our pre-training on ScanNet Data Efficient Benchmark. ScanNetV2 [9] consists of various indoor scenes formed as 3D point clouds and corresponding semantic annotations. The data is taken from 707 distinct spaces and covers 20 semantic classes. We follow the official data scarcity scenarios: *i)* Limited Annotations (LA) considers only a few labeled points in each scene, *ii)* and Limited Reconstructions(LR) considers the a few number of labeled scenes. We use O-CNN pre-trained by our approach on Matterport3D dataset for all the results submitted to official ScanNet.

Limited Annotations. Following the official configuration in the *3D Semantic label with Limited Annotations benchmark*, there are four different scales, including $\{20, 50, 100, 200\}$ labeled points per training scene. We show the quantitative comparison on the testing split in Table 1. We achieve state-of-the-art results on purely supervised fine-tuning setup and semi-supervised setup. It is worth noting that our results without using unlabeled data already outperforms the semi-supervised OTOC [26] on 50, 100, and 200 labeled points per scene, which further highlights the effectiveness of our pre-training approach.

Limited Reconstruction. Limited Reconstruction (LR) is constructed by limiting the number of scenes. The official subset is randomly sampled from 1201 scenes, where $\{1\%, 5\%, 10\%, 20\%\}$ of training data is subsampled (corresponding to 12, 60, 120, and 240 scenes). We compare our results on *3D Semantic label with Limited Reconstructions benchmark* in Table 2. When only 1% of the training data is available, all methods achieve similar performance. When $5\%, 10\%, 20\%$ of the training data is given, we achieve a superior mIoU compared to previous methods. Additionally, we adopt semi-supervised learning and examine the capability under the limited reconstruction scenario. We first trained our model on the given labeled data, and then we randomly sample a subset of unlabeled scenes to produce pseudo labels. The pseudo labels is generated by selecting the

Table 1. Quantitative comparisons on official ScanNet Limited Annotations(LA) track. For a fair comparisons, the results are separated based on whether the unlabeled data points are used in training.

Method	Semi	# of labeled points per scene			
		20	50	100	200
PointContrast [41]	–	55.0	61.4	63.6	65.3
CSC [21]	–	53.1	61.2	64.4	66.5
ViewPointBN [27]	–	54.8	62.3	65.0	66.9
Ours	–	<u>57.9</u>	<u>65.8</u>	**71.4**	**71.1**
OTOC [26]	✓	59.4	64.2	67.0	69.4
Ours	✓	**63.9**	**69.5**	<u>70.4</u>	<u>70.9</u>

Table 2. Quantitative comparisons on offical ScanNet Limited Reconstructions(LR) track.

Method	Semi	Percentage of labeled scene			
		1%	5%	10%	20%
PointContrast [41]	–	25.3	43.8	55.5	60.3
CSC [21]	–	**27.0**	46.0	57.5	61.2
ViewPointBN [27]	–	25.6	45.2	56.6	62.5
Ours	–	26.6	<u>46.7</u>	**61.2**	<u>64.0</u>
Ours	✓	26.3	**50.8**	60.8	**66.3**

top-most 20% confident predictions of each class. We can then combined the labeled, pseudo-labeled, and unlabeled data points to train our models.

From the comparisons in Table 1 and Table 2, we find our pre-trained models with simple supervised fine-tuning already outperform existing methods tailored for 3D data-efficiency. Combining our pre-training with semi-supervised sometimes can further improve our results. In cases that semi-supervised does not improve, we still achieve similar performance. As we only use the simplest semi-supervised techniques, a more tailored one could be more helpful. In sum, our pre-training provides a good starting point for future work to develop 3D data-efficient approach.

4.3   Pre-training on Different 3D Models

To verify the effectiveness of our pre-training strategy on different 3D models, we implement O-CNN [37] and SparseConv [14] for 3D scene semantic segmentation. SparseConv uses a sparse voxelized input representation and keeps the same level of sparsity throughout the model. O-CNN builds up the octree structure of 3D points representation. As the octree depth increases, the octree-based CNN layers extract the information with higher resolution. Both models work

Table 3. mIoU of different 3D segmentation model and different pre-training strategy. The results are evaluated on ScanNet validation set. The oracle pre-training supervision is the ground-truth semantic segmentation maps provided in the pre-training dataset.

Based model	Pre-training strategy	Oracle pre-training supervision	LA (points) 20	200
SparseConv	Trained from scratch	–	51.6	65.4
	Perspective pre-training	✓	55.4	66.0
	Perspective pre-training	–	56.2	65.7
	Panoramic pre-training	–	58.5	67.3
O-CNN	Trained from scratch	–	52.9	65.0
	Perspective pre-training	✓	56.9	66.2
	Perspective pre-training	–	56.2	65.3
	Panoramic pre-training	–	55.6	67.8

on occupied point sets only to ensure computational efficiency, and the networks are constructed by U-Net [14,33] architecture to produce point-wise predictions. The overall results are summarized in Table 3. On both SparseConv and O-CNN, all different variations of our pre-training approach improved mIoU significantly over training from scratch.

4.4   Shape Analysis Under Limited Data Scenario

Our approach discussed above presents a positive outcome on scene-level semantic segmentation under the limited data scenario. In this section, we apply our pre-trained model on object-level 3D tasks such as object classification and shape part segmentation. With the data scarcity scenario for shape analysis, the implementation details and results are presented in the following paragraph. For all the experiments in shape analysis, we use O-CNN pre-trained by our approach on Matterport3D, which can also show the generalizibility from scene-level to object-level of our pre-training.

Object Classification. Object classification is the fundamental task of shape analysis. Given the 3D shape represented as point clouds, the classification model intends to assign the category of the input objects. With our approaches, the 3D models are required to extract features by limited annotations.

Dataset. ModelNet40 [40] dataset contains 12,311 shapes from 40 object categories, and is split into 9,843 objects for training and 2,468 objects for testing. For the limited data scenario, we randomly sample the subset by ratio {1%, 5%, 20%} in each category from the training set, and evaluate on the original testing set. For semi-supervised learning, the unlabeled batches is sampled randomly from the remained data of the training set.

Results. We finetune the 3D models with panoramic images pre-trained for object classification. The results are shown in Table 4, which is divided by whether the

Table 4. Accuracy of 3D object classification in ModelNet40 [40] dataset under limited training data. The training data is random sampled by 1%, 5%, 20% and the results are evaluated on the same testing set.

Pre-training strategy	Semi-supervised	1%	5%	20%
Trained from scratch	–	60.2	79.6	86.5
	✓	61.1	80.4	87.6
Our pre-training	–	65.5	80.6	87.5
	✓	**69.0**	**82.9**	**89.4**

Table 5. mIoU of shape part segmentation in ShapeNet [40] dataset under sampled training data. We use category mIoU across all categories and instance mIoU across all shape instance as the evaluation metrics. The results are evaluated on ShapeNet testing set which split by [43].

Pre-training strategy	Category mIoU			Instance mIoU		
	1%	5%	20%	1%	5%	20%
Trained from scratch	63.3	67.5	73.9	67.9	72.2	76.7
Our pretraining	**64.1**	**74.1**	**76.3**	**68.5**	**76.8**	**78.9**

semi-supervised training is used. Considering the model trained without unlabeled data, our pre-training strategy can directly improve the accuracy compared with the model trained from scratch. In the cases that the models are trained from scratch, the improvements of using the unlabeled data are limited. Conversely, our pre-training with only supervised fine-tuning already outperform the results of semi-supervised with random weight initialization. Besides, the improvements by semi-supervised learning is much significant than the improvement from random initialization.

Shape Part Segmentation. Shape part segmentation is more complicated than classification for shape analysis. Part segmentation is expected to generate the dense prediction and assign the part category to each point in objects.

Dataset. Yi *et al.* [43] annotates a subset of ShapeNet [40] 3D models with semantic part labels. The annotated subset of ShapeNet contains 2 to 6 parts per category, and 50 distinct parts in total among 16 shape categories. For training on the limited scenario, we build the limited subset by random sampling as ratio {1%, 5%, 20%} from each shape category, and then evaluate on the original validation set. According to the setup of pre-processing in [37], the input points are condensed by triangle faces and built in octree structure.

Results. We finetune the segmentation models with panoramic images pre-training in each category separately, and evaluate the performance of part segmentation by mean IoU across all categories and mIoU across all object instances. The results are shown in Table 5. As the data split in [17, 38, 47] are not provided, we only compare the results on our own split. With the knowledge distilled from off-the-shelf image scene parser, the pre-trained model performs better category mIoU and instance mIoU across all limited scales (1%, 5% and 20%).

4.5 Ablation Study

Image Modality. To observe the effect of the distillation via different image modalities, we pre-train O-CNN [37] on perspective images and panoramic images, and the supervised fine-tune the O-CNN on ScanNet with limited annotations. The results are demonstrated in Table 3. For all the variations of our pre-training, we achieve significant better results than training from scratch. Note that the panoramic images pre-training rises the performance by 2.7% and 2.8% mIoU compared with baseline on 20, 200 labeled points of scene(55.6% vs 52.9% and 67.8% vs 65.0%). The results imply that models can learn non-trivial and informative representation from the strong and well-trained image scene parser via RGB-D images. The learned representation boosts the capability of scene understanding under the limited annotations.

This work can not conclude whether perspective pre-training or panoramic pre-training is better, as the number of their training data is different. The panoramic data used in this work leads us to better results, so we use panoramic pre-training in all the other experiments.

Pre-training by 2D Annotations and Pseudo Labels. In our pre-training strategy, we supervise 3D models by the pseudo-labels introduced in Sect. 3.2. To examine the difference between the information distilled from pseudo labels and ground-truth labels, we conduct our pre-training strategy on perspective images with ground-truth semantic labels and generated pseudo-labels. The ground-truth semantic labels are provided by SUNRGB-D [34], while the pseudo labels are predicted by a strong and well-trained image scene parser. With the 3D model trained on limited annotations scenario, the results are presented in Table 3. The perspective images attain higher performance than the model trained from scratch regardless of whether the ground-truth labels are provided. The results imply that the pseudo labels can also improve the knowledge distillation during pre-training. Additionally, with provided annotations of images, our pre-training strategy achieves better mIoU than the baseline by 4% mIoU on 20 points and 1% mIoU on 200 points. Overall speaking, the less constrained pseudo-labels pre-training can achieve comparable results to the ground-truth labels pre-training, which suggests that ground-truth semantic maps are not necessary in our pre-training approach.

Combining Pre-training with Semi-supervised Learning. We study the effectiveness of our pre-training when combining with semi-supervised learning to simultaneously learn from labeled and unlabeled data. We run semi-supervised learning with random initialized model weights and our pre-trained model weights, and then evaluate the performance on ScanNetv2 validation sets. The approach is examined on both limited annotation(LA) and limit reconstruction(LR).

For LA, the results are summarized in Table 6. In "trained from scratch" column, the semi-supervised learning slightly improves mIoU when unlabeled data is used during training. In contrast, our pre-training strategy increases the performance significantly when combining with the semi-supervised learning, particularly in the 20, 50 of labeled points per scene. Using unlabeled data with our pre-training strategy outperforms the models trained from scratch by 4.5% and 4.1%.(61.1% vs 56.6% and 66.6% vs 62.5%).

Table 6. Ablation studies of our pre-training with and without semi-supervised learning. We follow the limited training data from official. The results are presented as mIoU and compared on ScanNet validation set.

Method	Semi	# of labeled points per scene			
		20	50	100	200
Trained from scratch	–	54.2	61.1	64.1	65.9
Our pre-training	–	**55.9**	**63.4**	**70.2**	**69.4**
Trained from scratch	✓	56.6	62.5	65.3	66.0
Our pre-training	✓	**61.1**	**66.6**	**69.6**	**69.7**

Table 7. mIoU of limited scene variety on ScanNet validation set.

Model	Pre-training strategy	LR(%)			
		1	5	10	20
O-CNN	Trained from scratch	18.7	39.7	52.4	59.6
	Our pre-training	26.3	44.9	56.5	62.7
	Our pre-training + Semi-supervised	**27.5**	**49.4**	**59.2**	**64.8**

For LR, we follow Sect. 4.2 for semi-supervised training. As a result, Table 7 shows that our pre-training strategy achieves greater mIoU by 27.5%, 49.4%, 59.2% and 64.8% on {1%, 5%, 10%, 20%}. Consequently, unlabeled data enhances the ability of scene understanding on both limited annotations and limited reconstruction.

5   Conclusion

This work presents a new 3D deep models pre-training strategy. We use RGB-D images as the bridge to transfer the knowledge from a strong and well-trained 2D scene parsing network to 3D models. Our pre-training strategy is model agnostic, and we show its effectiveness on two popular 3D models architectures. On the official ScanNet data-efficient track, we establish new state-of-the-art results. Besides, we also show that the improvement by our pre-training is additive to other label-efficient techniques. We hope our pre-trained weights can serve as a stepping stone for future 3D approaches and encourage more exploration on how to make good use of 2D resources in 3D tasks.

Acknowledgements. This work is supported in part by Ministry of Science and Technology of Taiwan (MOST 110-2634-F-002-051). We would like to thank National Center for High-performance Computing (NCHC) for computational and storage resource.

References

1. Boulch, A.: Convpoint: continuous convolutions for point cloud processing. Comput. Graph. **88**, 24–34 (2020)
2. Caesar, H., Uijlings, J.R.R., Ferrari, V.: Coco-stuff: thing and stuff classes in context. In: CVPR, pp. 1209–1218 (2018)
3. Caron, M., Misra, I., Mairal, J., Goyal, P., Bojanowski, P., Joulin, A.: Unsupervised learning of visual features by contrasting cluster assignments. In: NeurIPS (2020)
4. Chang, A.X., Dai, A., Funkhouser, T.A., Halber, M., Nießner, M., Savva, M., Song, S., Zeng, A., Zhang, Y.: Matterport3d: Learning from RGB-D data in indoor environments. In: 3DV. pp. 667–676 (2017)
5. Chen, L., Papandreou, G., Kokkinos, I., Murphy, K., Yuille, A.L.: Deeplab: Semantic image segmentation with deep convolutional nets, atrous convolution, and fully connected crfs. IEEE Trans. Pattern Anal. Mach. Intell. **40**, 834–848 (2018)
6. Chen, T., Kornblith, S., Norouzi, M., Hinton, G.E.: A simple framework for contrastive learning of visual representations. In: ICML, pp. 1597–1607 (2020)
7. Chen, X., He, K.: Exploring simple siamese representation learning. In: CVPR, pp. 15750–15758 (2021)
8. Choy, C.B., Gwak, J., Savarese, S.: 4d spatio-temporal convnets: Minkowski convolutional neural networks. In: CVPR, pp. 3075–3084 (2019)
9. Dai, A., Chang, A.X., Savva, M., Halber, M., Funkhouser, T.A., Nießner, M.: Scannet: richly-annotated 3d reconstructions of indoor scenes. In: CVPR, pp. 2432–2443 (2017)
10. Deng, J., Dong, W., Socher, R., Li, L., Li, K., Fei-Fei, L.: Imagenet: a large-scale hierarchical image database. In: CVPR, pp. 248–255 (2009)
11. et al., Z.: Barlow twins: self-supervised learning via redundancy reduction. In: ICML (2021)
12. Genova, K., et al.: Learning 3d semantic segmentation with only 2d image supervision. In: 3DV (2021)
13. Graham, B.: Sparse 3d convolutional neural networks. In: Xie, X., Jones, M.W., Tam, G.K.L. (eds.) BMVC, pp. 150.1-150.9 (2015)
14. Graham, B., Engelcke, M., van der Maaten, L.: 3d semantic segmentation with submanifold sparse convolutional networks. In: CVPR, pp. 9224–9232 (2018)
15. Grandvalet, Y., Bengio, Y.: Semi-supervised learning by entropy minimization. In: NeurIPS, pp. 281–296 (2005)
16. Grill, J., et al.: Bootstrap your own latent - a new approach to self-supervised learning. In: NeurIPS (2020)
17. Hassani, K., Haley, M.: Unsupervised multi-task feature learning on point clouds. In: 2019 IEEE/CVF International Conference on Computer Vision, ICCV 2019, Seoul, Korea (South), October 27–November 2, 2019, pp. 8159–8170. IEEE (2019)
18. He, K., Chen, X., Xie, S., Li, Y., Dollár, P., Girshick, R.B.: Masked autoencoders are scalable vision learners. arXiv preprint arXiv:2111.06377 (2021)
19. He, K., Fan, H., Wu, Y., Xie, S., Girshick, R.B.: Momentum contrast for unsupervised visual representation learning. In: CVPR, pp. 9726–9735 (2020)
20. Hinton, G.E., Vinyals, O., Dean, J.: Distilling the knowledge in a neural network. arXiv preprint arXiv:1503.02531 (2015)

21. Hou, J., Graham, B., Nießner, M., Xie, S.: Exploring data-efficient 3d scene understanding with contrastive scene contexts. In: CVPR, pp. 15587–15597 (2021)
22. Hsiao, C., Sun, C., Chen, H., Sun, M.: Specialize and fuse: pyramidal output representation for semantic segmentation. In: ICCV (2021)
23. Kingma, D.P., Ba, J.: Adam: A method for stochastic optimization. In: Bengio, Y., LeCun, Y. (eds.) 3rd International Conference on Learning Representations, ICLR 2015, San Diego, CA, USA, 7–9 May, 2015, Conference Track Proceedings (2015)
24. Li, Y., Bu, R., Sun, M., Wu, W., Di, X., Chen, B.: Pointcnn: convolution on x-transformed points. In: Bengio, S., Wallach, H.M., Larochelle, H., Grauman, K., Cesa-Bianchi, N., Garnett, R. (eds.) NeurIPS, pp. 828–838 (2018)
25. Liu, Y.C., et al.: Learning from 2d: Contrastive pixel-to-point knowledge transfer for 3d pretraining. arXiv preprint arXiv:2104.04687 (2021)
26. Liu, Z., Qi, X., Fu, C.: One thing one click: a self-training approach for weakly supervised 3d semantic segmentation. In: CVPR, pp. 1726–1736 (2021)
27. Luo, L., Tian, B., Zhao, H., Zhou, G.: Pointly-supervised 3d scene parsing with viewpoint bottleneck. arXiv preprint arXiv:2109.08553 (2021)
28. Maturana, D., Scherer, S.A.: Voxnet: A 3d convolutional neural network for real-time object recognition. In: 2015 IEEE/RSJ International Conference on Intelligent Robots and Systems, IROS 2015, Hamburg, Germany, 28 September–2 October, 2015, pp. 922–928 (2015)
29. Qi, C.R., Su, H., Mo, K., Guibas, L.J.: Pointnet: deep learning on point sets for 3d classification and segmentation. In: CVPR, pp. 77–85 (2017)
30. Qi, C.R., Yi, L., Su, H., Guibas, L.J.: Pointnet++: deep hierarchical feature learning on point sets in a metric space. In: NeurIPS, pp. 5099–5108 (2017)
31. Ranftl, R., Bochkovskiy, A., Koltun, V.: Vision transformers for dense prediction. In: ICCV, pp. 12179–12188 (2021)
32. Riegler, G., Ulusoy, A.O., Geiger, A.: Octnet: learning deep 3d representations at high resolutions. In: CVPR, pp. 6620–6629 (2017)
33. Ronneberger, O., Fischer, P., Brox, T.: U-Net: convolutional networks for biomedical image segmentation. In: Navab, N., Hornegger, J., Wells, W.M., Frangi, A.F. (eds.) MICCAI 2015. LNCS, vol. 9351, pp. 234–241. Springer, Cham (2015). https://doi.org/10.1007/978-3-319-24574-4_28
34. Song, S., Lichtenberg, S.P., Xiao, J.: SUN RGB-D: a RGB-D scene understanding benchmark suite. In: CVPR, pp. 567–576 (2015)
35. Tarvainen, A., Valpola, H.: Mean teachers are better role models: Weight-averaged consistency targets improve semi-supervised deep learning results. In: ICLR (2017)
36. Thomas, H., Qi, C.R., Deschaud, J., Marcotegui, B., Goulette, F., Guibas, L.J.: Kpconv: flexible and deformable convolution for point clouds. In: ICCV, pp. 6410–6419 (2019)
37. Wang, P., Liu, Y., Guo, Y., Sun, C., Tong, X.: O-CNN: octree-based convolutional neural networks for 3d shape analysis. ACM Trans. Graph, 72:1–72:11 (2017)
38. Wang, P., Yang, Y., Zou, Q., Wu, Z., Liu, Y., Tong, X.: Unsupervised 3d learning for shape analysis via multiresolution instance discrimination. In: Thirty-Fifth AAAI Conference on Artificial Intelligence, AAAI 2021, Thirty-Third Conference on Innovative Applications of Artificial Intelligence, IAAI 2021, The Eleventh Symposium on Educational Advances in Artificial Intelligence, EAAI 2021, Virtual Event, 2–9 February 2021, pp. 2773–2781. AAAI Press (2021)
39. Wu, W., Qi, Z., Li, F.: Pointconv: Deep convolutional networks on 3d point clouds. In: CVPR. pp. 9621–9630 (2019)

40. Wu, Z., et al.: 3d shapenets: a deep representation for volumetric shapes. In: CVPR, pp. 1912–1920 (2015)
41. Xie, S., Gu, J., Guo, D., Qi, C.R., Guibas, L., Litany, O.: PointContrast: unsupervised pre-training for 3D point cloud understanding. In: Vedaldi, A., Bischof, H., Brox, T., Frahm, J.-M. (eds.) ECCV 2020. LNCS, vol. 12348, pp. 574–591. Springer, Cham (2020). https://doi.org/10.1007/978-3-030-58580-8_34
42. Yan, X., Zheng, C., Li, Z., Wang, S., Cui, S.: Pointasnl: robust point clouds processing using nonlocal neural networks with adaptive sampling. In: CVPR, pp. 5588–5597 (2020)
43. Yi, L., et al.: A scalable active framework for region annotation in 3d shape collections. ACM Trans. Graph. **35**, 210:1–210:12 (2016)
44. Yuan, Y., Chen, X., Wang, J.: Object-contextual representations for semantic segmentation. In: Vedaldi, A., Bischof, H., Brox, T., Frahm, J.-M. (eds.) ECCV 2020. LNCS, vol. 12351, pp. 173–190. Springer, Cham (2020). https://doi.org/10.1007/978-3-030-58539-6_11
45. Zhang, Z., Girdhar, R., Joulin, A., Misra, I.: Self-supervised pretraining of 3d features on any point-cloud. In: ICCV (2021)
46. Zhao, H., Shi, J., Qi, X., Wang, X., Jia, J.: Pyramid scene parsing network. In: CVPR, pp. 6230–6239 (2017)
47. Zhao, Y., Birdal, T., Deng, H., Tombari, F.: 3d point capsule networks. In: IEEE Conference on Computer Vision and Pattern Recognition, CVPR 2019, Long Beach, CA, USA, 16–20 June, 2019, pp. 1009–1018. Computer Vision Foundation/IEEE (2019)
48. Zhou, B., Zhao, H., Puig, X., Fidler, S., Barriuso, A., Torralba, A.: Scene parsing through ADE20K dataset. In: CVPR, pp. 5122–5130 (2017)
49. Zhu, Z., Xu, M., Bai, S., Huang, T., Bai, X.: Asymmetric non-local neural networks for semantic segmentation. In: ICCV, pp. 593–602 (2019)

Adaptive Spatial-BCE Loss for Weakly Supervised Semantic Segmentation

Tong Wu[1][iD], Guangyu Gao[1(✉)][iD], Junshi Huang[2(✉)][iD], Xiaolin Wei[2][iD],
Xiaoming Wei[2][iD], and Chi Harold Liu[1][iD]

[1] Beijing Institute of Technology, Beijing, China
guangyugao@bit.edu.cn
[2] Meituan, Beijing, China
huangjunshi@meituan.com

Abstract. For Weakly-Supervised Semantic Segmentation (WSSS)
with image-level annotation, mostly relies on the classification network
to generate initial segmentation pseudo-labels. However, the optimization target of classification networks usually neglects the discrimination between different pixels, like insignificant foreground and background regions. In this paper, we propose an adaptive Spatial Binary
Cross-Entropy (Spatial-BCE) Loss for WSSS, which aims to enhance
the discrimination between pixels. In Spatial-BCE Loss, we calculate
the loss independently for each pixel, and heuristically assign the optimization directions for foreground and background pixels separately. An
auxiliary self-supervised task is also proposed to guarantee the Spatial-
BCE Loss working as envisaged. Meanwhile, to enhance the network's
generalization for different data distributions, we design an alternate
training strategy to adaptively generate thresholds to divide the foreground and background. Benefiting from high-quality initial pseudo-
labels by Spatial-BCE Loss, our method also reduce the reliance on
post-processing, thereby simplifying the pipeline of WSSS. Our method
is validated on the PASCAL VOC 2012 and COCO 2014 datasets, and
achieves the new state-of-the-arts. Code is available at https://github.
com/allenwu97/Spatial-BCE.

Keywords: WSSS · Spatial-BCE · Pseudo-labels · Adaptive threshold

1 Introduction

Semantic segmentation aims to allocate labels to each pixel of an image, which
has made significant progress driven by the Deep Neural Networks (DNNs). However, fully-supervised semantic segmentation requires pixel-level annotations,
which are costly and time-consuming. Thus, researchers are motivated to use
cheaper alternatives, including bounding boxes [23,30], scribbles [26], points [3]
and image-level labels [31]. Among them, the image-level labels can be directly

S. Avidan et al. (Eds.): ECCV 2022, LNCS 13689, pp. 199–216, 2022.
https://doi.org/10.1007/978-3-031-19818-2_12

obtained from existing datasets, or when constructing large-scale datasets, web search engines can automatically provide images and corresponding category tags. Thus the image-level labels based methods have lowest cost and become the mainstream of WSSS.

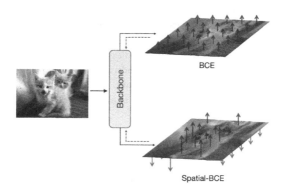

Fig. 1. The optimization directions of pixels when the network is trained by different loss function. The blue and red arrows correspond to the directions towards foreground and background, respectively. Best viewed in color. (Color figure online)

Most image-level labels based WSSS methods rely on Class Activation Maps (CAMs) [44] to generate the initial pseudo-labels for the training of semantic segmentation networks. Since the characteristics of the classification network, the CAMs tend to highlight the discriminative object regions, rather than the complete object regions, which deviates from the requirement of semantic segmentation. In this case, most previous WSSS works are motivated to extend the active regions of CAMs. One common solution is to obtain and fuse multiple highlight regions by multiple CAMs, including using multiple dilated convolutional blocks [36] and networks in different epochs [19]. These methods are intuitive, but easily to be over-activated or under-activated when faced with unconventional object sizes and shapes, and also their effects highly depend on hyper-parameters which harms the generalization.

In practice, the size of the foreground regions in the initial pseudo-label can be adjusted by the threshold, and a small threshold often refers to large foreground regions. However, classification networks usually ignore the feature discrimination between pixels, especially pixels in non-discriminative foreground regions and background regions. It is difficult and even unreasonable to divide precise target object boundaries by a global threshold overall images. Therefore, the crucial point to improve the quality of the initial pseudo-labels is enhancing the discrimination between foreground and background pixels, so that we can easily identify the object boundaries during inference.

For each positive category, the traditional BCE loss is calculated based on the average of the whole probability map, thus all pixels are optimized in the same direction, which reduces the discrimination between the foreground and background. Thus, we propose a novel Spatial Binary Cross-Entropy (Spatial-BCE) loss, which optimizes the foreground and background pixels in different directions, as shown in Fig. 1. Spatial-BCE Loss is designed to calculate the loss independently for each pixel. The optimization direction of each pixel depends on its probability in the prediction map and the threshold which divides the foreground and background. If its value is larger than the threshold, the parameters are optimized toward the foreground, otherwise, the parameters are optimized toward the background. With only image-level labels, the network lacks guidance on the division of foreground and background at the pixel level. Therefore, we introduce an auxiliary task to constrain the distribution of pixel probability, making Spatial-BCE Loss correctly assign the optimization direction for each pixel. Noticed that this auxiliary task is self-supervised, there is no requirement of the additional data. With this auxiliary task, the network which is trained by Spatial-BCE Loss, can correctly divide the foreground and background pixels and increase the discrimination between them. In that case, we can further obtain high-quality initial pseudo-labels with not only complete foreground regions but also precise boundaries.

In Spatial-BCE, we need a threshold to divide the foreground and background pixels. Although it can be a fixed threshold, a better way is to generate it through the network attentively for diverse inputs. Therefore, we design an alternate training strategy to alternately optimize this threshold. Previous methods determine such an optimal threshold through trial-and-error. However, it inevitably causes performance degradation when there is a lack of pixel-level labels for comparison required for trial-and-error in a real scenario of WSSS. Due to Spatial-BCE Loss allowing the network to generate the threshold adaptively, we can directly use it during inference. This training strategy makes us maintain stable performance even if without pixel-level labels as a comparison.

Our main contributions can be summarized as follows:

- We propose a novel Spatial-BCE Loss for WSSS, which can optimize the foreground and background pixels separately to increase the discrimination between them. With the help of an auxiliary self-supervised task, Spatial-BCE allows the network to generate high-quality initial labels.
- We design an alternate training strategy to allow the network to adaptively generate foreground and background dividing threshold. In that case, the network can generate the dividing threshold by itself during inference, reducing the impact of hyper-parameters on performance.
- Our method greatly improves the quality of the initial pseudo-labels, which in turn benefits the trained semantic segmentation network. Even if we remove post-processing that relies on additional networks or data, we still achieve competitive results. Under the same experimental configuration, we achieve the new state-of-the-arts of PASCAL VOC 2012 and MS-COCO 2014.

2   Related Work

2.1   Pseudo-Label Generation

For WSSS of image-level labels, the common pipeline is to generate pseudo-labels through a classification network, and then use the pseudo-labels to train a semantic segmentation network. Most methods use CAM as the initial segmentation pseudo-labels. Since CAM only highlights the discriminative regions, the previous works are motivated by expanding the focus regions of CAM. Hou *et al.* [17] proposed a self-erasing network, which forces the network to pay attention to the remaining part of the object by erasing discriminative regions. Erasing is intuitive and effective for expanding CAM, but it is difficult to give reliable termination conditions the iterative erasure. Wei *et al.* [36] expand the receptive field of discriminative features by adding dilated convolutions with different dilated rates. Different receptive fields can generate CAMs with different distributions of highlight regions. After the fusion of multiple CAMs, the purpose of expanding the highlight regions can be achieved. A similar method also includes the generation and fusion of multiple CAMs through different epoch parameters [19] and different layers [25]. This idea of merging multiple CAMs is mostly based on the designer's prior knowledge rather than guiding the network to learn to focus on the non-discriminative regions. Therefore, the extended range is relatively fixed and the robustness is low.

Most of the latest methods tend to guide the network to focus on non-discriminative regions autonomously by adding auxiliary tasks. SEAM [35] introduced consistency regularization to provide self-supervision for network learning; MCIS [33] explore the semantic relationships across images; Chang *et al.* [5] add sub-categories for each image through clustering, allowing the network to focus on more fine-grained features. EDAM [37] effectively combines semantic segmentation tasks and classification tasks, letting the classification network directly predict the mask of each category. EPS [24] use additional saliency maps as supervision during training, so that the predictions can have accurate object boundaries and discard co-occurring pixels. The above auxiliary tasks are based on the classification task, and the loss function of the classification task is BCE Loss. However, we found that BCE Loss has the disadvantage when it is used for WSSS. BCE Loss does not distinguish between foreground pixels and background pixels, which leads to the confusion of foreground and background in CAM. Therefore, we upgraded the BCE Loss to Spatial-BCE Loss, make the network can optimize the foreground and background separately, solve the problem from the roots.

2.2   Pseudo-Label Refinement

Besides improving the quality of the initial pseudo-labels, some works design additional networks to refine the initial pseudo-labels. SEC [20] proposed the three principles of seed, expansion, and constraint, which are followed by many subsequent works. DSRG [18] combines the seeded region growing with DNN, uses CAM as the initial cue, and continuously expands the seed regions. AffinityNet [2] generates a transition matrix by learning the similarity between pixels, and implements the semantic propagation by random walks. Ahn *et al.* [1] used the high-confidence foreground and high-confidence background to generate the supervision of the object boundaries, which further improves the accuracy when refining CAMs. Some works [13,19,24,33,36,37] use additional saliency maps as background cues to enhance the details of the object boundaries. Due to class-agnostic, these post-processings not only increase the complexity of the pipeline, but also easily introduce new noise. Our method can achieve competitive results without using additional saliency maps or training additional networks.

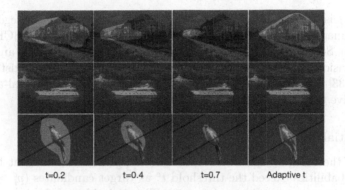

| t=0.2 | t=0.4 | t=0.7 | Adaptive t |

Fig. 2. Visualization of pseudo-labels from CAMs with different background threshold and our adaptive threshold. Best viewed in color. (Color figure online)

3 Approach

3.1 Overview

For an image with a specific image-level category, there is only part of pixels that strictly belong to such category, which we call target pixels, and the remained parts are the non-target pixels. In a multi-label classification task with traditional BCE Loss, the features of all pixels are averaged for the prediction of image probabilities belonging to certain categories. In this case, all pixels are assigned with the same optimization direction. Therefore, the discrimination of features between different pixels will be ignored, especially for the non-target pixels (e.g., pixels of background) and non-discriminative target pixels. The neglect of feature discrimination makes it difficult to get the precise object regions in the segmentation task. In Fig. 2, we present the pseudo-labels with various thresholds to distinguish the target pixels and the non-target pixels from the classification network. When setting a large threshold (0.7 or 0.4), the pseudo-labels only activate partial objects. To recall the rest target parts by reducing the threshold to 0.2, irrelative non-target regions surrounding the boundary are also activated.

Considering the feature discrimination of different pixels, we propose the Spatial-BCE Loss to optimize the prediction of pixels individually. Let's denote F the feature map of image I, and $P^c = Sigmoid(Linear^c(GAP(F)))$ represents the probability of I belonging to the cth category, where GAP is the Global Average Pooling. In the Spacial-BCE Loss, we generate the probability map $P^c = Sigmoid(Linear^c(F))$ to calculate the loss of cth category, where $P^c = \{p_i^c\} \in \mathcal{R}^{w \times h}$, and p_i^c is probability of ith pixel belonging to c^{th} category. Generally, the function of Spatial-BCE Loss is written as:

$$\mathcal{L}_{Spatial-BCE} = \sum_{c=1}^{C} \sum_{i=1}^{h \times w} (y^c R(p_i^c) - (1 - y^c) log(1 - p_i^c)) \qquad (1)$$

where $R(\cdot)$ is the pixel-wise re-factoring strategy for positive categories, and y^c is the image-level label of cth category. The details of Spatial-BCE loss are proposed in Sect. 3.2. To prevent the collapsing of Spatial BCE loss, an auxiliary self-supervision task is introduced in Sect. 3.3. Different from most existing works [13, 19, 24, 33] using the same threshold for all categories through trial-and-error, our adaptive threshold strategy is specified in Sect. 3.4.

3.2   Spatial-BCE Loss

Assuming the category-specific threshold t^c is pre-defined, we treat the pixels whose probabilities exceed the threshold t^c as target candidates ($p_i^c > t^c$), otherwise as non-target candidates ($p_i^c \leq t^c$). The main idea of $\mathcal{L}_{Spatial-BCE}$ is to divide the target candidates from non-target ones, and penalize the *uncertain* candidates by $R(\cdot)$. To this end, we propose three principles to design $R(\cdot)$:

1. The *uncertainty* of target candidates monotonically decreases with the increment of predicted probabilities, and reaches 0 if the probability is 1.
2. Likewise, the *uncertainty* of non-target candidates decreases monotonically as the predicted probabilities decrease and is 0 with the probability of 0.
3. The *uncertainty* of regions reach maximum if the predicted probability is surrounding the aforementioned threshold.

Based on principle 1&3, we define the uncertainty of target candidates as:

$$R_{tg}(p_i^c) = -(p_i^c - t^c)^2 + (1 - t^c)^2 \qquad (2)$$

where $p_i^c \in (t^c, 1)$. As a monotonically decreasing function over p_i^c, $R_{tg}(1) = 0$ (satisfying principle 1), and reaches the maximum value when p_i^c approaches t^c (satisfying principle 3). In practice, we use an α-balanced variable of $R_{tg}(p_i^c)$:

$$R_{tg}(p_i^c) = -\alpha(p_i^c - t^c)^2 + 1 \qquad (3)$$

where $\alpha = (1 - t^c)^{-2}$ is a variable to normalize the maximum value to be 1. Similarly, for pixels in non-target candidates ($p_i^c \in (0, t^c]$), following principle 2&3, we define the uncertainty of non-target candidates as:

$$R_{ntg}(p_i^c) = -\beta(p_i^c - t^c)^2 + 1 \qquad (4)$$

where $\beta = t^{c-2}$ is a normalized variable for non-target pixels.

Finally, the $R(p_i^c)$ can be written as a *differentiable* piecewise function:

$$R(p_i^c) = \begin{cases} R_{tg}(p_i^c) & p_i^c > t^c \\ R_{ntg}(p_i^c) & p_i^c \leq t^c \end{cases} \tag{5}$$

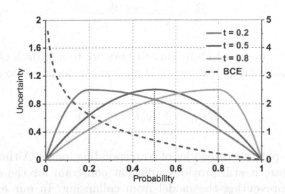

Fig. 3. The curve of uncertainty for the positive samples with different prediction probabilities. Best viewed in color. (Color figure online)

In Fig. 3, we visualize the pixel uncertainty of positive samples by BCE Loss and Spatial-BCE Loss respectively. The solid lines represent $R(p_i^c)$ with different t^c, and the dashed line refers to the $\log p_i^c$ which is the uncertainty metric in BCE Loss. Different from BCE Loss, which regards samples with low probability as the main optimization object, $R(p_i^c)$ turns to consider it as a confident non-target pixel. Furthermore, $R(p_i^c)$ increases the uncertainty of p_i^c when the value surrounds t^c, allowing the network to focus more on the hard-examples. Since the value of t^c is adaptively learned (sec Sect. 3.4), $R(p_i^c)$ has strong generalization when assigning optimization directions to pixels on different images.

3.3   Auxiliary Supervision

As we may observe in the experiment, the Spatial-BCE Loss can easily collapse to a trivial solution that the probabilities of all pixels are zero. Therefore, we import the distribution of target and non-target pixels as an extra constraint for the training of Spatial-BCE Loss. Since this pixel distribution is lacking in image-level labels, we design a self-supervised method without additional data to obtain it.

Given the image-level labels, we first train a classification network by BCE Loss to generate rough pseudo-labels. After using dense Conditional Random Field (dCRF) [21] for enhancement, we calculate Q^c for each image as:

$$Q^c = \frac{\sum_{i=1}^{h \times w} |l_i = c|}{\sum_{i=1}^{h \times w} |l_i \neq c|} \tag{6}$$

where l_i is the predicted category of ith pixel after dCRF, and Q^c represents Target and Non-Target (T/NT) proportion of the cth category. Note again that Q^c is calculated independently for each image.

With the initialized Q^c, we integrate our self-supervised task into the network as an extra constraint. We first predict the T/NT proportion of images by:

$$\widehat{P^c} = \frac{\sum_{i=1}^{h \times w} p_i^c}{\sum_{i=1}^{h \times w} (1 - p_i^c)} \tag{7}$$

Then, we urge the predicted $\widehat{P^c}$ by current network to approach Q^c. To this end, the Kullback-Leibler (KL) Divergence is used as the loss function:

$$\mathcal{D}_{KL} = \sum_{c=1}^{Pos} y_c \widehat{P^c} log(\frac{\widehat{P^c}}{Q^c}) \tag{8}$$

where Pos is the positive categories of corresponding image. Although the initial Q^c is not accurate, it still provides sufficient constrain for the distribution of pixels, thereby preventing the model from collapsing. In our experiment, we update Q^c of training images by the on-training network at every 8k iterations.

Finally, we use $\mathcal{L} = \mathcal{L}_{Spatial-BCE} + \mathcal{D}_{KL}$ as the whole losses. $\mathcal{D}_{KL}$ provides the constraint for the distribution of target and non-target pixels. $\mathcal{L}_{Spatial-BCE}$ converts the blurry soft predictions into polarized results, thus reducing the uncertainty of target and non-target pixels. The conjunction of $\mathcal{L}_{Spatial-BCE}$ and $\mathcal{D}_{KL}$ for joint training induces the network gradually generates more accurate pseudo-labels, as well as T/NT proportion.

3.4 Adaptive Threshold

We assume the category-specific threshold t^c is pre-defined in Sect. 3.2. In practice, rather than treating t^c as a fixed hyper-parameter, we generate t^c adaptively during the training process, and the algorithm is presented in Algorithm 1.

Specifically, we separate every γ iterations as a *training phase*. At the first θ iterations of each phase, we initialize the threshold t based on T/NT proportion Q for input images. In detail, we estimate the coarse proportion of target pixels in the image for each positive category c by $\frac{Q^c}{1+Q^c}$, and use the $\frac{Q^c}{1+Q^c}$-th percentile of predicted probabilities $\boldsymbol{P}^c$ as the image-specific category threshold t^c. In this case, the network is only updated by the gradient of loss w.r.t $\boldsymbol{P}$. This strategy provides a relatively reliable value for t_c and thus stabilizes the network training.

To chase a better threshold, we generate t_c by the network through convolution layers and update it via the Spatial-BCE loss in the rest $(\gamma - \theta)$ iterations of each *training phase*. Note that the gradient of loss w.r.t $\boldsymbol{P}$ is detached in this sub-phase, since simultaneous learning of t and $\boldsymbol{P}$ usually result in the instability of network training. As aforementioned, we update the T/NT proportion Q based on the on-training network parameters at the last iteration of each phase.

Pseudo-Label Generation. During inference, we directly use t generated by network to as the image-specific threshold of foreground and background. Generally, the pixel category l_i for specific image is determined by:

$$l_i = \begin{cases} \arg\max_c(S_i^c) & \text{if} \max(S_i^c) > 0 \\ \text{background} & \text{otherwise} \end{cases} \tag{9}$$

where $S_i^c = \{p_i^c - t^c, \forall c \in Pos\}$ is the de-biased probability set of pixels for c^{th} positive category. This strategy of adaptive threshold allows our method to select reliable dividing threshold (see Fig. 2), and thus alleviates the degradation of performance caused by artificial hyper-parameters.

Algorithm 1. Training with Adaptive Threshold

Input: Initial T/NT proportion Q; Hyper-parameter θ, γ
Output: Probability map P; Thresholds t
   for iter < Iteration **do**
     if $iter\%\gamma < \theta$ **then**
       $t = \frac{Q}{1+Q}$-th percentile of P
       $P = P.update()$
     else
       $t = t.update()$
       $P = P.detach()$
       if $iter\%\gamma == \gamma - 1$ **then**
         update Q
       end if
     end if
   end for

---

4 Experiments

4.1 Datasets and Evaluation Metric

We train and validate our method on PASCAL VOC 2012 [12] and MS-COCO 2014 [27]. The PASCAL VOC 2012 contains 21 categories (including a background category). Following the previous works, we use the augmented training set [15] containing $10,582$ images for training. The validation set and test set of PASCAL VOC 2012 contain $1,449$ and $1,456$ images respectively. The MS-COCO 2014 data set contains 81 categories including a background category, and the training set and validation set contain $82,081$ and $40,137$ images, respectively. We have excluded images that do not contain any foreground categories in MS-COCO 2014 as [9,24]. As some pixels in the ground-truth of MS-COCO overlap with multiple categories, we use the annotation of COCO-Stuff [4] as an alternative, which shares the same image set with MS-COCO.

We validate our method on the validation and test set of PASCAL VOC and the validation set of MS-COCO. The performance of the test set of PASCAL VOC is obtained by submitting the results to the official evaluation website. For all experiments, the mean IoU (mIoU) is used as the evaluation metric.

4.2   Implementation Details

We use ResNet38 [38] as the backbone of the classification network in our method. We randomly crop each image to the size of 368 × 368 and use RandomAug [10] as the data augmentation during training. During training, we also set learnable category-specific logit scales, which are initialized to $ln(10)$. A large logit scale can decrease the uncertainty that the pixel belongs to the foreground or background and speed up the convergence. During inference, we drop out logit scales to make prediction smoother, which is convenient for finding the optimal threshold. We perform multi-scales fusion when generating the initial pseudo-labels. After generating the pseudo-labels, we respectively apply IRN [1] and saliency maps generated by PoolNet [28] as post-processing. In different configurations, we use Deeplab-LargeFOV [6] and Deeplab-ASPP [7] as the semantic segmentation network, and their backbones are ResNet101 [16] and VGG16 [32] respectively. All backbones are pretrained on ImageNet [11].

We train our model on 4 T V100 GPUs with 16 GB memory. We use the SGD optimizer and set the initial learning rate to 10^{-2}. In the initialization phase, the learning rate (LR) will drop from 10^{-2} to 10^{-3}. After each iteration of Q^c, the LR will drop from 10^{-3} to 0. Other important hyper-parameters are set as follows: batch-size is 24, weight-decay is $5e^{-4}$, and momentum is 0.9.

4.3   Determining Thresholds

To determine the optimal threshold t^c, most of the previous methods need to traverse the range of $[0, 1]$ to generate pseudo-labels with different thresholds. The final threshold is obtained by calculating mIoU with the pixel-level ground-truth of the training set. Our Spatial-BCE loss allows the network to adaptively generate the t^c, as mentioned in Sect. 3.4. In that case, we can do inference without comparing with the pixel-level ground-truth, more strictly following the requirements of WSSS. In the subsequent experiments, we use * to indicate the results of using the adaptive threshold. For a fair comparison with previous methods, we also report the performance of manually choosing the optimal threshold.

4.4   Comparisons to State-of-the-Arts

In Table 1, we compare the quality of pseudo-labels of different methods. *Baseline* means method with pseudo-labels generated by ResNet38 based classification network using multi-scale inference through DA Layer [37], and the mIoU on training set of PASCAL VOC is 53.0%. The different ways to generate thresholds are mentioned in Sect. 4.3. After iteratively updating Q^c by 3 times, our initial pseudo-labels achieve mIoU of 65.3% with the adaptive thresholds, which is 9.7% higher than the previous SoTA of AdvCAM [22]. Since we have already used dCRF before updating Q^c, we also list results after dCRF for fair comparisons. It shows that our method is 4.2% higher than the previous SoTA, even if we do not need pixel-level labels during inference. When we manually choose the best threshold as previous works, our results can be 8.3% higher than the previous

Table 1. The mIoU (%) of the initial pseudo-labels (Init) and after refining by dCRF (+dCRF), on PASCAL VOC 2012 training set.

Method	Init	+dCRF
Baseline	53.0	60.2
Chang *et al.* [CVPR2020] [5]	50.9	55.3
SEAM[CVPR2020] [35]	55.4	56.8
AdvCAM[CVPR2021] [22]	55.6	62.1
Ours*	**65.3**	**66.3**
Ours	**68.1**	**70.4**

SoTA. The significant improvements on the initial pseudo-labels quality laid the foundation for us to simplify the post-processing process.

Table 2. Comparison to previous state-of-the-art approaches for WSSS on PASCAL VOC 2012 validation and test sets. $\mathcal{I}$: **Image-level labels**, $\mathcal{S}$: **Saliency maps**.

Method	Sup	Val	Test
IRNet[CVPR2019] [1]	$\mathcal{I}$	63.5	64.8
BES[ECCV2020] [8]	$\mathcal{I}$	65.7	66.6
CONTA[NeurIPS2020] [43]	$\mathcal{I}$	66.1	66.7
RRM[AAAI2020] [42]	$\mathcal{I}$	66.3	66.5
MBMNet[MM2020] [29]	$\mathcal{I}$	66.2	67.1
ECS-Net[ICCV2021] [34]	$\mathcal{I}$	66.6	67.6
AdvCAM[CVPR2021] [22]	$\mathcal{I}$	68.1	68.0
Ours*	$\mathcal{I}$	**68.5**	**69.7**
Ours	$\mathcal{I}$	**70.0**	**71.3**
OAA$^+$[ICCV2019] [19]	$\mathcal{I}, \mathcal{S}$	65.2	66.4
CIAN[AAAI2020] [14]	$\mathcal{I}, \mathcal{S}$	64.3	65.3
MCIS[ECCV2020] [33]	$\mathcal{I}, \mathcal{S}$	66.2	66.9
ICD[CVPR2020] [13]	$\mathcal{I}, \mathcal{S}$	67.8	68.0
AuxSegNet[ICCV2021] [39]	$\mathcal{I}, \mathcal{S}$	69.0	68.6
NSROM[CVPR2021] [41]	$\mathcal{I}, \mathcal{S}$	70.4	70.2
EDAM[CVPR2021] [37]	$\mathcal{I}, \mathcal{S}$	70.9	70.6
EPS[CVPR2021] [24]	$\mathcal{I}, \mathcal{S}$	71.0	71.8
Ours*	$\mathcal{I}, \mathcal{S}$	**70.5**	**71.6**
Ours	$\mathcal{I}, \mathcal{S}$	**71.8**	**73.4**

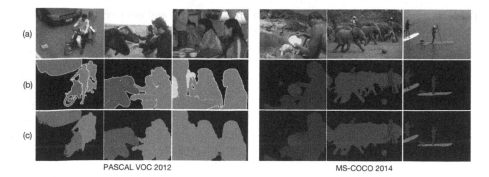

Fig. 4. Qualitative examples of segmentation results on the validation set of PAS-CAL VOC 2012 and MS-COCO 2014. (a) Original images, (b) Ground truth, (c) Our predictions. Best viewed in color. (Color figure online)

Table 2 shows the results on the PASCAL VOC 2012 validation and test set. The mainstream post-processing in previous methods is divided into two groups: methods with only image-level labels (such as AffinityNet [2], IRN [1]) or using additional saliency map. When there are only image-level labels, the post-processing relies on dCRF to construct the supervision for training extra network. Due to the training of extra network also requiring pixel-level ground-truth to finetune the hyper-parameters, we only use dCRF when we adaptively generate thresholds. It can be seen that under the premise of removing additional network and optimal threshold, our method can still achieve the new SoTA. When our method uses IRN as post-processing, the results are higher than the SoTA of the same post-processing by 1.9% and 3.3% on the validation set and test set, respectively.

Another mainstream post-processing method is using a pre-trained saliency detection network to generate saliency maps, and use saliency maps as the background cues, and the predictions as the foreground cues. Saliency map has accurate boundaries, but it can only highlight salient objects in images that are class-agnostic. To reduce the noise, we introduce the approach mentioned in EDAM [37], modifying the misjudgment regions in the saliency maps based on the predicted probability maps. In the end, our performance based on the saliency maps is 71.8% and 73.4% on the validation set and test set, also achieving the new SoTA of this experimental configuration.

Table 3. Comparison to previous SoTA approaches of weakly-supervised semantic segmentation on MS-COCO 2014 validation set.

Method	Backbone	Sup	Val
SEC[ECCV2016] [20]	VGG16	$\mathcal{I} + \mathcal{S}$	22.4
DSRG[CVPR2018] [18]	VGG16	$\mathcal{I} + \mathcal{S}$	26.0
ADL[TPAMI2020] [9]	VGG16	$\mathcal{I} + \mathcal{S}$	30.8
CONTA[NeurIPS2020] [43]	ResNet38	$\mathcal{I}$	32.8
SGAN[ACCESS2020] [40]	VGG16	$\mathcal{I} + \mathcal{S}$	33.6
AuxSegNet[ICCV2021] [39]	ResNet38	$\mathcal{I} + \mathcal{S}$	33.6
EPS[CVPR2021] [24]	VGG16	$\mathcal{I} + \mathcal{S}$	35.7
Ours*	VGG16	$\mathcal{I}$	**35.2**
Ours	VGG16	$\mathcal{I} + \mathcal{S}$	**38.6**

Table 3 shows the results of our method on MS-COCO 2014. On the MS-COCO dataset, we use VGG16 as the backbone, and use original CAMs as initial pseudo-labels. When we only use image-level labels and adaptively generate thresholds, we can achieve comparable results with previous SoTA which use extra saliency maps. If We use the same post-processing strategy which is mentioned above for saliency maps. On the validation set, our segmentation network achieves mIoU of 38.6%, which is 2.9% higher than EPS [23].

4.5 Ablation Studies

Iteratively Update of Q^c. Since the model generates more accurate T/NT proportion during the training process, we update Q^c at every 8k iterations to make Q^c gradually approaches ground-truth. In Table 4, we show the mIoU of the pseudo-labels after each phase. It can be seen that the improvement in the first iteration is most significant, and then the improvements gradually weaken. Considering the time cost, we terminate the training process after the third phase. Based on two different strategies for determining the threshold of pseudo-labels, we improve the performance of 6.1% and 10.2% respectively compared with the baseline. It is worth noting that our method does not depend on the specific network structure, in theory, any current model can be further finetuned by our method to achieve performance enhancement.

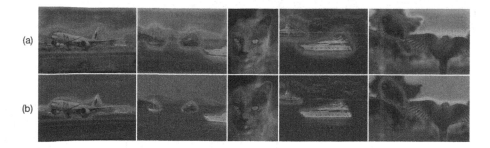

Fig. 5. Visualization of heatmaps on PASCAL VOC 2012. (a) Heatmap generated by our baseline. (b) Our final heatmap. Best viewed in color. (Color figure online)

Table 4. The mIoU on PASCAL VOC 2012 training set when each time Q^c is updated.

Iteration	Init	+dCRF
Baseline	53.0	60.2
Phase1*/Phase1	61.7/64.2	64.4/68.6
Phase2*/Phase2	64.5/67.5	66.1/70.7
Phase3*/Phase3	65.3/68.1	66.3/70.4

Contribution of Components. In Table 5, we measure the effect of different loss functions on the performance. After generating the initial Q^c based on the same results, we show the results of using only $\mathcal{D}_{KL}$ and $\mathcal{L}_{Spatial-BCE} + \mathcal{D}_{KL}$. When Q^c is updated after the first phase, the gap between the two sets is not obvious, only 1.2%. However, when Q^c is updated after the second phase, the gap increases to 3.8%. It can be concluded that without Spatial-BCE to optimize the distribution of predicted probability, $\mathcal{D}_{KL}$ will soon fall into the bottleneck, and the performance cannot be continuously improved.

In Table 6, we show the performance of pseudo-labels when the T/NT proportion is obtained in different ways. When t^c is determined by Q, the T/NT proportion of prediction is fixed, so the performance is difficult to improve. When the t^c is adaptively generated based on P, the network has the opportunity to optimize T/NT proportion of prediction, which leads to better results.

Table 5. The mIoU on PASCAL VOC 2012 training set when the network is trained by different loss function.

Loss	Init	Phase1	Phase2
$\mathcal{D}_{KL}$	60.2	63.0	63.7
$\mathcal{L}_{Spatial-BCE} + \mathcal{D}_{KL}$	60.2	64.2	67.5

Table 6. The mIoU on PASCAL VOC 2012 training set when the T/NT proportion is determined in different ways.

Method	Init	Phase1	Phase2
Fixed	60.2	59.7	60.5
Adaptive	60.2	61.7	64.5

Discrimination Between Foreground and Background. In Fig. 5, we compare heatmaps generated by our method with heatmaps generated by our baseline. It can be seen that after the fineturn by Spatial-BCE, the highlight regions of the heatmaps are more complete, and the activations of the background regions are significantly reduced. The initial heatmaps are also an active part of the background regions while highlighting the regions of the object. Our heatmap can be found that the activation degrees of the pixels on the boundary are nearly the same, which makes us completely divide the foreground regions. At the same time, near the boundary of the object, the activation degree of the pixels drops rapidly (the color changes from red to blue), which means that it hardly divides the background into the foreground when generating pseudo-labels, avoiding over-activation. This is the contribution of Spatial-BCE which increases the discrimination between the pixels of the foreground and background.

5   Conclusion

In this paper, we propose the novel loss function, Spatial-BCE Loss, to improve the quality of initial pseudo-labels for weakly-supervised semantic segmentation. Spatial-BCE Loss is re-factoring from the traditional BCE Loss, and is urged to assign the different optimization directions for the target and non-target pixels of each positive category. Through our alternate training, Spatial-BCE Loss can not only improve the feature discrimination between foreground and background pixels, but also allow the network to adaptively generate dividing threshold. The trained classification network can generate initial pseudo-labels without additional networks or data with accurate boundaries. Benefiting from high-quality initial pseudo-labels, we achieve new state-of-the-art of PASCAL VOC 2012 and MS-COCO 2014 datasets under various experimental configurations.

Acknowledgement. This work was supported by the National Natural Science Foundation of China under Grant No. 61972036, and in part by the Science and Technology on Optical Radiation Laboratory (China) under Grant No. 61424080213.

References

1. Ahn, J., Cho, S., Kwak, S.: Weakly supervised learning of instance segmentation with inter-pixel relations. In: Proceedings of the IEEE Conference on Computer Vision and Pattern Recognition, pp. 2209–2218 (2019)

2. Ahn, J., Kwak, S.: Learning pixel-level semantic affinity with image-level supervision for weakly supervised semantic segmentation. In: Proceedings of the IEEE Conference on Computer Vision and Pattern Recognition, pp. 4981–4990 (2018)

3. Bearman, A., Russakovsky, O., Ferrari, V., Fei-Fei, L.: What's the point: semantic segmentation with point supervision. In: Leibe, B., Matas, J., Sebe, N., Welling, M. (eds.) ECCV 2016. LNCS, vol. 9911, pp. 549–565. Springer, Cham (2016). https://doi.org/10.1007/978-3-319-46478-7_34

4. Caesar, H., Uijlings, J., Ferrari, V.: Coco-stuff: thing and stuff classes in context. In: Proceedings of the IEEE Conference on Computer Vision and Pattern Recognition, pp. 1209–1218 (2018)

5. Chang, Y.T., Wang, Q., Hung, W.C., Piramuthu, R., Tsai, Y.H., Yang, M.H.: Weakly-supervised semantic segmentation via sub-category exploration. In: Proceedings of the IEEE Conference on Computer Vision and Pattern Recognition, pp. 8991–9000 (2020)

6. Chen, L.C., Papandreou, G., Kokkinos, I., Murphy, K., Yuille, A.L.: Semantic image segmentation with deep convolutional nets and fully connected crfs. arXiv preprint arXiv:1412.7062 (2014)

7. Chen, L.C., Papandreou, G., Kokkinos, I., Murphy, K., Yuille, A.L.: Deeplab: semantic image segmentation with deep convolutional nets, atrous convolution, and fully connected crfs. IEEE Trans. Pattern Anal. Mach. Intell. **40**, 834–848 (2017)

8. Chen, L., Wu, W., Fu, C., Han, X., Zhang, Y.: Weakly supervised semantic segmentation with boundary exploration. In: Vedaldi, A., Bischof, H., Brox, T., Frahm, J.-M. (eds.) ECCV 2020. LNCS, vol. 12371, pp. 347–362. Springer, Cham (2020). https://doi.org/10.1007/978-3-030-58574-7_21

9. Choe, J., Lee, S., Shim, H.: Attention-based dropout layer for weakly supervised single object localization and semantic segmentation. IEEE Trans. Pattern Anal. Mach. Intell. **43**, 4256–4271 (2020)

10. Cubuk, E.D., Zoph, B., Shlens, J., Le, Q.V.: Randaugment: practical automated data augmentation with a reduced search space. In: Proceedings of the IEEE Conference on Computer Vision and Pattern Recognition Workshops, pp. 702–703 (2020)

11. Deng, J., Dong, W., Socher, R., Li, L.J., Li, K., Fei-Fei, L.: Imagenet: a large-scale hierarchical image database. In: Proceedings of the IEEE Conference on Computer Vision and Pattern Recognition, pp. 248–255 (2009)

12. Everingham, M., Van Gool, L., Williams, C.K.I., Winn, J., Zisserman, A.: The PASCAL Visual Object Classes Challenge 2012 (VOC2012) Results. http://www.pascal-network.org/challenges/VOC/voc2012/workshop/index.html

13. Fan, J., Zhang, Z., Song, C., Tan, T.: Learning integral objects with intra-class discriminator for weakly-supervised semantic segmentation. In: Proceedings of the IEEE Conference on Computer Vision and Pattern Recognition, pp. 4283–4292 (2020)

14. Fan, J., Zhang, Z., Tan, T., Song, C., Xiao, J.: Cian: cross-image affinity net for weakly supervised semantic segmentation. In: Proceedings of the AAAI Conference on Artificial Intelligence, pp. 10762–10769 (2020)

15. Hariharan, B., Arbeláez, P., Bourdev, L., Maji, S., Malik, J.: Semantic contours from inverse detectors. In: 2011 International Conference on Computer Vision. pp. 991–998 (2011)

16. He, K., Zhang, X., Ren, S., Sun, J.: Deep residual learning for image recognition. In: Proceedings of the IEEE Conference on Computer Vision and Pattern Recognition. pp. 770–778 (2016)

17. Hou, Q., Jiang, P., Wei, Y., Cheng, M.M.: Self-erasing network for integral object attention. In: Advances in Neural Information Processing Systems 31 (2018)
18. Huang, Z., Wang, X., Wang, J., Liu, W., Wang, J.: Weakly-supervised semantic segmentation network with deep seeded region growing. In: Proceedings of the IEEE Conference on Computer Vision and Pattern Recognition, pp. 7014–7023 (2018)
19. Jiang, P.T., Hou, Q., Cao, Y., Cheng, M.M., Wei, Y., Xiong, H.K.: Integral object mining via online attention accumulation. In: Proceedings of the IEEE International Conference on Computer Vision, pp. 2070–2079 (2019)
20. Kolesnikov, A., Lampert, C.H.: Seed, expand and constrain: three principles for weakly-supervised image segmentation. In: Leibe, B., Matas, J., Sebe, N., Welling, M. (eds.) ECCV 2016. LNCS, vol. 9908, pp. 695–711. Springer, Cham (2016). https://doi.org/10.1007/978-3-319-46493-0_42
21. Krähenbühl, P., Koltun, V.: Efficient inference in fully connected crfs with gaussian edge potentials. In: Advances in Neural Information Processing Systems 24 (2011)
22. Lee, J., Kim, E., Yoon, S.: Anti-adversarially manipulated attributions for weakly and semi-supervised semantic segmentation. In: Proceedings of the IEEE Conference on Computer Vision and Pattern Recognition, pp. 4071–4080 (2021)
23. Lee, J., Yi, J., Shin, C., Yoon, S.: Bbam: bounding box attribution map for weakly supervised semantic and instance segmentation. In: Proceedings of the IEEE Conference on Computer Vision and Pattern Recognition, pp. 2643–2652 (2021)
24. Lee, S., Lee, M., Lee, J., Shim, H.: Railroad is not a train: saliency as pseudo-pixel supervision for weakly supervised semantic segmentation. In: Proceedings of the IEEE Conference on Computer Vision and Pattern Recognition, pp. 5495–5505 (2021)
25. Lee, S., Lee, J., Lee, J., Park, C.K., Yoon, S.: Robust tumor localization with pyramid grad-cam. arXiv preprint arXiv:1805.11393 (2018)
26. Lin, D., Dai, J., Jia, J., He, K., Sun, J.: Scribblesup: scribble-supervised convolutional networks for semantic segmentation. In: Proceedings of the IEEE Conference on Computer Vision and Pattern Recognition, pp. 3159–3167 (2016)
27. Lin, T.Y., et al.: Microsoft coco: common objects in context. In: Proceedings of the European Conference on Computer Vision, pp. 740–755 (2014)
28. Liu, J.J., Hou, Q., Cheng, M.M., Feng, J., Jiang, J.: A simple pooling-based design for real-time salient object detection. In: Proceedings of the IEEE Conference on Computer Vision and Pattern Recognition, pp. 3917–3926 (2019)
29. Liu, W., Zhang, C., Lin, G., Hung, T.Y., Miao, C.: Weakly supervised segmentation with maximum bipartite graph matching. In: Proceedings of the 28th ACM International Conference on Multimedia. pp. 2085–2094 (2020)
30. Papandreou, G., Chen, L.C., Murphy, K.P., Yuille, A.L.: Weakly-and semi-supervised learning of a deep convolutional network for semantic image segmentation. In: Proceedings of the IEEE International Conference on Computer Vision, pp. 1742–1750 (2015)
31. Pinheiro, P.O., Collobert, R.: From image-level to pixel-level labeling with convolutional networks. In: Proceedings of the IEEE Conference on Computer Vision and Pattern Recognition, pp. 1713–1721 (2015)
32. Simonyan, K., Zisserman, A.: Very deep convolutional networks for large-scale image recognition. arXiv preprint arXiv:1409.1556 (2014)
33. Sun, G., Wang, W., Dai, J., Van Gool, L.: Mining cross-image semantics for weakly supervised semantic segmentation. In: Proceedings of the European Conference on Computer Vision, pp. 347–365 (2020)

34. Sun, K., Shi, H., Zhang, Z., Huang, Y.: Ecs-net: improving weakly supervised semantic segmentation by using connections between class activation maps. In: Proceedings of the IEEE International Conference on Computer Vision, pp. 7283–7292 (2021)
35. Wang, Y., Zhang, J., Kan, M., Shan, S., Chen, X.: Self-supervised equivariant attention mechanism for weakly supervised semantic segmentation. In: Proceedings of the IEEE Conference on Computer Vision and Pattern Recognition, pp. 12275–12284 (2020)
36. Wei, Y., Xiao, H., Shi, H., Jie, Z., Feng, J., Huang, T.S.: Revisiting dilated convolution: A simple approach for weakly-and semi-supervised semantic segmentation. In: Proceedings of the IEEE Conference on Computer Vision and Pattern Recognition, pp. 7268–7277 (2018)
37. Wu, T., et al.: Embedded discriminative attention mechanism for weakly supervised semantic segmentation. In: Proceedings of the IEEE Conference on Computer Vision and Pattern Recognition, pp. 16765–16774 (2021)
38. Wu, Z., Shen, C., Van Den Hengel, A.: Wider or deeper: revisiting the resnet model for visual recognition. Pattern Recognition, pp. 119–133 (2019)
39. Xu, L., Ouyang, W., Bennamoun, M., Boussaid, F., Sohel, F., Xu, D.: Leveraging auxiliary tasks with affinity learning for weakly supervised semantic segmentation. In: Proceedings of the IEEE International Conference on Computer Vision, pp. 6984–6993 (2021)
40. Yao, Q., Gong, X.: Saliency guided self-attention network for weakly and semi-supervised semantic segmentation. IEEE Access, pp. 14413–14423 (2020)
41. Yao, Y., et al.: Non-salient region object mining for weakly supervised semantic segmentation. In: Proceedings of the IEEE Conference on Computer Vision and Pattern Recognition, pp. 2623–2632 (2021)
42. Zhang, B., Xiao, J., Wei, Y., Sun, M., Huang, K.: Reliability does matter: an end-to-end weakly supervised semantic segmentation approach. In: Proceedings of the AAAI Conference on Artificial Intelligence, pp. 12765–12772 (2020)
43. Zhang, D., Zhang, H., Tang, J., Hua, X.S., Sun, Q.: Causal intervention for weakly-supervised semantic segmentation. In: Advances in Neural Information Processing Systems, pp. 655–666 (2020)
44. Zhou, B., Khosla, A., Lapedriza, A., Oliva, A., Torralba, A.: Learning deep features for discriminative localization. In: Proceedings of the IEEE Conference on Computer Vision and Pattern Recognition, pp. 2921–2929 (2016)

Dense Gaussian Processes for Few-Shot Segmentation

Joakim Johnander[1,4(✉)], Johan Edstedt[1], Michael Felsberg[1],
Fahad Shahbaz Khan[1,2], and Martin Danelljan[3]

[1] Department of Electrical Engineering, Linköping University, Linköping, Sweden
{joakim.johnander,johan.edstedt,michael.felsberg,fahad.khan}@liu.se
[2] Mohamed bin Zayed University of AI, Abu Dhabi, United Arab Emirates
[3] Computer Vision Lab, ETH Zürich, Zürich, Switzerland
[4] Zenseact AB, Göteborg, Sweden

Abstract. Few-shot segmentation is a challenging dense prediction task, which entails segmenting a novel query image given only a small annotated support set. The key problem is thus to design a method that aggregates detailed information from the support set, while being robust to large variations in appearance and context. To this end, we propose a few-shot segmentation method based on dense Gaussian process (GP) regression. Given the support set, our dense GP learns the mapping from local deep image features to mask values, capable of capturing complex appearance distributions. Furthermore, it provides a principled means of capturing uncertainty, which serves as another powerful cue for the final segmentation, obtained by a CNN decoder. Instead of a one-dimensional mask output, we further exploit the end-to-end learning capabilities of our approach to learn a high-dimensional output space for the GP. Our approach sets a new state-of-the-art on the PASCAL-5^i and COCO-20^i benchmarks, achieving an absolute gain of $+8.4$ mIoU in the COCO-20^i 5-shot setting. Furthermore, the segmentation quality of our approach scales gracefully when increasing the support set size, while achieving robust cross-dataset transfer. Code and trained models are available at https://github.com/joakimjohnander/dgpnet.

1 Introduction

Image few-shot segmentation (FSS) of semantic classes [28] has received increased attention in recent years. The aim is to segment novel *query* images based on only a handful annotated training samples, usually referred to as the *support set*. The FSS method thus needs to extract information from the support set in order to accurately segment a given query image. The problem is highly challenging, since the query image may present radically different views, contexts, scenes, and objects than what is represented in the support set.

Supplementary Information The online version contains supplementary material available at https://doi.org/10.1007/978-3-031-19818-2_13.

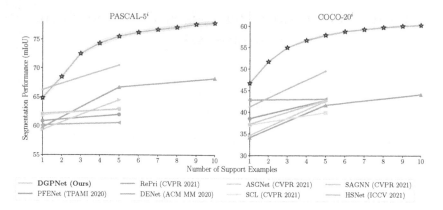

Fig. 1. Performance of the proposed DGPNet approach on the PASCAL-5^i and COCO-20^i benchmarks, compared to the state-of-the-art. We plot the mIoU (higher is better) for different number of support samples. For our approach, we show the mean and standard deviation over 5 experiments. Our GP-based method effectively leverages larger support sets, achieving substantial improvements in segmentation accuracy. Our method also excels in the extreme one-shot case, even outperforming all previously reported results on COCO-20^i.

The core component in any FSS framework is the mechanism that extracts information from the support set to guide the segmentation of the query image. However, the design of this module presents several challenges. First, it needs to aggregate detailed yet generalizable information from the support set, which requires a flexible representation. Second, the FSS method should effectively leverage larger support sets, achieving scalable segmentation performance when increasing its size. While perhaps trivial at first glance, this has proved to be a major obstacle for many state-of-the-art methods, as visualized in Fig. 1. Third, the method is bound to be queried with appearances not included in the support set. To achieve robust predictions even in such common cases, the method needs to assess the relevance of the information in the support images in order to gracefully revert to e.g. learned segmentation priors when necessary.

We address the aforementioned challenges by densely aggregating information in the support set using Gaussian Processes (GPs). Specifically, we use a GP to learn a mapping between dense local deep feature vectors and their corresponding mask values. The mask values are assumed to have a jointly Gaussian distribution with covariance based on the similarity between the corresponding feature vectors. This allows us to extract detailed relations from the support set, with the capability of modeling complex, non-linear mappings. As a non-parametric model [25], the GP further effectively benefits from additional support samples, since all given data is retained. As shown in Fig. 1, the segmentation accuracy of our approach improves consistently with the number of support samples. Lastly, the predictive covariance from the GP provides a principled measure of the uncertainty based on the similarity with local features in the support set.

Our FSS approach is learned end-to-end through episodic training, treating the GP as a layer in a neural network. This further allows us to learn the output space of the GP. To this end, we encode the given support masks with a neural network in order to achieve a multi-dimensional output representation. In order to generate the final masks, our decoder module employs the predicted mean query encodings, together with the covariance information. Our decoder is thus capable of reasoning about the uncertainty when fusing the predicted mask encodings with learned segmentation priors. Lastly, we further improve our FSS method by integrating dense GPs at multiple scales.

We perform comprehensive experiments on two benchmarks: PASCAL-5^i [28] and COCO-20^i [22]. Our proposed DGPNet outperforms existing methods for 5-shot by a large margin, setting a new state-of-the-art on both benchmarks. When using the ResNet101 backbone, our DGPNet achieves an absolute gain of 8.4 for 5-shot segmentation on the challenging COCO-20^i benchmark, compared to the best reported results in the literature. We further demonstrate the cross-dataset transfer capabilities of our DGPNet approach from COCO-20^i to PASCAL and perform detailed ablative studies to probe the effectiveness of our contributions.

2   Related Work

Few-Shot Segmentation. The earliest work in few-shot segmentation (FSS), by Shaban et al. [28], proposed a method for predicting the weights of a linear classifier based on the support set, which was further built upon in later works [4,15,29]. Instead of learning the classifier directly, Rakelly et al. [24] proposed to construct a global conditioning *prototype* from the support set and concatenate it to the query representation, with several subsequent works [6,22,35,46,48]. Recent works have strived to improve prototype-based approaches. Azad et al. [2] reduced the texture bias; Liu et al. [16] cross-referenced query and support features with a neural network and iteratively refined the predicted masks; Xie et al. [38] combined information at multiple feature levels with a graph neural network (with nodes being feature maps); and Wang et al. [33] proposed to model the prototype with a unimodal normal distribution. A major limitation of these methods is the *unimodality* assumption. Wu et al. [37] identified that additional support images may actually contaminate the unimodal target class representation, and proposed to weight the different support images. Zhang et al. [44] instead opted for more flexible target class models, comprising multiple feature vectors. Yang et al. [40], Liu et al. [17], and Li et al. [12] clustered feature vectors to create a multi-modal target class representation. However, clustering introduces extra hyperparameters, such as the number of clusters, as well as optimization difficulties. In this work, we propose a mechanism that increases the flexibility. The mechanism has few hyperparameters, primarily the choice of covariance function. More closely related to this work are methods that consider pointwise correspondences between the support and query set. These works have mostly focused on attention or attention-like mechanisms [10,31,34, 42,45,47]. The work of Min *et al.* [21] proposed a pointwise comparison that was then processed with hypercorrelation layers. In contrast with these methods, we construct a principled posterior over functions, which greatly aids the decoder.

Combining GPs and Neural Networks. While early work focused on combining GPs and neural networks in the standard supervised classification setting [5,27,36], there has recently been an increased interest in utilizing Gaussian processes in the context of *few-shot classification* [23,30]. These works employ the GP as the final output layer. This is problematic as the GP assumes the output to have a Gaussian distribution. In contrast, the output in classification has a *categorical* distribution. Thus, these works are forced to optimize proxies of either the predictive or marginal log-likelihood.

In this work, we instead adopt the GP as an internal layer and train a decoder to interpret the Gaussian output and make categorical predictions. As the GP is no longer used to make the final categorical predictions, we have the option to also *learn* the GP output space, further increasing its expressive power. Furthermore, we use a *dense* GP model to model the mapping from individual support features to corresponding encoded mask values. Compared to few-shot image classification, this leads to a high number of points and a high computational cost. We show that this can be addressed by downsampling and compensating the reduced resolution with a high-dimensional output space.

3   Method

3.1   Few-Shot Segmentation

Few-shot segmentation is a *dense* few-shot learning task [28]. The aim is to learn to segment objects from novel classes, given only a small set of annotated images. A single instance of this problem, referred to as an *episode*, comprises a small set of annotated samples, called the *support set*, and a set of samples on which prediction is to be made, the *query set*. Formally, we denote the support set as $\{(I_{Sk}, M_{Sk})\}_{k=1}^{K}$, comprising K image-mask pairs $I_{Sk} \in \mathbb{R}^{H_0 \times W_0 \times 3}$ and $M_{Sk} \in \{0,1\}^{H_0 \times W_0}$. A query image is denoted as $I_Q \in \mathbb{R}^{H_0 \times W_0 \times 3}$ and the aim is to predict its corresponding segmentation mask $M_Q \in \{0,1\}^{H_0 \times W_0}$.

To develop our approach, we first provide a general formulation for addressing the FSS problem, which applies to several recent methods, including prototype-based [12,15] and correlation-based [31,34] ones. Our formulation proceeds in three steps: feature extraction, few-shot learning, and prediction. In the first step, deep features are extracted from the given images,

$$\mathbf{x} = F(I) \in \mathbb{R}^{H \times W \times D} \ . \tag{1}$$

These features provide a more disentangled and invariant representation, which greatly aids the problem of learning from a limited number of samples.

The main challenge in FSS, namely how to most effectively leverage the annotated support samples, is encapsulated in the second step. As a general approach, we consider a learner module Λ that employs the support set in order to find a function f, which associates each query feature $x_Q \in \mathbb{R}^D$ with an *output* $y_Q \in \mathbb{R}^E$. Note that it is common to use a downsampled mask, setting $E = 1$, but we are not restricted to do so. The goal is to achieve an output y_Q that is strongly correlated with the ground-truth query mask, allowing it to act as a strong cue in the final prediction. Formally, we express this general formulation as,

$$f = \Lambda(\{\mathbf{x}_{Sk}, \mathbf{M}_{Sk}\}_k), \quad y_Q = f(x_Q) . \tag{2}$$

The learner Λ aggregates information in the support set $\{\mathbf{x}_{Sk}, \mathbf{M}_{Sk}\}_k$ in order to predict the function f. This function is then applied to query features (2). In the final step of our formulation, output from the function f on the query set is decoded by a separate network as $\widehat{M}_Q = U(\mathbf{y}_Q, \mathbf{x}_Q)$ to predict the segmentation $\widehat{M}_Q$. An overview of our formulation is shown in Fig. 2.

The general formulation in (2) encapsulates several recent approaches for few-shot segmentation. In particular, prototype-based methods, for instance PANet [35], are retrieved by letting Λ represent a mask-pooling operation. The function f then computes the cosine-similarity between the pooled feature vector and the input query features. In general, the design of the learner Λ represents the central problem in few-shot segmentation, since it is the module that extracts information from the support set. Next, we distinguish three key desirable properties of this module.

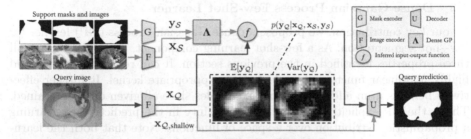

Fig. 2. Overview of our approach. Support and query images are fed through an encoder to produce deep features $\mathbf{x}_S$ and $\mathbf{x}_Q$ respectively. The support masks are fed through another encoder to produce $\mathbf{y}_S$ ($E = 1$ in this figure). Using Gaussian process regression, we infer the probability distribution of the query mask encodings $\mathbf{y}_Q$ given the support set and the query features (see Eqs. 5–7). We create a representation of this distribution and feed it through a decoder. The decoder then predicts a segmentation at the original resolution.

3.2 Key Properties of Few-Shot Learners

As discussed above, the core component in few-shot segmentation is the few-shot learner Λ. Much research effort has therefore been diverted into its design

[12,15,17,22,34,40]. To motivate our approach, we first identify three important properties that the learner should possess.

Flexibility of f. The intent in few-shot segmentation is to be able to segment a wide range of classes, unseen during training. The image feature distributions of different unseen classes are not necessarily linearly separable [1]. Prototypical few-shot learners, which are essentially linear classifiers, would fail in such scenarios. Instead, we need a mechanism that can learn and represent more complex functions f.

Scalability in Support Set Size K. An FSS method should be able to effectively leverage additional support samples and therefore achieve substantially better accuracy and robustness for larger support sets. However, many prior works show little to no benefit in the 5-shot setting compared to 1-shot. As shown by Li *et al.* [12] and Boudiaf *et al.* [4], it is crucial that new information is effectively incorporated into the model without *averaging* out useful cues.

Uncertainty Modeling. Since only a small number of support samples are available in FSS, the network needs to regularly handle unseen appearances in the query images. For instance, the query may include novel backgrounds, scenes, and objects. Since the function f predicted by the learner is not expected to generalize to such unseen scenarios, the network should instead utilize neighboring predictions or learned priors. However, this can only be achieved if f models and communicates the uncertainty of its prediction to the decoder.

3.3   Dense Gaussian Process Few-Shot Learner

As our key contribution, we propose a dense Gaussian Process (GP) learner for few-shot segmentation. As a few-shot learning component, the GP possesses all three properties identified in the previous section. It can represent flexible and highly non-linear functions by selecting an appropriate kernel. It further effectively benefits from additional support samples since all given data is retained. Third, the GP explicitly models the uncertainty in the prediction by learning a probabilistic distribution over a space of functions. Note that both the learning and inference steps of the GP are differentiable functions. While computational cost is a well-known challenge when deploying GPs, we demonstrate that even simple strategies can be used to keep the number of training samples at a tractable level for the FSS problem.

Our dense GP learner predicts a distribution of functions f from the input features x to an output y. First, the support feature maps are stacked and reshaped as a matrix $\mathbf{x}_S \in \mathbb{R}^{KHW \times D}$. The query feature maps are reshaped as $\mathbf{x}_Q \in \mathbb{R}^{HW \times D}$. Let $\mathbf{y}_S$ and $\mathbf{y}_Q$ be the corresponding outputs. The former is obtained from the support masks and the latter is to be predicted with the GP. The key assumption in the Gaussian process is that the support and query outputs $\mathbf{y}_S, \mathbf{y}_Q$ are jointly Gaussian according to,

$$\begin{pmatrix} \mathbf{y}_S \\ \mathbf{y}_Q \end{pmatrix} \sim \mathcal{N}\left(\begin{pmatrix} \mu_S \\ \mu_Q \end{pmatrix}, \begin{pmatrix} \mathbf{K}_{SS} & \mathbf{K}_{SQ} \\ \mathbf{K}_{SQ}^\top & \mathbf{K}_{QQ} \end{pmatrix} \right) . \tag{3}$$

For simplicity, we set the output prior means, μ_S and μ_Q, to zero. The covariance matrix K in (3) is defined by the input features at those points and a *kernel* $\kappa : \mathbb{R}^D \times \mathbb{R}^D \rightarrow \mathbb{R}$. In our experiments, we adopt the commonly used *squared exponential* (SE) kernel

$$\kappa(x^m, x^n) = \sigma_f^2 \exp(-\frac{1}{2\ell^2}\|x^m - x^n\|_2^2) , \qquad (4)$$

with scale parameter σ_f and length parameter ℓ. The kernel can be viewed as a similarity measure. If two features are similar, then the corresponding outputs are correlated.

Next, the posterior probability distribution of the query outputs is inferred (see Fig. 2). The rules for conditioning in a joint Gaussian give us [25]

$$\mathbf{y}_Q|\mathbf{y}_S, \mathbf{x}_S, \mathbf{x}_Q \sim \mathcal{N}(\boldsymbol{\mu}_{Q|S}, \boldsymbol{\Sigma}_{Q|S}) , \qquad (5)$$

where

$$\boldsymbol{\mu}_{Q|S} = \mathbf{K}_{SQ}^\top(\mathbf{K}_{SS} + \sigma_y^2\mathbf{I})^{-1}\mathbf{y}_S , \qquad (6)$$

$$\boldsymbol{\Sigma}_{Q|S} = \mathbf{K}_{QQ} - \mathbf{K}_{SQ}^\top(\mathbf{K}_{SS} + \sigma_y^2\mathbf{I})^{-1}\mathbf{K}_{SQ} . \qquad (7)$$

The measurements $\mathbf{y}_S$ are assumed to have been obtained with some additive i.i.d. Gaussian noise with variance σ_y^2. This corresponds to adding a scaled identity matrix $\sigma_y^2\mathbf{I}$ to the support covariance matrix $\mathbf{K}_{SS}$. The Eqs. 4–7 thus predict the distribution of the query outputs, represented by the mean value and the covariance. Note that the asymptotic complexity is $\mathcal{O}((KWH)^3)$. We therefore need to work on a sufficiently low resolution. Next, we introduce a way to incorporate more information into the predicted outputs while maintaining a low resolution.

3.4 Learning the GP Output Space

The decoder strives to transform the query outputs predicted by the few-shot learner into an accurate mask. This is a challenging task given the low resolution and the desire to generalize to classes unseen during offline training. We therefore explore whether additional information can be encoded into the outputs, $\mathbf{y}_Q$, in order to guide the decoder. This information could for instance include the shape of the mask or which of the object parts are at a given location. To this end, we train a mask-encoder to construct the outputs,

$$\mathbf{y}_{Sk} = G(M_{Sk}) . \qquad (8)$$

Here, G is a neural network with learnable parameters.

The aforementioned formulation allows us to learn the GP output space during the meta-training stage. To the best of our knowledge, this has not previously been explored in the context of GPs. There is some reminiscence to the attention mechanism in transformers [32]. The transformer queries, keys, and values correspond to our query features $\mathbf{x}_Q$, support features $\mathbf{x}_S$, and support outputs $\mathbf{y}_S$.

A difference is that in few-shot segmentation, there is a distinction between the features used for matching and the output – the features stem from the image whereas the output stems from the mask.

As mask encoder G, we employ a lightweight residual convolutional neural network. The network predicts multi-dimensional outputs for the support masks that are then reshaped into $\mathbf{y}_{\mathcal{S}} \in \mathbb{R}^{KHW \times E}$. These are fed into (6) when the posterior probability distribution of the query outputs is inferred. The two matrix-vector multiplications are transformed into matrix-matrix multiplications and the result $\boldsymbol{\mu}_{\mathcal{Q}|\mathcal{S}} \in \mathbb{R}^{HW \times E}$ is a matrix containing the multi-dimensional mean. The covariance in (7) is kept unchanged. This setup corresponds to the assumption that the covariance is isotropic over the output feature channels and that the different output feature channels are independent.

3.5   Final Mask Prediction

Next, we decode the predicted distributions of the query outputs in order to predict the final mask (see Fig. 2). Since the output of the GP is richer, including uncertainty and covariance information, we first need to consider what information to give to the final decoder network. We employ the following output features from the GP.

GP Output Mean. The multi-dimensional mean $\boldsymbol{\mu}_{\mathcal{Q}|\mathcal{S}}$ represents the best guess of $\mathbf{y}_{\mathcal{Q}}$. Ideally, it contains a representation of the mask $M_{\mathcal{Q}}$ to be predicted. The decoder works on 2D feature maps and $\boldsymbol{\mu}_{\mathcal{Q}|\mathcal{S}}$ is therefore reshaped as $\mathbf{z}_{\mu} \in \mathbb{R}^{H \times W \times E}$.

GP Output Covariance. The covariance $\boldsymbol{\Sigma}_{\mathcal{Q}|\mathcal{S}}$ captures both the uncertainty in the predicted query output $y_{\mathcal{Q}}$ and the correlation between different query outputs. The former lets the decoder network identify uncertain regions in the image, and instead rely on, e.g., learnt priors or neighbouring predictions. Moreover, local correlations can tell whether two locations of the image are similar. For each query output, we employ the covariance between it and each of its spatial neighbours in an $N \times N$ region. We represent the covariance as a 2D feature map $\mathbf{z}_{\Sigma} \in \mathbb{R}^{H \times W \times (N^2)}$ by vectorizing the N^2 covariance values in the $N \times N$ neighborhood. For more details, see Appendix E.

Shallow Image Features. Image features extracted from early layers of deep neural networks are of high resolution and permit precise localization of object boundaries [3]. We therefore store feature maps extracted from earlier layers of F and feed them into the decoder. These serve to guide the decoder as it transforms the low-resolution output mean into a precise, high-resolution mask.

We finally predict the query mask by feeding the above information into a decoder U,

$$\widehat{M_{\mathcal{Q}}} = U(\mathbf{z}_{\mu}, \mathbf{z}_{\Sigma}, \mathbf{x}_{\mathcal{Q},\text{shallow}}) \ . \tag{9}$$

Note that while it would be possible to sample from the inferred distribution of $\mathbf{y}_{\mathcal{Q}}$, we instead feed its parameters into the decoder that predicts the final mask. As a decoder U, we adopt DFN [43]. DFN processes its input at one scale at a time, starting with the coarsest scale. In our case, this is the concatenated mean and covariance. At each scale, the result at the previous scale is upsampled and processed together with any input at that scale. After processing the finest scaled input, the result is upsampled to the mask scale and classified with a linear layer and a `softmax` function.

3.6 FSS Learner Pyramid

Many computer vision tasks benefit from processing at multiple scales or multiple feature levels. While deep, high-level features directly capture the presence of objects or semantic classes, the mid-level features capture object parts [31]. In semantic segmentation, methods begin with high-level features and then successively add mid-level features while upsampling [18]. In object detection, a detection head is applied to each level in a feature pyramid [13]. Tian *et al.* [31] discuss and demonstrate the benefit of using both mid-level and high-level features for few-shot segmentation. We therefore adapt our framework to be able to process features at different levels.

We take features from our image encoder F extracted at multiple levels A. Let the support and query features at level $a \in A$ be denoted as $\mathbf{x}_S^a$ and $\mathbf{x}_Q^a$. From the mask encoder G we extract corresponding support outputs $\mathbf{y}_S^a$. We then introduce a few-shot learner Λ^a for each level a. For efficient inference, the support features and outputs are sampled on a grid such that the total stride compared to the original image is 32. The query features retain their resolution. Each few-shot learner then infers a posterior distribution over the query outputs $\mathbf{y}_Q^a$, parameterized by $\boldsymbol{\mu}_{Q|S}^a$ and $\boldsymbol{\Sigma}_{Q|S}^a$. These are then fed into the decoder, which processes them one scale at a time.

4 Experiments

We validate our proposed approach by conducting comprehensive experiments on two FSS benchmarks: PASCAL-5^i [28] and COCO-20^i [22].

4.1 Experimental Setup

Datasets. We conduct experiments on the PASCAL-5^i [28] and COCO-20^i [22] benchmarks. PASCAL-5^i is composed of PASCAL VOC 2012 [7] with additional SBD [8] annotations. The dataset comprises 20 categories split into 4 folds. For each fold, 15 categories are used for training and the remaining 5 for testing. COCO-20^i [22] is more challenging and is built from MS-COCO [14]. Similar to PASCAL-5^i, the COCO-20^i benchmark is split into 4 folds. For each fold, 60 base classes are used for training and the remaining 20 for testing.

Implementation Details. Following previous works, we employ ResNet-50 and ResNet-101 [9] backbones pre-trained on ImageNet [26] as image encoders. We let the dense GP work with feature maps produced by the third and the fourth residual module. In addition, we place a single convolutional projection layer that reduces the feature map down to 512 dimensions. As mask encoder, we use a light-weight CNN (see Appendix E). We use $\sigma_y^2 = 0.1$, $\sigma_f^2 = 1$, and $\ell^2 = \sqrt{D}$ in

Table 1. State-of-the-art comparison on the PASCAL-5^i and COCO-20^i benchmarks in terms of mIoU (higher is better). In each case, the best two results are shown in magenta and cyan font, respectively. Asterisk * denotes re-implementation by Liu *et al.* [15]. Our DGPNet achieves state-of-the-art results for 1-shot and 5-shot on both benchmarks. When using ResNet101, our DGPNet achieves absolute gains of 5.5 and 8.4 for 1-shot and 5-shot segmentation, respectively, on the challenging COCO-20^i benchmark, compared to the best reported results in the literature.

Backbone	Method	PASCAL-5^i		COCO-20^i	
		1-shot	5-shot	1-shot	5-shot
ResNet50	CANet [46] CVPR'19	55.4	57.1	40.8*	42.0*
	PGNet [45] ICCV'19	56.0	58.5	36.7*	37.5*
	RPMM [40] ECCV'20	56.3	57.3	30.6	35.5
	CRNet [16] CVPR'20	55.7	58.8	–	–
	DENet [15] MM'20	60.1	60.5	*42.8*	43.0
	LTM [42] MM'20	57.0	60.6	–	–
	PPNet [17] ECCV'20	51.5	62.0	25.7	36.2
	PFENet [31] TPAMI'20	60.8	61.9	–	–
	RePri [4] CVPR'21	59.1	66.8	34.0	42.1
	SCL [44] CVPR'21	61.8	62.9	–	–
	SAGNN [38] CVPR'21	62.1	62.8	–	-
	CWT [20] ICCV'21	56.4	63.7	32.9	41.3
	CMN [39] ICCV'21	62.8	63.7	39.3	43.1
	MLC [41] ICCV'21	62.1	66.1	33.9	40.6
	MMNet [37] ICCV'21	60.2	61.8	37.2	37.4
	HSNet [21] ICCV'21	*64.0*	*69.5*	39.2	*46.9*
	CyCTR [47] NeurIps'21	64.2	65.6	40.3	45.6
	DGPNet (Ours)	63.5±0.4	73.5±0.3	45.0±0.4	56.2±0.4
ResNet101	FWB [22] CVPR'19	56.2	60.0	21.2	23.7
	DAN [34] ECCV'20	58.2	60.5	24.4	29.6
	PFENet [31] TPAMI'20	60.1	61.4	38.5	42.7
	RePri [4] CVPR'21	59.4	65.6	–	–
	VPI [33] WACV'21	57.3	60.4	23.4	27.8
	ASGNet [12] CVPR'21	59.3	64.4	34.6	42.5
	SCL [44] CVPR'21	–	–	37.0	39.9
	SAGNN [38] CVPR'21	–	–	37.2	42.7
	CWT [20] ICCV'21	58.0	64.7	32.4	42.0
	MLC [41] ICCV'21	62.6	68.8	36.4	44.4
	HSNet [21] ICCV'21	66.2	*70.4*	*41.2*	*49.5*
	CyCTR [47] NeurIps'21	64.3	66.6	–	–
	DGPNet (Ours)	*64.8±0.5*	75.4±0.4	46.7±0.3	57.9±0.3

our GP. We train our models for 20k and 40k iterations à 8 episodes for PASCAL-5^i and COCO-20^i, respectively. On one NVIDIA A100 GPU, this takes up to 10 h. We use the AdamW [11, 19] optimizer with a weight decay factor of 10^{-3} and a cross-entropy loss with 1 to 4 background to foreground weighting. We use a learning rate of $5 \cdot 10^{-5}$ for all parameters save for the image encoder, which uses a learning rate of 10^{-6}. The learning rate is decayed with a factor of 0.1 when 10k iterations remain. We freeze the batch normalization layers of the image encoder. Episodes are sampled in the same way as we evaluate. We randomly flip images horizontally followed by resizing to a size of 384×384 for PASCAL-5^i and 512×512 for COCO-20^i.

Evaluation. We evaluate our approach on each fold by randomly sampling 5k and 20k episodes respectively, for PASCAL-5^i and COCO-20^i. This follows the work of Tian *et al.* [31] in which it is observed that the procedure employed by many prior works, using only 1000 episodes, yields fairly high variance. Additionally, following Wang *et al.* [35], Liu *et al.* [17], Boudiaf *et al.* [4], and Zhang *et al.* [44], our results in the state-of-the-art comparison are computed as the average of 5 runs with different seeds. We also report the standard deviation. Performance is measured in terms of mean Intersection over Union (mIoU). First, the IoU is calculated per class over all episodes in a fold. The mIoU is then obtained by averaging the IoU over the classes. As in the original work on FSS by Shaban *et al.* [28], we calculate the IoU on the original resolution of the images.

Table 2. Performance comparison (mIoU, higher is better), when performing cross-dataset evaluation from COCO-20^i to PASCAL. When using the same ResNet50 backbone, our approach achieves significant improvements for both 1-shot and 5-shot settings, with absolute gains of 5.8 and 11.3 mIoU over RePRI [4].

Backbone	Method	1-Shot	5-Shot
ResNet50	RPMM [40]	49.6	53.8
	PFENet [31]	61.1	63.4
	RePRI [4]	*63.1*	66.2
	HSNet [21]	61.6	*68.7*
	DGPNet (Ours)	**68.9±0.4**	**77.5±0.2**
ResNet101	HSNet [21]	64.1	70.3
	DGPNet (Ours)	**70.1±0.3**	**78.5±0.3**

4.2 State-of-the-Art Comparison

We compare our proposed DGPNet with state-of-the-art FSS approaches on the PASCAL-5^i and COCO-20^i benchmarks, see Table 1. Following prior works, we report results given a single support example, *1-shot*, and given five support examples, *5-shot*. On PASCAL-5^i with a ResNet50 backbone, recently proposed approaches – RePri [4], SCL [44], SAGNN [38], CWT [20], CMN [39], MLC [41], MMNet [37], HSNet [21], and CyCTR [47] – range from 56.4 mIoU to 64.2 mIoU. Our proposed DGPNet obtains a competitive performance of 63.5 mIoU. Where the powerful dense Gaussian process of DGPNet really shines, however, is when additional support examples are given. In the 5-shot setting, these recently proposed approaches range from 61.8 mIoU to 69.5 mIoU. Our proposed DGPNet obtains 73.5 mIoU, setting a new state-of-the-art for 5-shot few-shot segmentation on PASCAL-5^i.

On the challenging COCO-20^i benchmark, the previous best reported 1-shot result comes from the work of Liu *et al.* [15], 42.8 mIoU. In their work, however, no significant improvement is reported when additional support examples are given. Three other approaches – CMN [39], HSNet [21], and CyCTR [47] – obtain performance close to the work of Liu *et al.* [15], from 39.3 mIoU to 40.3 mIoU. Compared to the work of Liu *et al.* [15], CMN, HSNet, and CyCTR scale better with additional support examples, obtaining between 43.1 mIoU and 46.9 mIoU, with HSNet reporting the largest gain, 7.7 mIoU. Our proposed DGPNet obtains a 1-shot performance of 45.0 mIoU, setting a new state-of-the-art. As additional support examples are given, the gap to prior works increases. In the 5-shot setting, DGPNet obtains 56.2 mIoU, beating the previous best approach, HSNet, by 9.7 mIoU. Our per-fold results, including the standard deviation for each fold, are provided in Appendix A. Qualitative examples are found in Fig. 3 and Appendix F.

Scaling Segmentation Performance with Increased Support Size. As discussed earlier, the FSS approach is desired to effectively leverage additional support samples, thereby achieving superior accuracy and robustness for larger shot-sizes. We therefore analyze the effectiveness of the proposed DGPNet by increasing the number of shots on PASCAL-5^i and COCO-20^i. Figure 1 shows the results on the two benchmarks. We use the model trained for 5-shot in all cases, except for 1-shot. Compared to most existing works, our DGPNet effectively benefits from additional support samples, achieving remarkable improvement in segmentation accuracy with larger support sizes.

We provide a qualitative comparison between different support set sizes in Fig. 3. For this comparison, we use COCO-20^i. In general, the 1-shot setting is quite challenging for several classes due to major variations in where the object appears and what it looks like. If the single support sample is too different from the query sample, the model tends to struggle. With five support samples, the model performs far better. At five, the benefit of additional support samples seems to saturate, with ten samples often bringing only marginal improvements. An example is the train episode in Fig. 3, where minor improvements are obtained when going from five to ten support samples.

Cross-dataset Evaluation. In Table 2, we present the cross-dataset evaluation capabilities of our DGPNet from COCO-20^i to PASCAL. For this cross-dataset evaluation experiment, we followed the same protocol as in Boudiaf *et al.* [4], where the folds are constructed to ensure that there is no overlap with COCO-20^i training folds. When using the same ResNet50 backbone, our approach obtains

Support 1 Support 2-5 Support 6-10   Query   1-shot   5-shot   10-shot

Fig. 3. Right: Qualitative results of our approach given 1, 5, and 10 support samples. The 1-shot results are based on (see left) Support 1; the 5-shot results on Support 1 and Support 2–5; and the 10-shot Support 1, Support 2–5, and Support 6–10. The results are from COCO-20^i and human faces have been pixelized in the visualization, but the model makes predictions on the non-pixelized images.

1-shot and 5-shot mIoU of 68.9 and 77.5, respectively, with absolute gains of 5.8 and 11.3 over RePRI [4]. Further, DGPNet achieves segmentation mIoU of 70.1 and 78.5 in 1-shot and 5-shot setting, respectively, using ResNet101. For more details on the experiment, see Appendix A.

Table 3. Performance of different kernels on the PASCAL-5^i and COCO-20^i benchmarks. Notably, both the Exponential and SE kernels significantly outperform the linear kernel, confirming the need for a flexible learner. Measured in mIoU (higher is better). Best results are in bold.

Kernel	PASCAL-5^i		COCO-20^i		
	1-shot	5-shot	1-shot	5-shot	Mean Δ
Linear	58.6	61.8	37.1	45.3	0.0
Exponential	59.3	67.3	40.9	**51.8**	4.1
SE	**62.1**	**69.9**	**41.7**	51.2	**5.5**

Table 4. Analysis of learning the GP output space (GPO) and incorporating covariance (Cov). Performance in mIoU (higher is better). Best results are in bold.

Cov	GPO	PASCAL-5^i		COCO-20^i		
		1-shot	5-shot	1-shot	5-shot	Mean Δ
		62.1	69.9	41.7	51.2	0.0
✓		**62.5**	71.8	43.8	53.7	1.7
	✓	61.7	**72.7**	43.1	54.2	1.7
✓	✓	**62.5**	72.6	**44.7**	**55.0**	**2.5**

Table 5. Performance of different multilevel configurations of the dense GP few-shot learner on the PASCAL-5^i and COCO-20^i benchmarks. Measured in mIoU (higher is better). Best results are in bold.

Str. 16	Str. 32	PASCAL-5^i		COCO-20^i		
		1-shot	5-shot	1-shot	5-shot	Mean Δ
	✓	62.5	72.6	44.7	55.0	0.0
✓		60.4	69.3	40.7	51.1	−3.9
✓	✓	**63.9**	**73.6**	**45.3**	**56.4**	**1.4**

4.3 Ablation Study

Here, we analyze the impact of the key components in our proposed architecture. We first investigate the choice of kernel, κ. Then, we analyze the impact of the predictive covariance provided by the GP and the benefits of learning the GP output space. Last, we experiment with dense GPs at multiple feature levels. All experiments in this section are conducted with the ResNet50 backbone. Detailed experiments are provided in the Appendix. In each experiment, we also report the mean gain over a baseline (Mean Δ), where the mean is computed over the 1-shot and 5-shot performance on PASCAL-5^i and COCO-20^i.

Choice of Kernel. Going from a linear few-shot learner to a more flexible function requires an appropriate choice of kernel. We consider the *homogenous linear kernel* as our baseline. Note that the homogenous linear kernel is equivalent to Bayesian linear regression under appropriate priors [25]. To make our learner more flexible, we consider two additional choices of kernels, the *Exponential* and *Squared Exponential* (SE) kernels. The kernel equations are given in Appendix C. In Table 3, results on both the PASCAL and COCO benchmarks are presented. Both the Exponential and SE kernels greatly outperform the linear kernel, with the SE kernel leading to a mean gain of 5.5 in mIoU. These results show the benefit of a more flexible and scalable learner.

Incorporating Uncertainty and Learning the GP Output Space. We adopt the SE kernel and analyze the performance of letting the decoder process the predictive covariance provided by the dense GP. With the covariance in a

5×5 window, we obtain a gain of 1.7 mIoU. Then, we add the mask-encoder and learn the GP output space (GPO). This leads to a 1.7 improvement in isolation, or a 2.5 mIoU gain together with the covariance.

Multilevel Representations. Finally, we investigate the effect of introducing a multilevel hierarchy of dense GPs. Specifically we investigate using combinations of stride 16 and 32 features. The results are reported in Table 5. The results show the benefit of using dense GPs at different feature levels, leading to an average gain of 1.4 mIoU.

5 Conclusion

We have proposed a few-shot learner based on Gaussian process regression for the few-shot segmentation task. The GP models the support set in deep feature space and its flexibility permits it to capture complex feature distributions. It makes probabilistic predictions on the query image, providing both a point estimate and additional uncertainty information. These predictions are fed into a CNN decoder that predicts the final segmentation. The resulting approach sets a new state-of-the-art on 5-shot PASCAL-5^i and COCO-20^i, with absolute improvements of up to 8.4 mIoU. Our approach scales well with larger support sets during inference, even when trained for a fixed number of shots.

Acknowledgements. This work was partially supported by the Wallenberg Artificial Intelligence, Autonomous Systems and Software Program (WASP) funded by Knut and Alice Wallenberg Foundation; ELLIIT; and the ETH Future Computing Laboratory (EFCL), financed by a donation from Huawei Technologies. The computations were enabled by resources provided by the Swedish National Infrastructure for Computing (SNIC), partially funded by the Swedish Research Council through grant agreement no. 2018-05973.

References

1. Allen, K., Shelhamer, E., Shin, H., Tenenbaum, J.: Infinite mixture prototypes for few-shot learning. In: International Conference on Machine Learning, pp. 232–241. PMLR (2019)
2. Azad, R., Fayjie, A.R., Kauffmann, C., Ben Ayed, I., Pedersoli, M., Dolz, J.: On the texture bias for few-shot CNN segmentation. In: Proceedings of the IEEE/CVF Winter Conference on Applications of Computer Vision, pp. 2674–2683 (2021)
3. Bhat, G., Johnander, J., Danelljan, M., Khan, F.S., Felsberg, M.: Unveiling the power of deep tracking. In: Ferrari, V., Hebert, M., Sminchisescu, C., Weiss, Y. (eds.) ECCV 2018. LNCS, vol. 11206, pp. 493–509. Springer, Cham (2018). https://doi.org/10.1007/978-3-030-01216-8_30
4. Boudiaf, M., Kervadec, H., Masud, Z.I., Piantanida, P., Ben Ayed, I., Dolz, J.: Few-shot segmentation without meta-learning: a good transductive inference is all you need? In: Proceedings of the IEEE/CVF Conference on Computer Vision and Pattern Recognition, pp. 13979–13988 (2021)

5. Calandra, R., Peters, J., Rasmussen, C.E., Deisenroth, M.P.: Manifold Gaussian Processes for regression. In: Proceedings of the International Joint Conference on Neural Networks (2016). https://doi.org/10.1109/IJCNN.2016.7727626
6. Dong, N., Xing, E.P.: Few-shot semantic segmentation with prototype learning. In: British Machine Vision Conference 2018, BMVC 2018 (2019)
7. Everingham, M., Van Gool, L., Williams, C.K., Winn, J., Zisserman, A.: The pascal visual object classes (VOC) challenge. Int. J. Comput. Vision (2010). https://doi.org/10.1007/s11263-009-0275-4
8. Hariharan, B., Arbeláez, P., Bourdev, L., Maji, S., Malik, J.: Semantic contours from inverse detectors. In: Proceedings of the IEEE International Conference on Computer Vision (2011). https://doi.org/10.1109/ICCV.2011.6126343
9. He, K., Zhang, X., Ren, S., Sun, J.: Deep residual learning for image recognition. In: Proceedings of the IEEE Conference on Computer Vision and Pattern Recognition, pp. 770–778 (2016)
10. Hu, T., Yang, P., Zhang, C., Yu, G., Mu, Y., Snoek, C.G.: Attention-based multi-context guiding for few-shot semantic segmentation. In: Proceedings of the AAAI Conference on Artificial Intelligence, vol. 33, pp. 8441–8448 (2019)
11. Kingma, D.P., Ba, J.L.: Adam: a method for stochastic optimization. In: 3rd International Conference on Learning Representations, ICLR 2015 - Conference Track Proceedings (2015)
12. Li, G., Jampani, V., Sevilla-Lara, L., Sun, D., Kim, J., Kim, J.: Adaptive prototype learning and allocation for few-shot segmentation. In: Proceedings of the IEEE/CVF Conference on Computer Vision and Pattern Recognition, pp. 8334–8343 (2021)
13. Lin, T.Y., Dollár, P., Girshick, R., He, K., Hariharan, B., Belongie, S.: Feature pyramid networks for object detection. In: Proceedings of the IEEE Conference on Computer Vision and Pattern Recognition, pp. 2117–2125 (2017)
14. Lin, T.-Y., Maire, M., Belongie, S., Hays, J., Perona, P., Ramanan, D., Dollár, P., Zitnick, C.L.: Microsoft COCO: common objects in context. In: Fleet, D., Pajdla, T., Schiele, B., Tuytelaars, T. (eds.) ECCV 2014. LNCS, vol. 8693, pp. 740–755. Springer, Cham (2014). https://doi.org/10.1007/978-3-319-10602-1_48
15. Liu, L., Cao, J., Liu, M., Guo, Y., Chen, Q., Tan, M.: Dynamic extension nets for few-shot semantic segmentation. In: Proceedings of the 28th ACM International Conference on Multimedia, pp. 1441–1449 (2020)
16. Liu, W., Zhang, C., Lin, G., Liu, F.: CRNet: cross-reference networks for few-shot segmentation. In: Proceedings of the IEEE Computer Society Conference on Computer Vision and Pattern Recognition, pp. 4164–4172 (2020). https://doi.org/10.1109/CVPR42600.2020.00422
17. Liu, Y., Zhang, X., Zhang, S., He, X.: Part-aware prototype network for few-shot semantic segmentation. In: Vedaldi, A., Bischof, H., Brox, T., Frahm, J.-M. (eds.) ECCV 2020. LNCS, vol. 12354, pp. 142–158. Springer, Cham (2020). https://doi.org/10.1007/978-3-030-58545-7_9
18. Long, J., Shelhamer, E., Darrell, T.: Fully convolutional networks for semantic segmentation. In: Proceedings of the IEEE Conference on Computer Vision and Pattern Recognition, pp. 3431–3440 (2015)
19. Loshchilov, I., Hutter, F.: Decoupled weight decay regularization. In: International Conference on Learning Representations (2019). https://openreview.net/forum?id=Bkg6RiCqY7
20. Lu, Z., He, S., Zhu, X., Zhang, L., Song, Y.Z., Xiang, T.: Simpler is better: few-shot semantic segmentation with classifier weight transformer. In: Proceedings of the IEEE/CVF International Conference on Computer Vision, pp. 8741–8750 (2021)

21. Min, J., Kang, D., Cho, M.: Hypercorrelation squeeze for few-shot segmentation. In: Proceedings of the IEEE/CVF International Conference on Computer Vision, pp. 6941–6952 (2021)

22. Nguyen, K., Todorovic, S.: Feature weighting and boosting for few-shot segmentation. In: Proceedings of the IEEE International Conference on Computer Vision. vol. 2019-Octob, pp. 622–631 (2019). https://doi.org/10.1109/ICCV.2019.00071

23. Patacchiola, M., Turner, J., Crowley, E.J., Storkey, A.: Bayesian meta-learning for the few-shot setting via deep kernels. In: Advances in Neural Information Processing Systems (2020)

24. Rakelly, K., Shelhamer, E., Darrell, T., Efros, A., Levine, S.: Conditional networks for few-shot semantic segmentation. In: 6th International Conference on Learning Representations, ICLR 2018 - Workshop Track Proceedings (2018)

25. Rasmussen, C.E., Williams, C.K.I.: Gaussian Processes for Machine Learning (2006). https://doi.org/10.7551/mitpress/3206.001.0001

26. Russakovsky, O., Deng, J., Su, H., Krause, J., Satheesh, S., Ma, S., Huang, Z., Karpathy, A., Khosla, A., Bernstein, M., Berg, A.C., Fei-Fei, L.: ImageNet large scale visual recognition challenge. Int. J. Comput. Vision **115**(3), 211–252 (2015). https://doi.org/10.1007/s11263-015-0816-y

27. Salakhutdinov, R., Hinton, G.: Using deep belief nets to learn covariance kernels for Gaussian processes. In: Advances in Neural Information Processing Systems 20 - Proceedings of the 2007 Conference (2009)

28. Shaban, A., Bansal, S., Liu, Z., Essa, I., Boots, B.: One-shot learning for semantic segmentation. In: British Machine Vision Conference 2017, BMVC 2017 (2017). https://doi.org/10.5244/c.31.167

29. Siam, M., Oreshkin, B., Jagersand, M.: AMP: adaptive masked proxies for few-shot segmentation. In: Proceedings of the IEEE International Conference on Computer Vision, vol. 2019-Octob, pp. 5248–5257 (2019). https://doi.org/10.1109/ICCV.2019.00535

30. Snell, J., Zemel, R.: Bayesian few-shot classification with one-vs-each pólya-gamma augmented gaussian processes. In: International Conference on Learning Representations (2021). https://openreview.net/forum?id=lgNx56yZh8a

31. Tian, Z., Zhao, H., Shu, M., Yang, Z., Li, R., Jia, J.: Prior Guided Feature Enrichment Network for few-shot segmentation. IEEE Trans. Pattern Anal. Mach. Intell. (2020). https://doi.org/10.1109/tpami.2020.3013717

32. Vaswani, A., et al.: Attention is all you need. In: Advances in Neural Information Processing Systems, pp. 5998–6008 (2017)

33. Wang, H., Yang, Y., Cao, X., Zhen, X., Snoek, C., Shao, L.: Variational prototype inference for few-shot semantic segmentation. In: Proceedings of the IEEE/CVF Winter Conference on Applications of Computer Vision, pp. 525–534 (2021)

34. Wang, H., Zhang, X., Hu, Y., Yang, Y., Cao, X., Zhen, X.: Few-shot semantic segmentation with democratic attention networks. In: Vedaldi, A., Bischof, H., Brox, T., Frahm, J.-M. (eds.) ECCV 2020. LNCS, vol. 12358, pp. 730–746. Springer, Cham (2020). https://doi.org/10.1007/978-3-030-58601-0_43

35. Wang, K., Liew, J.H., Zou, Y., Zhou, D., Feng, J.: PANet: few-shot image semantic segmentation with prototype alignment. In: Proceedings of the IEEE International Conference on Computer Vision, vol. 2019-Octob, pp. 9196–9205 (2019). https://doi.org/10.1109/ICCV.2019.00929

36. Wilson, A.G., Hu, Z., Salakhutdinov, R., Xing, E.P.: Deep kernel learning. In: Proceedings of the 19th International Conference on Artificial Intelligence and Statistics, AISTATS 2016 (2016)

37. Wu, Z., Shi, X., Lin, G., Cai, J.: Learning meta-class memory for few-shot semantic segmentation. In: Proceedings of the IEEE/CVF International Conference on Computer Vision, pp. 517–526 (2021)
38. Xie, G.S., Liu, J., Xiong, H., Shao, L.: Scale-aware graph neural network for few-shot semantic segmentation. In: Proceedings of the IEEE/CVF Conference on Computer Vision and Pattern Recognition, pp. 5475–5484 (2021)
39. Xie, G.S., Xiong, H., Liu, J., Yao, Y., Shao, L.: Few-shot semantic segmentation with cyclic memory network. In: Proceedings of the IEEE/CVF International Conference on Computer Vision, pp. 7293–7302 (2021)
40. Yang, B., Liu, C., Li, B., Jiao, J., Ye, Q.: Prototype mixture models for few-shot semantic segmentation. In: Vedaldi, A., Bischof, H., Brox, T., Frahm, J.-M. (eds.) ECCV 2020. LNCS, vol. 12353, pp. 763–778. Springer, Cham (2020). https://doi.org/10.1007/978-3-030-58598-3_45
41. Yang, L., Zhuo, W., Qi, L., Shi, Y., Gao, Y.: Mining latent classes for few-shot segmentation. In: Proceedings of the IEEE/CVF International Conference on Computer Vision, pp. 8721–8730 (2021)
42. Yang, Y., Meng, F., Li, H., Wu, Q., Xu, X., Chen, S.: A new local transformation module for few-shot segmentation. In: Ro, Y.M., Cheng, W.-H., Kim, J., Chu, W.-T., Cui, P., Choi, J.-W., Hu, M.-C., De Neve, W. (eds.) MMM 2020. LNCS, vol. 11962, pp. 76–87. Springer, Cham (2020). https://doi.org/10.1007/978-3-030-37734-2_7
43. Yu, C., Wang, J., Peng, C., Gao, C., Yu, G., Sang, N.: Learning a Discriminative Feature Network for Semantic Segmentation. In: Proceedings of the IEEE Computer Society Conference on Computer Vision and Pattern Recognition (2018). https://doi.org/10.1109/CVPR.2018.00199
44. Zhang, B., Xiao, J., Qin, T.: Self-guided and cross-guided learning for few-shot segmentation. In: Proceedings of the IEEE/CVF Conference on Computer Vision and Pattern Recognition, pp. 8312–8321 (2021)
45. Zhang, C., Lin, G., Liu, F., Guo, J., Wu, Q., Yao, R.: Pyramid graph networks with connection attentions for region-based one-shot semantic segmentation. In: Proceedings of the IEEE International Conference on Computer Vision, vol. 2019-Octob, pp. 9586–9594 (2019). https://doi.org/10.1109/ICCV.2019.00968
46. Zhang, C., Lin, G., Liu, F., Yao, R., Shen, C.: CANET: class-agnostic segmentation networks with iterative refinement and attentive few-shot learning. In: Proceedings of the IEEE Computer Society Conference on Computer Vision and Pattern Recognition (2019). https://doi.org/10.1109/CVPR.2019.00536
47. Zhang, G., Kang, G., Yang, Y., Wei, Y.: Few-shot segmentation via cycle-consistent transformer. In: Advances in Neural Information Processing Systems 34 (2021)
48. Zhang, X., Wei, Y., Yang, Y., Huang, T.S.: Sg-one: similarity guidance network for one-shot semantic segmentation. IEEE Trans. Cybern. **50**(9), 3855–3865 (2020)

3D Instances as 1D Kernels

Yizheng Wu[1], Min Shi[1], Shuaiyuan Du[1], Hao Lu[1], Zhiguo Cao[1]([⊠]),
and Weicai Zhong[2]

[1] Key Laboratory of Image Processing and Intelligent Control, Ministry of Education
School of AIA, Huazhong University of Science and Technology, Wuhan, China
{yzwu21,min_shi,sydu,hlu,zgcao}@hust.edu.cn
[2] Huawei CBG Consumer Cloud Service Search and Maps BU, Shenzhen, China
zhongweicai@huawei.com

Abstract. We introduce a 3D instance representation, termed *instance kernels*, where instances are represented by one-dimensional vectors that encode the semantic, positional, and shape information of 3D instances. We show that instance kernels enable easy mask inference by simply scanning kernels over the entire scenes, avoiding the heavy reliance on proposals or heuristic clustering algorithms in standard 3D instance segmentation pipelines. The idea of instance kernel is inspired by recent success of dynamic convolutions in 2D/3D instance segmentation. However, we find it non-trivial to represent 3D instances due to the disordered and unstructured nature of point cloud data, *e.g.*, poor instance localization can significantly degrade instance representation. To remedy this, we construct a novel 3D instance encoding paradigm. First, potential instance centroids are localized as candidates. Then, a candidate merging scheme is devised to simultaneously aggregate duplicated candidates and collect context around the merged centroids to form the instance kernels. Once instance kernels are available, instance masks can be reconstructed via dynamic convolutions whose weights are conditioned on instance kernels. The whole pipeline is instantiated with a dynamic kernel network (DKNet). Results show that DKNet outperforms the state of the arts on both ScanNetV2 and S3DIS datasets with better instance localization. Code is available: https://github.com/W1zheng/DKNet.

Keywords: Instance kernel · Point cloud · Instance segmentation

1 Introduction

3D Instance segmentation aims to predict point-level instance labels [8, 12]. Standard approaches heavily rely on proposals [4,17,28] or heuristic clustering algorithms [2,12]. In this work, we show that instance masks can be reconstructed by

Y. Wu and M. Shi—Contributed equally.

Supplementary Information The online version contains supplementary material available at https://doi.org/10.1007/978-3-031-19818-2_14.

scanning a scene with *instance kernels*, a representation for 3D instances, which simultaneously encodes the positional, semantic, and shape information of 3D instances.

3D instance representation addresses two fundamental problems: i) how to localize an instance precisely and ii) how to aggregate features effectively to depict the instance. Unlike 2D instances that can be directly encoded via grid sampling [26] or dynamic kernel assigning [31], in the 3D domain, the disordered and unstructured nature of point cloud data renders difficulties for precise instance localization and reliable representation; and top-performing approaches [2,12,16] implicitly localize instances with centroid offsets [22], which only provides coarse information for instance representation, as shown in Fig. 1. Our instance kernel also draws inspiration from DyCo3D [9], which first applies dynamic convolution into 3D instance segmentation. However, DyCo3D is built upon the existing bottom-up segmentation pipeline [12], leaving the fundamental problems of instance encoding unsolved. To alleviate the difficulties above, we design a novel instance encoding paradigm that efficiently localizes different instances and encodes the semantic, positional, and shape information of instances into *instance kernels* for mask generation.

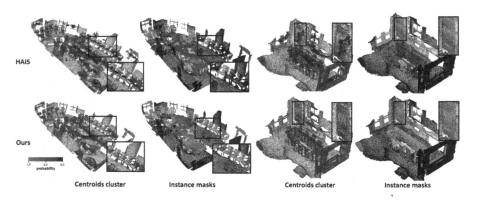

Fig. 1. Comparison of inferred centroid clusters and instance masks. Compared with HAIS [2], our DKNet generates more focused centroid clusters that can guide precise localization such that small and close instances can be discriminated and large instances have consistent predictions. Best viewed by zooming in and in color. (Color figure online)

We further incorporate the kernel encoding paradigm into a dynamic kernel network (DKNet) for 3D instance segmentation. To localize instances, as shown in Fig. 1, DKNet predicts a centroid map for instances and extracts centroids via a customized Non-Maximum Suppression (NMS) operator with local normalization. Observing that duplicated candidates may be predicted for a single instance (especially for large ones), we design an iterative aggregation mechanism to merge duplicated candidates guided by a predicted merging score map.

The score map indicates the probability whether each paired candidates should be merged. Afterwards, the merged instances are encoded into instance kernels by adaptively fusing the point features around the localized instance centroids. Finally, instance masks can be reconstructed with a few convolution layers, whose weights are conditioned on the generated instance kernels.

We evaluate DKNet on two popular 3D instance segmentation datasets, including ScanNetV2 [3] and S3DIS [1]. The results show that DKNet outperforms previous state-of-the-art approaches, ranking the first AP among published methods on the ScanNetV2 online leaderboard.[1] Thanks to the instance kernels and the specially designed instance localization pipeline, DKNet can better distinguish instances from dense areas than current top-performing approaches, as shown in Fig. 1. A series of ablation studies also demonstrate that the proposed instance localization and aggregation pipeline can greatly enhance the instance representation.

Our contributions are two-fold:

- We extend the idea of dynamic convolution into instance kernel, a comprehensive representation for 3D instances in point clouds;
- We propose a dynamic kernel network for 3D instance segmentation, with a novel instance kernel encoding paradigm;

2   Related Work

Here we briefly review the 3D instance segmentation approaches and kernel-based instance segmentation.

Proposal-Based 3D Instance Segmentation. Proposal-based approaches [29] assign instances with proposals and the instance masks are generated upon proposals. 3D-BoNet [28] directly predicts the bounding boxes, within instance masks are generated. 3D-MPA [4] samples proposals from predicted centroids; masks of proposals are then clustered to form the instance masks. GICN [17] simultaneously predicts the centroids and sizes of instances to obtain bounding box proposals. Predictions from proposal-based approaches show good objectness, while two major drawbacks exist: 1) the multi-stage training and the proposal generation process introduce large computational overhead; 2) the results highly rely on the proposals.

Propoal-Free 3D Instance Segmentation. Proposal-free approaches cluster points into instances in a bottom-up manner. SSTNet [16] models the entire scene by constructing a tree of superpoints and uses top-bottom traversal to aggregate nodes and form instance masks. PointGroup [12] clusters points using semantic labels and centroid offsets as clues. PE [30] encodes points into an

---

[1] http://kaldir.vc.in.tum.de/scannet_benchmark/semantic_instance_3d

embedding space where points from the same instances are close. Then cluster-ing are performed in this embedding space. Considering that clustering points into instances with various sizes in one shot is difficult, HAIS [2] proposes a novel hierarchical clustering pipeline to gradually refine the aggregation results. How-ever, even with the implicit guide of object signals, the objectness of predictions is still low, as shown in Fig. 1. Hence, directly adding kernel-based dynamic con-volution modules upon existing proposal-free approaches cannot bring out the best of kernel-based instance segmentation paradigm.

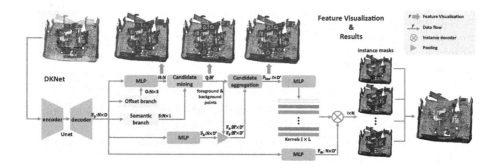

Fig. 2. Pipeline of the dynamic kernel network.

Kernel-Based Instance Segmentation. Kernel-based instance segmentation uses instance-aware kernels to scan the whole scene to reconstruct instance masks, the pivot of which is to represent or associate instances with different ker-nels. After obtaining the kernels, the common solutions are scanning the scene via dot product or dynamic convolution [11,19]. CondInst [24] predicts instance proposals by object detection and encodes the proposal features into kernels. K-Net [31] associates a fixed number of kernels with instances in certain regions and dynamically updates them. SOLOv2 [26] partitions the feature maps into grids and generates a kernel for each grid. However, representing instances as kernels in the 3D domain is non-trivial due to the disordered and unstructured nature of point cloud data. DyCo3D [9] first introduce the kernel-based paradigm in 3D instance segmentation, which is built upon existing bottom-up approach [12]. However, they focus on the concrete implementation of dynamic convolutions, and bypass the core of the kernel-based paradigm: how to encode instances into kernels? In this work, we further explore the underlying relation between the dis-criminative representation of instances and effective kernel-based segmentation, resulting in a novel localize-then-aggregate instance kernel encoding paradigm.

3   Dynamic Kernel Network for 3D Instance Segmentation

3.1   Overview

As illustrated in Fig. 2, at the core of our dynamic kernel network (DKNet) is to encode instances into discriminative instance kernels. The encoding process consists of three key stages: 1) processing raw point clouds with a UNet-like backbone and predicting point features, centroid offsets, and semantic masks; 2) localizing centroids for instances with a candidate mining branch; 3) merging duplicated candidates and collecting context around instance centroids to form instance kernels. Once the instance kernels are acquired, the instance masks can be obtained by processing the point cloud features with a few convolution layers, whose weights are conditioned on instance kernels.

3.2   Point-Wise Feature Extraction

Following recent proposal-free approaches [2,12,16], we adopt the backbone from PointGroup [12] for feature extraction. Given the raw point cloud $P \in \mathbb{R}^{N \times 6}$ with N points, a 3D UNet-like [23] backbone with sparse convolutions [5] outputs point features $F_p \in \mathbb{R}^{N \times D}$. F_p is then fed to a semantic branch which predicts semantic mask $S \in \mathbb{R}^{N \times C}$, and additionally, a centroid offset branch which infers the offset $O \in \mathbb{R}^{N \times 3}$ of each point to the corresponding instance centroid. The semantic branch is a Multi-Layer Perceptron (MLP) with $\mathtt{softmax}$ activation at the output layer. Cross-entropy loss and multi-class dice loss [20] are used to supervise the training of this branch. Similar to the semantic branch, the centroid offset branch maps F_p into offsets O. For each point P_i, $O_i \in \mathbb{R}^3$ is a vector pointing to the centroid of instance that covers this point. Further details of the backbone can be referred to the supplementary.

3.3   Finding Instances

To generate the kernel for each instance, we should first find all the instances. However, the top-performing proposal-free approaches predict centroid offsets as implicit object signals, which is rather coarse to precisely localize instances, as shown in Fig. 1. Hence, learning from proposal-based approaches, we propose a candidate mining branch to generate centroid maps, followed by a searching algorithm to localize the instance candidates.

As shown in Fig. 3, point features $F_p \in \mathbb{R}^{N \times D}$ and centroid offsets $O \in \mathbb{R}^{N \times 3}$ are jointly concatenated to form the input $F_c \in \mathbb{R}^{N \times (D+3)}$ for the centroid mining branch. Then F_c is fed into an MLP with $\mathtt{softmax}$ activation at the output layer to obtain sharp centroid heatmap $H \in \mathbb{R}^N$. Each element H_i indicates the probability of the ith point being an instance centroid.

During training, we place a 3D Gaussian kernel on every instance centroid to form a pseudo ground truth heatmap as $\hat{H}_i = \exp(-\alpha \cdot d_i^2 / r_i^2)$, where d_i denotes the distance between point i to the centroid of the instance covering it. r_i, which controls the variance of the Gaussian kernel, equals the maximum side length of the axis-aligned bounding box of the corresponding instance. Hence, the Gaussian kernels are geometry adaptive w.r.t. the size of different instances. α is set to 25 to keep the average of $\hat{H}$ around 0.1. To supervise the training, the loss function $\mathcal{L}_{center}$ for the candidate mining branch is defined as:

$$\mathcal{L}_{center} = \frac{1}{\sum_{i=1}^{N} \mathbb{I}(P_i)} \sum_{i=1}^{N} |H_i - \hat{H}_i| \cdot \mathbb{I}(P_i), \tag{1}$$

where $\mathbb{I}(P_i)$ is an indicator function that outputs 1 when i^{th} point belongs to an instance, otherwise outputs 0.

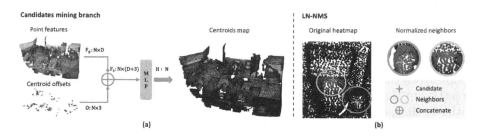

Fig. 3. Centroid mining branch. (a) The input of candidate mining branch; (b) The customized Non-Maximum Suppression with local normalization.

With the predicted heatmap H, we iteratively search the local maximum as instance candidates with a customized local normalized NMS (LN-NMS) strategy. During each iteration, the algorithm localizes the point with the highest centroid score among the foreground points; the centroid scores of other points in its neighbor with radius R are then normalized via the division of the maximum value in this R-radius neighbor. If the normalized centroid score is larger than a threshold T_θ, this point will be considered a candidate and all other points within its R-radius neighbor are suppressed and excluded in the next iteration, no matter whether the point is chosen as the candidate. We set $T_\theta = 0.5$ and the radius $R = 0.3m$ according to the average size of instances in ScanNet [3]. The iteration ends when no point remains or N_θ candidates have been found. N_θ is empirically set to 200. Finally, a candidate set $Q \in \mathbb{R}^{N'}$ can be collected, where N' denotes the number of candidates. Refer to supplementary for more details.

3.4 Representing Instances as Kernels

After localizing the instance centroids, we represent these candidates as instance kernels. We expect that one kernel is extracted for one instance, and the kernel should be discriminative. Therefore, we design a duplicated candidate aggregation strategy that simultaneously eliminates extra candidates and adaptively fuses features around candidates for instance representation.

Aggregating Duplicate Candidates. We judge whether two candidates should be aggregated based on the context of each candidate. For each raw candidate, we use the features from its "foreground points", "background points" to describe the context. The "foreground points" denote points with the same semantic label within a R-radius neighbor of each candidate, while the "background points" denote all the points with different semantic labels within a $2R$-radius neighbor of each candidate. To aggregate features from the foreground and background points, we first process point feature F_p with an MLP for dimensionality reduction. Then, the output features of the MLP w.r.t. "foreground points" and "background points" are averaged and respectively form the descriptive feature $F_n \in \mathbb{R}^{N' \times D'}$ and the background feature $F_b \in \mathbb{R}^{N' \times D'}$ for each candidate. As the above two features only encode semantic and shape information, we concatenate them with the shifted coordinate (add the raw coordinates with the centroid offset vectors) of each candidate as positional information to form the aggregation feature $F_a \in \mathbb{R}^{N' \times (2D'+3)}$ for duplicated candidates aggregation.

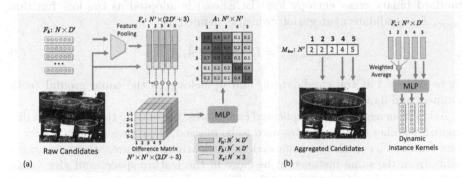

Fig. 4. Candidates aggregation module. (a) The process of predicting the merging score map; (b) The process of generating instance kernels.

As shown in Fig. 4, for each candidate, its aggregation feature $F_{a,i}$ is subtracted from the aggregation feature of all other candidates. By repeating this process for every instance, a candidate difference matrix reflecting how similar each pair of candidates will be generated. Taking the absolute values of the difference matrix as inputs, an MLP with `sigmoid` function outputs the merging score map $A \in \mathbb{R}^{N' \times N'}$, where A_{ij} indicates the probability that the ith and j^{th} candidates shall be merged because they can belong to the same instance.

Once the merging score map A is obtained, a simple greedy algorithm will be used to iteratively merge candidates. We first initialize an instance centroid map $M_{ins} \in \mathbb{R}^N$ where $M_{ins,i} = i$. M_{ins} records the indices of instance centroids that each candidate belongs to, and we define candidates with the same index as an instance *group*. Before aggregation, the instance centroids of candidates are themselves, and each candidate is an instance *group*. During each iteration, if A_{ij} is the maximum in A excluding diagonal elements, all candidates in the ith

and j^{th} instance *groups* will be merged. The instance centroid indices of these candidates will also be unified to the index of candidate with the highest centroid score among them. Since candidates within the same instance group can no longer be merged, we then update all merging scores between them to be 0. The iteration ends when all merging scores are below a predefined threshold, which is set to 0.5. The candidates with the same index are treated as a predicted instance. After aggregation, supposing I instance groups are generated, the centroid coordinates $C_{ins} \in \mathbb{R}^{I \times 3}$ of new instances are set to the coordinates of the center candidates, while the features of new instances $F_{ins} \in \mathbb{R}^{I \times D'}$ are obtained by a weighted average upon descriptive features of grouped candidates w.r.t. their neighbor sizes. The average can help dynamically aggregate the features of instances with different sizes. As shown in Fig. 4, if multiple candidates are predicted on one large instance, information from all the candidates will be propagated to the instance centroids to describe instances.

During training, it is easy to figure out which pairs of candidates are duplicated, and accordingly a ground truth merging map $\hat{A}$ can be generated. The standard binary cross entropy loss (BCELoss) is adopted as the loss function $\mathcal{L}_{aggre}$ for candidate aggregation, which is defined as:

$$\mathcal{L}_{aggre} = BCELoss(A, \hat{A}), \tag{2}$$

where $\hat{A}_{i,j} = 1$ if the candidate q_i and q_j belong to the same ground truth instance, and 0 otherwise.

Aside from aggregating duplicated candidates, by guiding the network to distinguish whether two candidates need to be merged, the representation ability of point feature F_p can also be enhanced, as the learning of aggregation encourages points from the same instance to be close in the feature space, and vice versa, similar to the idea of contrast learning [7].

Encoding Instance Kernels. After candidate aggregation, all instances in the scene are assigned with centroids, denoted by C_{ins}, and corresponding features, denoted by F_{ins}. F_{ins} is fed into an MLP to generate instance kernels $\mathcal{W} \in \mathbb{R}^{I \times L}$, where L is the length of instance kernels. In analogous to CondInst [24] and DyCo3D [9], the instance kernels are transformed into the weights for a few convolution layers in the instance decoding stage. Hence, L depends on the specific configurations of the convolution layers, which can be computed by Eq. 4.

3.5 Generating Masks with Instance Kernels

The instance kernels, denoted by $\mathcal{W}$, have encoded the positional, semantic, and shape characteristics of instances. To decode instances, the instance kernels are transformed into the weights with a few convolution layers, which are applied to augmented point cloud features to reconstruct instance masks.

To augment the point cloud features, an MLP further extracts the mask feature $F_m \in \mathbb{R}^{N \times D'}$ from the point feature F_p. To inject instance-aware positional information into F_m, inspired by DyCo3d [9], the offset between each point to the instance centroids are added to F_m before convolution. $E.g.$, for each point P_i, we compute its offset to the centroid of the kth instance as $Z_{k,i} = C_{ins,k} - X_i$. Then, the point decoding feature $F_d \in \mathbb{R}^{N \times (D'+3)}$ for the kth instance is generated by concatenating F_m and Z_k along the channel dimension. Although DyCo3D also generates instance masks via instance-specific kernels, the kernels are only applied to the points within the same semantic category. In contrast, the instance kernels in our approach scan the entire scene, which avoids the reliance of semantic prediction. Hence, the decoding process can correct some errors in semantic prediction. The bottom-up approaches [2,12] cannot correct such errors. The decoding process outputs the instance mask $M \in \mathbb{R}^{I \times N}$ by

$$M_i = Conv(F_d, \mathcal{W}_i), \quad i \in [1, I], \tag{3}$$

where $\mathcal{W}_i \in \mathbb{R}^L$ is transformed into the weights and biases via two 1×1 convolution layers. The first layer has 16 output channels with ReLU activation function and the second one has 1 output channel with sigmoid for mask decoding. To fit the number of parameters of convolution layers, the length of the instance kernel L can be computed by

$$L = (16+3) \times 16(weight) + 16(bias) + 16 \times 1(weight) + 1(bias) = 337. \tag{4}$$

The use of dynamic convolution is the same as [9,24]. Implementation details are depicted in the supplementary.

To supervise the generation of instance masks, we first match the predicted instances with actual instances with Hungarian algorithm [13] according to a cost matrix. Then, we apply the BCELoss and dice loss [20] for supervision. Supposing $M \in \mathbb{R}^{I \times N}$ with I instance masks is generated, and $\hat{M} \in \mathbb{R}^{G \times N}$ with G ground-truth instance masks is provided, the cost matrix $\mathcal{C} \in \mathbb{R}^{I \times G}$ for the Hungarian algorithm is obtained by:

$$\mathcal{C}_{i,j} = \|C_{ins,i} - C_{gt,j}\|_2 + \mathbb{I}(S_{ins,i} == S_{gt,j}), \tag{5}$$

where C_{ins} and C_{gt} are centroid coordinates of predicted an ground truth instances, respectively, and S_{ins} and S_{gt} are corresponding semantic labels. We determine the instance semantic label S_{ins} by voting within the predicted instance. With the cost matrix, one predicted instance mask is expected to be matched to a instance with the closet centroid and identical semantic label.

After the matching process, the predicted instance mask M is assigned with the ground truth instance masks $\hat{M} \in \mathbb{R}^{G \times N}$. Then, BCELoss $\mathcal{L}_{bce}$ and dice loss $\mathcal{L}_{dice}$ are computed by:

$$\mathcal{L}_{mask} = \frac{1}{I'} \sum_{k=1}^{I} (BCE(M_k, \hat{M}_k) + (1 - 2\frac{M_k \cdot \hat{M}_k}{|M_k| + |\hat{M}_k|})) \cdot \mathbb{I}(iou_k > 0.25), \tag{6}$$

where $\hat{M}_k$ denotes the ground truth instance mask for the kth instance, and M_k denotes the predicted one. iou_k denotes the Intersection-over-Union (IoU) between M_k and $\hat{M}_k$, and $\mathbb{I}$ is an indicator function. We add the constraints so that the loss will only be computed when instances are correctly matched and I' is defined by $I' = \sum_{k=1}^{I} \mathbb{I}(iou_k > 0.25)$.

Inference Post-Processing. During inference, to convert the soft instance masks M_k's into hard instance labels and filter out potentially wrong predictions, a simple yet effective two-stage refinement pipeline is proposed. First, some small duplicated fragments or noise is removed. In the second stage, superpoints [15] are applied to refine the shapes of generated instance masks. Unlike proposal-free approaches [2,9,12], our approach do not need NMS or ScoreNet in post-processing, which is efficient.

In the first stage, given the predicted soft instance mask $M \in \mathbb{R}^{I \times N}$, we first generate the raw instance label by selecting the label of the instance with the highest score in M. Then, we define a coverage score $S_c \in \mathbb{R}^I$ by:

$$S_{c,k} = N_{inter,k}/N_{intra,k}, \tag{7}$$

where $N_{inter,k}$ and $N_{intra,k}$ denote the number of "inter-point" and "intra-point" for the kth instance, respectively. The "inter-point" denotes the number of points being assigned to the kth instance in the raw instance label, while the "intra-point" denotes the number of points in the soft instance mask of the kth instance that are above a threshold $T_{m,k}$. $T_{m,k}$ is determined by the Otsu algorithm [21] that is adaptive to different instances. The cover score indicates the completeness and independence of each instance prediction. We then multiply the coverage score S_c with M to generate the refined soft instance mask. By taking the instance with the highest score in the refined mask, the final hard instance labels can be obtained. Refer to supplementary for more details on computing T_m.

In the second stage, we first aggregate the raw point clouds into superpoints [15]. Points within a certain superpoint should belong to the same instance. Hence, we unify the instance label in each superpoint to be the one that most points belong to. Aside from the instance label R, we need to assign a confidence score for each predicted instance that indicates the prediction quality for evaluation. This score is obtained by multiplying the average of instance score and semantic score of the "intra-point".

Table 1. Quantitative results on the ScanNetV2 test set. Refer to supplementary full results with all the 20 categories.

Approaches	mAP	AP@50	Bed	Booksh.	Cabinet	Chair	Curtain	Desk	Door	Otherfu.	Picture	Refrige.	Sofa	Table	Window
3D-BoNet [28]	25.3	48.8	67.2	59.0	30.1	48.4	62.0	30.6	34.1	25.9	12.5	43.4	49.9	51.3	43.9
3D-SIS [10]	16.1	38.2	43.2	24.5	19.0	57.7	26.3	3.3	32.0	24.0	7.5	42.2	69.9	27.1	23.5
MTML [14]	28.2	54.9	80.7	58.8	32.7	64.7	81.5	18.0	41.8	36.4	18.2	44.5	68.8	57.1	39.6
3D-MPA [4]	35.5	61.1	83.3	76.5	52.6	75.6	58.8	47.0	43.8	43.2	35.8	65.0	76.5	55.7	43.0
PointGroup [12]	40.7	63.6	76.5	62.4	50.5	79.7	69.6	38.4	44.1	55.9	47.6	59.6	75.6	55.6	51.3
GICN [17]	34.1	63.8	**89.5**	80.0	48.0	67.6	73.7	35.4	44.7	40.0	36.5	70.0	83.6	59.9	47.3
DyCo3D [9]	39.5	64.1	84.1	**89.3**	53.1	80.2	58.8	44.8	43.8	53.7	43.0	55.0	76.4	65.7	56.8
Occuseg [6]	48.6	67.2	75.8	68.2	57.6	84.2	50.4	52.4	56.7	58.5	45.1	55.7	79.7	56.3	46.7
PE [30]	39.6	64.5	77.3	79.8	53.8	78.6	79.9	35.0	43.5	54.7	54.5	64.6	76.1	55.6	50.1
SSTNet [16]	50.6	69.8	69.7	88.8	55.6	80.3	62.6	41.7	55.6	58.5	70.2	60.0	72.0	69.2	50.9
HAIS [2]	45.7	69.9	84.9	82.0	67.5	80.8	75.7	46.5	51.7	59.6	55.9	60.0	76.7	67.6	56.0
SoftGroup [25]	50.4	**76.1**	80.8	84.5	**71.6**	86.2	**82.4**	**65.5**	62.0	**73.4**	69.9	**79.1**	**84.4**	**76.9**	59.4
Ours	**53.2**	71.8	81.4	78.2	61.9	**87.2**	75.1	56.9	**67.7**	58.5	**72.4**	63.3	81.9	73.6	**61.7**

4   Experiments

In this section, we first compare the proposed Dynamic Kernel Network (DKNet) with other state-of-the-art approaches on two 3D instance segmentation benchmarks: ScanNetV2 [3] and S3DIS [1]. Then, we verify the effectiveness of different components in DKNet via a controlled ablation study.

4.1   Implementation Details

Training Details. For data preparation, the coordinates and colors are concatenated together to form 6D vectors for each point. The network is trained on a single RTX 3090 GPU with a batch size of 4 for 400 epochs. We use the AdamW [18] optimizer with an initial learning rate of 0.001, which is adjusted by a cosine scheduler [27] during training. Weight decay is set to $1e$-5. Following previous methods [12], we voxelize the point clouds with the size of 0.02 m for ScanNetV2 and 0.05 m for S3DIS.

The overall training loss combines the loss from the semantic prediction, offset prediction, candidate mining, candidate aggregation branch, and the mask generation process, which can be defined as:

$$\mathcal{L} = \mathcal{L}_{sem} + \mathcal{L}_{off} + \mathcal{L}_{center} + \mathcal{L}_{aggre} + \mathcal{L}_{mask}, \tag{8}$$

where $\mathcal{L}_{sem}$ and $\mathcal{L}_{off}$ are the losses for semantic segmentation and centroid offsets prediction, respectively.

Table 2. Object detection results on ScanNetV2 validation set. We report per-class mAP with an IoU of 25% and 50%. The IoU is computed on bounding boxes. We evaluate the performance of HAIS with the provided model. "Ours⁻" denotes the model without candidates aggregation part.

Approach	$AP@25$	$AP@50$
VoteNet [22]	58.6	33.5
3DSIS [10]	40.2	22.5
3D-MPA [4]	64.2	49.2
DyCo3D [9]	58.9	45.3
HAIS [2]	66.0	54.2
Ours⁻	65.4	57.9
Ours	**67.4**	**59.0**

Table 3. Quantitative results on S3DIS dataset. We report mCov, mWCov, mPre, and mRec. approaches with ‡ marks are evaluated on scenes in Area-5. The others are evaluated via 6-fold cross validation.

Approach	mCov	mWCov	mPre	mRec
PointGroup‡ [12]	–	–	61.9	62.1
DyCo3D‡ [9]	63.5	64.6	64.3	64.2
SSTNet‡ [16]	–	–	65.5	64.2
HAIS‡ [2]	64.3	**66.0**	**71.1**	65.0
Ours‡	**64.7**	65.6	70.8	**65.3**
OccuSeg [6]	–	–	72.8	60.3
GICN [17]	–	–	68.5	50.8
PointGroup [12]	–	–	69.6	69.2
SSTNet [16]	–	–	73.5	**73.4**
HAIS [2]	67.0	70.4	73.2	69.4
SoftGroup [25]	69.3	71.7	**75.3**	69.8
Ours	**70.3**	**72.8**	**75.3**	71.1

Datasets. We use ScanNetV2 and S3DIS for training and evaluation. ScanNetV2 includes 1,613 scenes with 20 different semantic categories. 1,201, 312 and 100 scenes are selected as the training, validation and test set, respectively. Note that the labels for test set are hidden for a fair comparison. Following official evaluation protocol, we use mean average precisions (mAPs) under different IoU thresholds as the evaluation metrics. $AP@25$ and $AP@50$ denote the average precision scores with IoU thresholds set to 25% and 50%. mAP denotes the average of all the APs with IoU thresholds ranging from 50% to 95% with a step size of 5%. S3DIS dataset consists of 271 scenes collected from 6 different areas with 13 different object categories. Following previous approaches [2,9,12,16], we train and evaluate our approach in two ways: 1) scenes from area-5 are used for testing while scenes in other areas are used for training; 2) 6-fold cross validation where each area is used in turn for testing. On S3DIS, with the threshold IoU set to 0.5, we report coverage (mCov), weighted coverage (mWCov), mean precision (mPrec), and mean recall (mRec) as evaluation metrics.

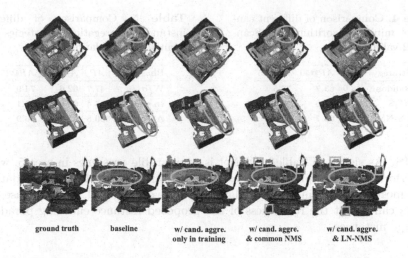

| ground truth | baseline | w/ cand. aggre. only in training | w/ cand. aggre. & common NMS | w/ cand. aggre. & LN-NMS |

Fig. 5. Qualitative results of our approaches on ScanNetV2 validation set. We highlight the key details with red marks. Best viewed by zooming in and in color.

4.2 Comparison with the State of the Arts

ScanNetV2. Comparisons with the state of the arts on the ScanNetV2 test set are shown in Table 1. Our approach achieves an mAP of 53.2%, outperforming previous state-of-the-art approaches. The proposed DKNet obtains significant improvement on small instances like chairs or pictures, and competitive results on large instances like beds or tables. We also notice that, compared with a recent well-designed bottom-up method SoftGroup [25], DKNet shows inferior $AP@50$. The plausible reason are two fold. First, we find the DKNet retains relative high AP under strict IoU thresholds (> 0.6), indicating that the predicted masks can well preserve the instance shapes. However, under lower IoU thresholds (< 0.6), DKNet becomes less advantageous. Second, our pure data-driven candidate merging process shows mistakes on some difficult scenes, such as bookshelves with vague boundaries, which can be better tackled by well-designed bottom-up clustering. These results suggest that DKNet can be improved with careful design to further boost the potential of the proposed kernel-based paradigm.

To evaluate the instance localization performance of 3D instance segmentation approaches, we compare different 3D instance approaches under object detection metrics on ScanNetV2 validation set in Table 2. Predicted masks are converted into axis-aligned bounding boxes following DyCo3D [9]. Our approach achieves the best performance in $AP@25$ of 67.4% and $AP@50$ of 59.0%, which demonstrate that kernels are extracted from solid instance localization results.

Table 4. Comparison of different candidate mining algorithms on ScanNetV2 validation set.

Strategy	AP	$AP@50$	$AP@25$
Random	48.0	63.7	73.5
NMS	49.7	64.6	75.7
LN-NMS	50.8	66.7	76.9

Table 5. Comparison of different instance aggregation strategies on ScanNetV2 validation set.

Phase	AP	$AP@50$	$AP@25$
W/o	47.7	62.6	74.9
In training	48.4	64.5	76.1
All phases	50.8	66.7	76.9

S3DIS. As shown in Table 3, DKNet is comparable on scenes in area-5, while outperforms other approaches in 3 out of 4 metrics on 6-fold cross validation. Since most categories in S3DIS are large ones like ceiling, wall, and bookcase, the results can reflect the robustness of the proposed instance encoding paradigm on large instances.

4.3   Qualitative Evaluation

We visualize the predicted masks in Fig. 5. As is shown in column 2, excluding candidate aggregation (baseline) leads to severe over-segmentation as multiple kernels will be generated for one instance. The over-segmentation can be effectively alleviated with the candidate aggregation, which will simultaneously guides the generation of instance-aware features. Comparing with the common NMS in GICN [17], the proposed LN-NMS can better localize instances with different sizes, while common NMS omits some small instances (garbage bins in row 2 and chairs in row 3).

4.4   Ablation Study

Here, we first compare different candidate mining and aggregation strategies. Then, we analyze how to better represent instances as kernels.

Candidate Mining. Here we verify the effectiveness of the proposed LN-NMS algorithm for candidate mining. As in Table 4, three different ways are compared. **Random** (row 1) denotes points above a threshold is randomly selected (at most 200) as instance candidates. **NMS** denotes the searching algorithm with common NMS used in GICN [17]. And the proposed **LN-NMS** algorithm shown in Sec. 3.3. Results in row 1 show that: 1) poor instance localization can significantly degrade instance segmentation performance; 2) even selecting points randomly in the heatmaps as candidates can yield competitive results, which demonstrates the robustness of the subsequent candidate aggregation module. Comparing results in row 2 with row 3, LN-NMS improves $AP@50$ by 2.1%.

Instance Candidates Aggregations. Comparison of different candidate aggregation strategies are shown in Table 5. "W/o" means no candidate aggregation are preformed. "In training" means only optimizing aggregation loss during training, while the raw candidates will not be aggregated for inference. "All phases" denotes our full approach. When adding the aggregation loss only for supervision (row 2), $AP@50$ increases by 1.9%; Motivated by the aggregation loss, there is an instance-clustering trend for the identical instances in the feature space, which is similar to contrast learning [7]. By aggregating the duplicated candidates, the full approach promotes the AP, $AP@50$ and $AP@25$ by 3.1%, 4.1% and 2.0% comparing with baseline, which demonstrates the effectiveness of candidate aggregation.

Generating Instance Kernels. With the raw instance candidates and merging map M_{ins} marking which candidates shall be aggregated, naturally, there are 2 different ways to represent each instance: 1) only using the feature from the candidate with the highest centroids scores; 2) aggregating features from all the merged candidates. As in Table 6, the latter way (default) obtains better results, which demonstrates that aggregating features from different candidates benefits the representation. In addition, we test an ideal representation in row 3 where features are collected from all the points within each instance. It can be observed that the performances of our instance representation approach are comparable with this ideal representation. This indicates that representative and discriminative instance contexts are obtained in our kernel generation process.

Table 6. Comparisons of different candidates aggregation approaches on ScanNetV2 validation set.

Strategy	AP	$AP@50$	$AP@25$
Candidates elimination	50.4	65.2	75.8
Candidates aggregation	50.8	66.7	76.9
Full instances	51.5	67.0	77.0

5 Conclusion

We introduce a 3D instance representation, termed *instance kernels*, which encodes the positional, semantic, and shape information of instances into a 1D vector. We find that the difficulty in representing 3D instances lies in precisely localizing instances and collecting discriminative features. Accordingly, we design a novel instance encoding paradigm that first mines centroids candidates for localization. Then, an aggregation process simultaneously eliminates duplicated candidates and gathers features around each instance for representation. We incorporate the instance kernel into a Dynamic Kernel Network (DKNet), which outperforms previous state-of-the-art approaches on public benchmarks.

Acknowledgment. This work was supported in part by the National Key R&D Program of China (No. 2018YFB1305504) and the DigiX Joint Innovation Center of Huawei-HUST.

References

1. Armeni, I., et al.: 3D semantic parsing of large-scale indoor spaces. In: Proceedings of IEEE Conference on Computer Vision Pattern Recognition (CVPR), pp. 1534–1543 (2016). https://doi.org/10.1109/CVPR.2016.170
2. Chen, S., Fang, J., Zhang, Q., Liu, W., Wang, X.: Hierarchical aggregation for 3D instance segmentation. In: Proceedings of IEEE International Conference on Computer Vision (ICCV), pp. 15447–15456 (2021). https://doi.org/10.1109/ICCV48922.2021.01518
3. Dai, A., Chang, A.X., Savva, M., Halber, M., Funkhouser, T., Nießner, M.: ScanNet: richly-annotated 3D reconstructions of indoor scenes. In: Proceedings of IEEE Conference on Computer Vision Pattern Recognition (CVPR), pp. 2432–2443 (2017). https://doi.org/10.1109/CVPR.2017.261
4. Engelmann, F., Bokeloh, M., Fathi, A., Leibe, B., Nießner, M.: 3D-MPA: multi-proposal aggregation for 3D semantic instance segmentation. In: Proceedings of IEEE Conference on Computer Vision Pattern Recognition (CVPR), pp. 9028–9037 (2020). https://doi.org/10.1109/CVPR42600.2020.00905
5. Graham, B., Engelcke, M., van der Maaten, L.: 3D semantic segmentation with submanifold sparse convolutional networks. In: Proceedings of IEEE Conference on Computer Vision Pattern Recognition (CVPR) (2018)
6. Han, L., Zheng, T., Xu, L., Fang, L.: OccuSeg: occupancy-aware 3d instance segmentation. In: Proceedings of IEEE Conference on Computer Vision Pattern Recognition (CVPR), pp. 2937–2946 (2020). https://doi.org/10.1109/CVPR42600.2020.00301
7. He, K., Fan, H., Wu, Y., Xie, S., Girshick, R.: Momentum contrast for unsupervised visual representation learning. In: Proceedings of IEEE Conference on Computer Vision Pattern Recognition (CVPR), pp. 9726–9735 (2020). https://doi.org/10.1109/CVPR42600.2020.00975
8. He, K., Gkioxari, G., Dollár, P., Girshick, R.: Mask R-CNN. In: Proceedings of IEEE International Conference on Computer Vision (ICCV), pp. 2980–2988 (2017). https://doi.org/10.1109/ICCV.2017.322
9. He, T., Shen, C., van den Hengel, A.: Dyco3d: robust instance segmentation of 3d point clouds through dynamic convolution. In: Proceedings of IEEE Conference on Computer Vision Pattern Recognition (CVPR), pp. 354–363 (2021)
10. Hou, J., Dai, A., Nießner, M.: 3D-sis: 3D semantic instance segmentation of RGB-D scans. In: Proceedings of IEEE Conference on Computer Vision Pattern Recognition (CVPR), pp. 4421–4430 (2019). https://doi.org/10.1109/CVPR.2019.00455
11. Jia, X., De Brabandere, B., Tuytelaars, T., Gool, L.V.: Dynamic filter networks. In: Proceedings of Advances in Neural Information Processing Systems (NeurIPS), pp. 667–675 (2016)
12. Jiang, L., Zhao, H., Shi, S., Liu, S., Fu, C.W., Jia, J.: PointGroup: dual-set point grouping for 3d instance segmentation. In: Proceedings of IEEE Conference on Computer Vision Pattern Recognition (CVPR), pp. 4866–4875 (2020). https://doi.org/10.1109/CVPR42600.2020.00492
13. Kuhn, H.W.: The Hungarian method for the assignment problem. Naval Research Logistics Quarterly (1955)
14. Lahoud, J., Ghanem, B., Oswald, M.R., Pollefeys, M.: 3D instance segmentation via multi-task metric learning. In: Proceedings of IEEE International Conference on Computer Vision (ICCV), pp. 9255–9265 (2019). https://doi.org/10.1109/ICCV.2019.00935

15. Landrieu, L., Simonovsky, M.: Large-scale point cloud semantic segmentation with superpoint graphs. In: Proceedings of IEEE Conference on Computer Vision Pattern Recognition (CVPR), pp. 4558–4567 (2018). https://doi.org/10.1109/CVPR.2018.00479

16. Liang, Z., Li, Z., Xu, S., Tan, M., Jia, K.: Instance segmentation in 3d scenes using semantic superpoint tree networks. In: Proceedings of IEEE International Conference on Computer Vision (ICCV), pp. 2763–2772 (2021). https://doi.org/10.1109/ICCV48922.2021.00278

17. Liu, S., Yu, S., Wu, S., Chen, H., Liu, T.: Learning gaussian instance segmentation in point clouds. arXiv Computer Research Repository (2020)

18. Loshchilov, I., Hutter, F.: Decoupled weight decay regularization. arXiv Computer Research Repository (2017)

19. Lu, H., Dai, Y., Shen, C., Xu, S.: Index networks. IEEE Trans. Pattern Anal. Mach. Intell. **44**(1), 242–255 (2022)

20. Milletari, F., Navab, N., Ahmadi, S.A.: V-Net: fully convolutional neural networks for volumetric medical image segmentation. In: International Conference on 3D Vision (3DV), pp. 565–571 (2016). https://doi.org/10.1109/3DV.2016.79

21. Otsu, N.: A threshold selection method from gray-level histograms. Man, and Cybernetics, IEEE Transactions on Systems (1979)

22. Qi, C.R., Litany, O., He, K., Guibas, L.: Deep hough voting for 3D object detection in point clouds. In: Proceedings of IEEE International Conference on Computer Vision (ICCV), pp. 9276–9285 (2019). https://doi.org/10.1109/ICCV.2019.00937

23. Ronneberger, O., Fischer, P., Brox, T.: U-Net: convolutional networks for biomedical image segmentation. In: Navab, N., Hornegger, J., Wells, W.M., Frangi, A.F. (eds.) MICCAI 2015. LNCS, vol. 9351, pp. 234–241. Springer, Cham (2015). https://doi.org/10.1007/978-3-319-24574-4_28

24. Tian, Z., Shen, C., Chen, H.: Conditional convolutions for instance segmentation. In: Vedaldi, A., Bischof, H., Brox, T., Frahm, J.-M. (eds.) ECCV 2020. LNCS, vol. 12346, pp. 282–298. Springer, Cham (2020). https://doi.org/10.1007/978-3-030-58452-8_17

25. Vu, T., Kim, K., Luu, T.M., Nguyen, X.T., Yoo, C.D.: SoftGroup for 3D instance segmentation on point clouds. In: Proceedings of IEEE Conference on Computer Vision Pattern Recognition (CVPR) (2022)

26. Wang, X., Zhang, R., Kong, T., Li, L., Shen, C.: Solov2: dynamic and fast instance segmentation. In: Proceedings of Advances in Neural Information Processing Systems (NeurIPS) (2020)

27. Xu, G., Cao, H., Dong, Y., Yue, C., Zou, Y.: Stochastic gradient descent with step cosine warm restarts for pathological lymph node image classification via pet/CT images. In: International Conference on Signal and Image Processing (ICSIP) (2020)

28. Yang, B., et al.: Learning object bounding boxes for 3D instance segmentation on point clouds. In: Proceedings of Advances in Neural Information Processing Systems (NeurIPS), pp. 6737–6746 (2019)

29. Yi, L., Zhao, W., Wang, H., Sung, M., Guibas, L.J.: GSPN: generative shape proposal network for 3d instance segmentation in point cloud. In: Proceedings of IEEE Conference on Computer Vision Pattern Recognition (CVPR), pp. 3947–3956 (2019). https://doi.org/10.1109/CVPR.2019.00407

30. Zhang, B., Wonka, P.: Point cloud instance segmentation using probabilistic embeddings. In: Proceedings of IEEE Conference on Computer Vision Pattern Recognition (CVPR), pp. 8883–8892 (2021)
31. Zhang, W., Pang, J., Chen, K., Loy, C.C.: K-Net: towards unified image segmentation. In: Proceedings of Advances in Neural Information Processing Systems (NeurIPS), pp. 10326–10338 (2021)

TransMatting: Enhancing Transparent Objects Matting with Transformers

Huanqia Cai[1,2], Fanglei Xue[1,2], Lele Xu[1,2(✉)], and Lili Guo[1,2]

[1] University of Chinese Academy of Sciences, Beijing, China
{caihuanqia19,xuefanglei19}@mails.ucas.ac.cn
[2] Key Laboratory of Space Utilization, Technology and Engineering Center for Space Utilization, Chinese Academy of Sciences, Beijing, China
{xulele,guolili}@csu.ac.cn

Abstract. Image matting refers to predicting the alpha values of unknown foreground areas from natural images. Prior methods have focused on propagating alpha values from known to unknown regions. However, not all natural images have a specifically known foreground. Images of transparent objects, like glass, smoke, web, etc., have less or no known foreground. In this paper, we propose a Transformer-based network, TransMatting, to model transparent objects with a big receptive field. Specifically, we redesign the trimap as three learnable tri-tokens for introducing advanced semantic features into the self-attention mechanism. A small convolutional network is proposed to utilize the global feature and non-background mask to guide the multi-scale feature propagation from encoder to decoder for maintaining the contexture of transparent objects. In addition, we create a high-resolution matting dataset of transparent objects with small known foreground areas. Experiments on several matting benchmarks demonstrate the superiority of our proposed method over the current state-of-the-art methods.

Keywords: Image matting · Vision Transformer · Deep learning

1 Introduction

Image matting is a technique to separate the foreground object and the background from an image by predicting a precise alpha matte as a result. It has been widely used in many applications, such as image and video editing, background replacement, and virtual reality [5,18,46]. Image matting assumes that every pixel in the image I is a linear combination of the foreground object F and the background B by an alpha matte α:

$$I = \alpha F + (1 - \alpha)B, \alpha \in [0, 1] \tag{1}$$

H. Cai and F. Xue—Contributed equally.

L. Xu—This work was supported by the National Natural Science Foundation of China under Grant 61901454.

Project page: https://github.com/AceCHQ/TransMatting.

S. Avidan et al. (Eds.): ECCV 2022, LNCS 13689, pp. 253–269, 2022.
https://doi.org/10.1007/978-3-031-19818-2_15

As only the image I is known in this equation, the image matting is an ill-posed problem. So many existing methods [5,18,22,29,38,43,46] take a trimap as an auxiliary input. The trimap segments the image into three parts: known foreground and background, and unknown area, indicated as white, black, and gray, separately.

Most traditional methods, including both the sampling-based methods [2,7,12,15,36,43] and propagation-based methods [5,17,18,38], utilize the known area samples to find candidate colors or propagate the known alpha value. They heavily rely on the information from known areas, especially the known foreground areas. Recently, learning-based methods directly predict alpha mattes by neural network learning from well-annotated datasets. Although these methods take a great improvement in image matting, they also need specific information from known areas to predict unknown areas. However, according to [25], more than 50% pixels in the unknown areas cannot be correlated to pixels in the known regions due to the limited reception field of deep learning methods. LFP-Net [25] proposes a Center-Surround Pyramid Pooling module to propagate the context feature from the known regions to the near unknown regions. However, not all natural images have a salient and opaque object as the known foreground [21]. Images of the glass, bonfires, plastic bags, etc., have salient foregrounds but with transparent or meticulous interiors; images of the web, smoke, water drops, etc., have non-salient foregrounds. The corresponding trimaps of these kinds of images will have very few or even no foreground areas. Most of the areas will be divided into the unknown regions. It is very challenging for existing models to learn long-range features with little known information. Furthermore, with the development of modern cameras, picture resolution is becoming higher and higher. However, the reception fields of existing models could not increase as the resolution of input images do, which makes the problem even worse.

To address this issue, we make the first attempt to introduce Vision Transformer (ViT) [9] to extract features with a large receptive field. The Transformer model is first proposed in natural language processing (NLP) and has achieved great performance in computer vision tasks, such as classification [9,27,41], segmentation [30,53], and detection [4,47]. It mainly consists of a multi-head self-attention (MHSA) module and a multi-layer perception module. The MHSA module could mine information in a global scope. Thus, the ViT model could learn global semantic features of the foreground object with high-level position relevance. To further help the model integrate the low-level appearance features (*e.g.,* texture) with high-level semantic features (*e.g.,* shape), a Multi-scale Global-guided Fusion (MGF) module is proposed. The MGF takes three adjacent scales of features as input, uses the non-background mask to guide the low-level feature, and employs the high-level feature to guide the information integration. With this new MGF module, only foreground features could be transmitted to the decoder, reducing the influence of background noises.

Since the DIM [46] concatenates the trimap and RGB image to feed into the network, almost all subsequent trimap-based methods follow this strategy. However, compared with the RGB image, the trimap is very sparse and has some high-level positional relevance [26]. Most areas in the trimap have the same value, making convolution neural networks with small kernels inefficient in extracting features. Inspired by the [cls] token in ViT, we propose a new form of trimap named the tri-token map. Three learnable tokens are used to indicate the foreground, background, and unknown categories. We denote them as tri-tokens. Based on these tri-tokens, we propose a Tri-token Guided Transformer Block (TGTB), which adds the query with the corresponding tri-tokens for introducing the trimap information into the self-attention mechanism. With this high-level position information, the Transformer module could identify which features are from the known areas and which are from the unknown areas.

Besides, there has not been any testbed for images with transparent or non-salient foreground objects. Previous datasets mainly focus on salient and opaque foregrounds, like animals [20] and portraits [26,37], which have significantly been investigated. To further help the community to dig into the transparent and non-salient cases, we collect 460 high-solution natural images with large unknown areas and manually label their alpha mattes.

Our main contributions can be summarized as follows:

1. We propose a TGTB module, introducing the Vision Transformer module to extract global semantic features with a big receptive field. We also redesign the trimap as a tri-token map to directly bring location information to the self-attention mechanism.
2. A MGF module is proposed to integrate multi-scale features, and the global information is well organized to guide the integration with low-level Transformer features.
3. We build a high-resolution matting dataset with 460 images of the transparent or non-salient foreground. The dataset will be released to promote the development of matting technology.
4. Experiments on three matting datasets demonstrate that the proposed Trans-Matting method outperforms the current SOTA methods, indicating the effectiveness of our proposed modules.

2   Related Works

In this section, we first briefly review matting from two perspectives: traditional methods and deep-learning methods. Then, we further give an overview of Vision Transformer models, as the Tri-token Guided Transformer Block (TGTB) is one of the main contributions of this work.

2.1   Traditional Matting

Traditional matting methods can be divided into two categories: sampling-based and propagation-based methods. These methods mainly rely on low-level features, like color, location, etc. The sampling-based methods [2,7,12,15,36,43] first predict the colors of the foreground and background by evaluating the similarity of colors between the known foreground, background, and unknown area in samples, and then predict alpha mattes. Various sampling techniques have been investigated, including color cluster sampling [36], edge sampling [15], ray casting [12], etc. The propagation-based methods [5,17,38] propagate the information from the known foreground and background to the unknown area by solving the sparse linear equation system [18], the Poisson equation system [13], etc., to obtain the best global optimal alpha.

2.2 Deep-Learning Matting

In recent decades, deep learning technologies have boomed in various fields of computer vision. The same goes for the image matting task. [40] combines the sampling and deep neural network to improve the accuracy of alpha matting prediction. The Indices matter method [29] proposes an index-guided method for up-sampling and down-sampling to make the detailed information in the prediction graph more complete. Based on providing a larger dataset Composition-1k [46], DIM utilizes an encoder-decoder model to directly predict alpha mattes, which effectively improves the accuracy. [39] introduces semantic classification information of the matting region and uses learnable weights and multi-class discriminators to revise the prediction results. [51] proposes a general matting framework, which is conducive to obtaining better results under the guidance of different qualities and forms. [26] further mines the information of the RGB map and trimap and fuses the global information from these maps for obtaining better alpha mattes. All of the above methods use trimap as guidance. Some trimap-free methods can predict alpha mattes without using trimap. However, the accuracy of these trimap-free methods still has a big gap compared to that of the trimap-guided ones [6,35,48,49], indicating that the trimap could help the model to capture information efficiently.

2.3 Vision Transformer

The Transformer is firstly proposed in [42] to model long-range dependencies for machine translation and has demonstrated impressive performance on NLP tasks. Inspired by this, numerous attempts have been made to adapt transformers for vision tasks, and promising results have been shown for vision fields such as image classification, objection detection, semantic segmentation, etc. In particular, ViT [9] divides the input image into patches with a size of 16 × 16 and feeds the patch sequences to the vanilla Transformer model. To help the training process and improve the performance, DeiT [41] proposes a teacher-student strategy, which includes a distillation token for the student to learn from the teacher. Later, Swin [27], PVT [44], Crossformer [45], and HVT [32] combine the Transformer and pyramidal structure to decrease the number of patches progressively for obtaining multi-scale feature maps. To reduce computing and memory complexity, Swin, HRFormer [52], and CrossFormer apply local-window self-attention in Transformer, which also shows superior or comparable performance compared to the counterpart CNNs. The powerful self-attention mechanism in Transformer shows great advantages over CNN by capturing global

Table 1. Comparison between different public matting datasets.

Image matting dataset	Total num	TT num	Resolution
DAPM [37]	2000	0	800 × 600
Composition-1 k [46]	481	86	1297 × 1082
Distinction-646 [34]	646	79	1727 × 1565
AIM-500 [21]	500	76	1260 × 1397
Transparent-460 (ours)	460	460	3820 × 3766

attention of the whole image. However, some researchers [23] argue that locality and globality are both essential for vision tasks. Therefore, various researchers have tried combining the locality of CNN with the globality of Transformer to improve performance further. LocalViT [23] brings depth-wise convolutions to vision transformer to combine self-attention mechanism with locality, and shows great improvement compared to the pure Transformer, like DeiT, PVT, and TNT [14].

3   Matting Dataset

According to the transparency of foregrounds, we could divide the images of matting into two types: 1) Transparent partially (TP): TP refers to that there are significant foreground and uncertainty areas, and the foreground areas can provide information for the prediction of uncertainty areas. For example, when the foreground is human, the opaque and unknown regions are the hair or clothes. 2) Transparent totally (TT): there are minor or non-salient foreground areas, and the entire image is semi-transparent or high transparent. These images include glass, plastic bags, fog, water drops, etc.

As illustrated in Table 1, we select four popular image matting datasets for comparison, including DAPM [37], Composition-1k, Distinctions-646 [34], and AIM-500 [21]. The DAPM dataset consists only of portraits with no translucent or transparent objects. The Composition-1k dataset contains multiple categories, while most images are portraits (227 out of 481, TP-type). The Distinctions-646 dataset also mainly consists of portraits (343 out of 646, TP-type) [26]. The AIM-500 dataset contains only 76 TT-type images (correspond to the Salient Transparent/Meticulous type and the Non-Salient type in the original dataset) but 424 TP-type images.

As we can see, the transparent objects in the above datasets only occupy a small portion. This may be because it is much more difficult to label transparent objects than other objects, limiting the progress of transparent objects in the matting field. In this work, we propose the first large-scale dataset targeting various high transparent objects called Transparent-460 dataset. Our Transparent-460 dataset includes 460 high-quality manually-annotated alpha mattes, where 410 images are for training and 50 for testing. Furthermore, to our best knowledge, the resolution of our Transparent-460 is the highest (the average resolution

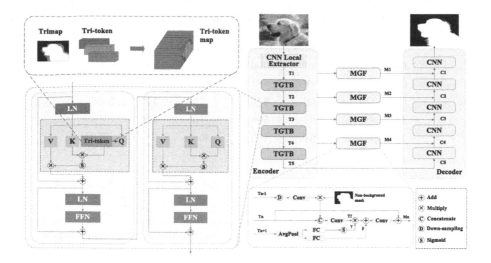

Fig. 1. The structure of our TransMatting.

is up to 3820 × 3766) among all datasets with high transparent objects. We believe this new matting dataset will greatly advance the matting research on objects with massive transparent areas.

4   Methodology

4.1   Motivation

By evaluating the results of some SOTA methods on TT and TP objects separately on the Composition-1k dataset (Table 2), we find that the results of TT, which denotes the total transparent objects, are much worse than TP, indicating that TT objects are the key to affecting the overall evaluation results. Furthermore, we find that most of the existing methods rely on the information of the foreground region for predicting the unknown region [1,5,17,19,22]. However, such methods will become useless or ineffective when facing images with no definite known regions. For example, [22] borrows features from both the known and unknown regions; when the unknown region is overwhelming in the images, the opacity propagation and the mattes prediction will face difficulties. Therefore, global information with a large or global receptive field and local features with inherent representation are needed to enhance the understanding and recognition capacity for objects with totally unknown regions. Although we can stack CNN layers to enlarge the receptive field, the information that covers the whole image is still hard to be obtained [25]. Besides, CNN also lacks global connectivity [23]. By contrast, the Transformer is good at modeling long-range connectivity with its attention mechanism.

Table 2. Performance of TT and TP objects on the Composition-1k dataset.

Methods	MSE↓			SAD↓		
	TT	TP	TT+TP	TT	TP	TT+TP
IndexNet [29]	22.87	8.9	13	110.3	18.08	45.8
GCAMatting [22]	15.89	6.2	9.1	85.72	13.68	35.3
MGMatting [51]	13.01	4.65	7.18	77.88	11.87	31.76
TransMatting (ours)	**7.49**	**3.4**	**4.58**	**59.37**	**10.35**	**24.96**

Moreover, most existing SOTA trimap-guided methods directly concatenate the trimap with RGB image as input. However, the huge gap between the two modalities of RGB image and trimap brings great difficulties in semantic feature extraction. At the same time, the trimap cannot effectively help the model focus on the region of interest. Therefore, a more efficient way to promote the guiding role of trimap is needed.

In short, to improve the performance of TT objects, more global and local features should be captured, and an effective guidance method for the trimap should be developed.

4.2 Baseline Structure

To extract both the local and global features, we combine CNN and the Transformer model as our encoder. Specifically, the first part, like [22,51] is the same as the first two stages of ResNet34-UNet (denoted as CNN Local Extractor in Fig. 1). The second part consists of a stack of our proposed Tri-token Guided Transformer Block (TGTB) based on the Swin Transformer [27]. As the decoder, we adopt the original ResNet34-UNet, a widespread network in the matting field [22,51].

4.3 Trimap Guided Methods

Almost all SOTA methods [3,22,26,39,46,50,51] use trimap as a guide and directly concatenate the RGB image and the annotated trimap as the model's input. However, the modalities of the RGB image and trimap are quite different. The RGB image scales from 0 to 255 and shows fine low-level features like texture, color similarity, etc. The trimap includes three values, containing high-level semantic information, like shape, location, etc., [26]. Thus, the direct concatenation between them is not the most efficient way to extract features.

Although trimap can explicitly indicate the region of interest, it is still hard to take full advantage of this information. To the best of our knowledge, we are the first to attempt to harmonize the RGB image and trimap rather than simply concatenating them. We insert a learnable trimap into the Transformer module to guide the model to concentrate on the valuable area, making the network learning more efficient and robust.

4.4   Tri-token

Inspired by the [cls] token in Vision Transformer, we design a new tri-token (shown in Fig. 1) structure, aiming to introduce the high-level semantic information directly into the self-attention mechanism to replace the inefficient concatenation methods. Given a vanilla $Trimap \in \mathbb{R}^{H \times W}$, we generate three **learnable** tri-tokens (denoted as $Token_i$, $i = \{0, 1, 2\}$) with different initialization to represent the known foreground, known background, and unknown areas, respectively. Every tri-token is a 1D vector, that is, $Token_i \in \mathbb{R}^C$. Then we replace every pixel in the trimap with the corresponding tri-token to generate the tri-token map, formulated as:

$$Trimap[Trimap == i] = Token_i, i = \{0, 1, 2\} \tag{2}$$

In this manner, the tri-token map can directly guide the self-attention process in the Transformer to pay more attention to the unknown areas for self-updating.

4.5   Tri-token Guided Transformer Block

Global connectivity is much more important for the prediction of total transparent objects. CNN does not have global attention, and its receptive field cannot cover the whole image [25], which leads to poor estimation for pixels outside receptive fields, while Transformer has global attention, and its receptive field can cover every pixel at the first layer.

The Transformer consists of multi-head self-attention (MHSA) and Multilayer Perceptron (MLP) blocks. The self-attention mechanism can be thought of as a mapping between a query and a collection of key-value pairs. The output is a weighted sum of the values, and the weights are assigned by the compatibility function between the query and the relevant key. This can be implemented by Scaled Dot-Product Attention [42], in which a softmax function is used to activate the dot products of query and all keys for obtaining the weights. MHSA means that more than one self-attention is performed in parallel.

Like [27,45,52], we use non-overlapping windows whose size is $M \times M$ to divide the feature maps. The MHSA is performed within each window. The formulations of vanilla attention and our tri-token attention in a specific window are shown as follows:

$$Attention(Q, K, V) = Softmax(QK^T/\sqrt{d})V \tag{3}$$

$$Tri\text{-}token\ Attention(Q, K, V) = Softmax((Q + Tri\text{-}token)K^T/\sqrt{d})V \tag{4}$$

where $Q, K, V \in R^{M^2 \times d}$ represent the query, key, and value in the attention mechanism, respectively. d is the query/key dimension. In the Tri-token Attention formulation, Q, K, and V are the same as that in the standard self-attention. The $Tri\text{-}token$ is our proposed learnable $trimap$ that adds to the query for forming a new tri-token query. In this way, our tri-token attention mechanism can selectively aggregate contexts and evaluate which region should be paid more attention to with the guidance of our learnable tri-tokens.

In this way, we combine the self-attention and tri-tokens to focus on more valuable regions by considering the relationship between non-background and background areas, and finally achieve the best performance. We use our tri-token attention every five blocks in each Tri-token Guided Transformer Block (TGTB).

4.6   Multi-scale Global-Guided Fusion Module

In the multi-scale feature pyramid structure, in-depth features contain more global information, while shallow features have rich local information like texture, color similarity, etc. Fusing these features is vital for accurately predicting alpha mattes for high transparent objects [35]. Although the direct sum operation can realize feature fusion, the details in the shallow features may attenuate the impact of the advanced semantics, resulting in some subtle regions missing [35]. To address this issue, we propose a Multi-scale Global-guided Fusion (MGF) module in the decoder process (see Fig. 1 for details), with both the non-background information and the advanced semantic features as guidance, to fuse the high-level semantic information and the lower ones effectively.

Specifically, we denote three adjacent features from shallow to deep as T_{n-1}, T_n, and T_{n+1}. The T_{n-1} is first down-sampled, then the Hadamard product is employed between the non-background mask and T_{n-1} to extract the low-level features of non-background, which helps to reduce the impact of complex background influence. This can guide the network to pay more attention to the foreground and unknown areas. After that, the T_{n-1} is concatenated with T_n, and a convolution layer is performed to align the channel of fused features. We mark this feature as T_f.

For the T_{n+1}, we first perform a global average pooling to generate channel-wise statistics and then use two fully connected (FC) layers to squeeze channels. As shown in Fig. 1, features output from the two FC layers are denoted as γ and β, separately. To fully capture channel-wise dependencies, we add a sigmoid function to activate γ and perform broadcast multiplication with T_f for channel re-weighting. After that, broadcast addition is performed between the channel-weighted feature and β. A convolution layer is used to fuse information from different groups. Notably, a skip connection from T_n is employed for obtaining the final fused features of MGF.

In short, considering that fusing low-level features directly may cause a negative impact on the advanced semantics [35], two techniques are proposed here. Firstly, the non-background mask is introduced into the fusion process to filter out the complex background information and further help to concentrate more attention on the foreground and unknown areas. Secondly, the global channel-wise attention from higher-level features is used for re-weighting and enhancing the important information in the fused features.

4.7   Loss Function

Following [51], we use three losses, including the alpha loss ($\mathcal{L}_\alpha$), Compositional loss [46] ($\mathcal{L}_{comp}$), and Laplacian loss [16] ($\mathcal{L}_{lap}$). As formulated below, their weights are set as 0.4, 1.2, and 0.16, respectively.

$$\mathcal{L}_{final} = 0.4 * \mathcal{L}_\alpha + 1.2 * \mathcal{L}_{comp} + 0.16 * \mathcal{L}_{lap} \qquad (5)$$

5   Experiments

In this section, we show our experimental settings and compare our evaluation results on the test set of Composition-1k [46], Distinction-646 [34], and our Transparent-460 datasets with other state-of-the-art methods.

5.1   Dataset

Composition-1k. contains 431 and 50 unique foreground objects and manually labeled alpha mattes as training and test sets, respectively. Every foreground object is composited with 100 (for training set) and 20 (for test set) background images from COCO [24] and Pascal VOC [10]. As a result, there are 43,100 images for training and 1,000 images for testing.

Distinction-646. comprises 646 distinct foreground objects. Similar to the Composition-1k, 50 objects are divided as the test set. Following the same composition rule, there are 59,600 and 1000 images for training and testing, respectively.

Our Transparent-460. mainly consists of transparent and non-salient objects as the foreground, like water drops, jellyfish, plastic bags, glass, crystals, etc. We collect 460 high-resolution images and carefully annotate them with Photoshop. Considering the transparent objects are very meticulous, we keep the original resolution of all collected images (up to 3820 × 3766 pixels on average). To our best knowledge, this is the first transparent object matting dataset in such a high resolution.

5.2   Evaluation Metrics

Following [3,16,26,29], we use four metrics for evaluation, including the Sum of Absolute Differences (SAD), Mean Squared Error (MSE), Gradient error (Grad.) and Connectivity error (Conn). It is notable that the unit of MSE value is set to 1e-3 for easy reading.

5.3   Implementation Details

We use PyTorch [33] to implement our proposed method. All the experiments are trained for 200,000 iterations. We initialize our network with ImageNet [8] pre-trained weights. The ablation experiments in Table 3, 4, 5 are done with 2

Table 3. The effectiveness of our proposed TGTB and MGF modules on the Composition-1k dataset.

TGTB	MGF	SAD ↓	MSE ↓	Grad. ↓	Conn.↓
		29.14	6.34	12.06	25.21
✓		27.45	5.66	11.77	24.30
	✓	27.21	5.57	11.23	23.25
✓	✓	**26.83**	**5.22**	**10.62**	**22.14**

NVIDIA Tesla V100 GPU with a batch size of 32. Moreover, to compare our method with the existing SOTA methods, we use a batch size of 64 with 4 NVIDIA Tesla V100 GPU to train our proposed method in Table 6, 7, 8. The Adam optimizer is utilized, and the initial learning rate is set to 1e-4 with the same learning rate decay strategy as [28,51]. For a fair comparison, we follow the data augmentation methods used in [22], like random crop, rotation, scaling, shearing, etc. Moreover, the trimaps for training are generated using dilation and erosion ways on alpha images by random kernel sizes from 1 to 30. Finally, we crop 512×512 patches on the center of the unknown area of alpha and composite them with the background from COCO. We use the same training conditions on the Composition-1k and Distinction-646 datasets.

5.4   Ablation Study

To evaluate the effectiveness of our new proposed modules of TGTB and MGF, and the performance with different hyper-parameters, we design the ablation study on the Composition-1k dataset.

Evaluate the Effectiveness of Our Proposed Modules. The quantitative results under the SAD, MSE, Gradient, and Connectivity errors with and without our proposed TGTB and MGF modules are illustrated in Table 3. As we can see, with the TGTB module, the four metrics listed above decrease to 27.45, 5.66, 11.77, and 24.30, respectively. The main reason is that our redesigned tri-token map is more suitable for propagating location information than simply concatenating to the input image. The MGF module could solely achieve similar performance, indicating that our proposed multi-scale feature fusion strategy can also help the decoder to make better use of the local and global information. When combined with the TGTB and MGF modules, the model achieves the best performance, indicating the effectiveness of the two new proposed modules.

Determine Where to Introduce Tri-tokens. There are four TGTB stages in our encoder model. Table 4 reports the performance with different positions to introduce tri-tokens. As the position goes deep, the feature map size decreases, making more position information lose. On the other hand, deep stages have learned more abstract semantic features, which is suitable for mutual learning with tri-tokens. As shown in Table 4, both shallow and deep stages benefit from tri-tokens, indicating that the tri-tokens in TGTB modules could guide the encoder to focus on the right regions.

Table 4. Ablation results on the Composition-1k dataset with different positions to introduce the proposed tri-tokens.

Position	SAD↓	MSE↓	Grad.↓	Conn.↓
1	31.68	7.24	14.20	27.42
4	29.50	6.20	13.18	25.23
1,2,3,4	**26.83**	**5.22**	**10.62**	**22.14**

Table 5. Ablation results on the Composition-1k dataset with local or (and) global features in the proposed MGF module.

Local	Global	SAD↓	MSE↓	Grad.↓	Conn.↓
		27.45	5.66	11.77	24.30
✓		27.16	5.34	11.03	22.60
	✓	27.39	5.46	11.43	23.20
✓	✓	**26.83**	**5.22**	**10.62**	**22.14**

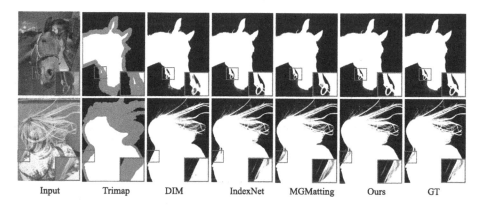

Input   Trimap   DIM   IndexNet   MGMatting   Ours   GT

Fig. 2. Visual comparison of our TransMatting against SOTA methods on the Composition-1k test set.

The Impact of Local and Global Features in MGF. Table 5 reports the effectiveness of our MGF module with and without local or global branches. The local branch is proposed to integrate T_{n-1} with the non-background mask, and the global branch is responsible for introducing global features from T_{n+1} to guide the feature flow. As we can see from Table 5, combining local and global branches could achieve the best performance compared with using one of them solely. The main reason is the effectiveness of our MGF in fusing local (texture, border) and global (semantic, location) features for modeling unknown regions.

5.5 Comparison with Prior Work

To evaluate our method's performance, we compare it with other state-of-the-art models on the following three datasets. Notably, we achieve the best performance on all three datasets.

Testing on Composition-1k. We show the quantitative and visual results on Table 6 and Fig. 2. Without any test-time augmentations, our proposed Trans-Matting outperforms other SOTA methods on all four evaluation metrics by only using the Composition-1k training set for training. As illustrated in Table 6, our

Table 6. The quantitative results on the Composition-1k test set [46]. † denotes results with test-time augmentation.

Methods	SAD↓	MSE↓	Grad.↓	Conn.↓
AlphaGAN [31]	52.4	30	38	53
DIM [29]	50.4	14	31.0	50.8
IndexNet [29]	45.8	13	25.9	43.7
AdaMatting [3]	41.7	10	16.8	-
ContextNet [16]	35.8	8.2	17.3	33.2
GCAMatting [22]	35.3	9.1	16.9	32.5
MGMatting [51]	31.5	6.8	13.5	27.3
TIMI-Net [26]	29.08	6.0	12.9	27.29
FBAMatting [11] †	25.8	5.2	10.6	20.8
TransMatting (ours)	**24.96**	**4.58**	**9.72**	**20.16**

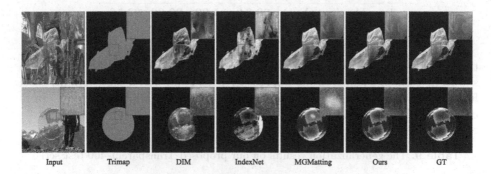

Input   Trimap   DIM   IndexNet   MGMatting   Ours   GT

Fig. 3. Visual comparison of our TransMatting against SOTA methods on our Transparent-460 test set.

model decreases the MSE and Grad metrics heavily: from 5.2, 10.6 to 4.58 and 9.72, respectively, indicating the effectiveness of our TransMatting.

Testing on Distinction-646. Table 7 compares the performance of our Trans-Matting with other state-of-the-art methods on Distinction-646. For a fair comparison, we follow the whole inference protocol in [34, 51] to calculate the metrics based on the whole image. Without any additional tuning, our method outperforms all the SOTA methods.

Testing on our Transparent-460. Based on their release codes, we train IndexNet and MGMatting methods on our dataset and compare them with ours in Table 8. Our Transparent-460 dataset mainly focuses on transparent and non-salient foregrounds, which is very difficult for existing image matting methods. Surprisingly, as illustrated in Table 8, our TransMatting achieves promising results with only a 4.02 MSE error. Furthermore, to evaluate the generalization performance of our model. We train our TransMatting on the Composition-1k

Table 7. The quantitative results on the Distinction-646 test set.

Methods	SAD↓	MSE↓	Grad.↓	Conn.↓
KNNMatting [5]	116.68	25	103.15	121.45
DIM [46]	47.56	9	43.29	55.90
HAttMatting [34]	48.98	9	41.57	49.93
GCAMatting [22]	27.43	4.8	18.7	21.86
MGMatting [51]	33.24	4.51	20.31	25.49
TransMatting (Ours)	**25.65**	**3.4**	**16.08**	**21.45**

Table 8. The quantitative results on our proposed Transparent-460 test set.

Methods	SAD↓	MSE↓	Grad.↓	Conn.↓
IndexNet [29]	573.09	112.53	140.76	327.97
MGMatting [51]	111.92	6.33	25.67	103.81
TransMatting (Ours)	**88.34**	**4.02**	**20.99**	**82.56**

training set and directly test it on the Transparent-460 test set. The results are shown in Table 9. Thanks to the big receptive field and well-designed multi-scale fusion module, our model reduces nearly half of the SAD, MSE, and Conn. errors compared to the SOTA methods.

Table 9. The generalization results on our proposed Transparent-460 test set.

Methods	SAD↓	MSE↓	Grad.↓	Conn.↓
DIM [46]	356.2	49.68	146.46	296.31
IndexNet [29]	434.14	74.73	124.98	368.48
MGMatting [51]	344.65	57.25	74.54	282.79
TIMI-Net [26]	328.08	44.2	142.11	289.79
TransMatting(Ours)	**192.36**	**20.96**	**41.8**	**158.37**

6   Conclusion

In order to generalize to transparent and non-salient foregrounds, matting algorithms must have the ability to mine long-range features and utilize the semantic features in trimap. In this paper, we propose a novel Transformer-based network by redesigning a tri-token map to introduce the trimap semantic features into the long-range dependencies of the self-attention mechanism. Furthermore, a multi-scale global-guided fusion module is proposed to take the global information and local non-background mask as a guide to fuse multi-scale features for better modeling the unknown regions in transparent objects. Experiments on the

Composition-1k, Distinctions-646, and our proposed Transparent-460 datasets demonstrate that our TransMatting outperforms the state-of-the-art methods.

References

1. Aksoy, Y., Aydın, T.O., Pollefeys, M.: Information-flow matting. arXiv preprint arXiv:1707.05055 (2017)
2. Berman, A., Dadourian, A., Vlahos, P.: Method for removing from an image the background surrounding a selected object, 17 Oct 2000, uS Patent 6,134,346
3. Cai, S., et al.: Disentangled image matting. In: Proceedings of the IEEE/CVF International Conference on Computer Vision, pp. 8819–8828 (2019)
4. Carion, N., Massa, F., Synnaeve, G., Usunier, N., Kirillov, A., Zagoruyko, S.: End-to-end object detection with transformers. In: Vedaldi, A., Bischof, H., Brox, T., Frahm, J.-M. (eds.) ECCV 2020. LNCS, vol. 12346, pp. 213–229. Springer, Cham (2020). https://doi.org/10.1007/978-3-030-58452-8_13
5. Chen, Q., Li, D., Tang, C.K.: Knn matting. IEEE Trans. Pattern Anal. Mach. Intell. **35**(9), 2175–2188 (2013)
6. Chen, Q., Ge, T., Xu, Y., Zhang, Z., Yang, X., Gai, K.: Semantic human matting. In: Proceedings of the 26th ACM international conference on Multimedia, pp. 618–626 (2018)
7. Chuang, Y.Y., Curless, B., Salesin, D.H., Szeliski, R.: A bayesian approach to digital matting. In: Proceedings of the 2001 IEEE Computer Society Conference on Computer Vision and Pattern Recognition, CVPR 2001, vol. 2, pp. II-II. IEEE (2001)
8. Deng, J., Dong, W., Socher, R., Li, L.J., Li, K., Fei-Fei, L.: Imagenet: A large-scale hierarchical image database. In: 2009 IEEE Conference on Computer Vision and Pattern Recognition, pp. 248–255. Ieee (2009)
9. Dosovitskiy, A., et al.: An image is worth 16x16 words: Transformers for image recognition at scale. arXiv preprint arXiv:2010.11929 (2020)
10. Everingham, M., Van Gool, L., Williams, C.K., Winn, J., Zisserman, A.: The pascal visual object classes (voc) challenge. Int. J. Comput. Vision **88**(2), 303–338 (2010)
11. Forte, M., Pitié, F.: f, b, alpha matting. arXiv preprint arXiv:2003.07711 (2020)
12. Gastal, E.S., Oliveira, M.M.: Shared sampling for real-time alpha matting. In: Computer Graphics Forum, vol. 29, pp. 575–584. Wiley Online Library (2010)
13. Grady, L., Schiwietz, T., Aharon, S., Westermann, R.: Random walks for interactive alpha-matting. In: Proceedings of VIIP, vol. 2005, pp. 423–429 (2005)
14. Han, K., Xiao, A., Wu, E., Guo, J., Xu, C., Wang, Y.: Transformer in transformer. In: Advances in Neural Information Processing Systems, vol. 34 (2021)
15. He, K., Rhemann, C., Rother, C., Tang, X., Sun, J.: A global sampling method for alpha matting. In: CVPR 2011, pp. 2049–2056. IEEE (2011)
16. Hou, Q., Liu, F.: Context-aware image matting for simultaneous foreground and alpha estimation. In: Proceedings of the IEEE/CVF International Conference on Computer Vision, pp. 4130–4139 (2019)
17. Lee, P., Wu, Y.: Nonlocal matting. In: CVPR 2011, pp. 2193–2200. IEEE (2011)
18. Levin, A., Lischinski, D., Weiss, Y.: A closed-form solution to natural image matting. IEEE Trans. Pattern Anal. Mach. Intell. **30**(2), 228–242 (2007)
19. Levin, A., Rav-Acha, A., Lischinski, D.: Spectral matting. IEEE Trans. Pattern Anal. Mach. Intell. **30**(10), 1699–1712 (2008)

20. Li, J., Zhang, J., Maybank, S.J., Tao, D.: End-to-end animal image matting. arXiv e-prints arXiv-2010 (2020)
21. Li, J., Zhang, J., Tao, D.: Deep automatic natural image matting. arXiv preprint arXiv:2107.07235 (2021)
22. Li, Y., Lu, H.: Natural image matting via guided contextual attention. In: Proceedings of the AAAI Conference on Artificial Intelligence, vol. 34, pp. 11450–11457 (2020)
23. Li, Y., Zhang, K., Cao, J., Timofte, R., Van Gool, L.: Localvit: Bringing locality to vision transformers. arXiv preprint arXiv:2104.05707 (2021)
24. Lin, T.-Y., et al.: Microsoft COCO: common objects in context. In: Fleet, D., Pajdla, T., Schiele, B., Tuytelaars, T. (eds.) ECCV 2014. LNCS, vol. 8693, pp. 740–755. Springer, Cham (2014). https://doi.org/10.1007/978-3-319-10602-1_48
25. Liu, Q., Xie, H., Zhang, S., Zhong, B., Ji, R.: Long-range feature propagating for natural image matting. In: Proceedings of the 29th ACM International Conference on Multimedia, pp. 526–534 (2021)
26. Liu, Y., et al.: Tripartite information mining and integration for image matting. In: Proceedings of the IEEE/CVF International Conference on Computer Vision, pp. 7555–7564 (2021)
27. Liu, Z., et al.: Swin transformer: Hierarchical vision transformer using shifted windows. In: Proceedings of the IEEE/CVF International Conference on Computer Vision, pp. 10012–10022 (2021)
28. Loshchilov, I., Hutter, F.: Sgdr: Stochastic gradient descent with warm restarts. arXiv preprint arXiv:1608.03983 (2016)
29. Lu, H., Dai, Y., Shen, C., Xu, S.: Indices matter: Learning to index for deep image matting. In: Proceedings of the IEEE/CVF International Conference on Computer Vision, pp. 3266–3275 (2019)
30. Lu, Z., He, S., Zhu, X., Zhang, L., Song, Y.Z., Xiang, T.: Simpler is better: few-shot semantic segmentation with classifier weight transformer. In: ICCV 2021, pp. 8741–8750. https://openaccess.thecvf.com/content/ICCV2021/html/Lu_Simpler_Is_Better_Few-Shot_Semantic_Segmentation_With_Classifier_Weight_Transformer_ICCV_2021_paper.html
31. Lutz, S., Amplianitis, K., Smolic, A.: Alphagan: Generative adversarial networks for natural image matting. arXiv preprint arXiv:1807.10088 (2018)
32. Pan, Z., Zhuang, B., Liu, J., He, H., Cai, J.: Scalable visual transformers with hierarchical pooling. arXiv e-prints arXiv-2103 (2021)
33. Paszke, A., et al.: Pytorch: An imperative style, high-performance deep learning library. In: Advances in Neural Information Processing Systems, vol. 32 (2019)
34. Qiao, Y., et al.: Attention-guided hierarchical structure aggregation for image matting. In: Proceedings of the IEEE/CVF Conference on Computer Vision and Pattern Recognition, pp. 13676–13685 (2020)
35. Qiao, Y., et al.: Multi-scale information assembly for image matting. In: Computer Graphics Forum, vol. 39, pp. 565–574. Wiley Online Library (2020)
36. Shahrian, E., Rajan, D., Price, B., Cohen, S.: Improving image matting using comprehensive sampling sets. In: Proceedings of the IEEE Conference on Computer Vision and Pattern Recognition, pp. 636–643 (2013)
37. Shen, X., Tao, X., Gao, H., Zhou, C., Jia, J.: Deep automatic portrait matting. In: Leibe, B., Matas, J., Sebe, N., Welling, M. (eds.) ECCV 2016. LNCS, vol. 9905, pp. 92–107. Springer, Cham (2016). https://doi.org/10.1007/978-3-319-46448-0_6
38. Sun, J., Jia, J., Tang, C.K., Shum, H.Y.: Poisson matting. In: ACM SIGGRAPH 2004 Papers, pp. 315–321 (2004)

39. Sun, Y., Tang, C.K., Tai, Y.W.: Semantic image matting. In: Proceedings of the IEEE/CVF Conference on Computer Vision and Pattern Recognition, pp. 11120–11129 (2021)

40. Tang, J., Aksoy, Y., Oztireli, C., Gross, M., Aydin, T.O.: Learning-based sampling for natural image matting. In: Proceedings of the IEEE/CVF Conference on Computer Vision and Pattern Recognition, pp. 3055–3063 (2019)

41. Touvron, H., Cord, M., Douze, M., Massa, F., Sablayrolles, A., Jégou, H.: Training data-efficient image transformers & distillation through attention. In: International Conference on Machine Learning, pp. 10347–10357. PMLR (2021)

42. Vaswani, A., et al.: Attention is all you need. In: Advances in Neural Information Processing Systems, vol. 30 (2017)

43. Wang, J., Cohen, M.F.: Optimized color sampling for robust matting. In: 2007 IEEE Conference on Computer Vision and Pattern Recognition, pp. 1–8. IEEE (2007)

44. Wang, W., et al.: Pyramid vision transformer: A versatile backbone for dense prediction without convolutions. In: Proceedings of the IEEE/CVF International Conference on Computer Vision, pp. 568–578 (2021)

45. Wang, W., Yao, L., Chen, L., Cai, D., He, X., Liu, W.: Crossformer: A versatile vision transformer based on cross-scale attention. arXiv e-prints arXiv-2108 (2021)

46. Xu, N., Price, B., Cohen, S., Huang, T.: Deep image matting. In: Proceedings of the IEEE Conference on Computer Vision and Pattern Recognition, pp. 2970–2979 (2017)

47. Yang, J., et al.: Focal self-attention for local-global interactions in vision transformers. arXiv preprint arXiv:2107.00641 (2021)

48. Yang, X., et al.: Smart scribbles for image matting. ACM Trans. Multimedia Comput. Commun. Appli. (TOMM) 16(4), 1–21 (2020)

49. Yang, X., Xu, K., Chen, S., He, S., Yin, B.Y., Lau, R.: Active matting. In: Advances in Neural Information Processing Systems, vol. 31 (2018)

50. Yu, H., Xu, N., Huang, Z., Zhou, Y., Shi, H.: High-resolution deep image matting. arXiv preprint arXiv:2009.06613 (2020)

51. Yu, Q., et al.: Mask guided matting via progressive refinement network. In: Proceedings of the IEEE/CVF Conference on Computer Vision and Pattern Recognition, pp. 1154–1163 (2021)

52. Yuan, Y., et al.: Hrformer: High-resolution transformer for dense prediction. arXiv preprint arXiv:2110.09408 (2021)

53. Zheng, S., et al.: Rethinking Semantic Segmentation from a Sequence-to-Sequence Perspective with Transformers. In: 2021 IEEE/CVF Conference on Computer Vision and Pattern Recognition (CVPR), pp. 6877–6886. IEEE. https://doi.org/10.1109/CVPR46437.2021.00681, https://ieeexplore.ieee.org/document/9578646/

MVSalNet: Multi-view Augmentation for RGB-D Salient Object Detection

Jiayuan Zhou[1], Lijun Wang[1(✉)], Huchuan Lu[1,2], Kaining Huang[3],
Xinchu Shi[3], and Bocong Liu[3]

[1] Dalian University of Technology, Dalian, China
zjy@mail.dlut.edu.cn, {ljwang,lhchuan}@dlut.edu.cn
[2] Peng Cheng Laboratory, Shenzhen, China
[3] Meituan, Beijing, China
{huangkaining,shixinchu,liubocong}@meituan.com

Abstract. RGB-D salient object detection (SOD) enjoys significant advantages in understanding 3D geometry of the scene. However, the geometry information conveyed by depth maps are mostly under-explored in existing RGB-D SOD methods. In this paper, we propose a new framework to address this issue. We augment the input image with multiple different views rendered using the depth maps, and cast the conventional single-view RGB-D SOD into a multi-view setting. Since different views captures complementary context of the 3D scene, the accuracy can be significantly improved through multi-view aggregation. We further design a multi-view saliency detection network (MVSalNet), which firstly performs saliency prediction for each view separately and incorporates multi-view outputs through a fusion model to produce final saliency prediction. A dynamic filtering module is also designed to facilitate more effective and flexible feature extraction. Extensive experiments on 6 widely used datasets demonstrate that our approach compares favorably against state-of-the-art approaches.

Keywords: RGB-D salient object detection · Multi-view augmentation · Multi-view fusion

1 Introduction

RGB-D salient object detection (SOD) aims to identify and segment the most conspicuous objects in the input scene considering both RGB images and the corresponding depth maps. With the rapid development of depth sensors, RGB-D SOD has found wide applications in surveillance [54], autonomous driving [43], and robotics [42], to name a few. Since additional depth information permits comprehensive understanding of the 3D geometry, RGB-D SOD is inherently more superior than its RGB based counterpart in handling challenging scenarios, including background clutter, illumination variation, etc., and therefore has attracted increasingly more attention from the community.

© The Author(s), under exclusive license to Springer Nature Switzerland AG 2022
S. Avidan et al. (Eds.): ECCV 2022, LNCS 13089, pp. 270 287, 2022.
https://doi.org/10.1007/978-3-031-19818-2_16

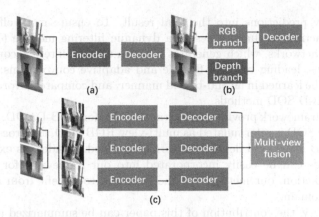

Fig. 1. Framework comparison. (a)(b) Existing RGB-D SOD methods mainly use the input depth map as an additional feature channel. (c) We leverage the 3D geometry of the input depth to perform multi-view saliency detection.

Since the depth map and RGB images are from two different modalities with significant cross-modal gap, it is not a trivial task to perform SOD by simultaneously utilizing the two input data modalities. As such, recent research efforts [5,30] mainly focus on cross-modal fusion between the input RGB and depth for SOD (cf. Fig. 1 (a) (b)). Although significant progress has been achieved, these existing methods mostly use the depth information as an additional input channel to provide low-level cues like edges, contours, and regions, while the essential 3D geometry information are under-explored. This drawback may potentially restrict the merits of existing RGB-D SOD, leading to unsatisfactory performance.

As we humans move freely in the 3D world, we can perceive the scene from different views, allowing more precise foreground detection even at adversarial cases. In fact, the human vision system are also binocular for more effective 3D perception. All these evidences indicate that multi-view perception enabled by 3D geometry can significantly benefit vision tasks.

Motivated by above observations, we propose a new framework to fully explore the geometry information for RGB-D SOD. Instead of using depth map as only low-level cues, we leverage the contained 3D geometry to render the input image under different views, which allows multi-view perception to be mimicked from a single static image. SOD can then be performed for each view independently and the generated single view predictions are eventually fused to produce the final saliency maps (As illustrated in Fig. 1 (c)). Since different views may capture different context of input scene and are complementary to each other, the saliency predictions aggregated from multiple views are shown to be more accurate and robust.

We implement the above idea by designing a multi-view saliency detection network (MVSalNet), which contains multiple saliency prediction streams for the augmented input views, and a multi-view fusion module to incorporate single-view saliency predictions into the final result. To ensure more effective deep feature extraction, we further design a dynamic filtering module (TDF) using transformer networks, which generates position-specific filters according to the input features, leading to more flexible and adaptive convolutions. The entire network can be learned in an end-to-end manner, and compares favorably against existing RGB-D SOD methods.

Our new framework provides an alternative idea for RGB-D SOD. Since single view RGB-D SOD is reformulated as multi-view RGB SOD, the cross-modal gap between RGB image and depth is naturally resolved. Besides, as existing RGB SOD methods can be easily incorporated into our framework for single-view saliency prediction, our method has the potential to benefit from advances in RGB SOD domain.

In summary, the contribution of this paper can be summarized as follows.

- We present a new framework for RGB-D SOD with multi-view augmentation, which can effectively leverage the geometry information carried in input depth maps.
- We design a multi-view saliency prediction network with dynamic filtering modules, which can not only enhance saliency prediction in each single view, but also enables cross-view prediction fusion, yielding more accurate SOD results.

Our method sets new state of the art on 6 benchmark datasets. Extensive evaluation has justified the effectiveness of our contribution.

2   Related Work

2.1   RGB-D Salient Object Detection

Traditional methods are mainly based on hand-crafted features, such as contrast [36], shape [7], compactness [8], background enclosure [16] and so on. As the representation ability of the hand-crafted features is limited, all the above models can not cope with complex scenes. While recently, deep learning-based methods have made significant progress [18,46,50] due to the powerful ability in discriminative feature representation. Based on the scope of this paper, we divide existing deep-based models into single-stream models [41,54] and multi-stream models [15,48]. The single-stream models directly fuse RGB images and depth maps to send to the network. For example, DANet [52] uses depth-enhanced dual attention to generate contrasted features for the decoder. For the multi-stream models, the frameworks employ parallel networks to extract and fuse multi-modal features with various strategies. For example, Zhang et al. [48] propose an asymmetric two-stream network and design a flow ladder module for RGB stream and a depth attention module for depth stream. Generally speaking, single-stream model is lighter and multi-stream model has better performance.

However, unlike the aforementioned methods in which depth cues are only treated as the direct input of the feature extractor. In this paper, we further exploit the use of depth information. As the depth information contains abundant geometric prior knowledge, we utilize the depth cues to rotate the corresponding RGB images. Then we get multi-view saliency results and fuse them to generate the final output. This results in two major benefits: 1) We generate multi-view RGB images to replace the original depth map, in this way, we explicitly eliminate the modal gap; 2) The noise in low-quality depth map is largely reduced as we use late-fusion [19] to fuse the multi-view saliency results.

2.2   Novel View Synthesis

In some tasks, new views can be synthesized as a data augmentation method. [55] formulates the 3D object detection problem as the detection of rotated bounding boxes in images from bird's eye view generated using the homography. [26] randomly manipulates the camera system, including its focal length, receptive field and location, to generate new training images with geometric shifts. [53] introduces a perspective-aware data augmentation that synthesizes new training examples with more diverse views by perturbing the existing ones in a geometrically consistent manner. Inspired by them, we propose to generate multi-view RGB images in RGB-D SOD.

2.3   Attention Mechanism and Transformer

Fully-convolutional networks [31] are mature architecture for dense prediction, they adopt convolution and subsampling as fundamental elements in order to learn multi-scale features that can leverage an appropriately large context. Furthermore, attention has already proven to be an effective architecture for learning strong models for natural language processing (NLP) [10,27]. There have been several works that adapt attention mechanisms to computer vision tasks and get competitive results, such as image classification [11], object detection [2], and panoptic segmentation [44]. This is likely because attention can capture long-range associations, which further lead to the trends that combine CNNs with transformers [2,28]. Notice the advantages of combining the two, we propose to leverage a transformer-based dynamic filtering module to generate adaptive kernels and get more effective features.

3   Method

In this section, we present a new paradigm for RGB-D saliency detection with multi-view augmentation. Figure 2 overviews the pipeline of our method. Given an input image **I** and its corresponding depth map **D**, we first render the RGB image from multiple novel views. Saliency detection is then independently performed under each of the newly rendered views as well as the original input view. Finally, we aggregate all the predicted saliency maps from different views to produce the output saliency prediction. We implement the above multi-view saliency detection and aggregation procedures through a multi-view saliency detection network (MVSalNet). In the following, we first elaborate on multi-view image synthesis in Sect. 3.1, and then describe the architecture of our proposed MVSalNet in Sect. 3.2.

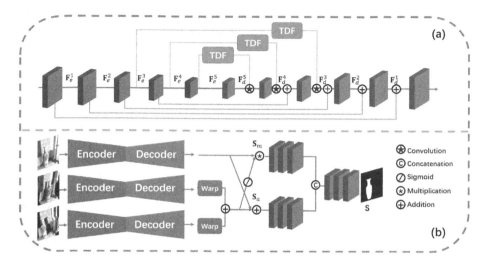

Fig. 2. (a) The details of the encoder-decoder branch in MVSalNet. (b) The overall architecture of MVSalNet. The network has three parallel encoder-decoder branches, and are fed multi-view RGB images, respectively. A multi-view fusion module is added at the end of the network to fuse the multi-view saliency results.

3.1 Multi-view Rendering

As opposed to prior RGB-D SOD methods that mainly use depth as an additional input feature, we propose to explore the 3D geometry information encoded in the depth maps for novel view synthesis, allowing single image RGB-D SOD to be conducted in a multi-view setting. To this end, we develop a multi-view rendering module to efficiently perform multi-view augmentation for the input image. Our basic principle is to reconstruct the 3D point cloud based on the input scene depth, which is then projected to a specific target novel view to render the RGB image.

Technically, given the depth value d of a pixel and its 2D coordinate $\mathbf{p}$ in the input image, its 3D point coordinate $\mathbf{P}$ can be computed. Given the relative motion between the input and a novel target view, we can further transform the 3D point $\mathbf{P}$ to the target view, and then project it onto the target image plane to obtain its corresponding pixel coordinate $\bar{\mathbf{p}}$ in the target image.

The above process establishes a position mapping from each pixel in the input view to its corresponding pixel in the novel target view, based on which we can synthesize the target view image using the input image, or inversely, warp the

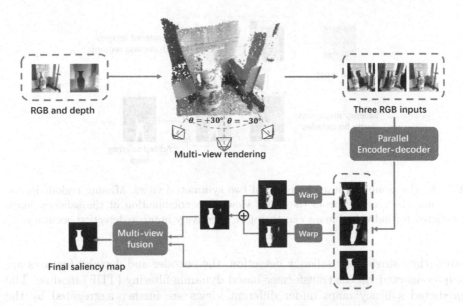

Fig. 3. Implementation of multi-view data augmentation.

predicted saliency map of the target view to the input view. For missing regions in the rendered images or saliency maps caused by occlusion, out-of-view, etc., we fill the missing values with 0.

Considering both efficiency and effectiveness, we augment the input image with two additional novel views. Although augmentation with more novel views may lead to better performance, it will also increase computational overhead. Our preliminary experiment further shows that using fixed relative motions for novel views performs more superior than random generated ones. Therefore, we restrict the rotation of the two novel views on the xy (horizontal) plane in the camera coordinate system. The rotation angles are empirically set to $\pm 30°$ around the z (vertical) axis, respectively. See Fig. 3 for an illustration. Since the position of the two novel views are symmetric w.r.t. the original view, the rendered images are complementary in the sense that missing regions in one view will be rendered in the other view (See Fig. 4). As a result, the two symmetric views can partially alleviate the impact of occluded or out-of-view regions during novel view rendering.

3.2 Multi-view Saliency Detection Network

We design a multi-view saliency detection network (MVSalNet) with multi-view augmented images as input. As shown in Fig. 2 (b), the MVSalNet can be divided into two parts, including the single-view saliency prediction module and multi-view fusion module. Since we augment the input image with two additional views, the single-view saliency prediction module contains three encoder-decoder networks, each of them operating in a specific input view. To further

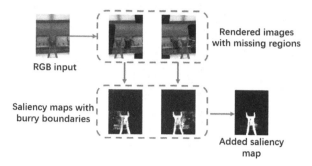

Fig. 4. The complementary property of two symmetric views. Missing regions in one view may be rendered from the other view. The combination of the saliency maps predicted for individual maps can therefore effectively improve detection accuracy.

strengthen single-view saliency detection, the encoder and decoder features are skip-connected via the transformer-based dynamic filtering (TDF) module. The predicted saliency maps under different views are further aggregated by the multi-view fusion module to produce the final output.

Single-View Encoder-Decoder. The three encoder-decoder networks under different views share the same architecture with untied network parameters. For the encoder, we adopt the ResNet-50 [20] backbone architecture, which produces a multi-scale feature pyramid denoted as $\{\mathbf{F}_e^i | i = 1, 2, \ldots, 5\}$ with i indicating the resolution index. The feature resolution becomes smaller as the layer goes deeper. The decoder then takes the coarsest-level feature $\mathbf{F}_e^i$ as input and progressively upsamples the intermediate feature maps $\{\mathbf{F}_d^i | i = 5, 4, \ldots, 1\}$ to the original input resolution. Short-cut connections are also added between encoder and decoder features of the same resolutions. Different from existing methods [40] that either use addition or concatenation to combine the corresponding features in the short-cut connections, we design a TDF module (as detailed below) which takes the encoder features $\mathbf{F}_e^3$-$\mathbf{F}_e^5$ and produce three position-specific dynamic filters. The generated filters are then applied to the corresponding decoder features $\mathbf{F}_d^3$-$\mathbf{F}_d^5$, respectively. Each decoder then independently predicts a saliency map for its input view.

Transformer-Based Dynamic Filtering Module. Figure 5 (a) overviews the network architecture of the proposed transformer-based dynamic filtering (TDF) module. For an input feature from the single-view saliency encoder, the TDF module aims to generate a position-specific dynamic filter which can then be applied to the corresponding features in the decoder. Due to its remarkable capabilities in modeling global correlation, we adopt transformer networks for the dynamic filter generation. To this end, we first partition the input encoder feature $\mathbf{F}_e^i$ into 1×1 patches, which are then fed into a linear layer to produce a set of 1D feature embeddings corresponding to each location. The embeddings are further processed by a window based multi-head self attention (W-MSA) block [29] followed by three shifted window based MSA (SW-MSA) blocks [29]

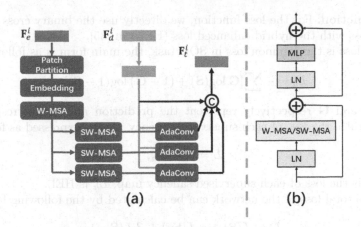

Fig. 5. (a) The structure of TDF. (b) The detail structure of W-MSA and SW-MSA.

(cf. Figure 5 (b)), producing three convolutional kernels for each spatial position on the input feature map. The generated convolutional kernels are then
applied to the corresponding feature $\mathbf{F}_d^i$ in the decoder through adaptive convolutions [23] with dilation rates of 1, 3, and 5, respectively. The final output $\mathbf{F}_t^i$
of the TDF module can be computed as:

$$\mathbf{F}_t^i = \mathcal{M}(< \mathcal{H}(\mathbf{F}_d^i), \mathcal{H}_a(\mathbf{F}_d^i; \mathbf{K}_1^i, \mathbf{K}_2^i, \mathbf{K}_3^i) >), \qquad (1)$$

where $\mathcal{H}$ denotes the standard 3×3 convolution; $\mathcal{H}_a$ denotes adaptive convolution
layer using the three generated position specific kernels $\mathbf{K}_1^i$-$\mathbf{K}_3^i$; $< \cdot, \cdot >$ indicates
channel-wise concatenation; and $\mathcal{M}$ is a linear transformation layer. As a result,
the obtained features are more effective for the decoding of saliency map.

Multi-view Fusion. Figure 2 (b) demonstrates the pipeline of our multi-view
fusion module. The single-view encoder-decoders predict the saliency map $\mathbf{S}_0$
for the input image and $\hat{\mathbf{S}}_1$, $\hat{\mathbf{S}}_2$ for the two augmented views. We first warp
the saliency predictions for two augmented views to the input view to obtain
the saliency maps $\mathbf{S}_1$ and $\mathbf{S}_2$, respectively. Considering their complimentary
property, we add the warped augmented view together as $\mathbf{S}_{1,2} = \mathbf{S}_1 + \mathbf{S}_2$ to tackle
occluded or out-of-view regions with missing values. To achieve more effective
multi-view fusion, we adopt both element-wise multiplication and addition to
combine the current and augmented view saliency maps:

$$\begin{aligned} \mathbf{S}_m &= \mathbf{S}_0 \odot \sigma(\mathbf{S}_{1,2}), \\ \mathbf{S}_a &= \mathbf{S}_0 + \mathbf{S}_{1,2}, \end{aligned} \qquad (2)$$

where $\sigma(\cdot)$ denotes sigmoid function. Intuitively, $\mathbf{S}_m$ is able to suppress false-
positive background noises while $\mathbf{S}_a$ allows to identify false-negative foreground
regions. We then concatenate both $\mathbf{S}_m$ and $\mathbf{S}_a$ along the channel dimension and
send their concatenation to an additional convolution layer to generate the final
saliency map $\mathbf{S}$ of the input image.

Loss Function. For the loss function, we directly use the binary cross entropy (BCE) loss with the hybrid enhanced loss (HEL) in [35].

BCE loss is the common loss in SOD task, the main form is as follows:

$$L_b = -\sum [\mathbf{G}\log(\mathbf{S}) + (1 - \mathbf{G})\log(1 - \mathbf{S})], \tag{3}$$

where $\mathbf{S}$ and $\mathbf{G}$ respectively represent the prediction and the corresponding ground truth. the loss of each supervised saliency map is expressed as follows:

$$L = L_b + L_h, \tag{4}$$

where L is the loss of each supervised saliency map, L_h is HEL.

So the total loss of the network can be calculated by the following formula:

$$L_t = L(\mathbf{S}) + \alpha \cdot L(\mathbf{S_0}) + \beta \cdot L(\mathbf{S}_{1,2}), \tag{5}$$

where L_t is the total loss of the network, α and β are the weight coefficients, we set $\alpha = \beta = 0.25$ in this paper.

4   Experiments

In this section, we conduct extensive experiments to verify the effectiveness of our method. First, we compare our model with other methods. Then we perform a series of ablation studies to evaluate each component of our framework.

4.1   Datasets and Evaluation Metrics

Datasets. We perform our experiments on six widely used RGB-D datasets for fair comparisons. LFSD [25] contains 100 image pairs. NJUD [24] contains 1985 image pairs. NLPR [36] contains 1000 image pairs. RGBD135 [6] contains 135 image pairs. STEREO [34] contains 1000 stereoscopic image pairs. DUTRGBD [38] 1200 image pairs. To guarantee fair comparisons, we follow the setting of [38]. On the DUTRGBD dataset, we choose the same 800 samples for training and 400 images for testing. For the other datasets, we follow the data partition of [3] to use 1485 samples from NJUD and 700 samples from NLPR to train and the remaining samples are used to test.

Evaluation Metrics. To comprehensively and fairly evaluate various methods, we employ five widely used metrics for evaluating, including F-measure [1], weighted F-measure [33], MAE [37], S-measure [12], and E-measure [13]. F-measure [1] reflects the performance of the binary predictions under different thresholds. Weighted F-measure is proposed to improved the existing metric F-measure, it defines a weighted precision and a weighted recall. MAE measures the average of the per-pixel absolute difference between the saliency maps and the ground truth. S-measure can evaluate the structural similarities. E-measure can jointly utilize image-level statistics and local pixel-level statistics for evaluating the binary saliency map.

Table 1. Quantitative RGB-D SOD results on DUTRGBD dataset. The best results are highlighted in red.

	Metric	TANet [4]	A2dele [39]	HDFNet [35]	JL-DCF [17]	UCNet [47]	DANet [52]	BTSNet [49]	Ours
DUTRGBD [38]	F_{max}	0.862	0.906	0.930	0.924	0.882	0.918	0.929	0.935
	F_{ada}	0.815	0.891	0.885	0.883	0.856	0.888	0.906	0.914
	F_β^ω	0.764	0.865	0.864	0.863	0.822	0.860	0.872	0.893
	MAE	0.067	0.042	0.041	0.043	0.056	0.043	0.039	0.034
	S_m	0.853	0.884	0.907	0.905	0.863	0.899	0.903	0.915
	E_m	0.901	0.929	0.938	0.938	0.906	0.937	0.942	0.951

4.2 Implementation Details

Parameter Setting. Three encoders of the proposed model are based on ResNet-50 [20], and only the convolutional layers in the corresponding classification networks are retained. During the training phase, we use the weight parameters pretrained on the ImageNet [9] to initialize the encoders.

Training Setting. During the training stage, we apply random horizontal flipping, random rotating as data augmentation for RGB images and depth images to improve generalization and avoid overfitting. And we employ random color jittering for RGB images. We use the momentum SGD optimizer with a weight decay of 5e-4, an initial learning rate of 5e-3, and a momentum of 0.9. Besides, we apply the CosineAnnealing strategy [32] with the minimum learning rate of 0. The input images are resized to 320×320. We train the model for 40 epochs on a NVIDIA GTX 1080 Ti GPU with a batch size of 4.

Testing Details. During the testing stage, we resize RGB and depth images to 320×320. The final prediction is rescaled to the original size for evaluation.

4.3 Comparisons

To demonstrate the effectiveness of the proposed method, we compare it with 14 state-of-the-art (SOTA) methods, including TANet [4], D3Net [14], A2dele [39], AFNet [45], CoNet [22], CPFP [51], JL-DCF [17], PCF [3], UCNet [47], HDFNet [35], DCF [21], BBSNet [15], DANet[52], BTSNet [49]. Quantitative results on the DUTRGBD dataset are shown in Table 1, while those on the rest five datasets are shown in Table 2. Our methods consistently outperforms all the other SOTAs across all the datasets in terms of different metrics.

Figure 6 shows sampled visualization results under challenging scenarios, including cluttered background (Row 3, 4, 10), complex objects (Row 2, 7, 8, 9), low-quality depth map (Row 1, 6), and small objects with misleading depth map (Row 5).

4.4 Ablation Study

In this section, we perform a series of ablation studies on the NLPR dataset [36] to further investigate the relative importance and specific contribution of each component in the proposed framework using as test dataset.

Table 2. Results ($\uparrow$:F_{max}, F_{ada}, F_β^ω, S_m, and E_m;$\downarrow$:MAE) of different RGB-D SOD methods across five datasets. The best results are highlighted in red.

	Metric	TANet [4]	D3Net [14]	A2dele [39]	AFNet [45]	CoNet [22]	CPFP [51]	JL-DCF [17]	PCF [3]	UCNet [47]	HDFNet [35]	DCF [21]	BBSNet [15]	BTSNet [49]	Ours
LFSD[25]	F_{max}	0.827	0.849	–	0.780	0.874	0.850	0.872	–	0.871	0.872	0.861	0.879	0.849	0.880
	F_{ada}	0.794	0.801	–	0.742	0.835	0.813	0.830	–	0.844	0.833	0.815	0.850	0.823	0.856
	F_β^ω	0.719	0.756	–	0.671	0.802	0.775	0.792	–	0.813	0.789	0.776	0.815	0.770	0.819
	MAE	0.111	0.099	–	0.133	0.077	0.088	0.082	–	0.072	0.088	0.087	0.073	0.098	0.072
	S_m	0.801	0.832	–	0.738	0.856	0.828	0.847	–	0.851	0.841	0.828	0.860	0.829	0.856
	E_m	0.851	0.860	–	0.810	0.892	0.867	0.885	–	0.896	0.885	0.865	0.902	0.874	0.906
NJUD[24]	F_{max}	0.888	0.903	0.888	0.804	0.900	0.890	0.914	0.887	0.906	0.921	0.920	0.922	0.927	0.922
	F_{ada}	0.844	0.840	0.873	0.768	0.780	0.837	0.881	0.844	0.885	0.887	0.898	0.894	0.901	0.902
	F_β^ω	0.805	0.833	0.844	0.696	0.848	0.828	0.866	0.803	0.867	0.877	0.886	0.879	0.884	0.886
	MAE	0.061	0.051	0.051	0.100	0.047	0.053	0.042	0.059	0.043	0.037	0.036	0.038	0.035	0.035
	S_m	0.878	0.895	0.868	0.772	0.895	0.878	0.902	0.877	0.894	0.909	0.908	0.915	0.918	0.910
	E_m	0.909	0.901	0.916	0.847	0.924	0.900	0.935	0.909	0.932	0.930	0.936	0.933	0.942	0.939
NLPR[36]	F_{max}	0.876	0.904	0.895	0.816	0.895	0.883	0.924	0.864	0.911	0.926	0.914	0.921	0.912	0.929
	F_{ada}	0.796	0.834	0.878	0.747	0.844	0.818	0.868	0.795	0.885	0.887	0.887	0.882	0.874	0.901
	F_β^ω	0.780	0.826	0.859	0.693	0.838	0.807	0.873	0.762	0.872	0.881	0.881	0.875	0.869	0.895
	MAE	0.041	0.034	0.028	0.058	0.031	0.038	0.023	0.044	0.026	0.024	0.022	0.024	0.027	0.021
	S_m	0.886	0.906	0.895	0.799	0.904	0.884	0.921	0.873	0.912	0.924	0.920	0.924	0.920	0.927
	E_m	0.916	0.934	0.945	0.884	0.933	0.920	0.953	0.916	0.952	0.955	0.958	0.952	0.949	0.959
RGBD135 [6]	F_{max}	0.853	0.917	0.893	0.775	0.908	0.882	0.931	–	0.931	0.932	0.903	0.934	0.929	0.934
	F_{ada}	0.795	0.876	0.868	0.730	0.866	0.829	0.899	–	0.916	0.908	0.870	0.901	0.899	0.908
	F_β^ω	0.740	0.831	0.838	0.641	0.845	0.787	0.892	–	0.901	0.897	0.844	0.879	0.873	0.903
	MAE	0.046	0.030	0.029	0.068	0.030	0.038	0.021	–	0.019	0.020	0.026	0.023	0.023	0.019
	S_m	0.858	0.904	0.883	0.770	0.906	0.872	0.929	–	0.927	0.929	0.897	0.926	0.917	0.931
	E_m	0.919	0.956	0.919	0.874	0.944	0.927	0.967	–	0.974	0.969	0.947	0.961	0.961	0.971
STEREO[34]	F_{max}	0.878	0.897	–	0.848	0.908	0.882	0.915	–	0.903	0.908	0.909	0.907	0.905	0.920
	F_{ada}	0.835	0.833	–	0.807	0.879	0.830	0.858	–	0.875	0.862	0.875	0.874	0.870	0.898
	F_β^ω	0.787	0.815	–	0.752	0.864	0.817	0.850	–	0.857	0.846	0.863	0.847	0.848	0.879
	MAE	0.060	0.054	–	0.075	0.038	0.051	0.041	–	0.041	0.044	0.040	0.043	0.044	0.035
	S_m	0.871	0.891	–	0.825	0.902	0.879	0.901	–	0.895	0.896	0.897	0.901	0.899	0.911
	E_m	0.916	0.911	–	0.887	0.939	0.907	0.932	–	0.939	0.928	0.937	0.933	0.932	0.946

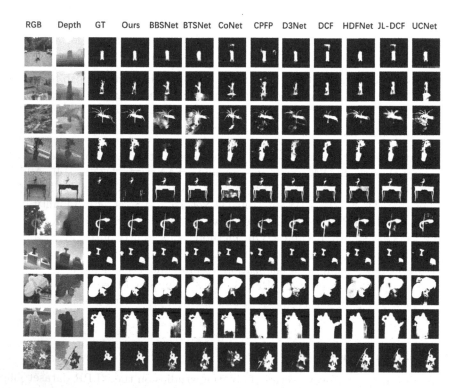

Fig. 6. Visual comparisons with SOTA RGB-D SOD models.

Multi-view Augmentation. To validate the effectiveness of our multi-view augmentation, we conduct several experiments. Results are shown in Table 3. "ED" means that we only use one encoder-decoder branch with RGB input. "EDaug" means that we use one encoder-decoder branch and add the RGB images from novel views to the training set to get extra training data. "3ED" means that we use three encoder-decoder branches with the same RGB input. "Ours-1view" means that we only generate one novel view, the two images pass through two parallel encoder-decoders. "Ours-depth" means that we use our proposed network and change the input of the original encoder-decoder branch to the concatenation of the original RGB image and depth map. It shows that the multi-view augmentation can effectively improve performance when used as extra input or just as extra data. Notice that "Ours-depth" uses more data and gets lower performance compared to "Ours", suggesting that our method can make full use of depth information, while a vanilla single-view encoder-decoder may fail to address the modal gap between RGB and depth. "Ours-15°", "Ours-45°", and "Ours-60°" mean that the rotation angles are set to ±15°, ±45°, and ±60°, respectively. Our preliminary experiments show that rotation angles larger than 60° will lead to degraded results. "Ours-random" means that we randomly choose the rotation angle between 0° and 60° in each side. "Ours-4views" means that we add two more views at ±60° and they share weight with ±30° branches, respectively. Compared with "Ours-1view" and "Ours-4views", we can see that "Ours" reach a good balance between efficiency and effectiveness.

Table 3. Ablation on the multi-view augmentation. The best results are highlighted in red.

Model	F_{max}	F_{ada}	F_β^ω	MAE	S_m	E_m
ED	0.919	0.860	0.855	0.029	0.910	0.939
EDaug	0.919	0.874	0.868	0.027	0.916	0.948
3ED	0.928	0.891	0.882	0.023	0.920	0.954
Ours-1view	0.922	0.882	0.876	0.024	0.917	0.951
Ours-depth	0.922	0.895	0.889	0.023	0.923	0.958
Ours-15°	0.929	0.894	0.889	0.021	0.924	0.957
Ours-45°	0.929	0.893	0.887	0.022	0.924	0.958
Ours-60°	0.929	0.889	0.884	0.022	0.923	0.955
Ours-random	0.928	0.894	0.890	0.021	0.925	0.958
Ours-4views	0.929	0.899	0.888	0.021	0.927	0.962
Ours	0.929	0.901	0.895	0.021	0.927	0.959

Dynamic Filtering Module. Ablations of the dynamic filtering module are reported in Table 4. "SK" means using normal skip connection without TDF. "no dilation" means that we only use one SW-MSA head with dilation rate of

Table 4. Ablation on dynamic filtering module. The best results are highlighted in red.

Model	F_{max}	F_{ada}	F_β^ω	MAE	S_m	E_m
SK	0.922	0.886	0.878	0.025	0.918	0.952
no dilation	0.926	0.892	0.885	0.022	0.921	0.957
dilation embed	0.927	0.896	0.886	0.022	0.924	0.957
DDPM	0.929	0.896	0.889	0.022	0.922	0.956
Ours	0.929	0.901	0.895	0.021	0.927	0.959

Table 5. Ablation on multi-view fusion. The best results are highlighted in red.

Model	F_{max}	F_{ada}	F_β^ω	MAE	S_m	E_m
Add	0.923	0.883	0.881	0.023	0.920	0.951
Supervise two	0.927	0.896	0.890	0.022	0.923	0.958
Ours	0.929	0.901	0.895	0.021	0.927	0.959

Table 6. Ablation on loss function. The best results are highlighted in red.

Model	F_{max}	F_{ada}	F_β^ω	MAE	S_m	E_m
$\alpha = \beta = 0$	0.925	0.893	0.883	0.024	0.920	0.955
$\alpha = \beta = 0.5$	0.928	0.901	0.895	0.021	0.926	0.960
$\alpha = \beta = 1$	0.928	0.894	0.886	0.022	0.923	0.957
Ours	0.929	0.901	0.895	0.021	0.927	0.959

1. "dilation embed" means that we use one SW-MSA head but embed features using dilated convolution with dilation rates of 1, 3, 5 in parallel and concatenate the outputs. "DDPM" means that we replace TDF module in our model with the CNN dynamic filtering branch in [35]. It shows that our proposed TDF module outperforms the counterparts. Besides, "DDPM" has 389M parameters, "Ours" has 83M parameters, the number of parameters is reduced by 78.6%.

Multi-view Fusion. Ablations of multi-view fusion are reported in Table 5. Among them, "add" means directly adding the three saliency maps in original view. "supervise two" denotes that the saliency maps of the two synthesized RGB images are supervised in original view respectively. It shows that our lightweight module can learn the complementary property between the saliency maps and generate accurate final result. In Fig. 7, we can see that S_0 mainly focuses on texture information, while $S_{1,2}$ captures abundant spatial structure information. The final result S is generated by fusing S_0 and $S_{1,2}$.

Loss Function. Ablations of the weight coefficients α and β in loss function are reported in Table 6. Experiments show that our model is robust to the hyper parameters α and β.

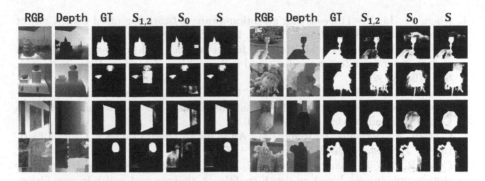

Fig. 7. Visual comparisons for showing the benefits of multi-view fusion. GT, $S_{1,2}$, and S_0 denotes ground truth of saliency map, the added saliency map from augmented views, and the saliency map from original view, respectively.

5 Conclusion

In this paper, we propose a new RGB-D salient object detection (SOD) framework to take full advantages of 3D geometry information contained in depth maps. Instead of using input depth maps as low-level cues, we render the input image from multiple different views and formulate SOD from a single static images to a multi-view setting. We further design a multi-view salient detection network (MVSalNet), which performs SOD independently for each individual view and fuses the output from multiple views to obtain the final prediction. The proposed method outperforms state-of-the-art RGB-D SOD approaches on 6 benchmark datasets with a considerable margin, which demonstrates the effectiveness of our contributions.

Acknowledgements. This paper is supported by National Natural Science Foundation of China (61725202, U1903215, 61906031, 61829102), and Fundamental Research Funds for Central Universities (DUT21RC(3)025).

References

1. Achanta, R., Hemami, S., Estrada, F., Susstrunk, S.: Frequency-tuned salient region detection. In: 2009 IEEE Conference on Computer Vision and Pattern Recognition, pp. 1597–1604. IEEE (2009)
2. Carion, N., Massa, F., Synnaeve, G., Usunier, N., Kirillov, A., Zagoruyko, S.: End-to-end object detection with transformers. In: Vedaldi, A., Bischof, H., Brox, T., Frahm, J.-M. (eds.) ECCV 2020. LNCS, vol. 12346, pp. 213–229. Springer, Cham (2020). https://doi.org/10.1007/978-3-030-58452-8_13
3. Chen, H., Li, Y.: Progressively complementarity-aware fusion network for rgb-d salient object detection. In: Proceedings of the IEEE Conference on Computer Vision and Pattern Recognition, pp. 3051–3060 (2018)

4. Chen, H., Li, Y.: Three-stream attention-aware network for rgb-d salient object detection. IEEE Trans. Image Process. **28**(6), 2825–2835 (2019)
5. Chen, H., Li, Y., Su, D.: Multi-modal fusion network with multi-scale multi-path and cross-modal interactions for rgb-d salient object detection. Pattern Recogn. **86**, 376–385 (2019)
6. Cheng, Y., Fu, H., Wei, X., Xiao, J., Cao, X.: Depth enhanced saliency detection method. In: Proceedings of International Conference on Internet Multimedia Computing and Service, pp. 23–27 (2014)
7. Ciptadi, A., Hermans, T., Rehg, J.M.: An in depth view of saliency. Georgia Institute of Technology (2013)
8. Cong, R., Lei, J., Zhang, C., Huang, Q., Cao, X., Hou, C.: Saliency detection for stereoscopic images based on depth confidence analysis and multiple cues fusion. IEEE Signal Process. Lett. **23**(6), 819–823 (2016)
9. Deng, J., Dong, W., Socher, R., Li, L.J., Li, K., Fei-Fei, L.: Imagenet: A large-scale hierarchical image database. In: 2009 IEEE Conference on Computer Vision and Pattern Recognition, pp. 248–255. IEEE (2009)
10. Devlin, J., Chang, M.W., Lee, K., Toutanova, K.: Bert: Pre-training of deep bidirectional transformers for language understanding. arXiv preprint arXiv:1810.04805 (2018)
11. Dosovitskiy, A., et al.: An image is worth 16 × 16 words: Transformers for image recognition at scale. arXiv preprint arXiv:2010.11929 (2020)
12. Fan, D.P., Cheng, M.M., Liu, Y., Li, T., Borji, A.: Structure-measure: A new way to evaluate foreground maps. In: Proceedings of the IEEE International Conference on Computer Vision, pp. 4548–4557 (2017)
13. Fan, D.P., Gong, C., Cao, Y., Ren, B., Cheng, M.M., Borji, A.: Enhanced-alignment measure for binary foreground map evaluation. arXiv preprint arXiv:1805.10421 (2018)
14. Fan, D.P., Lin, Z., Zhang, Z., Zhu, M., Cheng, M.M.: Rethinking rgb-d salient object detection: Models, data sets, and large-scale benchmarks. IEEE Trans. Neural Netw. Learn. Syst. **32**(5), 2075–2089 (2020)
15. Fan, D.-P., Zhai, Y., Borji, A., Yang, J., Shao, L.: BBS-Net: RGB-D salient object detection with a Bifurcated backbone strategy network. In: Vedaldi, A., Bischof, H., Brox, T., Frahm, J.-M. (eds.) ECCV 2020. LNCS, vol. 12357, pp. 275–292. Springer, Cham (2020). https://doi.org/10.1007/978-3-030-58610-2_17
16. Feng, D., Barnes, N., You, S., McCarthy, C.: Local background enclosure for rgb-d salient object detection. In: Proceedings of the IEEE Conference on Computer Vision and Pattern Recognition, pp. 2343–2350 (2016)
17. Fu, K., Fan, D.P., Ji, G.P., Zhao, Q.: Jl-dcf: Joint learning and densely-cooperative fusion framework for rgb-d salient object detection. In: Proceedings of the IEEE/CVF Conference on Computer Vision and Pattern Recognition, pp. 3052–3062 (2020)
18. Fu, K., Fan, D.P., Ji, G.P., Zhao, Q., Shen, J., Zhu, C.: Siamese network for rgb-d salient object detection and beyond. IEEE Trans. Pattern Anal. Mach. Intell. **44**, 5541–5559 (2021)
19. Han, J., Chen, H., Liu, N., Yan, C., Li, X.: Cnns-based rgb-d saliency detection via cross-view transfer and multiview fusion. IEEE Trans. Cybern. **48**(11), 3171–3183 (2017)
20. He, K., Zhang, X., Ren, S., Sun, J.: Deep residual learning for image recognition. In: Proceedings of the IEEE Conference on Computer Vision and Pattern Recognition, pp. 770–778 (2016)

21. Ji, W., et al.: Calibrated rgb-d salient object detection. In: Proceedings of the IEEE/CVF Conference on Computer Vision and Pattern Recognition, pp. 9471–9481 (2021)
22. Ji, W., Li, J., Zhang, M., Piao, Y., Lu, H.: Accurate RGB-D salient object detection via collaborative learning. In: Vedaldi, A., Bischof, H., Brox, T., Frahm, J.-M. (eds.) ECCV 2020. LNCS, vol. 12363, pp. 52–69. Springer, Cham (2020). https://doi.org/10.1007/978-3-030-58523-5_4
23. Jia, X., De Brabandere, B., Tuytelaars, T., Gool, L.V.: Dynamic filter networks. Adv. Neural. Inf. Process. Syst. **29**, 667–675 (2016)
24. Ju, R., Liu, Y., Ren, T., Ge, L., Wu, G.: Depth-aware salient object detection using anisotropic center-surround difference. Signal Process, Image Commun. **38**, 115–126 (2015)
25. Li, N., Ye, J., Ji, Y., Ling, H., Yu, J.: Saliency detection on light field. In: Proceedings of the IEEE Conference on Computer Vision and Pattern Recognition, pp. 2806–2813 (2014)
26. Lian, Q., Ye, B., Xu, R., Yao, W., Zhang, T.: Geometry-aware data augmentation for monocular 3d object detection. arXiv preprint arXiv:2104.05858 (2021)
27. Liu, Y., et al.: Roberta: A robustly optimized bert pretraining approach. arXiv preprint arXiv:1907.11692 (2019)
28. Liu, Y., Sun, G., Qiu, Y., Zhang, L., Chhatkuli, A., Van Gool, L.: Transformer in convolutional neural networks. arXiv preprint arXiv:2106.03180 (2021)
29. Liu, Z., et al.: Swin transformer: Hierarchical vision transformer using shifted windows. arXiv preprint arXiv:2103.14030 (2021)
30. Liu, Z., Zhang, W., Zhao, P.: A cross-modal adaptive gated fusion generative adversarial network for rgb-d salient object detection. Neurocomputing **387**, 210–220 (2020)
31. Long, J., Shelhamer, E., Darrell, T.: Fully convolutional networks for semantic segmentation. In: Proceedings of the IEEE Conference on Computer Vision and Pattern Recognition, pp. 3431–3440 (2015)
32. Loshchilov, I., Hutter, F.: Sgdr: Stochastic gradient descent with warm restarts. arXiv preprint arXiv:1608.03983 (2016)
33. Margolin, R., Zelnik-Manor, L., Tal, A.: How to evaluate foreground maps? In: Proceedings of the IEEE Conference on Computer Vision and Pattern Recognition, pp. 248–255 (2014)
34. Niu, Y., Geng, Y., Li, X., Liu, F.: Leveraging stereopsis for saliency analysis. In: 2012 IEEE Conference on Computer Vision and Pattern Recognition, pp. 454–461. IEEE (2012)
35. Pang, Y., Zhang, L., Zhao, X., Lu, H.: Hierarchical dynamic filtering network for RGB-D salient object detection. In: Vedaldi, A., Bischof, H., Brox, T., Frahm, J.-M. (eds.) ECCV 2020. LNCS, vol. 12370, pp. 235–252. Springer, Cham (2020). https://doi.org/10.1007/978-3-030-58595-2_15
36. Peng, H., Li, B., Xiong, W., Hu, W., Ji, R.: RGBD salient object detection: a benchmark and algorithms. In: Fleet, D., Pajdla, T., Schiele, B., Tuytelaars, T. (eds.) ECCV 2014. LNCS, vol. 8691, pp. 92–109. Springer, Cham (2014). https://doi.org/10.1007/978-3-319-10578-9_7
37. Perazzi, F., Krähenbühl, P., Pritch, Y., Hornung, A.: Saliency filters: Contrast based filtering for salient region detection. In: 2012 IEEE Conference on Computer Vision and Pattern Recognition, pp. 733–740. IEEE (2012)
38. Piao, Y., Ji, W., Li, J., Zhang, M., Lu, H.: Depth-induced multi-scale recurrent attention network for saliency detection. In: Proceedings of the IEEE/CVF International Conference on Computer Vision, pp. 7254–7263 (2019)

39. Piao, Y., Rong, Z., Zhang, M., Ren, W., Lu, H.: A2dele: Adaptive and attentive depth distiller for efficient rgb-d salient object detection. In: Proceedings of the IEEE/CVF Conference on Computer Vision and Pattern Recognition, pp. 9060–9069 (2020)
40. Ronneberger, O., Fischer, P., Brox, T.: U-net: Convolutional networks for biomedical image segmentation. In: International Conference on Medical Image Computing and Computer-Assisted Intervention, pp. 234–241. Springer (2015). https://doi.org/10.1007/978-3-662-54345-0_3
41. Shigematsu, R., Feng, D., You, S., Barnes, N.: Learning rgb-d salient object detection using background enclosure, depth contrast, and top-down features. In: Proceedings of the IEEE International Conference on Computer Vision Workshops, pp. 2749–2757 (2017)
42. Skoczeń, M., et al.: Obstacle detection system for agricultural mobile robot application using rgb-d cameras. Sensors 21(16), 5292 (2021)
43. Wan, T., et al.: Rgb-d point cloud registration based on salient object detection. IEEE Trans. Neural Netw. Learn. Syst. 33, 3547–3559 (2021)
44. Wang, H., Zhu, Y., Green, B., Adam, H., Yuille, A., Chen, L.-C.: Axial-DeepLab: Stand-alone axial-attention for panoptic segmentation. In: Vedaldi, A., Bischof, H., Brox, T., Frahm, J.-M. (eds.) ECCV 2020. LNCS, vol. 12349, pp. 108–126. Springer, Cham (2020). https://doi.org/10.1007/978-3-030-58548-8_7
45. Wang, N., Gong, X.: Adaptive fusion for rgb-d salient object detection. IEEE Access 7, 55277–55284 (2019)
46. Zhang, J., et al.: Uncertainty inspired rgb-d saliency detection. IEEE Trans. Pattern Anal. Mach. Intell. 44, 5761–5779 (2021)
47. Zhang, J., et al.: Uc-net: Uncertainty inspired rgb-d saliency detection via conditional variational autoencoders. In: Proceedings of the IEEE/CVF Conference on Computer Vision and Pattern Recognition, pp. 8582–8591 (2020)
48. Zhang, M., Fei, S.X., Liu, J., Xu, S., Piao, Y., Lu, H.: Asymmetric two-stream architecture for accurate rgb-d saliency detection. In: Vedaldi, A., Bischof, H., Brox, T., Frahm, J.-M. (eds.) ECCV 2020. LNCS, vol. 12373, pp. 374–390. Springer, Cham (2020). https://doi.org/10.1007/978-3-030-58604-1_23
49. Zhang, W., Jiang, Y., Fu, K., Zhao, Q.: Bts-net: Bi-directional transfer-and-selection network for rgb-d salient object detection. In: 2021 IEEE International Conference on Multimedia and Expo (ICME), pp. 1–6. IEEE (2021)
50. Zhang, Z., Lin, Z., Xu, J., Jin, W.D., Lu, S.P., Fan, D.P.: Bilateral attention network for rgb-d salient object detection. IEEE Trans. Image Process. 30, 1949–1961 (2021)
51. Zhao, J.X., Cao, Y., Fan, D.P., Cheng, M.M., Li, X.Y., Zhang, L.: Contrast prior and fluid pyramid integration for rgbd salient object detection. In: Proceedings of the IEEE/CVF Conference on Computer Vision and Pattern Recognition, pp. 3927–3936 (2019)
52. Zhao, X., Zhang, L., Pang, Y., Lu, H., Zhang, L.: A single stream network for robust and real-time rgb-d salient object detection. In: Vedaldi, A., Bischof, H., Brox, T., Frahm, J.-M. (eds.) ECCV 2020. LNCS, vol. 12367, pp. 646–662. Springer, Cham (2020). https://doi.org/10.1007/978-3-030-58542-6_39
53. Zhao, Y., Kong, S., Fowlkes, C.: Camera pose matters: Improving depth prediction by mitigating pose distribution bias. In: Proceedings of the IEEE/CVF Conference on Computer Vision and Pattern Recognition, pp. 15759–15768 (2021)

54. Zhu, C., Cai, X., Huang, K., Li, T.H., Li, G.: Pdnet: Prior-model guided depth-enhanced network for salient object detection. In: 2019 IEEE International Conference on Multimedia and Expo (ICME), pp. 199–204. IEEE (2019)
55. Zhu, M., Zhang, S., Zhong, Y., Lu, P., Peng, H., Lenneman, J.: Monocular 3d vehicle detection using uncalibrated traffic cameras through homography. arXiv preprint arXiv:2103.15293 (2021)

k-means Mask Transformer

Qihang Yu[1](✉), Huiyu Wang[1], Siyuan Qiao[2], Maxwell Collins[2], Yukun Zhu[2], Hartwig Adam[2], Alan Yuille[1], and Liang-Chieh Chen[2]

[1] Johns Hopkins University, Baltimore, USA
yucornetto@gmail.com
[2] Google Research, Mountain View, USA

Abstract. The rise of transformers in vision tasks not only advances network backbone designs, but also starts a brand-new page to achieve end-to-end image recognition (*e.g.*, object detection and panoptic segmentation). Originated from Natural Language Processing (NLP), transformer architectures, consisting of self-attention and cross-attention, effectively learn long-range interactions between elements in a sequence. However, we observe that most existing transformer-based vision models simply borrow the idea from NLP, neglecting the crucial difference between languages and images, particularly the extremely large sequence length of spatially flattened pixel features. This subsequently impedes the learning in cross-attention between pixel features and object queries. In this paper, we rethink the relationship between pixels and object queries, and propose to reformulate the cross-attention learning as a clustering process. Inspired by the traditional *k*-means clustering algorithm, we develop a **k**-means **Mask X**former (*k*MaX-DeepLab) for segmentation tasks, which not only improves the state-of-the-art, but also enjoys a simple and elegant design. As a result, our *k*MaX-DeepLab achieves a new state-of-the-art performance on COCO *val* set with 58.0% PQ, and Cityscapes *val* set with 68.4% PQ, 44.0% AP, and 83.5% mIoU without test-time augmentation or external dataset. We hope our work can shed some light on designing transformers tailored for vision tasks. Code and models are available at https://github.com/google-research/deeplab2.

Keywords: Segmentation · Transformer · *k*-means Clustering

1 Introduction

Transformers [89] are receiving a growing attention in the computer vision community. On the one hand, the transformer encoder, with multi-head self-attention as the central component, demonstrates a great potential for building powerful

Work done during an internship at Google.

Supplementary Information The online version contains supplementary material available at https://doi.org/10.1007/978-3-031-19818-2_17.

network architectures in various visual recognition tasks [32,70,93]. On the other hand, the transformer decoder, with multi-head cross-attention at its core, provides a brand-new approach to tackling complex visual recognition problems in an end-to-end manner, dispensing with hand-designed heuristics.

Recently, the pioneering work DETR [10] introduces the first end-to-end object detection system with transformers. In this framework, the pixel features are firstly extracted by a convolutional neural network [58], followed by the deployment of several transformer encoders for feature enhancement to capture long-range interactions between pixels. Afterwards, a set of learnable positional embeddings, named object queries, is responsible for interacting with pixel features and aggregating information through several interleaved cross-attention and self-attention modules. In the end, the object queries, decoded by a Feed-Forward Network (FFN), directly correspond to the final bounding box predictions. Along the same direction, MaX-DeepLab [92] proves the success of transformers in the challenging panoptic segmentation task [55], where the prior arts [21,54,100] usually adopt complicated pipelines involving hand-designed heuristics. The essence of this framework lies in converting the object queries to mask embedding vectors [49,87,97], which are employed to yield a set of mask predictions by multiplying with the pixel features.

The end-to-end transformer-based frameworks have been successfully applied to multiple computer vision tasks with the help of transformer decoders, especially the cross-attention modules. However, the working mechanism behind the scenes remains unclear. The cross-attention, which arises from the Natural Language Processing (NLP) community, is originally designed for language problems, such as neural machine translation [4,86], where both the input sequence and output sequence share a similar short length. This implicit assumption becomes problematic when it comes to certain vision problems, where the cross-attention is performed between object queries and spatially flattened pixel features with an exorbitantly large length. Concretely, usually a small number of object queries is employed (*e.g.*, 128 queries), while the input images can contain thousands of pixels for the vision tasks of detection and segmentation. Each object query needs to learn to highlight the most distinguishable features among the abundant pixels in the cross-attention learning process, which subsequently leads to slow training convergence and thus inferior performance [37,112].

In this work, we make a crucial observation that the cross-attention scheme actually bears a strong similarity to the traditional *k*-means clustering [72] by regarding the object queries as cluster centers with learnable embedding vectors. Our examination of the similarity inspires us to propose the novel **k**-means **Mask Xformer** (*k*MaX-DeepLab), which rethinks the relationship between pixel features and object queries, and redesigns the cross-attention from the perspective of *k*-means clustering. Specifically, when updating the cluster centers (*i.e.*, object queries), our *k*MaX-DeepLab performs a different operation. Instead of performing *softmax* on the large spatial dimension (image height times width) as in the original Mask Transformer's cross-attention [92], our *k*MaX-DeepLab performs *argmax* along the cluster center dimension, similar to the *k*-means pixel-cluster assignment step (with a *hard* assignment). We then update cluster centers by aggregating the pixel features based on the pixel-cluster assignment (computed by their feature affinity), similar to the k-means center-update step. In spite of being conceptually simple, the modification has a striking impact: on COCO *val* set [66], using the standard ResNet-50 [41] as backbone, our kMaX-DeepLab demonstrates a significant improvement of **5.2%** PQ over the original cross-attention scheme at a negligible cost of extra parameters and FLOPs. When comparing to state-of-the-art methods, our kMaX-DeepLab with the simple ResNet-50 backbone already outperforms MaX-DeepLab [92] with MaX-L [92] backbone by **1.9%** PQ, while requiring **7.9** and **22.0** times fewer parameters and FLOPs, respectively. Our kMaX-DeepLab with ResNet-50 also outperforms MaskFormer [24] with the strong ImageNet-22K pretrained Swin-L [70] backbone, and runs **4.4** times faster. Finally, our kMaX-DeepLab, using the modern ConvNeXt-L [71] as backbone, sets a new state-of-the-art performance on the COCO *val* set [66] with 58.0% PQ. It also outperforms other state-of-the-art methods on the Cityscapes *val* set [28], achieving 68.4% PQ, 83.5% mIoU, 44.0% AP, without using any test-time augmentation or extra dataset pretraining [66, 75].

2   Related Works

Transformers. Transformer [89] and its variants [2,8,26,39,57,74,94,106] have advanced the state-of-the-art in natural language processing tasks [30,31, 82] by capturing relations across modalities [4] or in a single context [25,89]. In computer vision, transformer encoders or self-attention modules are either combined with Convolutional Neural Networks (CNNs) [9,96] or used as standalone backbones [32,44,70,80,93]. Both approaches have boosted various vision tasks, such as image classification [7,19,32,44,64,70,80,93,101,105], image generation [42,77], object detection [10,43,80,83,96,112], video recognition [3,19,33,96], semantic segmentation [11,17,35,46,99,108,109,111,113], and panoptic segmentation [93].

Mask Transformers for Segmentation. Besides the usage as backbones, transformers are also adopted as task decoders for image segmentation. MaX-DeepLab [92] proposed **Mask Xformers** (MaX) for end-to-end panoptic segmentation. Mask transformers predict class-labeled object masks and are trained by Hungarian matching the predicted masks with ground truth masks. The essential component of mask transformers is the conversion of object queries to mask embedding vectors [49,87,97], which are employed to generate predicted masks. Both Segmenter [85] and MaskFormer [24] applied mask transformers to semantic segmentation. K-Net [107] proposed dynamic kernels for generating the masks. CMT-DeepLab [104] proposed to improve the cross-attention with an additional clustering update term. Panoptic Segformer [65] strengthened mask transformer with deformable attention [112], while Mask2Former [23] further boosted the performance with masked cross-attention along with a series of technical improvements including cascaded transformer decoder, deformable attention [112], uncertainty-based pointly supervision [56], *etc.*. These mask transformer methods generally outperform box-based methods [54] that decompose panoptic segmentation into multiple surrogate tasks (*e.g.*, predicting masks for each detected object bounding box [40], followed by fusing the instance segments ('thing') and semantic segments ('stuff') [14] with merging modules [60,62,67,78,100,103]). Moreover, mask transformers showed great success in the video segmentation problems [20,52,61].

Clustering Methods for Segmentation. Traditional image segmentation methods [1,72,110] typically cluster image intensities into a set of masks or superpixels with gradual growing or refinement. However, it is challenging for these traditional methods to capture high-level semantics. Modern clustering-based methods usually operate on semantic segments [13,15,18] and group 'thing' pixels into instance segments with various representations, such as instance center regression [22,50,63,76,88,93,102], Watershed transform [5,90], Hough-voting [6,59,91], or pixel affinity [36,47,51,69,84].

Recently, CMT-DeepLab [104] discussed the similarity between mask transformers and clustering algorithms. However, they only used the clustering update as a complementary term in the cross-attention. In this work, we further discover the underlying similarity between mask transformers and the *k*-means clustering algorithm, resulting in a simple yet effective *k*-means mask transformer.

3   Method

In this section, we first overview the mask-transformer-based segmentation framework presented by MaX-DeepLab [92]. We then revisit the transformer cross-attention [89] and the *k*-means clustering algorithm [72], and reveal their underlying similarity. Afterwards, we introduce the proposed **k-means Mask Xformer** (*k*MaX-DeepLab), which redesigns the cross-attention from a clustering perspective. Even though simple, *k*MaX-DeepLab effectively and significantly improves the segmentation performance.

3.1   Mask-Transformer-Based Segmentation Framework

Transformers [89] have been effectively deployed to segmentation tasks. Without loss of generality, we consider panoptic segmentation [55] in the following problem formulation, which can be easily generalized to other segmentation tasks.

Problem Statement. Panoptic segmentation aims to segment the image $I \in \mathbb{R}^{H \times W \times 3}$ into a set of non-overlapping masks with associated semantic labels:

$$\{y_i\}_{i=1}^{K} = \{(m_i, c_i)\}_{i=1}^{K}. \tag{1}$$

The K ground truth masks $m_i \in \{0,1\}^{H \times W}$ do not overlap with each other, *i.e.*, $\sum_{i=1}^{K} m_i \leq 1^{H \times W}$, and c_i denotes the ground truth class label of mask m_i.

Starting from DETR [10] and MaX-DeepLab [92], approaches to panoptic segmentation shift to a new end-to-end paradigm, where the prediction directly matches the format of ground-truth with N masks (N is a fixed number and $N \geq K$) and their semantic classes:

$$\{\hat{y}_i\}_{i=1}^N = \{(\hat{m}_i, \hat{p}_i(c))\}_{i=1}^N, \tag{2}$$

where $\hat{p}_i(c)$ denotes the semantic class prediction confidence for the corresponding mask, which includes 'thing' classes, 'stuff' classes, and the void class $\varnothing$.

The N masks are predicted based on the N object queries, which aggregate information from the pixel features through a transformer decoder, consisting of self-attention and cross-attention modules.

The object queries, updated by multiple transformer decoders, are employed as mask embedding vectors [49,87,97], which will multiply with the pixel features to yield the final prediction $\mathbf{Z} \in \mathbb{R}^{HW \times N}$ that consists of N masks. That is,

$$\mathbf{Z} = \underset{N}{\mathrm{softmax}}(\mathbf{F} \times \mathbf{C}^{\mathrm{T}}), \tag{3}$$

where $\mathbf{F} \in \mathbb{R}^{HW \times D}$ and $\mathbf{C} \in \mathbb{R}^{N \times D}$ refers to the final pixel features and object queries, respectively. D is the channel dimension of pixel features and object queries. We use underscript N to indicate the axis to perform softmax.

3.2   Relationship Between Cross-Attention and k-means Clustering

Although the transformer-based segmentation frameworks successfully connect object queries and mask predictions in an end-to-end manner, the essential problem becomes how to transform the object queries, starting from learnable embeddings (randomly initialized), into meaningful mask embedding vectors.

Cross-Attention.   The cross-attention modules are used to aggregate affiliated pixel features to update object queries. Formally, we have

$$\hat{\mathbf{C}} = \mathbf{C} + \underset{HW}{\mathrm{softmax}}(\mathbf{Q}^c \times (\mathbf{K}^p)^{\mathrm{T}}) \times \mathbf{V}^p, \tag{4}$$

where $\mathbf{C} \in \mathbb{R}^{N \times D}$ refers to N object queries with D channels, and $\hat{\mathbf{C}}$ denotes the updated object queries. We use the underscript HW to represent the axis for softmax on spatial dimension, and superscripts p and c to indicate the feature projected from the pixel features and object queries, respectively. $\mathbf{Q}^c \in \mathbb{R}^{N \times D}, \mathbf{K}^p \in \mathbb{R}^{HW \times D}, \mathbf{V}^p \in \mathbb{R}^{HW \times D}$ stand for the linearly projected features for query, key, and value. For simplicity, we ignore the multi-head mechanism and feed-forward network (FFN) in the equation.

As shown in Eq. (4), when updating the object queries, a *softmax* function is applied to the image resolution (HW), which is typically in the range of thousands of pixels for the task of segmentation. Given the huge number of pixels, it can take many training iterations to learn the attention map, which starts from a uniform distribution at the beginning (as the queries are randomly initialized). Each object query has a difficult time to identify the most distinguishable features among the abundant pixels in the early stage of training. This behavior is very different from the application of transformers to natural language processing tasks, *e.g.*, neural machine translation [4,86], where the input and output sequences share a similar short length. Vision tasks, especially segmentation problems, present another challenge for efficiently learning the cross-attention.

Discussion. Similar to cross-attention, self-attention needs to perform a *softmax* function operated along the image resolution. Therefore, learning the attention map for self-attention may also take many training iterations. An efficient alternative, such as axial attention [93] or local attention [70] is usually applied on high resolution feature maps, and thus alleviates the problem, while a solution to cross-attention remains an open question for research.

k-Means Clustering. In Eq. (4), the cross-attention computes the affinity between object queries and pixels (*i.e.*, $\mathbf{Q}^c \times (\mathbf{K}^p)^\mathrm{T}$), which is converted to the attention map through the spatial-wise softmax (operated along the image resolution). The attention map is then used to retrieve (and weight accordingly) affiliated pixel features to update the object queries. Surprisingly, we observe that the whole process is actually similar to the classic k-means clustering algorithm [72], which works as follows:

$$\mathbf{A} = \underset{N}{\mathrm{argmax}}(\mathbf{C} \times \mathbf{P}^\mathrm{T}), \tag{5}$$

$$\hat{\mathbf{C}} = \mathbf{A} \times \mathbf{P}, \tag{6}$$

where $\mathbf{C} \in \mathbb{R}^{N \times D}$, $\mathbf{P} \in \mathbb{R}^{HW \times D}$, and $\mathbf{A} \in \mathbb{R}^{N \times HW}$ stand for cluster centers, pixel features, and clustering assignments, respectively.

Comparing Eq. (4), Eq. (5), and Eq. (6), we notice that the k-means clustering algorithm is parameter-free and thus no linear projection is needed for query, key, and value. The updates on cluster centers are not in a residual manner. Most importantly, k-means adopts a *cluster-wise argmax* (*i.e.*, argmax operated along the cluster dimension) instead of the spatial-wise softmax when converting the affinity to the attention map (*i.e.*, weights to retrieve and update features).

This observation motivates us to reformulate the cross-attention in vision problems, especially image segmentation. From a clustering perspective, image segmentation is equivalent to grouping pixels into different clusters, where each cluster corresponds to a predicted mask. However, the cross-attention mechanism, also attempting to group pixels to different object queries, instead employs a different *spatial-wise softmax* operation from the *cluster-wise argmax* as in k-means. Given the success of k-means, we hypothesize that the cluster-wise argmax is a more suitable operation than the spatial-wise softmax regarding pixel clustering, since the cluster-wise argmax performs the hard assignment and efficiently reduces the operation targets from thousands of pixels (HW) to just a few cluster centers (N), which (we will empirically prove) speeds up the training convergence and leads to a better performance.

3.3 k-means Mask Transformer

Herein, we first introduce the crucial component of the proposed k-means Mask Transformer, $i.e.$, k-means cross-attention. We then present its meta architecture and model instantiation.

k-means Cross-Attention.   The proposed *k-means cross-attention* reformulates the cross-attention in a manner similar to k-means clustering:

$$\hat{\mathbf{C}} = \mathbf{C} + \underset{N}{\mathrm{argmax}}(\mathbf{Q}^c \times (\mathbf{K}^p)^\mathrm{T}) \times \mathbf{V}^p. \tag{7}$$

Comparing Eq. (4) and Eq. (7), the spatial-wise softmax is now replaced by the cluster-wise argmax. As shown in Fig. 1, with such a simple yet effective change, a typical transformer decoder could be converted to a kMaX decoder. Unlike the original cross-attention, the proposed k-means cross-attention adopts a different operation ($i.e.$, cluster-wise argmax) to compute the attention map, and does not require the multi-head mechanism [89]. However, the cluster-wise argmax, as a hard assignment to aggregate pixel features for the cluster center update, is not a differentiable operation, posing a challenge during training. We have explored several methods ($e.g.$, Gumbel-Softmax [48]), and discover that a simple deep supervision scheme turns out to be most effective. In particular, in our formulation, the affinity logits between pixel features and cluster centers directly correspond to the softmax logits of segmentation masks ($i.e.$, $\mathbf{Q}^c \times (\mathbf{K}^p)^\mathrm{T}$ in Eq. (7) corresponds to $\mathbf{F} \times \mathbf{C}^\mathrm{T}$ in Eq. (3)), since the cluster centers aim to group pixels of similar affinity together to form the predicted segmentation masks. This formulation allows us to add deep supervision to every kMaX decoder, in order to train the parameters in the k-means cross-attention module.

Meta Architecture.   Figure 2 shows the meta architecture of our proposed kMaX-DeepLab, which contains three main components: pixel encoder, enhanced pixel decoder, and kMaX decoder. The pixel encoder extracts the pixel features either by a CNN [41] or a transformer [70] backbone, while the enhanced pixel decoder is responsible for recovering the feature map resolution as well as enhancing the pixel features via transformer encoders [89] or axial attention [93]. Finally, the kMaX decoder transforms the object queries ($i.e.$, cluster centers) into mask embedding vectors from the k-means clustering perspective.

Model Instantiation.   We build kMaX based on MaX-DeepLab [92] with the official code-base [98]. We divide the whole model into two paths: the pixel path and the cluster path, which are responsible for extracting pixel features and cluster centers, respectively. Figure 3 details our kMaX-DeepLab instantiation with two example backbones.

Pixel Path.   The pixel path consists of a pixel encoder and an enhanced pixel decoder. The pixel encoder is an ImageNet-pretrained [81] backbone, such as ResNet [41], MaX-S [92] ($i.e.$, ResNet-50 with axial attention [93]), and ConvNeXt [71]. Our enhanced pixel decoder consists of several axial attention blocks [93] and bottleneck blocks [41].

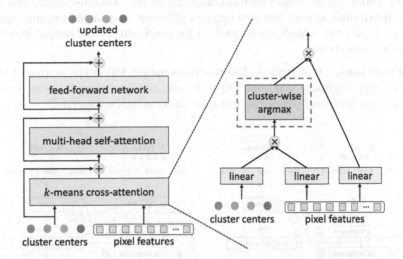

Fig. 1. To convert a typical transformer decoder into our *k*MaX decoder, we simply replace the original cross-attention with our *k*-means cross-attention (*i.e.*, with the only simple change *cluster-wise argmax* high-lighted in red) (Color figure online)

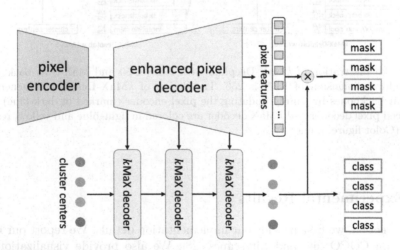

Fig. 2. The meta architecture of *k*-means Mask Transformer consists of three components: pixel encoder, enhanced pixel decoder, and *k*MaX decoder. The pixel encoder is any network backbone. The enhanced pixel decoder includes transformer encoders to enhance the pixel features, and upsampling layers to generate higher resolution features. The series of *k*MaX decoders transform cluster centers into (1) mask embedding vectors, which multiply with the pixel features to generate the predicted masks, and (2) class predictions for each mask.

Cluster Path.   The cluster path contains totally six kMaX decoders, which are evenly distributed among features maps of different spatial resolutions. Specifically, we deploy two kMaX decoders each for pixel features at output stride 32, 16, and 8, respectively.

Loss Functions.   Our training loss functions mostly follow the setting of MaX-DeepLab [92]. We adopt the same PQ-style loss, auxiliary semantic loss, mask-id cross-entropy loss, and pixel-wise instance discrimination loss [104].

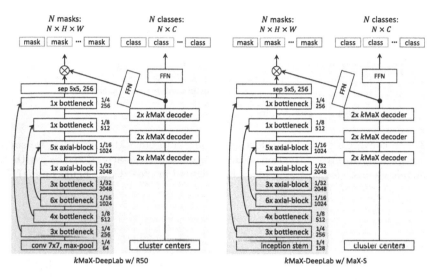

Fig. 3. An illustration of kMaX-DeepLab with ResNet-50 and MaX-S as backbones. The hidden dimension of FFN is 256. The design of kMaX-DeepLab is general to different backbones by simply updating the pixel encoder (marked in dark-blue). The enhanced pixel decoder and kMaX decoder are colored in light-blue and yellow, respectively (Color figure online)

4   Experimental Results

In this section, we first provide our implementation details. We report our main results on COCO [66] and Cityscapes [28]. We also provide visualizations to better understand the clustering process of the proposed kMaX-DeepLab. The ablation studies are provided in the appendix.

4.1   Implementation Details

The meta architecture of the proposed kMaX-DeepLab contains three main components: the pixel encoder, enhanced pixel decoder, and kMaX decoder, as shown in Fig. 2. We provide the implementation details of each component below.

Pixel Encoder. The pixel encoder extracts pixel features given an image. To verify the generality of *k*MaX-DeepLab across different pixel encoders, we experiment with ResNet-50 [41], MaX-S [92] (*i.e.*, ResNet-50 with axial attention [93] in the 3rd and 4th stages), and ConvNeXt [71].

Enhanced Pixel Decoder. The enhanced pixel decoder recovers the feature map resolution and enriches pixel features via self-attention. As shown in Fig. 3, we adopt one axial block with channels 2048 at output stride 32, and five axial blocks with channels 1024 at output stride 16. The axial block is a bottleneck block [41], but the 3×3 convolution is replaced by the axial attention [93]. We use one bottleneck block at output stride 8 and 4, respectively. We note that the axial blocks play the same role (*i.e.*, feature enhancement) as the transformer encoders in other works [10,24,104], where we ensure that the total number of axial blocks is six for a fair comparison to previous works [10,24,104].

Cluster Path. As shown in Fig. 3, we deploy six *k*MaX decoders, where each two are placed for pixel features (enhanced by the pixel decoders) with output stride 32, 16, 8, respectively. Our design uses six transformer decoders, aligning with the previous works [10,24,104], though some recent works [23,65] adopt more transformer decoders to achieve a stronger performance.

Training and Testing. We mainly follow MaX-DeepLab [92] for training settings. The ImageNet-pretrained [81] backbone has a learning rate multiplier 0.1. For regularization and augmentations, we adopt drop path [45], random color jittering [29], and panoptic copy-paste augmentation, which is an extension from instance copy-paste augmentation [34,38] by augmenting both 'thing' and 'stuff' classes. AdamW [53,73] optimizer is used with weight decay 0.05. The *k*-means cross-attention adopts cluster-wise argmax, which aligns the formulation of attention map to segmentation result. It therefore allows us to directly apply deep supervision on the attention maps. These auxiliary losses attached to each *k*MaX decoder have the same loss weight of 1.0 as the final prediction, and Hungarian matching result based on the final prediction is used to assign supervisions for all auxiliary outputs. During inference, we adopt the same mask-wise merging scheme used in [24,65,104,107] to obtain the final segmentation results.

COCO Dataset. If not specified, we train all models with batch size 64 on 32 TPU cores with 150k iterations (around 81 epochs). The first 5k steps serve as the warm-up stage, where the learning rate linearly increases from 0 to 5×10^{-4}. The input images are resized and padded to 1281×1281. Following MaX-DeepLab [92], the loss weights for PQ-style loss, auxiliary semantic loss, mask-id cross-entropy loss, instance discrimination loss are 3.0, 1.0, 0.3, and 1.0, respectively. The number of cluster centers (*i.e.*, object queries) is 128, and the final feature map resolution has output stride 4 as in MaX-DeepLab [92].

We have also experimented with doubling the number of object queries to 256 for *k*MaX-DeepLab with ConvNeXt-L, which however leads to a performance loss. Empirically, we adopt a **drop query** regularization, where we randomly drop half of the object queries (*i.e.*, 128) during each training iteration, and all queries (*i.e.*, 256) are used during inference. With the proposed drop query regularization, doubling the number of object queries to 256 consistently brings 0.1% PQ improvement under the large model regime.

Cityscapes Dataset. We train all models with batch size 32 on 32 TPU cores with 60k iterations. The first 5k steps serve as the warm-up stage, where learning rate linearly increases from 0 to 3×10^{-4}. The inputs are padded to 1025×2049. The loss weights for PQ-style loss, auxiliary semantic loss, mask-id cross-entropy loss, and instance discrimination loss are 3.0, 1.0, 0.3, and 1.0, respectively. We use 256 cluster centers, and add an additional bottleneck block in the pixel decoder to produce features with output stride 2.

4.2 Main Results

Our main results on the COCO [66] and Cityscapes [28] *val* set are summarized in Table 1 and Table 2, respectively.

Table 1. COCO *val* set results. Our FLOPs and FPS are evaluated with the input size 1200×800 and a Tesla V100-SXM2 GPU. †: ImageNet-22K pretraining. ⋆: Using 256 object queries with drop query regularization. ‡: Using COCO *unlabeled* set

method	backbone	params	FLOPs	FPS	PQ	PQTh	PQSt
MaskFormer [24]	ResNet-50 [41]	45M	181G	17.6	46.5	51.0	39.8
K-Net [107]	ResNet-50 [41]	–	–	–	47.1	51.7	40.3
CMT-DeepLab [104]	ResNet-50 [41]	–	–	–	48.5	–	–
Panoptic SegFormer [65]	ResNet-50 [41]	51M	214G	7.8	49.6	54.4	42.4
Mask2Former [23]	ResNet-50 [41]	44M	226G	8.6	51.9	57.7	43.0
kMaX-DeepLab	ResNet-50 [41]	57M	168G	22.8	**53.0**	**58.3**	**44.9**
MaX-DeepLab [92]	MaX-S [92]	62M	324G	-	48.4	53.0	41.5
CMT-DeepLab	MaX-S† [92]	95M	396G	8.1	53.0	57.7	45.9
kMaX-DeepLab	MaX-S† [92]	74M	240G	16.9	**56.2**	**62.2**	**47.1**
MaskFormer [24]	Swin-B (W12)† [70]	102M	411G	8.4	51.8	56.9	44.1
CMT-DeepLab [104]	Axial-R104† [104]	135M	553G	6.0	54.1	58.8	47.1
Panoptic SegFormer [65]	PVTv2-B5† [95]	105M	349G	–	55.4	61.2	46.6
Mask2Former [23]	Swin-B (W12)† [70]	107M	466G	–	56.4	62.4	47.3
kMaX-DeepLab	ConvNeXt-B† [71]	122M	380G	11.6	**57.2**	**63.4**	**47.8**
MaX-DeepLab [92]	MaX-L [92]	451M	3692G	–	51.1	57.0	42.2
MaskFormer [24]	Swin-L (W12)† [70]	212M	792G	5.2	52.7	58.5	44.0
K-Net [107]	Swin-L (W7)† [70]	–	–	–	54.6	60.2	46.0
CMT-DeepLab [104]	Axial-R104-RFN† [79]	270M	1114G	3.2	55.3	61.0	46.6
Panoptic SegFormer [65]	Swin-L (W7)† [70]	221M	816G	-	55.8	61.7	46.9
Mask2Former [23]	Swin-L (W12)† [70]	216M	868G	4.0	57.8	64.2	48.1
kMaX-DeepLab	ConvNeXt-L† [71]	232M	744G	6.7	57.9	64.0	48.6
kMaX-DeepLab⋆	ConvNeXt-L† [71]	232M	749G	6.6	58.0	64.2	48.6
kMaX-DeepLab‡	ConvNeXt-L† [71]	232M	744G	6.7	**58.1**	**64.3**	**48.8**

COCO *val* Set. In Table 1, we compare our *k*MaX-DeepLab with other transformer-based panoptic segmentation methods on COCO *val* set. Notably, with a simple ResNet-50 backbone, *k*MaX-DeepLab already achieves 53.0% PQ, surpassing *most* prior arts with stronger backbones. Specifically, *k*MaX-DeepLab outperforms MaskFormer [24] and K-Net [107], all with the ResNet-50 backbone as well, by a large margin of **6.5%** and **5.9%**, while maintaining a similar level of computational costs. Our *k*MaX-DeepLab with ResNet-50 even surpasses the largest variants of MaX-DeepLab [92] by **1.9%** PQ (while using **7.9×** fewer parameters and **22.0×** fewer FLOPs), and MaskFormer (while using **3.7×** fewer parameters and **4.7×** fewer FLOPs) by 0.3% PQ, respectively. With a stronger backbone MaX-S [92], *k*MaX-DeepLab boosts the performance to 56.2% PQ, outperforming MaX-DeepLab with the same backbone by **7.8%** PQ. Our *k*MaX-DeepLab with MaX-S backbone also improves over the previous state-of-art K-Net with Swin-L [70] by **1.6%** PQ. To further push the envelope, we adopt the modern CNN backbone ConvNeXt [71] and set new state-of-the-art results of 57.2% PQ with ConvNeXt-B and 58.0% PQ with ConvNeXt-L, outperforming K-Net with Swin-L by a significant margin of **3.4%** PQ.

When compared to more recent works (CMT-DeepLab [104], Panoptic Seg-Former [65], and Mask2Former [23]), *k*MaX-DeepLab still shows great performances without the advanced modules, such as deformable attention [112], cascaded transformer decoder [23], and uncertainty-based pointly supervision [56]. As different backbones are utilized for each method (*e.g.*, PVTv2 [95], Swin [70], and ConvNeXt [71]), we start with a fair comparison using the ResNet-50 backbone. Our *k*MaX-DeepLab with ResNet-50 achieves a significant better performance compared to CMT-DeepLab, Panoptic SegFormer and Mask2Former by a large margin of **4.5%**, **3.4%**, and **1.1%** PQ, respectively. Additionally, our model runs almost **3×** faster than them (since *k*MaX-DeepLab enjoys a simple design without deformable attention). When employing stronger backbones, *k*MaX-DeepLab with ConvNeXt-B outperforms CMT-DeepLab with Axial-R104, Panoptic SegFormer with PVTv2-B5, and Mask2Former with Swin-B (window size 12) by **3.1%**, **1.8%**, and **0.8%** PQ, respectively, while all models have a similar level of cost (parameters and FLOPs). When scaling up to the largest backbone for each method, *k*MaX-DeepLab outperforms CMT-DeepLab, and Panoptic SegFormer significantly by **2.7%** and **2.2%** PQ. Although we already perform better than Mask2Former with Swin-L (window size 12), we notice that *k*MaX-DeepLab benefits much less than Mask2Former when scaling up from base model to large model (+0.7% for *k*MaX-DeepLab but +1.4% for Mask2Former), indicating *k*MaX-DeepLab's strong representation ability and that it may overfit on COCO *train* set with the largest backbone. Therefore, we additionally perform a simple experiment to alleviate the over-fitting issue by generating pseudo labels [12] on COCO *unlabeled* set. Adding pseudo labels to the training data slightly improves *k*MaX-DeepLab, yielding a PQ score of **58.1%** (the drop query regularization is not used here and the number of object query remains 128).

Cityscapes *val* Set.   In Table 2, we compare our kMaX-DeepLab with other state-of-art methods on Cityscapes *val* set. Our reported PQ, AP, and mIoU results use the same panoptic model to provide a comprehensive comparison. Notably, kMaX-DeepLab with ResNet-50 backbone already surpasses most baselines, while being more efficient. For example, kMaX-DeepLab with ResNet-50 achieves **1.3%** PQ higher performance compared to Panoptic-DeepLab [22] (Xception-71 [27] backbone) with **20%** computational cost (FLOPs) reduced. Moreover, it achieves a similar performance to Axial-DeepLab-XL [93], while using **3.1×** fewer parameters and **5.6×** fewer FLOPs. kMaX-DeepLab achieves even higher performances with stronger backbones. Specifically, with MaX-S backbone, it performs on par with previous state-of-the-art Panoptic-DeepLab with SWideRNet [16] backbone, while using **7.2×** fewer parameters and **17.2×** fewer FLOPs. Additionally, even only trained with panoptic annotations, our kMaX-DeepLab also shows superior performance in instance segmentation (AP) and semantic segmentation (mIoU). Finally, we provide a comparison with the recent work Mask2Former [23], where the advantage of our kMaX-DeepLab becomes even more significant. Using the ResNet-50 backbone for a fair comparison, kMaX-DeepLab achieves **2.2%** PQ, **1.2%** AP, and **2.2%** mIoU higher performance than Mask2Former. For other backbone variants with a similar size, kMaX-DeepLab with ConvNeXt-B is **1.9%** PQ higher than Mask2Former with Swin-B (window size 12). Notably, kMaX-DeepLab with ConvNeXt-B already obtains a PQ score that is **1.4%** higher than Mask2Former with their best backbone. With ConvNeXt-L as backbone, kMaX-DeepLab sets a new state-of-the-art record of 68.4% PQ without any test-time augmentation or COCO [66]/Mapillary Vistas [75] pretraining.

Fig. 4. Visualization of kMaX-DeepLab (ResNet-50) pixel-cluster assignments at each kMaX decoder stage, along with the final panoptic prediction. In the cluster assignment visualization, pixels with same color are assigned to the same cluster and their features will be aggregated for updating corresponding cluster centers

Visualizations.   In Fig. 4, we provide a visualization of pixel-cluster assignments at each kMaX decoder and final prediction, to better understand the working mechanism behind kMaX-DeepLab. Another benefit of kMaX-DeepLab

Table 2. Cityscapes *val* set results. We only consider methods without extra data [66, 75] and test-time augmentation for a fair comparison. We evaluate FLOPs and FPS with the input size 1025×2049 and a Tesla V100-SXM2 GPU. Our instance (AP) and semantic (mIoU) results are based on the same panoptic model (*i.e.*, no task-specific fine-tuning). †: ImageNet-22K pretraining

method	backbone	params	FLOPs	FPS	PQ	AP	mIoU
Panoptic-DeepLab [22]	Xception-71 [27]	47M	548G	5.7	63.0	35.3	80.5
Axial-DeepLab [93]	Axial-ResNet-L [93]	45M	687G	–	63.9	35.8	81.0
Axial-DeepLab [93]	Axial-ResNet-XL [93]	173M	2447G	–	64.4	36.7	80.6
CMT-DeepLab [104]	MaX-S [92]	–	–	–	64.6	–	81.4
Panoptic-DeepLab [22]	SWideRNet-(1,1,4.5) [16]	536M	10365G	1.0	66.4	40.1	82.2
Mask2Former [23]	ResNet-50 [41]	–	–	–	62.1	37.3	77.5
Mask2Former [23]	Swin-B (W12)† [70]	–	–	–	66.1	42.8	82.7
Mask2Former [23]	Swin-L (W12)† [70]	–	–	–	66.6	43.6	82.9
SETR [109]	ViT-L† [32]	–	–	–	–	–	79.3
SegFormer [99]	MiT-B5 [99]	85M	1460G	2.5	–	–	82.4
Mask R-CNN [40]	ResNet-50 [41]	–	–	–	–	31.5	-
PANet [68]	ResNet-50 [41]	–	–	–	–	36.5	–
kMaX-DeepLab	ResNet-50 [41]	56M	434G	9.0	64.3	38.5	79.7
kMaX-DeepLab	MaX-S† [92]	74M	602G	6.5	66.4	41.6	82.1
kMaX-DeepLab	ConvNeXt-B† [71]	121M	858G	5.2	68.0	43.0	83.1
kMaX-DeepLab	ConvNeXt-L† [71]	232M	1673G	3.1	**68.4**	**44.0**	**83.5**

is that with the cluster-wise argmax, visualizations can be directly drawn as segmentation masks, as the pixel-cluster assignments are exclusive to each other with cluster-wise argmax. Noticeably, the major clustering update happens in the first three stages, which already updates cluster centers well and generates reasonable clustering results, while the following stages mainly focus on refining details. This coincides with our observation that 3 *k*MaX decoders are sufficient to produce good results. Besides, we observe that 1st clustering assignment tends to produce over-segmentation effects, where many clusters are activated and then combined or pruned in the later stages. Moreover, though there exist many fragments in the first round of clustering, it already surprisingly distinguishes different semantics, especially some persons are already well clustered, which indicates that the initial clustering is not only based on texture or location, but also depends on the underlying semantics. Another visualization is shown in Fig. 5, where we observe that *k*MaX-DeepLab behaves in a part-to-whole manner to capture an instance. More experimental results (*e.g.*, ablation studies, test set results) and visualizations are available in the appendix.

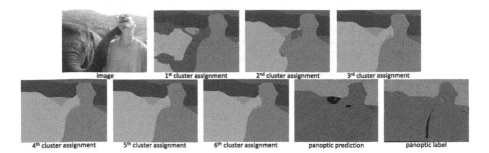

Fig. 5. Visualization of kMaX-DeepLab (ResNet-50) pixel-cluster assignments at each kMaX decoder stage, along with the final panoptic prediction. kMaX-DeepLab shows a behavior of recognizing objects starting from their parts to their the whole shape in the clustering process. For example, the elephant's top head, body, and nose are separately clustered at the beginning, and they are gradually merged in the following stages

5   Conclusion

In this work, we have presented a novel end-to-end framework, called k-means Mask Transformer (kMaX-DeepLab), for segmentation tasks. kMaX-DeepLab rethinks the relationship between pixel features and object queries from the clustering perspective. Consequently, it simplifies the mask-transformer model by replacing the multi-head cross attention with the proposed single-head k-means clustering. We have tailored the transformer-based model for segmentation tasks by establishing the link between the traditional k-means clustering algorithm and cross-attention. We hope our work will inspire the community to develop more vision-specific transformer models.

Acknowledgments. We thank Jun Xie for the valuable feedback on the draft. This work was supported in part by ONR N00014-21-1-2812.

References

1. Achanta, R., Shaji, A., Smith, K., Lucchi, A., Fua, P., Süsstrunk, S.: Slic superpixels compared to state-of-the-art superpixel methods. In: IEEE TPAMI (2012)
2. Ainslie, J., Ontanon, S., Alberti, C., Pham, P., Ravula, A., Sanghai, S.: Etc: Encoding long and structured data in transformers. In: EMNLP (2020)
3. Arnab, A., Dehghani, M., Heigold, G., Sun, C., Lučić, M., Schmid, C.: Vivit: A video vision transformer. In: ICCV (2021)
4. Bahdanau, D., Cho, K., Bengio, Y.: Neural machine translation by jointly learning to align and translate. In: ICLR (2015)
5. Bai, M., Urtasun, R.: Deep watershed transform for instance segmentation. In: CVPR (2017)
6. Ballard, D.H.: Generalizing the hough transform to detect arbitrary shapes. In: Pattern Recognition (1981)

7. Bello, I., Zoph, B., Vaswani, A., Shlens, J., Le, Q.V.: Attention augmented convolutional networks. In: ICCV (2019)
8. Beltagy, I., Peters, M.E., Cohan, A.: Longformer: The long-document transformer. arXiv:2004.05150 (2020)
9. Buades, A., Coll, B., Morel, J.M.: A non-local algorithm for image denoising. In: CVPR (2005)
10. Carion, N., Massa, F., Synnaeve, G., Usunier, N., Kirillov, A., Zagoruyko, S.: End-to-end object detection with transformers. In: Vedaldi, A., Bischof, H., Brox, T., Frahm, J.-M. (eds.) ECCV 2020. LNCS, vol. 12346, pp. 213–229. Springer, Cham (2020). https://doi.org/10.1007/978-3-030-58452-8_13
11. Chen, J., et al.: Transunet: Transformers make strong encoders for medical image segmentation. arXiv:2102.04306 (2021)
12. Chen, L.-C., et al.: Naive-student: leveraging semi-supervised learning in video sequences for urban scene segmentation. In: Vedaldi, A., Bischof, H., Brox, T., Frahm, J.-M. (eds.) ECCV 2020. LNCS, vol. 12354, pp. 695–714. Springer, Cham (2020). https://doi.org/10.1007/978-3-030-58545-7_40
13. Chen, L.C., Papandreou, G., Kokkinos, I., Murphy, K., Yuille, A.L.: Semantic image segmentation with deep convolutional nets and fully connected crfs. In: ICLR (2015)
14. Chen, L.C., Papandreou, G., Kokkinos, I., Murphy, K., Yuille, A.L.: Deeplab: Semantic image segmentation with deep convolutional nets, atrous convolution, and fully connected crfs. In: IEEE TPAMI (2017)
15. Chen, L.C., Papandreou, G., Schroff, F., Adam, H.: Rethinking atrous convolution for semantic image segmentation. arXiv:1706.05587 (2017)
16. Chen, L.C., Wang, H., Qiao, S.: Scaling wide residual networks for panoptic segmentation. arXiv:2011.11675 (2020)
17. Chen, L.C., Yang, Y., Wang, J., Xu, W., Yuille, A.L.: Attention to scale: Scale-aware semantic image segmentation. In: CVPR (2016)
18. Chen, L.-C., Zhu, Y., Papandreou, G., Schroff, F., Adam, H.: Encoder-decoder with atrous separable convolution for semantic image segmentation. In: Ferrari, V., Hebert, M., Sminchisescu, C., Weiss, Y. (eds.) ECCV 2018. LNCS, vol. 11211, pp. 833–851. Springer, Cham (2018). https://doi.org/10.1007/978-3-030-01234-2_49
19. Chen, Y., Kalantidis, Y., Li, J., Yan, S., Feng, J.: Aˆ 2-nets: Double attention networks. In: NeurIPS (2018)
20. Cheng, B., Choudhuri, A., Misra, I., Kirillov, A., Girdhar, R., Schwing, A.G.: Mask2former for video instance segmentation. arXiv:2112.10764 (2021)
21. Cheng, B., et al.: Panoptic-DeepLab. In: ICCV COCO + Mapillary Joint Recognition Challenge Workshop (2019)
22. Cheng, B., et al.: Panoptic-DeepLab: A Simple, Strong, and Fast Baseline for Bottom-Up Panoptic Segmentation. In: CVPR (2020)
23. Cheng, B., Misra, I., Schwing, A.G., Kirillov, A., Girdhar, R.: Masked-attention mask transformer for universal image segmentation. In: CVPR (2022)
24. Cheng, B., Schwing, A.G., Kirillov, A.: Per-pixel classification is not all you need for semantic segmentation. In: NeurIPS (2021)
25. Cheng, J., Dong, L., Lapata, M.: Long short-term memory-networks for machine reading. In: EMNLP (2016)
26. Child, R., Gray, S., Radford, A., Sutskever, I.: Generating long sequences with sparse transformers. arXiv:1904.10509 (2019)
27. Chollet, F.: Xception: Deep learning with depthwise separable convolutions. In: CVPR (2017)

28. Cordts, M., et al.: The cityscapes dataset for semantic urban scene understanding. In: CVPR (2016)
29. Cubuk, E.D., Zoph, B., Mane, D., Vasudevan, V., Le, Q.V.: Autoaugment: Learning augmentation policies from data. In: CVPR (2019)
30. Dai, Z., Yang, Z., Yang, Y., Carbonell, J.G., Le, Q., Salakhutdinov, R.: Transformer-xl: Attentive language models beyond a fixed-length context. In: ACL (2019)
31. Devlin, J., Chang, M.W., Lee, K., Toutanova, K.: BERT: Pre-training of deep bidirectional transformers for language understanding. In: NAACL (2019)
32. Dosovitskiy, A., et al.: An image is worth 16×16 words: Transformers for image recognition at scale. In: ICLR (2021)
33. Fan, H., et al.: Multiscale vision transformers. In: ICCV (2021)
34. Fang, H.S., Sun, J., Wang, R., Gou, M., Li, Y.L., Lu, C.: Instaboost: Boosting instance segmentation via probability map guided copy-pasting. In: ICCV (2019)
35. Fu, J., et al.: Dual attention network for scene segmentation. In: CVPR (2019)
36. Gao, N., et al.: Ssap: Single-shot instance segmentation with affinity pyramid. In: ICCV (2019)
37. Gao, P., Zheng, M., Wang, X., Dai, J., Li, H.: Fast convergence of detr with spatially modulated co-attention. In: ICCV (2021)
38. Ghiasi, G., et al.: Simple copy-paste is a strong data augmentation method for instance segmentation. In: CVPR (2021)
39. Gupta, A., Berant, J.: Gmat: Global memory augmentation for transformers. arXiv:2006.03274 (2020)
40. He, K., Gkioxari, G., Dollár, P., Girshick, R.: Mask r-cnn. In: ICCV (2017)
41. He, K., Zhang, X., Ren, S., Sun, J.: Deep residual learning for image recognition. In: CVPR (2016)
42. Ho, J., Kalchbrenner, N., Weissenborn, D., Salimans, T.: Axial attention in multidimensional transformers. arXiv:1912.12180 (2019)
43. Hu, H., Gu, J., Zhang, Z., Dai, J., Wei, Y.: Relation networks for object detection. In: CVPR (2018)
44. Hu, H., Zhang, Z., Xie, Z., Lin, S.: Local relation networks for image recognition. In: ICCV (2019)
45. Huang, G., Sun, Yu., Liu, Z., Sedra, D., Weinberger, K.Q.: Deep networks with stochastic depth. In: Leibe, B., Matas, J., Sebe, N., Welling, M. (eds.) ECCV 2016. LNCS, vol. 9908, pp. 646–661. Springer, Cham (2016). https://doi.org/10.1007/978-3-319-46493-0_39
46. Huang, Z., Wang, X., Huang, L., Huang, C., Wei, Y., Liu, W.: Ccnet: Criss-cross attention for semantic segmentation. In: ICCV (2019)
47. Hwang, J.J., et al.: SegSort: Segmentation by discriminative sorting of segments. In: ICCV (2019)
48. Jang, E., Gu, S., Poole, B.: Categorical reparameterization with gumbel-softmax. In: ICLR (2017)
49. Jia, X., De Brabandere, B., Tuytelaars, T., Gool, L.V.: Dynamic filter networks. In: NeurIPS (2016)
50. Kendall, A., Gal, Y., Cipolla, R.: Multi-task learning using uncertainty to weigh losses for scene geometry and semantics. In: CVPR (2018)
51. Keuper, M., Levinkov, E., Bonneel, N., Lavoué, G., Brox, T., Andres, B.: Efficient decomposition of image and mesh graphs by lifted multicuts. In: ICCV (2015)
52. Kim, D., et al.: TubeFormer-DeepLab: Video Mask Transformer. In: CVPR (2022)
53. Kingma, D.P., Ba, J.: Adam: A method for stochastic optimization. In: ICLR (2015)

54. Kirillov, A., Girshick, R., He, K., Dollár, P.: Panoptic feature pyramid networks. In: CVPR (2019)
55. Kirillov, A., He, K., Girshick, R., Rother, C., Dollár, P.: Panoptic segmentation. In: CVPR (2019)
56. Kirillov, A., Wu, Y., He, K., Girshick, R.: Pointrend: Image segmentation as rendering. In: CVPR (2020)
57. Kitaev, N., Kaiser, Ł., Levskaya, A.: Reformer: The efficient transformer. In: ICLR (2020)
58. LeCun, Y., Bottou, L., Bengio, Y., Haffner, P.: Gradient-based learning applied to document recognition. Proc. IEEE **86**(11), 2278–2324 (1998)
59. Leibe, B., Leonardis, A., Schiele, B.: Combined object categorization and segmentation with an implicit shape model. In: Workshop on statistical learning in computer vision, ECCV (2004)
60. Li, Q., Qi, X., Torr, P.H.: Unifying training and inference for panoptic segmentation. In: CVPR (2020)
61. Li, X., et al.: Video k-net: A simple, strong, and unified baseline for video segmentation. In: CVPR (2022)
62. Li, Y., et al.: Attention-guided unified network for panoptic segmentation. In: CVPR (2019)
63. Li, Y., et al.: Fully convolutional networks for panoptic segmentation. In: CVPR (2021)
64. Li, Y., et al.: Neural architecture search for lightweight non-local networks. In: CVPR (2020)
65. Li, Z., et al.: Panoptic segformer. In: CVPR (2022)
66. Lin, T.-Y., et al.: Microsoft COCO: common objects in context. In: Fleet, D., Pajdla, T., Schiele, B., Tuytelaars, T. (eds.) ECCV 2014. LNCS, vol. 8693, pp. 740–755. Springer, Cham (2014). https://doi.org/10.1007/978-3-319-10602-1_48
67. Liu, H., et al.: An end-to-end network for panoptic segmentation. In: CVPR (2019)
68. Liu, S., Qi, L., Qin, H., Shi, J., Jia, J.: Path aggregation network for instance segmentation. In: CVPR (2018)
69. Liu, Y., Yang, S., Li, B., Zhou, W., Xu, J., Li, H., Lu, Y.: Affinity derivation and graph merge for instance segmentation. In: Ferrari, V., Hebert, M., Sminchisescu, C., Weiss, Y. (eds.) ECCV 2018. LNCS, vol. 11207, pp. 708–724. Springer, Cham (2018). https://doi.org/10.1007/978-3-030-01219-9_42
70. Liu, Z., et al.: Swin transformer: Hierarchical vision transformer using shifted windows. In: ICCV (2021)
71. Liu, Z., Mao, H., Wu, C.Y., Feichtenhofer, C., Darrell, T., Xie, S.: A convnet for the 2020s. In: CVPR (2022)
72. Lloyd, S.: Least squares quantization in pcm. IEEE Trans. Inf. Theory **28**(2), 129–137 (1982)
73. Loshchilov, I., Hutter, F.: Decoupled weight decay regularization. In: ICLR (2019)
74. Luong, M.T., Pham, H., Manning, C.D.: Effective approaches to attention-based neural machine translation. In: EMNLP (2015)
75. Neuhold, G., Ollmann, T., Rota Bulo, S., Kontschieder, P.: The mapillary vistas dataset for semantic understanding of street scenes. In: ICCV (2017)
76. Neven, D., Brabandere, B.D., Proesmans, M., Gool, L.V.: Instance segmentation by jointly optimizing spatial embeddings and clustering bandwidth. In: CVPR (2019)
77. Parmar, N., et al.: Image transformer. In: ICML (2018)

78. Porzi, L., Bulò, S.R., Colovic, A., Kontschieder, P.: Seamless scene segmentation. In: CVPR (2019)
79. Qiao, S., Chen, L.C., Yuille, A.: Detectors: Detecting objects with recursive feature pyramid and switchable atrous convolution. In: CVPR (2021)
80. Ramachandran, P., Parmar, N., Vaswani, A., Bello, I., Levskaya, A., Shlens, J.: Stand-alone self-attention in vision models. In: NeurIPS (2019)
81. Russakovsky, O., et al.: Imagenet large scale visual recognition challenge. IJCV **115**, 211–252 (2015)
82. Shaw, P., Uszkoreit, J., Vaswani, A.: Self-attention with relative position representations. In: NAACL (2018)
83. Shen, Z., Zhang, M., Zhao, H., Yi, S., Li, H.: Efficient attention: Attention with linear complexities. In: WACV (2021)
84. Sofiiuk, K., Barinova, O., Konushin, A.: Adaptis: Adaptive instance selection network. In: ICCV (2019)
85. Strudel, R., Garcia, R., Laptev, I., Schmid, C.: Segmenter: Transformer for semantic segmentation. In: ICCV (2021)
86. Sutskever, I., Vinyals, O., Le, Q.V.: Sequence to sequence learning with neural networks. In: NeurIPS (2014)
87. Tian, Z., Shen, C., Chen, H.: Conditional convolutions for instance segmentation. In: Vedaldi, A., Bischof, H., Brox, T., Frahm, J.-M. (eds.) ECCV 2020. LNCS, vol. 12346, pp. 282–298. Springer, Cham (2020). https://doi.org/10.1007/978-3-030-58452-8_17
88. Uhrig, J., Rehder, E., Fröhlich, B., Franke, U., Brox, T.: Box2pix: Single-shot instance segmentation by assigning pixels to object boxes. In: IEEE Intelligent Vehicles Symposium (IV) (2018)
89. Vaswani, A., et al.: Attention is all you need. In: NeurIPS (2017)
90. Vincent, L., Soille, P.: Watersheds in digital spaces: an efficient algorithm based on immersion simulations. In: IEEE TPAMI (1991)
91. Wang, H., Luo, R., Maire, M., Shakhnarovich, G.: Pixel consensus voting for panoptic segmentation. In: CVPR (2020)
92. Wang, H., Zhu, Y., Adam, H., Yuille, A., Chen, L.C.: Max-deeplab: End-to-end panoptic segmentation with mask transformers. In: CVPR (2021)
93. Wang, H., Zhu, Y., Green, B., Adam, H., Yuille, A., Chen, L.-C.: Axial-DeepLab: stand-alone axial-attention for panoptic segmentation. In: Vedaldi, A., Bischof, H., Brox, T., Frahm, J.-M. (eds.) ECCV 2020. LNCS, vol. 12349, pp. 108–126. Springer, Cham (2020). https://doi.org/10.1007/978-3-030-58548-8_7
94. Wang, S., Li, B., Khabsa, M., Fang, H., Ma, H.: Linformer: Self-attention with linear complexity. arXiv:2006.04768 (2020)
95. Wang, W., et al.: Pvtv 2: Improved baselines with pyramid vision transformer. arXiv:2106.13797 (2021)
96. Wang, X., Girshick, R., Gupta, A., He, K.: Non-local neural networks. In: CVPR (2018)
97. Wang, X., Zhang, R., Kong, T., Li, L., Shen, C.: SOLOv2: Dynamic and fast instance segmentation. In: NeurIPS (2020)
98. Weber, M., et al.: DeepLab2: A TensorFlow Library for Deep Labeling. arXiv: 2106.09748 (2021)
99. Xie, E., Wang, W., Yu, Z., Anandkumar, A., Alvarez, J.M., Luo, P.: Segformer: Simple and efficient design for semantic segmentation with transformers. In: NeurIPS (2021)
100. Xiong, Y., et al.: Upsnet: A unified panoptic segmentation network. In: CVPR (2019)

101. Yang, C., et al.: Lite vision transformer with enhanced self-attention. In: CVPR (2022)
102. Yang, T.J., et al.: Deeperlab: Single-shot image parser. arXiv:1902.05093 (2019)
103. Yang, Y., Li, H., Li, X., Zhao, Q., Wu, J., Lin, Z.: Sognet: Scene overlap graph network for panoptic segmentation. In: AAAI (2020)
104. Yu, Q., et al.: Cmt-deeplab: Clustering mask transformers for panoptic segmentation. In: CVPR (2022)
105. Yu, Q., Xia, Y., Bai, Y., Lu, Y., Yuille, A.L., Shen, W.: Glance-and-gaze vision transformer. In: NeurIPS (2021)
106. Zaheer, M., et al.: Big bird: Transformers for longer sequences. In: NeurIPS (2020)
107. Zhang, W., Pang, J., Chen, K., Loy, C.C.: K-net: Towards unified image segmentation. In: NeurIPS (2021)
108. Zhao, H., et al.: PSANet: Point-wise spatial attention network for scene parsing. In: Ferrari, V., Hebert, M., Sminchisescu, C., Weiss, Y. (eds.) ECCV 2018. LNCS, vol. 11213, pp. 270–286. Springer, Cham (2018). https://doi.org/10.1007/978-3-030-01240-3_17
109. Zheng, S., et al.: Rethinking semantic segmentation from a sequence-to-sequence perspective with transformers. In: CVPR (2021)
110. Zhu, S.C., Yuille, A.: Region competition: Unifying snakes, region growing, and bayes/mdl for multiband image segmentation. In: IEEE TPAMI (1996)
111. Zhu, X., Cheng, D., Zhang, Z., Lin, S., Dai, J.: An empirical study of spatial attention mechanisms in deep networks. In: ICCV (2019)
112. Zhu, X., Su, W., Lu, L., Li, B., Wang, X., Dai, J.: Deformable detr: Deformable transformers for end-to-end object detection. In: ICLR (2021)
113. Zhu, Z., Xu, M., Bai, S., Huang, T., Bai, X.: Asymmetric non-local neural networks for semantic segmentation. In: CVPR (2019)

SegPGD: An Effective and Efficient Adversarial Attack for Evaluating and Boosting Segmentation Robustness

Jindong Gu[1,3]([✉]), Hengshuang Zhao[2,3], Volker Tresp[1], and Philip H. S. Torr[3]

[1] University of Munich, Munich, Germany
jindong.gu@outlook.com
[2] The University of Hong Kong, Hong Kong, China
[3] Torr Vision Group, University of Oxford, Oxford, UK

Abstract. Deep neural network-based image classifications are vulnerable to adversarial perturbations. The image classifications can be easily fooled by adding artificial small and imperceptible perturbations to input images. As one of the most effective defense strategies, adversarial training was proposed to address the vulnerability of classification models, where the adversarial examples are created and injected into training data during training. The attack and defense of classification models have been intensively studied in past years. Semantic segmentation, as an extension of classifications, has also received great attention recently. Recent work shows a large number of attack iterations are required to create effective adversarial examples to fool segmentation models. The observation makes both robustness evaluation and adversarial training on segmentation models challenging. In this work, we propose an effective and efficient segmentation attack method, dubbed SegPGD. Besides, we provide a convergence analysis to show the proposed SegPGD can create more effective adversarial examples than PGD under the same number of attack iterations. Furthermore, we propose to apply our SegPGD as the underlying attack method for segmentation adversarial training. Since SegPGD can create more effective adversarial examples, the adversarial training with our SegPGD can boost the robustness of segmentation models. Our proposals are also verified with experiments on popular Segmentation model architectures and standard segmentation datasets.

Keywords: Adversarial robustness · Semantic segmentation

1 Introduction

Due to their vulnerability to artificial small perturbations, the adversarial robustness of deep neural networks has received great attention [13,40]. A large amount of attack and defense strategies have been proposed for classification in

Supplementary Information The online version contains supplementary material available at https://doi.org/10.1007/978-3-031-19818-2_18.

past years [1,5,14,34,37,39,43,46,47,52,54,57]. As an extension of classification, semantic segmentation also suffers from adversarial examples [2,50]. Segmentation models applied in real-world safety-critical applications also face potential threats, *e.g.*, in self-driving systems [4,19,25,32,33,36] and in medical image analysis [10,12,30,35]. Hence, the adversarial robustness of segmentation has also raised great attention recently [2,8,20,24,27,38,42,49,49–51,53].

In terms of the attack methods, different from classification, the attack goal in segmentation is to fool all pixel classifications at the same time. An effective adversarial example of a segmentation model are expected to fool as many pixel classifications as possible, which requires the larger number of attack iterations [15,50]. The observation makes both robustness evaluation and adversarial training on segmentation models challenging. In this work, we propose an effective and efficient segmentation attack method, dubbed SegPGD. Besides, we provide a convergence analysis to show why the proposed SegPGD can create more effective adversarial examples than PGD under the same number of attack iterations.

The right evaluation of model robustness is an important step to building robust models. Evaluation with weak or inappropriate attack methods can give a false sense of robustness [3]. Recent work [51] evaluates the robustness of segmentation models under a similar setting to the one used in classification. This could be problematic given the fact that a large number of attack iterations are required to create effective adversarial examples of segmentation [50]. We evaluate the adversarially trained segmentation models in previous work with a strong attack setting, namely with a large number of attack iterations. We found the robustness can be significantly reduced. Our SegPGD can reduce the mIoU score further. For example, the mIoU of adversarially trained PSPNet [56] on Cityscapes dataset [9] can be reduced to near zero under 100 attack iterations.

As one of the most effective defense strategies, adversarial training was proposed to address the vulnerability of classification models, where the adversarial examples are created and injected into training data during training [13,29]. One promising way to boost segmentation robustness is to apply adversarial training to segmentation models. However, the creation of effective segmentation adversarial examples during training can be time-consuming. In this work, we demonstrate that our effective and efficient SegPGD can mitigate this challenge. Since it can create effective adversarial examples, the application of SegPGD as the underlying attack method of adversarial training can effectively boost the robustness of segmentation models. It is worth noting that many adversarial training strategies with single-step attacks have been proposed to address the efficiency of adversarial training in classification [1,37,43,47,57]. However, they do not work well on segmentation models since the adversarial examples created by single-step attacks are not effective enough to fool segmentation models. The contributions of our work can be summarised as follows:

- Based on the difference between classification and segmentation, we propose an effective and efficient segmentation attack method, dubbed SegPGD. Especially, we show its generalization to single-step attack SegFGSM.

- We provide a convergence analysis to show the proposed SegPGD can create more effective adversarial examples than PGD under the same number of attack iterations.
- We apply SegPGD as the underlying attack method for segmentation adversarial training. The adversarial training with our SegPGD achieves state-of-the-art performance on the benchmark.
- We conduct experiments with popular segmentation model structures (*i.e.,* PSPNet and DeepLabV3) on standard segmentation datasets (*i.e.,* PASCAL VOC and Cityscapes) to demonstrate the effectiveness of our proposals.

2   Related Work

Adversarial Robustness of Segmentation Models. The work [2] makes an extensive study on the adversarial robustness of segmentation models and demonstrates the inherent robustness of standard segmentation models. Especially, they find that adversarial examples in segmentation do not transfer well across different scales and transformations. Another work [50] also found that the adversarial examples created by their attack method do not transfer well across different network structures. The observations in the two works [2,50] indicate the standard segmentation models are inherently robust to transfer-based black-box method. The belief is broken by the work [15] where they propose a method to improve the transferability of adversarial examples and show the feasibility of transfer-based black-box method. In addition, the adversarial robustness of segmentation models has also been studied from other perspectives, such as universal adversarial perturbation [20,23], adversarial example detection [49], and backdoor attack [28]. Theses works also imply the necessity of building robust segmentation models to defend against potential threats. Along this direction, the work [25] shows self-supervised learning with more data can improve the robustness of standard models. However, the obtained model can be easily completely fooled with a strong attack [25]. A recent work [51] makes the first exploration to apply adversarial training to segmentation models. We find that the adversarially trained models is still vulnerable under strong attacks. The robust accuracy of their adversarial trained models can be significantly reduced under PGD with a large number of attack iterations. In this work, we propose an effective and efficient segmentation attack method, which be used in adversarial training to build robust segmentation models against strong attacks.

Adversarial Training of Classification Models. Adversarial training has been intensively studied on classification models [13,29]. When a multi-step attack is applied to create adversarial examples for adversarial training, the obtained model is indeed robust against various attack to some extent, as shown in [5,29,46,52]. However, adversarial training with multi-step attack can be very time consuming due to the adversarial example creation, which is N times longer than standard natural training [29,37]. To accelerate the adversarial training, single-step attack has also been explored therein. When standard single-step attack is applied during training, the obtained model is only robust to single-step attack [41]. One reason behind is that the gradient masking phenomenon of the model can be observed on the adversarial examples created by single-step attack. Besides, another challenge to apply single-step attack in adversarial training is the label leaking problem where the model show higher robust accuracy against single-step attack than clean accuracy [26]. The low defensive effectiveness of single-step attack and the low efficiency of multi-step attack pose a dilemma.

One way to address the dilemma is to overcome the challenges using advanced single-step attacks [21,22,41,44,46,47,55], which can address label leaking problem and avoid gradient masking phenomenon. Though it boosts the robustness of the classification models, however, single-step attack based adversarial training does work well on segmentation model due to the challenge to create effective segmentation adversarial examples with a single-step attack. Another way to address the dilemma is to simulate the robustness performance of multi-step attack-based adversarial training in an efficient way [5,37,57]. However, it is not clear how well the generalization of the methods above to segmentation is.

3   SegPGD for Evaluating and Boosting Segmentation

In semantic segmentation, given the segmentation model $f_{seg}(\cdot)$, the clean image $X^{clean} \in \mathbb{R}^{H \times W \times C}$ and its segmentation label $Y \in \mathbb{R}^{H \times W \times M}$, the segmentation model classifies all individual pixels of the input image $f_{seg}(X^{clean}) \in \mathbb{R}^{H \times W \times M}$. The notation (H, W) corresponds to the size of input image, C is the number of image channels, and M stands for the number of output classes. The goal of the attack is to create an adversarial example to mislead classifications of all pixels of an input image.

3.1   SegPGD: An Effective and Efficient Segmentation Attack

Formally, the goal of attack is defined to create the adversarial example X^{adv} to mislead all the pixel classifications of an image X^{clean}, i.e., $argmax(f_{seg}(X^{adv})_i) \neq argmax(Y_i)$ where $i \in [1, H \times W]$ corresponds to the index of a input pixel. One of the most popular attack method PGD [29] creates adversarial examples via multiple iterations in Eq. 1.

$$X^{adv_{t+1}} = \phi^{\epsilon}(X^{adv_t} + \alpha * sign(\nabla_{X^{adv_t}} L(f(X^{adv_t}), Y))), \tag{1}$$

where α, ϵ are the step size and the perturbation range, respectively. X^{adv_t} is the adversarial example after the t-th attack step, and the initial value is set to $X^{adv_0} = X^{clean} + \mathcal{U}(-\epsilon, +\epsilon)$, which corresponds to perturbation random initialization. The $\phi^{\epsilon}(\cdot)$ function clips its output into the range $[X^{clean} - \epsilon, X^{clean} + \epsilon]$. Besides, X^{adv_t} is always clipped into a valid image space. $sign(\cdot)$ is the sign function and $\nabla_a(b)$ is the matrix derivative of b with respect to a. $L(\cdot)$ stands for the cross-entropy loss function. In segmentation, the loss is

$$L(f_{seg}(X^{adv_t}), Y) = \frac{1}{H \times W} \sum_{i=1}^{H \times W} CE(f_{seg}(X^{adv_t})_i, Y_i) = \frac{1}{H \times W} \sum_{i=1}^{H \times W} L_i. \tag{2}$$

We reformulate the loss function into two parts in Eq. 3. The first term therein is the loss of the correctly classified pixels, while the second one is formed by the wrongly classified pixels.

$$L(f_{seg}(\boldsymbol{X}^{adv_t}), \boldsymbol{Y}) = \frac{1}{H \times W} \sum_{j \in P^T} L_j + \frac{1}{H \times W} \sum_{k \in P^F} L_k, \qquad (3)$$

where P^T is the set of correctly classified pixels, P^F corresponds to wrongly classified ones. The two sets make up all pixels, $i.e.$, $\#P^T + \#P^F = H \times W$.

The loss of the second term is often large since the wrongly classified pixels lead to large cross-entropy loss. When creating adversarial examples, the gradient of the second loss term can dominate. However, the increase of the second-term loss does not lead to better adversarial effect since the involved pixels have already been wrongly classified. To achieve high effective adversarial examples on segmentation, a large number of attack iterations are required so that the update towards increasing the first-term loss can be accumulated to mislead correctly classified pixels.

To tackle the issue above, considering the dense pixel classifications in segmentation, we propose the **Segmentation-specific PGD**, dubbed **SegPGD**, which can create more effective adversarial examples with the same number of attack iterations in Eq. 4.

$$L(f_{seg}(\boldsymbol{X}^{adv_t}), \boldsymbol{Y}) = \frac{1-\lambda}{H \times W} \sum_{j \subset P^T} L_j + \frac{\lambda}{H \times W} \sum_{k \in P^F} L_k, \qquad (4)$$

where two loss terms are weighted with $1 - \lambda$ and λ, respectively. Note that the selection of λ is non-trivial. It does not work well by simply setting $\lambda = 0$ where only correctly classified pixels are considered. In such a case, the previous wrongly classified pixels can become benign again after a few attack iterations since they are ignored when updating perturbations. The claim is also consistent with the previous observation [45, 48] that adversarial perturbation is also sensitive to small noise. Furthermore, setting λ to a fixed value in $[0, 0.5]$ does not always lead to better attack performance due to a similar reason. When most of pixel classifications are fooled after a few attack iterations, less weight on the wrongly classified pixels can make some of them benign again.

In this work, instead of manually specifying a fixed value to λ, we propose to set λ dynamically with the number of attack iterations. The intuition behind the dynamic schedule is that we mainly focus on fooling correct pixel classifications in the first a few attack iterations and then treat the wrong pixel classifications quasi equally in the last few iterations. By doing this, our SegPGD can achieve similar attack effectiveness with less iterations. We list some instances of our dynamic schedule as follows

$$\lambda(t) = \frac{t-1}{2T}, \qquad \lambda(t) = \frac{1}{2} * \log_2(1 + \frac{t-1}{T}), \qquad \lambda(t) = \frac{1}{2} * (2^{(t-1)/T} - 1), \qquad (5)$$

where t is the index of current attack iteration and T are the number of all attack iterations. Our experiments show that all the proposed instances are similarly

---

Algorithm 1. SegPGD: An Efficient and Effective Segmentation Attack

---

Require: segmentation model $f_{seg}(\cdot)$, clean samples $\boldsymbol{X}^{clean}$, perturbation range ϵ, step size α, attack iterations T

$\boldsymbol{X}^{adv_0} = \boldsymbol{X}^{clean} + \mathcal{U}(-\epsilon, +\epsilon)$ ▷ initialize adversarial example

for t ← 1 to T **do**   ▷ loop over attack iterations

    $P = f_{seg}(\boldsymbol{X}^{clean})$ ▷ make predictions

    $P^T, P^F \leftarrow P$ ▷ split predictions

    $\lambda(t) \leftarrow (t-1)/2T$ ▷ compute weight

    $L \leftarrow \lambda(t) * L(P^T, \boldsymbol{Y}) + (1-\lambda(t)) * L(P^F, \boldsymbol{Y})$ ▷ loss for example updates

    $\boldsymbol{X}^{adv_{t+1}} \leftarrow \boldsymbol{X}^{adv_t} + \alpha * sign(\nabla_{\boldsymbol{X}^{adv_t}} L)$ ▷ update adversarial examples

    $\boldsymbol{X}^{adv_{t+1}} \leftarrow \phi^\epsilon(\boldsymbol{X}^{adv_{t+1}})$ ▷ clip into ϵ-ball of clean image

end for

---

effective. In this work, we mainly use the first simple linear schedule. The pseudo code of our SegPGD with the proposed schedule is shown in Algorithm 1. Further discussion on the schedules to dynamically set λ are in Sect. 4.2.

Similarly, the loss function in Eq. 4 can also be applied in single-step adversarial attack, *e.g.*, FGSM [13]. In the resulted SegFGSM, only correctly classified pixels are considered in case of the proposed λ schedule. Since it only takes one-step update, the wrongly classified pixels is less likely to become benign. Hence, SegFGSM with the proposed λ schedule (*i.e.*, $\lambda = 1$) also shows superior attack performance than FGSM.

In this subsection, we propose a fast segmentation attack method, *i.e.*, SegPGD. It can be applied to evaluate the adversarial robustness of segmentation models in an efficient way. Besides, SegPGD can also be applied to accelerate the adversarial training on segmentation models.

3.2 Convergence Analysis of SegPGD

Problem Formulation. The goal of the attack is to create an adversarial example $\boldsymbol{X}^{adv}$ to maximize cross-entropy loss of all the pixel classifications. The adversarial example is constrained into ϵ-ball of the clean example $\boldsymbol{X}^{clean}$. The cross-entropy loss of i-th pixel is

$$L(\boldsymbol{X}, \boldsymbol{Y}_i) = CE(f_{seg}(\boldsymbol{X}^{adv_t})_i, \boldsymbol{Y}_i). \tag{6}$$

The process to create adversarial example for segmentation can be formulated into a constrained minimization problem

$$\min_{\boldsymbol{X}} \frac{1}{H \times W} \sum_{i=1}^{H \times W} g_i(\boldsymbol{X}) \quad s.t. \ \|\boldsymbol{X} - \boldsymbol{X}^{clean}\|_\infty < \epsilon \text{ and } \boldsymbol{X} \in [0,1], \tag{7}$$

where $g_i(\boldsymbol{X}) = -L(\boldsymbol{X}, \boldsymbol{Y}_i)$. The variable is constrained into concave region since both constraints are linear.

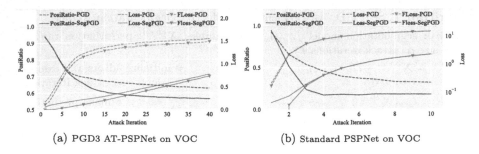

(a) PGD3 AT-PSPNet on VOC   (b) Standard PSPNet on VOC

Fig. 1. Convergence Analysis. SegPGD marked with blue solid lines achieve higher MisRatio than PGD under the same number of attack iterations. The loss of false classified pixels (**FLoss**) marked with triangle down dominate the overall loss (i.e. red lines without markers) during attacks. Compared to PGD, the FLoss in SegPDG makes up a smaller portion of the overall loss since SegPGD main focuses on correctly classified pixels in the first a few attack iterations. (Color figure online)

Projected Gradient Descent-based optimization method is often applied to solve the constrained minimization problem above [29]. The method first takes a step towards the negative gradient direction to get a new point while ignoring the constraint, and then correct the new point by projecting it back into the constraint set.

The gradient-descent step of PGD attack is

$$X^{t+1} = X^t - \alpha * sign(\nabla \sum_{i=1}^{H \times W} g_i(X^t)), \tag{8}$$

In contrast, the gradient-descent step of our SegPGD attack is

$$X^{t+1} = X^t - \alpha * sign(\nabla(\sum_{j \in P^T} (1 - \lambda(t))g_j(X^t) + \sum_{k \in P^F} \lambda(t)g_k(X^t))), \tag{9}$$

where α is the step size. The initial point is the original clean example X^{clean} or a random initialization $X^{clean} + \mathcal{U}(-\epsilon, +\epsilon)$.

Convergence Criterion. In classification task, the loss is directly correlated with attack goal. The larger the loss is, the more likely the input is to be misclassified. However, it does not hold in segmentation task. The large loss of segmentation not necessarily leads to more pixel misclassifications since the loss consists of losses of all pixel classifications. Once a pixel is misclassified, the increase of the loss on the pixel does not bring more adversarial effect. Hence, we propose a new convergence criterion for segmentation, dubbed MisRatio, which is defined as the ratio of misclassified pixels to all input pixels.

Convergence Analysis. In the first step to update adversarial examples, the update rule of our SegPGD can be simplified as

$$X^1 = X^0 + \alpha * sign(\sum_{j \in P^T} \nabla g_j(X^t)), \tag{10}$$

Algorithm 2. Segmentation Adversarial Training with SegPGD

Require: segmentation model $f_{seg}(\cdot)$, training iterations $\mathcal{N}$, perturbation range ϵ, step size α, attack iterations T

for i $\leftarrow$ 1 to $\mathcal{N}$ do

    X_1^{clean}, $X_2^{clean} \leftarrow X^{clean}$ ▷ split mini-batch

    $X_2^{adv} \leftarrow$ SegPGD$(f_{seg}(\cdot), X_2^{clean}, \epsilon, \alpha, i)$ ▷ create adversarial examples

    $L \leftarrow L(f_{seg}(X_1^{clean}), Y_1) + L(f_{seg}(X_2^{adv}), Y_2)$ ▷ loss for network updates

end for

For almost all misclassified pixels $k \in P^F$ of X^0, the k-th pixel of X^1 is still misclassified since natural misclassifications are not sensitive to small adversarial noise in general. The claim is also true with PGD update rule. Besides, our SegPGD can turn part of the pixels $k \in P^T$ of X^0 into misclassified ones of X^1. However, PGD is less effective to do so since the update direction also takes the misclassified pixels of X^0 into consideration. Therefore, our SegPGD can achieve higher MisRatio than PGD in the first step.

In all intermediate steps, both SegPGD and PGD leverage gradients of all pixels classification loss to update adversarial examples. The difference is that our SegPGD assign more weight to loss of correctly classified pixel classifications. The assigned value depends on the update iteration t. Our SegPGD focuses more on fooling correctly classified pixels at first a few iterations and then treat both quasi equally. By doing this, our SegPGD can achieve higher MisRatio than PGD under the same attack iterations.

In Fig. 1, we show the pixel classification loss and PosiRatio (=1 - MisRatio) in each attack iteration. Figure 1a shows the case to attack adversarially trained PSPNet on VOC (see more details in experimental section). SegPGD marked with blue solid lines achieve higher MissRatio than PGD under the same number of attack iterations. The loss of False classified pixels (FLoss) marked with triangle down dominate the overall loss (i.e. red lines without markers) during attacks. Compared to PGD, the FLoss in SegPDG makes up a smaller portion of the overall loss since SegPGD main focuses on correctly classified pixels in the first a few attack iterations. Note that the scale of loss does not matter since only the signs of input gradients are leveraged to create adversarial examples.

3.3 Segmentation Adversarial Training with SegPGD

Adversarial training, as one of the most effective defense methods, has been well studied in the classification task. In classification, the main challenge of applying adversarial training is computational cost. It requires multiple gradient propagation to produce adversarial images, which makes adversarial training slow. In fact, it can take 3–30 times longer to train a robust network with adversarial training than training a non-robust equivalent [37]. The segmentation task makes the adversarial training more challenging. More attack iterations are required to create effective adversarial examples for boosting segmentation robustness. *E.g.,* more than 100 attack iterations are required to fool segmentation [50].

In this work, we improve segmentation adversarial training by applying SegPGD as the underlying attack. As an effective and efficient segmentation attack method, SegPGD can create more effective adversarial examples than the popular PGD. By injecting the created adversarial examples into the training data, adversarial training with SegPGD can achieve a more robust segmentation model with the same computational cost. Following the previous work, the adversarial training procedure on segmentation is shown in Algorithm 2.

4   Experiment

In this section, we first introduce the experimental setting. Then, we show the effectiveness of SegPGD. Specifically, we show SegPGD can achieve similar attack effect with less attack iterations than PGD on both standard models and adversarially trained models. In the last part, we show that adversarial training with SegPGD can achieve more adversarially robust segmentation models.

4.1   Experimental Setting

Datasets. The popular semantic segmentation datasets, PASCAL VOC 2012 (VOC) [11] and Cityscapes (CS) [9], are adopted in experiments. VOC dataset contains 20 object classes and one class for background, with 1,464, 1,499, and 1,456 images for training, validation, and testing, respectively. Following the popular protocol [17], the training set is augmented to 10,582 images. Cityscapes dataset contains urban scene understanding images with 19 categories, which contains high-quality pixel-level annotations with 2,975, 500, and 1,525 images for training, validation, and testing, respectively.

Models. We choose popular semantic segmentation architectures PSPNet [56] and DeepLabv3 [7] for our experiments. The standard configuration of the model architectures is used as in [56]. By default, ResNet50 [18] is applied as a backbone for feature extraction in both segmentation models.

Adversarial Attack. We choose the popular single-step attack FGSM [13] and the popular multiple-step attack PGD [29] as our baseline attack methods. In this work, we focus on ℓ_∞-based perturbations. The maximum allowed perturbation value ϵ is set to $0.03 = 8/255$. The step size α is set to 0.03 for FGSM and 0.01 for PGD. The PGD with 3 attack iterations is denoted as PGD3. Besides, for evaluating the robustness of segmentation models, we also apply attack methods, such as CW attack [6], DeepFool [31] and ℓ_2-based BIM [26].

Metrics. The standard segmentation evaluation metric mIoU (in %) is used to evaluate the adversarial robustness of segmentation models. The mIoUs on both clean image and adversarial images are reported, respectively. The higher the mIoUs are, the more robust the model is.

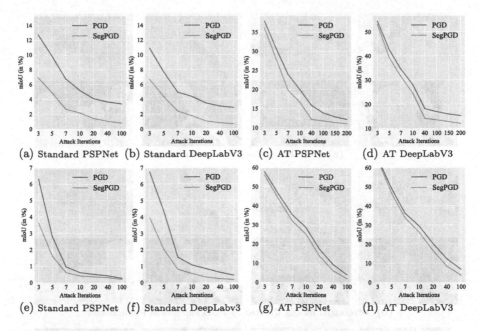

Fig. 2. SegPGD is more effective and efficient than PGD. SegPGD creates more effective adversarial examples with the same number of attack iterations and converges to a better minima than PGD. The subfigures (a-d) show the segmentation mIoUs on VOC, while the scores on Cityscapes are reported in the subfigures (e-h). AT PSPNet stands for the adversarially trained PSPNet.

4.2   Evaluating Segmentation Robustness with SegPGD

Quantitative Evaluation. We train PSPNet and DeepLabV3 on VOC and Cityscapes, respectively. Both standard training and adversarial training are considered in this experiment. PGD with 3 attack iterations is applied as the underlying attack method of adversarial training. This result in 8 models. We apply PGD and SegPGD on the 8 models. On each model, we report the final mIoU under attack with different attack iterations, *e.g.*, 20, 40 and 100. As shown in Fig. 2, the segmentation models show low mIoU on the adversarial examples created by our SegPGD. SegPGD achieve can converge faster to a better minima than PGD, which shows the high effectiveness and efficiency of SegPGD.

Qualitative Evaluation. For qualitative evaluations, we visualize the created adversarial examples and model's predictions on them. We take the adversarial examples created on standard PSPNet on VOC with 20 attack iterations as examples. As shown in Fig. 3, the adversarial perturbations created by both PGD and SegPGD are imperceptible to human vision. In other words, the created adversarial examples in Fig. 3b and 3b are not distinguishable from the counter-part clean images in Fig. 3a. The predicted masks on the adversarial examples by SegPGD have deviated more from the ground truth than the

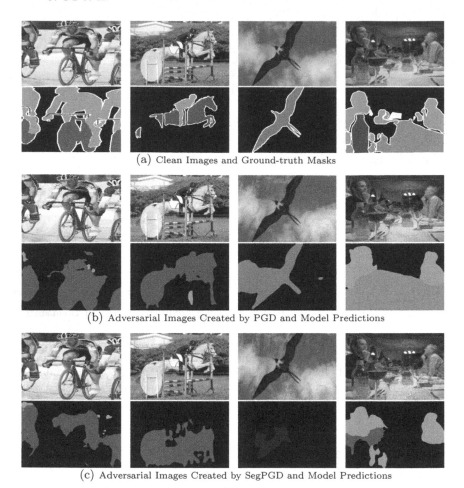

(a) Clean Images and Ground-truth Masks

(b) Adversarial Images Created by PGD and Model Predictions

(c) Adversarial Images Created by SegPGD and Model Predictions

Fig. 3. Visualizing of Adversarial Examples and Predictions on them. SegPGD create more effective adversarial examples than PGD.

ones corresponding to PGD. The visualization in Fig. 3 shows SegPGD creates more effective adversarial examples than PGD under the same number of attack iterations.

Comparison with other Segmentation Attack Methods. The segmentation attack methods have also been explored in related work. The work [20] aim to create adversarial perturbations that are always deceptive when added to any sample. Similarly, The work [23] creates universal perturbations to attack multiple segmentation models. Since more constraints are applied to universal perturbations, both types of universal adversarial perturbations are supposed to be less effective than the sample-specific ones. Another work [20] related to us proposes Dense Adversary Generation (DAG), which can be seen as a special

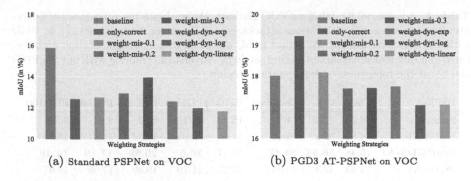

(a) Standard PSPNet on VOC   (b) PGD3 AT-PSPNet on VOC

Fig. 4. Schedules for weighting misclassified pixels. SegPGD with our *weight-dyn-linear* weighting schedules can better reduce the mIoU and achieve better attack effectiveness than the ones with baseline schedules.

case of our SegPGD along with other minor differences. DAG only considers the correctly classified pixels in each attack iteration, which is equivalent to set $\lambda = 0$ in our SegPGD. To further improve the attack effectiveness, the work [16] proposes multiple-layer attack (MLAttack) where the losses in feature spaces of multiple intermediate layers and the one in the final output layer are combined to create adversarial examples. SegPGD outperforms both DAG and MLAttack in terms of both efficiency and effectiveness, as shown in Appendix A.

Single-Step Attack. When a single attack iteration is applied, SegPGD is degraded to SegFGSM. In SegFGSM, only the loss of correctly classified pixels are considered in the case of the proposed λ schedule. We compare FGSM and SegFGSM and report the mIOU. Our SegFGSM outperforms FGSM on both standard models and adversarially trained models. See Appendix B for details.

Ablation on Weighting Schedules. In this work, we argue that the weight should be changed dynamically with the attack iterations. At the beginning of the attack, the update of adversarial example should focus more on fooling correctly classified pixels. In Eq. 5, we list three schedule instances, i.e., the *weight-dyn-linear*, *weight-dyn-exp*, and *weight-dyn-log* schedule respectively. We denote the case as *baseline* where the losses of all the pixels are equally treated. Another choice to weigh the loss of misclassified pixels is to use a constant λ, *e.g.*, 0.1, 0.2 or 0.3, which is denoted as *weight-mis-λ*. When the constant is set to zero, only correctly classified pixels are considered to compute the loss in all attack iterations, which is denoted as *only-correct*. We report the mIoU of segmentation under different weighting schedules in Fig. 4. As shown in the figure, SegPGD with our *weight-dyn-linear* weighting schedules can better reduce the mIoU and achieve better attack effectiveness than baselines. Given its simplicity, we apply the linear schedule rule in our SegPGD. We leave the exploration of more dedicated weighting schedules in future work.

Table 1. Adversarial Training on VOC Dataset. We evaluate the robustness of adversarially trained models with various attacks, especially under strong attacks (*e.g.,* PGD with 100 attack iterations). We report mIoU scores on different segmentation architectures and different adversarial training settings. Adversarial training with our SegPGD can boost the robustness of segmentation models.

PSPNet	Clean	CW	DeepFool	BIMl2	PGD10	PGD20	PGD40	PGD100
Standard	76.64	4.72	14.2	15.32	5.21	4.09	3.64	3.37
PGD3-AT	74.51	52.23	55.46	51.56	20.04	17.34	15.84	13.89
SegPGD3-AT	75.38	**56.52**	**59.47**	50.17	**26.6**	**20.69**	**17.19**	**14.49**
PGD7-AT	74.99	42.30	45.05	47.21	21.79	19.39	17.99	16.97
SegPGD7-AT	74.45	**48.79**	**51.44**	45.15	**25.73**	**22.05**	**20.61**	**19.23**
DeepLabv3	Clean	CW	DeepFool	BIMl2	PGD10	PGD20	PGD40	PGD100
Standard	77.36	5.24	13.57	14.76	4.36	3.46	3.05	2.85
PGD3-AT	75.03	57.10	60.23	36.83	28.16	20.77	18.12	16.91
SegPGD3-AT	75.01	**59.55**	**62.12**	**39.46**	26.29	**20.92**	**19.1**	**18.24**
PGD7-AT	73.45	48.51	48.87	43.13	26.23	21.15	20.06	19.10
SegPGD7-AT	74.46	**51.42**	**51.47**	42.91	**30.95**	**26.68**	**24.32**	**23.09**

4.3   Boosting Segmentation Robustness with SegPGD-AT

The setting of adversarial training in previous work [51] is adopted in this work. In the baseline, PGD is applied as the underlying attack method for adversarial training. In our approach, We apply SegPGD to create adversarial examples for adversarial training. For both standard training and adversarial training, we train for one more time and report the average results.

White-Box Attack. We evaluate the segmentation models with popular white-box attacks. The results are reported in Table 1 and Table 2. The mIoU of the standard segmentation model can be reduced to near zero. As expected, they are not robust at all to strong attack methods. Adversarial training methods boost the robustness of segmentation models to different degrees. Under the evaluation of all attack methods, adversarial training with our SegPGD achieves more robust segmentation performance than the one with PGD. Besides the popular segmentation attack methods, we also evaluate the adversarially-trained models with our SegPGD. The evaluation results also support our conclusion, which can be found in Appendix C.

We also compare our SegPGD-AT with the recently proposed segmentation adversarial training method DDCAT [51]. We load the pre-trained DDCAT models from their released the codebase and evaluate the model with strong attacks. We found that their models are vulnerable to strong attacks, *e.g.,* PGD100. For fair comparison, we compare the scores on our SegPGD3-AT with the ones on their models since three steps are applied to generate adversarial examples in both cases. Our model trained with SegPGD3-AT outperform the DDCAT by a large margin under strong attacks, *e.g.,* 10.98 (DDCAT) vs. 18.24 (ours) with

Table 2. Adversarial Training on Cityscapes Dataset. This table show that the boosting effect of adversarial training with our SegPGD still clearly holds on a different dataset. Besides, we show the previous adversarially trained baseline model can be reduced to near zero under strong attack, *i.e.*, PGD3-AT PSPNet under PGD100 attack. Our segPGD improves the robustness significantly.

PSPNet	Clean	CW	DeepFool	BIMl2	PGD10	PGD20	PGD40	PGD100
Standard	73.98	5.94	12.68	12.36	0.96	0.61	0.42	0.27
DDCAT [51]	76.64	4.72	14.2	15.32	5.21	4.09	3.64	3.37
PGD3-AT	71.28	35.21	36.84	32.22	28.79	17.3	9.29	3.95
SegPGD3-AT	71.01	**36.30**	**38.27**	**35.34**	**33.52**	**25.23**	**19.22**	**13.04**
PGD7-AT	69.85	27.78	28.44	27.87	26.00	24.75	23.86	22.8
SegPGD7-AT	70.21	**29.59**	**30.68**	**32.55**	**27.13**	**25.56**	**24.29**	**23.13**
DeepLabv3	Clean	CW	DeepFool	BIMl2	PGD10	PGD20	PGD40	PGD100
Standard	73.82	8.24	14.26	13.86	1.07	0.84	0.62	0.44
DDCAT [51]	76.64	4.72	14.2	15.32	5.21	4.09	3.64	3.37
PGD3-AT	71.45	36.72	38.98	36.78	29.52	20.23	12.22	6.74
SegPGD3-AT	71.04	**37.93**	37.63	34.54	**32.11**	**25.49**	**17.67**	**15.23**
PGD7-AT	69.91	28.87	29.63	30.58	25.64	24.48	22.87	21.24
SegPGD7-AT	69.93	**29.73**	**31.30**	**32.35**	**30.43**	**28.78**	**26.73**	**25.31**

DeepLabv3 architecture on VOC dataset under PGD100. More results can be found in Appendix D.

In our experiments, PGD-AT PSPNet on Cityscapes can be almost completely fooled under strong attack where the mIoU is 3.95 under PGD100 attack. Adversarial training with SegPGD boosts the robustness to 13.04. Although the improvement is large, there is still much space to improve.

Black-Box Attack. We also evaluate the segmentation robustness with black-box attacks. Different from white-box attacks, black-box attackers are supposed to have no access to the gradient of the target model. Following the previous work [51], we conduct experiments with transfer-based black-box attacks. We train PSPNet and DeepLabV3 on the same dataset. Then, we create adversarial examples on PSPNet with PGD100 or SegPGD100 and test the robustness of DeepLabV3 on these adversarial examples. The detailed results are reported in Appendix E. The DeepLabV3 models trained with different adversarial training methods are tested. The model trained with our SegPGD-AT shows the best performance against the transfer-based black-box attacks. The claim is also true when different attack methods are applied to create adversarial examples.

5   Conclusions

A large number of attack iterations are required to create effective segmentation adversarial examples. The requirement makes both robustness evaluation and adversarial training on segmentation challenging. In this work, we propose an effective and efficient segmentation-specific attack method, dubbed SegPGD. We first show SegPGD can converge better and faster than the baseline PGD. The effectiveness and efficiency of SegPGD is verified with comprehensive experiments on different segmentation architectures and popular datasets. Besides the evaluation, we also demonstrate how to boost the robustness of segmentation models with SegPGD. Specifically, we apply SegPGD to create segmentation adversarial examples for adversarial training. Given the high effectiveness of the created adversarial examples, the adversarial training with SegPGD improves the segmentation robustness significantly and achieves the state of the art. However, there is still much space to improve in terms of the effectiveness and efficiency of segmentation adversarial training. We hope this work can serve as a solid baseline and inspire more work to improve segmentation robustness.

Acknowledgement. This work is supported by the UKRI grant: Turing AI Fellowship EP/W002981/1, EPSRC/MURI grant: EP/N019474/1, HKU Startup Fund, and HKU Seed Fund for Basic Research. We would also like to thank the Royal Academy of Engineering and FiveAI.

References

1. Andriushchenko, M., Flammarion, N.: Understanding and improving fast adversarial training. In: NeurIPS (2020)
2. Arnab, A., Miksik, O., Torr, P.H.: On the robustness of semantic segmentation models to adversarial attacks. In: CVPR (2018)
3. Athalye, A., Carlini, N., Wagner, D.: Obfuscated gradients give a false sense of security: Circumventing defenses to adversarial examples. In: ICML (2018)
4. Bar, A., et al.: The vulnerability of semantic segmentation networks to adversarial attacks in autonomous driving: Enhancing extensive environment sensing. IEEE Signal Process. Mag. **38**(1), 42–52 (2020)
5. Cai, Q.Z., Du, M., Liu, C., Song, D.: Curriculum adversarial training. In: IJCAI (2018)
6. Carlini, N., Wagner, D.: Towards evaluating the robustness of neural networks. In: 2017 IEEE Symposium on Security and Privacy (sp), pp. 39–57. IEEE (2017)
7. Chen, L.C., Papandreou, G., Schroff, F., Adam, H.: Rethinking atrous convolution for semantic image segmentation. arXiv:1706.05587 (2017)
8. Cho, S., Jun, T.J., Oh, B., Kim, D.: Dapas: Denoising autoencoder to prevent adversarial attack in semantic segmentation. In: 2020 International Joint Conference on Neural Networks (IJCNN), pp. 1–8. IEEE (2020)
9. Cordts, M., et al.: The cityscapes dataset for semantic urban scene understanding. In: CVPR (2016)
10. Daza, L., Pérez, J.C., Arbeláez, P.: Towards robust general medical image segmentation. In: de Bruijne, M., et al. (eds.) MICCAI 2021. LNCS, vol. 12903, pp. 3–13. Springer, Cham (2021). https://doi.org/10.1007/978-3-030-87199-4_1
11. Everingham, M., Van Gool, L., Williams, C.K., Winn, J., Zisserman, A.: The pascal visual object classes (voc) challenge. International journal of computer vision (IJCV) (2010)

12. Full, P.M., Isensee, F., Jäger, P.F., Maier-Hein, K.: Studying robustness of semantic segmentation under domain shift in cardiac mri. In: International Workshop on Statistical Atlases and Computational Models of the Heart. pp. 238–249. Springer (2020)
13. Goodfellow, I.J., Shlens, J., Szegedy, C.: Explaining and harnessing adversarial examples. In: ICLR (2015)
14. Gu, J., Wu, B., Tresp, V.: Effective and efficient vote attack on capsule networks. arXiv preprint arXiv:2102.10055 (2021)
15. Gu, J., Zhao, H., Tresp, V., Torr, P.: Adversarial examples on segmentation models can be easy to transfer. arXiv preprint arXiv:2111.11368 (2021)
16. Gupta, P., Rahtu, E.: Mlattack: Fooling semantic segmentation networks by multi-layer attacks. In: German Conference on Pattern Recognition. pp. 401–413. Springer (2019)
17. Hariharan, B., Arbeláez, P., Girshick, R., Malik, J.: Hypercolumns for object segmentation and fine-grained localization. In: Proceedings of the IEEE conference on computer vision and pattern recognition. pp. 447–456 (2015)
18. He, K., Zhang, X., Ren, S., Sun, J.: Deep residual learning for image recognition. In: CVPR (2016)
19. He, X., Yang, S., Li, G., Li, H., Chang, H., Yu, Y.: Non-local context encoder: Robust biomedical image segmentation against adversarial attacks. In: Proceedings of the AAAI Conference on Artificial Intelligence. vol. 33, pp. 8417–8424 (2019)
20. Hendrik Metzen, J., Chaithanya Kumar, M., Brox, T., Fischer, V.: Universal adversarial perturbations against semantic image segmentation. In: ICCV (2017)
21. Jia, X., Zhang, Y., Wu, B., Ma, K., Wang, J., Cao, X.: Las-at: Adversarial training with learnable attack strategy. In: Proceedings of the IEEE/CVF Conference on Computer Vision and Pattern Recognition. pp. 13398–13408 (2022)
22. Jia, X., Zhang, Y., Wu, B., Wang, J., Cao, X.: Boosting fast adversarial training with learnable adversarial initialization. IEEE Trans. Image Process. (2022)
23. Kang, X., Song, B., Du, X., Guizani, M.: Adversarial attacks for image segmentation on multiple lightweight models. IEEE Access 8, 31359–31370 (2020)
24. Kapoor, N., et al.: From a fourier-domain perspective on adversarial examples to a wiener filter defense for semantic segmentation. In: 2021 International Joint Conference on Neural Networks (IJCNN), pp. 1–8. IEEE (2021)
25. Klingner, M., Bar, A., Fingscheidt, T.: Improved noise and attack robustness for semantic segmentation by using multi-task training with self-supervised depth estimation. In: Proceedings of the IEEE/CVF Conference on Computer Vision and Pattern Recognition Workshops, pp. 320–321 (2020)
26. Kurakin, A., Goodfellow, I., Bengio, S., et al.: Adversarial examples in the physical world. In: ICLR (2016)
27. Lee, H.J., Ro, Y.M.: Adversarially robust multi-sensor fusion model training via random feature fusion for semantic segmentation. In: 2021 IEEE International Conference on Image Processing (ICIP), pp. 339–343. IEEE (2021)
28. Li, Y., Li, Y., Lv, Y., Jiang, Y., Xia, S.T.: Hidden backdoor attack against semantic segmentation models. arXiv preprint arXiv:2103.04038 (2021)
29. Madry, A., Makelov, A., Schmidt, L., Tsipras, D., Vladu, A.: Towards deep learning models resistant to adversarial attacks. In: ICLR (2018)
30. Milletari, F., Navab, N., Ahmadi, S.A.: V-net: Fully convolutional neural networks for volumetric medical image segmentation. In: 3DV (2016)
31. Moosavi-Dezfooli, S.M., Fawzi, A., Frossard, P.: Deepfool: a simple and accurate method to fool deep neural networks. In: Proceedings of the IEEE Conference on Computer Vision and Pattern Recognition, pp. 2574–2582 (2016)

32. Nakka, K.K., Salzmann, M.: Indirect local attacks for context-aware semantic segmentation networks. In: Vedaldi, A., Bischof, H., Brox, T., Frahm, J.-M. (eds.) ECCV 2020. LNCS, vol. 12350, pp. 611–628. Springer, Cham (2020). https://doi.org/10.1007/978-3-030-58558-7_36

33. Nesti, F., Rossolini, G., Nair, S., Biondi, A., Buttazzo, G.: Evaluating the robustness of semantic segmentation for autonomous driving against real-world adversarial patch attacks. In: Proceedings of the IEEE/CVF Winter Conference on Applications of Computer Vision, pp. 2280–2289 (2022)

34. Park, G.Y., Lee, S.W.: Reliably fast adversarial training via latent adversarial perturbation. In: ICCV (2021)

35. Paschali, M., Conjeti, S., Navarro, F., Navab, N.: Generalizability vs. robustness: investigating medical imaging networks using adversarial examples. In: Frangi, A.F., Schnabel, J.A., Davatzikos, C., Alberola-López, C., Fichtinger, G. (eds.) MICCAI 2018. LNCS, vol. 11070, pp. 493–501. Springer, Cham (2018). https://doi.org/10.1007/978-3-030-00928-1_56

36. Rossolini, G., Nesti, F., D'Amico, G., Nair, S., Biondi, A., Buttazzo, G.: On the real-world adversarial robustness of real-time semantic segmentation models for autonomous driving. arXiv preprint arXiv:2201.01850 (2022)

37. Shafahi, A., et al.: Adversarial training for free! NeurIPS (2019)

38. Shen, G., Mao, C., Yang, J., Ray, B.: Advspade: Realistic unrestricted attacks for semantic segmentation. arXiv preprint arXiv:1910.02354 (2019)

39. Sriramanan, G., Addepalli, S., Baburaj, A., et al.: Towards efficient and effective adversarial training. In: NeurIPS (2021)

40. Szegedy, C., et al.: Intriguing properties of neural networks. In: ICLR (2014)

41. Tramèr, F., Kurakin, A., Papernot, N., Goodfellow, I., Boneh, D., McDaniel, P.: Ensemble adversarial training: Attacks and defenses. In: ICLR (2018)

42. Tran, H.-D., et al.: Robustness verification of semantic segmentation neural networks using relaxed reachability. In: Silva, A., Leino, K.R.M. (eds.) CAV 2021. LNCS, vol. 12759, pp. 263–286. Springer, Cham (2021). https://doi.org/10.1007/978-3-030-81685-8_12

43. Vivek, B., Babu, R.V.: Single-step adversarial training with dropout scheduling. In: CVPR (2020)

44. Vivek, B.S., Mopuri, K.R., Babu, R.V.: Gray-box adversarial training. In: Ferrari, V., Hebert, M., Sminchisescu, C., Weiss, Y. (eds.) ECCV 2018. LNCS, vol. 11219, pp. 213–228. Springer, Cham (2018). https://doi.org/10.1007/978-3-030-01267-0_13

45. Wang, D., Ju, A., Shelhamer, E., Wagner, D., Darrell, T.: Fighting gradients with gradients: Dynamic defenses against adversarial attacks. arXiv preprint arXiv:2105.08714 (2021)

46. Wang, J., Zhang, H.: Bilateral adversarial training: Towards fast training of more robust models against adversarial attacks. In: ICCV (2019)

47. Wong, E., Rice, L., Kolter, J.Z.: Fast is better than free: Revisiting adversarial training. In: ICLR (2020)

48. Wu, B., et al.: Attacking adversarial attacks as a defense. arXiv preprint arXiv:2106.04938 (2021)

49. Xiao, C., Deng, R., Li, B., Yu, F., Liu, M., Song, D.: Characterizing adversarial examples based on spatial consistency information for semantic segmentation. In: Ferrari, V., Hebert, M., Sminchisescu, C., Weiss, Y. (eds.) ECCV 2018. LNCS, vol. 11214, pp. 220–237. Springer, Cham (2018). https://doi.org/10.1007/978-3-030-01249-6_14

50. Xie, C., Wang, J., Zhang, Z., Zhou, Y., Xie, L., Yuille, A.: Adversarial examples for semantic segmentation and object detection. In: ICCV (2017)
51. Xu, X., Zhao, H., Jia, J.: Dynamic divide-and-conquer adversarial training for robust semantic segmentation. In: ICCV (2021)
52. Ye, N., Li, Q., Zhou, X.Y., Zhu, Z.: Amata: An annealing mechanism for adversarial training acceleration. In: AAAI (2021)
53. Yu, Y., Lee, H.J., Kim, B.C., Kim, J.U., Ro, Y.M.: Towards robust training of multi-sensor data fusion network against adversarial examples in semantic segmentation. In: ICASSP 2021–2021 IEEE International Conference on Acoustics, Speech and Signal Processing (ICASSP), pp. 4710–4714. IEEE (2021)
54. Zhang, D., Zhang, T., Lu, Y., Zhu, Z., Dong, B.: You only propagate once: Accelerating adversarial training via maximal principle. In: NeurIPS (2019)
55. Zhang, H., Wang, J.: Defense against adversarial attacks using feature scattering-based adversarial training. In: NeurIPS (2019)
56. Zhao, H., Shi, J., Qi, X., Wang, X., Jia, J.: Pyramid scene parsing network. In: CVPR (2017)
57. Zheng, H., Zhang, Z., Gu, J., Lee, H., Prakash, A.: Efficient adversarial training with transferable adversarial examples. In: CVPR (2020)

Adversarial Erasing Framework via Triplet with Gated Pyramid Pooling Layer for Weakly Supervised Semantic Segmentation

Sung-Hoon Yoon⑩, Hyeokjun Kweon⑩, Jegyeong Cho⑩, Shinjeong Kim⑩, and Kuk-Jin Yoon(✉)⑩

Korea Advanced Institute of Science and Technology, Daejeon, South Korea
{yoon307,0327june,j2k0618,aakseen,kjyoon}@kaist.ac.kr

Abstract. Weakly supervised semantic segmentation (WSSS) has employed Class Activation Maps (CAMs) to localize the objects. However, the CAMs typically do not fit along the object boundaries and highlight only the most-discriminative regions. To resolve the problems, we propose a Gated Pyramid Pooling (GPP) layer which is a substitute for a Global Average Pooling (GAP) layer, and an Adversarial Erasing Framework via Triplet (AEFT). In the GPP layer, a feature pyramid is obtained by pooling the CAMs at multiple spatial resolutions, and then be aggregated into an attention for class prediction by gated convolution. With the process, CAMs are trained not only to capture the global context but also to preserve fine-details from the image. Meanwhile, the AEFT targets an over-expansion, a chronic problem of Adversarial Erasing (AE). Although AE methods expand CAMs by erasing the discriminative regions, they usually suffer from the over-expansion due to an absence of guidelines on when to stop erasing. We experimentally verify that the over-expansion is due to rigid classification, and metric learning can be a flexible remedy for it. AEFT is devised to learn the concept of erasing with the triplet loss between the input image, erased image, and negatively sampled image. With the GPP and AEFT, we achieve new state-of-the-art both on the PASCAL VOC 2012 *val/test* and MS-COCO 2014 *val* set by 70.9%/71.7% and 44.8% in mIoU, respectively.

Keyword: Weakly supervised semantic segmentation

1 Introduction

Recently, semantic segmentation based on Deep Learning (DL) has been widely used in various applications such as autonomous driving and medi-

---

S.-H Yoon and H. Kweon—Equal contribution.

Supplementary Information The online version contains supplementary material available at https://doi.org/10.1007/978-3-031-19818-2_19.

cal imaging. However, since the semantic segmentation model requires pixel-level labels, a considerable amount of cost and time is consumed to generate labels. To reduce this burden and make DL-based semantic segmentation more practically applicable in general tasks, many Weakly Supervised Semantic Segmentation (WSSS) studies that utilize only weak supervision such as image-level labels [1,2,7,15,27,30,35,50,56,58], scribble [36,48], bounding boxes [13,24,31,40], and points [4] have been proposed. In this work, we focus on WSSS with image-level labels, an especially challenging task among weakly-supervised ones.

To learn semantic segmentation with image-level labels only, most existing WSSS approaches follow the steps: (1) localize the objects through Class Activation Maps(CAMs) [61], (2) refine the CAMs and generate pseudo-labels in a pixel-level, and (3) train the semantic segmentation model with the pseudo-labels. Although the CAMs can localize the objects to some extent, they are not precise at an object boundary and only highlight the most discriminative pattern.

As far as we know, most CNN-based classifiers in WSSS employ a Global Average Pooling (GAP) layer to aggregate the feature map and predict the existing classes in the image. However, since the GAP layer *averages* all the features, even including ones from object-irrelevant regions, CAMs usually ignore small segments and do not fit with the object boundary (*i.e.* impreciseness). To overcome this innate limitation of the GAP layer, BES [50] and Araslanov *et al.* [3] utilize softmaxed-CAMs as a pooling weight while making the class prediction. Instead, in this paper, we propose a Gated Pyramid Pooling (GPP) layer that not only captures the global context but also localizes fine-details. In the proposed GPP layer, CAMs are average-pooled with various bin sizes (*e.g.* 8×8 or 16×16) and form a spatial pyramid. Then, this pyramid of the pooled features is aggregated sequentially through a gating mechanism [47] in a coarse-to-fine manner for better localization. The final output of the aggregation is used as a pixel-level weight that decides to either encourage or discourage the contribution of CAMs for predicting the image-level class. Here, by building the pyramid features with different spatial resolutions and using them as weights to generate the class prediction, CAMs are trained to capture not only global context (from low-scale bins) but also localize fine-details (from high-scale bins).

In addition to the GPP layer that effectively resolves the impreciseness problem in CAMs, we utilize the concept of Adversarial Erasing (AE) method to further guide CAMs to be activated even on the less-discriminative regions. AE methods [19,27,33,51,59], one of the most actively studied strategy in WSSS, extend CAMs to whole object regions by erasing the most-discriminative regions of an image or intermediate feature. However, since there is no explicit guidance regarding when to stop the erasing, CAMs generated from the AE approach usually suffer from an over-expansion [27]. To benefit from AE methods while preventing the over-expansion problem, we propose an Adversarial Erasing Framework via Triplet (AEFT) that reformulates the AE methods as triplet learning with the GPP feature. Here, we experimentally verify that imposing relatively

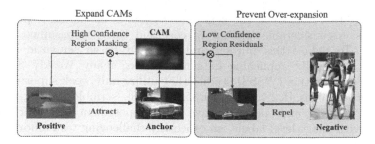

Fig. 1. Brief illustration of the proposed Adversarial Erasing Framework via Triplet (AEFT). An anchor image and a positive image denote the original image and the masked image according to CAMs, respectively. A negative image is sampled to do not have overlapping classes with the anchor image. In the feature space, we train the model to locate the positive image close to the anchor image while increasing the distance between the anchor image and the negative image.

rigid supervision (*e.g.* classification loss) on the AE framework leads the resulting CAMs to suffer from the over-expansion problem. Since we allow the framework to adjust its features according to the distance between them, this approach can be regarded as a more softened version compared to prior studies using rigid supervision [27,33,51,59]. For triplet learning, as shown in Fig. 1, we define the original image as an *anchor* image. After masking the high-confidence regions of the CAMs from the anchor image, the remained image is regarded as a *positive* image. Finally, the other image, which has no overlapping classes with the anchor image, is used as a *negative* image. In AEFT, we minimize the distance between the anchor and positive in the GPP feature space (*i.e. Attract*) while maximizing the distance between the anchor and negative feature (*i.e. Repel*). While *Attract* guides the CAMs to explore less-discriminative region, *Repel* prevents the over-expansion problem. Since the distance between the anchor and negative is often already far enough, we intentionally exclude the high-confidence region from the anchor to impose a harder but helpful constraint for AEFT.

In summary, we propose (1) the Gated Pyramid Pooling (GPP) layer to resolve the architectural limitation of classifier (or GAP) and (2) the Adversarial Erasing Framework via Triplet (AEFT) to effectively prevent the over-expansion via triplet, while preserving the benefits of AE (expanding the CAMs to less-discriminative regions). With the proposed GPP and AEFT, we achieve state-of-the-art WSSS performance by a large gap on both the PASCAL VOC 2012 and MS-COCO 2014 sets, using the image-level labels only.

2   Related Works

Earlier Works in WSSS. WSSS with image-level labels generally utilizes
CAMs to localize target objects on the images. However, as CAMs tend to only
focus on the most discriminative parts and do not fit along the object boundary, subsequent works in WSSS have tried to generate high-quality pseudo-labels
from the CAMs for training semantic segmentation. Many studies proposed to
refine the CAMs using pixel-level affinity [2,16,43] or region growing [20,25].
Much research targeted to enhance the quality of localization of CAMs by using
stochastic feature selection [29], attention map accumulation [22], and scale-
invariance [50]. Also, lots of methods employ additional constrains such as sub-
categorical classification [7], co-attention constraints [34,45], and complementary
patch loss [58]. Several studies [15,32,34,45,54,55] employ saliency map to indicate dominant foreground objects distinguished from the background. Despite
the efficacy, neither saliency module nor an external dataset was adopted in the
proposed method, in line with the objective of WSSS learning from only image-
level labels. Like the proposed method, BES [10] and Araslanov et al. [3] utilize
CAMs as weights for pooling when making class predictions. However, while
both methods used softmaxed-CAMs as weights for pooling, the improvement
of BES is marginal (within 1%), and Araslanov et al. [3] requires to define background constant. Unlike the previous methods, the proposed method not only
preserves global context but also captures fine-details through sign-preserving
gated convolution and pyramid pooling.

Adversarial Erasing. Adversarial Erasing (AE) [19,27,33,51,59] is widely
used strategy in WSSS. By erasing the most discriminative region from the
image, the AE method promotes the network to expand its CAMs to the less
discriminative object region. The first AE method is proposed by Wei et al. [51],
which is a recursive find-and-erase scheme. Zhang et al. [59] proposes an end-to-
end feature-level erasing framework with complementary branches. However, if
the initial classifier succeeded in completely erasing the object, the complementary classifier would suffer from the over-erasing. SeeNet [19] suggested using
the ternary thresholding method for mask generation process to relieve the
over-erasing, but it requires a pre-trained saliency detection module. In recent,
GAIN [33] and OC-CSE [27] proposed soft erasing approaches that generate
learnable masks. In these methods, the CAMs generation network is jointly
trained by a classification loss and auxiliary loss regarding the adversarial erasing
process. GAIN propose an attention mining loss to assure that the erased image
does not contain any objects. However, this self-guidance makes the framework
difficult to self-correct the over-expansion. In OC-CSE, the CAM of only one
class is selected for erasing, and then the guidance from the pre-trained ordinary classifier is used to prevent the erasing network from erasing the objects
of the not selected classes. But the usage of the pre-trained classifier limits the
performance of this method. All of the aforementioned AE methods are based
on imposing a classification loss on the erased image. In our view, forcing the

network to make the prediction from the erased image according to the binary classification label (exist or not) is the main reason for over-expansion. Instead of this "rigid" constrain, we aim to let the network understand the concept of *erasing* in the form of triplet learning. This is a more softened approach compared to prior AE-based studies while not harming the benefit of AE methods.

Deep Metric Learning. Deep metric learning has widely been used for resolving various computer vision tasks [11,12,18,21,38,39]. Generally, it aims to learn a metric that measures the semantic distance between instances. As a metric function, the embedding function is trained to map an instance to be close to the similar inputs than the dissimilar inputs. Contrastive loss [17] directly optimizes this goal by decreasing the distance between semantically close instances while increasing the distance between dissimilar instances. On the other hand, triplet loss [42] takes three inputs at once: anchor instance, positive instance, and negative instance. Then, the loss minimizes the distance between the anchor instance and the positive object while it maximizes the distance between the anchor instance and the negative instance. For semantic segmentation, deep metric learning is used to improve performance in supervised learning [49] or to overcome the lack of data in challenging cases such as weak supervision [23] and open-world scenario [6]. Though we borrow the concept of the triplet loss, as far as we know, our method is the first AE-based method that incorporates metric learning in WSSS.

3   Proposed Method

3.1   Overview

In this paper, we propose a Gated Pyramid Pooling (GPP) layer, which is a simple but effective replacement of the Global Average Pooling (GAP) layer widely used in WSSS. To fully utilize the outperforming localization ability of the GPP layer, we also devise a novel Adversarial Erasing Framework via Triplet (AEFT). The proposed framework mainly focuses on training the network to find less-discriminative regions while relieving the over-expansion problem. Note that our method only utilizes image-level labels.

3.2   CAMs Generation

Before discussing our main approaches, we briefly introduce the general process of generating CAMs. Let $f \in \mathbb{R}^{K \times h \times w}$ denote a feature map of the last convolution layer of the classifier, where K is the number of classes and h, w represent the spatial dimensions of the feature map, respectively. Then, an image-level class prediction p can be acquired by applying Global Average Pooling (GAP) on the feature map as $p = \sigma\left(\frac{1}{hw}\sum_{i,j} f(i,j)\right)$, where $f(i,j)$ denotes the feature vector at a location (i,j) and σ is a sigmoid function. By taking the Rectified

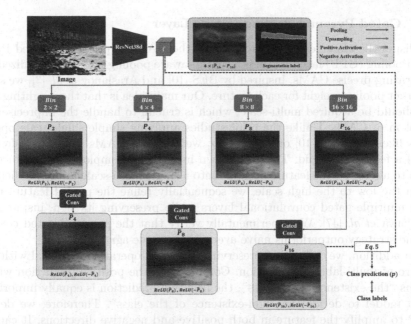

Fig. 2. Overview of the proposed Gated Pyramid Pooling (GPP). By applying pooling with different bin sizes to feature map f, we generate feature pyramid (P_2, P_4, P_8, P_{16}). Each component of the feature pyramid is interpolated to feature map size to apply gated convolution. Along with the feature pyramid, Gated features $(\hat{P}_4, \hat{P}_8, \hat{P}_{16})$ are visualized. *Red* and *green* color represent positive and negative activation, respectively. For simplicity only *train* class is visualized. (Color figure online)

Linear Unit (ReLU) to the feature map and normalizing it between 0 and 1 for each class, an activation map of kth class (A^k) is generated as follows:

$$A^k \leftarrow \frac{ReLU(f^k)}{max(ReLU(f^k))}. \qquad (1)$$

Here, we also apply bilinear upsampling on the CAMs to fit the spatial dimension of them with the input image.

Considering the formulation for the class prediction p, the GAP makes the feature contribute *equally* irrelevant to their location. As claimed by several works [3,10], the GAP increases a dependency on the context and misleads the classifier to learn erroneous correlations between image pixels and image-level class labels. Therefore, the resulting CAMs tend to be activated on highly-correlated background regions (*e.g.* railroad of *train* class, water of *boat* class) while ignoring the small objects. Since generating fine pixel-level pseudo-labels is crucial for WSSS, this is a critical disadvantage.

3.3   Gated Pyramid Pooling (GPP) Layer

To dispel the aforementioned problems of the GAP, we propose a Gated Pyramid Pooling (GPP) layer, which is a spatial-aware pooling method specialized for generating precise CAMs. Inspired by BES [10] and Araslanov et al. [3], we set a different pooling weight for each feature. Our main idea is that the weighting factor should be acquired multi-scale, which is crucial to handle the impreciseness problem of CAMs. Unlike the prior studies applying simple single-scale operations like softmax [3,10] on the CAMs, we pool the CAMs at multi-scale and build a feature pyramid. Then, as shown in Fig. 2, we employ a gating mechanism to aggregate the feature pyramid into a single multi-scale-aware prediction. From the low to the high scale, we sequentially refine the pooled feature map with multiple gated convolutional layers while preserving its sign, inspired by Takikawa et al. [47]. We experimentally verify that the proposed gated coarse-to-fine strategy outperforms naive averaging or scale-agnostic fusion.

In addition, we define sign-preserving attention operation $\mathcal{G}$ to deal with the nature of multi-label classification. Compared to the positive prediction which means "the existence of the class", the negative prediction is equally important for a model to decide the "non-existence of the class". Therefore, we devise GPP to amplify the feature in both positive and negative directions. It can be formulated by taking ReLU and concatenating two different features with 3×3 convolution layer $Conv_{3\times3}$. The process can be defined as follows:

$$\mathcal{G}(x,y) = \sigma(Conv_{3\times3}(ReLU(x)||ReLU(y))). \tag{2}$$

Let P_γ denote the averaged result of feature map f with $\gamma \times \gamma$ pooling. By applying pooling with different sizes ($\gamma \in \{2,4,8,16\}$) to f, we generate a feature pyramid (P_2, P_4, P_8, P_{16}) as in PSPNet [60]. Each component of the feature pyramid is upsampled to feature map f resolution. Then we obtain attention for positive (α) and negative (β) (where $\alpha, \beta \in \mathbb{R}^{2\times h\times w}$) maps as follows:

$$\alpha_n = \mathcal{G}_\alpha(\hat{P}_{2^n}, P_{2^{n+1}}), \quad \beta_n = \mathcal{G}_\beta(-\hat{P}_{2^n}, -P_{2^{n+1}}), \tag{3}$$

where σ is a sigmoid function and $n \in \{1,2,3\}$. And the gated feature $\hat{P}_{2^{n+1}}$ can be obtained as follows:

$$\hat{P}_{2^{n+1}} = \left(ReLU(\hat{P}_{2^n}) \odot \alpha_{n,1} + ReLU(P_{2^{n+1}}) \odot \alpha_{n,2}\right)/2$$
$$- \left(ReLU(-\hat{P}_{2^n}) \odot \beta_{n,1} + ReLU(-P_{2^{n+1}}) \odot \beta_{n,2}\right)/2, \tag{4}$$

where $\odot$ indicates element-wise product between tensors. Here, $\alpha_{n,1}$ and $\alpha_{n,2}$ are first and second channel of α_n, respectively. And $\hat{P}_{2^n}$ equals P_{2^n} only when $n = 1$. The final output of the Gated Pyramid Pooling (GPP) is $\hat{P}_{16} \in \mathbb{R}^{K\times h\times w}$. Here, K denotes the total number of classes.

$$p = \sigma(\frac{1}{hw} \sum \{ReLU(\hat{P}_{16}) \odot ReLU(f)\}$$
$$-\{ReLU(-\hat{P}_{16}) \odot ReLU(-f)\}), \tag{5}$$

By decoupling the feature f with the feature pyramid and aggregating it with the proposed Gated Pyramid Pooling (GPP), regions that are not related to objects are penalized while regions that are highly related are encouraged. Since GPP aggregates features from coarse level (*i.e.* small bin size pooling) to fine level (*i.e.* large bin size pooling) thoroughly, the generated CAMs not only fit along the object boundary but also localize the whole object region. A more detailed ablation study regarding GPP will be discussed in Sect. 4.

3.4 Adversarial Erasing Framework via Triplet (AEFT) for WSSS

The proposed GPP layer enables the model to generate CAMs with a higher localization quality than the GAP layer. However, this architectural improvement is still insufficient to acquire dense pseudo-labels for semantic segmentation. Additional guidance is required to make the CAMs cover the less-discriminative regions, which are difficult to be activated by a mere classification task.

In the field of WSSS, an Adversarial Erasing (AE) is one of the most widely used approaches to mitigate this problem. For AE, the most-discriminative regions of the CAMs are intentionally erased from the image. Then, the model is trained again to classify the erased image according to the original image-level classification labels. Continuously iterating this process make the model focus more on the less-discriminative regions, which were originally ignored, and thereby the resulting CAMs are also expanded. Because of its clear and intuitive strategy, plenty of WSSS studies [19,27,33,51,59] has been conducted based on the AE method. However, due to the lack of supervision for when to stop expanding, CAMs generated from the AE approach usually suffer from an over-expansion problem. To relieve such an over-expansion problem, a method using guidance from a pre-trained model [27] has been proposed recently. However, updating the guidance makes the training process unstable, and therefore the method has a limited performance due to the fixed classifier. As aforesaid, there are two main obstacles for the AE method to gently expand its CAMs while rejecting the undesired derailment. First, though the image-level classification labels are valuable supervision in WSSS, such supervision is often *too rigid* to follow and usually leads to the over-expansion problem when classifying the masked image/feature. Second, direct guidance from the AE branch to the CAMs makes the model be unstable in terms of the quality of the generated CAMs.

In this paper, we aim to train the model to understand the concept of *erasing* in a more flexible manner. To achieve this goal, we propose a novel Adversarial Erasing framework via Triplet (AEFT), the modified AE framework using a triplet loss between the images. In our framework, we make the representation of the masked image I_P embedded by the model to be close to that of the original input image I_A. To prevent the over-expansion, we maximize the feature-level distance between the I_A and the negative image I_N, an image that does not share any class with I_A. In other words, for the original image I_A, the masked image I_P and the negative image I_N are regarded as a positive sample and a negative sample, respectively. To avoid direct guidance to the CAMs, we utilize the feature space of the GPP layer as embedding space. Compared to using the rigid classification based on the binary label for the masked image in conventional works, the proposed metric-based approach helps the model flexibly adjust the distance between the features and the decision boundary.

Acquiring Masked Image. To erase the highly activated regions from input image I_A and obtain corresponding masked image I_P, we generate a foreground map A^{fg} from the CAMs of I_A as follows:

$$A^{fg}(i,j) = \max\{A^k(i,j) : k = 1, \ldots, K\}, \tag{6}$$

where A^k is an activation map of kth class and (i,j) denotes the pixel position. Then, according to the foreground map, we acquire the masked image I_P as follows:

$$I_P(i,j) = \begin{cases} 0, & \text{if } A^{fg}(i,j) \geq t_H \\ A^{fg}(i,j)I_A(i,j), & \text{otherwise.} \end{cases} \tag{7}$$

Note that we combine the hard-masking [19,51,59] and soft-masking [27,33] according to the threshold (which is denoted as t_H in Eq. 7). Since the regions with already higher activation are not the main target of learning in the proposed framework, we empirically find that this strategy is valid.

Adversarial Erasing via Metric Learning. As depicted in Fig. 3, each image should be represented as a feature vector in the embedding space for metric learning. We map the anchor (input) image I_A and the negative image I_N to the anchor embedding $e_A = \frac{1}{hw} \sum \hat{P}_{16,I_A}$ and the negative embedding $e_N = \frac{1}{hw} \sum \hat{P}_{16,I_N}$, respectively. Likewise, embedding for the masked image

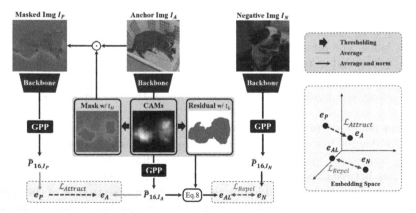

Fig. 3. Overview of the proposed Adversarial Erasing Framework via Triplet (AEFT). The weight of the networks is shared. In AEFT, we merged anchor image I_A, masked image I_P, and negative image I_N as a triplet set. We extract embedding from each member of the triplet set and impose loss relations based on their traits.

e_P is obtained in the same manner. By minimizing the distance between the anchor embedding and positive embedding, the model focuses more on the less-discriminative regions. However, though we soften the loss as a form of metric learning, it is still true that minimizing the distance between two embeddings pushes the model to keep exploring the regions even after the complete erasing. Therefore, in AEFT, we devise another constraint to inhibit the over-expansion using negative image I_N. In specific, we intentionally aggregate the features on low confidence regions according to CAMs to acquire 'the embedding for low confidence region of the anchor image' (e_{AL}). Then, we maximize the distance between e_{AL} and negative embedding e_N. Once the CAMs are over-expanded, the embedding from the low confidence region would include less information regarding the objects in the image. Then it would be difficult for the networks to separate such less-information embedding with the negative embedding. There-fore, intuitively, the expansion of CAMs is suppressed while maximizing the distance between e_{AL} and the negative embedding. The embedding for the low confidence region of the anchor image (e_{AL}) can be acquired as follows:

$$e_{AL}^k = \frac{1}{N^k} \sum_{(i,j)} \mathbb{1}(A^k(i,j) < t_L) \cdot \hat{P}_{16,I_A}(i,j), \text{ where } N^k = \sum_{(i,j)} \mathbb{1}(A^k(i,j) < t_L).$$

$$(8)$$

Here, k and (i,j) denote the class order and the pixel index, respectively. $\mathbb{1}$ is an indicator function that returns 1 if the statement is true, otherwise 0. The AEFT is composed of two metric losses: (1) $\mathcal{L}_{Attract}$ that minimizes the distance between anchor embedding and positive embedding and (2) $\mathcal{L}_{Repel}$ that maximizes the distance between the embedding for low confidence region of the anchor image and negative embedding. Each loss can be formulated as follows:

$$\mathcal{L}_{Attract} = ||e_A - e_P||^2, \tag{9}$$

$$\mathcal{L}_{Repel} = [-||e_{AL} - e_N||^2 + \epsilon]_+, \tag{10}$$

where $|| \cdot ||^2$ denotes mean squared error and ϵ denotes a fixed margin that constrains the maximum distance between embeddings. For the $\mathcal{L}_{Repel}$, we only consider positive distance ($[\cdot]_+$) to prevent unbounded embedding space. Each embedding is normalized before calculating the distance between them.

Our total loss function for training the AEFT is formulated as follows:

$$\mathcal{L}_{AEFT} = \mathcal{L}_{cls} + \lambda_1 \mathcal{L}_{Attract} + \lambda_2 \mathcal{L}_{Repel}, \tag{11}$$

where $\mathcal{L}_{cls}$ denotes the binary cross-entropy loss between the class prediction (p) and image-level labels.

In AEFT, we utilize features from the GPP layer to construct the embedding space for metric learning. The metric learning on the GPP feature can be inter-preted as implicit learning of the pooling weight, which is an effective way to handle the CAMs while not interrupting the backbone features of the classifier themselves. We experimentally verify that using the GPP feature is beneficial than using the CAMs or the intermediate features of the classifier. Also, compared with methods directly using GT classification labels for training the AE branch, the proposed AEFT shows superior performance in both qualitative and quantitative manners. In specific, thanks to the metric-based approach of AEFT, learning a semantic distance between the images enables CAMs to explore the less-discriminative regions while preventing the over-expansion problem.

4   Experimental Results

4.1   Dataset and Evaluation Metric

Evaluation of the proposed method is conducted on the PASCAL VOC 2012 dataset [14] and MS-COCO 2014 dataset [37]. COCO dataset is is more challenging in WSSS since it contains more classes (81) with small objects than PASCAL VOC 2012 (21). For VOC dataset, the proposed framework is trained with the augmented *train* set (10,582), and evaluated using both *val* (1,449) and *test* sets (1,456). For COCO dataset, the proposed method is trained with *train* set (80 k) and evaluated on *val* set (40 k). We use the mean Intersection over Union (mIoU) for evaluating our methods as similar to many other WSSS studies. As pointed out in Lee *et al.* [32], we utilize GT segmentation labels from COCO-Stuff dataset [5] for evaluation since the ground truth segmentation labels of the MS-COCO 2014 dataset have some overlaps between objects.

4.2   Implementation Details

The proposed network is implemented with PyTorch. We employ ResNet38 [53] as a backbone and the network is initialized with ImageNet [41] parameters. The data is augmented using horizontal flipping, color jittering [26], and cropping. The model is trained on 4 RTX 3090 GPUs with batch size 32. We use a poly learning rate [9] with the initial learning rate of 0.01 and the power of 0.9. For the semantic segmentation network, we use Deeplab [8] with ResNet38 backbone as in [2,27,35,46,58] for fair comparison. Margin (ϵ) between anchor and negative feature is set to 0.5. Weights of loss terms, λ_1 and λ_2, are set to 0.15 and 0.15, respectively. Our code is available at https://github.com/KAIST-vilab/AEFT.

4.3   Ablation Studies

To evaluate the proposed GPP layer, we conduct experiments with several different pooling bin sizes and feature aggregation methods. We set our baseline as a mere classifier with the GAP layer (achieves 48.4% in mIoU). As shown in Table 1, the larger the bin size for pooling, the higher performance (mIoU) of CAMs can be achieved. Furthermore, when averaging $\mathcal{A}$ of different feature pyramid is used as a weighting factor of Eq. 5, the performance is higher than using one of pooled feature pyramid. The results also show that the direction of aggregation is important, since using gated convolution in fine-to-coarse direction ($\mathcal{G}_I$) shows lower performance than naive averaging ($\mathcal{A}$), while the proposed

Table 1. Ablation study of the Gated Pyramid Pooling, evaluated on the PASCAL VOC *train* set. Pooled feature maps of bin sizes $\{2 \times 2, 4 \times 4, 8 \times 8, 16 \times 16\}$ are listed and used as an weighting factor of Eq. 5. Aggregation methods are noted as: $\mathcal{A}$ (averaged), $\mathcal{G}$ (gated convolution, from coarse to fine), and $\mathcal{G}_I$ (gated convolution, from fine to coarse). The performance is evaluated with the PASCAL VOC 2012 *train* set.

2×2	4×4	8×8	16×16	Aggregation	mIoU (%)
✓				-	49.9
	✓			-	51.6
		✓		-	52.9
			✓	-	53.1
✓	✓	✓	✓	$\mathcal{A}$	53.3
✓	✓	✓	✓	$\mathcal{G}_I$	51.3
✓	✓	✓	✓	$\mathcal{G}$	**54.2**

coarse-to-fine ($\mathcal{G}$) outperforms both. It supports our design intention: the global context and fine details are well preserved in both small and large bin sizes, and the coarse-to-fine aggregation can effectively exploit both information.

To clarify the source of improvements in AEFT, we conduct an ablation study as in Table 2. With the attraction loss ($\mathcal{L}_{Attract}$), the proposed AEFT achieves 55.0% in mIoU, while the repelling loss ($\mathcal{L}_{Repel}$) achieves 54.8%. Actually, we did not expect that the AEFT could increase the performance of the generated CAMs with the repelling loss only, which is designed to aid the attraction loss. Since the background regions around the foreground objects are sometimes activated by the CAMs, we interpret this result as the repelling loss successively penalizes such unwanted intrusion. It leads the framework to generate precise CAMs fit along the object boundary. When both loss functions are used, the proposed framework achieves 56.0% in mIoU, and these results represent that benefits from each loss function are synergistic. Furthermore, compare to using CAMs itself as a embedding space for triplet learning (*Direct*, 54.6%), using the GPP feature (*Indirect*, 56.0%) shows better performance. This result indirectly supports our hypothesis (the direct guidance from the AE branch to the CAMs makes the model be unstable). Moreover, if we maximize the distance between e_A and e_N instead of e_{AL} and e_N, the performance of AEFT decreases to 54.1%. This result implies that the distance between the two images without sharing class is already large enough, as we expected, and our strategy for using only the low-confidence region from the anchor image is effective in terms of the quality of the CAMs. Also, it is noteworthy that the gain of CRF is even larger in our method (↑7.5%) than vanilla CAMs (↑5.9%). It implies that the benefit of our method is not overlapped with that of CRF, which is advantageous in terms of generating high-quality pseudo-labels.

In addition, we experimentally verify that using rigid classification labels triggers an over-expansion problem in Adversarial Erasing (AE). To quantitatively compare the degree of over-expansion, we use *Precision* and *Recall* scores of the

Table 2. Ablation study of the proposed AEFT. *Direct*: global average pooled result of CAMs is used as embedding for metric, *Indirect*: global average pooled result of GPP is used. The performance (mIoU,%) is evaluated with the VOC 2012 *train* set.

Distance	$\mathcal{L}_{Attract}$	$\mathcal{L}_{Repel}$	CAMs	CAMs w/CRF
Indirect	✓		54.9	62.2
Indirect		✓	54.6	61.5
Direct	✓	✓	54.7	61.6
Indirect	✓	✓	**56.0**	**63.5**

generated CAMs. Here, *Precision* means that the true activation over the whole activation and *Recall* is the true activation over the GT. Although *Precision* and *Recall* are not direct metrics for the over-expansion of CAMs, they could be reasonable measure for quantitative comparison. As shown in Table 3, we compare the *Precision*, *Recall*, and mIoU performances from the various settings in our proposed AEFT. Here, we quantitatively verify that forcing the network to make the prediction from the erased image I_P according to the binary classification label (exist or not) leads to over-expansion. When we use the classification label for guiding the classifier to explore the less-discriminative regions as conventional AE-based WSSS methods (*Attract(Rigid)* in Table 3), the performance becomes lower than using the GPP layer alone (54.2%). Though using the rigid classification labels increases *Recall* by 0.5%, it causes over-expansion and thereby harms *Precision* by -1.4%. Instead of using the rigid labels, by minimizing the distance between the anchor embedding e_A and positive embedding e_P in a soft manner (*Attract(Soft)*), the proposed method increases *Recall* by 1.6% while not degrading the *Precision*. When we additionally employ repelling loss devised for preventing over-expansion (denoted as *Attract(Soft)+Repel*), both *Precision* and *Recall* are increased by 1.9% and 0.7%, respectively.

Table 3. Comparison of the precision, recall, and mIoU from the various settings in the AEFT. ***Attract (Rigid)***: uses rigid classification labels for the masked image I_P, ***Attract (Soft)***: minimizes the distance between the anchor e_A and positive e_P in a soft manner, ***Atrract (Soft)+Repel***: uses repelling loss $\mathcal{L}_{Repel}$ along with attraction loss $\mathcal{L}_{Attract}$ (our setting). The performance is evaluated on the VOC 2012 *train* set.

	Precision (%)	Recall (%)	mIoU (%)
GPP only	66.5	75.6	54.2
Attract (Rigid)	65.1 (−1.4)	76.1 (+0.5)	53.4 (−0.8)
Attract (Soft)	66.6 (+0.1)	77.2 (+1.6)	55.0 (+0.8)
Atrract (Soft)+Repel	68.4 (+1.9)	76.3 (+0.7)	56.0 (+1.8)

Table 4. Evaluation (mIoU,%) of the CAMs, the CAMs with CRF, and the CAMs with CRF and RW [2] on the PASCAL VOC 2012 *train* set.

Methods	seed	w/CRF	w/CRF, RW
CONTA [57] $_{NeurIPS20}$	56.2	65.4	66.1
EDAM [52] $_{CVPR21}$	52.8	58.2	68.1
AdvCAM [30] $_{CVPR21}$	55.6	62.1	68.0
ECS [46] $_{ICCV21}$	56.6	58.6	–
OC-CSE [27]$_{ICCV21}$	56.0	62.8	66.9
CDA [44] $_{ICCV21}$	58.4	–	66.4
PMM [35] $_{ICCV21}$	58.2	61.5	61.0
RIB [30] $_{NeurIPS}$	56.5	62.9	70.6
Ours	56.0	63.5	**71.0**

4.4 Comparison with State-of-the-Arts

By applying a commonly used Random Walk (RW) approach [2] as in [7,30, 46,50,58], we acquire further improved pixel-level pseudo labels for training the semantic segmentation model. As shown in Table 4, though the performance of CAMs of the proposed method is similar to the existing state-of-the-art, our method greatly benefits from CRF (about 7.5%). According to Kweon *et al.* [27], we can interpret this performance gain as a benefit from more precise CAMs that match along object boundaries while activating the whole object. The resulting performance of pseudo labels achieves 71.0% mIoU on PASCAL VOC 2012 *train* set. For a fair comparison with the current state-of-the-art, we train the Deeplab-LargeFOV [8] with the corresponding pseudo labels. The backbone of the segmentation model is ResNet38d. As shown in Table 5, the proposed AEFT achieves a state-of-the-art with 70.9% and 71.7% mIoU on PASCAL VOC 2012 *val* and *test* sets, respectively. Qualitative segmentation results of the proposed method can be found in the *Supplementary Material*, which depicts that the segmentation model can capture fine details as well, owing to the high quality pseudo labels used for training the model. We also evaluate our method in the MS-COCO 2014 dataset to show the superiority and versatility of the proposed framework. It achieves 44.8% on the MS-COCO *val* set, which is a new state-of-the-art, outperforming the other methods by a meaningful margin (1.0%).

Table 5. Performance (mIoU, %) comparison with other state-of-the-art WSSS methods on the PASCAL VOC 2012 and MS-MOCO 2014. Since we use neither saliency nor external dataset at all, we list the methods using image-level only in this table. **Bold** numbers represent the best results.

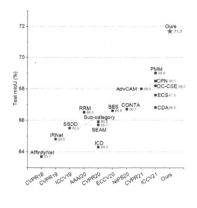

Methods	Backbone	VOC val	VOC test	COCO val
AffinityNet [2]$_{CVPR18}$	ResNet38	61.7	63.7	–
ICD [15]$_{CVPR20}$	ResNet101	64.1	64.3	–
IRNet [1]$_{CVPR19}$	ResNet50	63.5	64.8	32.6
SSDD [43]$_{ICCV19}$	ResNet38	64.9	65.5	–
SEAM [50]$_{CVPR20}$	ResNet38	64.5	65.7	31.9
Sub-category [7]$_{CVPR20}$	ResNet101	66.1	65.9	–
CONTA [57]$_{NIPS20}$	ResNet38	66.1	66.7	33.4
RRM [56]$_{AAAI20}$	ResNet101	66.3	66.5	–
BES [10]$_{ECCV20}$	ResNet101	65.7	66.6	–
CDA [44]$_{ICCV21}$	ResNet38	66.1	66.8	–
ECS [46]$_{ICCV21}$	ResNet38	66.6	67.6	–
AdvCAM [30]$_{CVPR21}$	ResNet101	68.1	68.0	–
OC-CSE [27]$_{ICCV21}$	ResNet38	68.4	68.2	36.4
CPN [58]$_{ICCV21}$	ResNet38	67.8	68.5	–
RIB [28]$_{NeurIPS21}$	ResNet101	68.3	68.6	43.8
PMM [35]$_{ICCV21}$	ResNet38	68.5	69.0	36.7
Ours	ResNet38	**70.9**	**71.7**	**44.8**

5   Conclusions

To address the problems in weakly supervised semantic segmentation (WSSS), we propose a Gated Pyramid Pooling (GPP) layer that replaces the GAP layer by using a feature pyramid and a novel Adversarial Erasing framework via Triplet (AEFT) that incorporates metric learning for suppressing the over-expansion problem in AE. Extensive ablation studies support that the proposed GPP layer outperforms the conventional GAP layer while effectively resolving the impreciseness problem of CAMs with the help of the feature pyramid. In addition, the proposed AEFT succeeds in relieving the over-expansion problem of AE by exploiting the triplet loss as a softer criterion compared to classification loss conventionally used. With the proposed GPP and AEFT, we achieve the state-of-the-art performance both on the PASCAL VOC 2012 and MS-COCO 2014 *val* set with a great margin, only utilizing image-level supervision.

Acknowledgements. This work was supported by Institute of Information and Communications Technology Planning & Evaluation(IITP) Grants funded by Korea Government (MSIT), No. 2020–0-00440, Development of Artificial Intelligence Technology that Continuously Improves Itself as the Situation Changes in the Real World, and No. 2014–3-00123, Development of High Performance Visual BigData Discovery Platform for Large-Scale Realtime Data Analysis.

References

1. Ahn, J., Cho, S., Kwak, S.: Weakly supervised learning of instance segmentation with inter-pixel relations. In: Proceedings of the IEEE Conference on Computer Vision and Pattern Recognition, pp. 2209–2218 (2019)
2. Ahn, J., Kwak, S.: Learning pixel-level semantic affinity with image-level supervision for weakly supervised semantic segmentation. In: Proceedings of the IEEE Conference on Computer Vision and Pattern Recognition, pp. 4981–4990 (2018)
3. Araslanov, N., Roth, S.: Single-stage semantic segmentation from image labels. In: Proceedings of the IEEE/CVF Conference on Computer Vision and Pattern Recognition, pp. 4253–4262 (2020)
4. Bearman, A., Russakovsky, O., Ferrari, V., Fei-Fei, L.: What's the point: semantic segmentation with point supervision. In: Leibe, B., Matas, J., Sebe, N., Welling, M. (eds.) ECCV 2016. LNCS, vol. 9911, pp. 549–565. Springer, Cham (2016). https://doi.org/10.1007/978-3-319-46478-7_34
5. Caesar, H., Uijlings, J., Ferrari, V.: Coco-stuff: Thing and stuff classes in context. In: Computer Vision and Pattern Recognition (CVPR), 2018 IEEE Conference on. IEEE (2018)
6. Cen, J., Yun, P., Cai, J., Wang, M.Y., Liu, M.: Deep metric learning for open world semantic segmentation. 2021 IEEE/CVF International Conference on Computer Vision (ICCV), pp. 15313–15322 (2021)
7. Chang, Y.T., Wang, Q., Hung, W.C., Piramuthu, R., Tsai, Y.H., Yang, M.H.: Weakly-supervised semantic segmentation via sub-category exploration. In: Proceedings of the IEEE/CVF Conference on Computer Vision and Pattern Recognition, pp. 8991–9000 (2020)
8. Chen, L., Papandreou, G., Kokkinos, I., Murphy, K., Yuille, A.L.: Semantic image segmentation with deep convolutional nets and fully connected crfs. In: Bengio, Y., LeCun, Y. (eds.) In: 3rd International Conference on Learning Representations, ICLR (2015), http://arxiv.org/abs/1412.7062
9. Chen, L.C., Papandreou, G., Kokkinos, I., Murphy, K., Yuille, A.L.: Deeplab: semantic image segmentation with deep convolutional nets, atrous convolution, and fully connected crfs. IEEE Trans. Pattern Anal. Mach. Intell. 40(4), 834–848 (2017)
10. Chen, L., Wu, W., Fu, C., Han, X., Zhang, Y.: Weakly supervised semantic segmentation with boundary exploration. In: Vedaldi, A., Bischof, H., Brox, T., Frahm, J.-M. (eds.) ECCV 2020. LNCS, vol. 12371, pp. 347–362. Springer, Cham (2020). https://doi.org/10.1007/978-3-030-58574-7_21
11. Chen, T., Kornblith, S., Norouzi, M., Hinton, G.: A simple framework for contrastive learning of visual representations. In: International conference on machine learning, pp. 1597–1607. PMLR (2020)
12. Chen, X., Fan, H., Girshick, R., He, K.: Improved baselines with momentum contrastive learning. arXiv preprint arXiv:2003.04297 (2020)
13. Dai, J., He, K., Sun, J.: Boxsup: Exploiting bounding boxes to supervise convolutional networks for semantic segmentation. In: Proceedings of the IEEE international conference on computer vision, pp. 1635–1643 (2015)
14. Everingham, M., Van Gool, L., Williams, C.K., Winn, J., Zisserman, A.: The pascal visual object classes (voc) challenge. Int. J. Comput. Vision 88(2), 303–338 (2010)
15. Fan, J., Zhang, Z., Song, C., Tan, T.: Learning integral objects with intra-class discriminator for weakly-supervised semantic segmentation. In: Proceedings of the IEEE/CVF Conference on Computer Vision and Pattern Recognition, pp. 4283–4292 (2020)

16. Fan, J., Zhang, Z., Tan, T., Song, C., Xiao, J.: Cian: Cross-image affinity net for weakly supervised semantic segmentation. In: Proceedings of the AAAI Conference on Artificial Intelligence, vol. 34, pp. 10762–10769 (2020)
17. Hadsell, R., Chopra, S., LeCun, Y.: Dimensionality reduction by learning an invariant mapping. In: 2006 IEEE Computer Society Conference on Computer Vision and Pattern Recognition (CVPR'06), vol. 2, pp. 1735–1742 (2006). https://doi.org/10.1109/CVPR.2006.100
18. He, K., Fan, H., Wu, Y., Xie, S., Girshick, R.: Momentum contrast for unsupervised visual representation learning. In: Proceedings of the IEEE/CVF Conference on Computer Vision and Pattern Recognition (CVPR) (2020)
19. Hou, Q., Jiang, P., Wei, Y., Cheng, M.M.: Self-erasing network for integral object attention. In: Advances in Neural Information Processing Systems, pp. 549–559 (2018)
20. Huang, Z., Wang, X., Wang, J., Liu, W., Wang, J.: Weakly-supervised semantic segmentation network with deep seeded region growing. In: Proceedings of the IEEE Conference on Computer Vision and Pattern Recognition, pp. 7014–7023 (2018)
21. Jia, C., et al.: Scaling up visual and vision-language representation learning with noisy text supervision. In: International Conference on Machine Learning. pp. 4904–4916. PMLR (2021)
22. Jiang, P.T., Hou, Q., Cao, Y., Cheng, M.M., Wei, Y., Xiong, H.K.: Integral object mining via online attention accumulation. In: Proceedings of the IEEE International Conference on Computer Vision, pp. 2070–2079 (2019)
23. Ke, T.W., Hwang, J.J., Yu, S.: Universal weakly supervised segmentation by pixel-to-segment contrastive learning. In: International Conference on Learning Representations (2021). https://openreview.net/forum?id=N33d7wjgzde
24. Khoreva, A., Benenson, R., Hosang, J., Hein, M., Schiele, B.: Simple does it: Weakly supervised instance and semantic segmentation. In: Proceedings of the IEEE Conference on Computer Vision and Pattern Recognition, pp. 876–885 (2017)
25. Kolesnikov, A., Lampert, C.H.: Seed, expand and constrain: three principles for weakly-supervised image segmentation. In: Leibe, B., Matas, J., Sebe, N., Welling, M. (eds.) ECCV 2016. LNCS, vol. 9908, pp. 695–711. Springer, Cham (2016). https://doi.org/10.1007/978-3-319-46493-0_42
26. Krizhevsky, A., Sutskever, I., Hinton, G.E.: Imagenet classification with deep convolutional neural networks. Commun. ACM **60**(6), 84–90 (2017)
27. Kweon, H., Yoon, S.H., Kim, H., Park, D., Yoon, K.J.: Unlocking the potential of ordinary classifier: Class-specific adversarial erasing framework for weakly supervised semantic segmentation. In: Proceedings of the IEEE/CVF International Conference on Computer Vision, pp. 6994–7003 (2021)
28. Lee, J., Choi, J., Mok, J., Yoon, S.: Reducing information bottleneck for weakly supervised semantic segmentation. Adv. Neural. Inf. Process. Syst. **34**, 27408–27421 (2021)
29. Lee, J., Kim, E., Lee, S., Lee, J., Yoon, S.: Ficklenet: Weakly and semi-supervised semantic image segmentation using stochastic inference. In: Proceedings of the IEEE Conference on Computer Vision and Pattern Recognition, pp. 5267–5276 (2019)
30. Lee, J., Kim, E., Yoon, S.: Anti-adversarially manipulated attributions for weakly and semi-supervised semantic segmentation. In: Proceedings of the IEEE/CVF Conference on Computer Vision and Pattern Recognition, pp. 4071–4080 (2021)

31. Lee, J., Yi, J., Shin, C., Yoon, S.: Bbam: Bounding box attribution map for weakly supervised semantic and instance segmentation. In: Proceedings of the IEEE/CVF Conference on Computer Vision and Pattern Recognition, pp. 2643–2652 (2021)
32. Lee, S., Lee, M., Lee, J., Shim, H.: Railroad is not a train: Saliency as pseudo-pixel supervision for weakly supervised semantic segmentation. In: Proceedings of the IEEE/CVF Conference on Computer Vision and Pattern Recognition, pp. 5495–5505 (2021)
33. Li, K., Wu, Z., Peng, K.C., Ernst, J., Fu, Y.: Tell me where to look: Guided attention inference network. In: Proceedings of the IEEE Conference on Computer Vision and Pattern Recognition, pp. 9215–9223 (2018)
34. Li, X., Zhou, T., Li, J., Zhou, Y., Zhang, Z.: Group-wise semantic mining for weakly supervised semantic segmentation. arXiv preprint arXiv:2012.05007 (2020)
35. Li, Y., Kuang, Z., Liu, L., Chen, Y., Zhang, W.: Pseudo-mask matters in weakly-supervised semantic segmentation. In: Proceedings of the IEEE/CVF International Conference on Computer Vision, pp. 6964–6973 (2021)
36. Lin, D., Dai, J., Jia, J., He, K., Sun, J.: Scribblesup: Scribble-supervised convolutional networks for semantic segmentation. In: Proceedings of the IEEE Conference on Computer Vision and Pattern Recognition, pp. 3159–3167 (2016)
37. Lin, T.-Y., et al.: Microsoft COCO: common objects in context. In: Fleet, David, Pajdla, Tomas, Schiele, Bernt, Tuytelaars, Tinne (eds.) ECCV 2014. LNCS, vol. 8693, pp. 740–755. Springer, Cham (2014). https://doi.org/10.1007/978-3-319-10602-1_48
38. Liu, W., Wen, Y., Yu, Z., Li, M., Raj, B., Song, L.: Sphereface: Deep hypersphere embedding for face recognition. In: Proceedings of the IEEE Conference on Computer Vision and Pattern Recognition, pp. 212–220 (2017)
39. van den Oord, A., Li, Y., Vinyals, O.: Representation learning with contrastive predictive coding. ArXiv abs/1807.03748 (2018)
40. Papandreou, G., Chen, L.C., Murphy, K.P., Yuille, A.L.: Weakly-and semi-supervised learning of a deep convolutional network for semantic image segmentation. In: Proceedings of the IEEE International Conference on Computer Vision, pp. 1742–1750 (2015)
41. Russakovsky, O., et al.: Imagenet large scale visual recognition challenge. Int. J. Comput. Vision 115(3), 211–252 (2015)
42. Schroff, F., Kalenichenko, D., Philbin, J.: Facenet: A unified embedding for face recognition and clustering. In: 2015 IEEE Conference on Computer Vision and Pattern Recognition (CVPR), pp. 815–823 (2015). https://doi.org/10.1109/CVPR.2015.7298682
43. Shimoda, W., Yanai, K.: Self-supervised difference detection for weakly-supervised semantic segmentation. In: Proceedings of the IEEE International Conference on Computer Vision, pp. 5208–5217 (2019)
44. Su, Y., Sun, R., Lin, G., Wu, Q.: Context decoupling augmentation for weakly supervised semantic segmentation. arXiv preprint arXiv:2103.01795 (2021)
45. Sun, G., Wang, W., Dai, J., Van Gool, L.: Mining cross-image semantics for weakly supervised semantic segmentation. arXiv preprint arXiv:2007.01947 (2020)
46. Sun, K., Shi, H., Zhang, Z., Huang, Y.: Ecs-net: Improving weakly supervised semantic segmentation by using connections between class activation maps. In: Proceedings of the IEEE/CVF International Conference on Computer Vision, pp. 7283–7292 (2021)
47. Takikawa, T., Acuna, D., Jampani, V., Fidler, S.: Gated-scnn: Gated shape cnns for semantic segmentation. In: Proceedings of the IEEE International Conference on Computer Vision, pp. 5229–5238 (2019)

48. Vernaza, P., Chandraker, M.: Learning random-walk label propagation for weakly-supervised semantic segmentation. In: Proceedings of the IEEE Conference on Computer Vision and Pattern Recognition, pp. 7158–7166 (2017)
49. Wang, W., Zhou, T., Yu, F., Dai, J., Konukoglu, E., Van Gool, L.: Exploring cross-image pixel contrast for semantic segmentation. In: Proceedings of the IEEE/CVF International Conference on Computer Vision (ICCV), pp. 7303–7313 (2021)
50. Wang, Y., Zhang, J., Kan, M., Shan, S., Chen, X.: Self-supervised equivariant attention mechanism for weakly supervised semantic segmentation. In: Proceedings of the IEEE/CVF Conference on Computer Vision and Pattern Recognition, pp. 12275–12284 (2020)
51. Wei, Y., Feng, J., Liang, X., Cheng, M.M., Zhao, Y., Yan, S.: Object region mining with adversarial erasing: A simple classification to semantic segmentation approach. In: Proceedings of the IEEE conference on computer vision and pattern recognition, pp. 1568–1576 (2017)
52. Wu, T., et al.: Embedded discriminative attention mechanism for weakly supervised semantic segmentation. In: Proceedings of the IEEE/CVF Conference on Computer Vision and Pattern Recognition, pp. 16765–16774 (2021)
53. Wu, Z., Shen, C., Van Den Hengel, A.: Wider or deeper: revisiting the resnet model for visual recognition. Pattern Recogn. **90**, 119–133 (2019)
54. Xu, L., Ouyang, W., Bennamoun, M., Boussaid, F., Sohel, F., Xu, D.: Leveraging auxiliary tasks with affinity learning for weakly supervised semantic segmentation. In: Proceedings of the IEEE/CVF International Conference on Computer Vision, pp. 6984–6993 (2021)
55. Yao, Y., et al.: Non-salient region object mining for weakly supervised semantic segmentation. In: Proceedings of the IEEE/CVF Conference on Computer Vision and Pattern Recognition, pp. 2623–2632 (2021)
56. Zhang, B., Xiao, J., Wei, Y., Sun, M., Huang, K.: Reliability does matter: An end-to-end weakly supervised semantic segmentation approach. In: Proceedings of the AAAI Conference on Artificial Intelligence, vol. 34, pp. 12765–12772 (2020)
57. Zhang, D., Zhang, H., Tang, J., Hua, X., Sun, Q.: Causal intervention for weakly-supervised semantic segmentation. In: Advances in Neural Information Processing Systems (2020)
58. Zhang, F., Gu, C., Zhang, C., Dai, Y.: Complementary patch for weakly supervised semantic segmentation. In: Proceedings of the IEEE/CVF International Conference on Computer Vision, pp. 7242–7251 (2021)
59. Zhang, X., Wei, Y., Feng, J., Yang, Y., Huang, T.S.: Adversarial complementary learning for weakly supervised object localization. In: Proceedings of the IEEE Conference on Computer Vision and Pattern Recognition, pp. 1325–1334 (2018)
60. Zhao, H., Shi, J., Qi, X., Wang, X., Jia, J.: Pyramid scene parsing network. In: Proceedings of the IEEE conference on computer vision and pattern recognition, pp. 2881–2890 (2017)
61. Zhou, B., Khosla, A., Lapedriza, A., Oliva, A., Torralba, A.: Learning deep features for discriminative localization. In: Proceedings of the IEEE conference on computer vision and pattern recognition, pp. 2921–2929 (2016)

Continual Semantic Segmentation via Structure Preserving and Projected Feature Alignment

Zihan Lin⬩, Zilei Wang(✉)⬩, and Yixin Zhang⬩

University of Science and Technology of China, Hefei, China
{myustc,zhyx12}@mail.ustc.edu.cn, zlwang@ustc.edu.cn

Abstract. Deep networks have been shown to suffer from catastrophic forgetting. In this work, we try to alleviate this phenomenon in the field of continual semantic segmentation (CSS). We observe that two main problems lie in existing arts. First, attention is only paid to designing constraints for encoder (*i.e.*, the backbone of segmentation network) or output probabilities. But we find that forgetting also happens in the decoder head and harms the performance greatly. Second, old and new knowledge are entangled in intermediate features when learning new categories, making existing practices hard to balance between plasticity and stability. On these bases, we propose a framework driven by two novel constraints to address the aforementioned problems. First, a structure preserving loss is applied to the decoder's output to maintain the discriminative power of old classes from two different granularities in embedding space. Second, a feature projection module is adopted to disentangle the process of preserving old knowledge from learning new classes. Extensive evaluations on VOC2012 and ADE20K datasets show the effectiveness of our approach, which significantly outperforms existing state-of-the-art CSS methods.

Keywords: Incremental learning · Semantic segmentation

1 Introduction

Semantic segmentation aims to assign every pixel a semantic category for a given image, which is a fundamental but challenging computer vision task. In recent years, the state-of-the-art methods [7,47,48] based on Fully Convolutional Network (FCN) [30] have achieved great success on large-scale benchmarks [9,15,49]. These models are designed to be trained in a one-shot mode with all data prepared in advance. When fitted in real-world applications, they will inevitably encounter situations where new categories need to be gradually learned. A naive way is to fine-tune the models on new data. But they often fail to preserve the performance of learned classes when updating themselves, which is called catastrophic forgetting [16] and becomes a main obstacle in practical applications.

S. Avidan et al. (Eds.): ECCV 2022, LNCS 13689, pp. 345–361, 2022.
https://doi.org/10.1007/978-3-031-19818-2_20

Incremental learning, which enables models to continuously learn new knowledge like human beings, is considered to be a promising solution to this challenge.

Class incremental learning (CIL) has been widely studied in image classification [14,22,26,29,39], but it only received attention very recently in semantic segmentation. Extending methods from classification, current CSS approaches have managed to make some progress, but two problems still remain unsolved. One of them is that existing practices add constraints on either output probabilities [3] or encoder [13,37] to prevent forgetting. But they fail to consider the decoder head, which is a component exclusively in segmentation model. We argue that forgetting phenomenon also exists in decoder and has a significant impact on performance (details in Sect. 5.5). On this demand, we aim to design a constraint specially for decoder to mitigate the forgetting. Another problem lies in the knowledge distillation (KD) [20] used by existing methods [12–14] to prevent changes in intermediate features (i.e., the output of encoder). Fixing the model activations is reasonable in classification since the ground truth of a certain image is consistent across all learning steps, but it's not the case in CSS. The semantic information of a given image might gradually increase, which is known as the background shift [3]. This requires the intermediate features to be updated accordingly to integrate the knowledge of upcoming classes, making the aforementioned practices fail in this situation. Simply mimicking the intermediate features of the old model would cause a conflict, where the model tries to be persistent on the old activations while struggling to learn the new knowledge, and ultimately results in degradation of performance.

In response to these problems, we propose a novel framework called SPPA, standing for **S**tructure **P**reserving and **P**rojected feature **A**lignment. It consists of two main components to tackle the forgetting of encoder and decoder, respectively. For encoder, we adopt a feature projection module to extract from intermediate features some low-dimensional representations of old knowledge, on which alignment is performed. This practice disentangles the preservation of old knowledge and the integration of new knowledge, which enables us to maintain the integrity of old knowledge without hindering the update process of new knowledge. For decoder, inter-class and intra-class structures are first modeled from its output, which reflect both coarse and fine-grained relations in embedding space. The structures are then explicitly maintained during incremental learning, which can effectively preserve the discriminability of old classes while being more flexible for new classes. The extensive experiments on Pascal-VOC 2012 [15] and ADE20K [49] demonstrate the effectiveness of our method. The contributions of this paper are summarized as

- We unveil the forgetting phenomena of decoder head in CSS, and propose to mitigate this by maintaining the class-related structures in embedding space at two different granularities.
- We manage to solve the potential conflict of feature KD in CSS by disentangle the preservation of old knowledge and the integration of new knowledge. Consequently provides a better trade-off between stability and plasticity.

– We integrate both modules into a unified framework, which is evaluated on two popular datasets with diverse experimental settings. The results reveal that our method outperforms previous state-of-the-art methods.

2   Related Works

Continual Learning. Deep neural networks have achieved great success in many fields, such as image classification [19], semantic segmentation [7] and object detection [40]. However, they often suffer from catastrophic forgetting [16] when situated in real scenarios where continuous streams of new data are involved. Continual learning that aims at tackling this obstacle has become an active field recently.

Continual learning for image classification has been extensively studied these years. Current techniques can be mainly divided into rehearsal methods, architectural methods and regularization methods. Rehearsal methods store a limited amount of raw images [2,22,39] and interleave them with new data to relieve forgetting. Some methods store intermediate features [10,18,25] instead, as they require less storage and contain richer information. Generation-based methods are derived from them that resort to generative networks to obtain images [8,42] or features [43] of old classes. Architectural methods either use a sub-network [32,33] to solve each independent task at a cost of limited scalability, or dynamically expand the network [28,44,45] for learning new tasks with growing complexity. Most of these methods require additional task labels and are restricted to multi-head [34] setup. Regularization methods design extra loss functions to maintain the previous knowledge, which can be divided into parameter regularization [26,46] and distillation-based methods. Parameter regularization prevents updates on the most important parameters for old tasks based on various metrics like fisher information matrix [26] or gradient magnitude [46]. Distillation-based methods are the most common currently. It was first introduced by LwF [29] to penalize the changes in output logits. LwM [12] additionally performs distillation on attention maps. [22] introduces a stronger constraint by punishing changes in feature vectors. Distillation on intermediate features is used in [14] to further reduce forgetting.

Continual Semantic Segmentation. Modern deep neural networks for semantic segmentation are mostly based on the fully convolutional network (FCN) [30]. Recent developments mainly try to exploit spatial information to improve accuracy. For example, encoder-decoder architecture [1,41] is used to prevent spatial information loss. Atrous convolution [35] enlarges the field of view to incorporate more spatial context information. Multi-scale information is further considered in ASPP [5] and PSPNet [47]. More recently, attention mechanisms have been used to model spatial dependences [17,24,48].

Despite their great success, segmentation networks inevitably suffer from catastrophic forgetting when used in online scenarios. [36] is the first one targeting for CSS. But it requires the labels of both old and current classes be provided,

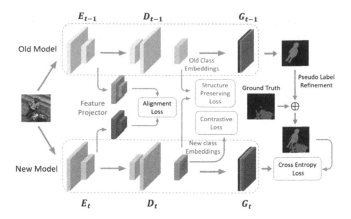

Fig. 1. Overview of the proposed framework. Alignment is performed on the representations generated by the feature projector to solve the potential conflict in feature distillation for encoder. The structure preserving loss is used to maintain discriminative structures for old classes, while the contrastive loss is used to optimize the structures of new classes. Finally, pseudo label is used to provide missing annotations for old classes.

which greatly limits its capability in real scenarios. [3] formalized background shift, which is a new obstacle specially for CSS. To overcome this, they modified traditional distillation loss and cross-entropy loss to an unbiased version considering the semantic inconsistency of background. More recently, some methods proposed to tackle forgetting in encoder. [37] designed a framework combining prototype distillation, feature sparcification and contrastive learning. However, a plasticity issue might arise when trying to punish any changes in prototypes. [13] proposed a multi-scale pooling distillation to preserve long and short range statistics in intermediate features. But it cannot solve the conflict of old and new knowledge stated before. What's more, all of the existing works never considered the forgetting of decoder. Our work is the first attempt to tackle forgetting in both encoder and decoder in a single framework, with two novel distillation approaches to achieve a better plasticity-stability trade-off than existing methods.

3   Problem Definition

In semantic segmentation, let $\mathcal{X} \in \mathbb{R}^{H \times W \times 3}$ be the input space, and $\mathcal{Y} \in \mathcal{C}^{H \times W}$ be the label space where $\mathcal{C}$ is a given category set. Given a dataset $\mathcal{T} = \{(\mathbf{x}_n, \mathbf{y}_n)\}$ where $(\mathbf{x}_n, \mathbf{y}_n) \subset \mathcal{X} \times \mathcal{Y}$. The goal is to produce a segmentation map $\hat{\mathbf{y}} \subset \mathcal{Y} \in \mathcal{C}^{H \times W}$ by assigning each pixel $x_i \in \mathbf{x}$ a class in $\mathcal{C}$. This is usually done using a deep neural network $M : \mathcal{X} \mapsto \mathbb{R}^{H \times W \times C}$ and the segmentation result is calculated as $\hat{\mathbf{y}} = \arg\max_c M(\mathcal{X})[h, w, c]$. Nowadays M is usually an encoder-decoder architecture made by a feature extractor E and a segmentation head

D (*i.e.*, $M = E \circ D$). In our work, we treat the final classifier as an individual module G, which makes $M = E \circ D \circ G$. We use $\mathbf{F}$ to represent the intermediate feature output by encoder $E(\mathbf{x})$, and $\mathbf{E}$ to represent the embedding output by decoder $D(\mathbf{F})$ before final classifier.

In conventional training pipeline, the complete training set $\mathcal{T}$ is available and the model is trained only once to learn all classes $\mathcal{C}$. While in continual learning, the training procedure is composed of multiple learning steps. At each step, a subset of the training set is provided together with a set of novel classes to be learned. More specifically, at the initial learning step $t = 0$, a standard supervised training is performed on a subset of training data $\mathcal{T}^0$ with labels of $\mathcal{C}^0$. Next moving to a more general step t, a set of novel classes $\mathcal{C}^t \in \mathcal{C}$ is introduced, expanding the learned label set to $\mathcal{C}^{0:t} = \mathcal{C}^{0:t-1} \cup \mathcal{C}^t$. A training set $\mathcal{T}^t$ with all $\mathcal{C}^{0:t-1}$ labeled as background and only $\mathcal{C}^t$ been labeled is provided to update the previous model $M_{t-1} : \mathcal{X} \mapsto \mathbb{R}^{H \times W \times \mathcal{C}^{0:t-1}}$ to $M_t : \mathcal{X} \mapsto \mathbb{R}^{H \times W \times \mathcal{C}^{0:t}}$. As in standard class incremental learning, we assume that the classes introduced in different steps are disjoint ($\mathcal{C}^i \cap \mathcal{C}^j = \emptyset$) except for the background class b.

4   Method

In this section, we will introduce the proposed modules one by one, following the structure depicted in Fig. 1 from left to right to make it easier to follow.

4.1   Projected Feature Alignment for Encoder

We first go through the feature distillation in existing works. [12,14,22] have attempted to perform distillation on intermediate features, which consists of a distance metric to minimize the differences between feature maps from old and new models. Due to the rich information contained in intermediate features, this approach often yields better performance than logits distillation. An Euclidean distance is usually applied, and the distillation loss can be formulated as

$$L = \|E_t(\mathbf{x}) - E_{t-1}(\mathbf{x})\|_2, \tag{1}$$

where $E_t(\mathbf{x})$, $E_{t-1}(\mathbf{x})$ indicate the feature of image $\mathbf{x}$ extracted by the encoder of step t and step $t-1$, respectively.

It's reasonable to keep the activations invariant in classification since the semantic information of an image will not change across all incremental learning steps. In CSS however, the semantic information might increase (*e.g.*, we want to learn a new class which was ignored and labeled as background before). This is known as the background shift phenomenon [3] exclusively in CSS. Ideally, the intermediate features should be updated to integrate new knowledge. It means that even if the current model does not forget the old knowledge, its intermediate features could still be quite different from the ones extracted by the old model. If we make the current model directly mimic the old one, a conflict arises between being unchanged in the activations and adapting to new classes at the same time.

Therefore, we try to disentangle the processes of retaining old knowledge and learning new classes for intermediate features. It is known that an auto-encoder can generate compact representations that contain sufficient information to reconstruct the original feature maps [21]. We leverage this property to extract low-dimensional representations of old knowledge from intermediate features. These representations are then used for alignment. It's worth noting that we do not reduce the spatial resolution like traditional auto-encoder since the segmentation task requires high spatial precision. We define the auto-encoder architecture as a projector P and a reconstructor R to avoid being confused with the encoder and the decoder in segmentation model. Before starting learning step t, we use the intermediate features of T^t extracted by E_{t-1} to train a projection module P_{t-1} from scratch using the reconstruction loss L_{recon}.

$$L_{recon} = \|E_{t-1}(\mathbf{x}) - R_{t-1}(P_{t-1}(E_{t-1}(\mathbf{x})))\|_2 \tag{2}$$

After it converges, the reconstructor R_{t-1} is discarded, the old model M_{t-1} and the trained P_{t-1} is used to initialize the current model (i.e., M_t and P_t). During the training of current step t, M_{t-1} and P_{t-1} are fixed, while P_t and M_t are updated. L1 distance is minimized between the output of P_{t-1} and P_t.

$$L_{ali} = \|P_{t-1}(E_{t-1}(\mathbf{x})) - P_t(E_t(\mathbf{x}))\|_1 \tag{3}$$

The design intuition is as follows: The output representations of P_{t-1} contain sufficient information to restore the old features $E_{t-1}(\mathbf{x})$ which stand for the old knowledge. If we can extract the same representations from the new feature maps $E_t(\mathbf{x})$ by P_t, it indicates that the new feature maps still hold the same old knowledge as that of the old. With this design, we make it possible to add constraints on the old knowledge without the potential impediment on learning new knowledge.

4.2   Class Structure Preserving for Decoder

Existing approaches utilizing pre-classifier embeddings commonly fall into adding constraints for feature vectors [22] or class prototypes [4,37]. However, when applied in CSS to prevent forgetting in decoder head, we find that these methods yield overly strong constraints on the output, which penalizes any possible changes with respect to its previous position in the embedding space. On the one hand, it might impede the model from finding a globally optimal position that benefits both old and new classes, thus hindering the learning of new classes. On the other hand, the model tries to be persistent in the absolute positions of old classes while integrating new classes. This might cause an optimization conflict, which results in uncontrollable small drifts of old classes in embedding space and ultimately leads to a chaotic distribution, as shown in Fig. 6.

Since it is undesirable to directly fix the embeddings of old classes, we turn to seeking a practice that can preserve the performance of old classes without sacrificing the freedom to learn new classes. It is known that the embeddings from a well-trained model form a space, in which the instances of the same class are close to each other and far from those of different class. This geometry in embedding space is the key to making each class linearly separable. Inspired by this, we propose to model the structures in embedding space and explicitly maintain them during incremental learning steps. In this work, we consider two structures of different granularity: the structure between different classes (inter-class structure) and the structure within a single class (intra-class structure).

Class prototypes (*i.e.*, mean of feature vectors of a class) are needed when modeling these two structures. Generally, they can be computed in a global, in-batch or in-image manner, which represents the class information from coarse-grained (global-level) to fine-grained (instance-level). In segmentation, the embeddings of the same class vary across images and, in our observation, form image-level mini-clusters. Adopting in-image prototypes to obtain a fine-grained relation in our case helps to reflect these differences at an instance level and slightly boosts the performance. Similar practices are also observed in other designs for segmentation, like [23] uses in-image prototypes for contrastive learning. The in-image prototype $\mathbf{p}_c$ of class c for an input image is defined as

$$\mathbf{p}_c = \frac{1}{|\mathbf{y}^* = c|} \sum_{\substack{e_i \in \mathbf{E} \\ i = c}} \mathbf{e}_i. \tag{4}$$

Since $\mathbf{E}$ has lower spatial resolution than input due to the network architecture, $\mathbf{y}$ should be resized to match the spatial dimension of $\mathbf{E}$ and is denoted as $\mathbf{y}^*$. $i = c$ means the position i is labeled as class c in $\mathbf{y}^*$. Because we don't have labels for old classes, the pseudo label $\hat{\mathbf{y}}^*$ output by the old model is used to obtain old class prototypes.

The inter-class structure is formulated as a distance matrix A between each in-image prototype of old classes obtained in the current batch, where A_{ij} is the cosine distance between prototypes i and j. The consistency constraint is then applied between A^t from M_t and A^{t-1} from M_{t-1}. $\bar{\mathbf{p}}_i$ is l2-normalized version of $\mathbf{p}_i$ and $\langle \bar{\mathbf{p}}_i, \bar{\mathbf{p}}_j \rangle$ calculates the cosine similarity between $\mathbf{p}_i$ and $\mathbf{p}_j$.

$$A_{ij} = 1 - \langle \bar{\mathbf{p}}_i, \bar{\mathbf{p}}_j \rangle, \quad i, j \in \mathcal{C}^{0:t-1} \tag{5}$$

$$L_{inter} = \|A^t - A^{t-1}\|_F \tag{6}$$

As for the intra-class structure, the embedding vectors of the same class are clustered closely with each other, so we leverage Euclidean distance to better reflect the small changes in the structure. Intuitively, the intra-class structure can be maintained by keeping the distance between embedding vectors and their prototypes. We further integrate the direction information to prevent the embedding vectors from rotating (See Fig. 2 right), making the constraints as fixing the *relative position* of embedding vectors with regard to their prototypes. The loss function for a class c is defined as

$$L_{intra}(c) = \frac{1}{|\hat{\mathbf{y}}^* = c|} \sum_{\substack{e_i \in \mathbf{E} \\ i = c}} \|(\mathbf{e}_i^{t-1} - \mathbf{p}_c^{t-1}) - (\mathbf{e}_i^t - \mathbf{p}_c^t)\|_2, \tag{7}$$

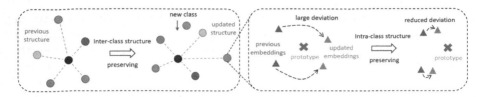

Fig. 2. Illustration of the structure preserving loss, which maintains inter-class structure and intra-class structure respectively.

in which $\mathbf{e}_i^{t-1}$, $\mathbf{p}_c^{t-1}$ is obtained from M_{t-1}. $\mathbf{e}_i^t$, $\mathbf{p}_c^t$ is obtained from M_t. ν is used to balance between inter-class term and intra-class term.

$$L_{str} = L_{inter} + \nu \sum_{c \in \mathcal{C}^{0:t-1}} L_{intra}(c). \tag{8}$$

Compared to previous approaches, the structure preserving loss allows the model to freely re-arrange the feature vectors in the embedding space as long as the discriminative structures are intact during this procedure. By maintaining both coarse and fine-grained structures, it effectively mitigates the forgetting of old classes while at the same time avoiding the optimization dilemma, thus providing a better stability-plasticity trade-off.

The proposed structure preserving loss mainly focuses on old classes. We want to obtain good initial structures in the embedding space when learning new classes. Then the good structures can be explicitly maintained in future steps by the structure preserving loss. We adopt contrastive learning to achieve this. The losses are extended from [22,37], and we confirm that a modern contrastive loss with similar effects can also be used. L_{comp} encourages the embedding vectors to be close to their prototypes, while L_{mar} is to ensure enough distance between different classes. By doing this, a structure with compact distributions within classes and wide boundaries between classes can be obtained, which is less prone to catastrophic forgetting. The loss functions are formulated as follows, with μ a hyper-parameter and Δ a pre-defined distance.

$$L_{comp}(c) = \frac{1}{|\mathbf{y}^* = c|} \sum_{\substack{\mathbf{e}_i \in \mathbf{E} \\ i=c}} \|\mathbf{e}_i^t - \mathbf{p}_c^t\|_2, \tag{9}$$

$$L_{mar} = \sum_i \sum_j \max\{0, \Delta - \langle \bar{\mathbf{p}}_i, \bar{\mathbf{p}}_j \rangle\}, \quad i, j \in \mathcal{C}^t, \tag{10}$$

$$L_{cont} = L_{mar} + \mu \sum_{c \in \mathcal{C}^t} L_{comp}(c). \tag{11}$$

4.3   Pseudo Label for Old Classes

Training the model directly on given labels using cross-entropy loss will aggravate catastrophic forgetting because previous classes are labeled as background. A common practice is to adopt the pseudo label [27] technique to provide missing annotations for old classes. At step t, we use the prediction of M_{t-1} on the current training set $\mathcal{T}^t$ to label the background regions. The pseudo labels are then refined by only accepting pixels with certainty above a threshold. Considering that the model cannot learn each class equally well. If the same threshold is applied to all classes, the poorly-learned classes might be overly rejected and unnecessarily sacrifice the integrity of pseudo label. It's better to let each class has its own threshold. We use entropy (denoted as u) as certainty measurement. An entropy threshold t_c is selected for each class to keep a fixed percentage of the raw pseudo label whose prediction entropy is below t_c.

$$y_i^{pseudo} = \begin{cases} y_i & \text{if } y_i \in \mathcal{C}^t, \\ \text{argmax}_{c \in \mathcal{C}^{0:t-1}} M_{t-1}(\mathbf{x}) & \text{elif } u < t_c, \\ 0 & \text{otherwise.} \end{cases} \quad (12)$$

4.4 Loss Function

Combining the losses introduced above, the overall loss is formulated as follows, with α, β, γ being hyper-parameters:

$$L = L_{ce} + \alpha L_{ali} + \beta L_{str} + \gamma L_{cont}. \quad (13)$$

5 Experiments

We compare the performance of our method against four strong CSS methods, including ILT [36], MiB [3], SDR [37], and PLOP [13]. All methods in comparison (including ours) do not use the replay strategy [39], where a small amount of previous data is stored and rehearsed. The theoretical performance is obtained by training the model in an offline manner with all data and labels (given as Upper Bound).

5.1 Experiments Setup

Datasets and Metrics. We evaluated our framework on Pascal-VOC 2012 [15] with 20 foreground classes and a background class, and ADE20K [49] with 150 foreground classes. We use mean Intersection over Union (mIoU) as the performance metric. After the whole training process, the performance on initial classes $\mathcal{C}^0$ is denoted as *old*. On classes learned in incremental steps $\mathcal{C}^{1:T}$, it is denoted as *new*. And on all classes $\mathcal{C}^{0:T}$, it is denoted as *all*.

Incremental Protocols. Two different settings of CSS are described in [3] concerning different split methods for training set. **Disjoint setting:** at step t, the given images contain pixels only belonging to either old classes or current classes $\mathcal{C}^{0:t-1} \cup \mathcal{C}^t$. **Overlapped setting:** at step t, images that contain at least one pixel belonging to the current classes $\mathcal{C}^t$ are provided. In both settings, Only labels for $\mathcal{C}^t$ is provided, and $\mathcal{C}^{0:t-1}$ are labeled as background. The main difference is that, the background in disjoint setting contains only seen classes $\mathcal{C}^{0:t-1}$. While in overlapped setting, it might contain old and future classes $\mathcal{C}^{0:t-1} \cup \mathcal{C}^{t+1:T}$. This is a more realistic setting for CSS because it is hard to guarantee that there will not be a demand to learn new concepts that were ignored previously. It is also more challenging due to severer semantic shifts.

Training Procedure. For fair comparison, we follow most of the settings in previous works [3,13]. We adopt the Deeplabv3 [6] with an output stride of 16. ResNet-101 [19] serves as the backbone and is pretrained on ImageNet [11]. The model is optimized using SGD with momentum, with a learning rate of $7e-3$ for the initial step and $7e-4$ for the following steps. The learning rate is decreased according to the polynomial decay rule of power 0.9. The model is trained with a batch size of 16 for 30 epochs on VOC2012 and 60 epochs on ADE20K. The input image is cropped to 512×512 for both training and validation. The data augmentation for training consists of a random scaling with a factor from 0.5 to 2.0 and a random left-right flip.

5.2   Evaluation on Pascal VOC2012

We present the results on Pascal-VOC 2012 in this section. Following [3,13,37], we use three experiment setups: adding one class for a step (19–1), adding five classes for a step (15–5), and adding one class sequentially for five steps (15–1). The results of disjoint and overlapped are reported in Table 1 and Table 2. The results of ours are averaged over three runs to increase statistical significance.

Single Step Addition of One Class (19–1). In this setup, the model first learns 19 classes and then the last class is added. Directly fine-tuning the model always leads to the worst performance. ILT, which is the first method targeting CSS, struggles to learn the new class but still suffers from forgetting. The methods from MiB start to explicitly consider the background shift in CSS and all of them have steady improvements. Our method achieves the best overall performance in both disjoint and overlapped settings. We notice that despite of only a single class being added, this setting is actually difficult since the training set is rather small and monotonous.

Table 1. Results of different methods on **overlapped** Pascal-VOC 2012. Best in **bold**.

Method	19–1 (2 tasks)			15–5 (2 tasks)			15–1 (6 tasks)		
	Old	New	All	Old	New	All	Old	New	All
Finetune	34.7	14.9	33.8	12.5	36.9	18.3	4.9	3.2	4.5
ILT [36]	67.1	12.3	64.4	66.3	40.6	59.9	4.9	7.8	5.7
MiB [3]	70.2	22.1	67.8	75.5	49.4	69.0	46.2	12.9	37.9
SDR [37]	69.1	32.6	67.4	75.4	52.6	69.9	44.7	21.8	39.2
PLOP [13]	75.3	**37.3**	73.5	75.7	51.7	70.0	65.1	21.1	54.6
Ours	**76.5**	36.2	**74.6**	**78.1**	**52.9**	**72.1**	**66.2**	**23.3**	**56.0**
Upper bound	77.6	76.7	77.5	79.0	72.8	77.5	79.0	72.8	77.5

Table 2. Results of different methods on **disjoint** Pascal-VOC 2012. Best in **bold**.

Method	19–1 (2 tasks)			15–5 (2 tasks)			15–1 (6 tasks)		
	Old	New	All	Old	New	All	Old	New	All
Finetune	35.2	13.2	34.2	8.4	33.5	14.4	5.8	4.9	5.6
ILT [36]	69.1	16.4	66.4	63.2	39.5	57.3	3.7	5.7	4.2
MiB [3]	69.6	25.6	67.4	71.8	43.3	64.7	35.1	13.5	29.7
SDR [37]	69.9	37.3	68.4	73.5	47.3	67.2	59.2	12.9	48.1
PLOP [13]	75.3	**38.8**	73.6	71.0	42.8	64.2	57.8	13.6	46.4
Ours	**75.5**	38.0	**73.7**	**75.3**	**48.7**	**69.0**	**59.6**	**15.6**	**49.1**
Upper bound	77.6	76.7	77.5	79.0	72.8	77.5	79.0	72.8	77.5

Single Step Addition of Five Classes (15–5). In this setup, the remaining 5 classes are added in a single step after learning the first 15 classes. It has the most severe semantic shift due to the most classes being added in one go. Such a setup requires the model to have enough plasticity to learn 5 classes in one go while being stable on old knowledge. In this scenario, our method outperforms all previous works by a large margin in both disjoint and overlapped settings. In the disjoint setting, our method improves by about 2% compared to SDR and about 5% compared to PLOP. In the overlapped setting, our method has another gain of about 2% compared to the second place. What's more, our method boosts the performance of old classes to only 1% lower than the upper bound, together with a notable gain in new class performance.

Multi-step Addition of Five Classes (15–1). This is similar to 15–5 except that the last 5 classes are added one by one, making it the most challenging because of multiple learning phases. As presented in Table 1 and Table 2, all methods suffer a great performance drop. ILT almost forgets all the knowledge. Even the second place, PLOP, has a significant drop compared to 15–5. Our method again achieves state-of-the-art performance on both the disjoint and overlapped setups, proving its ability under longer learning sequences.

Visual Results. We visualize the results of all methods on different setups to unveil more details. As shown in Fig. 3, our method consistently produces better segmentation results than all competitors. For example, in 19–1 our method successfully distinguishes all three classes while the competitors fail to recognize at least one class. The same situation can be observed in 15–5 and 15–1, in which our method generates a segmentation map closer to the ground truth and less confusion between classes is observed. We further provide a visualization of each step in 15–1 setup to present the knowledge shift across each step. As shown in Fig. 4, all the competitors suffer more or less from catastrophic forgetting. The change is especially great in step 5, in which most of the old classes are misclassified into the new class *train* learned in this step. But our method shows good robustness against forgetting, which presents constant segmentation results across each step.

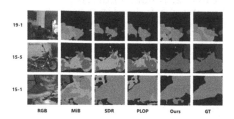

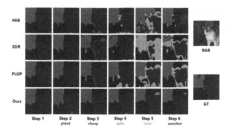

Fig. 3. Visual results of competing methods on different setups of overlapped VOC2012. Best viewed in color. (Color figure online)

Fig. 4. Visual results of competing methods on each step of 15–1 overlapped VOC2012. Best viewed in color. (Color figure online)

5.3   Evaluation on ADE20K

In this section, we evaluate our method on the ADE20K dataset. Following [3,13], we adopt three setups on the overlapped protocol: adding the last 50 classes in a single step (100–50), adding 50 classes each step (50–50), and adding 10 classes each step for the last 50 classes (100–10). ADE20K is much more complex than VOC2012, as we can only get 37.9% mIoU under the offline training. We compare the same methods as in Sect. 5.2 except for SDR since it does not provide their results on ADE20K.

The full results are presented in Table 3. ILT suffers from critical forgetting, while PLOP and MiB achieve much better results. Our method outperforms all previous methods in both 100–50 and 50–50 setups with an improvement of more than 2% compared to PLOP. When diving into longer learning sequences, our method shows higher performance on old classes, and achieves comparable overall performance compared to PLOP. These experiments show the effectiveness of our method on a large-scale dataset.

Table 3. Results of different methods on overlapped ADE20K in mIoU, best in **bold**

Method	100–50 (2 tasks)			50–50 (3 tasks)			100–10 (6 tasks)		
	old	New	All	Old	New	All	Old	New	All
ILT [36]	18.2	14.4	17.0	3.5	12.8	9.7	0.1	3.0	1.0
MiB [3]	40.5	17.1	32.7	45.5	21.0	29.3	38.2	11.1	29.2
PLOP [13]	41.8	14.8	32.9	48.8	20.9	30.4	40.4	**13.6**	**31.5**
Ours	**42.9**	**19.9**	**35.2**	**49.8**	**23.9**	**32.5**	41.0	12.5	**31.5**
Upper Bound	43.5	26.7	37.9	50.3	31.7	37.9	43.5	26.7	37.9

5.4   Ablation Study

We investigate the effects of key components in our framework using an ablation study on 15–1 overlapped VOC2012. Starting from the basic fine-tuning, we gradually add the proposed components upon it. The full results are shown in Table 4. It can be seen that the performance is boosted steadily as more components are added. Each module in our framework can bring benefits to both old and new classes. We conducted another experiment by removing a single component from the framework to better show their individual contribution to performance. Among all the components, the structure preserving loss contributes the most to performance (-4.7% if removed), and the second is the projected feature alignment (-2.1% if removed). This well demonstrates the effectiveness of the two proposed techniques.

Table 4. Performance contribution of each component on 15–1 overlapped VOC2012

L_{ce}	pseudo	L_{ali}	L_{str}	L_{cont}	Old	New	All
✓					4.9	3.2	4.5
✓	✓				23.5	6.6	19.5
✓	✓	✓			60.8	17.2	50.4
✓	✓	✓	✓		65.3	21.2	54.8
✓	✓	✓	✓	✓	**66.2**	**23.0**	**55.9**
✓	✓		✓	✓	63.7	22.0	53.8
✓	✓	✓		✓	61.7	17.8	51.2

5.5   Further Analysis

Forgetting in Decoder Head. To give a straightforward view of the forgetting phenomenon in decoder head, we utilize Canonical Correlation Analysis [38] to measure the model output similarity between adjacent incremental steps. Results are shown in Fig. 5. The X-axis indicates different layers of the network. The Y-axis represents learning steps. Lower value on step 1 indicates more forgetting. (a) only performs probability distillation and results in the most forgetting. (b) adds constraints on intermediate features, which alleviates the forgetting of encoder, but a clear gap can be observed in decoder output ($0.943 \rightarrow 0.765$). (c) further applies constraints on decoder as in our framework, which alleviates this forgetting and improves the overall performance.

Effect of Structure Preserving Loss. To demonstrate the effects of the structure preserving loss against the widely used prototype alignment, we performed an experiment by replacing structure preserving loss with the prototype alignment used in [37]. We select five old classes (*plane, bicycle, bird, boat, bottle*) and use t-SNE [31] to visualize their distributions in embedding space after training on new classes. To erase the randomness in initial learning phase, we use the same checkpoint learned the first 15 classes to start from. Results are shown in Fig. 6. Our structure preserving loss better maintains the discriminative power (*i.e.*, a clearer boundary) of old classes than prototype alignment.

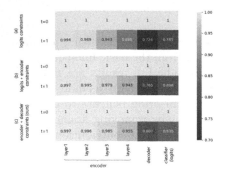

(a) Prototype Alignment   (b) Ours

Fig. 5. Visualization of the forgetting phenomena in segmentation network.

Fig. 6. Comparing the embedding space distribution results by prototype alignment and structure preserving loss.

Effect of Projected Feature Alignment. We conducted an experiment on 15–1 overlapped VOC by adding different feature space constraints on a simple baseline, which is made of L_{ce} + pseudo label. Results of *all* classes are shown in Table 5. Pixel-wise aligns the feature map directly in a pixel-wise manner. Pooled indicates the practice in [13]. Our proposed module successfully surpasses its contenders by a clear margin of 2.3%.

Table 5. Performance comparison of different feature space constraints

Baseline	Pixel-wise	Pooled [13]	Ours
18.9	33.7	48.2	**50.5**

6 Limitation and Conclusion

Continual learning is often applied to situations where computational and space overhead are important. Our method, which is exemplar-free, does not require any extra storage. But we do admit that there's some extra computational cost of our method. It is mainly from two aspects: The first is from the training of projector P, which takes roughly 10 min per incremental step. The second is from all the proposed modules, which makes the training about 10% slower than fine-tuning. We think this is within an acceptable range for most situations.

In this paper, we present a novel framework to deal with the forgetting in both encoder and decoder of segmentation network. In detail, the projected feature alignment module is designed to disentangle the preservation of old knowledge from integrating new knowledge. The structure preserving loss exploits inter-class and intra-class structures to maintain the discriminability of each class in the embedding space. Their effects are demonstrated by extensive experiments.

Though our method managed to achieve a better stability-plasticity trade-off, the cause of large performance gap to the upper bound for new classes still remains to be explored in the future work.

Acknowledgements. This work is supported by the National Natural Science Foundation of China under Grant No. 62176246 and No. 61836008.

References

1. Badrinarayanan, V., Kendall, A., Cipolla, R.: Segnet: a deep convolutional encoder-decoder architecture for image segmentation. IEEE Trans. Pattern Anal. Mach. Intell. **39**(12), 2481–2495 (2017)
2. Castro, F.M., Marín-Jiménez, M.J., Guil, N., Schmid, C., Alahari, K.: End-to-end incremental learning. In: Ferrari, V., Hebert, M., Sminchisescu, C., Weiss, Y. (eds.) ECCV 2018. LNCS, vol. 11216, pp. 241–257. Springer, Cham (2018). https://doi.org/10.1007/978-3-030-01258-8_15
3. Cermelli, F., Mancini, M., Bulo, S.R., Ricci, E., Caputo, B.: Modeling the background for incremental learning in semantic segmentation. In: ICCV (2020)
4. Chaudhry, A., Gordo, A., Dokania, P.K., Torr, P., Lopez-Paz, D.: Using hindsight to anchor past knowledge in continual learning. In: AAAI (2021)
5. Chen, L.C., Papandreou, G., Kokkinos, I., Murphy, K., Yuille, A.L.: Deeplab: semantic image segmentation with deep convolutional nets, atrous convolution, and fully connected crfs. IEEE Trans. Pattern Anal. Mach. Intell. **40**(4), 834–848 (2017)
6. Chen, L.C., Papandreou, G., Schroff, F., Adam, H.: Rethinking atrous convolution for semantic image segmentation. arXiv preprint arXiv:1706.05587 (2017)
7. Chen, L.-C., Zhu, Y., Papandreou, G., Schroff, F., Adam, H.: Encoder-decoder with atrous separable convolution for semantic image segmentation. In: Ferrari, V., Hebert, M., Sminchisescu, C., Weiss, Y. (eds.) ECCV 2018. LNCS, vol. 11211, pp. 833–851. Springer, Cham (2018). https://doi.org/10.1007/978-3-030-01234-2_49
8. Cong, Y., Zhao, M., Li, J., Wang, S., Carin, L.: Gan memory with no forgetting. In: NIPS (2020)
9. Cordts, M., et al.: The cityscapes dataset for semantic urban scene understanding. In: CVPR (2016)
10. De Lange, M., Tuytelaars, T.: Continual prototype evolution: Learning online from non-stationary data streams. In: ICCV (2021)
11. Deng, J., Dong, W., Socher, R., Li, L.J., Li, K., Fei-Fei, L.: Imagenet: A large-scale hierarchical image database. In: CVPR (2009)
12. Dhar, P., Singh, R.V., Peng, K.C., Wu, Z., Chellappa, R.: Learning without memorizing. In: CVPR (2019)
13. Douillard, A., Chen, Y., Dapogny, A., Cord, M.: Plop: Learning without forgetting for continual semantic segmentation. In: CVPR (2021)
14. Douillard, A., Cord, M., Ollion, C., Robert, T., Valle, E.: PODNet: pooled outputs distillation for small-tasks incremental learning. In: Vedaldi, A., Bischof, H., Brox, T., Frahm, J.-M. (eds.) ECCV 2020. LNCS, vol. 12365, pp. 86–102. Springer, Cham (2020). https://doi.org/10.1007/978-3-030-58565-5_6
15. Everingham, M., Winn, J.: The Pascal Visual Object Classes Challenge 2012 (voc2012) Development Kit. Tech. Rep, Pattern Analysis, Statistical Modelling and Computational Learning (2011)

16. French, R.M.: Catastrophic forgetting in connectionist networks. Trends Cogn. Sci. **3**(4), 128–135 (1999)
17. Fu, J., et al.: Dual attention network for scene segmentation. In: CVPR (2019)
18. Hayes, T.L., Kafle, K., Shrestha, R., Acharya, M., Kanan, C.: REMIND your neural network to prevent catastrophic forgetting. In: Vedaldi, A., Bischof, H., Brox, T., Frahm, J.-M. (eds.) ECCV 2020. LNCS, vol. 12353, pp. 466–483. Springer, Cham (2020). https://doi.org/10.1007/978-3-030-58598-3_28
19. He, K., Zhang, X., Ren, S., Sun, J.: Deep residual learning for image recognition. In: CVPR (2016)
20. Hinton, G., Vinyals, O., Dean, J.: Distilling the knowledge in a neural network. arXiv preprint arXiv:1503.02531 (2015)
21. Hinton, G.E., Salakhutdinov, R.R.: Reducing the dimensionality of data with neural networks. Science **313**(5786), 504–507 (2006)
22. Hou, S., Pan, X., Loy, C.C., Wang, Z., Lin, D.: Learning a unified classifier incrementally via rebalancing. In: ICCV (2019)
23. Hu, H., Cui, J., Wang, L.: Region-aware contrastive learning for semantic segmentation. In: ICCV (2021)
24. Huang, Z., Wang, X., Huang, L., Huang, C., Wei, Y., Liu, W.: Ccnet: Criss-cross attention for semantic segmentation. In: ICCV (2019)
25. Iscen, A., Zhang, J., Lazebnik, S., Schmid, C.: Memory-efficient incremental learning through feature adaptation. In: Vedaldi, A., Bischof, H., Brox, T., Frahm, J.-M. (eds.) ECCV 2020. LNCS, vol. 12361, pp. 699–715. Springer, Cham (2020). https://doi.org/10.1007/978-3-030-58517-4_41
26. Kirkpatrick, J., et al.: Overcoming catastrophic forgetting in neural networks. Proc. Natl. Acad. Sci. **114**(13), 3521–3526 (2017)
27. Lee, D.H., et al.: Pseudo-label: The simple and efficient semi supervised learning method for deep neural networks. In: Workshop on Challenges in Representation Learning, ICML (2013)
28. Li, X., Zhou, Y., Wu, T., Socher, R., Xiong, C.: Learn to grow: A continual structure learning framework for overcoming catastrophic forgetting. In: ICML (2019)
29. Li, Z., Hoiem, D.: Learning without forgetting. IEEE Trans. Pattern Anal. Mach. Intell. **40**(12), 2935–2947 (2017)
30. Long, J., Shelhamer, E., Darrell, T.: Fully convolutional networks for semantic segmentation. In: CVPR (2015)
31. Van der Maaten, L., Hinton, G.: Visualizing data using t-sne. J. Mach. Learn. Res. **9**(11) (2008)
32. Mallya, A., Davis, D., Lazebnik, S.: Piggyback: adapting a single network to multiple tasks by learning to mask weights. In: Ferrari, V., Hebert, M., Sminchisescu, C., Weiss, Y. (eds.) ECCV 2018. LNCS, vol. 11208, pp. 72–88. Springer, Cham (2018). https://doi.org/10.1007/978-3-030-01225-0_5
33. Mallya, A., Lazebnik, S.: Packnet: Adding multiple tasks to a single network by iterative pruning. In: CVPR (2018)
34. Maltoni, D., Lomonaco, V.: Continuous learning in single-incremental-task scenarios. Neural Netw. **116**, 56–73 (2019)
35. Mehta, S., Rastegari, M., Caspi, A., Shapiro, L., Hajishirzi, H.: ESPNet: efficient spatial pyramid of dilated convolutions for semantic segmentation. In: Ferrari, V., Hebert, M., Sminchisescu, C., Weiss, Y. (eds.) ECCV 2018. LNCS, vol. 11214, pp. 561–580. Springer, Cham (2018). https://doi.org/10.1007/978-3-030-01249-6_34
36. Michieli, U., Zanuttigh, P.: Incremental learning techniques for semantic segmentation. In: ICCVW (2019)

37. Michieli, U., Zanuttigh, P.: Continual semantic segmentation via repulsion-attraction of sparse and disentangled latent representations. In: CVPR (2021)
38. Raghu, M., Gilmer, J., Yosinski, J., Sohl-Dickstein, J.: Svcca: Singular vector canonical correlation analysis for deep learning dynamics and interpretability. In: NIPS (2017)
39. Rebuffi, S.A., Kolesnikov, A., Sperl, G., Lampert, C.H.: icarl: Incremental classifier and representation learning. In: CVPR (2017)
40. Ren, S., He, K., Girshick, R., Sun, J.: Faster r-cnn: Towards real-time object detection with region proposal networks. Adv. Neural. Inf. Process. Syst. **28**, 91–99 (2015)
41. Ronneberger, O., Fischer, P., Brox, T.: U-net: Convolutional networks for biomedical image segmentation. In: International Conference on Medical Image Computing and Computer-assisted Intervention (2015)
42. Shin, H., Lee, J.K., Kim, J., Kim, J.: Continual learning with deep generative replay. In: NIPS (2017)
43. Xiang, Y., Fu, Y., Ji, P., Huang, H.: Incremental learning using conditional adversarial networks. In: ICCV (2019)
44. Yan, S., Xie, J., He, X.: Der: Dynamically expandable representation for class incremental learning. In: CVPR (2021)
45. Yoon, J., Yang, E., Lee, J., Hwang, S.J.: Lifelong learning with dynamically expandable networks. arXiv preprint arXiv:1708.01547 (2017)
46. Zenke, F., Poole, B., Ganguli, S.: Continual learning through synaptic intelligence. In: ICML, pp. 3987–3995 (2017)
47. Zhao, H., Shi, J., Qi, X., Wang, X., Jia, J.: Pyramid scene parsing network. In: CVPR (2017)
48. Zhao, H., Zhang, Y., Liu, S., Shi, J., Loy, C.C., Lin, D., Jia, J.: PSANet: point-wise spatial attention network for scene parsing. In: Ferrari, V., Hebert, M., Sminchisescu, C., Weiss, Y. (eds.) ECCV 2018. LNCS, vol. 11213, pp. 270–286. Springer, Cham (2018). https://doi.org/10.1007/978-3-030-01240-3_17
49. Zhou, B., Zhao, H., Puig, X., Fidler, S., Barriuso, A., Torralba, A.: Scene parsing through ade20k dataset. In: CVPR (2017)

Interclass Prototype Relation for Few-Shot Segmentation

Atsuro Okazawa[⊠]

R&D Promotion Office AI Strategy Office, SoftBank Corp., Tokyo, Japan
atsuro.okazawa@g.softbank.co.jp

Abstract. Traditional semantic segmentation requires a large labeled image dataset and can only be predicted within predefined classes. Solving this problem of few-shot segmentation, which requires only a handful of annotations for the new target class, is important. However, with few-shot segmentation, the target class data distribution in the feature space is sparse and has low coverage because of the slight variations in the sample data. Setting the classification boundary that properly separates the target class from other classes is an impossible task. In particular, it is difficult to classify classes that are similar to the target class near the boundary. This study proposes the Interclass Prototype Relation Network (IPRNet), which improves the separation performance by reducing the similarity between other classes. We conducted extensive experiments with Pascal-5^i and COCO-20^i and showed that IPRNet provides the best segmentation performance compared with previous research.

Keywords: Semantic segmentation · Few-shot segmentation · Few-shot learning · Metric learning

1 Introduction

Recent advances in semantic segmentation have been brought about by advanced convolutional neural networks (CNNs) [15] and large labeled image datasets [6, 9, 18, 50]. However, semantic segmentation with fully supervised learning requires a substantial number of annotations per pixel and can be time-consuming to create. To solve this problem, few-shot segmentation that requires only a handful of annotations for a new target class is important. Few-shot segmentation aims to obtain generalization ability from known classes and adapt them to new target classes via a few shots, namely, support data. However, few-shot segmentation is not under the condition in which features can be extracted from a large amount of data with all variations (Fig. 1(a)), and the target class data distribution in the feature space is sparse and has low coverage (Fig. 1(b)). Therefore, there is an essential problem in that it is not possible to set the classification boundary

---

Supplementary Information The online version contains supplementary material available at https://doi.org/10.1007/978-3-031-19818-2_21.

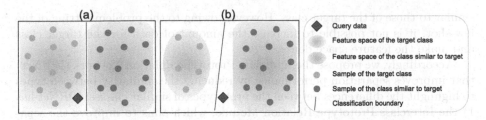

Fig. 1. Each shade is the area to which most of the samples in that class are mapped. The black line is the classification boundary calculated from the samples. (a) shows the area to be mapped from sufficient samples is given, not in the condition of few-shot. (b) shows the area mapped in the few-shot problem with a few samples. If query data is plotted near the boundary between the two classes you want to classify, we cannot detect the class that originally belonged to the target class in the case of few-shot with a narrow-mapped area. (Color figure online)

that separates the target class from other classes properly. In particular, it is difficult to classify classes that have features like those of the target class near the boundary.

To tackle this important problem without increasing the number of shots for the target class, it is important to differentiate the features between each class when learning generalization abilities from known classes.

Few-shot segmentation is an extension of the technology based on few-shot learning [11,28,31,32,36,39], and it tackles the more difficult task of predicting the label for each pixel, instead of predicting a single label for the entire image in few-shot learning. Few-shot learning has meta-learning [11,39] and metric learning [31,32,36] as the mainstream methods. Meta-learning was introduced by Shaban et al. [29] and metric learning by Snell et al. [8] for segmentation problems. In particular, the metric learning approach to the problem of few-shot segmentation has been actively studied in recent years and has been successful. Our research is related to few-shot segmentation using metric learning. The method of few-shot segmentation using metric learning has been pushed into a global descriptor called a prototype using supporting data [30]. The support data are a few samples with a target class. The prototype is a vector representation of the features for the target class and there are many studies based on the method of inference by comparing the prototype with the features of the query image. For better prototyping and proper comparison of support and query, there are several earlier studies. For example, there are studies that have introduced a mechanism to separate the foreground and background [19,38,42], and studies that have introduced a multi-scale architecture [16,34,45]. In these studies, the prototype extracted from the support data was appropriately compared with the query data, and training was performed based on the loss function between the query data and its ground truth.

However, in few-shot segmentation with only a few shot samples, as mentioned above, there are few variations of support data using the target class, and prototype generation is performed from a sparse feature space. Therefore, it is particularly difficult to obtain a prototype that can classify classes with features similar to those of the target class. However, owing to the problem setting of the few-shot, it is not possible to increase the amount of data for the target class and make the feature space dense.

Accordingly, we propose the Interclass Prototype Relation Network (IPRNet) that improves separation performance by reducing the similarity between types to highlight the differences between the prototypes of similar classes. IPRNet has 1) the Interclass Prototype Relation Module, which aims to improve the separation performance between similar classes by reducing the similarity between prototypes of each class, and 2) the Respective Classifier Module, which aims to improve the separation performance by integrating respective estimations of the target class and background. We hypothesized that these modules could improve the separation performance of the target and other classes. In this study, we verified this hypothesis using two experiments. First, we evaluated whether the performance would improve compared to earlier research with the best performance. Second, we conducted an ablation study to verify that the modules proposed in IPRNet are effective in classifying similar classes. The contributions of this study are as follows.

- We propose a novel few-shot segmentation method called IPRNet, which improves the separation performance between the target class and other classes that are especially similar to the target.
- We evaluate Pascal-5^i [29] and COCO-20^i [18] and show that the proposed method improves the mean intersection over union (mIoU) over the existing best-performing method.
- Through an ablation study, we verified whether the proposed method is effective for classifying similar classes.

Our code is at:
https://sb-biz.primedrive.jp/v2/access?key=PGru7XxrXU-tQe1ZzB2EpQ

2   Related Work

Few-shot segmentation mostly consists of few-shot learning-based technology that improves model generalization ability and semantic segmentation technology that solves pixel-level classification problems. We describe the existing research related to these constituent requirements and issues.

Semantic Segmentation. In semantic segmentation, deep neural networks based on convolutional neural networks (CNNs) [15] have been successful. Starting from fully convolutional networks [22], especially encoder-decoder structure proposed by Segnet [1] has become the basic network structure in recent semantic segmentation. Recently, a faster method Enet [27], and encoder-decoder structures that ensemble multi-scale features to express all frequency information have been proposed [3–5,43,47,48]. The latest research also proposed a convolution-free and resolution deterioration-free method [49] based on the transformer [35], which is a model that uses only the attention mechanism instead of CNNs [15].

Few-shot Learning. Few-shot learning focuses on the generalization ability of the model and enables learning for new class predictions using a few annotated samples. The mainstream existing methods are metric learning [31,32,36] and meta-learning [11,28,39]. The core idea of metric learning is the distance measurement, which is formulated as an optimization of the distance or similarity between the images and regions. Meta-learning focuses on achieving a high-speed learning ability by defining specific optimization functions and loss functions. Among these methods, the concept of a prototypical network [31] is widely adopted for few-shot segmentation, and it is possible to reduce the calculation cost significantly while maintaining high performance. Many methods focus on image classification, but recently few-shot segmentation has attracted attention.

Few-shot Segmentation. Few-shot segmentation is an extension of few-shot learning that addresses the more difficult task of predicting a label for each pixel rather than predicting a single label for the entire image. Few-shot meta-learning was introduced into the segmentation problem by Shaban et al. [29], and there is a lot of research on its enlargement [2,14,23,33]. Few-shot metric learning was successfully introduced by Snell et al. [8]. Many of the methods so far drop the problem into a 1-way classification problem in order to apply it to episodic learning [36] to acquire generalization ability. In previous research, they pushed the support data into a global descriptor to obtain a prototype that is the features of the class in the first step [30]. In the next step, the target object and background are separated by comparing the prototype with the query image [16,21,26,34,38,41,42,45,46]. In addition to these, research to absorb the difference in size, position, and orientation of the target object on the support image and the query image, and to compare them correctly have been conducted [10,20,37,40,44]. There is also a method to infer from the correlation between all positions of query data and support data [25]. However, most of them are solved by general 1-way classification problems, so only the relationship between the target class and the background can be considered. ASR [19] is a method that uses multiple latent class vectors, but the feature map channel is divided and assigned to each class. Therefore, if many classes are included, the number of channels assigned to one class will decrease and it will not work effectively. Existing methods do not fully consider the relationships between different classes, making proper classification difficult. In particular, it is the most difficult to separate from similar classes near the discriminant boundary, and to the best of our knowledge, no research has been conducted on this problem. This research proposes a novel IPRNet that focus on improvement between similar classes, which are particularly difficult to classify.

3   Problem Definition

The major difference between few-shot segmentation and general semantic segmentation is that the training and test set categories do not intersect. Particularly, at the inference stage, the test set had classes that were not found during the training. Specifically, given the train set $S_{train} = \{(I^{S/Q}, M^{S/Q})\}$ and the test set $S_{test} = \{(I^{S/Q}, M^{S/Q})\}$, the categories of the two sets do not intersect ($S_{train} \cap S_{test} = \phi$). Here, $I \in R_{H \times W \times 3}$ shows an RGB image and $M \in R_{H \times W}$ shows a segmentation mask. The subscripts S and Q represent support and query, respectively. We mimicked the first one-shot segmentation study [29] and applied training and testing to an episodic learning framework. In each episode, the input to the model consists of the query image I^Q and k samples (I_i^S, M_i^S), $i \in \{1, 2, \ldots, k\}$ from the support set. All support image and query image have the same class c. Training selects (I^Q, M^Q, I_i^S, M_i^S) of batch size b set from the train set $S_{train} = \{(I^{S/Q}, M^{S/Q})\}$ and estimates the query mask $\tilde{M}^Q$ to approximate ground truth mask M^Q.

4   Proposed Method

4.1   Design Guideline of Network Structure

As mentioned in the introduction, the target class, which has a few shots in few-shot segmentation, is difficult to classify similar classes because the data plotted in the feature space is sparse and has low coverage. To address this problem, we propose the Interclass Prototype Relation Network (IPRNet). IPRNet has two modules: the Interclass Prototype Relation Module (IPRM) and the Respective Classifier Module (RCM). These modules aim to improve the identification performance of similar classes by reducing the similarity between prototypes and extracting the differences between classes. An overview of the network is shown in Fig. 2.

4.2   Interclass Prototype Relation Network

This section describes the overall flow of IPRNet. First, the support and query images were fed into pretrained shared CNNs (pretrained by ImageNet [7]) to extract features. Next, we passed the support features with the support masks through the IPRM. Through IRPM, we obtain prototypes that represent feature vectors for each class, and the value L_r that indicates the similarity between each prototype. Then, for more accurate pixel-by-pixel matching, the matching between prototypes and the query feature is performed by calculating the cosine similarity in map process. The two similarity maps can be obtained by matching with the query feature for the target class prototype and the background prototype, respectively. This process was inspired by earlier research on MLC [42]. The input to the multi-scale network is the concatenation of the support features, the query feature, and two similarity maps. The output is a relation feature valid for classification and L_m which is a multi-scale loss introduced in PFENet [34]. The multi-scale network sets up the top-down structure of FPN-like [17] by using the feature enrichment module introduced in PFENet [34] to obtain multi-scale information. This structure enables fast multi-scale aggregation, by transferring features from finer to coarser and by easing feature interaction. Each scale yielded

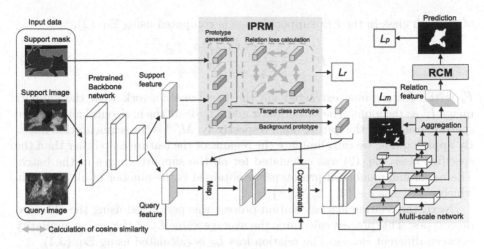

Fig. 2. The Overall architecture of the proposed method, Interclass Prototype Relation Network (IPRNet).

segmentation results for calculating the loss. The average loss value for each scale was L_m. Finally, the relation feature, a multi-scale information-intensive feature map, is generated by fusing all the different scales into a concatenated feature map by convolution.

The relation feature is the input to the RCM. The RCM performs a discriminative process to classify the foreground and background more and obtains the final inference result $\tilde{M}^Q$. Then, the loss L_p between the inference result and ground-truth mask M^Q is calculated. The loss function, which is the cost function of training, is given by Eq. (1).

$$Loss = w_1 L_r + w_2 L_m + w_3 L_p \tag{1}$$

w_1, w_2, and w_3 are the weight coefficients, which were trained with $w_1 = 0.4$, $w_2 = 0.2$, and $w_3 = 0.4$, respectively. Here, we describe the details of the proposed IPRM and RCM.

4.3   Interclass Prototype Relation Module

The Interclass Prototype Relation Module (IPRM) was proposed to reduce the similarity between classes. The IPRM has a prototype generation process and relation loss calculation process. The prototype generation process calculates the prototypes of all classes present in the batch and the support images. This process obtains a prototype, which is a global descriptor of a particular class in an image by taking as input the feature map extracted from the support image and the segmentation mask paired with it. We employed the masked average pooling strategy [30] to compute the prototype for each class. The prototype P_i^c

of the c th class in the i th support image is computed using Eq. (2).

$$P_i^c = \frac{\sum_{x,y} F_i^{x,y} \mathbb{1}[M_i^{x,y} = c]}{\sum_{x,y} \mathbb{1}[M_i^{x,y} = c]} \tag{2}$$

F_i is the feature map extracted via the backbone network with the support image I_i^S as the input. The subscripts x and y indicate the horizontal and vertical spatial positions of the feature map, respectively. $M_i^{x,y}$ is the segmentation mask. By applying this, we can eliminate the regions of the feature map other than the specified class. Eq. (2) was calculated for all the support images in the batch. The maximum number of prototypes n obtained is the number of all classes in which S_{train} has $c \in \{0, 1, 2, \ldots, n\}$.

Next, the relation loss calculation process was performed using the obtained prototypes. This process calculates the average value L_r of the cosine similarity between different classes. The relation loss L_r is calculated using Eq. (3,4).

$$L_r = \frac{\sum_{c_s}^n \sum_{c_t}^n Sim(P^{c_s}, P^{c_t}) \mathbb{1}[c_s \neq c_t]}{\sum_{c_s}^n \sum_{c_t}^n \mathbb{1}[c_s \neq c_t]} \tag{3}$$

$$Sim(P^{c_s}, P^{c_t}) = \frac{P^{c_s} \cdot P^{c_t}}{\|P^{c_s}\| \cdot \|P^{c_t}\|} \tag{4}$$

The c_s and c_t refer to class numbers and the difference between the prototypes of two different classes is measured by the cosine similarity expressed in the Eq. (4). The similarity between the prototypes of two different classes c_s, c_t computed in Eq. (4) is calculated for all combinations switching between pairs of prototypes to be measured using the Eq. (3). The average of these prototype similarity values are the relation loss L_r. The relation loss L_r is designed to improve the separation performance between each class by training the network such that the similarity between each class is low.

Further, among the prototypes calculated by the Eq. (2), the prototype of the target class and the prototype extracted from the background region is selected. The prototypes were compared with the query feature and their respective similarity maps were computed. These similarity maps were used as input to the multi-scale network.

4.4   Respective Classifier Module

An overview of the Respective Classifier Module (RCM) is shown in Fig. 3. The RCM takes as input the relation feature, which is the output of the multi-scale network shown in Fig. 2. It was designed to improve the separation performance between the target class of objects and the background by estimating each of them independently, reintegrating the results. The relation feature F_r is branched for foreground target class prediction and background prediction. Then, it is transformed into a probability distribution representation V_1 and V_0 through two convolutional layers, the activation layer, and the softmax layer respectively. V_1 indicates the probability distribution of a single object in the target class, and V_0 indicates the probability distribution of the background region only. The loss calculation between the probability distribution representation V and the ground-truth mask M^Q can be expressed by Eq. (5).

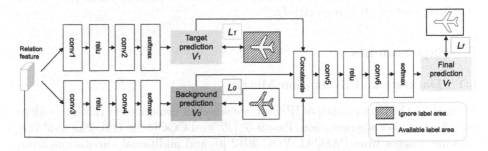

Fig. 3. Configuration diagram of the Respective Classifier Module (RCM).

$$L_c = \frac{-\sum_{x,y} D_c(M^Q) log(V_c^{x,y})}{\sum_{x,y} D_c(M^Q)} \qquad (5)$$

$$D_c(M^Q) = \begin{cases} c & \text{if } M^Q = c, \\ 255 & \text{if } M^Q \neq c. \end{cases} \qquad (6)$$

The loss is calculated by cross-entropy error. V_1, the calculated loss value L_1 is the loss function of a single object of the target class, and given probability distribution V_0, the calculated loss value L_0 is the loss function of the background region only. Equation (6) is a function that, according to the given class ID c, returns 255 pixel positions other than those corresponding to the class ID in the ground-truth mask M^Q. Two hundred and fifty-five means ignore the label, an area that is not used when calculating loss, thus eliminating the relationship between the background and foreground. This mechanism allows the RCM to acquire a discriminator through training that can make inferences about its pixels based on information about the object itself only.

The final estimation result V_f can be obtained by fusing the information from the concatenation of the foreground probability distribution V_1, background probability distribution V_0, and relation feature F_r with the two convolution layers, the activation layer, and the softmax layer. The cross-entropy error L_f was also obtained between the prediction result V_f and the ground-truth mask M^Q.

The final loss value L_n output by the RCM is obtained by the weighted addition of the loss value L_1 to the prediction result of the foreground object alone, the loss value L_0 to the prediction result of the background region only, and the loss value L_f to the final prediction result. This can be defined by the following Eq. (7).

$$L_p = \alpha L_1 + \beta L_0 + \gamma L_f \tag{7}$$

α, β, and γ are the weight coefficients, which were trained by setting $\alpha = 0.15$, $\beta = 0.15$, and $\gamma = 0.7$, respectively.

5   Experiments

5.1   Datasets and Evaluation Metric

To analyze the performance of IPRNet, we selected two datasets that are widely used for few-shot segmentation, Pascal-5^i [29] and COCO-20^i [18]. Pascal-5^i [29] contains images from PASCAL VOC 2012 [9] and additional annotations from SBD [12]. For training, a total of twenty class categories were evenly divided into four splits, and model training was performed using a cross-validation method. Specifically, three splits were selected as the training data during the training process, and the remaining splits were used for testing. During testing, one thousand support-query pairs were randomly sampled and evaluated [29]. COCO-20^i [18], unlike Pascal-5^i [29], is an incredibly challenging dataset. This is because it is a large dataset having 82,081 images, and many objects are included in the images of realistic scenes. Following the FWB [26], the eighty classes of COCO-20^i [18] were evenly divided into four splits, and the same cross-validation scheme was used. To obtain more stable results, we randomly sampled 20,000 pairs of during the testing [34]. As an evaluation metric, we used the mean Intersection-over-Union (mIoU), which is commonly used in semantic segmentation.

An ablation study was conducted to verify the influence of the proposed IPRM and RCM. To verify the separation performance between similar classes, we compared the per-class IoU results of IPRNet and the baseline without our IPRM and RCM.

5.2   Implementation Details

ResNet [13] is employed as the backbone network, and block2 and block3 are concatenated to generate the feature map [46]. The input support and query images were cropped to an image size of 400×400 pixels and fed into the backbone network. The initial learning rate was set to 0.05, momentum and weight decay to 0.9 and 0.0001, respectively, and the optimizer was trained with a batch size of thirty-two using the SGD optimizer. We also adapted the poly method [3], where the decay of the learning rate is achieved by multiplying by $(1 - current\_iter)^{power}$, and the $power$ is set to 0.9. The pretrained backbone network is frozen so that it does not learn class-specific representations of the training data. We implemented it using Pytorch experimented with an NVIDIA A10G GPU.

In the ablation study, for the IPRM deletion, we also modified a mechanism that uses only the conventional masked average pooling strategy [30] to acquire the target class prototype and background class prototype. The RCM is completely removed and replaced with a mechanism that calculates the loss between the output of the multi-scale network and the ground truth.

5.3   Experimental Results

The effectiveness of our method was evaluated on two benchmark datasets Pascal-5^i [29] and COCO-20^i [18]. We extensively experimented with the 1-shot and 5-shot on 1-way problems in various few-shot split settings using the widely used encoder networks ResNet-50 and ResNet-101. Here, an n-way k-shot implies that k samples are given for each class among the n classes. Extensive experiments show that the mIoU is improved over the conventional method in all the cases. The ablation study confirmed that removing the IPRM and RCM reduced the mIoU.

Pascal-5^i First, we describe the results for Pascal-5^i in Table 1. Our method has an improved mIoU in all conditions compared with HSNet [25], which has the highest mIoU among the conventional methods. In order of performance improvement, we first see a 1.7% mIoU improvement for ResNet-50 and 1.3% mIoU improvement for ResNet-101 in the 1-shot setting. These are followed by 0.7% mIoU improvement for ResNet-50 and 0.5% mIoU improvement for ResNet-101 in the 5-shot setting.

COCO-20^i Next, we describe the results for COCO-20^i in Table 2. Our method has improved mIoU in all conditions compared with HSNet [25], which has the highest mIoU among the conventional methods. In order of performance improvement, we first see a 6.1% mIoU improvement for ResNet-50 and 5.7% mIoU improvement for ResNet-101 in the 1-shot setting. This is followed by a 4.2% mIoU improvement for ResNet-50 and 3.8% mIoU improvement for ResNet-101 in the 5-shot setting.

Ablation Study. We conducted an ablation study to investigate the influence of the IPRM and RCM, which are the main components of our model. All ablation study experiments were the result of a 5-shot setup performed on the COCO-20^i dataset using the ResNet50 backbone. The results are presented in Table 3. The largest decrease in mIoU is seen in the case where both IPRM and RCM are removed by 4.4%, followed by the case where only IPRM is removed by 2.5%, and finally, the case where only RCM is removed 1.8%. The mIoU decreased by more than 0.7% when the IPRM was removed then when the RCM was removed.

Table 1. Performance on Pascal-5^i in IoU with per-splits results. Some results are from [16,21,25,34,38,42].

Backbone	Method	1shot					5shot				
		s-0	s-1	s-2	s-3	mean	s-0	s-1	s-2	s-3	mean
ResNet-50	PANet [38]	44.0	57.5	50.8	44.0	49.1	55.3	67.2	61.3	53.2	59.3
	PPNet [21]	48.6	60.6	55.7	46.5	52.8	58.9	68.3	66.8	58.0	63.0
	PFENet [34]	61.7	69.5	55.4	56.3	60.8	63.1	70.7	55.8	57.9	61.9
	ASGNet [16]	58.8	67.9	56.8	53.7	59.3	63.7	70.6	64.2	57.4	63.9
	MLC [42]	59.2	71.2	**65.6**	52.5	62.1	63.5	71.6	**71.2**	58.1	66.1
	HSNet [25]	64.3	70.7	60.3	60.5	64.0	**70.3**	73.2	67.4	**67.1**	69.5
	Ours	**65.2**	**72.9**	63.3	**61.3**	**65.7**	70.2	**75.6**	68.9	66.2	**70.2**
ResNet-100	FWB [26]	51.3	64.5	56.7	52.2	56.2	54.8	67.4	62.2	55.3	59.9
	PPNet [21]	52.7	62.8	57.4	47.7	55.2	60.3	70.0	69.4	60.7	65.1
	PFENet [34]	60.5	69.4	54.4	55.9	60.1	62.8	70.4	54.9	57.6	61.4
	ASGNet [16]	59.8	67.4	55.6	54.4	59.3	64.6	71.3	64.2	57.3	64.4
	MLC [42]	60.8	71.3	61.5	56.9	62.6	65.8	74.9	71.4	63.1	68.8
	HSNet [25]	67.3	72.3	62.0	**63.1**	66.2	**71.8**	74.4	67.0	**68.3**	70.4
	Ours	**67.8**	**74.6**	**65.7**	62.2	**67.5**	70.0	**75.9**	**71.8**	65.8	**70.9**

Table 2. Performance on COCO-20^i in IoU with per-splits results. Some results are from [21,25,34,42].

Backbone	Method	1shot					5shot				
		s-0	s-1	s-2	s-3	mean	s-0	s-1	s-2	s-3	mean
ResNet-50	PPNet [21]	28.1	30.8	29.5	27.7	29.0	39.0	40.8	37.1	37.3	38.5
	PFENet [34]	36.5	38.6	34.5	33.8	35.8	36.5	43.3	37.8	38.4	39.0
	MLC [42]	**48.0**	36.6	27.4	28.2	35.1	**54.0**	42.1	34.9	33.6	41.4
	HSNet [25]	36.3	43.1	38.7	39.2	39.2	43.3	51.3	48.2	45.0	46.9
	Ours	42.2	**48.9**	**45.5**	**44.6**	**45.3**	48.0	**55.7**	**50.7**	**50.1**	**51.1**
ResNet-100	FWB [26]	17.0	18.0	21.0	28.9	21.2	19.1	21.5	23.9	30.1	23.7
	PFENet [34]	34.3	33.0	32.3	30.1	32.4	38.5	38.6	38.2	34.3	27.4
	MLC [42]	**51.1**	38.7	28.5	31.6	37.5	**57.8**	47.1	37.8	37.6	45.1
	HSNet [25]	37.2	44.1	42.4	41.3	41.2	45.9	53.0	51.8	47.1	49.5
	Ours	42.9	**50.6**	**46.8**	**47.4**	**46.9**	50.7	**58.3**	**52.8**	**51.3**	**53.3**

IOU of Each Class. To verify whether our proposed method is effective for objects with similar features that are difficult to classify, we evaluated it using COCO-20^i is an incredibly challenging dataset that has many objects in the images of realistic scenes. The evaluation is perfomed by comparing IoUs per class with the baseline, which eliminates the IPRM and RCM, and our proposed IRPNet. The results are presented in Table 4. Significant improvements in IoU were as follows 1 person 20.7%, 61 dining table 20.2%, 73 fridge 15.4%, 66 remote 13.4%, 67 keyboard 12.3%, 27 handbag 11.6%, 29 suitcase 11.5%, 25 backpack 11.3%, 74 book 11.3%, 44 knife 10.3%; and an increase of more than 10% in IoU. The classes with a lower IoU were 34 kite −8.9%, 65 mouse −8.2%, 48 apple −4.6%, 59 potted plant −3.8%, 13 park meter −3.5%, 4 motorcycle −2.8%, 10 traffic light −1.8%, 26 umbrella −1.5%, 69 microwave −1.5%, 80 toothbrush -1.0%; and a decrease of over 1% in IoU. The following is a discussion.

Table 3. Influence of the IPRM and RCM on COCO-20^i in IoU with 5-shot ResNet-50 backbone condition.

IPRM	RCM	s-0	s-1	s-2	s-3	Mean
		42.0	52.1	46.4	46.4	46.7
✔		46.8	53.6	47.9	49.0	49.3
	✔	44.9	53.2	47.4	48.9	48.6
✔	✔	**48.0**	**55.7**	**50.7**	**50.1**	**51.1**

Table 4. COCO-20^i performance in all classes of IoU experimented with 5shot ResNet-50. Baseline means the result of eliminating the IPRM and RCM.

s-0	Baseline	Ours	s-1	Baseline	Ours	s-2	Baseline	Ours	s-3	Baseline	Ours
1 Person	30.3	**51.0**	2 Bicycle	52.9	**55.8**	3 Car	35.7	**38.5**	4 Motorcycle	**56.0**	53.2
5 Airplane	73.0	**76.0**	6 Bus	69.0	**72.2**	7 Train	72.1	**72.8**	8 Truck	**35.1**	34.7
9 Boat	40.9	**50.5**	10 T. light	**40.8**	39.0	11 Fire H.	77.4	**82.7**	12 Stop sign	76.5	**81.4**
13 Park meter	**60.1**	56.6	14 Bench	35.7	**38.3**	15 Bird	64.5	**69.0**	16 Cat	77.4	**82.0**
17 Dog	65.0	**73.6**	18 Horse	70.6	**74.7**	19 Sheep	75.2	**76.8**	20 Cow	73.0	**78.7**
21 Elephant	79.7	**83.0**	22 Bear	83.6	**85.6**	23 Zebra	75.9	**76.2**	24 Giraffe	72.6	**75.6**
25 Backpack	18.5	**29.8**	26 Umbrella	**60.0**	58.5	27 Handbag	21.2	**32.8**	28 Tie	17.8	**18.6**
29 Suitcase	42.7	**54.2**	30 Frisbee	69.6	**75.9**	31 Skis	31.3	**38.7**	32 Snowboard	37.4	**46.3**
33 Sports ball	41.3	**48.4**	34 Kite	**51.3**	42.4	35 B. bat	31.0	**35.1**	36 B. glove	48.3	**50.4**
37 Skateboard	**42.4**	**42.4**	38 Surfboard	64.7	**68.8**	39 T. racket	58.4	**65.3**	40 Bottle	28.4	**32.6**
41 W. glass	34.0	**37.6**	42 Cup	49.8	**56.6**	43 Fork	19.7	**22.7**	44 Knife	34.8	**45.1**
45 Spoon	14.2	**16.8**	46 Bowl	31.1	**31.9**	47 Banana	41.6	**45.5**	48 Apple	**39.4**	34.8
49 Sandwich	50.5	**52.8**	50 Orange	47.5	**49.5**	51 Broccoli	33.1	**36.8**	52 Carrot	23.1	**27.3**
53 Hot dog	61.7	**67.5**	54 Pizza	84.7	**87.5**	55 Donut	67.6	**70.8**	56 Cake	44.1	**51.9**
57 Chair	7.2	**14.2**	58 Couch	34.5	**37.0**	59 P. plant	**11.9**	8.1	60 Bed	53.7	**56.8**
61 D. table	27.5	**47.7**	62 Toilet	57.2	**63.6**	63 TV	44.5	**52.2**	64 Laptop	50.9	**55.4**
65 Mouse	**50.1**	41.9	66 Remote	45.4	**58.8**	67 Keyboard	21.0	**33.3**	68 Cellphone	56.2	**65.8**
69 Microwave	**34.9**	33.4	70 Oven	23.6	**29.7**	71 Toaster	**42.4**	41.8	72 Sink	30.8	**33.0**
73 Fridge	23.9	**39.3**	74 Book	10.2	**21.5**	75 Clock	59.6	**64.6**	76 Vase	41.7	**46.1**
77 Scissors	43.4	**43.8**	78 Teddy	60.6	**64.2**	79 Hairdrier	47.9	**49.4**	80 Toothbrush	**34.3**	33.3

5.4 Discussion

Comparison with State-of-the-Arts. We discuss the mIoU of our IPRNet and the latest and best performing HSNet [25]. By comparing the results of Pascal-5^i and COCO-20^i described in Table 1 and 2, we can see that the performance of the COCO-20^i condition is better than that of the Pascal-5^i condition, regardless of the backbone network and number of shots. We discuss that our mechanism of training to avoid similarity between prototypes is effective for this difficult problem; COCO-20^i, contains many more difficult objects to classify. Comparing the experimental results of 1-shot and 5-shot, the mIoU of 1-shot is higher regardless of the backbone network and dataset. This is because, as mentioned in the introduction, when the number of shots is smaller, the feature space covered by the support data is sparser and the classification performance was worse for classes near the classification boundary. Therefore, we can conclude that the proposed method is more effective.

Performance of Similar Classes that Are Difficult to Classify. For IPR-Net and the baseline without the IPRM and RCM, we compare and discuss IoU for each of the classes shown in Table 4. Two main cases of classes that are difficult to classify are considered because of the existence of similar classes.

The first is an object that is likely to be similar to many other classes with a wide range of variations, specifically morphological changes, pictorial changes through decoration, and combinations with complex backgrounds. Specifically, there are 1 person with IoU increased by 20.7% and 61 dining table with IoU increased by 20.2%. 1 Person has all types of shape changes due to a deformable body and pictorial changes due to decoration (Fig. 4(a)(b)). 61 Dining table is a complex combination of objects, most of which are placed on top of each other; therefore, the pixel boundaries of the objects are always shared with various occlusions (Fig. 4(c)(d)).

Second, the class is an object with a simple shape and few features. Specifically, 73 fridge IoU increased by 15.4%, 66 remote by 13.4%, 67 keyboard by 12.3%, 27 handbag by 11.6%, 29 suitcase by 11.5%, 25 backpack by 11.3%, 74 book by 11.3%, and 44 knife by 10.3%. For example, 73 fridge is a symmetrical rectangular object with a few prominent patterns or protrusions on its surface. Hence, it was difficult to classify them based on similar rectangles and backgrounds (Fig. 4(e)). 66 Remote is also a small rectangular body (Fig. 4(f)).

For all other classes with an improved IoU, there are many classes where the IoU has increased because the RCM effect is considered to have improved the separation performance from the background. However, compared to the classes where these performances have increased by over 10%, the degree of conformity between the two cases are considered to be low. Specifically, 45 spoon and 43 fork are considered not to have improved IoU by as much as 44 knife because the shape of the tip is more non-graphical and distinctive compared to 44 knife.

Performance of Characteristic Objects. Further, we consider the classes with the lowered IoU as proper nouns that have a complex shape or structure unique to that object that is unparalleled. Specifically, there are 80 toothbrush, 69 microwave, 26 umbrella, 10 traffic light, 4 motorcycle, 13 park meter, 59 potted plant, 48 apple, 65 mouse and 34 kite. For example, 34 kite is a discriminative proper nouns with no other similar concepts and limited use, so it is not often combined with several backgrounds (Fig. 4(g)). 9 Potted plants had complex shapes that could not be represented by rectangles or spheres (Fig. 4(h)). For these objects, we consider that how to extract the unique features of the object to be more important than acquiring the differences from other classes, and the training to reduce the similarity of prototypes between different classes, which is the aim of the IPRM, does not work effectively.

Fig. 4. Examples of the recognition result using our proposed method, baseline and HSNet [25]. The red shade is the area of the target class. It is the ground truth in the case of query, the recognition result in the cases of the baseline, HSNet [25] and Ours. (Color figure online)

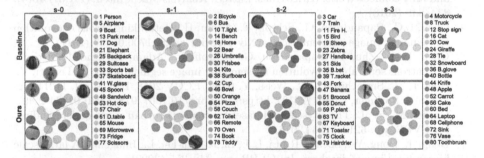

Fig. 5. Visualization results by t-SNE [24] of prototype obtained from query image of the baseline and IPRNet. They experiment with ResNet-50 5shot setting on COCO-20^i.

Qualitative Evaluation. Figure 5 shows the difference between the baseline and our IPRNet using t-SNE [24] for the prototype of the target class extracted from the query image. At the baseline, prototypes between different classes are adjacent or overlapping, and there is no clear classification. However, in IPRNet, the distance between each prototype is increased, and it can be observed that the prototypes are more separated.

6 Conclusion

This study proposes a novel IPRNet that introduces a mechanism to improve separation performance by reducing the similarity between different classes. Experiments show that mIoU is improved over the existing few-shot segmentation methods [21,25,34,42]. In addition, through an ablation study, we verified the separation performance between similar classes that are difficult to classify.

Acknowledgments. We would like to thank Mr. Katsushi Yamashita, Director, from R&D Promotion Office, SoftBank Corp. and all of R&D Promotion Office members. Mr. Yamashita's witty and helpful support helped us, especially. We would like to thank Mr. Shogo Hamano from Department of Mechano-Informatics, the University of Tokyo for his informative discussion.

References

1. Badrinarayanan, V., Kendall, A., Cipolla, R.: Segnet: A deep convolutional encoder-decoder architecture for image segmentation. In: IEEE Transactions on Pattern Analysis and Machine Intelligence (PAMI), pp. 2481–2495 (2017)
2. Boudiaf, B., Kervadec, H., Masud, Z.I., Piantanida, P., Ayed, I.B., Dolz, J.: Few-shot segmentation without meta-learning: A good transductive inference is all you need? In: CVPR, pp. 13979–13988 (2021)
3. Chen, L.C., Papandreou, G., Kokkinos, I., Murphy, K., Yuille, A.L.: Deeplab: Semantic image segmentation with deep convolutional nets, atrous convolution, and fully connected crfs. IEEE Trans. Pattern Analysis Mach. Intell. (PAMI) **40**(4), 834–848 (2018)
4. Chen, L.C., Papandreou, G., Schroff, F., Adam, H.: Rethinking atrous convolution for semantic image segmentation. arXiv preprint arXiv:1706.05587 (2017)
5. Chen, L.-C., Zhu, Y., Papandreou, G., Schroff, F., Adam, H.: Encoder-decoder with atrous separable convolution for semantic image segmentation. In: Ferrari, V., Hebert, M., Sminchisescu, C., Weiss, Y. (eds.) ECCV 2018. LNCS, vol. 11211, pp. 833–851. Springer, Cham (2018). https://doi.org/10.1007/978-3-030-01234-2_49
6. Cordts, M., et al.: The cityscapes dataset for semantic urban scene understanding. In: CVPR, pp. 3213–3223 (2016)
7. Deng, J., Dong, W., Socher, R., Li, L.J., Li, K., Fei-Fei, L.: Imagenet: A large-scale hierarchical image database. In: CVPR, pp. 248–255 (2009)
8. Dong, N., Xing, E.P.: Few-shot semantic segmentation with prototype learning. In: The British Machine Vision Conference (BMVC) (2017)
9. Everingham, M., Eslami, S.M.A., Gool, L.V., Williams, C.K.I., Winn, J.M., Zisserman, A.: The pascal visual object classes challenge: a retrospective. Int. J. Comput. Vision **111**, 98–136 (2014)
10. Fan, Z., et al.: Fgn: Fully guided network for few-shot instance segmentation. In: CVPR, pp. 9172–9181 (2020)
11. Finn, C., Abbeel, P., Levine, S.: Model-agnostic meta-learning for fast adaptation of deep networks. In: International Conference on Machine Learning(ICML), vol. 70, pp. 1126–1135 (2017)
12. Hariharan, B., Arbeláez, P., Girshick, R., Malik, J.: Simultaneous detection and segmentation. In: Fleet, D., Pajdla, T., Schiele, B., Tuytelaars, T. (eds.) ECCV 2014. LNCS, vol. 8695, pp. 297–312. Springer, Cham (2014). https://doi.org/10.1007/978-3-319-10584-0_20
13. He, K., Zhang, X., Ren, S., Sun, J.: Deep residual learning for image recognition. In: CVPR, pp. 770–778 (2016)
14. Hendryx, S.M., Leach, A.B., Hein, P.D., Morrison, C.T.: Meta-learning initializations for image segmentation. arXiv preprint arXiv:1912.06290 (2020)

15. LeCun, Y., Bernhard.E.Boser, Denker, J.S., Henderson, D., Howard, R.E., Hubbard, W.E., Jackel, L.D.: Backpropagation Applied to Handwritten Zip Code Recognition. Neural Comput. **1**(4), 541–551 (1989)
16. Li, G., Jampani, V., Sevilla-Lara, L., Sun, D., Kim, J., Kim, J.: Adaptive prototype learning and allocation for few-shot segmentation. In: CVPR, pp. 8334–8343 (2021)
17. Lin, T.Y., Dollár, P., Girshick, R., He, K., Hariharan, B., Belongie, S.: Feature pyramid networks for object detection. In: CVPR, pp. 2117–2125 (2017)
18. Lin, T.-Y., et al.: Microsoft COCO: common objects in context. In: Fleet, D., Pajdla, T., Schiele, B., Tuytelaars, T. (eds.) ECCV 2014. LNCS, vol. 8693, pp. 740–755. Springer, Cham (2014). https://doi.org/10.1007/978-3-319-10602-1_48
19. Liu, B., Ding, Y., Jiao, J., Ji, X., Ye, Q.: Anti-aliasing semantic reconstruction for few-shot semantic segmentation. In: CVPR, pp. 9747–9756 (2021)
20. Liu, W., Zhang, C., Lin, G., Liu, F.: Crnet: Cross-reference networks for few-shot segmentation. In: CVPR, pp. 4165–4173 (2020)
21. Liu, Y., Zhang, X., Zhang, S., He, X.: Part-aware prototype network for few-shot semantic segmentation. In: Vedaldi, A., Bischof, H., Brox, T., Frahm, J.-M. (eds.) ECCV 2020. LNCS, vol. 12354, pp. 142–158. Springer, Cham (2020). https://doi.org/10.1007/978-3-030-58545-7_9
22. Long, J., Shelhamer, E., Darrell, T.: Fully convolutional networks for semantic segmentation. In: CVPR, pp. 3431–3440 (2015)
23. Lu, Z., He, S., Zhu, X., Zhang, L., Song, Y.Z., Xiang, T.: Simpler is better: Few-shot semantic segmentation with classifier weight transformer. In: ICCV, pp. 8741–8750 (2021)
24. Maaten, L.V.D., Hinton, G.: Visualizing data using t-sne. J. Mach. Learn. Res. **9**(86), 2579–2605 (2008)
25. Min, J., Kang, D., Cho, M.: Hypercorrelation squeeze for few-shot segmentation. In: ICCV, pp. 6941–6952 (2021)
26. Nguyen, K.D.M., Todorovic, S.: Feature weighting and boosting for few-shot segmentation. In: ICCV, pp. 622–631 (2019)
27. Paszke, A., Chaurasia, A., Kim, S., Culurciello, E.: Enet: A deep neural network architecture for real-time semantic segmentation. arXiv preprint arXiv:1606.02147 (2016)
28. Ravi, S., Larochelle, H.: Optimization as a model for few-shot learning. In: International Conference on Learning Representations (ICLR) (2017)
29. Shaban, A., Bansal, S., Liu, Z., Essa, I., Boots, B.: One-shot learning for semantic segmentation. In: The British Machine Vision Conference (BMVC) (2017)
30. Siam, M., Oreshkin, B.N., Jagersand, M.: Amp: Adaptive masked proxies for few-shot segmentation. In: ICCV, pp. 5249–5258 (2019)
31. Snell, J., Swersky, K., Zemel, R.S.: Prototypical networks for few-shot learning. In: Neural Information Processing Systems(NeurIPS), vol. 30 (2017)
32. Sung, F., Yang, Y., Zhang, L., Xiang, T., Torr, P.H.S., Hospedales, T.M.: Learning to compare: Relation network for few-shot learning. In: CVPR. pp. 1199–1208 (2018)
33. Tian, P., Wu, Z., Qi, L., Wang, L., Shi, Y., Gao, Y.: Differentiable meta-learning model for few-shot semantic segmentation. The AAAI Conf. Artif. Intell. **34**(07), 12087–12094 (2020)
34. Tian, Z., Zhao, H., Shu, M., Yang, Z., Li, R., Jiaa, J.: Prior guided feature enrichment network for few-shot segmentation. IEEE Trans. Pattern Analysis Mach. Intell. (PAMI) **99** 1–1 (2020)
35. Vaswani, A., et al.: Attention Is All You Need. In: Neural Information Processing Systems(NeurIPS), vol. 30 (2017)

36. Vinyals, O., Blundell, C., Lillicrap, T., koray kavukcuoglu, Wierstra, D.: Matching networks for one shot learning. In: Neural Information Processing Systems(NeurIPS), vol. 29 (2016)
37. Wang, H., Zhang, X., Hu, Y., Yang, Y., Cao, X., Zhen, X.: Few-shot semantic segmentation with democratic attention networks. In: Vedaldi, A., Bischof, H., Brox, T., Frahm, J.-M. (eds.) ECCV 2020. LNCS, vol. 12358, pp. 730–746. Springer, Cham (2020). https://doi.org/10.1007/978-3-030-58601-0_43
38. Wang, K., Liew, J.H., Zou, Y., Zhou, D., Feng, J.: Panet: Few-shot image semantic segmentation with prototype alignment. In: ICCV, pp. 9197–9206 (2019)
39. Wang, Y.-X., Hebert, M.: Learning to learn: model regression networks for easy small sample learning. In: Leibe, B., Matas, J., Sebe, N., Welling, M. (eds.) ECCV 2016. LNCS, vol. 9910, pp. 616–634. Springer, Cham (2016). https://doi.org/10.1007/978-3-319-46466-4_37
40. Xie, G.S., Liu, J., Xiong, H., Shao, L.: Scale-aware graph neural network for few-shot semantic segmentation. In: CVPR, pp. 5475–5484 (2021)
41. Yang, B., Liu, C., Li, B., Jiao, J., Ye, Q.: Prototype mixture models for few-shot semantic segmentation. In: Vedaldi, A., Bischof, H., Brox, T., Frahm, J.-M. (eds.) ECCV 2020. LNCS, vol. 12353, pp. 763–778. Springer, Cham (2020). https://doi.org/10.1007/978-3-030-58598-3_45
42. Yang, L., Zhuo, W., Qi, L., Shi, Y., Gao, Y.: Mining latent classes for few-shot segmentation. In: ICCV, pp. 8721–8730 (2021)
43. Yu, F., Koltun, V.: Multi-scale context aggregation by dilated convolutions. In: International Conference on Learning Representations (ICLR) (2016)
44. Zhang, B., Xiao, J., Qin, T.: Self-guided and cross-guided learning for few-shot segmentation. In: CVPR, pp. 8312–8321 (2021)
45. Zhang, C., Lin, C., Liu, F., Guo, J., Wu, Q., Yao, R.: Pyramid graph networks with connection attentions for region-based one-shot semantic segmentation. In: ICCV, pp. 9587–9595 (2019)
46. Zhang, C., Lin, G., Liu, F., Yao, R., Shen, C.: Canet: Class-agnostic segmentation networks with iterative refinement and attentive few-shot learning. In: CVPR, pp. 5217–5226 (2019)
47. Zhao, H., Qi, X., Shen, X., Shi, J., Jia, J.: ICNet for real-time semantic segmentation on high-resolution images. In: Ferrari, V., Hebert, M., Sminchisescu, C., Weiss, Y. (eds.) ECCV 2018. LNCS, vol. 11207, pp. 418–434. Springer, Cham (2018). https://doi.org/10.1007/978-3-030-01219-9_25
48. Zhao, H., Shi, J., Qi, X., Wang, X., Jia, J.: Pyramid scene parsing network. In: CVPR, pp. 6230–6239 (2017)
49. Zheng, S., et al.: Rethinking semantic segmentation from a sequence-to-sequence perspective with transformers. In: CVPR, pp. 6881–6890 (2021)
50. Zhou, B., et al.: Semantic understanding of scenes through the ade20k dataset. Int. J. Comput. Vision **127**, 302–321 (2019)

Slim Scissors: Segmenting Thin Object from Synthetic Background

Kunyang Han[1,2]🆔, Jun Hao Liew[3]🆔, Jiashi Feng[3]🆔, Huawei Tian[4],
Yao Zhao[1,2](✉)🆔, and Yunchao Wei[1,2]🆔

[1] Institute of Information Science, Beijing Jiaotong University, Beijing, China
`yzhao@bjtu.edu.cn`
[2] Beijing Key Laboratory of Advanced Information Science and Network Technology,
Beijing, China
[3] ByteDance, Beijing, China
[4] People's Public Security University of China, Beijing, China
`https://kunyanghan.github.io/SlimScissors/`

Abstract. Existing interactive segmentation algorithms typically fail
when segmenting objects with elongated thin structures (*e.g.*, bicycle
spokes). Though some recent efforts attempt to address this challenge by
introducing a new synthetic dataset and a three-stream network design,
they suffer two limitations: 1) large performance gap when tested on real
image domain; 2) still requiring extensive amounts of user interactions
(clicks) if the thin structures are not well segmented. To solve them,
we develop **Slim Scissors**, which enables quick extraction of elongated
thin parts by simply brushing some coarse scribbles. Our core idea is
to segment thin parts by learning to compare the original image to a
synthesized background without thin structures. Our method is model-
agnostic and seamlessly applicable to existing state-of-the-art interactive segmentation models. To further reduce the annotation burden, we
devise a similarity detection module, which enables the model to automatically synthesize background for other similar thin structures from
only one or two scribbles. Extensive experiments on COIFT, HRSOD
and ThinObject-5K clearly demonstrate the superiority of Slim Scissors
for thin object segmentation: it outperforms TOS-Net by 5.9% IoU_{thin}
and 3.5% $\mathcal{F}$ score on the real dataset HRSOD.

Keywords: Interactive image segmentation · Thin object
segmentation

K. Han—Work done during an internship at ByteDance.

Supplementary Information The online version contains supplementary material
available at https://doi.org/10.1007/978-3-031-19818-2_22.

S. Avidan et al. (Eds.): ECCV 2022, LNCS 13689, pp. 379–395, 2022.
https://doi.org/10.1007/978-3-031-19818-2_22

1   Introduction

Interactive image segmentation denotes the task of extracting the object-of-interest given some user-hints, such as clicks [23, 24, 44, 48], scribbles [2, 3, 11, 12], polygons [1, 4, 27] or bounding boxes [35, 42, 43]. This task has received much attention over the past few years due to its wide application domains ranging from image editing, medical image analysis [41] to dataset annotation [4]. The segmentation process usually iterates between: (i) user providing feedback based on the errors of current segmentation, and (ii) the segmentation model updating its prediction accordingly; and terminates when the user is satisfied with the final segmentation result.

Existing deep learning-based interactive segmentation algorithms have shown exceptional performance in this task by typically requiring only a few clicks to obtain a high-quality segmentation mask [1, 26, 30, 43, 44, 48]. For example, the current state-of-the-art Inside-Outside Guidance (IOG) [48] only needs three clicks to attain >90% IoU accuracy. However, despite the overall good performance, these state-of-the-art methods can hardly be deployed to segment objects with elongated thin parts (*e.g.* bug legs and bicycle spokes), which is the main focus of this work. As a result, excessive user interactions for correcting those missing or wrongly segmented thin parts becomes unavoidable.

Recently, Liew *et al.* [22] found out that the failure in segmenting thin objects can be mainly attributed to two factors: (i) coarsely annotated dataset used for training interactive segmentation models (*e.g.*, PASCAL [10] and COCO [25]) and (ii) imbalanced distribution between thin and non-thin pixels, which cause difficulty in learning using standard pixel-wise cross-entropy loss. To tackle these challenges, they collected a large-scale synthetic dataset specifically tailored for segmentation of elongated thin objects, called ThinObject-5K. Based on this, they proposed a three-stream network named TOS-Net and demonstrated significant performance improvement over the baselines.

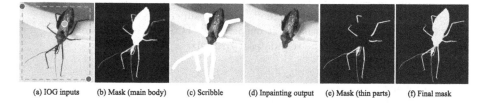

(a) IOG inputs   (b) Mask (main body)   (c) Scribble   (d) Inpainting output   (e) Mask (thin parts)   (f) Final mask

Fig. 1. Overall idea. Our Slim Scissors first takes the inside-outside guidance (IOG) [48] (a) to produce a coarse segmentation outlining the main body (b). Given some *coarse* scribbles around the thin structures (c), we construct a synthetic background image without the presence of thin parts (*e.g.*, bug legs and antennas) via image inpainting algorithm (d) and feed such (image, background) pair to a network for segmentation of thin parts (e). Finally, both features are fused to produce the final segmentation output (f).

Despite the outstanding performance of TOS-Net [22] in segmenting thin structures, it still exhibits the following drawbacks: (i) severe performance degradation when testing on real images due to the large domain gap between the synthetic training samples and real testing images (*e.g.* ~10% and ~20% IoU$_{thin}$ [22] on COIFT [31] and HRSOD [47] dataset, respectively); (ii) since TOS-Net follows the extreme clicking paradigm of DEXTR [30], it typically requires extensive user inputs (boundary clicks) for correction, especially when it comes to refining elongated thin parts, offering an unsatisfactory user experience.

To tackle this issue, we develop a new interactive segmentation paradigm, called **Slim Scissors** that enables fast extraction of elongated thin parts by simply drawing some *coarse* scribbles to cover them. Note that we employ *thick/coarse* scribbles whose radius is much larger than that of the thin structures (Figs. 1 and 6). As a result, scribbling along thin structures is fast. Compared to boundary clicking, coarse scribbling is not only much more convenient as users can quickly brush over the thin parts instead of carefully clicking on them, it also contains significantly richer information (*e.g.* shape prior) than sparse clicks.

The scribbles are then used to construct a synthetic background simulating the absence of thin parts. The underlying motivation is that the contrastive information in such an (image, background) pair can help enhance the signals of elongated thin parts, simplifying the learning process. For instance, segmenting thin spokes from a bicycle wheel becomes easier if able to compare the original image of bicycle wheel with a background image without the thin spokes. Based on this, we train a network that learns to segment thin objects based on such (image, background) pair input. Such design can be easily augmented with a generic interactive segmentation architecture to form an end-to-end trainable network. As shown in the experimental section, this simple design works surprisingly well.

As compared to boundary clicks, drawing *coarse* scribbles doubtlessly reduces the annotation burden by a significant margin. To further reduce the annotation cost, we propose a similarity detection module that automatically finds others similar thin structures guided by the current drawn scribbles. Taking Fig. 1 as an example, one only needs to coarsely draw one or two scribbles on the bug's legs while our similarity detection module generates coarse scribbles covering the remaining legs by computing per-pixel similarity with the user-provided ones. Such module can significantly reduce human efforts in drawing scribbles when many similar thin patterns appear in the same image.

Overall, the key contributions are as follows: 1) We develop **Slim Scissors**, a simple and effective tool that enables users to efficiently and accurately acquire elongated thin parts by simply drawing some coarse scribbles around them. The success of Slim Scissors can be attributed to the novel idea of segmenting from synthetic background. 2) We propose a similarity detection module to augment the Slim Scissors. This module significantly reduces the annotation burden when many similar thin patterns independently appear within the same image. 3) We perform extensive experiments on three publicly available benchmarks and show

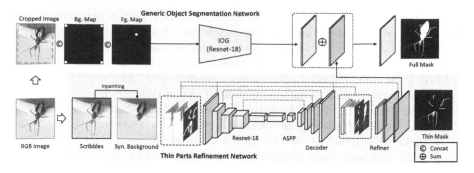

Fig. 2. Slim Scissors. Our segmentation network is composed of two parallel networks: (top) a Generic Object Segmentation (GOS) network that takes Inside-Outside Guidance (IOG) [48] as input to extract a coarse mask delineating the object main body; (bottom) a Thin Parts Refinement (TPR) network that first synthesizes a background without thin parts based on *coarsely* drawn scribbles. By comparing the (image, background) pair, our TPR networks produces a coarse prediction of thin parts, which is subsequently refined with a refiner to produce a mask of elongated thin structures only. Finally, features from both networks are fused to obtain the final segmentation mask. Note that we did not visualize the coarse mask prediction by IOG here for simplicity.

significant performance improvement upon the state-of-the-art on the two real datasets, COIFT [31] and HRSOD [47].

2   Related Works

Interactive Object Segmentation. Recently proposed interactive image segmentation algorithms are primarily driven by deep learning-based methods where significant improvements upon the traditional approaches have been demonstrated. Among them, Xu *et al.* [44] was the first to apply deep learning to this task. They proposed to apply Euclidean distance transformation to the user clicks and train a FCN based on such (image, clicks) pair. Since then, various forms of inputs representation [29,41], interaction types [1,8,20,27,30] have been explored. Some works [7,13,15,24,26] seek to better exploit useful cues from user-provided inputs while others [21,23] tackles the ambiguity in interactive segmentation given limited user hints. More recent works such as [16,19,37] focus on inference-time optimization to improve segmentation quality. Despite their overall good performance in segmenting general objects, these approaches typically fail when deployed to segment objects with elongated thin parts.

Interactive Thin Object Segmentation. Vicente *et al.* [39] tackled the shrinking bias problem in graph cut when segmenting elongated thin objects by proposing a connectivity prior. Jegalka and Bilmes [17] developed cooperative cuts that favors homogeneous boundaries by penalizing the number of types of label discontinuities. Mansilla *et al.* [31] introduced connectivity constraint on Oriented Image Foresting Transform (OIFT) and demonstrated considerable improvement on segmentation of thin objects. Dong *et al.* [9] proposed a sub-Markov random walk (subRW) algorithm and showed that segmentation of thin objects can be handled by adding label prior to subRW. More recently, Liew *et al.* [22] made the first attempt to extend the deep learning-based interactive segmentation methods to handle elongated thin objects. Nevertheless, this approach performs poorly on real image domain due to the large domain gap with the synthetic training data. Furthermore, the boundary clicking paradigm offers a poor user experience, making their methods less useful for practical application. In this work, we advocate the use of *coarse* scribble in replacement of boundary clicking to enable faster interaction and a novel idea of segmenting from synthetic background for more effective segmentation of thin structures.

3   Motivation

At the core of our approach is the novel idea of learning to segment thin parts from synthetic background where these backgrounds are synthesized based on the *coarsely* drawn scribbles highlighting the elongated thin structures. We begin by explaining the motivation before proceeding to explaining our approach.

Why Choose Scribbles for Interaction? Boundary clicking adopted by [22], in general, serve as an efficient means for interaction when conducting interactive segmentation of general objects. However, we consider that boundary clicking may not be an optimal choice for selecting thin objects as precise clicking on the thin parts can be extremely time-consuming. For example, as shown in Fig. 1(a), when segmenting the beetle's antennas, users need to be extremely careful when clicking such that the clicks are correctly located on the boundaries of the antennas, offering an overall poor user experience. Differently, taking *coarsely* drawn scribbles to cover elongated thin parts is arguably faster and much more convenient, as users only need to roughly brush over them. As shown in Figs. 1 and 6, the scribble size is much larger than that of the thin structure, easing the annotation process. This motivates us to employ coarse scribbles for selecting the interested thin parts for segmentation. It is also worth mentioning that there exist a large body of works which employ scribbles for interactive segmentation [2,3,5,11,12]. However, our approach completely differs from prior works in terms of both motivation and fashion. Specifically, previous works typically draw scribbles *within* the object to indicate the target-of-interest whereas we take *coarsely* drawn scribbles to cover elongated thin parts. An alternative to coarse scribbling is patch annotation over image grid [8]. However, we argue that patch annotation over fix-sized grid is less flexible.

Why Do We Need Synthetic Background? There are two reasons why synthetic background will be beneficial for thin object segmentation. First, the challenge of thin object segmentation task mainly lies in that many thin parts are a few pixels wide, making their features be easily overlaid by around background after several downsampling operations, which are common settings in popular CNN backbones. Intuitively, if we are able to acquire the background under the thin parts, the contrastive information will help lift the signals of the inconsistent regions between foreground and background and relieve the negative effect caused by the downsampling operations. Second, thin parts are often with rich contextual background, making the background of thin parts be easily and accurately synthesized even with very simple inpainting algorithms.

4   Slim Scissors

We propose Slim Scissors, a novel, generic, and effective solution for interactive thin object segmentation. As shown in Fig. 2, the overall network consists of two parallel subnetworks, *i.e.*, (i) a Generic Object Segmentation (GOS) network for obtaining a coarse segmentation mask delineating the object's main body; (ii) a Thin Parts Refinement (TPR) network for extracting the elongated thin structures based on the user-provided scribbles. To obtain the final segmentation output, we simply element-wise sum the features prior to the last layer of both subnetworks and pass to a 1×1 conv with sigmoid activation. Explicitly disentangling the processing of thin parts from the main body in this way not only encourages each subnetwork to better focus on different aspect of the object (main body *vs.* elongated thin structures), but also helps to alleviate the imbalanced thin and non-thin regions issue [22].

4.1   Generic Object Segmentation (GOS) Network

As our Slim Scissors is model-agnostic to GOS, it can be easily integrated with any generic interactive segmentation network to formulate an end-to-end trainable segmentation framework. In this work, we show an example instantiation based on Inside-Outside Guidance (IOG) [48], the current best performing interactive segmentation algorithm. In summary, IOG takes an inside (object's center) and two outside points (two symmetrical corners of bounding box enclosing the object) as input, outputting an object-centric crop based on the relaxed

Fig. 3. Examples of generated synthetic background images using cv2.inpaint. We challenge the reader to identify the missing thin structures by simply comparing the (image, background) pair (Beetle's legs and antennas, heron's neck and legs and morpho's antennas).

box. Concatenated with the foreground and background localization heatmaps derived from the inside and outside guidance, the cropped input is resized to 512×512 before passing to a coarse-to-fine network for segmentation. As a coarse-grained prediction usually suffices for segmenting the object's main body, we replace the ResNet-101 [14] backbone in IOG with a much lighter ResNet-18 [14] to reduce computational cost.

4.2   Thin Parts Refinement (TPR) Network

TPR provides a novel solution to segment thin parts by taking advantages of synthetic background, which consists of the following two stages during training.

Scribbles Generation. We ask the users to roughly draw some scribbles to identify the thin parts for segmentation. However, in practice, different users tend to have different scribbling behaviors, it is therefore almost infeasible to collect many interaction samples from real users for training. To simulate scribbles annotated by human annotators, we first extract the elongated thin structures for each object instance following [23]. During training, depending on the target object size, these thin regions are randomly dilated by $[0.5r, 1.5r]$ pixels to simulate coarse scribbling behavior (definition of r can be found in Eq. 5). The effects of dilations (coarseness of scribbles) will be studied in Sect. 6.3.

Background Synthesis. We propose to construct a synthetic background image based on the user-provided scribbles to simulate the absence of thin parts. In particular, we employ an image inpainting algorithm to inpaint the regions indicated by the scribbles. Since scribbles used to mark thin structures typically cover relatively small areas and are often with rich background context, a realistic-looking completed image can be easily obtained given existing inpainting algorithms. In this work, we opt for simplicity and use `cv2.inpaint`[1] from OpenCV [33]. Note that our proposed synthetic background idea is a universal one and more advanced inpainting methods such as [28,46] can be employed but it is out of the scope of this work. We will leave this to our future work. Figure 3 depicts some examples of generated synthetic background images where elongated thin structures are successfully removed.

Based on this, we concatenate the input image, its corresponding synthetic background and the scribbles mask as input for training the TPR network. Considering an RGB image I, the corresponding synthetic background B and scribbles mask S, the task of segmenting thin structures can be mathematically formulated as learning a mapping function $f(; \theta_{thin})$ that is parameterized by θ_{thin}:

$$M_{thin} = f(I, B, S; \theta_{thin}) \tag{1}$$

where M_{thin} denotes the mask of thin parts.

---

[1] https://docs.opencv.org/master/d7/d8b/group__photo__inpaint.html.

Network Architecture. Similarly, our TPR is realized with a coarse-to-fine network structure. More specifically, taking the cropped input from GOS sub-network as input, the first stage adopts a ResNet-18 [14]-based encoder-decoder structure, which appends an Atrous Spatial Pyramid Pooling (ASPP) module [6] at the end of the encoder to incorporate global context, followed by a decoder that sequentially upsamples and concatenates with low-level feature for subsequent convolution, producing a coarse mask and features of 32 channels.

Taking the concatenation of the original image, synthesized background, scribbles input, coarse mask prediction and features from the previous stage as input, the second stage first downsamples the concatenated inputs to 0.5× resolution to enlarge the receptive field for refinement, followed by two lightweight convolution layers. The output features is then bilinearly upsampled back to original resolution, concatenated with the (image, background, scribble) input before passing to another two convolution layers, followed by a sigmoid activation to produce M_{thin}. Please refer to our supplement for more details.

Training of Slim Scissors. We employ the same training strategy adopted in IOG [48] for training our GOS network, *i.e.* training with binary cross-entropy loss and applying side losses at each level of CoarseNet as deep supervision. Similarly, our TPR network (both decoder and refiner) is trained with standard binary cross-entropy loss, with thin structures extracted from [22] being the ground truth masks. In order to encourage the TPR network to mainly focus on refining the coarse predictions along elongated thin parts, we only consider those background pixels around the thin parts during training. In our experiments, the ratio of foreground and background pixels is set as 1:2.

5 Fast Slim Scissors

As explained in Sect. 3, coarse scribbling provides a more user-friendly way to advance the interactive thin parts segmentation process. Nevertheless, drawing scribbles for each and every thin part can still be time-consuming and redundant since most thin structures share similar appearance or texture (*e.g.*, bug legs). This observation motivates us to answer the following question: *in the scenario where many thin parts are with similar patterns, do users really need to brush them one by one?* The answer is **no**.

Similar Detection Module. To reduce the annotation burden in scribbling similar thin parts over and over again, we propose a similar detection module (SDM) to augment our Slim Scissors. SDM accepts one or two scribbles from the user and automatically mines other similar thin structures. The newly mined thin structures are often coarse yet sufficient to serve as new scribbles for guiding the subsequent background synthesis in Slim Scissors. In particular, given some user-marked scribbles, we first extract an initial segmentation of thin parts covered by those scribbles using our Slim Scissors, which help separate the scribbled regions into a set of foreground and background pixels. We next compute two distance maps by matching each pixel within the same image to both the foreground F and background sets B in the embedding feature space. Mathematically, for each pixel i, we define its matching as:

$$G_F(i) = \min_{j \in F} D(i, j) \tag{2}$$

$$G_B(i) = \min_{j \in B} D(i, j) \tag{3}$$

where $D(i, j)$ refers to the distance between pixel i and j in term of their corresponding embedding vector e_i and e_j, which is defined as [40,45]:

$$D(i, j) = \begin{cases} 1 - \frac{1}{1+\exp(\|e_i - e_j\|^2 + b_F)}, & \text{if } j \in F \\ 1 - \frac{1}{1+\exp(\|e_i - e_j\|^2 + b_B)}, & \text{if } j \in B \end{cases} \tag{4}$$

where b_F and b_B denote the learnable foreground and background bias.

Given the computed foreground and background matching maps, we concatenate them with the pixel-level embedding feature and pass to a lightweight decoder module for outputting a binary mask predicting other similar thin structures to the one(s) highlighted by current scribble(s). As a result, our Fast Slim Scissors only requires one or two scribbles from the user while attaining results similar to that of drawing scribbles on all the elongated thin structures.

In Fig. 4, we show how to apply our SDM to automatically generate more scribbles covering other similar thin parts. Specifically, in round one, the user first draws a coarse scribble (green scribble) on one of the beetle's leg (the most frequent thin parts). Our SDM then identifies the remaining legs and automatically generates coarse scribbles for them (orange scribbles). Similarly, in round two, the user draws another scribble on one of the missing parts, i.e., beetle's antennas, and uses SDM to locate another missing antennas within the image. In practice, such an iterative process will continue until all thin parts are covered by scribbles (either user drawn or generated by our SDM).

Overall Process. The overall interactive segmentation process of *Fast* Slim Scissors is summarized as follows: 1) The user first provides an inside (object's

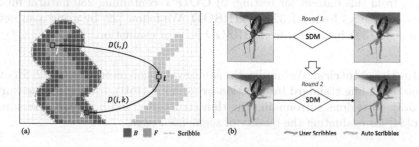

Fig. 4. Scribble generation process using our SDM. (Color figure online)

center) and two outside clicks (two symmetrical corners of a bounding box) [48] to indicate the object-to-segment; 2) Draw a coarse scribble on any of the thin parts; 3) SDM mines other similar thin structures by computing a global matching with the user-provided scribble(s) (Sect. 5); 4) Along with the previously drawn ones, these newly mined scribbles are passed to our Slim Scissors, producing an initial object segmentation mask including elongated thin structures (Sect. 4); 5) The user could either introduce more scribbles on the missing thin parts or remove the mislabeled thin regions; 6) Step 3 to 5 (step 3 is skipped if the user chose to erase the wrongly segmented thin parts) are repeated until the user is satisfied with the final segmentation output.

6 Experiment

6.1 Datasets and Settings

Implementation Details. We train our Slim Scissors on ThinObject-5K `train` split [22] with 4,743 images for 30 epochs. All networks are initialized from ResNet-18 [14] pre-trained on ImageNet [36]. We train our model using Adam optimizer [18] with batch size of 25 and weight decay of 10^{-4}. The base learning rate of the pre-trained and newly initialized weights are set to 10^{-4} and 10^{-3}, respectively. We employ a linear decay learning rate scheduler with 5 epochs of linear warmup. We randomly resize and rotate images with a scale factor from 0.75 to 1.25 and rotation angle from $-20°$ to $20°$. We also apply random horizontal flipping during training. For similar detection experiments, we created a subset from ThinObject-5K `train` split, where each image fulfills the following criteria: 1) each image must contain at least two separated thin regions; 2) the thin regions should belong to the same semantic category (*e.g.*, bug's legs and antennas are considered two different categories). The resulting subset contains 2,128 images. We train our SDM on this subset for 45 epochs with batch size of 16 using a constant learning rate. All other hyperparameters remain the same.

Datasets. We evaluate on the following three benchmarks: 1) **ThinObject-5K**, a large-scale synthetic dataset from [22] that is composed of different objects with elongated thin structures (*e.g.*, ants, racket, harps *etc.*). We use 500 testing images from this dataset for testing; 2) **COIFT** containing 280 natural images of birds and bugs from [31,32]; 3) **HRSOD**. We follow [22] by using a subset of 280 images (305 instances) from HRSOD [47] for evaluation.

Evaluation Metrics. We employ the same evaluation metrics as in [22] for evaluation, including standard Intersection-over-Union (IoU), IoU_{thin} for measuring IoU only on regions surrounding the thin structures, and boundary measure $\mathcal{F}$ from [34] for evaluating the quality of segmented edges.

6.2 Comparison with the State-of-the-Art Methods

We first compare our Slim Scissors with prior methods on ThinObject-5K `test` split, COIFT [31] and HRSOD [47] dataset. For our Slim Scissors, we simulate user-drawn scribbles by performing morphological dilation on the thin parts extracted from [22], where kernel radius r is adaptively determined based on the image resolution. The underlying idea is that thin parts in a higher resolution image appear to be thicker, requiring sufficiently coarse scribbles for annotation. Specifically, let the range of kernel radius be $[r_{min}, r_{max}]$, the maximum target object size in the three benchmarks be N_{max}, given a test object with pixel number N, the kernel radius is set to

$$r = \left\lfloor r_{min} + N \times \frac{r_{max}}{N_{max}} \right\rfloor, \tag{5}$$

where $r_{min} = 3$ and $r_{max} = 240$. The results are summarized in Table 1.

Table 1. Quantitative results on ThinObject-5K `test` set, COIFT and HRSOD dataset. [†] take scribble and synthetic background as input.

Method	Training Set	ThinObject-5K [22]			COIFT [31]			HRSOD [47]		
		IoU	IoU$_{thin}$	$\mathcal{F}$	IoU	IoU$_{thin}$	$\mathcal{F}$	IoU	IoU$_{thin}$	$\mathcal{F}$
DEXTR [22]	PASCAL-10K	61.8	43.5	49.0	70.6	36.8	74.4	69.5	35.4	66.1
DEXTR [22]	ThinObject-5K	88.8	74.0	89.3	88.2	68.3	93.4	82.5	57.7	84.8
f-BRS [38]	ThinObject-5K	90.9	80.6	89.4	87.8	63.6	89.1	78.1	49.7	74.1
f-BRS[†] [38]	ThinObject-5K	92.5	84.0	92.6	89.9	67.1	94.6	86.4	63.8	87.7
TOS-Net [22]	ThinObject-5K	94.3	86.5	94.8	92.0	76.4	95.3	86.4	65.1	87.9
Slim Scissors (ours)	ThinObject-5K	91.1	80.5	93.5	93.2	80.8	97.0	87.7	71.0	91.4

We first notice a significant performance gap between the synthetic and real image datasets for the state-of-the-art TOS-Net [22], suggesting that TOS-Net has possbily overfitted on the artefacts present in synthetic image. On the other hand, the performance gap of Slim Scissors is much smaller. More importantly, when tested on real image benchmarks, our Slim Scissors outperforms TOS-Net by a significant margin (4.4% IoU$_{thin}$ and 1.7% $\mathcal{F}$ score on COIFT and 5.9% IoU$_{thin}$ and 3.5% $\mathcal{F}$ on HRSOD dataset), demonstrating the better generalization capabilities of our Slim Scissors for practical application. Some qualitative comparison can be found in Fig. 5.

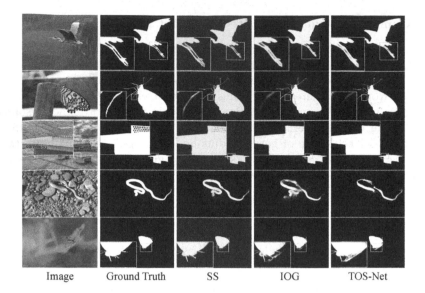

Image   Ground Truth   SS   IOG   TOS-Net

Fig. 5. Qualitative comparison between IOG [48], TOS-Net [22] and our SS.

6.3   Ablation Study

We perform ablation experiments to quantitatively justify the effectiveness of each proposed component. The results are summarized in Table 2.

Synthetic Background and Scribble Input. To study the effectiveness of synthetic background for segmentation of thin structures, we construct two baselines that take only image *or* image with synthetic background as input for TPR. As shown in Tab. 2 (top), the performance is significantly boosted when using synthetic background, and can be further improved when using coarse scribbles.

Dual Path Architecture. As shown in Tab. 2 (bottom), removing TPR (3rd row) reduces to IOG [48] except for using lighter ResNet-18 and being trained on ThinObject-5K dataset, where the performance severely degrades. However, as the large performance gap is partly due to different network input, we also train a baseline that additionally accepts synthetic background and scribbles input (2nd row) for fair comparison. Dual path design performs the best, suggesting explicitly separating the processing of thin parts from the object's main body is generally beneficial for this task.

Table 2. Ablation study. **TPR input** and **dual path structure**.

TPR input	COIFT [31]			HRSOD [47]		
	IoU	IoU$_{thin}$	$\mathcal{F}$	IoU	IoU$_{thin}$	$\mathcal{F}$
+BG +scribble	93.2	80.8	97.0	87.7	71.0	91.4
+BG	92.9	79.0	96.7	86.6	68.6	90.3
image only	91.2	76.1	94.3	83.2	60.5	85.0

Dual Branch	BG + Scribble	ThinObject-5K [22]			COIFT [31]		
		IoU	IoU$_{thin}$	$\mathcal{F}$	IoU	IoU$_{thin}$	$\mathcal{F}$
✓	✓	91.1	80.5	93.5	93.2	80.8	97.0
	✓	91.0	78.7	93.5	93.5	77.9	98.4
		89.0	75.7	88.9	91.3	73.1	94.4

Robustness to Scribble Size. Lastly, we study the robustness of our Slim Scissors to different scribble size. Thicker scribbles enable faster scribbling but may lead to poorer inpainting result which subsequently affect the segmentation quality; thinner scribbles encode stronger spatial prior but at the cost of being slower. As shown in Fig. 6, as expected, the segmentation performance gradually degrades when using larger r. Nevertheless, we can see that our approach is fairly robust on overall. Even with $1.5 \times r$ scheme, our Slim Scissors still consistently outperforms TOS-Net on both benchmarks, well demonstrating its practicality with various scribble size. We also visualize some examples in Fig. 6, where we can see $1 \times r$ is sufficient to enable users for fast scribbling over thin parts.

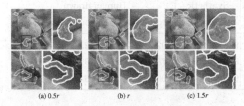

(a) 0.5r   (b) r   (c) 1.5r

r	COIFT [31]			HRSOD [47]		
	IoU	IoU$_{thin}$	$\mathcal{F}$	IoU	IoU$_{thin}$	$\mathcal{F}$
$0.5\times$	93.2	81.4	96.9	87.6	71.7	91.3
$1\times$	93.2	80.8	97.0	87.7	71.0	91.4
$1.5\times$	93.0	80.1	96.9	87.4	69.8	91.0

Fig. 6. Robustness against different scribble coarseness. r denotes the radius of morphological kernel used to dilate the scribbles for evaluation (Eq. 5).

6.4   Automated Scribble Recommendation

To validate the effectiveness of the similar detection module (SDM) in boosting the annotation process, we perform additional experiments to evaluate our fast Slim Scissors. The goal of SDM is to automatically acquire similar thin parts with the given scribbles. We perform evaluation on a subset of COIFT, and choose the objects (*e.g.* birds and bugs) that are with at least two or more similar thin parts, resulting in 268 images and 2063 independent thin parts in total. At the beginning, for each image, we randomly select a thin part and draw a scribble to cover it. SDM is then used to detect the remaining ones. In the following rounds, we progressively conduct additional interactions (*e.g.*, adding new scribbles to the missing parts or removing the false regions detected by SDM) until all the thin parts are well covered by scribbles. As shown in Fig. 7, our fast Slim Scissors reduces human efforts by more than a half ($\sim$1000 scribbles) and achieves competitive performance to that of a fully annotated counterpart (79.9%+ *vs.* 81.7%).

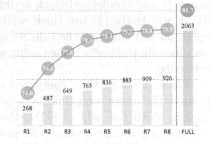

Fig. 7. The trend of IoU$_{thin}$ and the corresponding annotation burden of Fast Slim Scissors on a subset of COIFT.

6.5   User Study

Impact of Synthetic Scribbles. We conducted a user study to examine the impact of synthetic scribbles. Specifically, we recruited 9 participants and split them into 3 groups where each group is asked to annotate the entire COIFT dataset (280 images) with their best efficiency. In other words, each image is annotated 3 times and annotators can choose their preferred scribble sizes. As shown in Fig. 8, the performance on human-drawn scribbles are close to the synthetic ones, demonstrating the robustness of our approach to user variance.

COIFT	Synthetic		Human-drawn			
	SS	TOS-Net	Group1	Group2	Group3	Average
IoU	93.2	92.0	92.5	92.5	92.5	92.5 ± 0.00
IoU$_{thin}$	80.8	76.4	78.8	78.8	79.1	78.9 ± 0.14
$\mathcal{F}$	97.0	96.2	96.2	96.3	96.4	96.3 ± 0.08

Fig. 8. User study. Left: Top 3 easiest (top) and hardest (bottom) samples with their annotation time. Right: user study performance on COIFT dataset.

Compatibility with Real User Input. We also conducted another user study to examine the effectiveness of the proposed approach with real human user input. We collected 20 images from COIFT dataset [31] and manually chose two sets of scribble thickness for each test image, simulating thin and thick scribbles. Among these 20 images, there are 7 birds, 7 butterflies and 6 others (mainly insects). Some hardest and easiest samples are shown in Fig. 8. We recruited 5 participants where each participant is asked to draw scribbles along the thin parts under 4 different settings: (1) use only the pre-defined thin scribble; (2) use only the pre-defined thick scribble; (3) use thin scribbles with the help of SDM and lastly (4) use thick scribbles with SDM. The total time (seconds) spent on scribbling all the thin regions and IoU$_{thin}$ are reported in Table 3. We also include the performance using synthetic scribbles (dilation rate of $1 \times r$) as reference.

We first notice that, compared to thin scribbles, employing thick scribbles reduces the annotation time by nearly half with only slight performance degradation. With the help of SDM, the annotation time is further reduced while maintaining similar accuracy. In practice, when SDM is employed, much fewer scribbles are needed, we can therefore afford to use thinner scribbles for better accuracy. Lastly, the small performance gap between synthetic and real human-drawn scribbles demonstrates the robustness of our method to user variance.

Table 3. User study. Total time (s) spent on scribbling all the thin regions and IoU$_{thin}$ in 20 images using different annotation rules.

Annotation rule	User1		User2		User3		User4		User5		Average	
	IoU$_{thin}$	Time	IoU$_{thin}$	Time	IoU$_{thin}$	Time	IoU$_{thin}$	Time	IoU$_{thin}$	Time	IoU$_{thin}$	Time
Thin scribbles	73.3	178	73.9	219	75.7	218	75.5	251	75.7	210	74.8	215
Thick scribbles	72.4	112	72.5	143	74.1	108	74.0	149	73.8	86.3	73.4	120
Thin scribbles + SDM	73.8	65.2	74.5	76.2	75.6	90.5	75.5	73.6	76.0	59.2	75.1	72.9
Thick scribbles + SDM	72.9	55.7	73.8	51.4	72.7	44.8	74.0	53.4	74.6	33.2	73.6	47.7

7   Conclusion

This work addresses the task of interactively segmenting objects with elongated thin structures, such as bicycles with thin spokes. The underlying core idea is to synthesize a background image without thin parts such that the network could learn to compare the (image, background) pair for extraction of thin structures. Inspired by this, we devise an effective interactive thin object segmentation technique called Slim Scissors that only requires *coarsely*-drawn scribbles covering thin parts, which is arguably easier than precise clicking used in previous literature. We also show that it can be seamlessly combined with existing state-of-the-art interactive segmentation networks to form an end-to-end trainable segmentation framework. To reduce the annotation burden, we also propose a fast Slim Scissors variant, which augments Slim Scissors with a similar detection module for mining additional scribbles covering other similar thin structures in the same image. We demonstrate the effectiveness of our approach on three publicly available benchmarks and achieve the new state-of-the-art.

Limitations. Our proposed SDM occasionally falsely detects irrelevant thin parts (*e.g.*, another bird's legs) in the presence of multiple objects. How to reduce the false positives will be investigated in our future work.

Acknowledgments. This work was supported in part by the National Key R&D Program of China (No. 2021ZD0112100), the National NSF of China (No. U1936212, No. 62120106009, No. 61972405), the Fundamental Research Funds for the Central Universities (No. K22RC00010).

References

1. Acuna, D., Ling, H., Kar, A., Fidler, S.: Efficient interactive annotation of segmentation datasets with Polygon-RNN++. In: CVPR (2018)
2. Bai, X., Sapiro, G.: A geodesic framework for fast interactive image and video segmentation and matting. In: ICCV (2007)
3. Boykov, Y.Y., Jolly, M.P.: Interactive graph cuts for optimal boundary & region segmentation of objects in nd images. In: ICCV (2001)
4. Castrejon, L., Kundu, K., Urtasun, R., Fidler, S.: Annotating object instances with a Polygon-RNN. In: CVPR (2017)

5. Chen, B., Ling, H., Zeng, X., Gao, J., Xu, Z., Fidler, S.: ScribbleBox: interactive annotation framework for video object segmentation. In: Vedaldi, A., Bischof, H., Brox, T., Frahm, J.-M. (eds.) ECCV 2020. LNCS, vol. 12358, pp. 293–310. Springer, Cham (2020). https://doi.org/10.1007/978-3-030-58601-0_18

6. Chen, L.C., Papandreou, G., Kokkinos, I., Murphy, K., Yuille, A.L.: Deeplab: Semantic image segmentation with deep convolutional nets, atrous convolution, and fully connected crfs. TPAMI (2018)

7. Chen, X., Zhao, Z., Yu, F., Zhang, Y., Duan, M.: Conditional diffusion for interactive segmentation. In: ICCV (2021)

8. Dang, V.N., et al.: Vessel-captcha: an efficient learning framework for vessel annotation and segmentation. In: Medical Image Analysis (2021)

9. Dong, X., Shen, J., Shao, L., Van Gool, L.: Sub-markov random walk for image segmentation. TIP (2015)

10. Everingham, M., Van Gool, L., Williams, C.K., Winn, J., Zisserman, A.: The pascal visual object classes (VOC) challenge. In: IJCV (2010)

11. Grady, L.: Random walks for image segmentation. TPAMI (2006)

12. Gulshan, V., Rother, C., Criminisi, A., Blake, A., Zisserman, A.: Geodesic star convexity for interactive image segmentation. In: CVPR (2010)

13. Hao, Y., et al.: Edgeflow: Achieving practical interactive segmentation with edge-guided flow (2021)

14. He, K., Zhang, X., Ren, S., Sun, J.: Deep residual learning for image recognition. In: CVPR (2016)

15. Hu, Y., Soltoggio, A., Lock, R., Carter, S.: A fully convolutional two-stream fusion network for interactive image segmentation. In: Neural Networks (2019)

16. Jang, W.D., Kim, C.S.: Interactive image segmentation via backpropagating refinement scheme. In: CVPR (2019)

17. Jegelka, S., Bilmes, J.: Cooperative cuts for image segmentation. Tech. rep., Technical Report 2010–0003, University of Washington (2010)

18. Kingma, D.P., Ba, J.: Adam: A method for stochastic optimization (2015)

19. Kontogianni, T., Gygli, M., Uijlings, J., Ferrari, V.: Continuous adaptation for interactive object segmentation by learning from corrections. In: Vedaldi, A., Bischof, H., Brox, T., Frahm, J.-M. (eds.) ECCV 2020. LNCS, vol. 12361, pp. 579–596. Springer, Cham (2020). https://doi.org/10.1007/978-3-030-58517-4_34

20. Le, H., Mai, L., Price, B., Cohen, S., Jin, H., Liu, F.: Interactive boundary prediction for object selection. In: Ferrari, V., Hebert, M., Sminchisescu, C., Weiss, Y. (eds.) Computer Vision – ECCV 2018. LNCS, vol. 11218, pp. 20–36. Springer, Cham (2018). https://doi.org/10.1007/978-3-030-01264-9_2

21. Li, Z., Chen, Q., Koltun, V.: Interactive image segmentation with latent diversity. In: CVPR (2018)

22. Liew, J.H., Cohen, S., Price, B., Mai, L., Feng, J.: Deep interactive thin object selection. In: WACV (2021)

23. Liew, J.H., Cohen, S., Price, B., Mai, L., Ong, S.H., Feng, J.: MultiSeg: Semantically meaningful, scale-diverse segmentations from minimal user input. In: ICCV (2019)

24. Liew, J.H., Wei, Y., Xiong, W., Ong, S.H., Feng, J.: Regional interactive image segmentation networks. In: ICCV (2017)

25. Lin, T.-Y.: Microsoft COCO: common objects in context. In: Fleet, D., Pajdla, T., Schiele, B., Tuytelaars, T. (eds.) ECCV 2014. LNCS, vol. 8693, pp. 740–755. Springer, Cham (2014). https://doi.org/10.1007/978-3-319-10602-1_48

26. Lin, Z., Zhang, Z., Chen, L.Z., Cheng, M.M., Lu, S.P.: Interactive image segmentation with first click attention. In: CVPR (2020)

27. Ling, H., Gao, J., Kar, A., Chen, W., Fidler, S.: Fast interactive object annotation with Curve-GCN. In: CVPR (2019)
28. Liu, G., Reda, F.A., Shih, K.J., Wang, T.-C., Tao, A., Catanzaro, B.: Image inpainting for irregular holes using partial convolutions. In: Ferrari, V., Hebert, M., Sminchisescu, C., Weiss, Y. (eds.) ECCV 2018. LNCS, vol. 11215, pp. 89–105. Springer, Cham (2018). https://doi.org/10.1007/978-3-030-01252-6_6
29. Majumder, S., Yao, A.: Content-aware multi-level guidance for interactive instance segmentation. In: CVPR (2019)
30. Maninis, K.K., Caelles, S., Pont-Tuset, J., Van Gool, L.: Deep extreme cut: From extreme points to object segmentation. In: CVPR (2018)
31. Mansilla, L.A., Miranda, P.A.: Oriented image foresting transform segmentation: Connectivity constraints with adjustable width. In: SIBGRAPI Conference on Graphics, Patterns and Images (SIBGRAPI) (2016)
32. Mansilla, L.A., Miranda, P.A., Cappabianco, F.A.: Oriented image foresting transform segmentation with connectivity constraints. In: ICIP (2016)
33. OpenCV: Open source computer vision library (2015)
34. Perazzi, F., Pont-Tuset, J., McWilliams, B., Gool, L.V., Gross, M., Sorkine-Hornung, A.: A benchmark dataset and evaluation methodology for video object segmentation. In: CVPR (2016)
35. Rother, C., Kolmogorov, V., Blake, A.: Grabcut: Interactive foreground extraction using iterated graph cuts. In: ACM ToG (2004)
36. Russakovsky, O., et al.: Imagenet large scale visual recognition challenge. IJCV (2015)
37. Sofiiuk, K., Petrov, I., Barinova, O., Konushin, A.: f-BRS: Rethinking backpropagating refinement for interactive segmentation. In: CVPR (2020)
38. Sofiiuk, K., Petrov, I., Barinova, O., Konushin, A.: f-brs: Rethinking backpropagating refinement for interactive segmentation. In: CVPR (2020)
39. Vicente, S., Kolmogorov, V., Rother, C.: Graph cut based image segmentation with connectivity priors. In: CVPR (2008)
40. Voigtlaender, P., Chai, Y., Schroff, F., Adam, H., Leibe, B., Chen, L.C.: FEELVOS: Fast end-to-end embedding learning for video object segmentation. In: CVPR (2019)
41. Wang, G., et al.: DeepIGeoS: a deep interactive geodesic framework for medical image segmentation. TPAMI (2018)
42. Wu, J., Zhao, Y., Zhu, J.Y., Luo, S., Tu, Z.: Milcut: A sweeping line multiple instance learning paradigm for interactive image segmentation. In: CVPR (2014)
43. Xu, N., Price, B., Cohen, S., Yang, J., Huang, T.: Deep GrabCut for object selection. In: BMVC (2017)
44. Xu, N., Price, B., Cohen, S., Yang, J., Huang, T.S.: Deep interactive object selection. In: CVPR (2016)
45. Yang, Z., Wei, Y., Yang, Y.: Collaborative video object segmentation by foreground-background integration. In: Vedaldi, A., Bischof, H., Brox, T., Frahm, J.-M. (eds.) ECCV 2020. LNCS, vol. 12350, pp. 332–348. Springer, Cham (2020). https://doi.org/10.1007/978-3-030-58558-7_20
46. Yu, J., Lin, Z., Yang, J., Shen, X., Lu, X., Huang, T.S.: Free-form image inpainting with gated convolution. In: ICCV, pp. 4471–4480 (2019)
47. Zeng, Y., Zhang, P., Zhang, J., Lin, Z., Lu, H.: Towards high-resolution salient object detection. In: ICCV (2019)
48. Zhang, S., Liew, J.H., Wei, Y., Wei, S., Zhao, Y.: Interactive object segmentation with inside-outside guidance. In: CVPR (2020)

Abstracting Sketches Through Simple Primitives

Stephan Alaniz[1,2](✉) , Massimiliano Mancini[1] , Anjan Dutta[3] ,
Diego Marcos[4] , and Zeynep Akata[1,2,5]

[1] University of Tübingen, Tübingen, Germany
stephan.alaniz@uni-tuebingen.de
[2] MPI for Informatics, Saarbrücken, Germany
[3] University of Surrey, Guildford, UK
[4] Wageningen University, Wageningen, The Netherlands
[5] MPI for Intelligent Systems, Tübingen, Germany

Abstract. Humans show high-level of abstraction capabilities in games
that require quickly communicating object information. They decompose the message content into multiple parts and communicate them in
an interpretable protocol. Toward equipping machines with such capabilities, we propose the Primitive-based Sketch Abstraction task where the
goal is to represent sketches using a fixed set of drawing primitives under
the influence of a budget. To solve this task, our Primitive-Matching
Network (PMN), learns interpretable abstractions of a sketch in a self
supervised manner. Specifically, PMN maps each stroke of a sketch to
its most similar primitive in a given set, predicting an affine transformation that aligns the selected primitive to the target stroke. We learn
this stroke-to-primitive mapping end-to-end with a distance-transform
loss that is minimal when the original sketch is precisely reconstructed
with the predicted primitives. Our PMN abstraction empirically achieves
the highest performance on sketch recognition and sketch-based image
retrieval given a communication budget, while at the same time being
highly interpretable. This opens up new possibilities for sketch analysis,
such as comparing sketches by extracting the most relevant primitives
that define an object category. Code is available at https://github.com/
ExplainableML/sketch-primitives.

Keywords: Sketch abstraction · Sketch analysis

A. Dutta is with the Institute for People-Centred AI at the University of Surrey. S.
Alaniz, M. Mancini and Z. Akata are with the Cluster of Excellence Machine Learning
at the University of Tübingen.

Supplementary Information The online version contains supplementary material
available at https://doi.org/10.1007/978-3-031-19818-2_23.

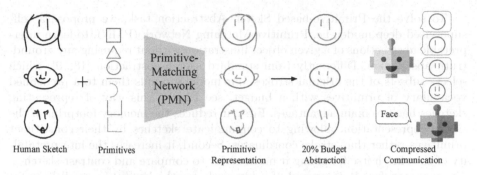

Fig. 1. Primitive-based sketch abstraction task. Our Primitive-Matching Network (PMN) takes human sketches and replaces their strokes with simple shapes from a set of 7 drawing primitives to create an abstract representation of the sketch. We can further compress the sketch by sub-selecting primitive strokes to meet a desired information budget. When communicating a sketch with a limited budget, our sketch abstractions retain the original semantics to perform well on downstream tasks.

1   Introduction

Consider the game *Pictionary*[1], where one player picks an object, e.g. a face, and draws the object in an iterative manner, e.g. using a large circle for the head, small lines for eyes and an arc for the mouth, until the other players guess the object correctly. The goal is to represent an object by decomposing it into parts that characterize this object using as few parts as possible such that another player can recognize it as fast as possible. The inherent human ability [10] that makes playing this game with multiple players possible is the ability to identify the most distinctive parts of the object and ground them into an interpretable communication protocol for the other players. In other words, humans are capable of a high level of abstraction when thinking about, recognizing and describing objects to other.

Inspired by this observation, we propose Primitive-based Sketch Abstraction as a new representation learning task, where the goal is to represent free-form drawings, i.e. sketches, by means of a fixed set of simple primitives. Sketches are an excellent tool for this task as they capture the essential parts of an object while removing the potentially adversarial texture and color information. However, humans have different drawing styles and skills, influenced by their upbringing and culture [33]. This causes different participants to draw the same instance of a real object in different ways (e.g. see Fig. 1, left). We argue, however, that there exists a fundamental representation of each object class. As demonstrated in [1,10,27] when a participant draws an object of their imagination using a fixed dictionary of shapes providing a heavily abstracted representation of the object, another participant still guesses the object correctly.

---

[1] https://en.wikipedia.org/wiki/Pictionary.

To solve the Primitive-based Sketch Abstraction task, we propose a self-supervised deep model, i.e. Primitive-Matching Network (PMN), to learn interpretable abstractions of a given object illustration without requiring any ground-truth abstraction. Differently from standard sketch-abstraction [18,19], which selects subsets of the original strokes, our model grounds them to a predefined vocabulary of primitives with a budget, see Fig. 1. This way of representing sketches has two main advantages. First, it reduces the memory footprint of the sketch representation, allowing to communicate sketches by their constituent primitives rather than stroke coordinates. Second, it increases the interpretability of the sketch itself, making it much easier to compare and contrast sketches, e.g. a *human face* is composed of a big circle for the head, two small lines for the eyes and one arc for the mouth whereas a *cat face* is similar to a *human face* but has triangles on top of the head for its ears.

Our PMN model replaces each stroke of a sketch with a single drawing primitive. This is achieved by mapping each stroke to its most similar primitive in a given set, and predicting an affine transformation that aligns the selected primitive to the target stroke. We train PMN by comparing the distance-transform of target strokes and their primitive-based version. At test time, given a sketch, we can efficiently choose a set of primitives and their spatial transformations, such that the generated sketch is fully composed of primitive shapes while being as similar as possible to the original one. Experiments on sketch recognition and fine-grained sketch-based image retrieval tasks, show that the PMN abstraction achieves the highest performance given a communication budget (i.e. number of bytes necessary to communicate the sketch). Moreover, we show how we can use our abstraction to compare sketches, extracting the most relevant primitives and patterns that define an object category.

To summarize, our contributions are: i) we propose the task of Primitive-based Sketch Abstraction, where the goal is to produce interpretable sketch representations by means of predefined drawing primitives; ii) we propose the first method for this task, Primitive-Matching Network, which learns to match strokes to primitives using as supervision a reconstruction loss over their distance transforms; iii) we show that PMN provides reliable sketch representations, communicating more information with a lower budget when compared with standard sketch abstraction methods, and eases sketch analysis.

2   Related Works

Sketch Abstraction. The goal of sketch abstraction [2,19] is to simplify the original strokes (or segments) from sketches without altering their semantic meaning. Abstracting sketches allows to communicate their information more effectively and efficiently, highlighting the most important traits of a sketch without corrupting its content [2]. This is used in many applications, ranging from sketch-based image retrieval from edge-maps, to controllable sketch synthesis at various abstraction levels. Previous approaches addressed this problem through reinforcement learning, learning to remove sketch parts while preserving some desired features (e.g. semantic category, attributes) [18,19]. Differently from previous works, we do not abstract sketches by removing strokes, but we ground them to a set of drawing primitives. This allows us to not only simplify the sketch representation itself, but to easily perform comparisons and analyses across sketches in a more straight forward manner than with stroke-based abstraction methods.

Sketch Applications. The release of the TU-Berlin [8] and QuickDraw [15] datasets attracted the attention of the research community towards sketch classification. Early works addressed the task with maximum margin classifiers over hand-crafted features [16,31]. Advent of large-scale sketch datasets led to the development of deep learning models for this task that even surpassed human performance [45]. Recent approaches explored deep and hand-crafted features [14], multi-graph transformers [42], coarse-to-fine hierarchical features [43], and learned tokenization schemes [25].

Another popular application of sketches is *sketch-based image retrieval* (SBIR), where the goal is to match free-hand sketches with corresponding natural images, both at category [23,37] and at instance level [3,21]. Existing approaches for this task bridge the domain gap between photos and sketches by means of two branch architectures focusing on each modality independently [6,7,9], and even applying attention-based objectives [20,35] or self-supervised ones [21]. Recently, [4] proposed to perform retrieval online, while the human is drawing. [29,30] perform SBIR by matching keyshapes to patches of sketch and contour images for SBIR, e.g. through S-HELO [28] descriptors. In this work, we do not directly address sketch recognition and SBIR, but we use them to quantitatively analyze the compression/quality of our abstract sketch representations.

Reconstruction with Primitives. One way of simplifying a complicated shape is to build an approximation using simple primitives. This is a central aspect of how humans understand the environment [5] and has been applied to vector-like bitmap images [13,24,40], CAD sketches [11,22,32], and 3D shape reconstruction from sketches [34] or images [17,46]. Interestingly, also many lossy image compression methods represent an image as a combination of predefined primitives [36,39]. One closely related work [40] focuses on diagram-like sketches, using shape proposals and an SVM classifier to assign the best-matching primitive. [41] represents sketches and edge maps of real images through lines and arcs for sketch-based image retrieval. Differently from these approaches, we are not restricted to specific domains [40], or primitives [41]. PMN is generic and can be applied to any sketch, and any set of drawing primitives.

3   Abstracting Sketches by Drawing Primitives

Given a sketch, our goal is to obtain an abstract representation by replacing its strokes with a set of drawing primitives (e.g. squares, circles, lines). Formally, we have a training set $\mathcal{T} = \{s^k\}_{k=1}^K$ of sketches, where $s^k \in \mathcal{S}$ is a sketch in the set of possible drawings $\mathcal{S}$. Following previous works [19], we assume that

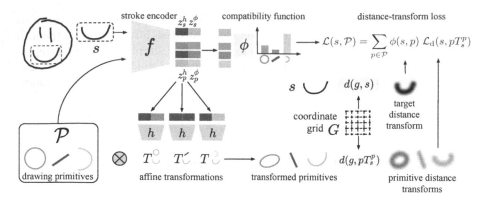

Fig. 2. PMN model architecture. Given an input stroke (top left) and a set of primitives (bottom left), PMN encodes them into a shared embedding space using f. The embeddings are split in two parts, one for h to compute the affine transformations aligning primitives to the target stroke, and one to compute the compatibility between primitives and the strokes with ϕ. From a coordinate grid G, we compute a distance transform function of the stroke and the transformed primitives. We then use distance transforms and the compatibility scores to build the self-supervised objective of PMN.

each sketch $\mathbf{s}^k$ (henceforth $\mathbf{s}$ for readability) is composed of a set of strokes (i.e. $\mathbf{s} = \{s_1, \ldots, s_n\}$), and that each stroke is defined as a sequence of two-dimensional points of length $m(s_i)$. Additionally, we assume to have a set $\mathcal{P} \subset \mathcal{S}$ of drawing primitives that we want to use to represent our sketches, where each primitive is also a sequence of points. Note that no constraint is imposed on the primitives composing $\mathcal{P}$. At test time, given a sketch $\mathbf{s}$, our goal is to re-draw each stroke $s \in \mathbf{s}$ with a primitive $p \in \mathcal{P}$. This requires two steps: first, we need to map each stroke s_i to its closest primitive $p_i \in \mathcal{P}$. Second, we need to compute the affine transform parameters making the primitive p_i better fit the original stroke s_i. In the following, we describe how we achieve these goals.

3.1   Learning to Match Strokes and Primitives

There are two main challenges in matching an arbitrary stroke s with a primitive $p \in \mathcal{P}$. First, we have no ground-truth pairs available, thus we have no direct information on which primitive p is the most similar to the stroke s. Second, even if we find the best primitive, we need still to align it to the stroke. As a simple example, if we take two straight lines of equal length, a perfect match in case they are parallel, they result in a bad match if they are orthogonal to each other. We overcome these issues by i) applying an affine transformation to the primitives in $\mathcal{P}$ and ii) comparing the original strokes and the transformed primitives through their *distance transform*.

Aligning Strokes and Primitives. We need to transform a primitive in such a way that it better matches a given target stroke. To this end, we instantiate two functions, a stroke encoder $f : \mathcal{S} \to \mathbb{R}^d$, mapping a stroke (or primitive) to a d-dimensional embedding, and an alignment function $h : \mathbb{R}^d \times \mathbb{R}^d \to \mathrm{Aff}(\mathbb{R}^2)$, predicting the affine transformation that best aligns two strokes given their encoded representations. With h, we compute a transformation matrix T_s^p as:

$$T_s^p = h(z_p, z_s) \tag{1}$$

where $z_y = f(y)$ is the feature vector of the encoded sketch/primitive y, and T_s^p the transformation aligning the primitive p to the stroke s.

Distance Transform Loss. Our goal is to find replacements for human strokes from a set of primitive shapes such that the visual difference is minimal. Given a stroke s, which is represented as a sequence of $m(s)$ connected points, i.e. $s = \{x_1, \ldots, x_m\}$ and given a coordinate $g \in G$, with G being a sampled coordinate grid, we can define the influence of the stroke at g as:

$$d(g, s) = \max_{i \in \{1, \ldots, m(s)-1\}, \, r \in [0,1]} \exp\left(-\gamma \|g - r\, x_i - (1 - r)x_{i+1}\|^2\right). \tag{2}$$

Computing $d(g, s)$ for every coordinate in G we obtain a distance map, also called *distance transform* [26]. Note that in Eq. (2) we do not use directly the distance transform but its exponentially inverted version. This allows us to highlight the map on points closer to the stroke, with γ acting as a smoothing factor. We can interpret this map as a visual rendering of the particular stroke, where the intensity of each pixel (coordinate) g decreases with the distance of g to the stroke itself. Considering a stroke s and a primitive p, we can then define the distance transform loss as:

$$\mathcal{L}_d(s, p|h) = \sum_{g \in G} \|d(g, s) - d(g, pT_s^p)\|. \tag{3}$$

With Eq. (3), we are defining a reconstruction loss that sidesteps possible mismatches in the number of points contained in s and p as well as the need of matching points across the two strokes. For simplicity, we normalize the coordinates of each point in s and p to the range $[-1, 1]$ before applying the loss and we consider G as a set of linearly spaced coordinates in $[-1.5, 1.5]$.

Exploiting Stroke Similarities. Up to now we have discussed how we can align one primitive to a target stroke by means of the affine transformation computed by h and how we can train h by comparing distance transforms. However, during inference we want to replace s with the best matching primitive selected from the set $\mathcal{P}$. With the current formulation, this could be done by replacing s with the primitive $p \in \mathcal{P}$ for which the loss $\mathcal{L}_d(s, p|h)$ has the lowest value.

While straightforward, this solution entails two issues. First, during inference we would need to compute the distance transform $d(g, p)$ for each $g \in G$ and $p \in \mathcal{P}$. Computing this map for each primitive is costly and would increase the inference time of the model. Second, if we do not consider how well a primitive p matches a stroke s, we may have misleading training signals for h. To clarify, let us consider a simple example, where s is a full circle and p a simple straight line. In such case, the loss $\mathcal{L}_d(s, p|h)$ would be high even if h predicts the best possible alignment. This means that the loss would be dominated by primitives that, such as p, cannot represent the stroke s, making h focus on an ill-posed problem rather than on matching compatible primitive-stroke pairs.

To address both issues, we inject the compatibility between a stroke and a primitive in the loss function. With this aim, we modify the stroke encoder as $f : \mathcal{S} \rightarrow \mathbb{R}^{2d}$ and, given an input y, we divide its embedding into two d-dimensional parts $z_y = [z_y^h, z_y^\phi] = f(s)$, where z_y^h will be the part used to compute the alignment function through h and z_y^ϕ will be used to compute the similarity between strokes/primitives. Given this embedding function, we calculate the relative similarity between a target stroke s and a primitive p as:

$$\phi(s, p) = \frac{\exp(\bar{z}_s^{\phi\mathsf{T}} \bar{z}_p^\phi / \kappa)}{\sum_{q \in \mathcal{P}} \exp\left(\bar{z}_s^{\phi\mathsf{T}} \bar{z}_q^\phi / \kappa\right)} \tag{4}$$

where κ is a temperature value, and $\bar{z}_y^\phi$ is the L2-normalized version of z_y^ϕ. Note that while ϕ needs to be invariant to the particular poses of s and p to score their compatibility, h in Eq. (1) needs to capture their pose to better align them. These conflicting objectives are what lead us to split the output of f in two parts. With the compatibility scores, we can define our final loss as:

$$\mathcal{L}(s, \mathcal{P}|h, f) = \sum_{p \in \mathcal{P}} \phi(s, p) \, \mathcal{L}_d(s, pT_s^p). \tag{5}$$

With this formulation, the lower the compatibility $\phi(s, p)$ between a primitive p and the stroke s, the lower the weight of the distance transform loss between p and s. Notably, the lowest value of $\mathcal{L}(s, \mathcal{P}|h, f)$ is achieved when i) the transformation matrices computed through h align all primitives to the target stroke in the best way (w.r.t. the distance transforms), and ii) the primitives with the highest compatibility scores are the ones that better match the target stroke. Thus, minimizing $\mathcal{L}(s, \mathcal{P}|h, f)$ forces h to output correct transformation matrices and f to encode similar strokes close in the second half of the embedding space, fulfilling both our goals. We name the full model composed of f, h and ϕ our *Primitive Matching Network* (PMN). Figure 2 shows the PMN pipeline.

3.2   Replacing Strokes with Primitives

After learning f and h, we can replace strokes with primitives at test time. In particular, since computing the distance transform for each possible primitive is costly, we can directly use f and ϕ to select the best matching primitive for a given stroke. Specifically, given a stroke s of a sketch $\mathbf{s}$, we replace it by:

$$\hat{p} = \arg \max_{p \in \mathcal{P}} \phi(s, p) \tag{6}$$

where $\hat{p}$ is the best-matching primitive. Given the primitive $\hat{p}$, we can now compute the corresponding alignment matrix as $T_s^{\hat{p}}$ from Eq.(1), and the abstracted sketch $\hat{s}$ as:

$$\hat{s} = \{\hat{p_1}^{\mathsf{T}} T_{s_1}^{\hat{p_1}}, \cdots, \hat{p_n}^{\mathsf{T}} T_{s_n}^{\hat{p_n}},\} \tag{7}$$

where $n = m(s)$ is the number of strokes in s. We highlight that our formulation is agnostic to the number of strokes in a sketch, the shape and number of primitives in $\mathcal{P}$, and the number of points composing each stroke.

4   Experiments

In this section, we present our experimental results. We first discuss our experimental setting (Sect. 4.1) and show results on sketch classification (Sect. 4.2) and fine-grained sketch-based image retrieval (Sect. 4.3) under a limited communication budget. Finally, we study the impact of the primitives (Sect. 4.4) and show qualitative analysis on the abstract representations (Sect. 4.5).

4.1   Experimental Setting

Datasets and Benchmarks. Following previous works on sketch abstraction [18,19] we test our model on sketch classification using Quickdraw [12], and on fine-grained sketch-based image retrieval (FG-SBIR) on ShoeV2 [44] and ChairV2 [44].

For **Quickdraw**, we follow [19] and select, 630k sketches from nine semantic categories (cat, chair, face, fire-truck, mosquito, owl, pig, purse and shoe). In this benchmark, we train a classifier on the original training sketches (details in the Supplementary), testing it on sketches abstracted using PMN or the competing methods given a specific budget (details below). We measure the performance as classification accuracy of the pretrained classifier given the abstracted inputs.

ShoeV2. comprises 5982 training and 666 testing image-sketch pairs of various shoes. For this task, we train a Siamese network [35] on the original training sketches with the same architecture of [4,19], replacing the standard triplet loss

Quickdraw   ChairV2   ShoeV2

Fig. 3. Message content. Human-drawn sketches (top) are split into messages of three points each (gray-coded). A primitive from the PMN representation (bottom) encodes more semantic information while requiring the same bytes for a single message. (Color figure online)

with a contrastive objective[2]. We measure the image-retrieval accuracy (top-10) of this network on test sketches abstracted using either PMN or one of competing methods, with the abstraction conditioned on a given budget.

ChairV2 contains 952 training and 323 testing pairs of chairs. For this dataset, we follow the FG-SBIR evaluation protocol described for ShoeV2.

Implementation Details. We train two neural networks f and h as described in Sect. 3. The stroke encoder f is a 6-layer Transformer [38], each with 8 self-attention heads. In all datasets, sketch data is represented as a list of points with their 2D coordinates and a binary label denoting whether the human is drawing or lifting the pen. We use the latter label to identify strokes. We feed as input to f the sequence of 2D-points of a stroke, together with an extra token used to obtain the final stroke embedding. We implement h as a 3-layer MLP that takes the concatenated embedding z_p^h and z_s^h as input. We use 7 predefined primitives $\mathcal{P}$, as shown in Fig. 1, as they can represent a wide variety of human strokes. We restrict T_s^p to be a composite transformation of rotation, anisotropic scale, rotation in sequence, since we found it to be flexible enough to represent a wide variety of hand-drawn strokes. The translation is directly taken from the coordinates of s and not predicted. The Supplementary contains more details about the transformation. Hyperparameters $\gamma = 6$ and $\kappa = 0.2$ are the same on all datasets and chosen by performing grid-search on a validation set.

Budget Computation. To quantify the level of abstraction of our primitive representation, we adopt a similar evaluation procedure as in [18]. Instead of measuring classification accuracy of full sketches, the evaluation is done at different subsets of the full sketch given a budget, amounting to different levels of sketch compression. Concretely, we test at budgets of 10%, 20% and 30% of the original sketch's information content to focus on the high compression regime. To compute the budget, we follow the same procedure of [19], considering a single message as made of three stroke points, i.e. three sets of 2D coordinates (see Fig. 3). Note that each message contains information equivalent to six floating points values and a categorical value indicating to which stroke the points belong. This is the same amount of information as a single primitive of our proposed model, defined as a 2D-translation, 2D-scale, 2D-rotation for its transformation and a categorical label indicating which of the 7 primitives is used. When evaluating the budget on human sketches, each message corresponds to three points of a hand-drawn stroke while, when using our abstract representation, each message is a single primitive. Given a $N\%$ budget, we calculate the number of messages that can be communicated as the $N\%$ of the total messages forming the sketch. In Fig. 3, we illustrate what constitutes a message for human-drawn sketches and our primitive representations.

---

[2] We found the contrastive objective to stabilize and speed up the training without sacrificing retrieval accuracy.

Compared Methods. There are two ways in which a sketch can be abstracted. The first is by keeping the input representation untouched (i.e. original hand-draw strokes) but ranking the messages based on their importance for preserving the sketch content. Given a budget, we can then select only the most important subset of the messages. This is the approach of standard sketch abstraction methods [18,19]. We categorize this strategy as *Selection*-based abstraction.

The second strategy is orthogonal and simplifies the sketch by grounding strokes to shapes in a fixed vocabulary, as in our PMN. This strategy does not define any ranking for the messages, but achieves abstraction by changing the stroke itself. We categorize this strategy under the name *Shape*-based abstraction. In the experiments, we consider both type of approaches.

Selection-Based. For this category, we consider two state-of-the-art methods: *Deep Sketch Abstraction* (DSA) [19] and *Goal-Driven Sequential-data Abstraction* (GDSA) [18]. DSA and GDSA are reinforcement learning methods that learn to order messages based on the performance on a downstream task. Specifically, DSA models the importance of each stroke by means of a classification (retrieval) rank-based reward, encouraging the target class (photo instance) to be highly ranked at all communication steps. GDSA is a more general strategy, applicable to various type of data. It directly uses the accuracy on the downstream task as reward function for the reinforcement learning agent, enforcing that the performance is preserved when the number of messages increases.

Shape-Based. Since PMN is the first approach, we did not find other competitors in the literature addressing the same abstraction problem. As additional baseline we consider Shape Words (SW) [41], proposed in the context of sketch-based image retrieval. SW uses an heuristic algorithm to split the original strokes into multiple parts, fitting either a line or an arc to each part through Least Squares. Since SW cannot use arbitrary primitives, we use the same set of the original paper, i.e. lines and arcs. When PMN and SW are applied alone, the message order is the same on which the original strokes were drawn.

Shape+Selection-Based. Since the two type of approaches are orthogonal, it is interesting to test if they can benefit each other. For this purpose, we also test other two models, combining GDSA with SW and our PMN.

4.2   Sketch Classification

In Table 1, we report the classification accuracy for both our PMN and the competitors on Quickdraw for budgets 10%, 20%, 30%, and 100% as reference. From the experiments, we can see that methods SW and PMN, based on shape abstraction, outperform by a margin DSA and GDSA, based on message selection. This is a direct consequence of using shapes as messages rather than original stroke parts, since the former can communicate much more semantic information in a

Table 1. Classification accuracy on Quickdraw at budgets of 10%, 20% and 30% evaluated with a classifier trained on the original human-drawn sketches.

Abstraction method		Budget (%)			
Type	Name	10	20	30	100
Selection	DSA [19]	20.12	43.85	64.04	**97.20**
	GDSA [18]	26.88	51.65	71.60	
Shape	SW [41]	51.21	68.20	75.60	78.30
	PMN	67.08	83.69	89.15	91.78
Selection + Shape	SW+GDSA	62.70	74.87	77.61	78.30
	PMN+GDSA	**77.22**	**87.79**	**90.23**	91.78

single message. We see the largest gain at low budgets, e.g. at a 10% budget, DSA achieves 20.12% accuracy, and GDSA 26.88%, whereas SW reaches 51.21% and PMN obtains 67.08%, outperforming the rest significantly. This shows how PMN is better than SW at preserving the content of the sketch. This is a consequence of the higher flexibility in terms of 1) shapes that PMN can use and 2) precision of the alignment procedure, guided by the distance transform loss rather than Least Squares on heuristically selected points. The trend is similar at 20% and 30% budgets, at which point PMN achieves an accuracy of 89.18% against 75.60% of SW and 71.60% of GDSA. Notably, abstracting strokes with PMN is not lossless and the data distribution is different from the classifier's training data such that the accuracy at 100% of PMN (91.78%) is lower than using human sketches (97.20%). On the up side, this allows PMN to reach an accuracy close to the upper bound of the original sketches at already 30% budget showing that PMN well retains the semantic of the original dataset.

Finally, if we couple a selection-based method (GDSA) with a shape-based ones, we see a consistent improvement of performance, with an improvement of 10% (77.22% accuracy) at 10% budget for PMN+GDSA over simple PMN. Despite the improvement, SW+GDSA achieves lower performance than PMN alone at every budget (e.g. 62.70% at 10%), showing again how the abstraction of PMN is more precise than SW one.

4.3   Sketch-based Image Retrieval

In Table 2, we show the results of our PMN abstractions and the competing methods in the fine-grained sketch-based image retrieval (FG-SBIR) task for the ShoeV2 (left) and ChairV2 (right) datasets. Similarly to classification, we report the results at three different budgets: 10%, 20% and 30%.

FG-SBIR has different challenges from sketch classification, since the communicated representation should precisely capture the specific characteristics of an instance rather than the shared ones of object categories. Despite our PMN abstraction smooths the specific details of strokes when grounding them to drawing primitives, it still preserves the most recognizable characteristics of an

Table 2. Fine-grained sketch-based image-retrieval (FG-SBIR) results (top-10 accuracy) on ShoeV2 and ChairV2 at budgets of 10%, 20% and 30% evaluated with a retrieval network trained on the original sketch-image pairs.

Abstraction method		ShoeV2, Budget (%)				ChairV2, Budget (%)			
Type	Name	10	20	30	100	10	20	30	100
Selection	DSA [19]	10.96	18.32	26.88	**75.22**	16.72	31.58	45.20	**86.99**
	GDSA [18]	14.86	21.32	31.08		20.74	33.13	47.68	
Shape	SW [41]	15.47	25.53	29.13	29.82	28.79	45.82	48.92	51.27
	PMN	29.58	48.50	54.35	56.55	53.87	70.59	73.99	75.92
Selection	SW+GDSA	19.96	28.97	29.27	29.82	35.60	47.98	50.77	51.27
+Shape	PMN+GDSA	**36.18**	**50.45**	**55.10**	56.55	**63.15**	**73.68**	**75.23**	75.92

instance given a specific budget. Overall, the results are consistent with the ones on sketch classification, with PMN achieving the best results in both datasets and for each level of abstraction. For instance, at 10% budget, PMN achieves a retrieval accuracy of 29.58% on ShoeV2 and 53.87% on ChairV2, surpassing by a comfortable margin SW (i.e. 15.47% on ShoeV2, 28.79% on ChairV2) and selection-based models (e.g. GDSA, 14.86% on ShoeV2, 20.74% on ChairV2).

As a direct consequence of the inherent challenges of this FG-SBIR (requiring more detailed information), we see that the higher the budget, the higher the the gap between PMN and the competitors. With 30% budget, PMN achieves 54.35% accuracy on ShoeV2 and 73.99% on ChairV2 while SW best result is 29.13% on ShoeV2 and 49.92% on ChairV2 and GDSA achieves 31.08% on ShoeV2 and 47.68% on ChairV2. SW shows an opposite trend, with the performance gap with selection-based methods becoming smaller as the budget increases, performing lower than GSDA on ShoeV2 for a 30% budget. These results highlight that PMN makes a more precise use of the available primitives, achieving the best trade-off between compression and distinctiveness of the sketch representation.

As expected, coupling PMN with GDSA leads to the best results overall (e.g. 36.18% on ShoeV2 and 63.15% on ChairV2 at 10%), with the performance of PMN alone consistently surpassing the ones of SW+GDSA (e.g. 19.96% on ShoeV2 and 35.60% on ChairV2 at 10%), highlighting that while selection and shape-based methods are complementary, it is fundamental that the latter precisely reconstructs the input, something achieved by PMN and not by SW.

4.4   Ablation Study

In Table 3, we analyze the importance of the primitive shapes by evaluating the PMN model with different subsets of primitives for Quickdraw and ChairV2. We use the PMN model trained with all seven primitives and at test time only provide a subset of them. We start with the most commonly used shape, the arc, and add one primitive at a time in the order of their usage frequency in the Quickdraw dataset. While the arc alone does not provide enough flexibility in Quickdraw, with only two primitives, arc and circle, our PMN model achieves a

Table 3. Results of PMN with different primitives on Quickdraw (acc.) and ChairV2 (top-10 acc.). Primitives added one at a time in order of usage in Quickdraw.

| Primitives | Quickdraw (Classification) | | | | ChairV2 (FG-SBIR) | | | |
| | Usage (%) | Budget (%) | | | Usage (%) | Budget(%) | | |
		10	20	30		10	20	30
⌒	22.72	45.99	70.58	79.82	39.17	41.79	62.53	69.34
+ ○	20.97	62.48	79.68	86.15	10.63	42.72	63.77	69.65
+ −	19.60	62.04	79.92	86.63	15.89	43.03	64.08	69.65
+ ∟	15.84	64.07	81.26	87.81	13.81	43.96	64.70	69.96
+ △	8.67	64.98	82.64	88.85	7.57	44.58	65.94	70.27
+ □	6.20	66.93	83.59	89.12	5.79	49.22	69.34	73.37
+ ⊔	6.00	67.08	83.69	89.15	7.12	53.87	70.59	73.99

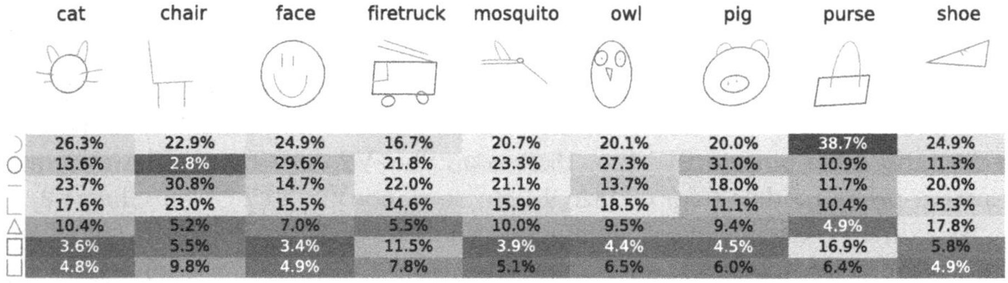

	cat	chair	face	firetruck	mosquito	owl	pig	purse	shoe
⌒	26.3%	22.9%	24.9%	16.7%	20.7%	20.1%	20.0%	38.7%	24.9%
○	13.6%	2.8%	29.6%	21.8%	23.3%	27.3%	31.0%	10.9%	11.3%
−	23.7%	30.8%	14.7%	22.0%	21.1%	13.7%	18.0%	11.7%	20.0%
∟	17.6%	23.0%	15.5%	14.6%	15.9%	18.5%	11.1%	10.4%	15.3%
△	10.4%	5.2%	7.0%	5.5%	10.0%	9.5%	9.4%	4.9%	17.8%
□	3.6%	5.5%	3.4%	11.5%	3.9%	4.4%	4.5%	16.9%	5.8%
⊔	4.8%	9.8%	4.9%	7.8%	5.1%	6.5%	6.0%	6.4%	4.9%

Fig. 4. Primitive analysis on quickdraw. Primitive representation of each Quickdraw class and the usage frequency of each primitive per class (in %).

higher classification accuracy than both GDSA and SW with 62.48% (vs. 51.21% in SW) at a 10% budget and 86.15% (vs. 75.60% in SW) at a 30% budget. To put these results into perspective, SW is able to represent line, circle and arc shapes, so even without using lines the PMN model can better fit shapes and reconstruct sketches while retaining their semantics. This is particularly evident for ChairV2, where, even by using only arcs, PMN surpasses SW at all budgets (e.g. 41.79% vs 28.79% at 10% budget).

As more shapes are added, there are diminishing returns in increasing classification accuracy for Quickdraw. However, every primitive contributes to the performance our model achieves. The triangle, square and U-shape stand out to provide a significant improvement despite their relatively low usage of 8.67%, 6.20% and 6.00% respectively. Interestingly, on ChairV2 we see a more monotonic increase in the performance (e.g. from 44.58% to 49.22% when adding squares at 10% budget). This is expected, since FG-SBIR requires a more precise reconstruction of the original sketch, thus having more primitives helps in better modeling the specificity of each stroke, improving the overall results.

As a final note, while there are many options on which primitives to include, these results validate the choice of these seven primitives. Nonetheless, depending on the dataset and use case, other choices could be considered.

4.5   Qualitative Analysis

How are objects represented through primitives? An interesting aspect of PMN is that we can now compare strokes across different sketches, extracting possible patterns. To show one possible application, in Fig. 4, we analyze the use of primitives when reconstructing Quickdraw classes. We show a representative abstracted sample of each class and the distribution of the primitives per class.

When inspecting the primitive distributions, we observe that the most used primitives are arcs and circles. As shown in our ablation study (cf. Table 3), using these two primitives alone can already cover a lot of variation on human strokes. Common use cases for arcs include the ears in animals, smiles in faces and handles in purses. Circles most frequently represent heads in animals and faces and firetrucks' wheels. The body of the firetruck and the purse are often represented by rectangles. These correlations can be observed when comparing the average distribution of primitives per class, e.g. more frequent use of line and corner in chairs or rectangle and arc in purses than in other classes.

Figure 4 also shows a limitation of our PMN model. PMN tries to match one primitive to each human stroke. However, when a stroke cannot be easily represented by a primitive, PMN may provide inaccurate representations. This is the case of the shoe class, where the main part of the shoe is usually drawn in a single stroke with a closed L-shape. In this case, PMN approximates this L-shape with a triangle (17.8% of shoe primitives are triangles, more than in any other class) that, despite driving the semantic of the sketch, provides a less accurate abstraction. In the future it would be interesting to address such cases by either learning to split/merge strokes and their parts into simpler shapes or by learning the primitives $\mathcal{P}$ together with PMN.

Representations at Different Budgets. We inspect some example qualitative results of our model in Fig. 5, showing sketch abstractions with varying compression budgets. We can see that partitioning the original strokes into three-point sub-strokes results in unrecognisable sketches even if GDSA is used to optimize the selection order. On the other hand, both SW and our PMN preserve the semantic much better given the same budget levels (e.g. shoe, bottom right). However, the additional flexibility allowed by PMN results in a much more faithful abstraction than SW, as exemplified by the body and ladder of the firefighting truck (top left), which are both represented by unnaturally rounded shapes by SW. Even when SW and PMN select the same shapes, PMN better aligns them to the original stroke, as can be seen from the circle used as seat of the chair (top right), or the arc used as left-ear of the cat (bottom left). This confirms the advantage of our self-supervised alignment objective w.r.t. the less flexible Least Squares solution of SW.

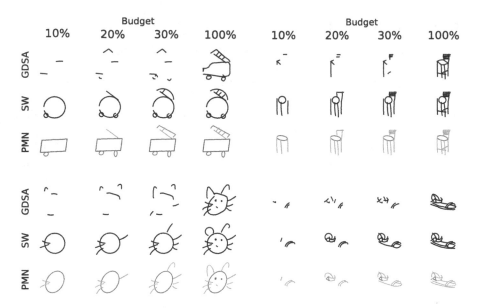

Fig. 5. Qualitative example of sketches at different budgets. We show example sketches of Quickdraw (left), ChairV2 (top right) and ShoeV2 (bottom right) at 10%, 20%, 30% budgets when using GDSA, SW, PMN. Primitive color legend: ⊃ ○ − ⌐ △ □ ⌐. (Color figure online)

5   Conclusion

Motivated by how humans abstract object representations in interpretable messages when playing communication games, in this paper we proposed a new representation learning task, Primitive-based Sketch Abstraction, where the goal is to represent a sketch with a given set of simple drawing primitives. To address this task we proposed a model, Primitive-Matching Network, that maps each stroke of a sketch to its closest drawing primitive and predicts the affine transformation to align them. We overcome the lack of annotation for stroke abstractions by developing a self-supervised objective using the distance transforms of the primitives and the target strokes. Experiments show that our model surpasses standard sketch abstraction methods on sketch classification and sketch-based image retrieval at a given budget. Differently from hand-drawn strokes, our PMN abstraction is highly interpretable and leads to new types of sketch analyses, comparing sketches by means of their primitives.

Acknowledgments. This work has been partially funded by the ERC (853489 - DEXIM), by the DFG (2064/1 - Project number 390727645), and as part of the Excellence Strategy of the German Federal and State Governments.

References

1. Barquero, B., Logie, R.: Imagery constraints on quantitative and qualitative aspects of mental synthesis. Europ. J. Cogn. Psychol. **11**(3), 315–333 (1999)
2. Berger, I., Shamir, A., Mahler, M., Carter, E., Hodgins, J.: Style and abstraction in portrait sketching. ACM Transactions on Graphics (TOG) 32(4), 1–12 (2013)
3. Bhunia, A.K., Chowdhury, P.N., Sain, A., Yang, Y., Xiang, T., Song, Y.Z.: More photos are all you need: Semi-supervised learning for fine-grained sketch based image retrieval. In: CVPR (2021)
4. Bhunia, A.K., Yang, Y., Hospedales, T.M., Xiang, T., Song, Y.: Sketch less for more: On-the-fly fine-grained sketch-based image retrieval. In: CVPR (2020)
5. Biederman, I.: Recognition-by-components: a theory of human image understanding. Psychol. Rev. **94**(2), 115 (1987)
6. Cao, Y., Wang, C., Zhang, L., Zhang, L.: Edgel index for large-scale sketch-based image search. In: CVPR (2011)
7. Cao, Y., Wang, H., Wang, C., Li, Z., Zhang, L., Zhang, L.: Mindfinder: interactive sketch-based image search on millions of images. In: ACM MM (2010)
8. Eitz, M., Hildebrand, K., Boubekeur, T., Alexa, M.: Sketch-based image retrieval: Benchmark and bag-of-features descriptors. IEEE Trans. Visualization Comput. Graph. **17**(11), 1624–1636 (2010)
9. Federici, M., Dutta, A., Forré, P., Kushman, N., Akata, Z.: Learning Robust Representations via Multi-View Information Bottleneck. In: ICLR (2020)
10. Finke, R.A., Slayton, K.: Explorations of creative visual synthesis in mental imagery. Memory Cogn. **16**(3), 252–257 (1988)
11. Ganin, Y., Bartunov, S., Li, Y., Keller, E., Saliceti, S.: Computer-aided design as language. In: NeurIPS (2021)
12. Ha, D., Eck, D.: A neural representation of sketch drawings. In: ICLR (2018)
13. Hammond, T., Paulson, B.: Recognizing sketched multistroke primitives. ACM Trans. Interact. Intell. Syst. (TiiS) **1**(1), 1–34 (2011)
14. Jia, Q., Fan, X., Yu, M., Liu, Y., Wang, D., Latecki, L.J.: Coupling deep textural and shape features for sketch recognition. In: ACM MM (2020)
15. Jongejan, J., Rowley, H., Kawashima, T., Kim, J., Fox-Gieg, N.: The Quick, Draw! - A.I. Experiment. https://quickdraw.withgoogle.com (2016)
16. Li, Y., Song, Y., Gong, S.: Sketch recognition by ensemble matching of structured features. In: BMVC (2013)
17. Liu, Z., Freeman, W.T., Tenenbaum, J.B., Wu, J.: Physical primitive decomposition. In: ECCV (2018)
18. Muhammad, U.R., Yang, Y., Hospedales, T.M., Xiang, T., Song, Y.: Goal-driven sequential data abstraction. In: ICCV (2019)
19. Muhammad, U.R., Yang, Y., Song, Y., Xiang, T., Hospedales, T.M.: Learning deep sketch abstraction. In: CVPR (2018)
20. Pang, K., Li, K., Yang, Y., Zhang, H., Hospedales, T.M., Xiang, T., Song, Y.Z.: Generalising fine-grained sketch-based image retrieval. In: CVPR (2019)
21. Pang, K., Yang, Y., Hospedales, T.M., Xiang, T., Song, Y.Z.: Solving mixed-modal jigsaw puzzle for fine-grained sketch-based image retrieval. In: CVPR (2020)
22. Para, W., Bhat, S., Guerrero, P., Kelly, T., Mitra, N., Guibas, L.J., Wonka, P.: Sketchgen: Generating constrained cad sketches. NeurIPS (2021)
23. Parui, S., Mittal, A.: Similarity-invariant sketch-based image retrieval in large databases. In: Fleet, D., Pajdla, T., Schiele, B., Tuytelaars, T. (eds.) ECCV 2014. LNCS, vol. 8694, pp. 398–414. Springer, Cham (2014). https://doi.org/10.1007/978-3-319-10599-4_26

24. Reddy, P., Gharbi, M., Lukac, M., Mitra, N.J.: Im2vec: Synthesizing vector graphics without vector supervision. In: CVPR (2021)
25. Ribeiro, L.S.F., Bui, T., Collomosse, J., Ponti, M.: Sketchformer: Transformer-based representation for sketched structure. In: Proceedings of the IEEE/CVF conference on computer vision and pattern recognition, pp. 14153–14162 (2020)
26. Rosenfeld, A., Pfaltz, J.L.: Distance functions on digital pictures. Pattern Recogn. **1**(1), 33–61 (1968)
27. Roskos-Ewoldsen, B., Intons-Peterson, M.J., Anderson, R.E.: Imagery, creativity, and discovery: A cognitive perspective. Elsevier (1993)
28. Saavedra, J.M.: Sketch based image retrieval using a soft computation of the histogram of edge local orientations (s-helo). In: ICIP (2014)
29. Saavedra, J.M.: Rst-shelo: sketch-based image retrieval using sketch tokens and square root normalization. Multimedia Tools Appl. **76**(1), 931–951 (2017)
30. Saavedra, J.M., Barrios, J.M., Orand, S.: Sketch based image retrieval using learned keyshapes (lks). In: BMVC (2015)
31. Schneider, R.G., Tuytelaars, T.: Sketch classification and classification-driven analysis using fisher vectors. ACM Trans. Graphics (TOG) 33(6), 1–9 (2014)
32. Seff, A., Zhou, W., Richardson, N., Adams, R.P.: Vitruvion: A generative model of parametric cad sketches. In: International Conference on Learning Representations (2021)
33. Segall, M.H., Campbell, D.T., Herskovits, M.J.: The influence of culture on visual perception. Bobbs-Merrill (1966)
34. Smirnov, D., Bessmeltsev, M., Solomon, J.: Deep sketch-based modeling of man-made shapes. In: ICLR (2021)
35. Song, J., Yu, Q., Song, Y.Z., Xiang, T., Hospedales, T.M.: Deep spatial-semantic attention for fine-grained sketch-based image retrieval. In: ICCV (2017)
36. Taubman, D., Marcellin, M.: JPEG2000 image compression fundamentals, standards and practice: image compression fundamentals, standards and practice, vol. 642. Springer Science & Business Media (2012)
37. Tolias, G., Chum, O.: Asymmetric feature maps with application to sketch based retrieval. In: CVPR (2017)
38. Vaswani, A., et al.: Attention is all you need. In: NeurIPS (2017)
39. Wallace, G.K.: The jpeg still picture compression standard. IEEE Trans. Consum. Electron. **38**(1), 30–44 (1992)
40. Wu, J., Wang, C., Zhang, L., Rui, Y.: Offline sketch parsing via shapeness estimation. In: AAAI (2015)
41. Xiao, C., Wang, C., Zhang, L., Zhang, L.: Sketch-based image retrieval via shape words. In: ACM ICMR (2015)
42. Xu, P., Joshi, C.K., Bresson, X.: Multigraph transformer for free-hand sketch recognition. IEEE TNNLS (2021)
43. Yang, L., Pang, K., Zhang, H., Song, Y.Z.: Sketchaa: Abstract representation for abstract sketches. In: ICCV (2021)
44. Yu, Q., Song, Y.Z., Xiang, T., Hospedales, T.M.: Sketchx! - shoe/chair fine-grained sbir dataset (2017)
45. Yu, Q., Yang, Y., Liu, F., Song, Y.Z., Xiang, T., Hospedales, T.M.: Sketch-a-net: A deep neural network that beats humans. IJCV (2017)
46. Zeng, H., Joseph, K., Vest, A., Furukawa, Y.: Bundle pooling for polygonal architecture segmentation problem. In: CVPR (2020)

Multi-scale and Cross-scale Contrastive Learning for Semantic Segmentation

Theodoros Pissas[1,2]([✉]) [ID], Claudio S. Ravasio[1,2] [ID], Lyndon Da Cruz[3,4] [ID], and Christos Bergeles[2] [ID]

[1] Wellcome/EPSRC Centre for Interventional and Surgical Sciences, University College London (UCL), London, UK
rmaptpi@ucl.ac.uk
[2] School of Biomedical Engineering and Imaging Sciences, King's College London (KCL), London, UK
[3] Moorfields Eye Hospital, London, UK
[4] Institute of Ophthalmology, University College London, London, UK

Abstract. This work considers supervised contrastive learning for semantic segmentation. We apply contrastive learning to enhance the discriminative power of the multi-scale features extracted by semantic segmentation networks. Our key methodological insight is to leverage samples from the feature spaces emanating from multiple stages of a model's encoder itself requiring neither data augmentation nor online memory banks to obtain a diverse set of samples. To allow for such an extension we introduce an efficient and effective sampling process, that enables applying contrastive losses over the encoder's features at multiple scales. Furthermore, by first mapping the encoder's multi-scale representations to a common feature space, we instantiate a novel form of supervised local-global constraint by introducing cross-scale contrastive learning linking high-resolution local features to low-resolution global features. Combined, our multi-scale and cross-scale contrastive losses boost performance of various models (DeepLabv3, HRNet, OCRNet, UPerNet) with both CNN and Transformer backbones, when evaluated on 4 diverse datasets from natural (Cityscapes, PascalContext, ADE20K) but also surgical (CaDIS) domains. Our code is available at https://github.com/RViMLab/MS_CS_ContrSeg.

Keywords: Contrastive learning · Segmentation · Mutli-scale · Cross-scale

1 Introduction

Supervised deep learning has driven remarkable progress in semantic segmentation, catalyzed by advances in convolutional network architecture design and

---

L. Cruz and C. Bergeles—The last two authors contributed equally.

Supplementary Information The online version contains supplementary material available at https://doi.org/10.1007/978-3-031-19818-2_24.

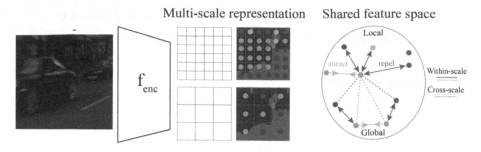

Fig. 1. Key idea: We leverage the multiscale representation provided by semantic segmentation encoders to propose a supervised-contrastive learning loss that is applied both at multiple scales and across them, within a shared feature space.

the availability of large-scale pixel-level annotated datasets. Regarding the former, the standard paradigm involves convolutional encoders [12,30,33] to extract non-linear embeddings from images followed by a decoder that maps them to a task-specific output space. For semantic segmentation, the seminal work of fully convolutional networks [28] demonstrated end-to-end learning of both these components by simply supervising the decoder with per-pixel classification losses. Further, considerable research has focused on designing inductive biases in convolutional architectures that enable complex, both local and global, visual relations across the input image to be encoded [5,6,33,41,42], leading to impressive results on challenging visual domains [9,47].

Recently, contrastive learning has been revisited as a tool for shaping network feature spaces under desired constraints over their samples, while avoiding representation collapse [34]. It has achieved strong results in unsupervised representation learning, and has established that contrastive pretraining over global [8,11,13,40] or dense [36] features can give rise to encoders that are on par with their supervised counterparts when fine-tuned on downstream tasks. This progress has motivated the incorporation of contrastive loss functions and training methodologies for the tasks of supervised classification and segmentation.

For the latter, recent works have shown that dense supervised contrastive learning applied over the final encoder layer can boost semantic segmentation performance when labelled examples are scarce [46], or in semi-supervised learning [1]. When using the full datasets, the method of [35] showed significant overall gains when enhanced with a class-wise memory bank maintaining a large diverse set of features during training. While recognizing and delineating objects at multiple scales is essential for segmentation, these methods directly regularize only the feature space of the final encoder layer thus operating at a single scale.

We instead apply multi-scale contrastive learning at multiple model layers. This direct feature-space supervision on early convolution/attention layers, usually learnt by back-propagating gradients all the way through many layers starting from the output space, can allow them to capture more complex relations between image regions, complementary to the usual function of early layers as local texture or geometry feature extractors. Intuitively, we treat the network as a function that maps an image to a multi-scale representation, distributed across its different stages, and endow it with class-aware, feature-space constraints.

A second central element of contrastive approaches is the sampling of positive and negative samples to respectively attract and repel same-class embeddings. The usual sample generation mechanisms for unsupervised learning are augmentations, which as extensively discussed in [8,11,39], is a non-trivial and lengthy tuning step. Similarly, maintaining a class-wise memory bank, while providing a large set of examples, introduces hyper-parameters such as its size, update frequency and sample selection heuristic that can be dataset dependent [35].

We propose a simpler alternative to those options, by collecting samples from the feature spaces of multiple layers. Essentially, we leverage internal, intermediate information from the encoder that is available simply through the feedforward step of the network without the need to resort to external steps such as data augmentation or online storing of samples via memory banks. Specifically, to find diverse *views* of the data, we sample them from within the model's multi-scale feature spaces and link them both within and across scales (Fig. 1). Conclusively, our contributions are the following:

- A batch-level hyper-parameter-free anchor sampling process that balances the contributions from frequent and rare classes while allowing efficient application of multiple contrastive loss terms.
- The introduction of supervised contrastive loss terms at multiple scales, and a novel cross-scale loss that enforces local-global consistency between high-resolution local and low-resolution global features emanating from different stages of the encoder.
- A model-agnostic overall method that can be successfully applied to a variety of strong baselines and state-of-the-art methods, comprising both transformer and CNN-based backbones, on 4 challenging datasets leading to improved performance.

Notably, on ADE20K, our method improves OCRNet [42] by **2.3%** and UPerNet with Swin-T and Swin-S, by **1.4%** and **1.3%**, respectively (for single scale evaluation). Further, on Cityscapes, our approach achieves an improvement of **1.1%** for both HRNet [33] and UPerNet with ResNet-101. Finally, on a challenging surgical scene understanding dataset, CaDIS, we outperform the state-of-art [25], especially improving rare classes by **1.2%**.

2   Related Works

We outline connections between several research areas and our method and discuss its differences with existing methods.

Context Aggregation for Semantic Segmentation: An intrinsic property of convolutional encoders is that information is aggregated from each pixel's local neighborhood, the size of which can be expanded at the sacrifice of spatial feature resolution. The latter, however, is crucial for segmentation. The simple and effective aggregation approach of [28] underpinned extensive research into more sophisticated context-aggregation mechanisms such as dilated-convolution-based approaches [5–7,41], attention [32,42], feature pyramid networks [38], or maintaining a high resolution representation throughout the network [33]. Recently, transformer and hybrid architectures have also been proposed. By design, these enable long-range context aggregation at the cost of increased computational requirements for training and inference [26,31]. Broadly, these approaches only support information aggregation from within a single input image. On the contrary, our method links features from different images during training, while being compatible with any architecture with hierarchical structure.

Contrastive Learning is a feature learning paradigm under which, given a known assignment of samples into classes, the objective is to minimize the distance between samples of the same class while maximizing the distance between samples of different classes. From an information-theoretic perspective, minimizing an instantiation of this objective, termed the *InfoNce* loss [24], maximizes a lower bound of the mutual information (MI) between same class features. The same loss was used to learn a pretext task for unsupervised representation learning leading to impressive downstream task results [8,11,36,37,40], while in [19] it was shown to perform on par with standard cross-entropy for classification.

Supervised Contrastive Learning for Segmentation: Concurrently with this work, improvements on strong baselines were demonstrated for segmentation by employing a **single** supervised contrastive loss term as an auxiliary loss applied after the inner-most layer of the encoder [15,35]. We instead explore **multiple** contrastive loss terms computed over different encoder layers, directly regularizing their feature spaces. Both [15,35] employ a memory bank used to maintain an extended set of pixel-embeddings or class-wise averaged region-embeddings. Instead, we opt for a simpler memory-less approach that avoids the need to tune memory size and memory update frequency, and instead collect samples from within and across different encoder layers. Further, while [15,35] focused on ResNet and HRNet, we demonstrate effective application of our method on a wider set of architectures and backbones, also exploring UPer-Net and transformers. Finally, in [46], a similar contrastive loss term was used as a pretraining objective for DeepLabv3 leading to significant gains in performance when labelled examples are scarce. However, it provided small benefits when using the complete datasets, and required long $(300 - 600K$ steps) 2-phase training schedules, which is close to 3-6× the training steps for our approach. Notably, in [46], contrasted feature maps had to spatially downsampled, for efficiency, which we circumvent by employing a balanced sampling approach.

Local-Global Representation Learning: Contrastive learning with local and global representations of images has been studied for unsupervised pretraining. [14] proposes training an encoder by maximizing the MI between local and global representations to force the latter to compactly describe the shared information across the former. The method of [44] is similar; the maximization of MI between features encoding different local windows of a sentence, and global sentence embeddings, is used as a pretraining objective in natural language processing leading to improved downstream transfer. Moreover, in [2], the InfoNCE loss is optimized over local and global features computed for two augmented views of an image, while in [4] it is applied on medical images separately over global features, and local features. In the latter, computation of the local loss term assumes the presence of common anatomical structures in medical scans to a relatively fixed position, an assumption invalid for datasets with diverse structures and scenes. There, strict prior knowledge on the location of an object/region is unrealistic. Finally, in [40] with the goal of pretraining for object detection, multiple InfoNCE objectives are optimized: globally, by extracting a feature vector for the whole image, locally by extracting multiple feature vectors each describing a single random patch of the same image, and across local-global levels by forcing whole-image and patch-level features of the same image to be similar.

Our approach is supervised, allowing to directly align the negative/positive assignment with the task of segmentation. We do not employ augmentation to obtain views of the data as in [2,4], as tuning it is laborious [8,39]. Instead, we leverage each dataset's spatial class distribution as a source for diverse samples. Finally, our local-global loss term is better aligned to segmentation than [40]: instead of enforcing scale invariance of globally pooled encoder features, our loss is computed over dense features from multiple layers.

3 Method

We now provide an introduction to the InfoNCE loss function and the proposed multi-scale and cross-scale losses, as well as a sampling procedure to perform balanced and computationally feasible contrastive learning.

3.1 Preliminaries

Let $I \in \mathbb{R}^{H \times W \times 3}$ be an image, with H its height and W its width. Semantic segmentation frameworks usually comprise a backbone encoder and a segmentation head. The backbone encoder, $f_{enc} : \mathbb{R}^{H \times W \times 3} \mapsto \mathbb{R}^{h \times w \times c}$, maps the input image to a c-dimensional feature space. The segmentation head, $f_{seg} : \mathbb{R}^{h \times w \times c} \mapsto [0,1]^{H \times W \times N_c}$, decodes the features $F = f_{enc}(I)$ to produce the output per-pixel segmentation $\hat{Y} = f_{seg}(F)$. These two modules are usually trained with a *cross-entropy* loss $\mathcal{L}_{ce}(\hat{Y}, Y)$, where Y denotes the ground truth per-pixel labels. Both f_{enc}, f_{seg} can be learned end-to-end with only supervision of f_{seg}. Under this training paradigm, the learning signal is restricted to unary classification errors.

An extension to this paradigm is to produce gradients that consider the distance of encoded image regions relative to other regions (not necessarily from the same image) in feature space. To directly (rather than implicitly via the decoder's gradients) shape the encoder's latent space, a supervised contrastive loss term [19] can be used. The loss forces features from image regions belonging to the same (different) class to be pushed close (apart). In this work, to identify the classes in each image, we use the labels downsampled to the spatial dimensions of the feature space, which we denote by $\hat{Y}$. Given this class assignment, the InfoNCE [24] loss can be computed over a set of feature vectors, which is usually termed *anchor* or *query* set; we denote this set by $\mathcal{A}$. For $\forall z_i \in \mathcal{A}$, there exist the sets of positives and negatives denoted by $\mathcal{P}_i$ and $\mathcal{N}_i$ respectively, which in the supervised setting that we examine, are identified according to the labels. The availability of dense rather than global features and labels allows a simple way to identify positive samples, referred also as "views", without the need to craft a set of appearance and geometric perturbations as done in other unsupervised [8,11] and supervised approaches [46]. Thus, instead of requiring a dataset-specific data augmentation pipeline, we exploit the natural occurrences of same or different class-pixels across the scene and across different images.

$$\mathcal{L}_c(\mathcal{A}) = \frac{1}{|\mathcal{A}|} \sum_{i \in \mathcal{A}} \frac{1}{|\mathcal{P}(i)|} \sum_{j \in \mathcal{P}(i)} \mathcal{L}(z_i, z_j), \tag{1}$$

where

$$\mathcal{L}(z_i, z_j) = -\log \frac{\exp(z_i \cdot z_j / \tau)}{\exp(z_i \cdot z_j / \tau) + \sum_{n \in \mathcal{N}(i)} \exp(z_i \cdot z_n / \tau)} \tag{2}$$

This formulation is identical to the supervised contrastive loss for classification proposed in [19]. There, however, the choices of $\mathcal{P}_i$ and $\mathcal{N}_i$ are straightforward as each image in a batch has only a single global label. Crucially, as is standard practice in contrastive learning over convolutional feature maps [8,11], the loss is not directly computed over the encoder features F, but rather over a non-linear projection using a small FCN $f_{proj} : \mathbb{R}^{h \times w \times c} \mapsto \mathbb{R}^{h \times w \times d}$ such that $Z = f_{proj}(F)$. For the rest, z_i refers to a d-dimensional feature vector from the i-th spatial position of Z. We now motivate and describe the proposed anchor sampling process and the multi- and cross-scale loss terms.

3.2   Fully-dense Contrastive Learning

In the general case, $\mathcal{A}$ can consist of all feature vectors from all spatial positions of F, the set of which is hereby denoted as Ω_F. For each element z_i of $\mathcal{A}$ and knowing the downsampled labels $\tilde{Y}$, we select as $\mathcal{N}_i = \{z_j \in \mathcal{A} : j \neq i, \tilde{Y}(j) \neq \tilde{Y}(i)\}$ and $\mathcal{P}_i = \{z_j \in \mathcal{A} : j \neq i, \tilde{Y}(j) = \tilde{Y}(i)\}$. The computational complexity of this operation is quadratic in the spatial dimensions of the feature vector, i.e., $\mathcal{O}(h^2 w^2)$. As most semantic segmentation methods, e.g. [7,33,42], require a small output stride for f_{enc}, it can become prohibitively expensive. Additionally, the quadratic complexity of this choice becomes even more prohibitive when introducing contrastive losses at multiple layers. Further, it is a well studied property of contrastive losses that the number [8,11] and the hardness [17,24,27] of the negatives affect the learned representations. Therefore, minimizing (1) over Ω_F can become trivial due to the consideration of many easy samples.

3.3   Anchor-set Sampling

An alternative is to sample from Ω_F while maintaining a balanced number of feature vectors from each present class. We generate $\mathcal{A}$ across a batch of images,

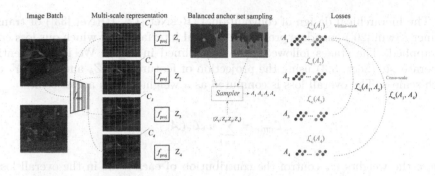

Fig. 2. Method overview: (1) An input batch is mapped to a multi-scale representation $F_1, ...F_4$, each having varying spatial dimensions and channels $C_1, ..., C_4$, using the convolutional encoder. (2) Each dense feature is projected using a separate f_{proj} that preserves spatial dimensions but maps all features to a common d-dimensional space. (3) Each projected feature map Z_s undergoes a label-based balanced sampling process (Sect. 3.2) to produce an anchor set $\mathcal{A}_s$. (4) The feature vectors in $\mathcal{A}$ and their class assignments are used to calculate the contrastive loss of (1) within each scale. Further, pairs of anchor sets from two different scales are utilized to compute a cross-scale loss (Sect. 3.5).

comparably to [35]. The intuition behind this is to further expand the diversity of samples considered by not restricting semantic relations between samples to span a single image. Rather, we allow samples to extend across different images.

Specifically, we sample K anchors per class present in the batch, equally divided among instances of that class. The sets of positives and negatives of each anchor are populated with elements of $\mathcal{A}$. This sampling process can be described as $\mathcal{A} \sim \{i \in \cup_{b=1}^{B} \Omega_F^{(b)}\}$, such that $\mathcal{A}$ has at most A_{max} samples, with B being the batch size. Importantly and contrary to [35,46], K is selected on-the-fly rather than as a hyper-parameter, and is the number of samples from the class with the least occurrences in the batch. This is motivated by the observation that classes of small objects or regions will only occupy a small fraction of Ω_F, even more so due to the spatial stride of network encoders. This heuristic enables a balanced contribution of semantic classes in the loss and removes the requirement for tuning K. Additionally, it reduces the computational cost of each contrastive term enabling the multi-scale and cross-scale extensions described next.

3.4 Multi-scale Contrastive Loss

Having obtained an efficient way to compute the loss of (1), we extend it to multiple scales. This extension regularizes the feature space of different network layers, by pushing same-class features closer, and maximizes the MI of features and their semantic labels. Importantly, while applying a pixel-space classification loss would also achieve the latter, the way we generate the anchor set allows us to attract features of the same class *from across different images*.

The hierarchical design of convolutional (ResNet [12], HRNet [33]) or transformer (Swin [20]) encoders provides a natural interface over which our loss can be applied. The stages followed are also outlined in Fig. 2. We independently generate $\mathcal{A}_s$ (Sect. 3.3) using the projection of features F_s, Z_s and labels $\tilde{Y}_s$ at each scale s. The overall loss is computed as a weighted sum across scales:

$$\mathcal{L}_{cms} = \sum_{s=1}^{S} w_s \mathcal{L}_c(\mathcal{A}_s) \tag{3}$$

where the weights w_s control the contribution of each scale in the overall loss, and S is the number of different scales. As described in Sect. 3.3 generating $\mathcal{A}_s$ involves randomization, thus it is ensured that the image regions involved in computing the above loss at each scale are independent.

3.5   Cross-scale Contrastive Loss

We further push same-class features from across different scales closer together. Specifically, we push high-resolution local features to be close to lower resolution global features. Given that global features encapsulate high level semantic concepts of the encoded image, guided by Lc_{ce}, we require that local features also encode those concepts by forcing them to lie close by in the projector's feature space. Intuitively, this enables local features describing parts of objects/regions to be predicative of their global structure of the object and vice versa.

Importantly, we note that directly requiring that the features be similar across scales would be very hard to satisfy without causing collapse. The use of separate small non-linear convolutional projector at each scale (Fig. 2) provides a compromise between a hard contrastive constraint on the encoder's features and the lower dimensional common space spanned by f_{proj} wherein the cross-scale loss is calculated. Therefore, we compute it using the independently generated anchor sets $\mathcal{A}_s$ and $\mathcal{A}_{s'}$, from scales s and s':

$$\mathcal{L}_{ccs} = \sum_{(s,s')} w_{s,s'} \mathcal{L}_c(\mathcal{A}_s, \mathcal{A}_{s'}) \tag{4}$$

Negatives and positives for samples of one scale are collected from the anchor set of the other. Concretely, for each element $z_i \in \mathcal{A}_s$ we specify $\mathcal{N}_i = \{z_j \in \mathcal{A}_{s'} : j \neq i, \tilde{Y}_{s'}(j) \neq \tilde{Y}_s(i)\}$ and $\mathcal{P}_i = \{z_j \in \mathcal{A}_{s'} : j \neq i | j, \tilde{Y}_{s'}(j) = \tilde{Y}_s(i)\}$. Finally, the gradients derived from this loss are backpropagated to both involved anchor sets thus instantiating a form of bidirectional local-global consistency for learning the encoder. Combining all the above terms, our complete objective is:

$$\mathcal{L}_{total} = \mathcal{L}_{ce} + \lambda_{cms} * \mathcal{L}_{cms} + \lambda_{ccs} * \mathcal{L}_{ccs} \tag{5}$$

Table 1. (a), (c): Ablations for each component of our contrastive loss on **Cityscapes val** and **ADE20K**, respectively, reporting the mean and std deviation of each variant across 4 random seeds. (b), (d): Comparisons with auxiliary losses on **Cityscapes** (results taken from [35]) and **ADE20K**, respectively.

(a)

Model	Backbone	$\mathcal{L}_c$	$\mathcal{L}_{cms}$	$\mathcal{L}_{ccs}$	Scale pairs	mean±std
HRNet	HR48v2				-	79.1 ± 0.2
HRNet	HR48v2	✓			-	80.0 ± 0.1
HRNet	HR48v2	✓			-	80.7 ± 0.2
HRNet	HR48v2	✓	✓		1	81.2 ± 0.1
HRNet	HR48v2	✓	✓		2	81.4 ± 0.1

(Network / Loss / mIoU (ss))

(b)

Model	Loss	mIoU (ss)
HRNet	CE	79.1
HRNet	AFF [18]	78.7
HRNet	RMI [45]	79.9
HRNet	Lovasz [3]	80.3
HRNet	ours	**81.5**

(c)

Model	Backbone	$\mathcal{L}_c$	$\mathcal{L}_{cms}$	$\mathcal{L}_{ccs}$	Scale pairs	mean±std
UPerNet	Swin-T				-	44.5 ± 0.4
UPerNet	Swin-T	✓			-	45.1 ± 0.2
UPerNet	Swin-T	✓	✓		2	45.8 ± 0.1

(Network / Loss / mIoU (ss))

(d)

Model	Loss	mIoU (ss)
OCRNet	CE	44.5
OCRNet	Lovasz	44.7
OCRNet	ours	**46.8**
SwinT	CE	44.5
SwinT	Lovasz	45.2
SwinT	ours	**45.9**

4 Experiments

4.1 Datasets

We benchmark our approach on the following challenging datasets from natural and surgical image domains using the mIoU as the main performance metric:

ADE20K [47] comprises 20210 and 2000 and train and val images, respectively, capturing 150 semantic classes from natural scenes.

Cityscapes [9] consists of 5000 images of urban scenes on which 19 classes are pixel-level labelled. We train, and evaluate, on the *train*, and *val* sets, respectively. We also report performance on the server-withheld *test* set.

Pascal-Context [23] comprises 4998 and 5105 and train and val images, respectively, capturing 59 classes from natural scenes.

CaDIS [10] comprises 25 cataract surgery videos and 4671 pixel-level annotated frames with labels for *anatomies, instruments* and *miscellaneous objects*. We experiment on tasks 2 and 3 defined in [10], comprising 17, 25 classes respectively. We follow [25] and also report the average of the per-class IoUs for anatomies, surgical tools and rare classes (present in less than 15% of the images) (Fig. 3).

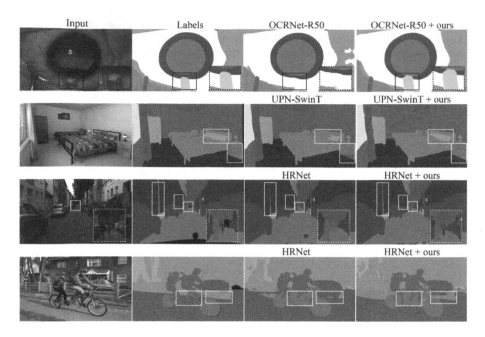

Fig. 3. Qualitative results: Qualitative comparisons across all 4 datasets. More results are provided in the supplementary material.

4.2   Ablations and Comparisons to State-of-the-art

For ablations, we train for $40K$ steps with a batch size of 8 and for $160K$ steps with batch size of 16 on Cityscapes and ADE20K, respectively.

Ablation of Loss Components: First, we conduct an ablation study to demonstrate the importance of each component of our proposed loss on Cityscapes using HRNet, reported in Table 1(a). We repeat this ablation on ADE20K using UPerNet with a Swin-T backbone, shown in Table 1(c). We report the average mIoU across 4 runs with different random seeds to demonstrate the stability of the ranking of methods with respect to initialization and training stochasticity. **Comparisons to other auxiliary losses**: Tables 1(b), 1(d) compare the proposed loss to other auxiliary losses on Cityscapes and ADE20K, respectively.

Comparisons to Contrastive Learning Approaches: Table 2 compares our approach to concurrent contrastive learning methods for semantic segmentation.

Performance Improvements: In Tables 3(a), 3(b), 4 and 5, we report performance improvements obtained by training a variety of models with our loss (Eq. 5) (referred to as "ours") on 4 datasets. We refer to cross-entropy as "CE" in tables. We experiment with DeepLabv3, HRNet, OCRNet and UPerNet with ResNet-101 and Swin (T,S,B,L) Transformer backbones. To train models using our proposed loss we leave all other settings the same as for the referenced or implemented baselines (i.e. without our loss) to enable fair comparisons.

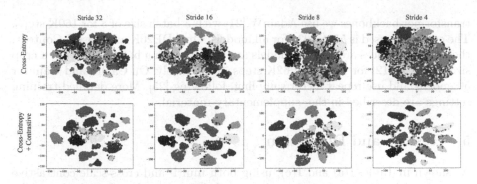

Fig. 4. TSNE [22] visualisation of the feature spaces of HRNet, on Cityscapes, trained without (top) and with (bottom) our proposed contrastive loss. Color indicates each sample's class.

Table 2. Comparison with concurrent supervised contrastive learning methods for semantic segmentation on **Cityscapes val** with single scale evaluation. Our proposed loss matches the performance of those methods without using a memory bank and relying only on anchors sampled from multiscale features.

Network	Loss			mIoU	
Model	Function	Memory	$\mathcal{A}$ Selection	@40K	Best
HRNet(ours)	$\mathcal{L}_{cms}$, $\mathcal{L}_{ccs}$	-	Multi-scale Balanced	**81.5**	**82.2**
HRNet[35]	$\mathcal{L}_c$	Pixel	Hard-example	80.5	-
HRNet[35]	$\mathcal{L}_c$	Region+Pixel	Hard-example	81.0	**82.2**
HRNet[15]	$\mathcal{L}_c$+aux	Region	Averaging	-	81.9

4.3 Implementation Details

Contrastive Losses: All instantiations of the loss of Eq. (1) utilize a temperature $\tau = 0.1$. A distinct projector comprising 2 $Conv$1x1-$ReLU$-BN layers and a linear mapping with $d = 256$, is attached to features of each scale for multi-scale and cross-scale loss variants. We utilize the $C2$, $C3$, and $C5$ features, with output strides of 4, 8 and 8, respectively, for models with a dilated ResNet backbone. When using HRNet as the backbone, we utilize features from all 4 scales, with output strides of $4, 8, 16, 32$. The cross-scale loss is applied for scale pairs $(4, 32)$ and $(4, 16)$. When using UPerNet we attach our loss on all 4 scales of the backbone on Cityscapes and on the feature pyramid net (FPN) outputs on ADE20K, The weights w_s in Eq. (3) are set to $1, 0.7, 0.4, 0.1$ for feature maps of strides $4, 8, 16, 32$ respectively in all experiments. The two latter choices are supported by ablations provided in the supplementary material.

Training Settings: For experiments on all datasets/models, to enable fair comparisons we closely follow the training settings specified in [6,20,33,42] and their respective official implementations. Unless otherwise stated, on Cityscapes for CNN backbones we use SGD with a batch size of 12 for 120K steps while for transformer backbones we use AdamW [21] with a batch size of 8 for 160K steps. The crops size used is 512×1024 for all models. On ADE20K and Pascal-Context, the only differences are that all models are trained with a batch size of 16, a crop size of 512×512, for 160K and 62K steps, respectively. On CaDIS and train for 50 epochs and use repeat factor sampling, following [25]. More detailed training settings are described in the supplementary material.

5 Results and Discussion

As shown in Tables 1(a) and 1(c), using both multi- and cross-scale contrastive terms provides the highest improvement relative to the baseline. Additionally, as shown in Tables 1(b) and 1(d), our loss outperforms other auxiliary losses.

Regarding **other supervised contrastive losses**, on **Cityscapes**, where we are able to provide a comparison between all methods, our loss is competitive,

Table 3. (a): Comparisons on **ADE20K** with single/multi-scale evaluation (ss/ms). If no reference is provided the value is obtained by our implementation and OHEM refers to the loss of [29]. (b): Comparisons on **CaDIS** tasks 2 and 3 that define 17 and 25 classes respectively. †: Imagenet-22K pretraining.

(a) ADE20K

Model	Backbone	#Params (M)	Source	Loss	mIoU ss/ms	Improvement
DeepLabv3	R101	63	[40]	CE	/44.1	
DeepLabv3	R101	63	-	ours	**44.2/45.6**	
DeepLabv3	R101	63	[15]	$\mathcal{L}_c$+Mem+OHEM	-/46.8	
OCRNet	HR48v2	71	[42]	OHEM	44.5/45.5	
OCRNet	HR48v2	71	-	ours	**46.8/47.4**	
UPerNet	R101	86	[38]	CE	42.0/42.7	
UPerNet	R101	86	-	ours	**43.8/45.3**	
UPerNet	Swin-T	60	-	CE	44.7/45.5	
UPerNet	Swin-T	60	[20]	CE	44.5/45.8	
UPerNet	Swin-T	60	-	ours	**45.9/46.6**	
UPerNet	Swin-S	81	-	CE	48.1/49.2	
UPerNet	Swin-S	81	[20]	CE	47.6/49.5	
UPerNet	Swin-S	81	-	ours	**48.9/50.0**	
UPerNet	Swin-B†	121	[20]	CE	50.1/51.6	
UPerNet	Swin-B†	121	-	ours	**51.3/52.2**	
UPerNet	Swin-L†	234	[20]	CE	52.0/53.5	
UPerNet	Swin-L†	234	-	ours	**52.9**/53.3	− 0.2)

(b) CaDIS

	Method		mIoU		Anatomies		Tools		Rare	
Model	Backbone	Loss	Task2	Task3	Task2	Task3	Task2	Task3	Task2	Task3
OCRNet	R50	CE	81.02	76.24	**90.80**	90.87	74.65	70.82	74.46	67.58
OCRNet	R50	+ours	**81.47**	**77.67**	90.66	**90.91**	**75.58**	**72.79**	**77.63**	**73.09**
Improvement					(−0.14)					
OCRNet	R50	Lovasz	82.36	77.77	**90.63**	90.59	76.89	72.96	77.52	71.44
OCRNet	R50	+ours	**82.56**	**78.25**	90.59	**90.76**	**77.41**	**73.11**	**78.55**	72.67
Improvement					(−0.04)					

Table 4. (a): Comparisons with strong baselines on **Cityscapes val** with single scale evaluation (ss). If no reference is provided the value is obtained by our implementation. The denoted improvements are relative to the baseline with the highest mIoU between our implementation and previously reported results. (b): Comparison on **Cityscapes test** with all models trained on **train+val**.

(a) Cityscapes val

Network			Loss	mIoU	(ss)
Model	Backbone	Source			Improvement
PSPNet	R101	[16]	Metric-learning	78.2	
DeepLabv3	R101	[6]	CE	77.8	
DeepLabv3	R101	-	CE	78.2	
DeepLabv3	R101	-	ours	79.0	
HRNet	HR48v2	-	CE	81.0	
HRNet	HR48v2	[33]	CE	81.1	
HRNet	HR48v2	-	ours	82.2	
HRNet	HR48v2	[35]	L_c+Mem	82.2	
OCR	HR48v2	-	CE	81.2	
OCR	HR48v2	[42]	CE	81.6	
OCR	HR48v2	-	ours	81.9	
UPerNet	R101	-	CE	78.0	
UPerNet	R101	-	ours	79.1	
UPerNet	Swin-T	-	CE	79.2	
UPerNet	Swin-T	-	ours	79.9	
UPerNet	Swin-S	-	CE	80.9	
UPerNet	Swin-S	-	ours	81.7	
UPerNet	Swin-B	-	CE	82.0	
UPerNet	Swin-B	-	ours	82.6	

(b) Cityscapes test

Method	Loss	mIoU	iIoU	IoU Cat	iIoU Cat
HRNet[33]	CE	81.6	61.8	92.1	82.2
HRNet	ours	81.9	62.9	92.2	83.4
Improvement		(+0.3)	(+1.1)	(-0.1)	(+1.2)
OCRNet [42]	CE	82.5	61.7	92.1	81.6
OCRNet	ours	82.4	63.6	92.2	83.7
Improvement		(-0.1)	(+1.9)	(+0.1)	(+2.1)

matching [35] and outperforming [15], despite not using a memory bank, hard-example mining or region averaging (Table 2). On **ADE20K**, [15] only report results with DeepLabv3 and achieve higher improvement over the baseline than ours (+**2.7**% vs +**1.5**%), albeit using both memory, OHEM and an intermediate auxiliary loss. Importantly, we showcase improvements for a wider range of models on this dataset. On **Pascal-Context**, we marginally outperform [35] when using HRNet and are outperformed when using OCRNet.

Notably, our loss improves performance for various CNN and Transformer-based models across datasets: On the challenging **ADE20K** dataset, (Table 3(a)), we obtain an improvement for OCRNet by +**2.3**% (ss) and +**1.9**% (ms), for UPerNet by +**1.8**% (ss) and +**2.6**% (ms) and for DeepLabv3 by +**1.5**% (ms). We also improve the state-of-the-art model of [20] using UPerNet with Swin Transformer for increasing backbone sizes (Swin-T,-S,-B, -L) by +**1.4**%, +**1.3**%, +**1.2**% and +**0.9**% (ss). Overall, with only a single-scale input, CNN models and Swin variants trained with our loss achieve performance close to that of the baseline when the latter employs the computationally expensive (several seconds per image) multiscale test-time augmentation.

On **Cityscapes-val** (Table 4), we improve both HRNet and UPerNet with ResNet101 by +**1.1**%, DeepLabv3 by +**0.8**% and UperNet with Swin-T and Swin-S by +**0.7**% and +**0.8**% respectively. Our approach with HRNet and OCR-Net, did not significantly improve mIoU on the **test** set but in both cases outperforms baselines reported in [33,42] in terms of **iIoU** and **iIoU-Cat** (Table 4(b))

Table 5. Comparisons on **Pascal-context val** with multi-scale evaluation (ms).

Network			Loss	mIoU (ms)	
Model	Backbone	Source			Improvement
HRNet	HR48v2	-	CE	53.5	
HRNet	HR48v2	[33]	CE	54.0	
HRNet	HR48v2	-	Ours	**55.3**	
HRNet	HR48v2	[35]	$\mathcal{L}_c$+Mem	55.1	
OCRNet	HR48v2	-	CE	55.8	
OCRNet	HR48v2	[42]	CE	56.2	
OCRNet	HR48v2	-	Ours	56.5	
OCRNet	HR48v2	[35]	$\mathcal{L}_c$+Mem	**57.2**	

by notable margins. This can be attributed to our sampling approach, that balances loss contributions from all present classes (Sect. 3.3) and the fact that our loss enhances the discriminative power of features at multiple scales, as exemplified in Fig. 4.

On **Pascal-Context**, adding our loss to HRNet leads to +**1.3**% and a small improvement for OCRNet (Table 5) while doing so on the method of [25], on **CaDIS** results in state-of-the-art performance, especially favouring the rarest of classes, respectively for tasks 2 and 3, by +**1.0**% and +**1.2**%, when combined with the Lovasz loss, and by +**3.1**% and +**5.2**% when combined with CE. While the mean mIoU is a standard metric for assessing semantic segmentation, focusing it over the rarest classes is crucial to assess long-tailed class performance. This is especially important in the surgical domain where collecting data of rarely appearing tools, under real surgery conditions, can be particularly difficult.

6   Conclusion

We presented an effective method for supervised contrastive learning both at multiple feature scales and across them. Overall, we showcased significant gains for most of the strong CNN and Transformer-based models we experimented with on 4 datasets. Notably, our approach achieved maximal gains on the challenging ADE20K dataset, which contains a large number semantic concepts (150 classes), where the recognition component of segmentation greatly benefits from the class-aware clustered feature spaces of our method.

Acknowledgements:. This work was supported by an ERC Starting Grant [714562].

References

1. Alonso, I., Sabater, A., Ferstl, D., Montesano, L., Murillo, A.C.: Semi-supervised semantic segmentation with pixel-level contrastive learning from a class-wise memory bank. In: Proceedings of the IEEE International Conference on Computer Vision (2021)

2. Bachman, P., Hjelm, R.D., Buchwalter, W.: Learning representations by maximizing mutual information across views. Adv. Neural. Inf. Process. Syst. **32**, 1–11 (2019)
3. Berman, M., Triki, A.R., Blaschko, M.B.: The lovász-softmax loss: a tractable surrogate for the optimization of the intersection-over-union measure in neural networks. In: Proceedings of the IEEE Conference on Computer Vision and Pattern Recognition (CVPR), June 2018
4. Chaitanya, K., Erdil, E., Karani, N., Konukoglu, E.: Contrastive learning of global and local features for medical image segmentation with limited annotations. Adv. Neural. Inf. Process. Syst. **33**, 1–13 (2020)
5. Chen, L.C., Papandreou, G., Kokkinos, I., Murphy, K., Yuille, A.L.: Semantic image segmentation with deep convolutional nets and fully connected CRFs (2016)
6. Chen, L.C., Papandreou, G., Schroff, F., Adam, H.: Rethinking Atrous convolution for semantic image segmentation (2017)
7. Chen, L.-C., Zhu, Y., Papandreou, G., Schroff, F., Adam, H.: Encoder-decoder with Atrous separable convolution for semantic image segmentation. In: Ferrari, V., Hebert, M., Sminchisescu, C., Weiss, Y. (eds.) ECCV 2018. LNCS, vol. 11211, pp. 833–851. Springer, Cham (2018). https://doi.org/10.1007/978-3-030-01234-2_49
8. Chen, T., Kornblith, S., Norouzi, M., Hinton, G.: A simple framework for contrastive learning of visual representations. In: International Conference on Machine Learning, pp. 1597–1607. PMLR (2020)
9. Cordts, M., et al.: The cityscapes dataset for semantic urban scene understanding. CoRR abs/1604.01685 (2016). https://arxiv.org/abs/1604.01685
10. Grammatikopoulou, M., et al.: CaDIS: cataract dataset for image segmentation (2020)
11. He, K., Fan, H., Wu, Y., Xie, S., Girshick, R.: Momentum contrast for unsupervised visual representation learning. arXiv preprint arXiv:1911.05722 (2019)
12. He, K., Zhang, X., Ren, S., Sun, J.: Deep residual learning for image recognition. In: Proceedings of the IEEE Conference on Computer Vision and Pattern Recognition, pp. 770–778 (2016)
13. Hénaff, O.J., Koppula, S., Alayrac, J.B., Oord, A., Vinyals, O., Carreira, J.: Efficient visual pretraining with contrastive detection. In: International Conference on Computer Vision (2021)
14. Hjelm, R.D., et al.: Learning deep representations by mutual information estimation and maximization. In: ICLR (2019)
15. Hu, H., Cui, J., Wang, L.: Region-aware contrastive learning for semantic segmentation. In: Proceedings of the IEEE/CVF International Conference on Computer Vision (ICCV), pp. 16291–16301, October 2021
16. Hwang, J.J., et al.: SegSrt: Segmentation by discriminative sorting of segments. In: Proceedings of the IEEE International Conference on Computer Vision, pp. 7334–7344 (2019)
17. Kalantidis, Y., Sariyildiz, M.B., Pion, N., Weinzaepfel, P., Larlus, D.: Hard negative mixing for contrastive learning. CoRR abs/2010.01028 (2020). https://arxiv.org/abs/2010.01028
18. Ke, T.-W., Hwang, J.-J., Liu, Z., Yu, S.X.: Adaptive affinity fields for semantic segmentation. In: Ferrari, V., Hebert, M., Sminchisescu, C., Weiss, Y. (eds.) ECCV 2018. LNCS, vol. 11205, pp. 605–621. Springer, Cham (2018). https://doi.org/10.1007/978-3-030-01246-5_36
19. Khosla, P., et al.: Supervised contrastive learning. arXiv preprint arXiv:2004.11362 (2020)

20. Liu, Z., et al.: Swin transformer: hierarchical vision transformer using shifted windows. In: Proceedings of the IEEE/CVF International Conference on Computer Vision (ICCV), pp. 10012–10022, October 2021
21. Loshchilov, I., Hutter, F.: Decoupled weight decay regularization. In: ICLR (2019)
22. Van der Maaten, L., Hinton, G.: Visualizing data using T-SNE. J. Mach. Learn. Res. **9**(11), 1–27 (2008)
23. Mottaghi, R., et al.: The role of context for object detection and semantic segmentation in the wild. In: IEEE Conference on Computer Vision and Pattern Recognition (CVPR) (2014)
24. van den Oord, A., Li, Y., Vinyals, O.: Representation learning with contrastive predictive coding. CoRR abs/1807.03748 (2018). https://arxiv.org/abs/1807.03748
25. Pissas, T., Ravasio, C.S., Da Cruz, L., Bergeles, C.: Effective semantic segmentation in cataract surgery: what matters most? In: de Bruijne, M., et al. (eds.) MICCAI 2021. LNCS, vol. 12904, pp. 509–518. Springer, Cham (2021). https://doi.org/10.1007/978-3-030-87202-1_49
26. Ranftl, R., Bochkovskiy, A., Koltun, V.: Vision transformers for dense prediction. In: Proceedings of the IEEE/CVF International Conference on Computer Vision (ICCV), pp. 12179–12188, October 2021
27. Saunshi, N., Plevrakis, O., Arora, S., Khodak, M., Khandeparkar, H.: A theoretical analysis of contrastive unsupervised representation learning. In: Chaudhuri, K., Salakhutdinov, R. (eds.) Proceedings of the 36th International Conference on Machine Learning. Proceedings of Machine Learning Research, vol. 97, pp. 5628–5637. PMLR, 09–15 June 2019. https://proceedings.mlr.press/v97/saunshi19a.html
28. Shelhamer, E., Long, J., Darrell, T.: Fully convolutional networks for semantic segmentation. IEEE Trans. Pattern Anal. Mach. Intell. **39**(4), 640–651 (2017). https://doi.org/10.1109/TPAMI.2016.2572683
29. Shrivastava, A., Gupta, A., Girshick, R.: Training region-based object detectors with online hard example mining. In: Proceedings of the IEEE Conference on Computer Vision and Pattern Recognition (CVPR), June 2016
30. Simonyan, K., Zisserman, A.: Very deep convolutional networks for large-scale image recognition. In: Proceedings of the International Conference on Learning Representations (2015)
31. Strudel, R., Garcia, R., Laptev, I., Schmid, C.: Segmenter: transformer for semantic segmentation. In: Proceedings of the IEEE/CVF International Conference on Computer Vision (ICCV), pp. 7262–7272, October 2021
32. Takikawa, T., Acuna, D., Jampani, V., Fidler, S.: Gated-SCNN: gated shape CNNs for semantic segmentation. In: Proceedings of the IEEE/CVF International Conference on Computer Vision (ICCV), October 2019
33. Wang, J., et al.: Deep high-resolution representation learning for visual recognition. TPAMI. **43**, 3349–3364 (2019)
34. Wang, T., Isola, P.: Understanding contrastive representation learning through alignment and uniformity on the hypersphere. In: III, H.D., Singh, A. (eds.) Proceedings of the 37th International Conference on Machine Learning. Proceedings of Machine Learning Research, vol. 119, pp. 9929–9939. PMLR, 13–18 July 2020. https://proceedings.mlr.press/v119/wang20k.html
35. Wang, W., Zhou, T., Yu, F., Dai, J., Konukoglu, E., Van Gool, L.: Exploring cross-image pixel contrast for semantic segmentation. In: Proceedings of the IEEE/CVF International Conference on Computer Vision (ICCV), pp. 7303–7313, October 2021

36. Wang, X., Zhang, R., Shen, C., Kong, T., Li, L.: Dense contrastive learning for self-supervised visual pre-training. In: Proceedings of IEEE Conference on Computer Vision and Pattern Recognition (CVPR) (2021)
37. Wu, Z., Xiong, Y., Yu, S.X., Lin, D.: Unsupervised feature learning via non-parametric instance discrimination. In: Proceedings of the IEEE Conference on Computer Vision and Pattern Recognition (CVPR), June 2018
38. Xiao, T., Liu, Y., Zhou, B., Jiang, Y., Sun, J.: Unified perceptual parsing for scene understanding. In: Ferrari, V., Hebert, M., Sminchisescu, C., Weiss, Y. (eds.) ECCV 2018. LNCS, vol. 11209, pp. 432–448. Springer, Cham (2018). https://doi.org/10.1007/978-3-030-01228-1_26
39. Xiao, T., Wang, X., Efros, A.A., Darrell, T.: What should not be contrastive in contrastive learning. In: International Conference on Learning Representations (2021). https://openreview.net/forum?id=CZ8Y3NzuVzO
40. Xie, E., et al.: DetCO: unsupervised contrastive learning for object detection. In: Proceedings of the IEEE/CVF International Conference on Computer Vision (ICCV), pp. 8392–8401, October 2021
41. Yu, F., Koltun, V., Funkhouser, T.: Dilated residual networks. In: Computer Vision and Pattern Recognition (CVPR) (2017)
42. Yuan, Y., Chen, X., Wang, J.: Object-contextual representations for semantic segmentation. In: Vedaldi, A., Bischof, H., Brox, T., Frahm, J.-M. (eds.) ECCV 2020. LNCS, vol. 12351, pp. 173–190. Springer, Cham (2020). https://doi.org/10.1007/978-3-030-58539-6_11
43. Zhang, H., et al.: ResNeST: Split-attention networks. arXiv preprint arXiv:2004.08955 (2020)
44. Zhang, Y., He, R., Liu, Z., Lim, K.H., Bing, L.: An unsupervised sentence embedding method by mutual information maximization. In: EMNLP (2021)
45. Zhao, S., Wang, Y., Yang, Z., Cai, D.: Region mutual information loss for semantic segmentation. In: Wallach, H., Larochelle, H., Beygelzimer, A., d'Alché-Buc, F., Fox, E., Garnett, R. (eds.) Advances in Neural Information Processing Systems, vol. 32. Curran Associates, Inc. (2019). https://proceedings.neurips.cc/paper/2019/file/a67c8c9a961b4182688768dd9ba015fe-Paper.pdf
46. Zhao, X., et al.: Contrastive learning for label efficient semantic segmentation. In: Proceedings of the IEEE/CVF International Conference on Computer Vision (ICCV), pp. 10623–10633, October 2021
47. Zhou, B., Zhao, H., Puig, X., Fidler, S., Barriuso, A., Torralba, A.: Semantic understanding of scenes through the ade20k dataset. Int. J. Comput. Vision **127**, 302–321 (2018)

One-Trimap Video Matting

Hongje Seong[1], Seoung Wug Oh[2], Brian Price[2], Euntai Kim[1],
and Joon-Young Lee[2(✉)]

[1] Yonsei University, Seoul, Korea
{hjseong,etkim}@yonsei.ac.kr
[2] Adobe Research, San Jose, CA, USA
{seoh,bprice,jolee}@adobe.com

Abstract. Recent studies made great progress in video matting by extending the success of trimap-based image matting to the video domain. In this paper, we push this task toward a more practical setting and propose One-Trimap Video Matting network (OTVM) that performs video matting robustly using only one user-annotated trimap. A key of OTVM is the joint modeling of trimap propagation and alpha prediction. Starting from baseline trimap propagation and alpha prediction networks, our OTVM combines the two networks with an alpha-trimap refinement module to facilitate information flow. We also present an end-to-end training strategy to take full advantage of the joint model. Our joint modeling greatly improves the temporal stability of trimap propagation compared to the previous decoupled methods. We evaluate our model on two latest video matting benchmarks, Deep Video Matting and VideoMatting108, and outperform state-of-the-art by significant margins (MSE improvements of 56.4% and 56.7%, respectively). The source code and model are available online: https://github.com/Hongje/OTVM.

Keywords: Video matting · Trimap propagation · Alpha prediction

1 Introduction

Video matting is the task of predicting accurate alpha mattes from a video. This is an essential step in video editing applications requiring an accurate separation of the foreground and the background layers such as video composition. For each video frame I, it aims to divide the input color into three components: the foreground color, the background color, and the alpha matte. Formally, for a given pixel, it can be written as, $I = \alpha F + (1 - \alpha)B$, where F and B are

---

H. Seong—This work was done during an internship at Adobe Research.

Supplementary Information The online version contains supplementary material available at https://doi.org/10.1007/978-3-031-19818-2_25.

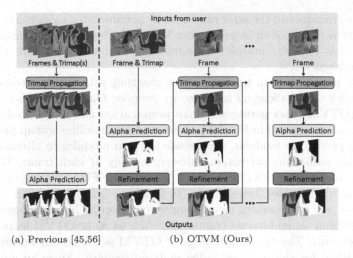

(a) Previous [45,56]   (b) OTVM (Ours)

Fig. 1. The previous decoupled approach *vs*. Our joint modeling. (a) Previous video matting methods [45,56] have two decoupled stages: first generate all missing trimaps and then predict the alpha mattes given the trimaps. This approach does not have any interaction between trimap propagation and alpha matting. Since the trimap propagation with no understanding of alpha mattes is often not stable, they require *multiple user-annotated trimaps* to prevent drifting at trimap propagation. (b) In our OTVM, trimap propagation and alpha prediction modules interact each other and the refinement step updates the predictions from the two modules. Our trimap propagation module memorizes and utilizes all information (RGB, trimap, alpha, latent features) to propagate trimaps. It results in accurate and robust predictions, enabling us to perform *one-trimap video matting*.

the foreground and background color, and $\alpha \in [0,1]$ represents the alpha value. Here, only 3 values (I) are known, and the remaining 7 values (F, B, and α) are unknown. Given the ill-posed nature of the problem, traditional methods utilize trimaps as additional inputs that indicate pixels that are either solid foreground, solid background, or uncertain. The trimap provides a clue for the target object and effectively improves the stability of the alpha prediction.

Leveraging the latest progress in trimap-based image matting [14,28,52] and mask propagation [34,39,54], recent studies [45,56] successfully developed learning-based video matting techniques. They decouple video matting into two stages, trimap propagation and alpha prediction. They re-purpose the latest mask propagation network [34] to propagate the given trimaps throughout the video, then design alpha prediction networks that take multiple trimaps as input and predict the alpha matte (Fig. 1(a)).

While the decoupled approach effectively simplifies the task, it has a critical limitation as illustrated in Fig. 2. By the nature of the trimap, the unknown region of one frame may be changed into foreground or background and vice versa at a different frame. Therefore, if we propagate a trimap based on visual correspondences without the knowledge for alpha matte [5,34], it may produce inaccurate trimaps and the error can be easily accumulated as shown in Fig. 2(c), leading to the failure of alpha prediction. With this challenge, the existing decoupled methods require multiple user-annotated trimaps to prevent drifting at trimap propagation.

In this paper, we aim to tackle video matting with a single trimap input. To cope with the challenging scenario, we propose One-Trimap Video Matting network (OTVM) that performs trimap propagation and alpha prediction as a joint task, as illustrated in Fig. 1(b). Starting from baseline trimap propagation and alpha prediction modules, we cascade the two modules to alternate trimap propagation and alpha prediction auto-regressively at each frame. We employ the space-time memory (STM) network [34] and the FBA matting network [14] as the baseline modules, respectively. To facilitate information flow within the pipeline, we add a refinement module and re-engineer STM accordingly. In addition, we present an end-to-end training pipeline to make OTVM learn the joint task successfully. The major advantage of OTVM is robust trimap propagation that is critical for the practical video matting scenario. Since an alpha matte contains richer information than a trimap, we are able to update the trimap after alpha prediction and this update step prevents error accumulation in the trimap, resulting in robust trimap propagation and accurate alpha prediction.

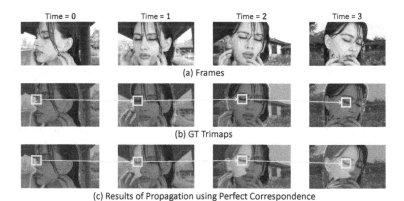

Fig. 2. **The intrinsic challenge of trimap propagation**. As shown in (b), the same part of an instance can have varying trimap labels at different times, *e.g.*., the label of the left eyelid has changed from the unknown to the foreground. If trimap propagation is conducted using visual correspondences without the knowledge for alpha matte, then it may produce inaccurate trimaps and the error can be easily accumulated as in (c).

OTVM produces accurate video mattes even in the challenging one-trimap scenario. We demonstrate that OTVM outperforms previous matting methods with large margins on two latest video matting benchmarks: 56.4% improvement on Deep Video Matting (DVM) [45] and 56.7% improvement on VideoMatting108 [56] in terms of MSE. We also conduct extensive analysis experiments and show that the proposed joint modeling and learning scheme are crucial for achieving robust and accurate video matting results.

2   Related Work

Image Matting. The image matting task was introduced in [36]. Unlike the image segmentation task that predicts a binary alpha value, the image matting problem aims to predict a high-precision alpha value in a continuous range. Therefore, the matting problem is harder to solve, and most existing works are addressed under some conditions. The most common condition is assuming a human-annotated trimap is given. The trimap is annotated into three different regions: definitely foreground region, definitely background region, and unknown region. The trimap serves to not only reduce the difficulty of the matting problem but also allows some user control over the results.

The traditional sampling-based methods [10,15,19,41,49] determine alpha values in unknown regions by combining sampled foreground and background pixels. Another traditional approach is the propagation-based [3,4,17,22,24–26], which propagates foreground and background pixels into unknown regions based on affinity scores. Recently, deep learning-based methods achieved great success in image matting [7,14,21,28,33,52] by training networks with a trimap input.

Some approaches try to find good alternatives to human-annotated trimap. Portrait matting [23,42,57] can extract an alpha matte without any external input, but they are only applicable for human subjects. Background matting [29, 38] proposes to take complete background information instead of trimap input and predicts high-resolution alpha matte. However, this method is hard to extend to general video matting because it can work only with a near-static background. Mask guided matting [55] proposes to replace the trimap with a coarse binary mask that is more accessible. All image matting methods mentioned above can be extended to video matting by applying frame-by-frame, but the constraint must be met for each frame (*e.g..*, trimap for every frame).

Video Matting. Early video matting methods largely extended traditional image matting methods by either extending the propagation temporally [1,8,27, 40] or by sampling in other frames [40,46]. While there was some work that would generate trimaps automatically by deriving them from segmentations [16,47] or using interpolation [9] or propagation [46], these were computed independently from the video matting method. Bai *et al..* [2] propagates trimaps based on predicted alpha mattes. Tang *et al..* [48] would use the alpha matte of one frame to help predict the trimap of the next frame, but did so by using the alpha matte to compute a binary segmentation from which a new trimap would be computed.

Recently, large-scale video matting datasets have been proposed [29,45,56] and they fueled video matting researches based on deep learning. Along with a large benchmark dataset, Sun *et al..* [45] proposed a two-stage deep learning-based method by decoupling video matting into trimap generation and alpha matting. They mainly focused on the alpha matting stage to achieve temporally coherent prediction by learning temporal feature alignment [50] and fusion [51]. Concurrently, Zhang et al.. [56] also approach the problem with a similar motivation. To be specific, [56] proposed a temporal aggregation module based on the guided contextual attention block [28] to maintain temporal consistency during the alpha prediction. They also released a large-scale benchmark dataset, VideoMatting108 [56].

However, both latest learning-based methods [45,56] overlooked the challenge in trimap propagation. Their temporally consistent results can be achieved only if the trimap is accurately generated for every frame. To prevent this issue, additional trimaps need to be manually annotated for several frames which makes the solutions less practical. Although there are trimap-free methods [30,44], those work only on human video and cannot generate user-desired results.

Toward robust and practical trimap-based video matting, we revisit the traditional video matting studies [2,48], where propagating a trimap based on the predicted alpha matte greatly improves the temporal coherency. Inspired by those studies while continuing the success of learning-based video matting, we propose OTVM that jointly learns trimap-based alpha prediction and alpha-based trimap propagation. Our method shows that high-quality and temporally consistent alpha prediction in video is achievable using only a single trimap input.

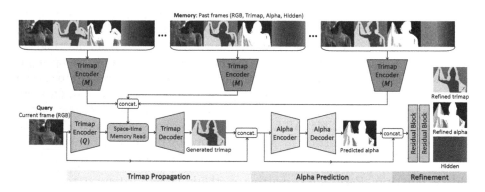

Fig. 3. Overall architecture of OTVM. Our trimap propagation network is inspired by space-time memory networks [34]. The network predicts a trimap based on the information from the previous frames and predictions that are embedded by the trimap memory encoder. From the given or generated trimap, we initially predict the alpha matte using our alpha prediction network inspired by [14]. Then we refine the generated trimap and predicted alpha matte via two light-weighted residual blocks. The refined trimap, alpha matte, and hidden features (dimension = 16) are fed to the trimap memory encoder so that they can be used for the next frames as a new memory. The framework is trained in an end-to-end manner. We further illustrate the details of each module in the supplementary material.

3   Method

The overall architecture of OTVM is illustrated in Fig. 3. Our method aims to perform video matting robustly with only a single trimap. From the given user-provided trimap at the first frame, we sequentially predict a trimap and an alpha matte for every frame in a video sequence. Starting from the user-provided trimap, we first predict the initial alpha matte by feeding the RGB frame and trimap to our alpha prediction network. Then, a lightweight refinement module is followed to correct errors in alpha matte and trimap, resulting in a refined alpha matte and trimap. The refinement module also produces a hidden latent feature map, and all the outputs from the refinement module are encoded as memory by the trimap memory encoder for trimap propagation. On the next frame, our trimap propagation module predicts the trimap by reading relevant information from the memory. This procedure – alpha prediction, refinement, and trimap propagation – is repeated until the end of the video sequence.

Note that recent works [45,56] also generate trimaps and predict alpha mattes, but our approach is completely different from those. In the existing works, the trimap propagation and alpha matting are totally decoupled. They tried to propagate trimap naively (*i.e.*. without consideration of the challenge introduced in Fig. 2), resulting in inaccurate trimap propagation. Therefore, multiple ground-truth (GT) trimaps should be provided to achieve good results. In contrast, our OTVM can extract accurate alpha matte results even with a single human-annotated trimap, thanks to our joint modeling.

3.1   Alpha Prediction with Trimap

Given an RGB image and (either a propagated or user-provided) trimap, we predict an alpha matte. Here, we opt for the state-of-the-art image-based matting network, FBA [14], to simplify the problem. The alpha matting network is an encoder-decoder architecture. The alpha encoder first takes a concatenation of the RGB and trimap along the channel dimension as input. Then, the resulting pyramidal features of the alpha encoder are fed into the alpha decoder that produces an alpha matte. To exploit the advantage of the coupled network trained end-to-end, we directly use the soft trimap from the trimap propagation module without binarization when a propagated trimap is given.

We can use any alpha matting network, however, we empirically observe that advanced alpha networks (*e.g.*. video alpha networks) make our framework complex and hinder end-to-end training given limited training data for video matting, while the latest image-based alpha networks work surprisingly well as long as a reliable trimap and end-to-end training are provided. Therefore, we take the simple image-based alpha prediction model and focus on developing the joint framework that can reliably propagate the trimaps.

3.2   Alpha-Trimap Refinement

In our video matting setting, which takes only one GT trimap, naively propagating the trimap may result in severe drifting as depicted in Fig. 2. To take the advantage of our coupled framework, we have an additional refinement module following the alpha network to provide refined information to the trimap propagation module afterward. The refinement module is light-weighted as it is composed of two residual blocks. The refinement module takes all available information for the current frame: an input RGB frame, the generated trimap, the predicted alpha matte, and the alpha decoder's latent features. Then, the module produces an updated trimap and a refined alpha matte along with unconstrained hidden features. The hidden features are intended for information that cannot be expressed in the form of trimap and alpha. These features are learnable through end-to-end training. All the outputs of the refinement module will be used for trimap propagation.

3.3   Trimap Propagation with Alpha

To propagate trimaps, we repurpose a state-of-the-art video segmentation network, space-time memory network (STM) [34], with important modifications. In the original STM [34], input images and corresponding masks in the past frames are set to the memory, while the image at the current frame is set to the query. Then, the memory and query are embedded through two independent ResNet50 [20] encoders. The embedded memory and query features are fed to the space-time memory read module. In the module, dense matching is performed and then a value of memory is retrieved based on the matching similarity. The decoder takes the retrieved memory value and query feature and then outputs an object mask. This approach can effectively exploit rich features of the intermediate frames and achieve state-of-the-art performance in video binary segmentation.

STM [34] simply can be extended from binary mask to trimap by increasing the input channel dimension of the memory encoder and the output channel dimension of the decoder. However, there is a fundamental limitation to apply STM directly for trimap propagation. In the binary mask, the foreground region and background region can be estimated by propagating from past binary masks. The trimap, however, cannot be estimated only with propagation because the unknown regions are frequently changed by the view of the foreground object (see Fig. 2) and trimap-only supervision does not provide a consistent clue for the changes. To address this problem, we additionally impose rich cues for generating the trimap into the memory encoder. Since the trimap has been determined by the alpha matte, it effectively helps to learn for trimap generation. We additionally impose a hidden feature extracted from the refinement module. With the hidden features, any errors can be easily propagated backward at training time, resulting in stable training. By imposing those into the memory encoder, we significantly reduce errors that occurred by drifting of the unknown regions.

3.4   End-to-End Training

To make OTVM work, it is critical to train the model end-to-end because each module depends on each other's outputs. However, video matting data are extremely difficult to annotate and existing video supervisions are not sufficient to train the model directly. As a practical solution, we train each module stage-wise then fine-tune the whole network in an end-to-end manner. First, we initialize the trimap propagation and the alpha matting modules with the pretrained weights of off-the-shelf STM [34] and FBA [14], respectively. Specifically, both pretrained models leverage ImageNet [37]. In addition, the STM is trained using image segmentation datasets [6,13,18,31,43] and video object segmentation datasets [35,53]. The FBA is trained on the Adobe Image Matting (AIM) dataset [52]. After the initialization, we pretrain our network modules in three stages (from Stage 1 to Stage 3) on the AIM dataset [52] and then finetune the whole model end-to-end on either VideoMatting108 [56] or DVM [45], depending on the target evaluation benchmark (Stage 4).

Stage 1: Training the alpha matting module and trimap propagation module separately. As two modules depend on each other, if we train the alpha matting module and the trimap propagation module simultaneously from scratch, this can lead the model to either poor convergence or simply memorizing training data (*i.e.*, overfitting). It is because both modules cannot learn meaningful features from almost randomly initialized input data which is as the output of other modules. Therefore, we first separately train two modules without the connections between two. Specifically, we train the alpha matting module with GT trimaps and train the trimap propagation module without taking inputs of an alpha matte and hidden features.

Stage 2: Training the alpha matting and refinement modules with propagated trimaps. We train the alpha matting model and refinement modules together while the trimap propagation module is frozen. This stage enables the refinement module to take a soft and noisy trimap as input and learn to predict accurate trimap and alpha matte.

Stage 3: Training the trimap propagation module. In the trimap propagation module, we activate all input layers for alpha matte and hidden features. Then, we train the trimap propagation module while the parameters for the remainders – alpha prediction and refinement – are frozen. In this stage, we leverage not only the loss from the predicted trimap but also the losses from the alpha prediction. This enables the trimap propagation module to predict a more reliable trimap for estimating the alpha matte. While we are not updating the alpha network and refinement module in this stage, we can leverage the gradients from their losses for updating the trimap propagation module.

Stage 4: End-to-end training. Finally, we train the whole network end-to-end using a video matting dataset. With the stage-wise pretraining, we can effectively leverage both image and video data, and achieve stable performance improvement at the end-to-end training.

3.5   Training Details

Data preparation. During training, we randomly sampled three temporally ordered foreground and background frames from each video sequence. If an image dataset (*e.g.*., AIM [52]) is used, we simulate three video frames from a pair of foreground and background images by applying three different random affine transforms into both foreground and background images. The random affine transforms include horizontal flipping, rotation, shearing, zooming, and translation. For each foreground and background frame, we randomly crop patches into 320×320, 480×480, or 640×640, centered on pixels in the unknown regions. And then we resize the cropped patches into 320×320. Additionally, we employ several augmentation strategies on both foreground and background frames: histogram matching between foreground and background colors, motion blur, Gaussian noise, and JPEG compression. Then we composite foreground and background on-the-fly to generate an input frame. The GT trimaps are generated by dilating the GT alpha matte with a random kernel size from 1×1 to 26×26.

Loss functions. We set objective functions for all outputs of the models, except for the hidden features. For both initially predicted and refined trimaps, we use the cross-entropy loss to compare with the GT. For the first frame where the GT trimap is provided as the input, we only apply the loss to the refined trimap. Ideally, there should be no change after refinement. We find penalizing any change after the refinement is helpful to prevent it from corrupting already accurate trimap. For the alpha predictions, we leverage the temporal coherence loss [45] and image matting losses used in FBA [14]. Different from some previous methods that only compute alpha losses on unknown regions (*e.g.*. [52]), we compute our losses on every pixel.

In addition to trimap and alpha losses, we also employ losses for the foreground and background color predictions. We estimate foreground and background colors from the alpha decoder and refinement module following [14]. We minimize all foreground and background losses used in [14], and additionally employ temporal coherence loss on both foreground and background. For the foreground color, we compute the losses only where an alpha value is greater than 0 because the exact foreground color is available only in those regions. More detailed explanations of loss functions are given in the supplementary material.

Other training details. We opt for RAdam optimizer [32] with a learning rate of 1e-5. We drop the learning rate to 1e-6 once at 90% iteration for each training stage. We freeze all the batch normalization layers in the networks. We used a mini-batch size of 4 and trained with four NVIDIA GeForce 1080Ti GPUs. At the first pretraining stage, we trained the alpha matting model about 100,000 iterations and we trained the trimap propagation model about 400,000 iterations. We trained about 50,000 iterations at each of the second and third training stages. Finally, we trained about 80,000 iterations at the last end-to-end training stage.

3.6   Inference Details

We used full-resolution inputs to achieve high-quality alpha matte results. For the memory management in the trimap propagation module, we generally follow STM [34] that stores the first and the previous frames to the memory by default, and additionally saves new memory periodically. We add the intermediate frames to the memory for every 10 frames. To avoid GPU memory overflow, we store only the last three intermediate frames and discard old frames.

4   Experiments

4.1   Evaluation Datasets and Metrics

We present experimental results and analysis on two latest benchmarks, Video-Matting108 [56] and DVM [45].

VideoMatting108 Dataset [56] includes 28 foreground video sequences paired with background video sequences in the validation set. The evaluation is conducted with three different trimap settings: narrow, medium, and wide. The groundtruth trimaps are generated by discretizing the groundtruth alpha mattes into the trimaps, followed by dilating the unknown regions with different kernel sizes; 11×11 for narrow, 25×25 for medium, and 41×41 for wide. This benchmark contains long video sequences and the average length of the videos is about 850 frames. Therefore, predicting alpha mattes with only a single trimap at the first frame is challenging in this dataset.

DVM Dataset. [45] provides 12 foreground videos and 4 background videos for validation. The validation set is constructed with an additional 50 foreground images from the AIM dataset [52]. In total, 62 foregrounds are composited with every 4 background videos, resulting in 248 test videos. Following the DVM paper [45], we discard 106 foreground images from the AIM dataset [52] and use the remaining 325 foreground images during our training because there are overlaps in the evaluation.

Evaluation Metrics. For a fair comparison, we follow the evaluation metrics from two large-scale video matting benchmarks [45,56]. To evaluate on Video-Matting108 [56], we compute SSDA (average sum of squared difference), MSE (mean squared error), MAD (mean absolute difference), dtSSD (mean squared difference of direct temporal gradients), and MESSDdt (mean squared difference between the warped temporal gradient) [12]. To evaluate on DVM [45], we compute SAD (sum of absolute difference), MSE, Grad (gradient error), Conn (connectivity error), dtSSD, and MESSDdt. For all computed metrics, lower is better.

In the original works [45,56], the metrics are computed only on the unknown regions of the GT trimaps. We follow this rule for fair comparisons with the existing methods. However, using only the unknown region cannot capture the errors in the foreground and background regions that occurred by inaccurate trimap propagation, which is important for evaluating the performance of end-to-end video matting methods. Therefore, we present *the modified versions of the metrics* that compute the scores on the full-frames, suffixed with "-V". The modified metrics are used for our analysis and ablation experiments.

4.2   Analysis Experiments

To validate our hypotheses experimentally, we conduct a set of analysis experiments. We use the one-trimap setting, where GT trimap is given only at the first frame. All these analysis experiments are conducted on the VideoMatting108 benchmark with the medium trimap setting.

Effectiveness of the Joint Modeling. We first validate the importance of the joint modeling of trimap propagation and alpha prediction in video matting. For this purpose, we design a simple baseline model for video matting, STM+FBA, that cascades STM [34] for trimap propagation and FBA [14] for alpha prediction from the propagated trimap. Note that we do not use OTVM because our proposals (*i.e.*, trimap and alpha refinement and using hidden features) are only applicable for the joint modeling. To show the effect of joint modeling, we train the baseline model with two different training strategies. One is obtained by training trimap propagation (STM) and alpha prediction (FBA) separately (*i.e.*, decoupled), and the other is by training both modules jointly (*i.e.*, joint). As shown in Table 1(a), the joint modeling greatly improves the alpha matte quality in the practical one-trimap scenario.

Efficacy of the Stage-Wise Training. We evaluate the importance of the proposed stage-wise training in OTVM and summarize the result in Table 1(b). When we activate all input layers for alpha matte and hidden features and end-to-end train OTVM from the pretraining stage (*i.e.*, joint), the model marginally surpasses the STM+FBA (joint). If our stage-wise training is applied, OTVM significantly outperforms STM+FBA (joint). The results demonstrate the superiority of our stage-wise training.

Efficacy of Each Training Stage. In this experiment, we use OTVM and validate our training strategy. The result is summarized in Table 1(c). To learn the trimap propagation model with the hidden features of the refinement module, we applied the last training stage (*i.e.*, Stage 4: end-to-end training) for all cases. As shown in the table, each pretraining stage consistently improves performance. The results demonstrate that our training strategy, stage-wise pretraining then end-to-end finetuning, is effective.

Table 1. Analysis experiments on VideoMatting108 validation set. For all experiments, we use 1-trimap setting where GT trimap is given only at the first frame. "-V" denotes the error has been computed in all regions of the frames (see Sec. 4.1).

(a) **Joint modeling.**

Model	Training method	SSDA-V	MSE-V	MAD-V	dtSSD-V	MESSDdt-V
STM+FBA	decoupled	83.61	10.62	22.12	36.31	3.45
	joint	**75.36**	**9.40**	**21.01**	**29.64**	**2.74**

(b) **Stage-wise training.**

Model	Training method	SSDA-V	MSE-V	MAD-V	dtSSD-V	MESSDdt-V
OTVM	joint	72.07	9.29	20.81	30.20	2.79
	joint + stage-wise	**54.67**	**2.61**	**13.02**	**29.87**	**1.78**

(c) **Ablation on training stages.** Each training stage is described in Sec. 3.4.

Train stages				SSDA-V	MSE-V	MAD-V	dtSSD-V	MESSDdt-V
Stage 1	Stage 2	Stage 3	Stage 4					
			✓	87.31	11.16	23.35	33.29	3.15
✓			✓	76.55	9.68	23.10	31.64	3.14
✓	✓		✓	75.33	9.54	22.30	31.44	3.06
✓	✓	✓	✓	**54.67**	**2.61**	**13.02**	**29.87**	**1.78**

(d) **Ablation on modules.** The modules are depicted in Fig. 3.

Refinement module (output)		Trimap module (input)		SSDA-V	MSE-V	MAD-V	dtSSD-V	MESSDdt-V	Time (sec/frame)
Alpha	Trimap	Alpha	Hidden						
				83.56	12.17	23.59	30.11	2.80	0.799
✓				83.51	11.26	22.49	31.78	2.89	0.951
✓	✓			78.70	9.80	21.51	29.97	2.80	0.946
✓	✓	✓		77.58	10.55	22.42	31.04	2.98	0.952
✓	✓	✓		62.47	4.27	15.06	29.98	1.85	0.955
✓	✓	✓	✓	**54.67**	**2.61**	**13.02**	**29.87**	**1.78**	0.964

(e) **Different image matting backbones.**

Backbone	Model	SSDA-V	MSE-V	MAD-V	dtSSD-V	MESSDdt-V
DIM [52]	STM+DIM	102.77	12.58	28.35	47.13	5.14
	OTVM	**92.30**	**11.04**	**25.98**	**39.27**	**4.21**
GCA [28]	STM+GCA	100.22	12.25	27.33	41.72	4.36
	OTVM	**84.73**	**10.04**	**24.23**	**36.02**	**3.46**

(f) **Trimap propagation.** "-T" is presented to estimate trimap quality and denotes that the unknown region in GT trimap has been modified (see Sec. 4.2).

Method	Precision-T	Recall-T	Average
Decoupled STM [45,56]	96.98	93.58	95.28
OTVM	**98.17**	**95.92**	**97.05**

Effectiveness of the Proposed modules. We thoroughly evaluate our proposed modules in Table 1(d). As shown in the table, the alpha matte and trimap refinements lead to complementary performance improvement. More importantly, feeding the alpha matte and hidden features of the refinement module to the trimap propagation module significantly improves performance with small extra computation. The result demonstrates the importance of reliable trimap propagation in the practical video matting setting. It also shows that our joint learning framework and the proposed modifications effectively address the problem.

Applicability to Different Image Matting Networks. Table 1(e) compares the results with other image matting backbones. We tested DIM [52] and GCA [28]. As shown in the table, OTVM consistently leads to performance improvements over the corresponding baselines.

Accuracy of Trimap Propagation. We study the effect of OTVM on trimap propagation. In Table 1(f), we compare OTVM with recent video matting methods [45,56] that naively propagate trimap using a single STM [34]. Ideally, the unknown area in a trimap needs to cover the entire soft matte area in the GT matte while being tight enough not to be trivial. To measure the quality of trimaps, we present two metrics: (1) *Precision-T*, precision of the estimated unknown area compared with the widely dilated GT unknown (dilation kernel size of 41×41) and (2) *Recall-T*, recall of the estimated unknown area compared with the minimum GT unknown (*i.e.*, no dilation of the GT unknown regions). As shown in Table 1(f), OTVM significantly outperforms the state-of-the-art approach and achieves high precision and high recall.

More Analysis in the Supplementary Material. We present additional results on the input of the trimap encoder, a visual analysis of the hidden feature, the effect of the refinement module, an analysis of runtime and GPU memory consumption, an analysis of temporal stability, and quantitative results with image matting metrics (*i.e.*. without "-V") in the supplementary material.

4.3   Comparison with State-of-the-Art Methods

VideoMatting108 [56]. Table 2(a) shows quantitative evaluation results on the VideoMatting108 validation set with medium trimap setting. Results with other trimap settings (*i.e.*., narrow and wide) are provided in the supplementary material. In the experiment, we measured errors only on the unknown regions for a fair comparison with [56]. We report two results depending on whether GT trimap is used for all frames or for the first frame. For testing other methods on 1-trimap setting, we used the baseline trimap propagation model we trained to generate trimaps. For a fair comparison with TCVOM [56], which uses GCA [28] for a backbone network, we additionally show FBA backbone results. When a single trimap is used, OTVM significantly outperforms all other methods, demonstrating the superiority of our approach.

Table 2. Comparison with state-of-the-art methods on public benchmarks. The trimap setting indicates how many GT trimaps are given as input, *i.e..*, "full-trimap" for all frames, "20/40-frame" for every 20/40th frames, "1-trimap" for only at the first frame.

(a) **Comparison on VideoMatting108 validation set**. Results for other methods are directly copied from [56]. † denotes our reproduced results using our training setup.

Trimap Setting	Methods	SSDA	MSE	MAD	dtSSD	MESSDdt
full-trimap	DIM [52]	61.85	9.99	44.38	34.55	2.82
	IndexNet [33]	58.53	9.37	43.53	33.03	2.33
	GCA [28]	55.82	8.20	40.85	31.64	2.15
	TCVOM (GCA) [56]	50.41	7.07	37.65	27.28	1.48
	TCVOM (FBA)† [56]	**39.76**	**4.01**	**28.68**	**22.93**	**1.06**
1-trimap	STM + TCVOM (GCA) [56]	77.23	22.15	57.40	32.18	2.97
	STM + TCVOM (FBA)† [56]	69.96	19.80	51.21	29.76	2.72
	STM + FBA† [14]	70.63	20.18	51.20	31.00	2.86
	OTVM	**50.51**	**8.58**	**37.16**	**28.28**	**1.63**

(b) **Comparison on DVM validation set**. Results for other methods are directly copied from [45].

Trimap Setting	Methods	SAD	MSE	Grad	Conn	dtSSD	MESSDdt
full-trimap	DIM [52]	54.55	0.030	35.38	55.16	23.48	0.53
	IndexNet [33]	53.68	0.028	27.52	54.44	19.50	0.49
	Context-Aware [21]	51.78	0.027	28.57	49.46	19.37	0.50
	GCA [28]	47.49	0.022	26.37	45.23	18.36	0.33
	DVM [45]	**40.91**	**0.014**	**19.02**	**40.58**	**15.11**	**0.25**
20-frame	DVM [45]	43.66	0.016	26.39	42.23	16.34	0.28
	OTVM	**37.90**	**0.013**	**19.13**	**36.48**	**14.76**	**0.22**
40-frame	DVM [45]	52.85	0.026	-	-	19.23	-
	OTVM	**38.24**	**0.014**	**19.29**	**36.83**	**14.82**	**0.22**
1-trimap	DVM [45]	65.33	0.039	-	-	35.46	-
	OTVM	**41.02**	**0.017**	**20.17**	**39.79**	**14.85**	**0.25**

DVM [45]. We conduct quantitative evaluation on the DVM benchmark and Table 2(b) shows the results. We computed errors only on the unknown regions according to the official metric. Table 2(b) shows that OTVM largely surpasses DVM in 1-trimap setting. Furthermore, OTVM using only 1-trimap (SAD 41.02) achieves comparable performance with DVM using full-trimap (SAD 40.91).

Qualitative Results on Real-World Videos. Figure 4 shows qualitative results on a real-world video. We compare OTVM with the cascaded STM and FBA model, denoted by STM+FBA. As shown in the figure, the STM+FBA model cannot sharply separate foreground and unknown regions on the object boundary. The cascade baseline model fails to predict accurate trimaps in the challenging scenes, resulting in poor alpha mattes. In contrast, OTVM predicts trimaps reliably, resulting in accurate alpha mattes. More qualitative results are provided in the supplementary material. In addition, we provide full-frame results online: https://youtu.be/qkda4fHSyQE.

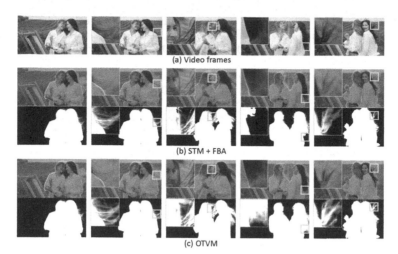

Fig. 4. Qualitative comparison on a real-world video sequence. As shown in zoomed trimap, OTVM generates more accurate and fine trimaps, *e.g.*, sharply separates hairs in the foreground region, but STM roughly predicts the regions as unknown. Therefore, we can extract high-quality alpha matte results.

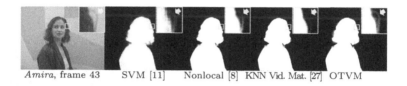

Amira, frame 43   SVM [11]   Nonlocal [8] KNN Vid. Mat. [27] OTVM

Fig. 5. Comparison with non-deep learning methods.

Comparison with Non-deep Learning Methods. Figure 5 shows a comparison with [8,11,27] on the *Amira* benchmark [9]. The benchmark [9] does not provide the GT alpha matte and we took the results of the previous methods from [11,27]. As shown in the figure, SVM [11] result is noisy. Both Nonlocal [8] and KNN Video Matting [27] fail to predict hair strand details. In contrast, OTVM predicts the precise alpha matte.

4.4 Limitations

Since our framework takes only a single user-annotated trimap, not only the input trimap quality but also the rich cues of the object in the frame are important. In Fig. 6(a), OTVM struggles to generate accurate trimaps if the user-annotated frame contains almost no object information, resulting in failure to predict the alpha matte. In Fig. 6(b), although the object is presented in the annotated frame, we may struggle to predict precise alpha mattes if there is no strong signal for the foreground object in the given trimap.

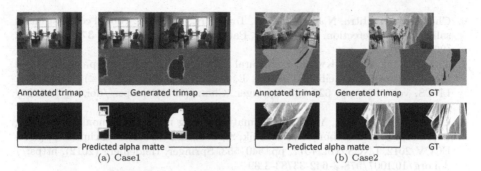

Fig. 6. Limitations. We indicate error areas with green boxes. (a) The user-annotated trimap contains almost no object information. (b) There is no strong signal for the foreground in the given trimap. (Color figure online)

5 Conclusion

In this paper, we present a new video matting framework that only needs a single user-annotated trimap. In contrast to the recent decoupled methods that focus on alpha prediction given the trimaps, we propose a coupled framework, OTVM, that performs trimap propagation and alpha prediction jointly. OTVM with one user-annotated trimap significantly outperforms the previous works in the same setting and even achieves comparable performance with the previous works using full-trimaps as input. OTVM is simple yet effective and works robustly in the practical one-trimap scenario. We hope that our research motivates follow-up studies and leads to practical video matting solutions.

Acknowledgements. This research was supported in part by the Yonsei Signature Research Cluster Program of 2022 (2022-22-0002). This research was also supported in part by the KIST Institutional Program (Project No. 2E31051-21-204).

References

1. Apostoloff, N., Fitzgibbon, A.: Bayesian video matting using learnt image priors. In: CVPR (2004)
2. Bai, X., Wang, J., Simons, D.: Towards temporally-coherent video matting. In: Gagalowicz, A., Philips, W. (eds.) MIRAGE 2011. LNCS, vol. 6930, pp. 63–74. Springer, Heidelberg (2011). https://doi.org/10.1007/978-3-642-24136-9_6
3. Chen, Q., Li, D., Tang, C.K.: KNN matting. IEEE Trans. Pattern Anal. Mach. Intell. **35**(9), 2175–2188 (2013)
4. Chen, X., Zou, D., Zhiying Zhou, S., Zhao, Q., Tan, P.: Image matting with local and nonlocal smooth priors. In: CVPR, pp. 1902–1907 (2013)
5. Cheng, H.K., Tai, Y.W., Tang, C.K.: Rethinking space-time networks with improved memory coverage for efficient video object segmentation. In: NeurIPS (2021)

6. Cheng, M.M., Mitra, N.J., Huang, X., Torr, P.H., Hu, S.M.: Global contrast based salient region detection. IEEE Trans. Pattern Anal. Mach. Intell. **37**(3), 569–582 (2014)
7. Cho, D., Tai, Y.-W., Kweon, I.: Natural image matting using deep convolutional neural networks. In: Leibe, B., Matas, J., Sebe, N., Welling, M. (eds.) ECCV 2016. LNCS, vol. 9906, pp. 626–643. Springer, Cham (2016). https://doi.org/10.1007/978-3-319-46475-6_39
8. Choi, I., Lee, M., Tai, Y.-W.: video matting using multi-frame nonlocal matting Laplacian. In: Fitzgibbon, A., Lazebnik, S., Perona, P., Sato, Y., Schmid, C. (eds.) ECCV 2012. LNCS, vol. 7577, pp. 540–553. Springer, Heidelberg (2012). https://doi.org/10.1007/978-3-642-33783-3_39
9. Chuang, Y.Y., Agarwala, A., Curless, B., Salesin, D., Szeliski, R.: Video matting of complex scenes. In: SIGGRAPH (2002)
10. Chuang, Y.Y., Curless, B., Salesin, D.H., Szeliski, R.: A Bayesian approach to digital matting. In: CVPR. IEEE (2001)
11. Eisemann, M., Wolf, J., Magnor, M.A.: Spectral video matting. In: VMV, pp. 121–126. Citeseer (2009)
12. Erofeev, M., Gitman, Y., Vatolin, D.S., Fedorov, A., Wang, J.: Perceptually motivated benchmark for video matting. In: BMVC, pp. 99–1 (2015)
13. Everingham, M., Van Gool, L., Williams, C.K., Winn, J., Zisserman, A.: The pascal visual object classes (VOC) challenge. Int. J. Comput. Vision **88**(2), 303–338 (2010)
14. Forte, M., Pitié, F.: f, b, alpha matting. arXiv preprint arXiv:2003.07711 (2020)
15. Gastal, E.S.L., Oliveira, M.M.: Shared sampling for real-time alpha matting. Comput. Graph. Forum **29**(2), 575–584 (2010)
16. Gong, M., Wang, L., Yang, R., Yang, Y.H.: Real-time video matting using multi-channel Poisson equations. In: Graphics Interface (2010)
17. Grady, L., Schiwietz, T., Aharon, S., Westermann, R.: Random walks for interactive alpha-matting. In: Proceedings of VIIP, vol. 2005, pp. 423–429 (2005)
18. Hariharan, B., Arbeláez, P., Bourdev, L., Maji, S., Malik, J.: Semantic contours from inverse detectors. In: ICCV, pp. 991–998. IEEE (2011)
19. He, K., Rhemann, C., Rother, C., Tang, X., Sun, J.: A global sampling method for alpha matting. In: CVPR, pp. 2049–2056. IEEE (2011)
20. He, K., Zhang, X., Ren, S., Sun, J.: Deep residual learning for image recognition. In: CVPR, pp. 770–778 (2016)
21. Hou, Q., Liu, F.: Context-aware image matting for simultaneous foreground and alpha estimation. In: ICCV, pp. 4130–4139 (2019)
22. Jian, S., Jia, J., Tang, C.K., Shum, H.Y.: Poisson matting. ACM Trans. Graph. **23**, 315–321 (2004)
23. Ke, Z., Sun, J., Li, K., Yan, Q., Lau, R.W.: ModNet: real-time trimap-free portrait matting via objective decomposition. In: AAAI (2022)
24. Lee, P., Wu, Y.: Nonlocal matting. In: CVPR, pp. 2193–2200. IEEE (2011)
25. Levin, A., Lischinski, D., Weiss, Y.: A closed-form solution to natural image matting. IEEE Trans. Pattern Anal. Mach. Intell. **30**(2), 228–242 (2007)
26. Levin, A., Rav-Acha, A., Lischinski, D.: Spectral matting. IEEE Trans. Pattern Anal. Mach. Intell. **30**(10), 1699–1712 (2008)
27. Li, D., Chen, Q., Tang, C.K.: Motion-aware KNN Laplacian for video matting. In: ICCV (2013)
28. Li, Y., Lu, H.: Natural image matting via guided contextual attention. In: AAAI (2020)

29. Lin, S., Ryabtsev, A., Sengupta, S., Curless, B.L., Seitz, S.M., Kemelmacher-Shlizerman, I.: Real-time high-resolution background matting. In: CVPR, pp. 8762–8771 (2021)

30. Lin, S., Yang, L., Saleemi, I., Sengupta, S.: Robust high-resolution video matting with temporal guidance. In: WACV (2022)

31. Lin, T.-Y., et al.: Microsoft COCO: common objects in context. In: Fleet, D., Pajdla, T., Schiele, B., Tuytelaars, T. (eds.) ECCV 2014. LNCS, vol. 8693, pp. 740–755. Springer, Cham (2014). https://doi.org/10.1007/978-3-319-10602-1_48

32. Liu, L., et al.: On the variance of the adaptive learning rate and beyond. In: ICLR, April 2020

33. Lu, H., Dai, Y., Shen, C., Xu, S.: Indices matter: learning to index for deep image matting. In: ICCV, pp. 3266–3275 (2019)

34. Oh, S.W., Lee, J.Y., Xu, N., Kim, S.J.: Video object segmentation using space-time memory networks. In: ICCV, October 2019

35. Pont-Tuset, J., Perazzi, F., Caelles, S., Arbeláez, P., Sorkine-Hornung, A., Van Gool, L.: The 2017 Davis challenge on video object segmentation. arXiv preprint arXiv:1704.00675 (2017)

36. Porter, T., Duff, T.: Compositing digital images. In: SIGGRAPH, pp. 253–259 (1984)

37. Russakovsky, O., et al.: ImageNet large scale visual recognition challenge. Int. J. Comput. Vision 115(3), 211–252 (2015)

38. Sengupta, S., Jayaram, V., Curless, B., Seitz, S.M., Kemelmacher-Shlizerman, I.: Background matting: the world is your green screen. In: CVPR, pp. 2291–2300 (2020)

39. Seong, H., Oh, S.W., Lee, J.Y., Lee, S., Lee, S., Kim, E.: Hierarchical memory matching network for video object segmentation. In: ICCV, pp. 12889–12898 (2021)

40. Shahrian, E., Price, B., Cohen, S., Rajan, D.: Temporally consistent and spatially accurate video matting. In: Eurographics (2014)

41. Shahrian, E., Rajan, D., Price, B., Cohen, S.: Improving image matting using comprehensive sampling sets. In: CVPR, pp. 636–643 (2013)

42. Shen, X., Tao, X., Gao, H., Zhou, C., Jia, J.: Deep automatic portrait matting. In: Leibe, B., Matas, J., Sebe, N., Welling, M. (eds.) ECCV 2016. LNCS, vol. 9905, pp. 92–107. Springer, Cham (2016). https://doi.org/10.1007/978-3-319-46448-0_6

43. Shi, J., Yan, Q., Xu, L., Jia, J.: Hierarchical image saliency detection on extended CSSD. IEEE Trans. Pattern Anal. Mach. Intell. 38(4), 717–729 (2015)

44. Sun, J., Ke, Z., Zhang, L., Lu, H., Lau, R.W.: ModNet-V: improving portrait video matting via background restoration. arXiv preprint arXiv:2109.11818 (2021)

45. Sun, Y., Wang, G., Gu, Q., Tang, C.K., Tai, Y.W.: Deep video matting via spatio-temporal alignment and aggregation. In: CVPR, pp. 6975–6984 (2021)

46. Lee, S.-Y., Yoon, J.-C., Lee, I.K.: Temporally coherent video matting. Graph. Models 72, 25–33 (2010)

47. Tang, Z., Miao, Z., Wan, Y.: Temporally consistent video matting based on bilayer segmentation. In: ICME (2010)

48. Tang, Z., Miao, Z., Wan, Y., Zhang, D.: Video matting via opacity propagation. Visual Comput. 28, 47–51 (2012)

49. Wang, J., Cohen, M.F.: Optimized color sampling for robust matting. In: CVPR. IEEE (2007)

50. Wang, X., Chan, K.C., Yu, K., Dong, C., Change Loy, C.: EDVR: video restoration with enhanced deformable convolutional networks. In: CVPRW (2019)

51. Woo, S., Park, J., Lee, J.-Y., Kweon, I.S.: CBAM: convolutional block attention module. In: Ferrari, V., Hebert, M., Sminchisescu, C., Weiss, Y. (eds.) ECCV 2018. LNCS, vol. 11211, pp. 3–19. Springer, Cham (2018). https://doi.org/10.1007/978-3-030-01234-2_1

52. Xu, N., Price, B., Cohen, S., Huang, T.: Deep image matting. In: CVPR, pp. 2970–2979 (2017)

53. Xu, N., et al.: YouTube-VOS: sequence-to-sequence video object segmentation. In: Ferrari, V., Hebert, M., Sminchisescu, C., Weiss, Y. (eds.) ECCV 2018. LNCS, vol. 11209, pp. 603–619. Springer, Cham (2018). https://doi.org/10.1007/978-3-030-01228-1_36

54. Xu, N., et al.: Youtube-vos: A large-scale video object segmentation benchmark. arXiv preprint arXiv:1809.03327 (2018)

55. Yu, Q., et al.: Mask guided matting via progressive refinement network. In: CVPR, pp. 1154–1163 (2021)

56. Zhang, Y., et al.: Attention-guided temporal coherent video object matting. In: ACM MM (2021)

57. Zhu, B., Chen, Y., Wang, J., Liu, S., Zhang, B., Tang, M.: Fast deep matting for portrait animation on mobile phone. In: ACM MM, pp. 297–305 (2017)

D²ADA: Dynamic Density-Aware Active Domain Adaptation for Semantic Segmentation

Tsung-Han Wu[1][✉], Yi-Syuan Liou[1], Shao-Ji Yuan[1], Hsin-Ying Lee[1],
Tung-I Chen[1], Kuan-Chih Huang[1], and Winston H. Hsu[1,2]

[1] National Taiwan University, Taipei City, Taiwan
tsunghan@cmlab.csie.ntu.edu.tw
[2] Mobile Drive Technology, Taipei City, Taiwan

Abstract. In the field of domain adaptation, a trade-off exists between the model performance and the number of target domain annotations. Active learning, maximizing model performance with few informative labeled data, comes in handy for such a scenario. In this work, we present **D²ADA**, a general active domain adaptation framework for semantic segmentation. To adapt the model to the target domain with minimum queried labels, we propose acquiring labels of the samples with high probability density in the target domain yet with low probability density in the source domain, complementary to the existing source domain labeled data. To further facilitate labeling efficiency, we design a dynamic scheduling policy to adjust the labeling budgets between domain exploration and model uncertainty over time. Extensive experiments show that our method outperforms existing active learning and domain adaptation baselines on two benchmarks, GTA5 → Cityscapes and SYNTHIA → Cityscapes. With less than 5% target domain annotations, our method reaches comparable results with that of full supervision.

Keywords: Active learning · Domain adaptation · Semantic segmentation

1 Introduction

Semantic segmentation is vital for many intelligent systems, such as self-driving cars and robotics. Over the past ten years, supervised deep learning methods [5–8,22,41,45,54,56] have achieved great success assisted by rich labeled datasets. However, obtaining large-scale datasets with manual pixel-by-pixel annotation is still costly in terms of time and effort. Thus, several prior works utilized *domain adaptation* techniques to transfer the knowledge learned from the whole labeled

Supplementary Information The online version contains supplementary material available at https://doi.org/10.1007/978-3-031-19818-2_26.

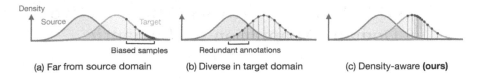

Density

Source   Target

Biased samples   Redundant annotations

(a) Far from source domain   (b) Diverse in target domain   (c) Density-aware **(ours)**

Fig. 1. Different Exploration Techniques in the ADA. (a) [40,58] proposed acquiring labels of target samples that are far from the source domain by the trained domain discriminator. However, the selected biased samples are inconsistent with real target distribution. (b) [13,31,39] proposed selecting diverse samples in the target domain with clustering techniques for label acquisition. Nonetheless, redundant annotations exist on samples that are similar to the existing labeled source domain dataset. (c) We propose a **density-aware** ADA strategy that acquires labels for samples that are representative in the target domain yet scarce the source domain, which is better than only considering either the source domain (a) or the target domain (b).

source domain, such as simulated game environments, to real-world unlabeled target domain.

In the field of domain adaptation, a trade-off exists between model performance and the amount of target domain annotations. With sufficient target domain manual annotations, supervised learning method can reach high performance (e.g. 71.3 mIoU on Cityscapes shown in Table 1). However, without any target labeling, the performance of unsupervised domain adaptation (UDA) methods [4,10,12,20,23,24,26,37,42,43,53,55,59–61] are still far below that of full supervision (e.g. 57.5 mIoU on GTA [32] → Cityscapes [11] reported by [55]). Compared to the above two extreme cases, which are impractical in real-world applications, a more reasonable manner is to strike a balance between model performance and the cost of labeling efforts.

Active learning, maximizing model performance with few informative labeled data, comes in handy for such a scenario. In the past few years, several works utilized model uncertainty [14,34,36,38,44,46] or data diversity [2,21,28,35,50] as the indicator to select valuable samples for labeling. The core concept of *uncertainty-based methods* is to acquire labels for data close to the model decision boundary, and the main idea of *data diversity approaches* is to query labels for a batch of samples that are far away from each other in the feature space.

Recently, some studies leveraged uncertainty and diversity active learning methods for domain adaptation [13,31,39,40,58], named Active Domain Adaptation (ADA); however, these methods have two significant defects. First of all, existing techniques for exploring target domain distributions is not efficient. Figure 1 (a, b) shows two typical approaches. As observed in (a), selecting samples far from source domain distributions [40,58] might lead to outliers or biased data [29]. As shown in (b), labeling diverse data in the target domain [13,31,39] might cause unnecessary annotations. In Fig. 3, we showed that the methods with 5% annotations still performed worse than ours with 2.5x fewer labels.

Furthermore, in the field of active learning, label acquisition strategies will be executed for multiple rounds [35,44]; yet, existing methods [13,31,39,40,58], combining uncertainty and diversity, did not notice the uncertainty measurement is less informative in the first few rounds in the ADA problem. As shown in Fig. 2 (a, b, c), model uncertainty might fail to detect high-confident but erroneous target domain regions (e.g. mispredict the pavement road as the sidewalk) under severe domain shift in earlier rounds. Thus, heavily relying on uncertainty cues led to poor results under low labeling budgets (see Table 2 and Fig. 3).

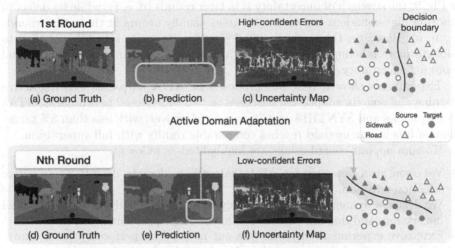

Fig. 2. Analysis of Uncertainty Criterion in the ADA. Uncertainty-based active learning methods acquire labels of data close to the decision boundary (red background). Since its inability to detect high-confident error under severe domain shift, several representative samples in the target domain far from the decision boundary will not be selected (yellow background) in earlier rounds, resulting in low label efficiency under few labeling budgets as shown in (a, b, c). However, as the two domains gradually align by fine-tuning with acquired labels, the number of high-confident but erroneous regions are rapidly reduced. Meanwhile, uncertainty measurement is able to capture the low-confident error with an accurate model in later rounds, as shown in (d, e, f). Hence, we propose a dynamic scheduling policy to pay more attention on domain exploration in earlier stages and rely more on model uncertainty later (Sect. 3.3). (Color figure online)

To address the two problems, we present an **D**ynamic **D**ensity-aware **A**ctive **D**omain **A**daptation (**D²ADA**) framework for semantic segmentation. To select the most informative target domain data for labeling, we propose a novel *density-aware selection method* to select data with the largest domain gaps. In this work, we use the term, **domain density**, to describe the prevalence of an observed sample in a specific domain. Then, we acquire the labels of regions with the largest density difference between the source and the target domain. As our intuition in Fig. 1 (c), since the model has already performed well on rich labeled source domain data, labeling few samples with high target domain density but low source domain density is sufficient for models to overcome the domain shift. Compared to prior works in Fig. 1 (a, b), the superiority of our method is demonstrated in Fig. 3. Also, it is mathematically proved to reduce the generalization bound of domain adaptation.

To maximize the advantages of domain exploration and model uncertainty, we develop a *dynamic scheduling policy* to adjust the labeling budgets between the two selection strategies over time. Since the key idea of the exploration technique is to reduce the domain gap, it greatly improves the performance in the first few rounds when the domain gap is large but is less effective later when the two domains are almost aligned as shown in Table 2. In contrast, as discussed in Fig. 2, the strength of uncertainty is in later rounds (d, e, f), while its defect of ignoring high-confident but erroneous areas usually occurs in the earlier rounds (a, b, c). Based on the analysis, we designed a dynamic scheduling policy to allocate more labeling budgets for our density selection in earlier rounds and more for uncertainty selection later.

Experiments showed that our proposed **D²ADA** surpasses existing active learning and domain adaptation baselines on two widely used benchmarks, GTA5 → Cityscapes and SYNTHIA → Cityscapes. Moreover, with less than 5% target domain labels, our method reaches comparable results with full supervision.

To sum up, our contributions are highlighted as follows,

- We open up a new way for ADA that utilizes domain density as an active selection criterion.
- We design a dynamic scheduling policy to adjust the budgets between model uncertainty and domain exploration in the ADA problem.
- Extensive experiments demonstrate our method outperforms current state-of-the-art active learning and domain adaptation methods.

2   Related Work

We introduce domain adaptation methods for semantic segmentation with none or few target labels and briefly review the progress of deep active learning.

2.1   Unsupervised Domain Adaptation for Semantic Segmentation

Many researchers have delved into unsupervised domain adaptation (UDA) methods for semantic segmentation. One widely-used approach is to align the distribution of the source and target domain through adversarial training [4,12,24,42,43]. Another popular method is to apply self-training techniques [47,55,57,59–61] by regularizing model uncertainty and assigning pseudo labels to reliable target domain data. However, the performance of distribution alignment is still not satisfying [28], and self-training often suffers from label noise [55]. These defects make existing UDA methods hard to reach the performance of full supervision on labeled target domain data.

2.2   Domain Adaptation for Semantic Segmentation with Few Target Labels

To reduce the performance gap between UDA methods and supervision on the whole labeled target domain data, some recent works have explored the utilization of weak and few target domain annotations.

[30] introduced a weakly-supervised domain adaptation (WDA) framework for semantic segmentation, where image-level annotations are provided initially, and few manual or pseudo pixel-level labels are gradually added.

Some considered the semi-supervised domain adaptation (SSDA) scenario, where few pixel-level annotations are available during training. [48] attempted to align the features of the source and target domain globally and semantically. [9] reduced the domain shift by mixing domains and knowledge distillation.

Unlike the above settings, where few labels were initially given, others used active learning strategies to acquire few target annotations from humans. For example, [28] chose diverse target domain images dissimilar to multiple source domain anchors for labeling. [36] utilized the inconsistency map generated by the disagreement of two classifiers to acquire few uncertain target pixel labels.

Different from those prior works focusing on uncertain or diverse samples, our density-aware selection method acquires labels of few but vital samples that can bridge the gap between the source and target domains.

2.3 Active Learning for Domain Adaptation

Active learning can reduce the annotation effort by gradually selecting the most valuable unlabeled data to be labeled.

Conventional active learning strategies can be roughly divided into two categories: uncertainty-based methods and diversity-aware approaches. The main idea of uncertainty-based methods is to acquire labels of data close to the model decision boundary. For example, [44,46] utilized the softmax entropy, confidence, or margin of network outputs as the selection criterion. [14,15] proposed a better measurement of uncertainty with MC-dropout. On the other hand, the key concept of diversity approaches is querying annotations of a batch of samples far away from each other in the feature space. Prior works used clustering [2] or greedy selection [21,35,50] to achieve the diversity in a query batch.

In addition, several recent works have studied active domain adaptation scenarios that consider both uncertainty and diversity. [40,58] used the trained classifier and domain discriminator to select highly uncertain samples far from the source domain distribution. [13,31,39] utilized clustering to select highly uncertain samples that are diverse in target domain distribution. Very recently, [52] proposed an energy-based approach and [51] presented the concept of regional impurity to address this problem.

To the best of our knowledge, we propose the first density-aware active domain adaptation strategy to acquire labels of representative samples with high density in the target domain but low density in the source domain. Also, we developed the first dynamic scheduling policy between domain exploration and model uncertainty for active domain adaptation.

3   Method

3.1   Overview

In the Active Domain Adaptation (ADA) problem, we have a data pool of C-category semantic segmentation from two domains, comprising rich labeled data D_S in the source domain and some data D_T in the target domain. D_T can be divided into D_T^L and D_T^U, where D_T^L contains the data in the target domain that has been labeled and D_T^U stores the remaining unlabeled data.

Given the initial D_T without any annotations, i.e., $D_T^L = \emptyset$, the ADA algorithm iteratively acquires annotations of few data in D_T^U to maximize the performance improvement over the test data in the target domain. The label acquisition process will be performed several rounds and the labeling budgets in single round is a fixed number of pixels.

Our **D**ynamic **D**ensity-aware **A**ctive **D**omain **A**daptation (**D²ADA**) framework can be divided into 3 steps: (1) Train an initial model on $D_S \cup D_T$ with UDA methods as a warmup step. (2) Perform the density-aware selection to determine representative regions for each category according to the domain density and balance the budgets across classes. (Sect. 3.2). (3) Dynamically determine the labeling budgets for domain exploration and model uncertainty in each round. Then, for the top-ranked regions in D_T^U, acquire ground truth labels and move them to D_T^L. Finally, fine-tune the model on $D_S \cup D_T^L$ in a supervised manner and go back to step (2) for a new round (Sect. 3.3). The complete and detailed algorithm can be referred in the supplementary material.

3.2   Density-Aware Selection

To efficiently label the most informative target domain data, unlike prior domain exploration techniques causing biased samples (Fig. 1(a)) or redundant labeling (Fig. 1(b)), our density-aware approach (Fig. 1(c)) selects data representative in the target distribution but uncommon in the source distribution. To achieve this, we introduce the concept of **domain density** to calculate the prevalence of data in a single domain and estimate the categorical domain gaps between the two domains. By selecting data with the largest domain gaps, we can adapt the model to the target domain with the least possible amount of queried annotations.

In the following section, we first introduce the definition of domain density and provide ways to estimate it for each category. Then, we elaborate on our density difference metric and its theoretical foundation. Finally, we describe the motivation and operation of our class-balanced selection.

Domain Density Estimation. Domain density is defined as a measurement of how common a region R is predicted as a certain class in a particular domain. Following prior works [3,19,25,38], we divide an image into multiple regions using the widely-used SLIC [1] algorithm. To derive the domain density, we first calculate the feature z of a region R as the average feature over all the pixels within R. Similarly, a category $c \in [C] = \{1, 2, \cdots C\}$ is assigned to a region based on the predicted probability averaged across all pixels in the region. Given a domain S, a feature vector z, and a category c, we define the **source domain density**

$$d_S = p_S(z|c) \tag{1}$$

as the probability that z is classified as c in S. To estimate $p_S(z|c)$, we construct the density estimators from all the observed (z, c) pairs in the source domain dataset. The target domain follows the manner similarly.

In our implementation, a set of Gaussian Mixture Models (GMMs) are served as the density estimators. The construction of GMMs can be efficiently completed by offline and parallel execution, which takes about 0.01 s per region with an 8-core CPU personal computer. More discussions and details are left in the supplementary material.

Density Difference as Metric. The key idea of our density-aware method is to select unlabeled target regions that contribute most to the domain gap so that we can adapt the model from the existing labeled source domain to the target domain efficiently. In practice, we use the difference between the target and the source domain density of a single region to evaluate the domain gap.

For the i-th unlabeled target region, we first obtain its source domain density d_S^i and target domain density d_T^i by feeding (z_i, c_i) to the pre-constructed density estimators. Then, we introduce a metric to rank all regions in D_T^U:

$$\pi_i = \log(\frac{d_T^i}{d_S^i}), \tag{2}$$

where $i = 1, 2, \cdots, |\#\text{regions} \in D_T^U|$. The ratio at the log scale is the log-likelihood difference of the two domain densities [16]. According to Eq. 2, regions with larger π are more prevalent in the target domain and less observable in the source domain. Therefore, we select regions with large π values to replenish the insufficient knowledge of the source domain and to maximize the exploration of the target domain. For each category c, we sorted all unlabeled regions based on their importance scores. Then, a fixed number of top-ranked regions for each category will be selected for labeling until running out of annotation budgets.

Theoretical Foundation. To provide the theoretical foundations for our proposed density difference metric, we leverage the proposition presented in [27]

$$\ell_{test} \le \ell_{train} + \frac{M}{\sqrt{2}} \sqrt{D_{\text{KL}}(p_T(c, z) \| p_S(c, z))}, \tag{3}$$

where ℓ_{test} and ℓ_{train} denote the testing and training loss respectively. p_S and p_T are the joint probability distributions of the category c and the extracted feature z in the source and the target domain distribution respectively. M is a constant term and $D_{\text{KL}}(\|)$ is the KL divergence of the two distributions.

As shown in Eq. 3, to provide a tighter generalization bound for ℓ_{test}, we aim to minimize the KL of the two joint probability distributions as follows:

$$D_{\text{KL}}(p_T(c, z) \| p_S(c, z)) = D_{\text{KL}}(p_T(c)\&\| p_S(c))$$
$$+ \mathbb{E}_{p_T(c)}[D_{\text{KL}}(p_T(z|c) \| p_S(z|c))]. \tag{4}$$

The proof of Eq. 4 is provided in the supplementary material.

Since $p_T(c)$ and $p_S(c)$ are similar in most cases, we can obtain better generalization to the target domain by minimizing the expected value of the KL divergence from $p_S(z|c)$ to $p_T(z|c)$. Additionally, to deal with the label imbalance problem and to make the model perform well for each category for semantic segmentation, we opt to reduce the categorical KL term, i.e., $D_{KL}(p_T(z|c) \parallel p_S(z|c))$.

Although it is impossible to obtain the precise KL term by sampling all data points in the high dimensional continuous feature space, we could leverage the Monte Carlo simulation process to estimate the KL of the two distributions [18]. Given a sufficient number of N observed region features $\{z_i\}_{i=1}^N$ predicted to a certain category c, the KL divergence of the two conditional probability distributions can be approximated as:

$$D_{KL}(p_T(z|c) \parallel p_S(z|c)) \approx \frac{1}{N} \sum_{i=1}^{N} \log(\frac{p_T(z_i|c)}{p_S(z_i|c)}) = \frac{1}{N} \sum_{i=1}^{N} \log(\frac{d_T^i}{d_S^i}) = \frac{1}{N} \sum_{i=1}^{N} \pi_i,$$
(5)

where π is our designed metric to rank all regions in the same class (see Eq. 2).

Equation 3, 4, 5 bridge the theory and our proposed metric. By providing annotations for data that contribute the most to the KL divergence (large π value) for each category, the domain shift between the source and target domain could be reduced, making the generalization bound of ℓ_{test} tighter.

Class-Balanced Selection. With our proposed density difference metric, the labeling budget in each round can be equally divided into C categories. Nonetheless, we demonstrate that ADA can be achieved more efficiently.

We empirically observe the required budget for domain adaptation varies across classes, and thus it is inefficient to allocate the same budget to each class. For example, UDA methods can reach the performance close to fully supervised learning in some classes like "vegetarian" and "building". However, it is difficult for them to generalize to others like "train" or "sidewalk" (see Table 1). Hence, we propose a class-balanced selection to fully utilize the labeling budget.

Intuitively, we would like to spend more budgets on hard categories, while fewer selection budgets for already well-aligned ones. According to Eq. 5, the average π scores of a particular category can be regarded as an approximation of categorical KL divergence from the source domain to the target domain. Therefore, we use this indicator to balance the selection budget across classes. Suppose the total labeling budget of our density selection in each round is B^d pixels, for a certain category c, the allocated budget $B^{d,c}$ is as follows:

$$B^{d,c} = \frac{w_c}{\sum_{i \in [C]} w_i} \cdot B^d, \quad w_c = \sigma(D_{KL}(p_T(z|c) \parallel p_S(z|c))). \tag{6}$$

In the Eq. 6, σ is a normalization function to make the weighting term falls within a reasonable range. The equation ensures that more labels are assigned to classes with larger domain shift, while less are distributed to well-aligned ones.

3.3   Dynamic Scheduling Policy

As described in Sect. 1, prior ADA practices acquired labels considering both uncertainty and domain exploration equally over time. However, these practices, not aware of the defects of both methods under different stages, caused poor labeling efficiency, especially under low labeling budgets in earlier rounds.

As shown in the comparison of Fig. 2 (a, b, c) and (d, e, f), uncertainty indicator is less effective under severe domain shift but provides informative cues when more target domain labeled data are acquired. Conversely, since the design concept of domain exploration is to conquer the domain shift, it is effective in the earlier rounds in ADA but the growth rapidly slows down in the later rounds as shown in Table 2 (a, c). As a result, we develop a dynamic scheduling policy to benefit from both selection strategies and facilitate labeling efficiency.

Let B be the constant labeling budgets of each active selection round. In the n-th round, B can be divided to B_n^d for the density-aware approach and B_n^u for an uncertainty-based method. Our budget allocation arrangements for the two approaches are as follows:

$$\lambda = \alpha \cdot 2^{-\beta(n-1)},$$
$$B_n^d = \lambda B, \; B_n^u = (1 - \lambda)B,$$

(7)

where α is the balance coefficient between density-aware and the uncertainty-based method, and β is the decay rate of the labeling budget of the density-aware approach. The reason for using half decay is to reflect the observation of the rapid domain shift reduction, which is discussed in Sect. 4.4. The above formula ensures that the density-aware method is heavily relied on when the domain gaps are large in the first few rounds, and the uncertainty-based method is mainly used when the gaps are diminished in the later rounds. The effectiveness of the designed policy is shown in Table 2. In our implementation, we use the conventional softmax entropy [44] as the uncertainty measurement.

After deciding the labeling budgets B_n^d and B_n^u for density-aware and uncertainty methods with our dynamic scheduling policy in each round, we select regions with the two selection strategies. Then, we acquire ground truth labels for these regions and move them from the unlabeled target set D_T^U to the labeled target set D_T^L. Finally, we fine-tune the model with $D_S \cup D_T^L$ under full supervision and then execute the next active selection round.

4   Experiments

4.1   Experimental Settings

We elaborate the datasets, networks and active learning protocol in our experiments. The implementation details are reported in the supplementary material.

Datasets. We evaluated various active learning and domain adaptation methods on two widely-used domain adaptive semantic segmentation benchmarks: GTA5 → Cityscapes and SYNTHIA → Cityscapes. Cityscapes [11] is a real-world self-driving dataset containing 2975 labeled training images, 500 labeled validation images and 1225 test images. GTA5 [32] is a synthetic dataset consisting of 24966 annotated images, which shares 19 semantic categories with Cityscapes. SYNTHIA dataset [33] has 9400 synthetic annotated images, which share 16 semantic categories with Cityscapes. For a fair comparison, we evaluated all methods on the Cityscapes validation split.

Models. We used DeepLabV3+ [8], a semantic segmentation model, to evaluate all active learning strategies. To fairly compare our method with existing domain adaptation approaches, we report the results on both DeepLabV2 [6] and DeepLabV3+ [8]. The two models are based on the ResNet-101 [17] backbone.

Active Learning Protocol. For all experiments, we first adopted [42] to train the network with adversarial training as a warmup step. Then we performed N-round active selection comprising of the following steps: (1) Select a small number of regions with B pixels from the unlabeled Cityscapes training split D_T^U based on different active selection strategies. (2) Acquire the labels of these selected data and add them to D_T^L. (3) Fine-tune the model with $D_S \cup D_T^L$ in a supervised manner. In our setting, we choose $N = 5$, $B = 1\%$ of the total number of pixels in Cityscapes training split for both tasks.

4.2   Comparison with Active Learning Baselines

We compared **D^2ADA** with 8 other active learning strategies, including random region selection (RAND), uncertainty-based methods (CONF [44], MAR [44], and ENT [44]), hybrid methods (BADGE [2] and ReDAL [50]), and existing Active Domain Adaptation baselines (AADA [40] and CLUE [31]). The implementation details are described in the supplementary material.

The experimental results are shown in Fig. 3. In each subplot, the x-axis indicates the percentage of total target domain annotated points and the y-axis is the corresponding mIoU score achieved by the network.

The improvement of our proposed **D^2ADA** is significant compared to prior active learning strategies. Under 5% labels, **D^2ADA** surpassed the second-placed approach by 0.55 mIoU in GTA → Cityscapes, whereas the gap between the second and fourth place was only 0.2. Similarly, in SYNTHIA → Cityscapes, **D^2ADA** outperformed second place by 0.98 mIoU, while the difference between the second and fourth place was 0.39. Moreover, our method consistently prevailed over other methods by more than 0.5 mIoU under all cases. The results verify our method makes huge progress in active domain adaptation problems.

On the GTA5 → Cityscapes, our proposed **D^2ADA** surpasses uncertainty-based active learning methods (ENT, CONF, and MAR) for over 3% mIoU at the first round. This suggests that density as a selection metric clearly benefits ADA over uncertainty against domain shift in the first few rounds.

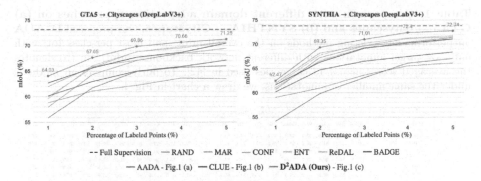

Fig. 3. Comparison with different active learning baselines. On two widely-used benchmarks, our D^2ADA outperforms 8 existing active learning strategies, including uncertainty-based methods (MAR, CONF, ENT), hybrid approaches (ReDAL, BADGE), and ADA practices. The result suggested that compared to prior domain exploration methods shown in Fig. 1 (a, b), our density-aware method (c) achieves significant improvement. More explanations and observations are reported in Sect. 4.2.

Compared with existing ADA methods (AADA, CLUE) or hybrid active learning approaches (ReDAL, BADGE), our method also achieves better performance under any tasks and with any amount of budget. On the two tasks, AADA even performs worse than random selection in the early stage. We conjecture that the selected biased samples might be inconsistent with the real target distribution as shown in Fig. 1 (a). As for CLUE, the methods with 5% annotations still performed worse than our method with merely 2.5x fewer labels (2% annotations) on the two tasks. We infer the main reason might be that they only considered choosing diverse target domain data but did not avoid the selection that were likely to appear in the source distribution as observed in Fig. 1 (b).

4.3   Comparison with Domain Adaptation Methods

We compared our D^2ADA with various domain adaptation methods, including unsupervised domain adaptations (UDAs) [10,20,23,26,37,42,43,49,53,55,59,60] weakly-supervised domain adaptation (WDA) [30], semi-supervised domain adaptation (SSDA) [9,48] and active domain adaptations (ADA) [28,36,51,52][1].

Table 1 (a, b) shows the result of GTA5 → Cityscapes and SYNTHIA → Cityscapes respectively. Compared with the current state-of-the-art UDA method [55], our method achieves an advantage of more than 10% mIoU with only 5% annotation effort. Furthermore, the performance of our method can achieve over 97% of the result of full supervision on target labeled data on both DeepLabV2 and DeepLabV3+ models. On the two tasks, compared with other label-efficient approaches (WDA, SSDA, ADA), our method not only performs

---

[1] We report the official results of the UDA, WDA, SSDA and ADA methods and train the DeepLabV2 and DeepLabV3+ ourselves.

Table 1. Comparison with different domain adaptation approaches on (a) GTA5 → Cityscapes and (b) SYNTHIA → Cityscapes. Our proposed D²ADA outperforms any existing methods on the overall mIoU and most per-class IoU with only 5% target annotations with different network backbones. mIoU* in (b) denotes the averaged scores across 13 categories used in [42]. To fairly compare all methods under the same number of annotations, we draw a chart in Fig. 4.

(a) GTA5 → Cityscapes

	Method	Road	SW	Build	Wall	Fence	Pole	TL	TS	Veg.	Terrain	Sky	PR	Rider	Car	Truck	Bus	Train	Motor	Bike	mIoU
UDA	AdaptSeg [42]	86.5	36.0	79.9	23.4	23.3	35.2	14.8	14.8	83.4	33.3	75.6	58.5	27.6	73.7	32.5	35.4	3.9	30.1	28.1	42.4
	ADVENT [43]	89.9	36.5	81.2	29.2	25.2	28.5	32.3	22.4	83.9	34.0	77.1	57.4	27.9	83.7	29.4	39.1	1.5	28.4	23.3	43.8
	SIMDA [49]	90.6	44.7	84.8	34.3	28.7	31.6	35.0	37.6	84.7	43.3	85.3	57.0	31.5	83.8	42.6	48.5	1.9	30.4	39.0	49.2
	CBST [60]	91.8	53.5	80.5	32.7	21.0	34.0	28.9	20.4	83.9	34.2	80.9	53.1	24.0	82.7	30.3	35.9	16.0	25.9	42.8	45.9
	CRST [61]	91.0	55.4	80.0	33.7	21.4	37.3	32.9	24.5	85.0	34.1	80.8	57.7	24.6	84.1	27.8	30.1	26.9	26.0	42.3	47.1
	LTIR [20]	92.9	55.0	85.3	34.2	31.1	34.9	40.7	34.0	85.2	40.1	87.1	61.0	31.1	82.5	32.3	42.9	0.3	36.4	46.1	50.2
	FDA [53]	92.5	53.3	82.4	26.5	27.6	36.4	40.6	38.9	82.3	39.8	78.0	62.6	34.4	84.9	34.1	53.1	16.9	27.7	46.4	50.5
	TPLD [37]	94.2	60.5	82.8	36.6	16.6	39.3	29.0	25.5	85.6	44.9	84.4	60.6	27.4	84.1	37.0	47.0	31.2	36.1	50.3	51.2
	IAST [26]	93.8	57.8	85.1	39.5	26.7	26.2	43.1	34.7	84.9	32.9	88.0	62.6	29.0	87.3	39.2	49.6	23.2	34.7	39.6	51.5
	ProDA [55]	87.8	56.0	79.7	46.3	44.8	45.6	53.5	53.5	88.6	45.2	82.1	70.7	39.2	88.8	45.5	59.4	1.0	48.9	56.4	57.5
	CAMDA [59]	91.3	46.0	84.5	34.4	29.7	32.6	35.8	36.4	84.5	43.2	83.0	60.0	32.2	83.2	35.0	46.7	0.0	33.7	42.2	49.2
	DPL-Dual [10]	92.8	54.4	86.2	41.6	32.7	36.4	49.0	34.0	85.8	41.3	86.0	63.2	34.2	87.2	39.3	44.5	18.7	42.6	43.1	53.3
	BAPA-Net [23]	94.4	61.0	88.0	26.8	39.9	38.3	46.1	55.3	87.8	46.1	89.4	68.8	40.0	90.2	60.4	59.0	0.0	45.1	54.2	57.4
WDA	WDA [30] (Point)	94.0	62.7	86.3	36.5	32.8	38.4	44.9	51.0	86.1	43.4	87.7	66.4	36.5	87.9	44.1	58.8	23.2	35.6	55.9	56.4
SSDA	ASS [48] (+50 city)	94.3	63.0	84.5	26.8	28.0	38.4	35.5	48.7	87.1	39.2	88.8	62.2	16.3	87.6	23.2	39.2	7.2	24.4	58.1	50.1
ADA	LabOR [36] (V2, 2.2%)	96.6	77.0	89.6	47.8	50.7	48.0	56.6	63.5	89.5	57.8	91.6	72.0	47.3	91.7	62.1	61.9	48.9	47.9	65.3	66.6
	MADA [28] (V3+, 5%)	95.1	69.8	88.5	43.3	48.7	45.7	53.3	59.2	89.1	46.7	91.5	73.9	50.1	91.2	60.6	56.9	48.4	51.6	68.7	64.9
	RIPU [51] (V3+, 5%)	97.0	77.3	90.4	54.6	53.2	47.7	55.9	64.1	90.2	59.2	93.2	75.0	54.8	92.7	73.0	79.7	68.9	55.5	70.3	71.2
	D²ADA (V2, 5%)	96.3	73.6	89.3	50.0	52.3	48.0	56.9	64.7	89.3	53.9	92.3	73.9	52.9	91.8	69.7	78.9	62.7	57.7	71.1	69.7
	D²ADA (V3+, 5%)	97.0	77.8	90.0	46.0	55.0	52.7	58.7	65.8	90.4	58.9	92.1	75.7	54.4	92.3	69.0	78.0	68.5	59.1	72.3	71.3
Target Only	DeepLabV2 [6]	97.4	79.5	90.3	51.1	52.4	49.0	57.5	68.0	90.5	58.1	93.1	75.1	53.9	92.7	72.0	80.2	65.0	58.1	71.1	71.3
	DeepLabV3+ [8]	97.6	81.3	91.1	49.8	57.6	53.8	59.6	69.1	91.2	60.5	94.4	76.7	55.6	93.3	75.8	79.9	72.9	57.7	72.2	73.2

(b) SYNTHIA → Cityscapes

	Method	Road	SW	Build	Wall*	Fence*	Pole*	TL	TS	Veg.	Sky	PR	Rider	Car	Bus	Motor	Bike	mIoU	mIoU*
UDA	AdaptSeg [42]	79.2	37.2	78.8	-	-	-	9.9	10.5	78.2	80.5	53.5	19.6	67.0	29.5	21.6	31.3	-	45.9
	ADVENT [43]	85.6	42.2	79.7	8.7	0.4	25.9	5.4	8.1	80.4	84.1	57.9	23.8	73.3	36.4	14.2	33.0	41.2	48.0
	SIMDA[49]	83.0	44.0	80.3	-	-	-	17.1	15.8	80.5	81.8	59.9	33.1	70.2	37.3	28.5	45.8	-	52.1
	CBST [60]	68.0	29.9	76.3	10.8	1.4	33.9	22.8	29.5	77.6	78.3	60.6	28.3	81.6	23.5	18.8	39.8	42.6	48.9
	CRST [61]	67.7	32.2	73.9	10.7	1.6	37.4	22.2	31.2	80.8	80.5	60.8	29.1	82.8	25.0	19.4	45.3	43.8	50.1
	LTIR [20]	92.6	53.2	79.2	-	-	-	1.6	7.5	78.6	84.4	52.6	20.0	82.1	34.8	14.6	39.4	-	49.3
	FDA [53]	79.3	35.0	73.2	-	-	-	19.9	24.0	61.7	82.6	61.4	31.1	83.9	40.8	38.4	51.1	-	52.5
	TPLD [37]	80.9	44.3	82.2	19.9	0.3	40.6	20.5	30.1	77.2	80.9	60.6	25.5	84.8	41.1	24.7	43.7	47.3	53.5
	IAST [26]	81.9	41.5	83.3	17.7	4.6	32.3	30.9	28.8	83.4	85.0	65.5	30.8	86.5	38.2	33.1	52.7	49.8	57.0
	ProDA [55]	87.8	45.7	84.6	37.1	0.6	44.0	54.6	37.0	88.1	84.4	74.2	24.3	88.2	51.1	40.5	45.6	55.5	62.0
	CAMDA [59]	82.5	42.2	81.3	-	-	-	18.3	15.9	80.6	83.5	61.4	33.2	72.9	39.3	26.6	43.9	-	52.4
	DPL-Dual [10]	87.5	45.7	82.8	13.3	0.6	33.2	22.0	20.1	83.1	86.0	56.6	21.9	83.1	40.3	29.8	45.7	47.0	54.2
	BAPA-Net [23]	91.7	53.8	83.9	22.4	0.8	34.9	30.5	42.8	86.6	88.2	66.0	34.1	86.6	51.3	29.4	50.5	53.3	61.2
WDA	WDA [30] (Point)	94.9	63.2	85.0	27.3	24.2	34.9	37.3	50.8	84.4	88.2	60.6	36.3	86.4	43.2	36.5	61.3	57.2	63.7
SSDA	ASS [48] (+50 city)	94.1	63.9	87.6	-	-	-	18.1	37.1	87.5	89.7	64.6	37.0	87.4	38.6	23.2	59.6	-	60.7
ADA	MADA [28] (V3+, 5%)	96.5	74.6	88.8	45.9	43.8	46.7	52.4	60.5	89.7	92.2	74.1	51.2	90.9	60.3	52.4	69.4	68.1	73.3
	RIPU[51] (V3+, 5%)	97.0	78.9	89.9	47.2	50.7	48.5	55.2	63.9	91.1	93.0	74.4	54.1	92.9	79.9	55.3	71.0	71.4	76.7
	D²ADA (V2, 5%)	96.4	76.3	89.1	42.5	47.7	48.0	55.6	66.5	89.5	91.7	75.1	55.2	91.4	77.0	58.0	71.8	70.6	76.3
	D²ADA (V3+, 5%)	96.7	76.8	90.3	48.7	51.1	54.2	58.3	68.0	90.4	93.4	77.4	56.4	92.5	77.5	58.9	73.3	72.7	77.7
Target Only	DeepLabV2 [6]	97.4	79.5	90.3	51.1	52.4	49.0	57.5	68.0	90.5	93.1	75.1	53.9	92.7	80.2	58.1	71.1	72.5	77.5
	DeepLabV3+ [8]	97.6	81.3	91.1	49.8	57.6	53.8	59.6	69.1	91.2	94.4	76.7	55.6	93.3	79.9	57.7	72.2	73.8	78.4

better on the overall mIoU score but also gains improvement on small objects, such as "traffic sign (TS)" or "bicycle", and difficult categories, like "fence" and "pole".

Since the labeling budgets used in prior label-efficient methods differ, Fig. 4 makes a fair comparison of these methods. The x-axis indicates the percentage of total target domain annotated points and the y-axis is the corresponding mIoU score achieved by the network. Obviously, our approach achieves higher performance over existing methods with the same number of labeled pixels.

Furthermore, according to statistics, our selected regions contain only 3.1 different categories on average, with 50% regions less than two categories. Thus, compared with prior works selecting scattered pixels [36] or small regions [51], our method, selecting superpixel-level regions, are more friendly for annotators to label (by drawing few polygons) and would require much less labeling effort.

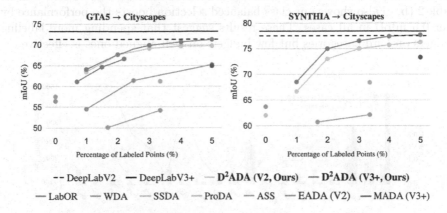

Fig. 4. Comparison with various label-efficient domain adaptation methods. We compare our method with the state-of-the-art UDA method, ProDA [55] as well as various label-efficient approaches, including WDA [30], SSDA [9], ASS [48], MADA [28], LabOR [36] and EADA [52]. On two tasks and networks, our proposed D^2ADA achieves the best result under the same number of labels.

Table 2. Ablation studies. The three columns on the right show the model performance when 1%, 3%, and 5% target annotations are acquired by different active selection strategies. The results validate the effectiveness of our density-aware method, class-balanced selection, and dynamic scheduling policy as the discussion in Sect. 4.4. It also demonstrates uncertainty perform worse with low labeling budgets (1%, 3%).

	Components				mIoU wrt. labels		
	Uncertainty	Density Aware	Class Balance	Dynamic Scheduling	1%	3%	5%
(a)	✓				59.97	68.79	70.70
(b)		✓			62.75	68.81	69.82
(c)		✓	✓		**64.03**	69.38	70.69
(d)	✓	✓	✓		63.30	69.66	71.15
(e)	✓	✓	✓	✓	**64.03**	**69.86**	**71.25**

4.4 Ablation Studies

We also conduct experiments to validate the effectiveness of all the proposed components with DeepLabV3+ backbone on GTA5 → Cityscapes. Extensive experiments about hyper-parameter settings and other in-depth analyses are left in the supplementary material.

Table 2 demonstrates the effectiveness of our density-aware selection, class-balanced selection, and dynamic scheduling policy. First, from the comparison of (a, b), with low budgets (1%, 3%), our density-aware selection solves the defect of uncertainty metrics, echoing our motivations stated in Fig. 2.

Second, as shown in Fig. 5, due to the class-balanced selection, our method queries more labels on minor classes but fewer labels on well-learned classes. Table 2 (b, c) also shows our class-balanced selection boosts the performance by over 0.5 mIoU in all cases. These results suggest that spending more labeling budgets on hard categories but few budgets on well-aligned ones is effective.

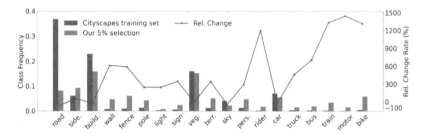

Fig. 5. The label distribution of our selected regions. Following [51], we draw histograms to compare the label distribution of the original target domain data (blue) and our 5% selected regions (red). The purple line chart shows the relative changes from the target dataset to our selection. Evidently, our class-balanced selection acquires more labels on minor classes, like 12× more labels on "train", "motor", and "bike". (Color figure online)

Third, as shown in Table 2 (a, e) and (c, e), given any budget, our dynamic scheduling policy surpasses the density or uncertainty methods alone. Also, the comparison of (d) and (e) validates our policy outperforms the "equally distributed" budget allocation under all situations. The results indicate that our dynamic scheduling strategy can take advantage of the two methods, further improving the labeling efficiency.

5   Conclusion

We propose **D²ADA**, a novel active domain adaptation method for semantic segmentation. By labeling samples with high probability density in the target domain yet with low probability density in the source domain, models can conquer the domain shift with limited annotation effort. Furthermore, we design the first dynamic scheduling policy to balance labeling budgets between domain exploration and model uncertainty for the ADA problem.

Acknowledgement. This work was supported in part by the Ministry of Science and Technology, Taiwan, under Grant MOST 110-2634-F-002-051, Mobile Drive Technology Co., Ltd (MobileDrive), and Industrial Technology Research Institute (ITRI). We are grateful to the National Center for High-performance Computing.

References

1. Achanta, R., Shaji, A., Smith, K., Lucchi, A., Fua, P., Süsstrunk, S.: SLIC superpixels compared to state-of-the-art superpixel methods. IEEE Trans. Pattern Anal. Mach. Intell. **34**(11), 2274–2282 (2012)
2. Ash, J.T., Zhang, C., Krishnamurthy, A., Langford, J., Agarwal, A.: Deep batch active learning by diverse, uncertain gradient lower bounds. In: ICLR (2020)
3. Casanova, A., Pinheiro, P.O., Rostamzadeh, N., Pal, C.J.: Reinforced active learning for image segmentation. In: International Conference on Learning Representations (2020). https://openreview.net/forum?id=SkgC6TNFvr
4. Chang, W.L., Wang, H.P., Peng, W.H., Chiu, W.C.: All about structure: adapting structural information across domains for boosting semantic segmentation. In: Proceedings of the IEEE/CVF Conference on Computer Vision and Pattern Recognition, pp. 1900–1909 (2019)
5. Chen, L.C., Papandreou, G., Kokkinos, I., Murphy, K., Yuille, A.L.: DeepLab: semantic image segmentation with deep convolutional nets, Atrous convolution, and fully connected CRFs. IEEE Trans. Pattern Anal. Mach. Intell. **40**(4), 834–848 (2017)
6. Chen, L.C., Papandreou, G., Kokkinos, I., Murphy, K.P., Yuille, A.L.: Semantic image segmentation with deep convolutional nets and fully connected CRFs. CoRR abs/1412.7062 (2015)
7. Chen, L.C., Papandreou, G., Schroff, F., Adam, H.: Rethinking Atrous convolution for semantic image segmentation. arXiv preprint arXiv:1706.05587 (2017)
8. Chen, L.-C., Zhu, Y., Papandreou, G., Schroff, F., Adam, H.: Encoder-decoder with atrous separable convolution for semantic image segmentation. In: Ferrari, V., Hebert, M., Sminchisescu, C., Weiss, Y. (eds.) ECCV 2018. LNCS, vol. 11211, pp. 833–851. Springer, Cham (2018). https://doi.org/10.1007/978-3-030-01234-2_49
9. Chen, S., Jia, X., He, J., Shi, Y., Liu, J.: Semi-supervised domain adaptation based on dual-level domain mixing for semantic segmentation. In: Proceedings of the IEEE/CVF Conference on Computer Vision and Pattern Recognition, pp. 11018–11027 (2021)
10. Cheng, Y., Wei, F., Bao, J., Chen, D., Wen, F., Zhang, W.: Dual path learning for domain adaptation of semantic segmentation. In: Proceedings of the IEEE/CVF International Conference on Computer Vision, pp. 9082–9091 (2021)
11. Cordts, M., et al.: The cityscapes dataset for semantic urban scene understanding. In: Proceedings of the IEEE Conference on Computer Vision and Pattern Recognition, pp. 3213–3223 (2016)
12. Du, L., et al.: SSF-DAN: separated semantic feature based domain adaptation network for semantic segmentation. In: Proceedings of the IEEE/CVF International Conference on Computer Vision, pp. 982–991 (2019)
13. Fu, B., Cao, Z., Wang, J., Long, M.: Transferable query selection for active domain adaptation. In: Proceedings of the IEEE/CVF Conference on Computer Vision and Pattern Recognition, pp. 7272–7281 (2021)

14. Gal, Y., Ghahramani, Z.: Dropout as a bayesian approximation: Representing model uncertainty in deep learning. In: International Conference on Machine Learning, pp. 1050–1059 (2016)
15. Gal, Y., Islam, R., Ghahramani, Z.: Deep Bayesian active learning with image data. In: International Conference on Machine Learning, pp. 1183–1192 (2017)
16. Hastie, T., Tibshirani, R., Friedman, J.: The Elements of Statistical Learning: Data Mining, Inference, and Prediction. SSS, vol. 2. Springer, New York (2009). https://doi.org/10.1007/978-0-387-84858-7
17. He, K., Zhang, X., Ren, S., Sun, J.: Deep residual learning for image recognition. In: Proceedings of the IEEE Conference on Computer Vision and Pattern Recognition, pp. 770–778 (2016)
18. Hershey, J.R., Olsen, P.A.: Approximating the Kullback Leibler divergence between gaussian mixture models. In: 2007 IEEE International Conference on Acoustics, Speech and Signal Processing-ICASSP 2007, vol. 4, pp. 4-317. IEEE (2007)
19. Kasarla, T., Nagendar, G., Hegde, G.M., Balasubramanian, V., Jawahar, C.: Region-based active learning for efficient labeling in semantic segmentation. In: 2019 IEEE Winter Conference on Applications of Computer Vision (WACV), pp. 1109–1117. IEEE (2019)
20. Kim, M., Byun, H.: Learning texture invariant representation for domain adaptation of semantic segmentation. In: Proceedings of the IEEE/CVF Conference on Computer Vision and Pattern Recognition, pp. 12975–12984 (2020)
21. Kirsch, A., van Amersfoort, J., Gal, Y.: BatchBald: efficient and diverse batch acquisition for deep Bayesian active learning. In: Advances in Neural Information Processing Systems, pp. 7026–7037 (2019)
22. Lin, G., Milan, A., Shen, C., Reid, I.: RefineNet: multi-path refinement networks for high-resolution semantic segmentation. In: Proceedings of the IEEE Conference on Computer Vision and Pattern recognition, pp. 1925–1934 (2017)
23. Liu, Y., Deng, J., Gao, X., Li, W., Duan, L.: BAPA-Net: boundary adaptation and prototype alignment for cross-domain semantic segmentation. In: Proceedings of the IEEE/CVF International Conference on Computer Vision, pp. 8801–8811 (2021)
24. Luo, Y., Zheng, L., Guan, T., Yu, J., Yang, Y.: Taking a closer look at domain shift: category-level adversaries for semantics consistent domain adaptation. In: Proceedings of the IEEE/CVF Conference on Computer Vision and Pattern Recognition, pp. 2507–2516 (2019)
25. Mackowiak, R., Lenz, P., Ghori, O., Diego, F., Lange, O., Rother, C.: CEREALS - cost-effective region-based active learning for semantic segmentation. In: British Machine Vision Conference 2018, BMVC 2018, Newcastle, UK, 3–6 September 2018, p. 121. BMVA Press (2018). https://bmvc2018.org/contents/papers/0437.pdf
26. Mei, K., Zhu, C., Zou, J., Zhang, S.: Instance adaptive self-training for unsupervised domain adaptation. In: Vedaldi, A., Bischof, H., Brox, T., Frahm, J.-M. (eds.) ECCV 2020. LNCS, vol. 12371, pp. 415–430. Springer, Cham (2020). https://doi.org/10.1007/978-3-030-58574-7_25
27. Nguyen, A.T., Tran, T., Gal, Y., Torr, P., Baydin, A.G.: KL guided domain adaptation. In: International Conference on Learning Representations (2022). https://openreview.net/forum?id=0JzqUlIVVDd
28. Ning, M., et al.: Multi-anchor active domain adaptation for semantic segmentation. In: Proceedings of the IEEE/CVF International Conference on Computer Vision, pp. 9112–9122 (2021)

29. Ovadia, Y., et al.: Can you trust your model's uncertainty? Evaluating predictive uncertainty under dataset shift. Adv. Neural. Inf. Process. Syst. **32**, 1–11 (2019)
30. Paul, S., Tsai, Y.-H., Schulter, S., Roy-Chowdhury, A.K., Chandraker, M.: Domain adaptive semantic segmentation using weak labels. In: Vedaldi, A., Bischof, H., Brox, T., Frahm, J.-M. (eds.) ECCV 2020. LNCS, vol. 12354, pp. 571–587. Springer, Cham (2020). https://doi.org/10.1007/978-3-030-58545-7_33
31. Prabhu, V., Chandrasekaran, A., Saenko, K., Hoffman, J.: Active domain adaptation via clustering uncertainty-weighted embeddings. In: Proceedings of the IEEE/CVF International Conference on Computer Vision, pp. 8505–8514 (2021)
32. Richter, S.R., Vineet, V., Roth, S., Koltun, V.: Playing for data: ground truth from computer games. In: Leibe, B., Matas, J., Sebe, N., Welling, M. (eds.) ECCV 2016. LNCS, vol. 9906, pp. 102–118. Springer, Cham (2016). https://doi.org/10.1007/978-3-319-46475-6_7
33. Ros, G., Sellart, L., Materzynska, J., Vazquez, D., Lopez, A.M.: The Synthia dataset: a large collection of synthetic images for semantic segmentation of urban scenes. In: Proceedings of the IEEE Conference on Computer Vision and Pattern Recognition, pp. 3234–3243 (2016)
34. Roth, D., Small, K.: Margin-based active learning for structured output spaces. In: Fürnkranz, J., Scheffer, T., Spiliopoulou, M. (eds.) ECML 2006. LNCS (LNAI), vol. 4212, pp. 413–424. Springer, Heidelberg (2006). https://doi.org/10.1007/11871842_40
35. Sener, O., Savarese, S.: Active learning for convolutional neural networks: a core-set approach. In: International Conference on Learning Representations (2018). https://openreview.net/forum?id=H1aIuk-RW
36. Shin, I., Kim, D.J., Cho, J.W., Woo, S., Park, K., Kweon, I.S.: Labor: labeling only if required for domain adaptive semantic segmentation. In: Proceedings of the IEEE/CVF International Conference on Computer Vision, pp. 8588–8598 (2021)
37. Shin, I., Woo, S., Pan, F., Kweon, I.S.: Two-phase pseudo label densification for self-training based domain adaptation. In: Vedaldi, A., Bischof, H., Brox, T., Frahm, J.-M. (eds.) ECCV 2020. LNCS, vol. 12358, pp. 532–548. Springer, Cham (2020). https://doi.org/10.1007/978-3-030-58601-0_32
38. Siddiqui, Y., Valentin, J., Nießner, M.: ViewAL: active learning with viewpoint entropy for semantic segmentation. In: Proceedings of the IEEE/CVF Conference on Computer Vision and Pattern Recognition, pp. 9433–9443 (2020)
39. Singh, A., et al.: Improving semi-supervised domain adaptation using effective target selection and semantics. In: Proceedings of the IEEE/CVF Conference on Computer Vision and Pattern Recognition, pp. 2709–2718 (2021)
40. Su, J.C., Tsai, Y.H., Sohn, K., Liu, B., Maji, S., Chandraker, M.: Active adversarial domain adaptation. In: Proceedings of the IEEE/CVF Winter Conference on Applications of Computer Vision, pp. 739–748 (2020)
41. Sun, K., Xiao, B., Liu, D., Wang, J.: Deep high-resolution representation learning for human pose estimation. In: Proceedings of the IEEE/CVF Conference on Computer Vision and Pattern Recognition, pp. 5693–5703 (2019)
42. Tsai, Y.H., Hung, W.C., Schulter, S., Sohn, K., Yang, M.H., Chandraker, M.: Learning to adapt structured output space for semantic segmentation. In: Proceedings of the IEEE Conference on Computer Vision and Pattern Recognition, pp. 7472–7481 (2018)
43. Vu, T.H., Jain, H., Bucher, M., Cord, M., Pérez, P.: Advent: adversarial entropy minimization for domain adaptation in semantic segmentation. In: Proceedings of the IEEE/CVF Conference on Computer Vision and Pattern Recognition, pp. 2517–2526 (2019)

44. Wang, D., Shang, Y.: A new active labeling method for deep learning. In: 2014 International Joint Conference on Neural Networks (IJCNN), pp. 112–119 (2014). https://doi.org/10.1109/IJCNN.2014.6889457
45. Wang, J., et al.: Deep high-resolution representation learning for visual recognition. IEEE Trans. Pattern Anal. Mach. Intell. **43**, 3349–3364 (2020)
46. Wang, K., Zhang, D., Li, Y., Zhang, R., Lin, L.: Cost-effective active learning for deep image classification. IEEE Trans. Circuits Syst. Video Technol. **27**(12), 2591–2600 (2016)
47. Wang, Y., Peng, J., Zhang, Z.: Uncertainty-aware pseudo label refinery for domain adaptive semantic segmentation. In: Proceedings of the IEEE/CVF International Conference on Computer Vision, pp. 9092–9101 (2021)
48. Wang, Z., et al.: Alleviating semantic-level shift: a semi-supervised domain adaptation method for semantic segmentation. In: Proceedings of the IEEE/CVF Conference on Computer Vision and Pattern Recognition Workshops, pp. 936–937 (2020)
49. Wang, Z., et al.: Differential treatment for stuff and things: a simple unsupervised domain adaptation method for semantic segmentation. In: Proceedings of the IEEE/CVF Conference on Computer Vision and Pattern Recognition, pp. 12635–12644 (2020)
50. Wu, T.H., et al.: ReDAL: region-based and diversity-aware active learning for point cloud semantic segmentation. In: Proceedings of the IEEE/CVF International Conference on Computer Vision, pp. 15510–15519 (2021)
51. Xie, B., Yuan, L., Li, S., Liu, C.H., Cheng, X.: Towards fewer annotations: active learning via region impurity and prediction uncertainty for domain adaptive semantic segmentation. In: Proceedings of the IEEE/CVF Conference on Computer Vision and Pattern Recognition, pp. 8068–8078 (2022)
52. Xie, B., Yuan, L., Li, S., Liu, C.H., Cheng, X., Wang, G.: Active learning for domain adaptation: an energy-based approach. In: Thirty-Sixth AAAI Conference on Artificial Intelligence (AAAI-2022) (2022)
53. Yang, Y., Soatto, S.: FDA: Fourier domain adaptation for semantic segmentation. In: Proceedings of the IEEE/CVF Conference on Computer Vision and Pattern Recognition, pp. 4085–4095 (2020)
54. Yuan, Y., Chen, X., Wang, J.: Segmentation transformer: object-contextual representations for semantic segmentation. In: Vedaldi, A., Bischof, H., Brox, T., Frahm, J.-M. (eds.) ECCV 2020. LNCS, vol. 12351, pp. 173–190. Springer, Cham (2020). https://doi.org/10.1007/978-3-030-58539-6_11
55. Zhang, P., Zhang, B., Zhang, T., Chen, D., Wang, Y., Wen, F.: Prototypical pseudo label denoising and target structure learning for domain adaptive semantic segmentation. In: Proceedings of the IEEE/CVF Conference on Computer Vision and Pattern Recognition, pp. 12414–12424 (2021)
56. Zhao, H., Shi, J., Qi, X., Wang, X., Jia, J.: Pyramid scene parsing network. In: Proceedings of the IEEE Conference on Computer Vision and Pattern Recognition, pp. 2881–2890 (2017)
57. Zheng, Z., Yang, Y.: Rectifying pseudo label learning via uncertainty estimation for domain adaptive semantic segmentation. Int. J. Comput. Vision **129**(4), 1106–1120 (2021)
58. Zhou, F., Shui, C., Yang, S., Huang, B., Wang, B., Chaib-draa, B.: Discriminative active learning for domain adaptation. Knowl. Based Syst. **222**, 106986 (2021)
59. Zhou, Q., et al.: Context-aware mixup for domain adaptive semantic segmentation. arXiv preprint arXiv:2108.03557 (2021)

60. Zou, Y., Yu, Z., Vijaya Kumar, B.V.K., Wang, J.: Unsupervised domain adaptation for semantic segmentation via class-balanced self-training. In: Ferrari, V., Hebert, M., Sminchisescu, C., Weiss, Y. (eds.) ECCV 2018. LNCS, vol. 11207, pp. 297–313. Springer, Cham (2018). https://doi.org/10.1007/978-3-030-01219-9_18

61. Zou, Y., Yu, Z., Liu, X., Kumar, B., Wang, J.: Confidence regularized self-training. In: Proceedings of the IEEE/CVF International Conference on Computer Vision, pp. 5982–5991 (2019)

Learning Quality-aware Dynamic Memory for Video Object Segmentation

Yong Liu[1], Ran Yu[1], Fei Yin[1], Xinyuan Zhao[2], Wei Zhao[2], Weihao Xia[3], and Yujiu Yang[1(✉)]

[1] Tsinghua Shenzhen International Graduate School, Tsinghua University, Beijing, China
{liu-yong20,yu-r19}@mails.tsinghua.edu.cn, yang.yujiu@sz.tsinghua.edu.cn
[2] Huawei Technologies, Shenzhen, China
[3] University College London, London, UK

Abstract. Recently, several spatial-temporal memory-based methods have verified that storing intermediate frames and their masks as memory are helpful to segment target objects in videos. However, they mainly focus on better matching between the current frame and the memory frames without explicitly paying attention to the quality of the memory. Therefore, frames with poor segmentation masks are prone to be memorized, which leads to a segmentation mask error accumulation problem and further affect the segmentation performance. In addition, the linear increase of memory frames with the growth of frame number also limits the ability of the models to handle long videos. To this end, we propose a **Q**uality-aware **D**ynamic **M**emory **N**etwork (QDMN) to evaluate the segmentation quality of each frame, allowing the memory bank to selectively store accurately segmented frames to prevent the error accumulation problem. Then, we combine the segmentation quality with temporal consistency to dynamically update the memory bank to improve the practicability of the models. Without any bells and whistles, our QDMN achieves new state-of-the-art performance on both DAVIS and YouTube-VOS benchmarks. Moreover, extensive experiments demonstrate that the proposed Quality Assessment Module (QAM) can be applied to memory-based methods as generic plugins and significantly improves performance. Our source code is available at https://github.com/workforai/QDMN.

Keywords: Video object segmentation · Memory bank

Y. Liu—This work was done during an internship at Huawei Technologies.

Supplementary Information The online version contains supplementary material available at https://doi.org/10.1007/978-3-031-19818-2_27.

1   Introduction

Given a video and the first frame's annotations of single or multiple objects, semi-supervised video object segmentation (Semi-VOS or One-shot VOS) aims at segmenting these objects in subsequent frames. Semi-VOS is one of the most challenging tasks in computer vision with many potential applications, including interactive video editing, augmented reality, and autonomous driving.

Unlike other segmentation tasks that aim to look for the relationship between features and specific categories, the critical problem of Semi-VOS lies in how to make full use of the spatial-temporal information to recognize the target objects. Consequently, the methods that perform matching with historical reference frames have received tremendous attention in recent years. Some works [40,50,52] utilize the first frame and the previous adjacent frame as referencesPer Springer style, both city and country names must be present in affiliations. Accordingly, we have inserted the city and country name " Guayaquil, Ecuador" in affiliation. Please check and confirm the inserted city and country names are correct. If not, please provide us with the correct city and country.. Due to limited reference information, these approaches tend to fail miserably under challenging scenarios, e.g., the target objects disappear for a while or are drastically deformed. To excavate more information, the Space-Time Memory Network (STM) [29] utilizes a memory network to memorize intermediate frames and their segmentation masks as references, which has been proved effective and has served as the current mainstream framework. Many approaches [6,7,14,21,35,36,42,46] further develop the feature extraction and memory readout process of STM and have achieved excellent performance.

However, these methods mainly focus on optimizing the matching process while ignoring the impact of the matching target, i.e., memory bank, on the segmentation results. Specifically, previous methods select memory frames in a straightforward way, i.e., storing at fixed frame intervals. This approach has two weaknesses: (1) Frames with poor segmentation results may be memorized and provide an erroneous reference for subsequent frames, which leads to an error accumulation problem. As shown in the first row of Fig. 1, if there are inaccurately segmented masks in the memory bank, the segmentation quality of

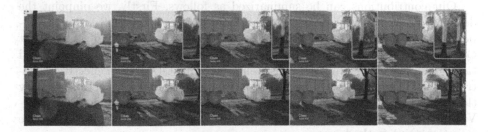

Fig. 1. Visual comparison of memory frames of different qualities. The first row shows the memory frames of MiVOS [6]. The second row shows the memory frames of our method. The yellow box area illustrates the error accumulation.

subsequent frames will be greatly degraded. Such an observation inspires us to pay more attention to the design of the memory bank. Since the matching-based approaches rely on a memory bank to identify the target objects, the memory bank's quality (especially the correctness) is very important. (2) In existing methods, the size of the memory bank would infinitely expand with the growth of frame number, which makes the models incapable of handling long videos and greatly limits their practicality.

Therefore, the way of designing the memory bank is a significant issue for spatial-temporal memory-based methods. Generally speaking, we believe that the design of the memory bank should meet the following principles: (1) *Accuracy: In a one-shot scenario, the memory bank should be composed of the annotated frame and frames that are segmented as accurately as possible to obtain correct supervision information.* (2) *Temporal consistency: Considering the continuity of motion, the state of objects in adjacent frames tends to be similar. In other words, the masks of adjacent frames are of great reference to the current frame.* Based on these two principles, we can selectively store frames with more reference information as memory and dynamically update the memory bank to handle videos of arbitrary length.

To this end, we propose a Quality-aware Dynamic Memory Network (QDMN), which introduces a simple but effective structure called Quality Assessment Module (QAM) in this paper to evaluate each frame's segmentation result and decide whether a frame can be added to the memory bank as a reference. Being aware of the segmentation quality limits the impact of noise and provides the accuracy credentials for dynamically updating the memory bank. Besides, since the objects in adjacent frames share a similar status to the current target, we introduce a temporal regularization to penalize the outdated memory. Extensive experiments demonstrate that the dynamic updating strategy of the memory bank designed according to the principles of accuracy and temporal consistency is reasonable and effective. By designing a high-quality memory bank and introducing temporal consistency, our method achieves new state-of-the-art performance on both DAVIS [33] and Youtube-VOS [47] benchmark without any bells and whistles. Furthermore, we also verify that memory-based methods can gain significant improvement by simply applying our QAM as a generic plugin for video object segmentation tasks.

Our contributions can be summarized as follows. Firstly, we pinpoint the design of the memory bank as the Achilles heel of the Semi-VOS task and propose the strategy for designing a high-quality memory bank. Secondly, we present QDMN for Semi-VOS, which can selectively memorize high-quality frames and take advantage of the temporal consistency. Thirdly, QDMN can effectively control the number of memory frames to avoid memory explosion. Experiments show that our method surpasses the existing methods on both DAVIS and YouTube-VOS datasets. Furthermore, QAM can be used as a generic plugin to improve memory-based methods.

2   Related Work

Propagation-Based Methods. Propagation-based methods [1,8,9,15,39,41, 49] treat semi-supervised video object segmentation as a mask propagation task. MaskTrack [31] concatenates the previous adjacent frame's segmentation mask with the current image as input and online fine-tunes the network. AGSS-VOS [24] proposes an attention-guided decoder to combine the instance-specific branch and instance-agnostic branch. Based on mask confidence and mask concentration, SAT [3] selectively propagates the entire image or local region to the next frame. The propagation-based method takes advantage of the strong prior provided by the previous adjacent frame. It can better deal with the appearance change of the target object, but it has fatal shortcomings in the problem of occlusion and error accumulation.

Detection-Based Methods. Detection-based methods divide the Semi-VOS task into three subtasks: detection, tracking and segmentation. DyeNet [20] utilizes RPN [34] to generate proposals and applies the re-identification module to perform matching. PReMVOS [26] uses Mask RCNN [12] to obtain coarse masks and performs optical flow, re-identification to achieve good performance. Huang *et al.* [16] and Sun *et al.* [38] integrate segmentation into tracking with a dynamic template bank. Detection-based methods rely heavily on the detectors, which dramatically limits the performance of such methods.

Matching-Based Methods. Matching-based methods perform matching between reference frames and the current frame to identify target objects, which has raised great attention for excellent performance and robustness. PML [4] proposes a pixel-level embedding network with the nearest neighbor classifier. FEELVOS [40] and CFBI [50] perform global and local matching with the first frame and the previous adjacent frame, respectively. AOT [51] associates multiple target objects into the same embedding space by employing an identification mechanism. STM [29] leverages the memory network to memorize intermediate frames as references, which has been proved effective and has served as the current mainstream framework. Based on STM, KMN [35] and RMNet [46] propose to perform local-to-local matching instead of non-local. SwiftNet [42] and AFB-URR [23] reduce memory duplication redundancy by calculating the similarity between query and memory. LCM [14] emphasizes the importance of the first frame and the previous adjacent frame. STCN [7] improves the feature extraction and performs reasonable matching by decoupling the image and masks. Following the memory-based idea, there are still many variants of STM, such as JOINT [27], EGMN [25], MiVOS [6], DMN-AOA [22], HMMN [36], and so on.

Although these methods have achieved great performance, they mainly focus on better matching the current frame with the memory frames. In other words, previous works dedicate to optimizing the matching process while neglecting the importance of matching with the correct object. Besides, they do not take into account that the size of the memory bank grows linearly with the length of the video, which greatly impacts the application of the models in real scenarios due to the hardware memory limitation.

3   Method

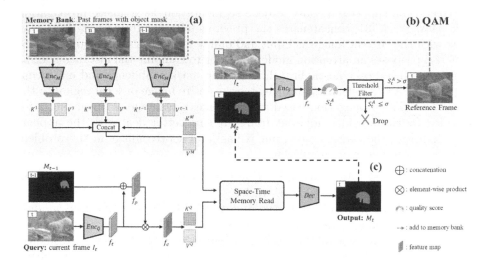

Fig. 2. Overview of QDMN. (a) is the feature extraction of the reference frames in the memory bank. (b) QAM is the module used to evaluate whether the current frame can be added to the memory bank. (c) is the pipeline for predicting the segmentation result of the current frame I_t.

3.1   Overview

The overall architecture of our QDMN is shown in Fig. 2. Similar to STM [29], during video processing, the current frame (t-th frame) is considered as the query, and the past reference frames with segmentation masks are considered as the memory. The query and memory are encoded into pairs of key and value maps through visual encoders and corresponding convolution layers. To highlight the temporal consistency of video, the query feature f_t is first enhanced with the prior mask to obtain the enhanced feature f_e. Then the enhanced feature is encoded into pairs of key K^Q and value V^Q through corresponding convolution layers. The Space-Time Memory Read block performs pixel-level matching between K^Q and the memory key K^M. The relative matching similarity is used to address the memory value V^M, and the corresponding values are combined to the decoder for segmentation. Finally, the Quality Assessment Module (QAM) evaluates the quality of the segmentation result and decides whether the query frame can become a memory frame.

Score: 0.64   Score: 0.92   Score: 0.39   Score: 0.81

Fig. 3. Illustrations of segmentation masks with different quality scores. The three rows represent the ground truth, segmentation results, and the quality scores predicted by QAM, respectively.

3.2 Quality Assessment Module

Designing the memory bank is a significant issue for memory network-based methods. For existing strategy, frames with erroneous masks may be memorized, which leads to an error accumulation problem. To alleviate this problem and ensure the accuracy of the memory bank, inspired by [17,18], we propose the Quality Assessment Module (QAM) to evaluate the segmentation quality and decide whether a frame can be added to the memory bank as a reference.

QAM is a simple structure but effective module composed of a score encoder, four convolution layers, and two MLP layers. It takes the query image $\mathbf{I}_t$ and its segmentation mask $\mathbf{M}_t$ as input and outputs the predicted quality scores. Since the feature extraction process of the score encoder Enc_s is the same as that of the memory encoder Enc_M (both takes images with segmentation masks as input), we directly use the memory encoder as the score encoder, which helps to save calculations and parameters. Specifically, the structure of the score encoder Enc_s and the memory encoder Enc_M is the same, and the parameters are shared. The QAM first takes the query image $\mathbf{I}_t \in \mathbb{R}^{3 \times H \times W}$ and its segmentation mask $\mathbf{M}_t \in \mathbb{R}^{1 \times H \times W}$ into the score encoder to obtain the score feature map $f_s \in \mathbb{R}^{C \times H/16 \times W/16}$, where $H \times W$ are resolutions of the input image. Then, f_s is input to the convolution layers and fully connected layers to learn the segmentation quality score $\mathbf{S}_t^A$ for the current frame. The process of segmentation quality assessment can be expressed as:

$$f_s = Enc_s(\mathbf{I}_t \oplus \mathbf{M}_t); \quad \mathbf{S}_t^A = Fc(Conv(f_s)), \tag{1}$$

where $\oplus$ denotes the concatenation operation. t is the index of the current frame. $Conv$ and Fc denote convolution and fully connected layers with sigmoid non-linear function, respectively.

During training, the target value of the quality score is defined as mask IoU between the segmentation mask and ground truth. The specific calculation process is as follows:

$$loss = \frac{1}{N} \sum_{i=1}^{N} (S_i^A - maskIoU(M_i, GT_i))^2, \tag{2}$$

where S_i^A represents the quality score of the segmentation result for i-th object, M_i indicates the segmentation result, GT_i is the ground truth. N indicates the total number of objects.

Since QAM evaluates the segmentation quality for each object individually, we take the average of all object scores in one frame as the quality score of this frame. In addition, considering that the segmentation difficulty varies for different video scenes, we normalize the quality scores of all frames in a video to better measure the relative quality of the segmentation results, which helps to memorize more helpful information under challenging scenarios. Specifically, the final quality score of each frame is its initial predicted score divided by the score of the first frame. Formally, the process can be expressed as:

$$\bar{S}_t^A = \frac{\frac{1}{N} \sum_{i=1}^{N} S_{t_i}^A}{\bar{S}_1^A}, \tag{3}$$

where N represents the total number of objects in the t-th frame, $\bar{S}_t^A$ indicates the quality score of the segmentation result in frame t, $\bar{S}_1^A$ represents the quality score of the first frame.

Figure 3 shows some visualization results of the quality assessment, the first two columns are the same video, and the last two columns represent another video. We can observe that the driver is considered part of the car in the first column, which is a bad case. The pink zebra in the third column is not recognized, and the orange zebra is matched with similar background objects.

For the hard case, our QAM identifies these suboptimal results well, which shows that the segmentation accuracy of a frame is consistent with its quality score. Extensive experiments also verify this. With QAM, the memory bank can selectively memorize frames whose quality scores are higher than the memory threshold σ, that is, frames with accurate segmentation masks. In this way, even if a frame is poorly segmented owing to fast object motion or other factors, it will not affect the subsequent frames or cause error accumulation.

Algorithm: 1 Pseudocode of Dynamic Memory Bank

Input : memory bank $Memory$, video frames sequence $\{I_t\}$ of length L

1: $t = 2$ # the ground truth mask of the first frame is given
2: $j = 1$ # the relative index of memory frames
3: **while** $t \leq L$ **do**
4:    **if** $S_t^A \geq \sigma$ **then**
5:       # to filter the inaccurately segmented frames
6:       $j = j + 1$
7:       **if** $len(Memory) \leq \beta$ **then**
8:          $Memory.add(\{j : [I_t, M_t, S_t^A]\})$ # store I_t, M_t, S_t^A to the j position in memory.
9:       **else**
10:          $S_{min}^R, id_{min} = inf, inf$
11:          **for** k in $Memory.keys()$ **do**   # k is the relative index of the frame in the memory bank
12:             $S_k^C = exp(k - j)$
13:             $S_k^R = S_k^A + S_k^C$
14:             **if** $S_k^R < S_{min}^R$ **then**
15:                $S_{min}^R = S_k^R, id_{min} = k$
16:             **end if**
17:          **end for**
18:          $Memory.del(id_{min})$ # remove the memory frame with the lowest reference score S^R
19:          $Memory.add(\{j : [I_t, M_t, S_t^A]\})$
20:       **end if**
21:    **end if**
22:    $t = t + 1$
23: **end while**

3.3 Dynamically Updated Memory Bank

The infinite increase of the memory frames with the growth of frame number greatly limits the practicability of the model in the real-world scenario. Thus, it is necessary to limit the size of the memory bank and update it dynamically to adapt to new scenarios.

Due to the temporal consistency of video, the appearance of the target objects in adjacent frames is similar. The masks of adjacent frames are more instructive for the segmentation of current frame. Combining the above analysis and considering accuracy, we suggest dynamically updating the memory bank in accordance with these two principles (Algorithm.1). Specifically, when the memory bank reaches a certain storage limit, we will dynamically update the memory bank to handle different video scenes. For quantifying the temporal consistency and measuring the distance between each memory frame and the current frame, we compute the temporal consistency score $\mathbf{S}^C$ as:

$$\mathbf{S}_k^C = e^{-|t-k|}, \tag{4}$$

where k is the index of each memory frame, t is the index of the current frame.

Based on the accuracy score $\mathbf{S}^A$ and the temporal consistency score $\mathbf{S}^C$, the reference score of each memory frame in the memory bank can be calculated by $\mathbf{S}_k^R = \bar{\mathbf{S}}_k^A + \mathbf{S}_k^C$. By removing the memory frames with the lowest reference score, the memory bank is dynamically updated to handle different video scenarios and prevent the memory explosion problem.

3.4   Prior Enhancement Strategy

In addition to considering temporal consistency when designing a memory bank, we further utilize the prior provided by the previous adjacent frame to enhance temporal information. We adopt a similar module structure to SCM [54] to introduce the prior information from the previous adjacent frame. Instead of introducing spatial constraint in the decoder like SCM, we utilize the prior information in the query encoding process to better learn the target object's appearance feature and avoid over-reliance on the prior information.

Specifically, in the query encoding process, the segmentation mask of the previous adjacent frame $\mathbf{M}_{t-1} \in \mathbb{R}^{1 \times H \times W}$ is downsampled and concatenated with the query's embedding $f_t \in \mathbb{R}^{C \times H/16 \times W/16}$. Then the resultant feature goes through convolution and non-linear function to fuse information between channels, through which a prior feature map $f_p \in \mathbb{R}^{1 \times H/16 \times W/16}$ is produced. Finally, we perform an element-wise product between f_p and f_t to get the enhanced feature $f_e \in \mathbb{R}^{C \times H/16 \times W/16}$. Formally, the process can be expressed as the following equation:

$$f_e = Conv(f_t \oplus \mathbf{M}_{t-1}) \otimes f_t. \tag{5}$$

Furthermore, we find that it is better to provide weak prior (mentioned above) than strong prior (masks of the previous frame have a great influence on the feature of the current frame). We found two primary reasons through experiments: the first one is that the prior information may be noisy, and providing a strong prior may lead to error accumulation; the second one is that providing strong prior makes the model overly dependent on it, which weakens its ability to extract features and identify objects. Table 6 shows the disadvantages of providing strong prior under challenging scenarios. In Sect. 5.3, we will describe the specific approach of providing strong prior.

3.5   Memory Read and Decoder

In the Space-Time Memory Read block [29], soft weights are first computed by measuring the similarities between query key K^Q and memory key K^M. Then the memory value V^M is retrieved by a weighted summation with the soft weights and concatenated with query value V^Q to get the output y. This operation can be summarized as:

$$y_i = V_i^Q \oplus \frac{1}{Z} \sum_{\forall j} \mathcal{D}(K_i^Q, K_j^M) V_j^M, \tag{6}$$

where i and j are the index of the query and the memory location, $Z = \sum_{\forall j} \mathcal{D}(K_i^Q, K_j^M)$ is the normalizing factor. $\mathcal{D}$ denotes the similarity measure (in our experiment is dot product).

Our decoder stays close to that of [29,55]. The decoder takes the output y of the Space-Time Memory Read block as input and predicts the object masks. It consists of an ASPP layer [2], a residual block, and two upsample blocks that upscale the feature map to the initial image size.

4   Implementation Details

Following the training strategy in MiVOS [6], we first pretrain our model on static image datasets [5, 19, 37, 43, 53] and then perform main training on YouTube-VOS and DAVIS datasets. Besides, we also experiment with the synthetic dataset BL30K proposed in MiVOS, which is not used unless otherwise specified. During pretraining, each image is expanded into a pseudo video of three frames by random affine, horizontal flip, color and brightness augmentation. We randomly pick three frames in chronological order (with a ground-truth mask for the first frame) from a video to form a training sample in the main training. The range of random sampling varies with the training process. In the intermediate period of training, the sampling range is set larger to improve the robustness of the model, while at the end of the training, it is set smaller to narrow the gap between training and inference. Our models are trained end-to-end with two 32GB Tesla V100 GPUs with the Adam optimizer in PyTorch. The batch size is set to 28 during pretraining and 16 during main training. We adopt ResNet-50 [13] as backbone for all encoders. Bootstrapped cross-entropy loss [6] is used for segmentation, and MSE loss is used for quality score evaluation. The initial learning rate is 2e-5. During inference, we choose the memory threshold σ of 0.8 by default. Ablation studies are conducted on a single 1080Ti GPU and DAVIS 2017 validation set in default.

5   Experiments

5.1   Comparisons with State-of-the-Art Methods

DAVIS 2016 [32] is a single object benchmark for video object segmentation. As shown in Table 1, QDMN trained without synthetic dataset still outperforms most previous methods (**91.0** $\mathcal{J}\&\mathcal{F}$). With synthetic training data, QDMN surpasses all existing methods and achieves the performance of **92.0** $\mathcal{J}\&\mathcal{F}$.

DAVIS 2017 [33] is a multiple objects extension of DAVIS 2016. In the Table 1, QDMN achieves an average score of **84.6** and **85.6** for training without synthetic data and with synthetic data, respectively. What's more, we also test our model on the challenging DAVIS 2017 testing split set. It achieves the best performance (**81.9**) compared to all previous methods.

YouTube-VOS [47] is a large-scale benchmark for video object segmentation. As shown in Table 2, without synthetic training data, our QDMN also achieves state-of-the-art performance (**83.0**). If we use synthetic data for training, the overall score of QDMN will be boosted to **83.8**.

Qualitative Results. The qualitative comparison between baseline and our QDMN are shown in Fig. 4. We show the performance on two challenging scenarios, *i.e.*, occlusion scenes and similar objects. Both STM [29] and MiVOS [6] have lost targets in the occlusion scene. STM lost targets in the scene with similar objects, while MiVOS identified other objects incorrectly. In contrast, our method can achieve satisfactory performance in challenging scenarios.

Table 1. Comparison with other methods on DAVIS dataset. '*' indicates using synthetic training dataset [6].

Method	DAVIS2016			DAVIS2017 val			DAVIS2017 test-dev		
	$\mathcal{J}$	$\mathcal{F}$	$\mathcal{J}\&\mathcal{F}$	$\mathcal{J}$	$\mathcal{F}$	$\mathcal{J}\&\mathcal{F}$	$\mathcal{J}$	$\mathcal{F}$	$\mathcal{J}\&\mathcal{F}$
RANet [44]	86.6	87.6	87.1	63.2	68.2	65.7	53.4	56.2	55.3
FEELVOS [40]	81.1	82.2	81.7	69.1	74.0	71.5	55.2	60.5	57.8
RGMP [28]	81.5	82.0	81.8	64.8	68.6	66.7	51.3	54.4	52.8
DMVOS [45]	88.0	87.5	87.8	-	-	-	-	-	-
STM [29]	88.7	89.9	89.3	79.2	84.3	81.8	69.3	75.2	72.2
KMN [35]	89.5	91.5	90.5	80.0	85.6	82.8	74.1	80.3	77.2
CFBI [50]	88.3	90.5	89.4	79.1	84.6	81.9	71.1	78.5	74.8
GIEL [11]	-	-	-	80.2	85.3	82.7	72.0	78.3	75.2
SwiftNet [42]	90.5	90.3	90.4	78.3	83.9	81.1	-	-	-
RMNet [46]	88.9	88.7	88.8	81.0	86.0	83.5	71.9	78.1	75.0
SSTVOS [10]	-	-	-	79.9	85.1	82.5	-	-	-
LCM [14]	89.9	91.4	90.7	80.5	86.5	83.5	74.4	81.8	78.1
MiVOS [6]	87.8	90.0	88.9	80.5	85.8	83.1	72.6	79.3	76.0
MiVOS* [6]	89.7	92.4	91.0	81.7	87.4	84.5	74.9	82.2	78.6
JOINT [27]	-	-	-	80.8	86.2	83.5	-	-	-
RPCMVOS [48]	87.1	94.0	90.6	81.3	86.0	83.7	75.8	82.6	79.2
DMN-AOA [22]	-	-	-	81.0	87.0	84.0	74.8	81.7	78.3
HMMN [36]	89.6	92.0	90.8	81.9	87.5	84.7	74.7	82.5	78.6
STCN [7]	90.8	92.5	91.6	82.2	88.6	85.4	72.7	79.6	76.1
AOT-L [51]	89.7	92.3	91.0	80.3	85.7	83.0	75.3	82.3	78.8
QDMN (Ours)	**90.2**	**91.7**	**91.0**	**81.8**	**87.3**	**84.6**	**74.2**	**81.2**	**77.7**
QDMN* (Ours)	**90.7**	**93.2**	**92.0**	**82.5**	**88.6**	**85.6**	**78.1**	**85.4**	**81.9**

5.2   Generic Plugins

To further prove the effectiveness of our proposed QAM, we apply it as a general plugin to other methods. The results on the DAVIS2017 validation set are shown in Table 3 (the baseline performance is our re-implementation results). It can be seen that with QAM, the performance of these methods has been significantly boosted. Besides, QAM is easy to be deployed on other methods, and we hope that the QAM would shed light on the studies of related fields that need to memorize reference information.

5.3   Ablation Study

The effectiveness of QAM. To demonstrate the effectiveness of the QAM, we conduct specific analyses from three dimensions.

Table 2. Evaluation on YouTube-VOS 2018 val set. Seen and Unseen denote whether the categories exist in the training set. $\mathcal{G}$ is averaged overall score.

Methods	Seen		Unseen		$\mathcal{G}$
	$\mathcal{J}$	$\mathcal{F}$	$\mathcal{J}$	$\mathcal{F}$	
STM [29]	79.7	84.2	72.8	80.9	79.4
AFB-URR [23]	78.8	83.1	74.1	82.6	79.6
GCM [21]	72.6	75.6	68.9	75.7	73.2
KMN [35]	81.4	85.6	75.3	83.3	81.4
G-FRTM [30]	68.6	71.3	58.4	64.5	65.7
SwiftNet [42]	77.8	81.8	72.3	79.5	77.8
GIEL [11]	80.7	85.0	75.0	81.9	80.6
SSTVOS [10]	80.9	-	76.6	-	81.8
RMNet [46]	82.1	85.7	75.7	82.4	81.5
LCM [14]	82.2	86.7	75.7	83.4	82.0
MiVOS [6]	80.0	84.6	74.8	82.4	80.4
MiVOS* [6]	81.1	85.6	77.7	86.2	82.6
JOINT [27]	81.5	85.9	78.7	86.5	83.1
HMMN [36]	82.1	87.0	76.8	84.6	82.6
DMN-AOA [22]	82.5	86.9	76.2	84.2	82.5
STCN [7]	81.9	86.5	77.9	85.7	83.0
AOT-L [51]	82.5	87.5	77.9	86.7	83.7
QDMN (Ours)	**82.0**	**86.8**	**77.5**	**85.5**	**83.0**
QDMN* (Ours)	**82.7**	**87.5**	**78.4**	**86.4**	**83.8**

Table 3. Applying QAM as general plugin. w/QAM indicates that whether the QAM is deployed on this method.

Methods	w / QAM	$\mathcal{J}$	$\mathcal{F}$	$\mathcal{J}\&\mathcal{F}$
STM [29]		78.8	84.2	81.5
	✓	81.0	86.2	**83.6**[†]
KMN [35]		79.7	85.5	82.6
	✓	81.9	87.4	**84.7**[†]
STCN [7]		81.5	87.7	84.6
	✓	82.5	88.7	**85.6**[†]

(1) Accuracy of the predicted scores. We perform a histogram visualization of the distribution of the ground truth mask IoU and prediction scores at 0.05 intervals (Fig. 5). When multiple objects are in a frame, the average is taken. We can see that the quality score and ground truth mask IoU are positively correlated, which verifies the accuracy of the scores predicted by QAM.

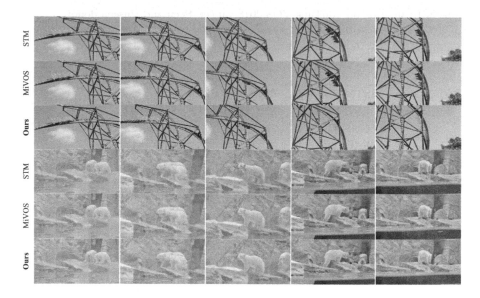

Fig. 4. Visual comparison of QDMN with baseline methods. Each row demonstrates five frames sampled from a video sequence.

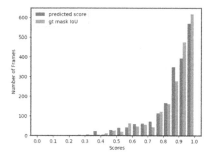

Fig. 5. Distribution of the prediction score and the ground truth mask IoU.

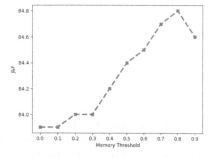

Fig. 6. The quantitative results of different memory threshold σ.

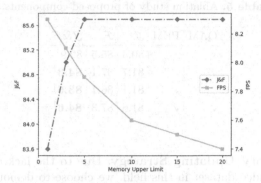

Fig. 7. The performance for different memory upper limit.

(2) Memory Threshold. We test different memory thresholds σ on DAVIS 2017 test-dev set, and the results are shown in Fig. 6. We can see that it will hurt the segmentation effect if the threshold is set too high or too low. The reason is that if the threshold σ is too high, only a few intermediate frames will be memorized, leading to losing a lot of helpful information; if the σ is too low, the model may memorize some incorrect noise information. Besides, the performance is worst when the memory threshold is 0 (at this time, QAM does not filter poor segmentation masks), which proves the motivation of the QAM is correct.

(3) Applying QAM only at inference stage. To further prove that filtering out inaccurately segmented frames has a beneficial effect on segmentation, we construct experiments that adding QAM only at the inference stage. Specifically, for QAM, we load its parameters trained in QDMN. For other parts, we load the weights of the initial model (trained without QAM). As shown in Table 4, the performance of all vanilla models has been improved after adding QAM, which shows the importance of filtering poorly segmented frames.

Table 4. The effect of adding QAM only in *the inference stage*

Methods	$\mathcal{J}\&\mathcal{F}_{(\sigma=0)}$	$\mathcal{J}\&\mathcal{F}_{(\sigma=0.8)}$
STM	81.5	82.5↑
KMN	82.6	83.4↑
MiVOS	82.7	83.5↑

Component Analysis. We analyze the effectiveness of our modules in Table 5. PE represents the prior enhancement strategy introduced to highlight temporal consistency. As shown in the table, both the QAM and PE bring remarkable performance improvement.

Table 5. Ablation study of proposed components.

QAM	PEM	$\mathcal{J}$	$\mathcal{F}$	$\mathcal{J}\&\mathcal{F}$
		80.3	85.5	82.9
✓		81.7	87.1	**84.3**$^{\uparrow}$
	✓	81.1	86.1	**83.6**$^{\uparrow}$
✓	✓	81.8	87.3	**84.6**$^{\uparrow}$

Dynamic Memory Updating Strategy. Due to the lack of a widely used large-scale long video dataset in this field, we choose to demonstrate the effectiveness of our proposed memory bank dynamic updating strategy by compressing the upper limit of the memory. As shown in Fig. 7, The segmentation effect remains unaffected even at low memory upper limit, and the speed is improved as a result of our memory bank design strategy. The similar phenomenon is observed on the YouTube-VOS set, which illustrates the effectiveness of our dynamic updating strategy.

Besides, we also perform analysis on long videos (without annotations). We find that previous memory network methods store up to about 70 frames and the memory explosion occurs, which greatly limits the practicability. But QDMN can handle videos of arbitrary length by setting upper memory limit and dynamically updating the memory. What's more, the FPS of previous methods will drop from 14 to about 2 before memory explodes, while the FPS of QDMN will stay around 7 after the initial drop (assuming the upper memory limit is 25).

Table 6. Ablation study of different enhancement strategy. "Weak" means providing weak prior (PE). "Strong" means providing strong location prior.

Strategy	DAVIS			YouTube-VOS		
	$\mathcal{J}$	$\mathcal{F}$	$\mathcal{J}\&\mathcal{F}$	$\mathcal{J}$	$\mathcal{F}$	$\mathcal{G}$
Weak	81.8	87.3	84.6	79.8	86.2	**83.0**
Strong	82.4	87.9	**85.2**	77.5	83.8	80.7

Enhancement Strategy. For PE, we directly concatenate the prior mask with the deepest layer feature of the current frame to provide a weak prior. In contrast, we also try to provide a strong prior. Specifically, we extract the feature of the prior mask and fuse it with the middle layer features of the current frame. After convolution and downsampling, the fused features are added to the deepest layer features of the current frame. Compared with the current enhancement strategy, this approach can significantly enhance the influence of the prior mask. However, although this approach works well in common scenarios, the performance drops significantly under challenging situations, as shown in Table 6. The reason for this phenomenon is that the strong prior makes the model overly dependent on it, which weakens the model's ability to recognize objects.

Speed Analysis. We also experiment with the impact of the proposed modules on the inference speed. With our modules, the FPS of baseline has changed from **8.6** to **7.8** on DAVIS2017 val set. The increased running time brought by QAM and PE is nearly negligible (no more than 10%), mainly because we directly use the feature extracted by the memory encoder for quality assessment.

6   Conclusion

In this paper, we propose that the design of the memory bank should follow the principles of accuracy and temporal consistency. To support this, we introduce a Quality-aware Dynamic Memory Network (QDMN) for semi-supervised video object segmentation, which selectively memorizes accurately segmented intermediate frames as references and emphasizes video temporal consistency. Without bells and whistles, our QDMN achieves new state-of-the-art performance on the popular benchmark YouTube-VOS and DAVIS with almost no additional inference time. Furthermore, the QAM also has a remarkable improvement for other approaches as a general plugin.

Acknowledgments. This research was supported in part by the National Natural Science Foundation of China under Grant No. U1903213, the Shenzhen Key Laboratory of Marine IntelliSense and Computation (NO. ZDSYS20200811142605016.)

References

1. Caelles, S., Maninis, K., Pont-Tuset, J., Leal-Taixé, L., Cremers, D., Gool, L.V.: One-shot video object segmentation. In: CVPR, pp. 5320–5329 (2017)
2. Chen, L., Papandreou, G., Kokkinos, I., Murphy, K., Yuille, A.L.: DeepLab: semantic image segmentation with deep convolutional nets, atrous convolution, and fully connected CRFs. TPAMI **40**, 834–848 (2018)
3. Chen, X., Li, Z., Yuan, Y., Yu, G., Shen, J., Qi, D.: State-aware tracker for real-time video object segmentation. In: CVPR, pp. 9381–9390 (2020)
4. Chen, Y., Pont-Tuset, J., Montes, A., Gool, L.V.: Blazingly fast video object segmentation with pixel-wise metric learning. In: CVPR, pp. 1189–1198 (2018)
5. Cheng, H.K., Chung, J., Tai, Y., Tang, C.: CascadePSP: toward class-agnostic and very high-resolution segmentation via global and local refinement. In: CVPR, pp. 8887–8896 (2020)
6. Cheng, H.K., Tai, Y., Tang, C.: Modular interactive video object segmentation: Interaction-to-mask, propagation and difference-aware fusion. arXiv preprint arXiv:2103.07941 (2021)
7. Cheng, H.K., Tai, Y., Tang, C.: Rethinking space-time networks with improved memory coverage for efficient video object segmentation. arXiv preprint arXiv:2106.05210 (2021)
8. Cheng, J., Tsai, Y., Hung, W., Wang, S., Yang, M.: Fast and accurate online video object segmentation via tracking parts. In: CVPR, pp. 7415–7424 (2018)

9. Cheng, J., Tsai, Y., Wang, S., Yang, M.: SegFlow: joint learning for video object segmentation and optical flow. In: ICCV, pp. 686–695 (2017)

10. Duke, B., Ahmed, A., Wolf, C., Aarabi, P., Taylor, G.W.: SSTVOS: sparse spatiotemporal transformers for video object segmentation. In: CVPR, pp. 5912–5921 (2021)

11. Ge, W., Lu, X., Shen, J.: Video object segmentation using global and instance embedding learning. In: CVPR, pp. 16836–16845 (2021)

12. He, K., Gkioxari, G., Dollár, P., Girshick, R.B.: Mask R-CNN. In: ICCV, pp. 2980–2988 (2017)

13. He, K., Zhang, X., Ren, S., Sun, J.: Deep residual learning for image recognition. In: CVPR, pp. 770–778 (2016)

14. Hu, L., Zhang, P., Zhang, B., Pan, P., Xu, Y., Jin, R.: Learning position and target consistency for memory-based video object segmentation. arXiv preprint arXiv:2104.04329 (2021)

15. Hu, Y., Huang, J., Schwing, A.G.: MaskRNN: instance level video object segmentation. In: NIPS, pp. 325–334 (2017)

16. Huang, X., Xu, J., Tai, Y., Tang, C.: Fast video object segmentation with temporal aggregation network and dynamic template matching. In: CVPR, pp. 8876–8886 (2020)

17. Huang, Z., Huang, L., Gong, Y., Huang, C., Wang, X.: Mask scoring R-CNN. In: CVPR, pp. 6409–6418 (2019)

18. Jiang, B., Luo, R., Mao, J., Xiao, T., Jiang, Y.: Acquisition of localization confidence for accurate object detection. In: Ferrari, V., Hebert, M., Sminchisescu, C., Weiss, Y. (eds.) Computer Vision – ECCV 2018. LNCS, vol. 11218, pp. 816–832. Springer, Cham (2018). https://doi.org/10.1007/978-3-030-01264-9_48

19. Li, X., Wei, T., Chen, Y.P., Tai, Y., Tang, C.: FSS-1000: a 1000-class dataset for few-shot segmentation. In: CVPR, pp. 2866–2875 (2020)

20. Li, X., Loy, C.C.: Video object segmentation with joint re-identification and attention-aware mask propagation. In: Ferrari, V., Hebert, M., Sminchisescu, C., Weiss, Y. (eds.) ECCV 2018. LNCS, vol. 11207, pp. 93–110. Springer, Cham (2018). https://doi.org/10.1007/978-3-030-01219-9_6

21. Li, Yu., Shen, Z., Shan, Y.: Fast video object segmentation using the global context module. In: Vedaldi, A., Bischof, H., Brox, T., Frahm, J.-M. (eds.) ECCV 2020. LNCS, vol. 12355, pp. 735–750. Springer, Cham (2020). https://doi.org/10.1007/978-3-030-58607-2_43

22. Liang, S., Shen, X., Huang, J., Hua, X.S.: Video object segmentation with dynamic memory networks and adaptive object alignment. In: ICCV, pp. 8065–8074 (2021)

23. Liang, Y., Li, X., Jafari, N.H., Chen, J.: Video object segmentation with adaptive feature bank and uncertain-region refinement. In: NIPS (2020)

24. Lin, H., Qi, X., Jia, J.: AGSS-VOS: attention guided single-shot video object segmentation. In: ICCV, pp. 3948–3956 (2019)

25. Lu, X., Wang, W., Danelljan, M., Zhou, T., Shen, J., Van Gool, L.: Video object segmentation with episodic graph memory networks. In: Vedaldi, A., Bischof, H., Brox, T., Frahm, J.-M. (eds.) ECCV 2020. LNCS, vol. 12348, pp. 661–679. Springer, Cham (2020). https://doi.org/10.1007/978-3-030-58580-8_39

26. Luiten, J., Voigtlaender, P., Leibe, B.: PReMVOS: proposal-generation, refinement and merging for video object segmentation. In: ACCV, pp. 565–580 (2018)

27. Mao, Y., Wang, N., Zhou, W., Li, H.: Joint inductive and transductive learning for video object segmentation. arXiv preprint arXiv:2108.03679 (2021)

28. Oh, S.W., Lee, J., Sunkavalli, K., Kim, S.J.: Fast video object segmentation by reference-guided mask propagation. In: CVPR, pp. 7376–7385 (2018)

29. Oh, S.W., Lee, J., Xu, N., Kim, S.J.: Video object segmentation using space-time memory networks. In: ICCV, pp. 9225–9234 (2019)
30. Park, H., Yoo, J., Jeong, S., Venkatesh, G., Kwak, N.: Learning dynamic network using a reuse gate function in semi-supervised video object segmentation. In: CVPR, pp. 8405–8414 (2021)
31. Perazzi, F., Khoreva, A., Benenson, R., Schiele, B., Sorkine-Hornung, A.: Learning video object segmentation from static images. In: CVPR, pp. 3491–3500 (2017)
32. Perazzi, F., Pont-Tuset, J., McWilliams, B., Gool, L.V., Gross, M.H., Sorkine-Hornung, A.: A benchmark dataset and evaluation methodology for video object segmentation. In: CVPR, pp. 724–732 (2016)
33. Pont-Tuset, J., Perazzi, F., Caelles, S., Arbelaez, P., Sorkine-Hornung, A., Gool, L.V.: The 2017 DAVIS challenge on video object segmentation. arXiv preprint arXiv:1704.00675 (2017)
34. Ren, S., He, K., Girshick, R.B., Sun, J.: Faster R-CNN: towards real-time object detection with region proposal networks. In: NIPS, pp. 91–99 (2015)
35. Seong, H., Hyun, J., Kim, E.: Kernelized memory network for video object segmentation. In: Vedaldi, A., Bischof, H., Brox, T., Frahm, J.-M. (eds.) ECCV 2020. LNCS, vol. 12367, pp. 629–645. Springer, Cham (2020). https://doi.org/10.1007/978-3-030-58542-6_38
36. Seong, H., Oh, S.W., Lee, J., Lee, S., Lee, S., Kim, E.: Hierarchical memory matching network for video object segmentation. arXiv preprint arXiv:2109.11404 (2021)
37. Shi, J., Yan, Q., Xu, L., Jia, J.: Hierarchical image saliency detection on extended CSSD. TPAMI. **38**, 717–729 (2016)
38. Sun, M., Xiao, J., Lim, E.G., Zhang, B., Zhao, Y.: Fast template matching and update for video object tracking and segmentation. In: CVPR, pp. 10788–10796 (2020)
39. Tsai, Y., Yang, M., Black, M.J.: Video segmentation via object flow. In: CVPR, pp. 3899–3908 (2016)
40. Voigtlaender, P., Chai, Y., Schroff, F., Adam, H., Leibe, B., Chen, L.: FEELVOS: fast end-to-end embedding learning for video object segmentation. In: CVPR, pp. 9481–9490 (2019)
41. Voigtlaender, P., Leibe, B.: Online adaptation of convolutional neural networks for video object segmentation. In: BMVC (2017)
42. Wang, H., Jiang, X., Ren, H., Hu, Y., Bai, S.: SwiftNet: real-time video object segmentation. In: CVPR, pp. 1296–1305 (2021)
43. Wang, L., et al.: Learning to detect salient objects with image-level supervision. In: CVPR, pp. 3796–3805 (2017)
44. Wang, Z., Xu, J., Liu, L., Zhu, F., Shao, L.: RANet: ranking attention network for fast video object segmentation. In: ICCV, pp. 3977–3986 (2019)
45. Wen, P., et al.: DMVOS: discriminative matching for real-time video object segmentation. In: ACMMM, pp. 2048–2056 (2020)
46. Xie, H., Yao, H., Zhou, S., Zhang, S., Sun, W.: Efficient regional memory network for video object segmentation. arXiv preprint arXiv:2103.12934 (2021)
47. Xu, N., et al.: Youtube-VOS: a large-scale video object segmentation benchmark. arXiv preprint arXiv:1809.03327 (2018)
48. Xu, X., Wang, J., Li, X., Lu, Y.: Reliable propagation-correction modulation for video object segmentation. In: AAAI, pp. 2946–2954 (2022)
49. Xu, Y., Fu, T., Yang, H., Lee, C.: Dynamic video segmentation network. In: CVPR, pp. 6556–6565 (2018)

50. Yang, Z., Wei, Y., Yang, Y.: Collaborative video object segmentation by foreground-background integration. In: Vedaldi, A., Bischof, H., Brox, T., Frahm, J.-M. (eds.) ECCV 2020. LNCS, vol. 12350, pp. 332–348. Springer, Cham (2020). https://doi.org/10.1007/978-3-030-58558-7_20
51. Yang, Z., Wei, Y., Yang, Y.: Associating objects with transformers for video object segmentation. arXiv preprint arXiv:2106.02638 (2021)
52. Yang, Z., Wei, Y., Yang, Y.: Collaborative video object segmentation by multi-scale foreground-background integration. In: IEEE TPAMI (2021)
53. Zeng, Y., Zhang, P., Lin, Z.L., Zhang, J., Lu, H.: Towards high-resolution salient object detection. In: ICCV, pp. 7233–7242 (2019)
54. Zhang, P., Hu, L., Zhang, B., Pan, P.: Spatial constrained memory network for semi-supervised video object segmentation. In: CVPR Workshops (2020)
55. Zhou, Z., et al.: Enhanced memory network for video segmentation. In: ICCV Workshops, pp. 689–692 (2019)

Learning Implicit Feature Alignment Function for Semantic Segmentation

Hanzhe Hu[1]([✉]), Yinbo Chen[2], Jiarui Xu[2], Shubhankar Borse[3], Hong Cai[3], Fatih Porikli[3], and Xiaolong Wang[2]

[1] Peking University, Beijing, China
huhz@pku.edu.cn
[2] University of California, San Diego, USA
[3] Qualcomm AI Research, San Diego, USA

Abstract. Integrating high-level context information with low-level details is of central importance in semantic segmentation. Towards this end, most existing segmentation models apply bilinear up-sampling and convolutions to feature maps of different scales, and then align them at the same resolution. However, bilinear up-sampling blurs the precise information learned in these feature maps and convolutions incur extra computation costs. To address these issues, we propose the Implicit Feature Alignment function (IFA). Our method is inspired by the rapidly expanding topic of implicit neural representations, where coordinate-based neural networks are used to designate fields of signals. In IFA, feature vectors are viewed as representing a 2D field of information. Given a query coordinate, nearby feature vectors with their relative coordinates are taken from the multi-level feature maps and then fed into an MLP to generate the corresponding output. As such, IFA implicitly aligns the feature maps at different levels and is capable of producing segmentation maps in arbitrary resolutions. We demonstrate the efficacy of IFA on multiple datasets, including Cityscapes, PASCAL Context, and ADE20K. Our method can be combined with improvement on various architectures, and it achieves state-of-the-art computation-accuracy trade-off on common benchmarks. Code is available at https://github.com/hzhupku/IFA.

Keywords: Semantic segmentation · Implicit neural representation · Feature alignment

H. Hu and Y. Chen—Equal contribution.

Supplementary Information The online version contains supplementary material available at https://doi.org/10.1007/978-3-031-19818-2_28.

S. Avidan et al. (Eds.): ECCV 2022, LNCS 13689, pp. 487–505, 2022.
https://doi.org/10.1007/978-3-031-19818-2_28

1   Introduction

Semantic Segmentation is one of the most fundamental and challenging tasks in computer vision. It aims at classifying each pixel in the image into a semantic category. Its wide applications include scene understanding, image editing, augmented reality, and autonomous driving. Most of these applications not only require the segmentation model to predict high resolution and high-quality masks but also demand high efficiency in speed and memory cost, especially when running online or on edge devices.

Most current approaches for semantic segmentation are built upon the Fully Convolutional Network [27]. At the heart of these approaches, is to integrate high-level context information with low-level details during segmentation. Empirically, deeper features with coarse resolution correspond to higher semantic information while shallow features in lower layers contain more local details. To aggregate different levels of information, state-of-the-art approaches such as Feature Pyramid Network (FPN) [20] and DeepLab V3+ [3] utilize bilinear up-sampling followed by convolutions to align the low resolution deep features with the high resolution shallow features. However, the bilinear up-sampling can blur the precise context learned in deep features and the convolutions are not optimal for speed and memory efficiency, especially for high resolution segmentation where the resolution difference between high-level context and low-level details can be large.

To perform efficient and precise feature alignment, we will require a representation that can flexibly query a location in any resolution and output the corresponding feature values. This formulation corresponds to the implicit neural

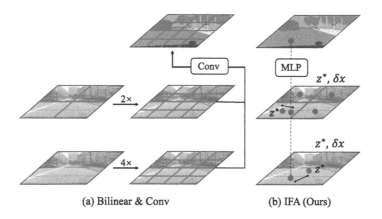

(a) Bilinear & Conv   (b) IFA (Ours)

Fig. 1. Implicit Feature Alignment function (IFA). (a) Prior works transform feature maps to the same resolution for alignment, where bilinear up-sampling blurs the precise information and convolutions can be inefficient. (b) IFA decodes directly from the original feature maps for arbitrary coordinates. An MLP takes as input the multi-level features around the query coordinate with their relative coordinates and outputs the aggregated information.

representation [18,32,35,39] proposed for high-quality 3D shape reconstruction, where a 3D object is represented as a neural network that maps a 3D coordinate to its occupancy in the object [28] or its signed distance to the object surface [32]. This idea is also migrated to the 2D domain [4,39], where neural functions are proposed as continuous image representations that allow the image to be decoded in an arbitrary resolution. Instead of decoding RGB values, can we use implicit neural representation to perform feature alignment?

In this paper, we propose a novel Implicit Feature Alignment function (IFA) to efficiently and precisely aggregate features from different levels for semantic segmentation. By forwarding an image to a ConvNet, the features can be viewed as latent codes evenly distributed in spatial dimensions (shown as green dots in Fig. 1). Intuitively, each latent code will represent a field of information. IFA will then decode the output segmentation map at every coordinate independently and parallelly. It takes as inputs the latent codes around the queried coordinate from different levels and the relative coordinates to the latent codes, then outputs the aggregated feature at the queried coordinate for classification, as illustrated in Fig. 1. Take the FPN [20] model as an example, the original design is to use bilinear up-sampling and convolutions to align multi-layer features. IFA can be a replacement here to align and aggregate the multi-level features. Instead of bilinear up-sampling the features and aligning them in a fixed resolution, IFA allows the features to be learned as precisely representing continuous fields of information. The information in different levels are functions of continuous coordinates, which leads to a precise feature alignment in a resolution-free manner, and we can query the coordinate in arbitrary output resolutions with IFA for semantic segmentation.

We demonstrate the effectiveness of IFA on multiple semantic segmentation architectures and multiple datasets including Cityscapes [8], PASCAL Context [31] and ADE20K [59]. We replace feature alignment approaches with IFA in different methods, and IFA outperforms the original methods in all cases. IFA shows state-of-the-art computation-accuracy trade-off on all experimented datasets. For performing high resolution image segmentation, we also experiment with reducing the high-level feature map size while maintaining the low-level feature resolution, which improves efficiency and reduces memory cost. IFA has shown a much larger gain on aligning features with larger resolution differences. This not only shows the effectiveness of IFA on precise feature alignment, but also reveals its potential on efficient high resolution semantic segmentation.

To sum up, our contributions are summarized as follows:

- We propose a novel implicit neural representation IFA for efficient and precise alignment among multi-level feature maps for semantic segmentation.
- Our IFA can be incorporated with multiple state-of-the-art semantic segmentation models and show improvement in all cases.
- We achieve state-of-the-art computation-accuracy trade-off on benchmarks including Cityscapes, PASCAL Context and ADE20K.

2 Related Work

Semantic Segmentation. With the success of deep neural networks [12,21,38], semantic segmentation has achieved great progress. Based on Fully Convolutional Network (FCN) [27], many works have been proposed. To produce high-resolution semantic segmentation map, spatial and semantic information are both indispensable. There are mainly two lines of research for learning the two kinds of information in semantic segmentation.

The first stream lies in that the final output of the network contains both spatial and semantic information. Many state-of-the-art methods follow this line to design segmentation head to capture contextual information. From the local perspective, DeepLabV3 [2] employs multiple atrous convolutions with different dilation rates to capture contextual information, while PSPNet [56] utilizes pyramid pooling over sub-regions to harvest information. While from the global perspective, Wang *et al.* [45] apply the idea of self-attention from transformer [42] into vision problems and propose the non-local module where correlations between all pixels are calculated to guide the dense contextual information aggregation. Recently, Transformer based models [7,40,46,58] have achieved great progress in semantic segmentation, while they also suffer from heavy parameters and computation cost.

The other stream dissipates information along outputs of different layers of the network. Hence, the success relies on the feature alignment among the outputs. Our method focuses on this direction.

Feature Alignment. A common knowledge in semantic segmentation is that outputs from shallower network layers contain more low-level spatial details, while outputs from deeper network layers possess more high-level semantic information. How to effectively align those features has become a vital problem in this stream of study. Many methods focus on fusing multi-level feature maps for high-resolution spatial details and rich semantics. U-Net [34] adds several expanding paths to the contracting path to enable precise localization with the context. CARAFE [43] introduces a context-aware feature upsampling method, where features inside a predefined region centered at each location are reassembled via a weighted combination. Semantic FPN [20] applies FPN [26] structure to semantic segmentation where multi-level features are aligned by several up-sampling stages consisted of convolution layers and bilinear up-sampling. On top of FPN, SFNet [24] proposes the flow alignment module to broadcast high-level context to high-resolution details. AlignSeg [17] explicitly learns the transformation offsets and adaptively aggregate contextual information for better alignment. However, aforementioned methods for alignment usually take up expensive computations and tend to be inefficient for real-time applications.

Implicit Neural Representation. In recent methods in 3D reconstruction, shape, object and scene can be represented by multi-layer perceptron (MLP) that maps coordinates to signals, known as implicit neural representations (INR) [10,29,30,32] since the parameters of the 3D representation are not explicitly encoded by point cloud, mesh or voxel. For example, DeepSDF [32] learns a set

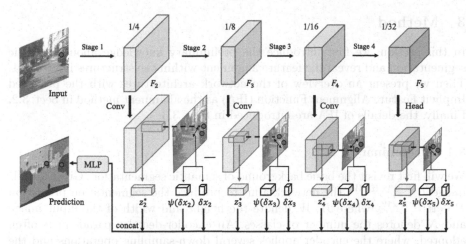

Fig. 2. Overview of our proposed Implicit Feature Alignment function (IFA).
The general architecture consists of an encoder part (in blue) and a decoder part (in orange). IFA aligns multi-level feature maps from different stages of the encoder. Each feature map is projected to the same dimension via a convolution layer. We could also project the last feature with a segmentation head like ASPP [2]. We view the features in feature maps as latent codes evenly distributed in the 2D space. Given a query coordinate x_q, we first find its nearest latent codes $\{z_i^*\}_{i=2}^5$ for each feature map i and use x_i^* to denote the coordinate of z_i^*. We then concatenate these latent codes $\{z_i^*\}_{i=2}^5$ and relative coordinates $\{\delta x_i = (x_q - x_i^*)\}_{i=2}^5$, and pass the concatenated vector into an MLP that directly predicts the segmentation label of point x_q. The red point refers to the query coordinate x_q, while the green point denotes the nearest coordinate x_i^* from x_q on feature F_i.

of continuous signed distance functions for shape representation. Later, NeRF [30] provides a more flexible way for synthesizing novel views of complex scenes.

Although implicit nerual representation has achieved great progress in 3D tasks, it is relatively under-explored for 2D tasks. [5] performs 2D shape generation from latent space for simple digits. [39] replaces ReLU with periodic activation functions inside MLP of implicit neural representation to model natural images in high quality. Recently, LIIF [4] applies implicit neural representation to model continuous image representation and UltraSR [47] further improves the accuracy by adding spatial encoding for implicit function on 2D images. CRM [36] performs image segmentation refinement by using implicit neural representations. Implicit PointRend [6] focuses on instance segmentation with point supervision, where implicit function is used to generate different parameters of the point head for each object. Different from these works, we focus on utilizing INR to perform implicit alignment of multi-level features.

3   Method

In this section, we first introduce the preliminary knowledge about semantic segmentation and reveal the feature alignment within the structures in Sect. 3.1. Then we present an overview of the network architectures with the proposed Implicit Feature Alignment Function (IFA) as the alignment method in Sect. 3.2. Finally, the details of IFA are introduced in Sect. 3.3.

3.1   Preliminary

We will first revisit the basic background of semantic segmentation. Given a RGB image $I \in \mathbb{R}^{3 \times H \times W}$, the network aims to produce the segmentation prediction $P \in \mathbb{R}^{N \times H \times W}$, where H, W denote the height and width of the input image and N denotes the number of classes. An encoder-decoder paradigm is often adopted, where the encoder applies several down-sampling operations and the decoder employs up-sampling modules to recover the original size. To capture rich information, state-of-the-art methods propose to aggregate features from different levels to capture both local details and high-level semantic information. Following the setting in FPN [20], different levels of features $\{F_i\}_{i=2}^5$ are extracted from different network stages, where a larger i denotes a deeper stage. $F_i \in \mathbb{R}^{C_i \times H_i \times W_i}$ is a C_i dimensional feature map defined on a spatial grid with size of $H_i \times W_i$ ($H_i = \frac{H}{2^i}$, $W_i = \frac{W}{2^i}$). To compensate the information loss during consecutive down-sampling operations, FPN aims to fuse different levels of feature for better representations. Originally proposed for object detection [26], FPN fuses high-level feature maps with low-level features in a top-down strategy step by step. At each step, FPN fuses high-level feature map with low-level one through several 2× bilinear up-sampling operations with convolution layers.

3.2   Overall Framework

Network Architecture. Fig. 2 demonstrates an overview of our network architecture, the general architecture can be described as an encoder part (in blue) and a decoder part (in orange). For a given input image, a bottom-up encoder will encode the image to the feature maps in different levels. A typical example of the encoder is ResNet [12], which generates four feature maps $\{F_i\}_{i=2}^5$ from different stages. For the decoder part, while FPN builds a top-down pathway with bilinear up-sampling, our method instead takes features $\{F_i\}_{i=2}^5$ from the encoder as inputs and decodes the output at every coordinate independently and parallelly, forming a point-independent prediction manner.

Supervision. Following standard practice in previous state-of-the-art works [13–15,23,24,53,56], we add the auxiliary supervision for improving the performance, as well as making the network easier to optimize. Specifically, the output of the third stage of the backbone is further fed into an auxiliary layer to produce an auxiliary prediction, which is supervised with the auxiliary loss. We apply standard cross entropy loss to supervise the auxiliary output and employ OHEM loss [37] to supervise the main output.

3.3 Implicit Feature Alignment Function

In this subsection, we will first introduce an implicit feature function defined on a single feature map. Then we present the details of the position encoding used in our method. Finally, we extend the implicit feature function to the IFA for multi-level features.

Implicit Feature Function. One of the main challenges of aggregating information from multi-level feature maps comes from their different resolutions. Up-sampling modules are usually necessary to align them within the same resolution. Our key idea is to define continuous feature maps (i.e. fields of features), which can be decoded at arbitrary coordinates, so that they are aligned in a continuous field and no up-sampling is required.

To define a continuous feature map M, we introduce the implicit feature function. It is inspired by recent works of implicit neural representations [4, 18] for 3D reconstruction and image super-resolution. Implicit feature function defines a decoding function f_θ (typically an MLP) over a discrete feature map to get the continuous feature map M. Given the discrete feature map, feature vectors are viewed as latent codes evenly distributed in the 2D space, each of them is assigned with a 2D coordinate. The feature value of M at x_q is defined by

$$M(x_q) = f_\theta(z^*, x_q - x^*), \tag{1}$$

where z^* is the nearest latent code from x_q and x^* is the coordinate of latent code z^*. To summarize, with the decoding function f_θ, we can define a continuous feature map M over a discrete feature map. In practice, f_θ is jointly learned with the feature encoder so that the features are learned to precisely represent continuous fields of information.

Position Encoding. As discussed in previous works [30,47], although neural networks can be treated as universal function approximators, the learning power gets limited when directly operated on xy coordinates due to its inferiority at representing high-frequency signals. This is consistent with the discovery of a recent work [33] that neural networks are biased towards low-frequency signals and are insensitive to high-frequency signals. Therefore, instead of directly feeding the coordinates to the network, we first encode them with the position encoding function. Formally, the encoding function we use is:

$$\psi(x) = (\sin(\omega_1 x), \cos(\omega_1 x), ..., \\ \sin(\omega_L x), \cos(\omega_L x)), \tag{2}$$

where the frequency ω_l are initialized as $\omega_l = 2e^l, l \in \{1, ..., L\}$ and can be fine-tuned during training, and the encoding function expands the 2D coordinates into the $2L$-dimensional encoding. We also perform experiments on position encoding functions using only sin or cos function, and Eq. 2 performs the best (see Sect. 4.3 for details). The final definition of implicit feature function is:

$$M(x_q) = f_\theta(z^*, \psi(x_q - x^*), x_q - x^*), \tag{3}$$

where the relative coordinates together with their position encodings are fed into the implicit function.

Feature Alignment. A direct way to perform feature alignment is to define implicit feature functions and convert each feature map in different levels to a continuous feature map, so that their features can be queried at arbitrary coordinates for alignment. In this subsection, we show that this can be simplified to a more efficient method.

Take aligning the feature maps $\{F_i\}_{i=2}^5$ as an example, we extend the implicit feature function to implicit feature alignment function (IFA), which directly defines a continuous feature map M over multi-level discrete feature maps in different resolutions. Specifically, we define the the value of M at x_q as

$$M(x_q) = f_\theta\big(\{z_i^*\}_{i=2}^5, \{\psi_i(\delta x_i), \delta x_i\}_{i=2}^5\big), \qquad (4)$$
$$\delta x_i = x_q - x_i^*,$$

where i denotes the index of feature level, z_i^* is the nearest latent code from x_q at level i and x_i^* is the coordinate of z_i^*. We implement f_θ as concatenating all its input vectors and passing it into an MLP. Intuitively, each latent code still represents a field of feature that can be decoded by relative coordinate, f_θ can decode the field for each level and model the interaction across different levels at the same time.

The alignment among features produced from different stages is also shown in the lower half of Fig. 2. For a query coordinate, we obtain the nearest latent codes from each level of features, noted as $\{z_i^*\}_{i=2}^5$, relative coordinates δx_i together with the corresponding encoded ones, noted as $\{\psi(\delta x_i)\}_{i=2}^5$. After concatenating the latent codes, relative coordinates and the encoded relative coordinates, we feed them into the MLP of the decoding function f_θ. Given a output resolution, we decode the segmentation map by querying every pixel location independently and parallelly. Therefore, IFA aligns the features in a resolution-free manner and allows decoding to arbitrary resolutions.

Besides FPN, the proposed IFA can be easily applied into other semantic segmentation models that require multi-level feature aggregation, such as DeepLab V3+ [3] and HRNet [44].

4   Experiments

4.1   Datasets and Evaluation Metrics

Cityscapes. The Cityscapes dataset [8] is tasked for urban scene understanding, which contains 30 classes and only 19 classes of them are used for scene parsing evaluation. The dataset contains 5000 finely annotated images and 20000 coarsely annotated images. The size of the images is 2048×1024 pixels. The finely annotated 5,000 images are split into 2975, 500 and 1525 images for training, validation and testing respectively. We only use finely annotated part in our experiments.

PASCAL Context. The PASCAL Context is a dataset [31] is a challenging scene parsing dataset which contains 59 semantic classes and 1 background class. The training set and test set consist of 4, 998 and 5, 105 images respectively.

ADE20K. The ADE20K dataset [59] is a large scale scene parsing benchmark which contains dense labels of 150 stuff/object categories. The annotated images are divided into 20K, 2K and 3K for training, validation and testing, respectively.

Evaluation Metric. The mean of class-wise Intersection over Union (mIoU) is used as the evaluation metric. Number of float-point operations (FLOPs) and number of parameters are also adopted for efficiency evaluations.

4.2   Implementation Details

We use ResNet pretrained on ImageNet [21] as our backbone. For Cityscapes dataset, we use stochastic gradient descent (SGD) optimizer with initial learning rate 0.01, weight decay 0.0005 and momentum 0.9. We adopt the 'poly' learning rate policy, where the initial learning rate is multiplied by $(1 - \frac{iter}{max\_iter})^{0.9}$. We adopt the crop size as 769×769, batch size as 16 and training iterations as 18k. For PASCAL Context dataset, we set the initial learning rate as 0.001, weight decay as 0.0001, crop size as 513×513, batch size as 16 and training iterations as 30K. For ADE20K dataset, we set the initial learning rate as 0.004, weight decay as 0.0001, crop size as 480×480, batch size as 16 and training iterations as 150K.

4.3   Results and Ablations

In this subsection, we conduct extensive experiments on the val set of Cityscapes dataset with different settings for our proposed IFA. For all the experiments in this subsection, we use ResNet-50 as the backbone and down-sampling rate as 32 if not specified. All compared methods are evaluated by single-scale inference.

Table 1. Performance comparisons of different aligning methods within the FPN structure on Cityscapes val set. GFLOPs calculations adopt 1024×1024 images as input.

Method	mIoU(%)	#Params	GFLOPs
Bilinear Up-sampling	76.52	27.7M	183.4
Nearest Neighbor	76.32	27.7M	183.4
Deconvolution	72.89	29.5M	304.4
Up-sampling Module	77.19	31.0M	219.1
CARAFE [43]	76.80	29.0M	190.5
AlignSeg [17]	78.50	49.7M	348.6
IFA (Ours)	78.02	27.8M	186.9

Table 2. Performance of IFA on different segmentation models on Cityscapes `val` set.

Method	mIoU(%)
DeepLab V3+	76.69
DeepLab V3+ (IFA)	**77.57**
PSPNet	73.64
PSPNet (IFA)	**74.42**
HRNet-W18	77.60
HRNet-W18 (IFA)	**78.10**
HRNet-W48-OCR	85.80
HRNet-W48-OCR (IFA)	**86.10**

Aligning Method. We first compare IFA against commonly used aligning methods, i.e. bilinear up-sampling, nearest up-sampling, deconvolution and up-sampling module (bilinear+convolution), and state-of-the-art methods including CARAFE [43] and AlignSeg [17]. In particular, we use FPN as the decoder, where the original aligning method is the up-sampling module. We then replace it with other aligning methods. As shown in Table 1, our proposed IFA performs the best over other baseline methods. While up-sampling module also achieves high performance, its overhead is much higher than the proposed IFA. IFA achieves better results than up-sampling module with 85% of its computation. Moreover, although AlginSeg obtains slightly better results, it takes up almost twice as many parameters and computation cost as ours. Hence, IFA achieves a better trade-off between computational cost and accuracy.

Extension to Other Models. Since our proposed IFA targets at aligning features from different levels, we can directly apply it into other segmentation models involving feature alignment such as DeepLab V3+ [3], PSPNet [56] and HRNet [44]. In particular, DeepLab V3+ aggregates low-level feature F_2 and high-level feature F_5 by simple bilinear up-sampling, which can be replaced by IFA. PSPNet aggregate features of different scales produced with different pooling strides by bilinear up-sampling as well. And HRNet also aggregates features of four different scales by bilinear up-sampling. Hence, we plug IFA into these models to replace bilinear up-sampling and perform alignment. The results are presented in Table 2. IFA improves DeepLab V3+ by 0.9%, PSPNet by 0.8%, and HRNet-W18 by 0.5%. Furthermore, IFA also boost the performance of HRNet-W48-OCR [52] by 0.3%, indicating the strong generalization ability of IFA for different segmentation models.

Resolution Difference. Commonly in FPN based methods, the largest scale difference between two feature maps is 8 times (between F_2 and F_5). The larger the scale difference is, the harder it is to align feature maps. Moreover, as the improvement of high-resolution image collection tools, we will obtain a large amount of super high resolution data for training and testing. However, with

Table 3. Effect of resolution difference on the feature maps of the FPN model. 'Stride' denotes the down-sampling rate of the network and 'Diff' denotes the scale different between F_2 and F_5. Results are reported on Cityscapes val set.

Method	Stride	Diff	IFA	mIoU(%)	Gain(%)
FPN	32	8		77.19	0.9
			✓	**78.02**	
	64	16		76.52	1.1
			✓	**77.69**	
	128	32		74.88	1.6
			✓	**76.40**	
DeepLab V3+	32	8		76.69	0.9
			✓	**77.57**	
	64	16		75.20	1.1
			✓	**76.23**	
	128	32		70.01	2.1
			✓	**72.18**	

higher-resolution images as input, current methods could be unable to fit into training machines due to limited memory capacity.

To demonstrate that our proposed IFA can better align feature maps with large scale difference, we alternate the stride of FPN and DeepLab V3+ models and apply IFA. In particular, we add average pooling operation after the first stage of the backbone to further downsample the feature maps, which simultaneously increase both the down-sampling rate of the network and the scale difference between the highest resolution feature (F_2) and the lowest resolution feature (F_5). The results of experiments on FPN is shown in Table 3. As the scale difference gets larger, the performance gain over the baseline becomes larger as well, demonstrating the capability of IFA to align feature map with large scale difference. And the results of experiments on DeepLab V3+ is also shown in Table 3. Similar conclusion can be obtained.

Visualizations of the effect of IFA. We further provide comparisons of visualization results on val set of Cityscapes dataset in Fig. 3. IFA considerably resolves category ambiguities within large objects and produces more precise boundaries of small objects, by effectively aggregating low-level and high-level feature maps. Hence, low-level spatial details and high-level semantic information can be precisely aligned to produce a more accurate prediction.

Visualizations of the feature maps. To better demonstrate the effect of IFA, we visualize feature maps from the first stage and final stage of the encoder and the decoded feature from IFA. Bright areas denote the existence of objects. The visualizations are generated by averaging the features along the channel dimension. As shown in Fig. 4, IFA can effectively leverage spatial details from

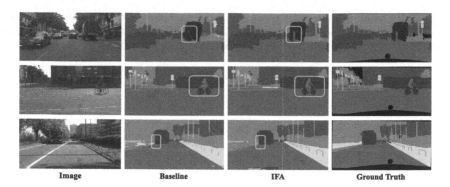

Image   Baseline   IFA   Ground Truth

Fig. 3. Visualization results on Cityscapes `val` set. From left to right: input image, predictions made by the FPN baseline, predictions made by the FPN with the proposed IFA and groundtruth map. Yellow squares denote the challenging regions that can be resolved by our proposed IFA.

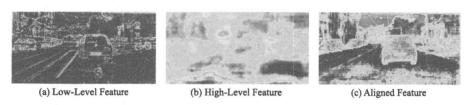

(a) Low-Level Feature   (b) High-Level Feature   (c) Aligned Feature

Fig. 4. Visualizations of feature maps. (a) Feature map from the first stage of the encoder. (b) Feature map from the last stage of the encoder. (c) Aligned feature from our proposed IFA.

the low-level feature and semantic information from the high-level feature, thus output a comprehensive feature representation.

4.4 Comparisons with State-of-the-Arts

In this subsection, we compare our method with other state-of-the-art methods on three benchmark datasets including Cityscapes, PASCAL Context and ADE20K. Specifically, we choose ResNet-101 as the encoder, FPN as the decoder and replace the up-sampling module in FPN with our proposed IFA. Moreover, to further improve the performance, we add an ASPP [2] module at the end of the encoder, only performing contextual learning on the last feature map of the pyramid (F_5). Similar to other state-of-the-art methods, we also use the multi-scale and flipping strategies for testing to achieve better results. For convenience, we use IFA to represent FPN with IFA.

Cityscapes. We train the proposed method using both training and validation set of Cityscapes dataset and make the evaluation on the **test** set by submitting our test results to the official evaluation server. Model parameters and computation FLOPs are also listed for comparison. From Table 4, it can be observed

Table 4. Comparisons with state-of-art on the Cityscapes `test` set. 'IFA*' denotes IFA with ASPP module. All methods use multi-scale inference. The GFLOPs is calculated with a image size of 1024×1024.

Method	Backbone	mIoU(%)	#Params	GFLOPs
RefineNet [25]	Dilated ResNet-101	73.6	-	-
PSPNet [56]	Dilated ResNet-101	78.4	68.1M	1104.4
AAF [19]	Dilated ResNet-101	79.1	-	-
DFN [51]	Dilated ResNet-101	79.3	96.7M	1185.6
PSANet [57]	Dilated ResNet-101	80.1	89.2M	1205.7
DenseASPP [48]	DenseNet-161	80.6	39.9M	640.1
ANNet [61]	Dilated ResNet-101	81.3	66.5M	1120.5
CPNet [49]	ResNet-101	81.3		
CCNet [16]	Dilated ResNet-101	81.4	69.8M	1190.0
DANet [9]	Dilated ResNet-101	81.5	69.7M	1335.9
STLNet [60]	Dilated ResNet-101	**82.3**	81.39M	535.9
SETR [58]	ViT-Large	81.1	318.3M	2352.0
SegFormer [46]	MiT-B5	82.2	84.7M	730.5
BiSeNet [50]	ResNet-18	77.7	15.6M	130.2
BiSeNet [50]	ResNet-101	78.9	54.2M	255.1
SFNet [24]	ResNet-18	79.5	15.9M	165.4
SFNet [24]	ResNet-101	81.8	55.0M	459.9
IFA (Ours)	ResNet-18	79.3	12.3M	93.3
IFA* (Ours)	ResNet-18	79.8	16.8M	98.0
IFA (Ours)	ResNet-101	81.2	46.7M	262.8
IFA* (Ours)	ResNet-101	**82.0**	64.3M	281.4

that our proposed IFA achieves competitive performance on Cityscapes `test` set with less computation cost. In particular, our proposed IFA achieves competitive result (81.2) compared with SFNet (81.8) with only 57% of its computational overhead. And with ASPP module bringing extra 7% computation, our method (82.0) surpasses SFNet (81.8) while only requiring 61% of its computation. Moreover, although recent Transformer based model SegFormer achieves slightly better results (82.2) than ours (82.0), it takes extra 32% parameters and 160% computation cost. The results demonstrate that our method achieved a better computation-accuracy trade-off compared with other state-of-the-art methods. It's also worth noting that SegFormer backbone is pre-trained with stronger data augmentation from the recipe of DeiT [41], while our backbone is pre-trained with standard data augmentation.

PASCAL Context. We also conduct experiments on the PASCAL Context dataset. We report the results under 60 classes without the background. Table 5 presents the results on the PASCAL Context `test` set. Our method achieves

Table 5. Comparisons with state-of-art on the PASCAL Context **test** set. The results are reported under 60 classes (w/o background). 'D-' denotes the dilated version of the backbone and 'IFA*' denotes IFA with ASPP module. All methods use multi-scale inference. GFLOPs is calculated with a image size of 480 × 480.

Method	Backbone	mIoU(%)	GFLOPs
FCN-8s [27]	VGG-16	37.8	-
DeepLab V2 [1]	D-ResNet-101	45.7	-
RefineNet [25]	ResNet-152	47.3	-
EncNet [54]	D-ResNet-101	51.7	-
DANet [9]	D-ResNet-101	52.6	296.4
ANNet [61]	D-ResNet-101	52.8	248.3
EMANet [22]	D-ResNet-101	53.1	212.3
SETR [58]	ViT-Large	**55.8**	519.5
SFNet [24]	ResNet-101	53.8	103.0
IFA (Ours)	ResNet-101	53.0	59.1
IFA* (Ours)	ResNet-101	**53.8**	63.5

competitive performance compared with state-of-the-art methods, with less computation cost. In particular, our method surpasses most of the previous methods with much less computation cost and achieves competitive results with SFNet, with only 61% of its overhead.

ADE20K. We also carry out experiments on the ADE20K dataset. Performance results are reported in Table 6. Our method achieves competitive result compared with CNN based CPNet, with only 22% of its computational cost.

Table 6. Comparisons with state-of-art on the ADE20K **val** set. 'D-' denotes the dilated version of the backbone, 'IFA*' denotes IFA with ASPP module. All methods use multi-scale inference. GFLOPs is calculated with a image size of 512 × 512.

Method	Backbone	mIoU(%)	GFLOPs
PSPNet [56]	D-ResNet-101	43.29	280.3
PSANet [57]	D-ResNet-101	43.77	305.4
CFNet [55]	D-ResNet-101	44.89	-
CCNet [16]	D-ResNet-101	45.22	301.2
APCNet [11]	D-ResNet-101	45.38	-
CPNet [49]	D-ResNet-101	46.27	314.3
SETR [58]	ViT-Large	50.28	591.0
SegFormer [46]	MiT-B5	**51.80**	184.6
SFNet [24]	ResNet-101	44.67	119.4
IFA (Ours)	ResNet-101	45.23	67.1
IFA* (Ours)	ResNet-101	**45.98**	72.0

5   Conclusion

In this paper, we focus on the feature alignment problem in popular semantic segmentation models involving feature aggregation operations. Hence, we present the Implicit Feature Alignment function (IFA) to perform precise feature alignment among multi-level features. IFA let multi-level features be learned as representing a continuous field of feature and represents the context from different levels as a function of continuous coordinates, which leads to a precise feature alignment in a resolution-free manner. Extensive experiments demonstrate the effectiveness of each component of IFA. Our IFA achieves competitive results on three benchmark datasets, *i.e.,* Cityscapes, PASCAL Context and ADE20K. Importantly, our method obtains a better trade-off between segmentation accuracy and computational cost than previous methods.

Acknowledgement. This work was supported, in part, by gifts from Qualcomm.

References

1. Chen, L.C., Papandreou, G., Kokkinos, I., Murphy, K., Yuille, A.L.: DeepLab: semantic image segmentation with deep convolutional nets, Atrous convolution, and fully connected CRFs. IEEE Trans. Pattern Anal. Mach. Intell. **40**(4), 834–848 (2017)
2. Chen, L.C., Papandreou, G., Schroff, F., Adam, H.: Rethinking atrous convolution for semantic image segmentation. arXiv preprint arXiv:1706.05587 (2017)
3. Chen, L.-C., Zhu, Y., Papandreou, G., Schroff, F., Adam, H.: Encoder-decoder with atrous separable convolution for semantic image segmentation. In: Ferrari, V., Hebert, M., Sminchisescu, C., Weiss, Y. (eds.) Computer Vision – ECCV 2018. LNCS, vol. 11211, pp. 833–851. Springer, Cham (2018). https://doi.org/10.1007/978-3-030-01234-2_49
4. Chen, Y., Liu, S., Wang, X.: Learning continuous image representation with local implicit image function. In: Proceedings of the IEEE/CVF Conference on Computer Vision and Pattern Recognition, pp. 8628–8638 (2021)
5. Chen, Z., Zhang, H.: Learning implicit fields for generative shape modeling. In: Proceedings of the IEEE/CVF Conference on Computer Vision and Pattern Recognition, pp. 5939–5948 (2019)
6. Cheng, B., Parkhi, O., Kirillov, A.: Pointly-supervised instance segmentation. In: Proceedings of the IEEE/CVF Conference on Computer Vision and Pattern Recognition, pp. 2617–2626 (2022)
7. Cheng, B., Schwing, A., Kirillov, A.: Per-pixel classification is not all you need for semantic segmentation. Adv. Neural. Inf. Process. Syst. **34**, 1–12 (2021)
8. Cordts, M., et al.: The cityscapes dataset for semantic urban scene understanding. In: Proceedings of the IEEE Conference on Computer Vision and Pattern Recognition, pp. 3213–3223 (2016)
9. Fu, J., et al.: Dual attention network for scene segmentation. In: Proceedings of the IEEE Conference on Computer Vision and Pattern Recognition, pp. 3146–3154 (2019)
10. Genova, K., Cole, F., Vlasic, D., Sarna, A., Freeman, W.T., Funkhouser, T.: Learning shape templates with structured implicit functions. In: Proceedings of the IEEE/CVF International Conference on Computer Vision, pp. 7154–7164 (2019)

11. He, J., Deng, Z., Zhou, L., Wang, Y., Qiao, Y.: Adaptive pyramid context network for semantic segmentation. In: Proceedings of the IEEE/CVF Conference on Computer Vision and Pattern Recognition, pp. 7519–7528 (2019)
12. He, K., Zhang, X., Ren, S., Sun, J.: Deep residual learning for image recognition. In: Proceedings of the IEEE Conference on Computer Vision and Pattern Recognition, pp. 770–778 (2016)
13. Hu, H., Cui, J., Wang, L.: Region-aware contrastive learning for semantic segmentation. In: Proceedings of the IEEE/CVF International Conference on Computer Vision, pp. 16291–16301 (2021)
14. Hu, H., Ji, D., Gan, W., Bai, S., Wu, W., Yan, J.: Class-wise dynamic graph convolution for semantic segmentation. In: Vedaldi, A., Bischof, H., Brox, T., Frahm, J.-M. (eds.) ECCV 2020. LNCS, vol. 12362, pp. 1–17. Springer, Cham (2020). https://doi.org/10.1007/978-3-030-58520-4_1
15. Hu, H., Wei, F., Hu, H., Ye, Q., Cui, J., Wang, L.: Semi-supervised semantic segmentation via adaptive equalization learning. Adv. Neural. Inf. Process. Syst. **34**, 22106–22118 (2021)
16. Huang, Z., Wang, X., Huang, L., Huang, C., Wei, Y., Liu, W.: CCNeT: Criss-cross attention for semantic segmentation. arXiv preprint arXiv:1811.11721 (2018)
17. Huang, Z., Wei, Y., Wang, X., Shi, H., Liu, W., Huang, T.S.: AlignSeg: feature-aligned segmentation networks. IEEE Trans. Pattern Anal. Mach. Intell. **44**, 550–557 (2021)
18. Jiang, C., et al.: Local implicit grid representations for 3d scenes. In: Proceedings of the IEEE/CVF Conference on Computer Vision and Pattern Recognition, pp. 6001–6010 (2020)
19. Ke, T.-W., Hwang, J.-J., Liu, Z., Yu, S.X.: Adaptive affinity fields for semantic segmentation. In: Ferrari, V., Hebert, M., Sminchisescu, C., Weiss, Y. (eds.) ECCV 2018. LNCS, vol. 11205, pp. 605–621. Springer, Cham (2018). https://doi.org/10.1007/978-3-030-01246-5_36
20. Kirillov, A., Girshick, R., He, K., Dollár, P.: Panoptic feature pyramid networks. In: Proceedings of the IEEE/CVF Conference on Computer Vision and Pattern Recognition, pp. 6399–6408 (2019)
21. Krizhevsky, A., Sutskever, I., Hinton, G.E.: ImageNet classification with deep convolutional neural networks. In: Advances In Neural Information Processing Systems, pp. 1097–1105 (2012)
22. Li, X., Zhong, Z., Wu, J., Yang, Y., Lin, Z., Liu, H.: Expectation-maximization attention networks for semantic segmentation. In: The IEEE International Conference on Computer Vision (ICCV), October 2019
23. Li, X., et al.: Improving semantic segmentation via decoupled body and edge supervision. In: Vedaldi, A., Bischof, H., Brox, T., Frahm, J.-M. (eds.) ECCV 2020. LNCS, vol. 12362, pp. 435–452. Springer, Cham (2020). https://doi.org/10.1007/978-3-030-58520-4_26
24. Li, X., et al.: Semantic flow for fast and accurate scene parsing. In: Vedaldi, A., Bischof, H., Brox, T., Frahm, J.-M. (eds.) ECCV 2020. LNCS, vol. 12346, pp. 775–793. Springer, Cham (2020). https://doi.org/10.1007/978-3-030-58452-8_45
25. Lin, G., Milan, A., Shen, C., Reid, I.: RefineNet: multi-path refinement networks for high-resolution semantic segmentation. In: Proceedings of the IEEE Conference on Computer Vision and Pattern Recognition, pp. 1925–1934 (2017)
26. Lin, T.Y., Dollár, P., Girshick, R., He, K., Hariharan, B., Belongie, S.: Feature pyramid networks for object detection. In: Proceedings of the IEEE Conference on Computer Vision and Pattern Recognition, pp. 2117–2125 (2017)

27. Long, J., Shelhamer, E., Darrell, T.: Fully convolutional networks for semantic segmentation. In: Proceedings of the IEEE Conference on Computer Vision and Pattern Recognition, pp. 3431–3440 (2015)
28. Mescheder, L., Oechsle, M., Niemeyer, M., Nowozin, S., Geiger, A.: Occupancy networks: Learning 3d reconstruction in function space. In: Proceedings of the IEEE/CVF Conference on Computer Vision and Pattern Recognition, pp. 4460–4470 (2019)
29. Michalkiewicz, M., Pontes, J.K., Jack, D., Baktashmotlagh, M., Eriksson, A.: Implicit surface representations as layers in neural networks. In: Proceedings of the IEEE/CVF International Conference on Computer Vision, pp. 4743–4752 (2019)
30. Mildenhall, B., et al.: NeRF: representing scenes as neural radiance fields for view synthesis. In: Vedaldi, A., Bischof, H., Brox, T., Frahm, J.-M. (eds.) ECCV 2020. LNCS, vol. 12346, pp. 405–421. Springer, Cham (2020). https://doi.org/10.1007/978-3-030-58452-8_24
31. Mottaghi, R., et al.: The role of context for object detection and semantic segmentation in the wild. In: Proceedings of the IEEE Conference on Computer Vision and Pattern Recognition, pp. 891–898 (2014)
32. Park, J.J., Florence, P., Straub, J., Newcombe, R., Lovegrove, S.: DeepSDF: learning continuous signed distance functions for shape representation. In: Proceedings of the IEEE/CVF Conference on Computer Vision and Pattern Recognition, pp. 165–174 (2019)
33. Rahaman, et al.: On the spectral bias of neural networks. In: International Conference on Machine Learning, pp. 5301–5310. PMLR (2019)
34. Ronneberger, O., Fischer, P., Brox, T.: U-Net: convolutional networks for biomedical image segmentation. In: Navab, N., Hornegger, J., Wells, W.M., Frangi, A.F. (eds.) MICCAI 2015. LNCS, vol. 9351, pp. 234–241. Springer, Cham (2015). https://doi.org/10.1007/978-3-319-24574-4_28
35. Saito, S., Huang, Z., Natsume, R., Morishima, S., Kanazawa, A., Li, H.: PIFu: pixel-aligned implicit function for high-resolution clothed human digitization. In: Proceedings of the IEEE/CVF International Conference on Computer Vision, pp. 2304–2314 (2019)
36. Shen, T., et al.: High quality segmentation for ultra high-resolution images. In: Proceedings of the IEEE/CVF Conference on Computer Vision and Pattern Recognition, pp. 1310–1319 (2022)
37. Shrivastava, A., Gupta, A., Girshick, R.: Training region-based object detectors with online hard example mining. In: Proceedings of the IEEE Conference on Computer Vision and Pattern Recognition, pp. 761–769 (2016)
38. Simonyan, K., Zisserman, A.: Very deep convolutional networks for large-scale image recognition. arXiv preprint arXiv:1409.1556 (2014)
39. Sitzmann, V., Martel, J., Bergman, A., Lindell, D., Wetzstein, G.: Implicit neural representations with periodic activation functions. Adv. Neural. Inf. Process. Syst. **33**, 1–12 (2020)
40. Strudel, R., Garcia, R., Laptev, I., Schmid, C.: Segmenter: Transformer for semantic segmentation. In: Proceedings of the IEEE/CVF International Conference on Computer Vision, pp. 7262–7272 (2021)
41. Touvron, H., Cord, M., Douze, M., Massa, F., Sablayrolles, A., Jégou, H.: Training data-efficient image transformers & distillation through attention. In: International Conference on Machine Learning, pp. 10347–10357. PMLR (2021)
42. Vaswani, A., et al.: Attention is all you need. In: Advances in Neural Information Processing Systems, pp. 5998–6008 (2017)

43. Wang, J., Chen, K., Xu, R., Liu, Z., Loy, C.C., Lin, D.: CARAFE: content-aware reassembly of features. In: 2019 IEEE/CVF International Conference on Computer Vision, ICCV 2019, Seoul, Korea (South), October 27–2 November 2019, pp. 3007–3016. IEEE (2019)
44. Wang, J., et al.: Deep high-resolution representation learning for visual recognition. IEEE Trans. Pattern Anal. Mach. Intell. **43**, 3349–3364 (2020)
45. Wang, X., Girshick, R., Gupta, A., He, K.: Non-local neural networks. In: Proceedings of the IEEE Conference on Computer Vision and Pattern Recognition, pp. 7794–7803 (2018)
46. Xie, E., Wang, W., Yu, Z., Anandkumar, A., Alvarez, J.M., Luo, P.: SegFormer: simple and efficient design for semantic segmentation with transformers. Adv. Neural. Inf. Process. Syst. **34**, 1–14 (2021)
47. Xu, X., Wang, Z., Shi, H.: UltraSR: spatial encoding is a missing key for implicit image function-based arbitrary-scale super-resolution. arXiv preprint arXiv:2103.12716 (2021)
48. Yang, M., Yu, K., Zhang, C., Li, Z., Yang, K.: DenseASPP for semantic segmentation in street scenes. In: Proceedings of the IEEE Conference on Computer Vision and Pattern Recognition, pp. 3684–3692 (2018)
49. Yu, C., Wang, J., Gao, C., Yu, G., Shen, C., Sang, N.: Context prior for scene segmentation. In: Proceedings of the IEEE/CVF Conference on Computer Vision and Pattern Recognition, pp. 12416–12425 (2020)
50. Yu, C., Wang, J., Peng, C., Gao, C., Yu, G., Sang, N.: BiSeNet: bilateral segmentation network for real-time semantic segmentation. In: Ferrari, V., Hebert, M., Sminchisescu, C., Weiss, Y. (eds.) ECCV 2018. LNCS, vol. 11217, pp. 334–349. Springer, Cham (2018). https://doi.org/10.1007/978-3-030-01261-8_20
51. Yu, C., Wang, J., Peng, C., Gao, C., Yu, G., Sang, N.: Learning a discriminative feature network for semantic segmentation. In: Proceedings of the IEEE Conference on Computer Vision and Pattern Recognition, pp. 1857–1866 (2018)
52. Yuan, Y., Chen, X., Wang, J.: Object-contextual representations for semantic segmentation. In: Vedaldi, A., Bischof, H., Brox, T., Frahm, J.-M. (eds.) ECCV 2020. LNCS, vol. 12351, pp. 173–190. Springer, Cham (2020). https://doi.org/10.1007/978-3-030-58539-6_11
53. Zhang, F., et al.: ACFNet: attentional class feature network for semantic segmentation. In: Proceedings of the IEEE International Conference on Computer Vision, pp. 6798–6807 (2019)
54. Zhang, H., et al.: Context encoding for semantic segmentation. In: Proceedings of the IEEE Conference on Computer Vision and Pattern Recognition, pp. 7151–7160 (2018)
55. Zhang, H., Zhang, H., Wang, C., Xie, J.: Co-occurrent features in semantic segmentation. In: Proceedings of the IEEE/CVF Conference on Computer Vision and Pattern Recognition, pp. 548–557 (2019)
56. Zhao, H., Shi, J., Qi, X., Wang, X., Jia, J.: Pyramid scene parsing network. In: Proceedings of the IEEE Conference on Computer Vision and Pattern Recognition, pp. 2881–2890 (2017)
57. Zhao, H., et al.: PSANet: point-wise spatial attention network for scene parsing. In: Ferrari, V., Hebert, M., Sminchisescu, C., Weiss, Y. (eds.) ECCV 2018. LNCS, vol. 11213, pp. 270–286. Springer, Cham (2018). https://doi.org/10.1007/978-3-030-01240-3_17

58. Zheng, S., et al.: Rethinking semantic segmentation from a sequence-to-sequence perspective with transformers. In: IEEE Conference on Computer Vision and Pattern Recognition, CVPR 2021, virtual, 19–25 June 2021, pp. 6881–6890. Computer Vision Foundation/IEEE (2021)
59. Zhou, B., Zhao, H., Puig, X., Fidler, S., Barriuso, A., Torralba, A.: Scene parsing through ade20k dataset. In: Proceedings of the IEEE Conference on Computer Vision and Pattern Recognition, pp. 633–641 (2017)
60. Zhu, L., Ji, D., Zhu, S., Gan, W., Wu, W., Yan, J.: Learning statistical texture for semantic segmentation. In: Proceedings of the IEEE/CVF Conference on Computer Vision and Pattern Recognition, pp. 12537–12546 (2021)
61. Zhu, Z., Xu, M., Bai, S., Huang, T., Bai, X.: Asymmetric non-local neural networks for semantic segmentation. In: Proceedings of the IEEE International Conference on Computer Vision, pp. 593–602 (2019)

Quantum Motion Segmentation

Federica Arrigoni[1($\boxtimes$)], Willi Menapace[2], Marcel Seelbach Benkner[3],
Elisa Ricci[2,4], and Vladislav Golyanik[5]

[1] Politecnico di Milano, Milan, Italy
federica.arrigoni@polimi.it
[2] University of Trento, Trento, Italy
[3] University of Siegen, Siegen, Germany
[4] Bruno Kessler Foundation, Trento, Italy
[5] MPI for Informatics, Saarbrücken, Germany

Abstract. Motion segmentation is a challenging problem that seeks to identify independent motions in two or several input images. This paper introduces the first algorithm for motion segmentation that relies on adiabatic quantum optimization of the objective function. The proposed method achieves on-par performance with the state of the art on problem instances which can be mapped to modern quantum annealers.

Keywords: Motion segmentation · Quantum approach · Synchronization

1 Introduction

Quantum computer vision is an emerging field. Recently, several classical problems were reformulated to enable quantum optimization, including recognition [15,44] and matching tasks [10,54]. Promising results were shown so far, thus encouraging further research. Among the two existing paradigms for quantum computing, *i.e.*, gate-based and adiabatic quantum computing (AQC), experimental realizations of AQC are already applicable to real-world problems, provided that the objective is given as a *quadratic unconstrained binary optimization* (QUBO) problem. Thus, quantum annealing (QA)—which refers to not perfectly adiabatic implementations of AQC [22,27]—is an experimental and promising technology for finding solutions to combinatorial problems leveraging quantum mechanics [21,24]. QA optimises objectives *without relaxation* and obtains *globally-optimal or low-energy solutions* with high probabilities. Note that these important properties are hardly present in traditional methods, hence it is crucial to identify problems benefiting from this new class of machines.

In [10] an AQC algorithm for *permutation synchronization* is proposed, which finds cycle-consistent matches across a set of images or shapes, where the matches

---

Supplementary Information The online version contains supplementary material available at https://doi.org/10.1007/978-3-031-19818-2_29.

are given as permutation matrices. The recent survey [3] discusses many synchronization problems already studied in the literature (*e.g.*, rotation synchronization for structure from motion [16] or pose synchronization for point-set registration [26]). Notwithstanding, permutation synchronization [10] is the only one that has been solved via quantum optimization so far.

| Ground truth | Our method | MODE [5] | SYNCH [4] | Xu et al. [62] |

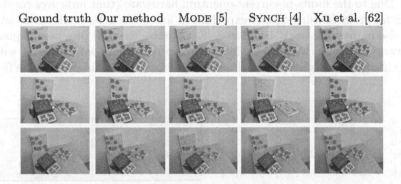

Fig. 1. Qualitative results on sample images from the new Q-MSEG dataset, where each color (symbol) represents a distinct planar motion. On average, the accuracies of our QuMoSEG-v1, MODE [5], SYNCH [4] and Xu et al. [62] are 0.97, 0.93, 0.93 and 0.89, respectively, on problems with 96 qubits. In the shown example, our approach outperforms the competitors. See Table 1 for further details.

This paper advances the state of the art in quantum computer vision by bringing a new synchronization problem, *i.e.*, *motion segmentation*, into an AQC-admissible form; see Fig. 1 for exemplary results. The task of motion segmentation [52] is to classify points in multiple images into different motions, which is relevant in such applications as dynamic 3D reconstruction [45] or autonomous driving [50]. Observe that quantum formulations do not make sense for all problems, but for those, *e.g.*, that include *combinatorial* optimisation objectives, which are usually $\mathcal{NP}$-hard. Motion segmentation is identified to have a combinatorial structure, and, hence, is a meaningful candidate to leverage the advantages of the quantum processor. Bringing motion segmentation into an AQC is challenging as only problems in a QUBO form are admitted. Thus, we primarily focus on how to formulate motion segmentation as a QUBO.

Our work adopts the synchronization formulation of motion segmentation from [4], which we carefully convert to a QUBO problem. This gives rise to the first variant of our quantum approach, named QuMoSEG-v1 (from "**QU**antum **MO**tion **SEG**mentation"): it works well in many practical scenarios but it can not manage large-scale problems since it is based on a *dense* matrix. For this reason, we also develop an alternative method based on a *sparse* matrix which can solve larger problems, resulting in QuMoSEG-v2: its derivation, however, requires additional assumptions, *i.e.*, the knowledge of the number of points per motion. In summary, our primary contributions are:

1) A new approach to motion segmentation that employs AQC (Sect. 3);
2) A new real dataset (Q-MSEG) for motion segmentation (Sect. 5).

In our extensive experiments, our approach achieves competitive accuracy (close to or higher than competing methods) on problem instances which are mappable to the AQC of the latest generation, and demonstrates its high robustness to noise. Due to the limits of current quantum hardware (that improves constantly and is far from maturity), our experiments are limited to small-scale data, as done also in previous work [10]. However, it is expected that progress in quantum hardware, alongside with the ability to solve combinatorial problems without approximation, will give practical advantages for large-scale data in the future.

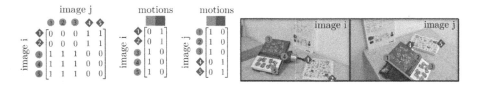

Fig. 2. Matrix representation of motion segmentation.

Our derivations share only a few similarities with Birdal *et al.* [10]: In fact, bringing motion segmentation into a QUBO form requires more analytical steps compared to previous work on quantum synchronization [10] because binary matrices are less constrained than permutation matrices. Moreover, our formulation for synchronization requires linearly-many variables in the number of input points, hence we can handle more points compared to [10]. Our source code and data are publicly available[1].

2   Background

Our work is inspired by Arrigoni and Pajdla [4] (reviewed in Sect. 2.1), where a convenient matrix representation is introduced for motion segmentation from pairwise correspondences. In Sect. 3 we will show how to rewrite such framework in terms of a QUBO, in order to enable adiabatic quantum optimization. In this respect, we report some preliminary notions on quantum computing in Sect. 2.2.

2.1   Motion Segmentation

The objective of the motion segmentation problem is to group key-points in multiple images according to a number of motions. We use the following notation: n is the number of images; p_i is the number of key-points in image i; $p = \sum_{i=1}^{n} p_i$ is the total amount of key-points; d is the number of motions

---

[1] See the project page https://4dqv.mpi-inf.mpg.de/QuMoSeg/.

(known by assumption). We focus here on motion segmentation from pairwise correspondences [4,5], which can be addressed in two steps: 1) motion segmentation is addressed on different *pairs* of images independently, which in turn can be done via multi-model fitting [6,39,40]; 2) the results derived in the first step are globally combined, thus producing the required *multi-frame* segmentation.

As shown in [4], motion segmentation can be seen as a "synchronization" of binary matrices. Indeed, the result of motion segmentation in two images i and j can be represented as a matrix $Z_{ij} \in \{0,1\}^{p_i \times p_j}$ as follows:

- $[Z_{ij}]_{h,k} = 1$ if point h in image i and point k in image j belong to the same motion;
- $[Z_{ij}]_{h,k} = 0$ otherwise.

The (known) binary matrix Z_{ij} is referred to as the *partial segmentation* or *relative segmentation* of the pair (i,j). It is a *local* representation of segmentation, since it reveals which points in two different images belong to the same motion, but it does not reveal which motion it is with respect to other pairs. Similarly, our desired output can be represented as a matrix $X_i \in \{0,1\}^{p_i \times d}$ as follows:

- $[X_i]_{h,k} = 1$ if point h in image i belongs to motion k;
- $[X_i]_{h,k} = 0$ otherwise.

The binary matrix X_i is called the *total segmentation* or *absolute segmentation* of image i. Observe that the number of rows is equal to the number of points while the number of columns is equal to the number of motions. Note also that it is a *global* representation of segmentation since it reveals the membership of all points with respect to an absolute order of motions. The notions of absolute and relative segmentations are illustrated in Fig. 2.

Remark 1. Let $\mathbf{m}_i$ be a vector of length d such that $[\mathbf{m}_i]_h$ counts how many points in image i belong to motion h. Then the columns in X_i sum to $\mathbf{m}_i$:

$$\mathbf{1}_{p_i}^\mathsf{T} X_i = \mathbf{m}_i^\mathsf{T}, \tag{1}$$

where $\mathbf{1}$ is a vector of ones (with length given as subscript). Note also that the product $X_i^\mathsf{T} X_i$ is a $d \times d$ diagonal matrix:

$$X_i^\mathsf{T} X_i = \mathrm{diag}(\mathbf{m}_i). \tag{2}$$

These simple properties will be exploited later.

The connection between relative and absolute segmentations [4] is given by:

$$Z_{ij} = X_i X_j^\mathsf{T}. \tag{3}$$

Recall that the left side in the above equation is known whereas the right side is unknown. In general, there are multiple image pairs giving rise to an equation of the form (3), which can be conveniently represented as the edge set $\mathcal{E}$ of a graph $\mathcal{G} = (\mathcal{V}, \mathcal{E})$, where $\mathcal{V} = \{1, \dots, n\}$ denotes the vertex set. In other terms, each vertex represents an image and an edge is present between two vertices if and only if the relative segmentation of that image pair is available. Thus motion segmentation can be cast to the task of recovering $X_1, \dots, X_n$ starting from Z_{ij} with $(i,j) \in \mathcal{E}$, such that (3) is satisfied. This is also called *synchronization* [3].

2.2   Adiabatic Quantum Optimization

Modern AQC can solve *quadratic unconstrained binary optimization* (QUBO) problems of the form

$$\min_{\mathbf{y} \in \mathcal{B}^k} \mathbf{y}^\top Q \mathbf{y} + \mathbf{s}^\top \mathbf{y}, \tag{4}$$

where $\mathcal{B}^k$ denotes the set of binary vectors of length k, $Q \in \mathbb{R}^{k \times k}$ is a real symmetric matrix and $\mathbf{s} \in \mathbb{R}^k$. Note that optimization is performed over binary variables. QUBO problems are $\mathcal{NP}$-hard. An optimization problem with hard constraints of the form $A_i \mathbf{y} = \mathbf{b}_i$ can be converted to a QUBO with soft constraints [10], in which linear terms weighted by multipliers λ_i rectify Q and $\mathbf{s}$:

$$\min_{\mathbf{y} \in \mathcal{B}^k} \mathbf{y}^\top Q \mathbf{y} + \mathbf{s}^\top \mathbf{y} + \sum_i \lambda_i \| A_i \mathbf{y} - \mathbf{b}_i \|^2. \tag{5}$$

As (4) does not easily allow including high-level constraints (*e.g.*, on a matrix rank), most computer vision problems cannot be easily posed in a QUBO form. Latest research thus focuses on finding such formulations [10,55].

AQC interprets $\mathbf{y}$ as a measurement result of k qubits, and optimization of (4) is performed on AQC not in the binary vector space but a "lifted", 2^k-dimensional space of k qubits, taking advantage of quantum-mechanical effects like superposition and entanglement. In contrast to a classical bit which can be either in state 0 or 1 at a time, a qubit $|q\rangle = \alpha|0\rangle + \beta|1\rangle$ can take any state fulfilling $\alpha, \beta \in \mathbb{C}$ and $|\alpha|^2 + |\beta|^2 = 1$. Once a QUBO form is known, it is first passed to a *minor embedding* algorithm such as Cai *et al.* [13]. Its purpose it to find a mapping of a QUBO problem (4) defined in terms of qubits—which in the following we call *logical* (*i.e.*, mathematical models)—to an AQC with *physical* qubits (*i.e.*, hardware realizations of the mathematical models). This step is necessary for most problems except the smallest ones, as the qubit connectivity pattern encoded in Q is not natively supported by the hardware [20] and several repeated physical qubits, building a *chain*, are required to represent a single logical qubit during quantum annealing.

After the initialisation in a problem-independent state, AQC is transitioning from the initial solution to the solution of the target problem in the QUBO form, *i.e.*, one says that the system evolves its state (or an *annealing* is taking place) according to the rules of quantum mechanics [24,42]. The notion *adiabatic* refers to how this transition happens in the ideal case, namely obeying the adiabatic theorem of quantum mechanics [12]. The remaining steps of an AQC algorithm are: 1) Sampling; 2) Unembedding; 3) Bitstring selection and 4) Solution interpretation [55]. QA is probabilistic in nature, and a globally-optimal measurement can be obtained with specific success probabilities. Thus, multiple annealings are required to reach a satisfactory result (QUBO *sampling*). The number of repetitions can vary by orders of magnitude depending on the probability to measure an optimal solution, the problem size and the minor embedding. Each sample is measured and unembedded, *i.e.*, returned in terms of the logical qubit measurements. Next, one or several samples are chosen as the final solution(s), and the most common criterion is the lowest energy (*i.e.*, minimal cost) over all samples. An interested reader can further refer to McGeoch [42].

3   Our Quantum Approach

At the core of our approach is a QUBO formulation of motion segmentation. This permits—for the first time in the literature—to solve the segmentation task via quantum optimization. Bringing it into a QUBO form is not straightforward and requires more analytical steps compared to previous work on quantum synchronization [10], as will be clarified later. We propose two methods: the first one is based on a dense matrix, and, hence, can require an increased embedding size (Sect. 3.1); the second one, instead, is based on a sparse matrix but relies on additional assumptions (Sect. 3.2). Both variants are principled (*i.e.*, are equivalent to synchronization) and can solve real-world problems (see Sect. 5).

3.1   QuMoSeg-v1

As explained in Sect. 2.1, motion segmentation can be posed as computing absolute segmentations (*i.e.*, $X_1, \ldots, X_n$) starting from pairwise segmentations Z_{ij} with $(i, j) \in \mathcal{E}$ such that $Z_{ij} = X_i X_j^\mathsf{T}$. In the presence of noise, the task is to solve the following optimization problem:

$$\min_{X_1, \ldots X_n} \sum_{(i,j) \in \mathcal{E}} \|Z_{ij} - X_i X_j^\mathsf{T}\|_F^2,$$

$$\text{s.t. } \mathrm{vec}(X_i) \in \mathcal{B}^{p_i}, \quad X_i \mathbf{1}_d = \mathbf{1}_{pi} \quad \forall i = 1, \ldots, n, \tag{6}$$

where $\mathcal{B}^k$ denotes the set of binary vectors of length k, and $\mathrm{vec}(\cdot)$ denotes the vectorization operator that transforms a matrix into a vector by stacking the columns one under the other. Recall that X_i has size $p_i \times d$, so $\mathrm{vec}(X_i)$ has length dp_i and should be a binary vector. The constraint $X_i \mathbf{1}_d = \mathbf{1}_{pi}$ means that each row sums to 1. Indeed, each row in X_i has exactly one entry equal to 1 (whereas all other entries are zero), meaning that each point should belong to exactly one motion. The cost in (6) measures, for each image pair, the discrepancy (in the Frobenius norm sense) between the input relative segmentation (*i.e.*, Z_{ij}) and the relative segmentation derived from the sought absolute segmentations (*i.e.*, $X_i X_j^\mathsf{T}$). It is also known as the *consistency error* [3].

Proposition 1. *Problem* (6) *is equivalent to*

$$\max_{X_1, \ldots X_n} \sum_{(i,j) \in \mathcal{E}} \mathrm{trace}(X_i^\mathsf{T}(2Z_{ij} - \mathbf{1}_{p_i \times p_j})X_j)$$

$$\text{s.t. } \mathrm{vec}(X_i) \in \mathcal{B}^{p_i}, \quad X_i \mathbf{1}_d = \mathbf{1}_{pi} \quad \forall i = 1, \ldots, n \tag{7}$$

where $\mathbf{1}_{p_i \times p_j}$ denotes a $p_i \times p_j$ matrix of ones.

Proof. For simplicity of exposition, we drop constraints and focus on the cost function itself. By computation and exploiting (2), we obtain:

$$
\begin{aligned}
||Z_{ij} - X_i X_j^\mathsf{T}||_F^2 &= \mathrm{trace}(Z_{ij}^\mathsf{T} Z_{ij}) + \mathrm{trace}(X_j X_i^\mathsf{T} X_i X_j^\mathsf{T}) - 2\,\mathrm{trace}(Z_{ij}^\mathsf{T} X_i X_j^\mathsf{T}) = \\
&\quad \mathrm{trace}(Z_{ij}^\mathsf{T} Z_{ij}) + \mathrm{trace}(\mathrm{diag}(\mathbf{m}_i)\,\mathrm{diag}(\mathbf{m}_j)) - 2\,\mathrm{trace}(X_i^\mathsf{T} Z_{ij} X_j) = \\
&= \mathrm{trace}(Z_{ij}^\mathsf{T} Z_{ij}) + \mathbf{m}_i^\mathsf{T} \mathbf{m}_j - 2\,\mathrm{trace}(X_i^\mathsf{T} Z_{ij} X_j).
\end{aligned}
\tag{8}
$$

Note that the first term is constant, for it depends on the input Z_{ij} only, hence it can be ignored in the optimization. As for the second term, using (1), we get:

$$
\mathbf{m}_i^\mathsf{T} \mathbf{m}_j = \mathrm{trace}(\mathbf{m}_i \mathbf{m}_j^\mathsf{T}) = \mathrm{trace}(X_i^\mathsf{T} \mathbf{1}_{p_i} \mathbf{1}_{p_j}^\mathsf{T} X_j) = \mathrm{trace}(X_i^\mathsf{T} \mathbf{1}_{p_i \times p_j} X_j).
\tag{9}
$$

Hence, the optimization in (6) is equivalent to (7). □

We now rewrite (7) in a compact form via the following notation, where all the measures/unknowns are grouped into block-matrices $X \in \mathbb{R}^{p \times d}$ and $Z \in \mathbb{R}^{p \times p}$:

$$
X = \begin{bmatrix} X_1 \\ X_2 \\ \dots \\ X_n \end{bmatrix}, \quad
Z = \begin{bmatrix} 0 & Z_{12} & \dots & Z_{1n} \\ Z_{21} & 0 & \dots & Z_{2n} \\ \dots & & & \dots \\ Z_{n1} & Z_{n2} & \dots & 0 \end{bmatrix}.
\tag{10}
$$

Proposition 2. *Problem* (6) *is equivalent to*

$$
\max_X \mathrm{vec}(X)^\mathsf{T} (I_{d \times d} \otimes (2Z - \mathbf{1}_{p \times p}))\,\mathrm{vec}(X),
\tag{11}
$$
$$
s.t. \ \mathrm{vec}(X) \in \mathcal{B}^{dp}, \quad X\mathbf{1}_d = \mathbf{1}_p.
$$

Proof. Let us define $W = 2Z - \mathbf{1}_{p \times p}$. Using (10), we get:

$$
\sum_{(i,j) \in \mathcal{E}} \mathrm{trace}(X_i^\mathsf{T} (2Z_{ij} - \mathbf{1}_{p_i \times p_j}) X_j) = \mathrm{trace}(X^\mathsf{T} W X).
\tag{12}
$$

The above equation can be further simplified by exploiting properties of the trace operator[2] and the Kronecker product[3] denoted by $\otimes$, resulting in:

$$
\mathrm{trace}(X^\mathsf{T} W X) = \mathrm{vec}(X)^\mathsf{T} \mathrm{vec}(W X) = \mathrm{vec}(X)^\mathsf{T} (I_{d \times d} \otimes W)\,\mathrm{vec}(X),
\tag{13}
$$

where $I_{d \times d}$ denotes the $d \times d$ identity matrix. Hence, the objective function (11) is the same as (7), which in turn is equivalent to (6), as shown in Proposition 1. As for constraints, it is easy to see that $\mathrm{vec}(X_i) \in \mathcal{B}^{p_i}$ and $X_i \mathbf{1}_d = \mathbf{1}_{p_i}$ translate into $\mathrm{vec}(X) \in \mathcal{B}^{dp}$ and $X\mathbf{1}_d = \mathbf{1}_p$, when considering all the unknowns simultaneously as stored in the block-matrix X. Hence we get the thesis. □

---

[2] For any matrices A, B of proper dimensions we have: $\mathrm{trace}(A^\mathsf{T} B) = \mathrm{vec}(A)^\mathsf{T} \mathrm{vec}(B)$.
[3] For any matrices A, B, Y of proper dimensions, the Kronecker product [38] satisfies: $\mathrm{vec}(AYB) = (B^\mathsf{T} \otimes A)\,\mathrm{vec}(Y)$.

Corollary 1. *Problem* (6) *can be mapped into a QUBO problem* (5)

$$\min_{\mathbf{y} \in \mathcal{B}^k} \mathbf{y}^\mathsf{T} Q \mathbf{y} + \lambda_1 \|A\mathbf{y} - \mathbf{b}\|^2, \tag{14}$$

where $Q = -I_{d\times d} \otimes (2Z - \mathbf{1}_{p\times p})$, $\mathbf{y} = \text{vec}(X)$, $A = (\mathbf{1}_d^\mathsf{T} \otimes I_{p\times p})$, $\mathbf{b} = \mathbf{1}_p$.

Proof. Problem (11) is the maximization of a quadratic cost function with binary variables and linear constraints, hence definitions of Q and $\mathbf{y}$ are immediate. The linear constraints $X\mathbf{1}_d = \mathbf{1}_p$ can be easily mapped into the canonical form $A\mathbf{y} = \mathbf{b}$ via properties of Kronecker product (see footnote 3) and vectorization.   □

Remark 2. $Q = -I_{d\times d} \otimes (2Z - \mathbf{1}_{p\times p})$ is symmetric and its size is $dp \times dp$, where d is the number of motions and p is the total amount of points over all images. Note also that the size of the optimization variable $\mathbf{y}$ is dp, so it scales linearly with the number of points (assuming $d \ll p$, which is usually the case in practice).

To summarize, the synchronization formulation for motion segmentation can be cast to a QUBO, thus enabling adiabatic quantum optimization. This gives rise to the first variant of our approach, which is called QuMoSeg-v1.

3.2   QuMoSeg-v2

Note that the block-matrix Z storing all partial segmentations is *sparse*, *i.e.*, most of its entries are zero (see Fig. 2 for an example of a partial segmentation). However, the matrix $2Z - \mathbf{1}_{p\times p}$ (which appears in the definition of Q in (14)) is *dense* (it has only -1 or +1 as possible entries). This may result in increased embedding size, which is undesirable in practice. This observation motivates the need for an alternative method based on a sparse matrix, which is explored here. This comes at the price of having additional assumptions, as shown below.

Proposition 3. *Let us assume that the amount of points per motion is known in each image, namely* $\mathbf{m}_i$ *is known* $\forall i = 1, \ldots, n$, *where the d-length vector* $\mathbf{m}_i$ *is defined in Remark 1. Then, Problem* (6) *is equivalent to*

$$\max_{X_1, \ldots X_n} \sum_{(i,j) \in \mathcal{E}} \text{trace}(X_i^\mathsf{T} Z_{ij} X_j), \tag{15}$$

$$s.t. \ \text{vec}(X_i) \in \mathcal{B}^{p_i}, \quad X_i \mathbf{1}_d = \mathbf{1}_{p_i}, \quad \mathbf{1}_{p_i}^\mathsf{T} X_i = \mathbf{m}_i^\mathsf{T} \quad \forall i = 1, \ldots, n.$$

Proof. The starting point is Eq. (8), which is copied here as a reference:

$$\|Z_{ij} - X_i X_j^\mathsf{T}\|_F^2 = \text{trace}(Z_{ij}^\mathsf{T} Z_{ij}) + \mathbf{m}_i^\mathsf{T} \mathbf{m}_j - 2\,\text{trace}(X_i^\mathsf{T} Z_{ij} X_j). \tag{16}$$

As already observed, the first term is constant, for it depends on Z_{ij} only; hence, it can be ignored in the optimization. Also the second term is constant, since $\mathbf{m}_i$ is known for each image i *by assumption*. Thus, the objective function (6) is equivalent to (15). Observe that we should add extra constraints to take into account our additional assumptions, which force the amount of points per motion to be equal to some predefined values in every image, namely $1_{p_i}^\mathsf{T} X_i = \mathbf{m}_i^\mathsf{T}$ (see also Remark 1). Hence we get the thesis.   □

Remark 3. Note that the knowledge of the number of points per motion was indeed essential in the proof of Proposition 3. In other terms, the synchronization problem (6) **is not equivalent** to (15) without such an assumption.

Proposition 4. *If the amount of points per motion is known for each image, then Problem (6) is equivalent to*

$$\max_X \text{vec}(X)^\mathsf{T} (I_{d\times d} \otimes Z) \, \text{vec}(X),$$
$$s.t. \ \text{vec}(X) \in \mathcal{B}^{dp}, \quad X1_d = 1_p, \quad KX = M, \tag{17}$$

where K and M are defined as follows:

$$K = \text{diag}(1_{p_1}^\mathsf{T}, 1_{p_2}^\mathsf{T}, \ldots, 1_{p_n}^\mathsf{T}), \quad M = \begin{bmatrix} \mathbf{m}_1^\mathsf{T} \ \mathbf{m}_2^\mathsf{T} \ \ldots \ \mathbf{m}_n^\mathsf{T} \end{bmatrix}^\mathsf{T}. \tag{18}$$

Proof. Problem (15) can be easily turned into the form (17) following the same reasoning as in the proof of Proposition 2. Concerning the additional constraints, it is easy to see that $KX = M$ is just a compact way of storing all the equations of the form $1_{p_i}^\mathsf{!} X_i = \mathbf{m}_i^\mathsf{!}$ simultaneously.   ⊔

Corollary 2. *If the amount of points per motion is known for each image, then Problem (6) can be mapped into a QUBO problem of the form (5), namely:*

$$\min_{\mathbf{y}\in\mathcal{B}^{dp}} \mathbf{y}^\mathsf{T} P \mathbf{y} + \lambda_2 ||A\mathbf{y} - \mathbf{b}||^2 + \lambda_3 ||E\mathbf{y} - \mathbf{f}||^2, \tag{19}$$

where $P = -I_{d\times d} \otimes Z$, $\mathbf{y} = \text{vec}(X)$, $A = (1_d^\mathsf{T} \otimes I_{p\times p})$, $\mathbf{b} = 1_p$, $\mathbf{f} = \text{vec}(M)$, and $E = (I_{d\times d} \otimes K)$.

Proof. Problem (17) is the maximization of a quadratic cost with binary variables and linear constraints, hence definitions of P and $\mathbf{y}$ are immediate. The linear constraints $X1_d = 1_p$ and $KX = M$ can be mapped into the forms $A\mathbf{y} = \mathbf{b}$ and $E\mathbf{y} = \mathbf{f}$ via properties of Kronecker product(see footnote 3) and vectorization.

□

In summary, we derived an alternative QUBO formulation for motion segmentation with a *sparse* matrix: P matrix defined in (19) is sparse as it inherits the same sparsity pattern as Z. This was possible **under simplified assumptions**, *i.e.*, the knowledge of the amount of points per motion in all images. This gives rise to the second variant of our approach, which is called QuMoSeg-v2.

Execution on AQC. Once the QUBO for the target data is known, namely (14) for QUMOSEG-V1 or (19) for QUMOSEG-V2, motion segmentation can be solved via adiabatic quantum optimization, as described in Sect. 2.2.

4   Related Work

After having introduced QUMOSEG, we next review the most related methods.

Synchronization refers to recovering elements of a group (associated to vertices in a graph) starting from a (redundant) set of pairwise ratios (associated to edges in the graph). Popular synchronization problems involve *rotations* and *rigid motions* [7,9,16,23,36,49,56–58], which are at the core of tasks such as camera motion estimation in SfM [46], point cloud registration [26] and SLAM [14]. Other synchronization problems concern *homographies* (which are related to image mosaicking [53]) and *affine transformations* (which were used to solve for global color correction [51]). Although synchronization has a well-established theory for the case where unknowns/measures belong to a group, specific routines can be possibly developed when the variables do not belong to a group and have a poorer structure. A notable example are *partial permutations*, which appear in the context of multi-view matching [8,17,41,47,64]. Another example are *binary matrices*, which are related to part segmentation in point clouds [29] and motion segmentation in images [4] (the basis of our method).

Motion Segmentation aims at detecting moving objects in a scene given multiple images by grouping all the key-points moving in the same way [52]. Existing techniques fall into three main categories, depending on the assumptions made on the inputs. Some methods assume extracted key-points and work with *unknown* correspondences [30,61]. Other algorithms assume some *local* information about correspondences in addition to the key-points, namely matches between image pairs [4,5]. The third category assumes known *global* information about the correspondences, namely multi-frame trajectories [31,34,37,48,62,63]. Our method belongs to the second category [4,5]. Accordingly, our technique splits motion segmentation into multiple two-frame sub-problems and then finds a global consistency among the partial results. Our method is related to Arrigoni and Pajdla [4], as we adapt their model and the data structures. Their approach relies on a similar formulation as (17) but solves the problem over real variables instead of binary ones, ending up with an approximate solution based on a spectral decomposition. In this paper, instead, we derive a QUBO formulation from scratch and solve it *without relaxation* on an AQC.

Quantum Computer Vision. Several quantum techniques are available for computer vision tasks, such as recognition and classification [15,44], object tracking [35], transformation estimation [25], point set and shape alignment [43,55], graph matching [54] and permutation synchronization [10]. Most of these methods are designed for an AQC. In [44], a binary matrix factorization is applied to feature extraction from facial images, while in [35] redundant detections are removed in multi-object detection with the help of an AQC. Another method classifies multi-spectral images with quantum SVM [15]. A quantum approach for correspondence problems on point sets [25] recovers rotations between pairs of point sets, which are approximated as sums of basis matrix elements. The qKC method described in [43] employs both classical and quantum kernel-based losses for point set matching. In contrast to [25], qKC is designed for a circuit-based quantum computer. QGM is the first approach for matching small graphs using AQC [54]. Q-Match [55] can non-rigidly match 3D shapes with up to 500 points and it overcomes the hardware limitations of the modern AQC (*e.g.*, the qubit connectivity pattern) by an iterative optimization scheme.

The quantum method most closely related to ours is QSync [10], as it also uses the framework of synchronization but solves a *different* problem. There are substantial differences between our task and permutation synchronization [10]: QSync operates on permutation matrices (*i.e.*, both rows and columns sum to 1), whereas the binary matrices in our case are less constrained (only rows sum up to 1). Although this might seem a minor difference, the term $\text{trace}(X_j X_i^\mathsf{T} X_i X_j^\mathsf{T})$ in QSync is constant and can be ignored in (8); the absence of "columns sum to 1" constraint means that we either i) require additional assumptions (QuMoSeg-v2) or ii) resort to different computations resulting in a new (dense) matrix (QuMoSeg-v1). In QSync, the number of variables scales quadratically with the number of points, whereas our method is more efficient per construction (*e.g.*, it uses matrices with $2k$ entries if there are two motions, see Remark 2). Hence, we can handle larger problems: five images of two motions with ten points each, *i.e.*, 100 points (see Sect. 5) *vs* five points in four images [10].

5   Experiments

In this section, we report experimental results on synthetic scenarios, a small dataset derived from the Hopkins benchmark [59] and a new real dataset. We evaluate our method on D-Wave's AQC of the latest generation, *i.e.*, D-Wave Advantage4.1 (Adv4.1), which is an AQC of the Pegasus architecture. It contains $\approx 5k$ physical qubits organized in cells of 24 qubits; each qubit is coupled to 15 other qubits; the total number of couplers is $\approx 40k$ [11]. Adv4.1 operates at temperatures below $17mK$ and is accessed remotely via Ocean [18]. We run all experiments with $20\mu s$ annealing time (no pause). The total QPU runtime of our experiments including overheads amounts to over 15 minutes (over 10^6 obtained samples). We sample $1k$ times and take the lowest-energy sample as the result.

Evaluation Methodology. Since our method is the first quantum approach to segmentation with two-frame correspondences, we compare it with traditional approaches (*i.e.*, not operating on AQC), namely Mode [5] and Synch [4], whose code is available online [2]. Both our approach and the competitors take the same input, namely a set of pairwise segmentations represented as a $dp \times dp$ block-matrix Z – see (10) – and they compute the absolute segmentations either in the form of a $p \times d$ matrix X or a vector $\mathbf{y} = \text{vec}(X)$ of length dp. In order to compare a given solution $\mathbf{y} \in \mathcal{B}^{dp}$ with the ground-truth solution named

$\mathbf{y}_{gt} \in \mathcal{B}^{dp}$, we use the *accuracy* μ^4, as done also in [10]. It is defined as the number of correct entries over the total amount of entries, namely

$$\mu = 1 - \mathcal{H}(\mathbf{y}_{gt}, \mathbf{y})/dp, \tag{20}$$

where $\mathcal{H}$ denotes the Hamming distance.

Experiment with Synthetic Noise. We consider 20 synthetic configurations with 3 or 4 images, 2 motions and 16 points per image, resulting in 96 or 128 qubits. In order to create the ground-truth segmentation X, each point is assigned to a motion which is randomly chosen among the two available ones. All the analyzed methods can solve these problems with accuracy $\mu = 1$. Next, we derive pairwise segmentations from (3) and we systematically inject into them increasing amounts of noise: in each pairwise segmentation we switch the motion of a percentage of points ranging from 0% to 50% (meaning that half of the points are corrupted). The results of this experiment are shown in Fig. 3a, which reports the accuracy (averaged over 20 problem instances) for the analyzed methods. QUMOSEG-V1 is certainly on par with the best traditional approach (namely MODE [5]). Concerning QUMOSEG-V2, it is worth noting that is starts from a similar formulation to SYNCH [4], but it solves the problem *without* relaxation: working over binary variables without relaxation in combination with AQC significantly improves the results. There are no significative differences between the two variants of our method on the smallest scenario with 96 qubits.

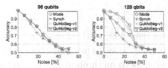 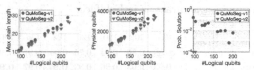

(a) Accuracy (max value 1) for several methods on synthetic data versus input noise.

(b) Average maximum chain length (left), average number of physical qubits (middle) and probability of finding a solution (right) versus number of logical qubits in Q-MSEG dataset.

Fig. 3. Results of synthetic experiments (a) and of real experiments on Q-MSEG (b).

Experiments on Q-MSEG Dataset. Due to the lack of small-scale datasets usable for our scenario (*i.e.*, whose size is manageable by current AQC), we generate a new dataset for motion segmentation with ground-truth annotations, which comprises six images depicting three planar objects captured from diverse viewpoints. We focus on planar motions (where the underlying model is the homography) since they can be recovered from a lower amount of points than general motions[5] (where the corresponding model is the fundamental matrix),

---

[4] Other measures can be considered with similar results, such as the misclassification error, which is widely adopted in motion segmentation.

[5] At least four points are needed to estimate a homography, whereas at least seven points are required for the fundamental matrix [28].

hence allowing more flexibility in the experimental setting. Each object is man-ually annotated with respectively 10, 11 and 12 keypoints, selected on highly-textured locations. Following [10,60], we extract features corresponding to each keypoint using an AlexNet model [33] pretrained on ImageNet. Several motion segmentation problems are derived from the dataset: we consider 14 different choices for the number of points/images/motions (see supplementary), and we sample 20 problem instances for each configuration, resulting in 280 problem instances in total. For each problem instance, we perform feature matching on every image pair via nearest neighbor search, where the cosine distance is used as the distance metric. Then, different motions are identified in each image pair by fitting multiple homographies to correspondences [40]. This defines a set of pairwise segmentations which represents the input to MODE [5], SYNCH [4] and our quantum methods. In order to enrich the evaluation, we evaluate here also the method by Xu et al. [62] (with public code [1]), although it makes differ-ent assumptions on the input (i.e., it requires multi-frame trajectories, which were computed by us following the same procedure as in [4,5]). For complete-ness, we also optimize the objective functions of our approach with a simulated annealing (SA) solver [19,32]: while in general it performs on par with QA for smaller problems, it provides an indication on the QPU's solution accuracy for the largest problems we test, which future QPU generations can potentially reach and outperform.

Table 1. Average accuracy (1 is the best) for several methods on our Q-MSEG dataset. The highest accuracy is in boldface.

# Qubits/Bin. Var.:	96	102	120	126	128	136	160	168	180	190	200	216	220	243
Xu et al. [62]	0.89	0.89	0.94	0.75	0.96	0.97	0.86	0.86	0.97	0.88	0.96	0.77	0.83	0.74
MODE [5]	0.93	0.93	0.96	0.93	**0.97**	0.97	**0.98**	0.99	**0.98**	**0.99**	**0.99**	0.93	1	**0.94**
SYNCH [4]	0.93	0.94	0.95	0.95	0.84	0.92	0.97	1	0.89	0.95	0.90	**0.94**	0.99	0.92
QUMOSEG-V1	0.97	0.97	0.97	0.96	0.95	**0.98**	**0.98**	0.99	**0.98**	**0.99**	**0.99**	0.64	–	–
QUMOSEG-V2	0.96	0.97	0.95	0.94	0.89	0.89	0.88	0.85	0.74	0.75	0.79	0.59	0.75	0.58
QUMOSEG-V1, SA	0.97	0.97	0.97	0.96	0.95	**0.98**	**0.98**	1	**0.98**	**0.99**	**0.99**	0.68	0.98	0.72
QUMOSEG-V2, SA	**0.98**	**0.99**	**0.99**	1	0.96	**0.98**	**0.98**	1	0.94	0.97	**0.99**	0.80	1	0.59

Results are given in Table 1 which reports, for each configuration, the mean accuracy μ (over 20 problem instances) for all the analyzed methods. Results show that there is no clear winner, since none of the methods outperforms all oth-ers in all cases. In particular, QUMOSEG-V1 is better than the state of the art on small-scale problems (i.e., 96–126 qubits) and comparable or better than the best traditional method (i.e., MODE [5]) on medium-scale problems (i.e., 136–200 qubits). On the largest cases, instead, either the performances of QUMOSEG-V1 significantly drop (i.e., 216 qubits) or it was unable to find a minor embedding and, hence, a solution (i.e., 220–243 qubits). Concerning QUMOSEG-V2, we can observe that—although working under easier assumptions—it is, in general, worse than QuMoSeg-v1. This might be caused by the difficulty of satisfying the additional (simplified) constraints in practice (as they are treated as soft instead of hard). Note that the performances of our approaches are largely affected by the limitation of current QPU. In this respect, the on-par or higher performance of SA (see Table 1) suggests that our QUBO approach to motion segmentation is sound and that future QPU generations are expected to reach or surpass that result. Qualitative results from an example with 96 qubits are shown in Fig. 1: in this case, QuMoSeg-v1 outperforms existing methods.

Figure 3b visualises how the expected number of physical qubits and the maximum chain length are increasing with the increasing problem size. It also reports the probability of finding a solution in a sampling, which is calculated as the portion of optimal solutions (as lowest energy solutions) among all solutions. Note that QuMoSeg-v1 has a non-zero probability to find an optimal solution for all problems except the largest embeddable one with 216 qubits, whereas QuMoSeg-v2 has a non-zero probability only on the two smallest cases.

Experiments on Hopkins Benchmark. Starting from the popular Hopkins155 dataset [59], we created small problems (with 120–240 qubits) by sampling a subset of images/points from the *cars2_06_g23* sequence (see our supplement for details). For each configuration, 20 instances were created, resulting in 400 examples in total. We used the fundamental matrix model to produce the pairwise segmentations which are given as input to MODE [5], SYNCH [4] and our quantum methods. We also consider the method by Xu et al. [62] and the simulated annealing (SA) solver [19,32], as done previously. Results are given in Table 2, showing that we outperform Xu et al. [62] (for number of qubits <198) and obtain comparable performance to classic methods [4,5] in several cases. Results for SA show that the global optima of our QUBOs match expected solutions (and that the obtained energy landscapes are as expected), even though the solutions cannot be recovered by QA in all cases due to hardware limitations. In particular, for small problems, the simulated annealing performs on par with QA (as expected), whereas large problems can be solved with the SA only.

Table 2. Accuracy (1.0 is the best) for several methods on sub-problems sampled from the Hopkins dataset [59]. The highest accuracy is in boldface.

# Qubits/Bin. Var. :	120	126	132	138	144	156	162	168	174	180	186	192	198	204	210	216	222	228	234	240
Xu et al. [62]	0.80	0.78	0.81	0.79	0.83	0.81	0.84	0.81	0.85	0.89	0.88	0.94	0.96	0.96	0.97	**1**	0.98	**1**	**0.99**	**1**
MODE [5]	0.89	0.91	0.90	0.93	0.92	0.94	0.95	0.95	0.96	0.95	0.97	0.98	0.98	0.98	0.98	0.98	**0.99**	0.98	0.98	**0.99**	0.99
SYNCH [4]	0.87	0.93	0.95	0.96	**0.99**	0.96	**0.99**	**0.99**	0.96	0.99	**1**	**1**	0.99	**1**	**1**	0.70	0.97	**1**	**0.99**	0.65
QuMoSeg-v1	0.92	0.89	0.93	0.93	0.93	0.93	0.95	0.94	0.96	0.95	0.96	0.97	0.96	-	-	-	-	-	-	-
QuMoSeg-v2	0.91	0.92	0.91	0.92	0.94	0.89	0.91	0.89	0.90	0.88	0.88	0.89	0.88	0.89	0.88	-	-	-	-	-
QuMoSeg-v1, SA	0.93	0.90	0.92	0.94	0.93	0.94	0.95	0.96	0.96	0.96	0.98	0.98	0.98	0.98	0.99	0.99	0.98	0.98	**0.99**	0.99
QuMoSeg-v2, SA	**0.96**	**0.97**	**0.98**	**0.98**	**0.99**	**0.99**	**0.99**	0.97	**0.99**	**1**	**1**	**1**	**1**	**1**	**1**	**1**	**1**	**0.99**	**1**	**0.99**	**1**

Discussion. In our experiments, QuMoSeg achieves on-par performance[6] with non-quantum state of the art, showing the viability of a quantum approach for segmentation. Among the two proposed variants, QuMoSeg-v1 is the most accurate and QuMoSeg-v2 can solve larger problems. Moreover, our approach in a combination with SA shows highly promising results on both Q-MSEG and the data we generated from Hopkins. The characteristics of the current quantum hardware starkly influence the performance of QuMoSeg: Problems with >120 points cannot be minor-embedded, and the largest mappable problems require maintaining too long qubit chains, which impedes the optimum search. As many other quantum algorithms, QuMoSeg will benefit from improved quantum hardware, both in terms of accuracy and solvable problem sizes. Indeed, it is expected that the possibility to solve combinatorial problems without approximation will give practical advantages for large-scale problems in the future.

6 Conclusion

We propose the first motion segmentation approach for an adiabatic quantum computer, which shows highly promising results and reaches state-of-the-art accuracy on a wide range of problems. We hope that the demonstrated progress encourages more work on quantum computer vision in the future.

Acknowledgements. This work was partially supported by the PRIN project LEGO-AI (Prot. 2020TA3K9N).

References

1. https://alex-xun-xu.github.io/ProjectPage/CVPR_18/
2. https://github.com/federica-arrigoni/ICCV_19
3. Arrigoni, F., Fusiello, A.: Synchronization problems in computer vision with closed-form solutions. Int. J. Comput. Vision **128**, 26–52 (2020)
4. Arrigoni, F., Pajdla, T.: Motion segmentation via synchronization. In: IEEE International Conference on Computer Vision Workshops (ICCVW) (2019)
5. Arrigoni, F., Pajdla, T.: Robust motion segmentation from pairwise matches. In: Proceedings of the International Conference on Computer Vision (2019)
6. Barath, D., Matas, J.: Multi-class model fitting by energy minimization and mode-seeking. In: Ferrari, V., Hebert, M., Sminchisescu, C., Weiss, Y. (eds.) ECCV 2018. LNCS, vol. 11220, pp. 229–245. Springer, Cham (2018). https://doi.org/10.1007/978-3-030-01270-0_14
7. Bernard, F., Thunberg, J., Gemmar, P., Hertel, F., Husch, A., Goncalves, J.: A solution for multi-alignment by transformation synchronisation. In: Proceedings of the IEEE Conference on Computer Vision and Pattern Recognition (2015)
8. Bernard, F., Thunberg, J., Swoboda, P., Theobalt, C.: HiPPI: higher-order projected power iterations for scalable multi-matching. In: Proceedings of the International Conference on Computer Vision (2019)

---

[6] Note that this reflects the current situation in the field: indeed, other quantum methods [10,55] do not outperform classical methods in all scenarios too.

9. Birdal, T., Arbel, M., Simsekli, U., Guibas, L.J.: Synchronizing probability measures on rotations via optimal transport. In: Proceedings of the IEEE Conference on Computer Vision and Pattern Recognition, pp. 1566–1576 (2020)

10. Birdal, T., Golyanik, V., Theobalt, C., Guibas, L.: Quantum permutation synchronization. In: Proceedings of the IEEE Conference on Computer Vision and Pattern Recognition (2021)

11. Boothby, K., Bunyk, P., Raymond, J., Roy, A.: Next-Generation Topology of D-Wave Quantum Processors. arXiv e-prints 2003.00133 (2020)

12. Born, M., Fock, V.: Beweis des adiabatensatzes. Z. Phys. **51**(3), 165–180 (1928)

13. Cai, J., Macready, W.G., Roy, A.: A practical heuristic for finding graph minors. arXiv e-prints 1406.2741 (2014)

14. Carlone, L., Tron, R., Daniilidis, K., Dellaert, F.: Initialization techniques for 3D SLAM: a survey on rotation estimation and its use in pose graph optimization. In: Proceedings of the IEEE International Conference on Robotics and Automation (2015)

15. Cavallaro, G., Willsch, D., Willsch, M., Michielsen, K., Riedel, M.: Approaching remote sensing image classification with ensembles of support vector machines on the d-wave quantum annealer. In: IEEE International Geoscience and Remote Sensing Symposium (IGARSS) (2020)

16. Chatterjee, A., Govindu, V.M.: Efficient and robust large-scale rotation averaging. In: Proceedings of the International Conference on Computer Vision (2013)

17. Chen, Y., Guibas, L., Huang, Q.: Near-optimal joint object matching via convex relaxation. In: Proceedings of the International Conference on Machine Learning, pp. 100–108 (2014)

18. D-Wave Systems Inc: D-wave ocean software documentation (2021). https://docs.ocean.dwavesys.com/en/stable/. Accessed 05 Mar 2022

19. D-Wave Systems Inc: dwave-neal documentation (2021). https://docs.ocean.dwavesys.com/_/downloads/neal/en/latest/pdf/. Accessed 6 Mar 2022

20. Dattani, N., Szalay, S., Chancellor, N.: Pegasus: the second connectivity graph for large-scale quantum annealing hardware. arXiv e-prints (2019)

21. Denchev, V.S., Boixo, S., Isakov, S.V., Ding, N., Babbush, R., Smelyanskiy, V., Martinis, J., Neven, H.: What is the computational value of finite-range tunneling? Phys. Rev. X **6**, 031015 (2016)

22. D-wave: What is quantum annealing? (2021). https://docs.dwavesys.com/docs/latest/c_gs_2.html. Accessed 05 Mar 2022

23. Eriksson, A., Olsson, C., Kahl, F., Chin, T.J.: Rotation averaging and strong duality. In: Proceedings of the IEEE Conference on Computer Vision and Pattern Recognition, pp. 127–135 (2018)

24. Farhi, E., Goldstone, J., Gutmann, S., Lapan, J., Lundgren, A., Preda, D.: A quantum adiabatic evolution algorithm applied to random instances of an np-complete problem. Science **292**(5516), 472–475 (2001)

25. Golyanik, V., Theobalt, C.: A quantum computational approach to correspondence problems on point sets. In: Proceedings of the IEEE Conference on Computer Vision and Pattern Recognition (2020)

26. Govindu, V.M., Pooja, A.: On averaging multiview relations for 3D scan registration. IEEE Trans. Image Process. **23**(3), 1289–1302 (2014)

27. Grant, E.K., Humble, T.S.: Adiabatic quantum computing and quantum annealing. In: Oxford Research Encyclopedia of Physics (2020)

28. Hartley, R., Zisserman, A.: Multiple View Geometry in Computer Vision, 2nd edn. Cambridge University Press, Cambridge (2004)

29. Huang, J., et al.: MultiBodySync: multi-body segmentation and motion estimation via 3d scan synchronization. In: Proceedings of the IEEE Conference on Computer Vision and Pattern Recognition, pp. 7108–7118 (2021)
30. Ji, P., Li, H., Salzmann, M., Dai, Y.: Robust motion segmentation with unknown correspondences. In: Fleet, D., Pajdla, T., Schiele, B., Tuytelaars, T. (eds.) ECCV 2014. LNCS, vol. 8694, pp. 204–219. Springer, Cham (2014). https://doi.org/10.1007/978-3-319-10599-4_14
31. Ji, P., Salzmann, M., Li, H.: Shape interaction matrix revisited and robustified: efficient subspace clustering with corrupted and incomplete data. In: Proceedings of the International Conference on Computer Vision, pp. 4687–4695 (2015)
32. Kirkpatrick, S., Gelatt, C.D., Vecchi, M.P.: Optimization by simulated annealing. Science **220**(4598), 671–680 (1983)
33. Krizhevsky, A., Sutskever, I., Hinton, G.E.: ImageNet classification with deep convolutional neural networks. In: Advances in Neural Information Processing Systems, vol. 25 (2012)
34. Lai, T., Wang, H., Yan, Y., Chin, T.J., Zhao, W.L.: Motion segmentation via a sparsity constraint. IEEE Trans. Intell. Transp. Syst. **18**(4), 973–983 (2017)
35. Li, J., Ghosh, S.: Quantum-soft QUBO Suppression for accurate object detection. In: Vedaldi, A., Bischof, H., Brox, T., Frahm, J.-M. (eds.) ECCV 2020. LNCS, vol. 12374, pp. 158–173. Springer, Cham (2020). https://doi.org/10.1007/978-3-030-58526-6_10
36. Li, X., Ling, H.: PoGO-Net: pose graph optimization with graph neural networks. In: Proceedings of the International Conference on Computer Vision, pp. 5895–5905 (2021)
37. Li, Z., Guo, J., Cheong, L.F., Zhou, S.Z.: Perspective motion segmentation via collaborative clustering. In: Proceedings of the International Conference on Computer Vision, pp. 1369–1376 (2013)
38. Liu, S., Trenkler, G.: Hadamard, Khatri-Rao, Kronecker and other matrix products. Int. J. Inf. Syst. Sci. **4**(1), 160–177 (2008)
39. Magri, L., Fusiello, A.: T-Linkage: A continuous relaxation of J-Linkage for multi-model fitting. In: Proceedings of the IEEE Conference on Computer Vision and Pattern Recognition. pp. 3954–3961 (June 2014)
40. Magri, L., Fusiello, A.: Multiple structure recovery via robust preference analysis. Image Vis. Comput. **67**, 1–15 (2017)
41. Maset, E., Arrigoni, F., Fusiello, A.: Practical and efficient multi-view matching. In: Proceedings of IEEE International Conference on Computer Vision, pp. 4568–4576 (2017)
42. McGeoch, C.C.: Adiabatic Quantum Computation and Quantum Annealing: Theory and Practice. Morgan & Claypool, Burlington (2014)
43. Noormandipour, M., Wang, H.: Matching Point Sets with Quantum Circuit Learning. arXiv e-prints 2102.06697 (2021)
44. O'Malley, D., Vesselinov, V.V., Alexandrov, B.S., Alexandrov, L.B.: Nonnegative/binary matrix factorization with a d-wave quantum anneale. PLoS ONE **13**(12), e0206653 (2018)
45. Ozden, K.E., Schindler, K., Van Gool, L.: Multibody structure-from-motion in practice. IEEE Trans. Pattern Anal. Mach. Intell. **32**(6), 1134–1141 (2010)
46. Ozyesil, O., Voroninski, V., Basri, R., Singer, A.: A survey of structure from motion. Acta Numer **26**, 305–364 (2017)
47. Pachauri, D., Kondor, R., Singh, V.: Solving the multi-way matching problem by permutation synchronization. In: Advances in Neural Information Processing Systems, vol. 26, pp. 1860–1868 (2013)

48. Rao, S., Tron, R., Vidal, R., Ma, Y.: Motion segmentation in the presence of outlying, incomplete, or corrupted trajectories. Pattern Anal. Mach. Intell. **32**(10), 1832–1845 (2010)

49. Rosen, D.M., Carlone, L., Bandeira, A.S., Leonard, J.J.: SE-Sync: a certifiably correct algorithm for synchronization over the special Euclidean group. Int. J. Robot. Res. **38**(2–3), 95–125 (2019)

50. Sabzevari, R., Scaramuzza, D.: Multi-body motion estimation from monocular vehicle-mounted cameras. IEEE Trans. Rob. **32**(3), 638–651 (2016)

51. Santellani, E., Maset, E., Fusiello, A.: Seamless image mosaicking via synchronization. In: ISPRS Annals of Photogrammetry, Remote Sensing and Spatial Information Sciences, IV-2, pp. 247–254 (2018)

52. Saputra, M.R.U., Markham, A., Trigoni, N.: Visual SLAM and structure from motion in dynamic environments: a survey. ACM Comput. Surv. **51**(2), 37:1–37:36 (2018)

53. Schroeder, P., Bartoli, A., Georgel, P., Navab, N.: Closed-form solutions to multiple-view homography estimation. In: IEEE Workshop on Applications of Computer Vision (WACV), pp. 650–657 (2011)

54. Seelbach Benkner, M., Golyanik, V., Theobalt, C., Moeller, M.: Adiabatic quantum graph matching with permutation matrix constraints. In: International Conference of 3D Vision (3DV) (2020)

55. Seelbach Benkner, M., Lähner, Z., Golyanik, V., Wunderlich, C., Theobalt, C., Moeller, M.: Q-match: iterative shape matching via quantum annealing. In: International Conference on Computer Vision (ICCV) (2021)

56. Singer, A.: Angular synchronization by eigenvectors and semidefinite programming. Appl. Comput. Harmon. Anal. **30**(1), 20–36 (2011)

57. Thunberg, J., Bernard, F., Goncalves, J.: Distributed methods for synchronization of orthogonal matrices over graphs. Automatica **80**, 243–252 (2017)

58. Tron, R., Daniilidis, K.: Statistical pose averaging with non-isotropic and incomplete relative measurements. In: Fleet, D., Pajdla, T., Schiele, B., Tuytelaars, T. (eds.) ECCV 2014. LNCS, vol. 8693, pp. 804–819. Springer, Cham (2014). https://doi.org/10.1007/978-3-319-10602-1_52

59. Tron, R., Vidal, R.: A benchmark for the comparison of 3-d motion segmentation algorithms. In: Proceedings of the IEEE Conference on Computer Vision and Pattern Recognition, pp. 1–8. IEEE (2007)

60. Wang, Q., Zhou, X., Daniilidis, K.: Multi-image semantic matching by mining consistent features. In: Computer Vision and Pattern Recognition (CVPR) (2018)

61. Wang, Y., Liu, Y., Blasch, E., Ling, H.: Simultaneous trajectory association and clustering for motion segmentation. IEEE Signal Process. Lett. **25**(1), 145–149 (2018)

62. Xu, X., Cheong, L.F., Li, Z.: 3d rigid motion segmentation with mixed and unknown number of models. IEEE Trans. Pattern Anal. Mach. Intell. **43**, 1–6 (2019)

63. Yan, J., Pollefeys, M.: A general framework for motion segmentation: independent, articulated, rigid, non-rigid, degenerate and non-degenerate. In: Leonardis, A., Bischof, H., Pinz, A. (eds.) ECCV 2006. LNCS, vol. 3954, pp. 94–106. Springer, Heidelberg (2006). https://doi.org/10.1007/11744085_8

64. Zhou, X., Zhu, M., Daniilidis, K.: Multi-image matching via fast alternating minimization. In: Proceedings of the International Conference on Computer Vision, pp. 4032–4040 (2015)

Instance as Identity: A Generic Online Paradigm for Video Instance Segmentation

Feng Zhu[1,2,3] (ID), Zongxin Yang[4] (ID), Xin Yu[3] (ID), Yi Yang[4] (ID),
and Yunchao Wei[5,6(✉)] (ID)

[1] Baidu Research, Beijing, China
Feng.Zhu@student.uts.edu.au
[2] ReLER, Centre for Artificial Intelligence, University of Technology Sydney,
Sydney, Australia
[3] Australian Artificial Intelligence Institute, University of Technology Sydney,
Sydney, Australia
[4] CCAI, College of Computer Science and Technology, Zhejiang University,
Hangzhou, China
[5] Institute of Information Science, Beijing Jiaotong University, Beijing, China
[6] Beijing Key Laboratory of Advanced Information Science and Network, Beijing,
China
yunchao.wei@bjtu.edu.cn

Abstract. Modeling temporal information for both detection and tracking in a unified framework has been proved a promising solution to video instance segmentation (VIS). However, how to effectively incorporate the temporal information into an online model remains an open problem. In this work, we propose a new online VIS paradigm named Instance As Identity (IAI), which models temporal information for both detection and tracking in an efficient way. In detail, IAI employs a novel identification module to predict identification number for tracking instances explicitly. For passing temporal information cross frame, IAI utilizes an association module which combines current features and past embeddings. Notably, IAI can be integrated with different image models. We conduct extensive experiments on three VIS benchmarks. IAI outperforms all the online competitors on YouTube-VIS-2019 (ResNet-101 41.9 mAP) and YouTube-VIS-2021 (ResNet-50 37.7 mAP). Surprisingly, on the more challenging OVIS, IAI achieves SOTA performance (20.3 mAP). Code is available at https://github.com/zfonemore/IAI.

Keyword: Video instance segmentation

F. Zhu—Work done during an internship at Baidu.

Supplementary Information The online version contains supplementary material available at https://doi.org/10.1007/978-3-031-19818-2_30.

1   Introduction

Instance Segmentation [5,8,9,12,16,26,44] is an important problem in the computer vision community, which aims to detect and segment objects of specific classes in an image. Benefiting from the rapid growth of deep learning techniques, this problem has achieved great progress in recent years. Most recently, Video instance segmentation (VIS) [37], an advanced version of instance segmentation that aims to simultaneously detect, segment, and track different objects of specific categories in videos, is attracting the attention of many researchers due to its wide application prospects, such as augmented reality and video editing.

The main challenge of VIS lies in how to assemble instance segmentation and tracking into a unified framework. Some latest progresses, e.g., MaskProp [2] and VisTR [35], achieved this target by handling a clip. However, these clip-based approaches fail to perform online inference and thus cannot be applied in real-world applications that requires real-time processing. To address this issue, some online solutions [6,21,37,39] following a tracking-by-detection paradigm are also proposed recently. Although such a paradigm can adopt temporal features for tracking, it does not utilize prior object information (e.g. appearance information and position information) to detect the corresponding ones in following frames. In view of this, we aim to design a novel online pipeline to fully exploit temporal object information and encode it for both detection and tracking processes.

To this end, we propose a new solution named Instance As Identity (IAI), as shown in Fig. 1(a). Within IAI, we first detect instances in the initial frame and assign IDentities (IDs) to them. Then, in the next frame, we directly predict IDs of instances. For those instances that fail to match any previous instances, we assign new unique IDs to them. By conducting this process on each frame, all the instances can be smoothly detected and tracked in an online manner.

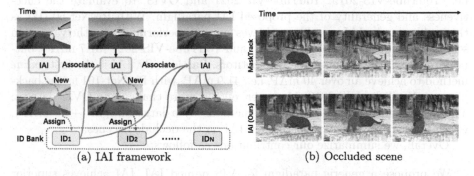

(a) IAI framework   (b) Occluded scene

Fig. 1. (a) The illustration of the IAI paradigm. In the video, the IAI first detects a car in the initial frame and assigns it ID1. In the second frame, IAI associates the first car with ID1, and recognizes the second car as a new instance. IAI assigns ID2 to the new car. In the next frames, IAI associates these two cars using ID1 and ID2. (b) Comparison of MaskTrack and IAI on occluded scenes from OVIS dataset.

To be specific, our IAI is achieved by a novel identification module and an efficient multi-object association module. The identification module consists of a new designed identification head and an identification bank. Particularly, the identification head can dynamically detect new instances and assign IDs, and the identification bank encodes the IDs and masks of objects into ID embeddings for propagating across frames. To construct an efficient multi-object association module, we propose an effective Hybrid Association Block (HAB), which adopts transformer and memory to propagate features for tracking and utilizes a classification projector to encode backbone features. It should be noted that our HAB is very different from the association module proposed in [42], which only works for the class-agnostic scenarios that the mask of first frame should be correctly provided by human and does not meet the requirements of VIS (i.e., one VIS model should be equipped with the ability of automatically performing instance segmentation and classification by itself). Through these two modules (i.e., identification and multi-object association), our IAI successfully achieves multiple object association at once for both detection and tracking.

To the best of our knowledge, IAI is the first VIS paradigm to use ID to unify detection and tracking in an online way. Besides, our IAI is pretty flexible and could be easily integrated with existing image segmentation models. Surprisingly, as shown in Fig. 1(b), we find that IAI shows a strong capability on handling object occlusion, which is a key problem in VIS. We attribute this robustness to the design of our identification module and association module. First, because our identification module encodes multiple object information into one identification embedding, the enriched surrounding information of each object helps model to separate different instance on occlude scenes. Second, the global memory in our association module helps the model to acquire object information which is absent in long-term occlusion.

We conduct extensive experiments on three challenging VIS benchmarks, i.e., YouTube-VIS-2019, YouTube-VIS-2021 and OVIS, to evaluate the effectiveness and generality of the proposed IAI paradigm. With ResNet-50 [17] as the backbone, the IAI paradigm achieves superior performance on the validation sets of YouTube-VIS-2019 (38.6 mAP) and YouTube-VIS-2021 (37.7 mAP), outperforming all the online model competitors. Particularly, IAI is the first *online* method to achieve an over 40 mAP, i.e., 41.9 mAP, with ResNet-101 as the backbone on YouTube-VIS-2019. Moreover, on the more challenging OVIS dataset, our method outperforms SOTA VIS methods by a large margin (+4.9 mAP), which further proves the robustness of IAI on the occluded scenes.

Overall, we summarize our contributions as follows:

- We propose a generic paradigm for VIS named IAI. IAI achieves superior performance on VIS benchmarks and outperforms all the online methods.
- We propose a novel identification module that can re-identify the previous instance and recognize a new instance, which is the first time in VIS to track instances explicitly using IDs.
- We propose a new hybrid association block as our association module, which combines backbone features with memory ID embeddings.

2 Related Work

VIS is highly related to several tasks, such as image instance segmentation and semi-supervised video object segmentation. In this section, we provide a brief overview of recent studies in VIS and related fields.

Image Instance Segmentation. Image instance segmentation algorithms are mainly built on either two-stage frameworks or one-stage frameworks. Though rapid progress has been witnessed in instance segmentation, the classical two-stage architecture Mask R-CNN [16] is still the most popular framework to date. Many state-of-the-art works are extended on the basis of Mask R-CNN. Mask R-CNN first predicts bounding-box proposals through a regional proposal network and then produces instance masks using the cropped features for proposals. As for one-stage algorithms, CondInst [32] is a good representative, which outperforms many state-of-the-art instance segmentation algorithms. CondInst adopts a dynamic instance-wise mask head to produce instance masks, thus avoiding ROI operations and enabling mask prediction with higher resolution features.

Semi-Supervised Video Object Segmentation. Semi-supervised video object segmentation (VOS) [29,30] targets at segmenting the given objects with the annotated object masks of the first frame in a video. Many semi-supervised VOS approaches rely on fine-tuning the first frame at test time. Some recent works [11,34,38] propose methods without fine-tuning to achieve a better runtime. STM [28] leverages a memory network to perform long-term propagation. CFBI [41,43] utilizes the feature embedding from the target foreground object and its corresponding background collaboratively. AOT [42] proposes a novel identification mechanism for multi-object association and utilizes a Long Short-Term Transformer to propagate information from memory frames.

Video Instance Segmentation. The VIS task consists of classification, segmentation, and tracking of instances in a video. Along with the YouTube-VIS 2019 datasets, [37] proposes a representative algorithm MaskTrack R-CNN. MaskTrack R-CNN employs a tracking branch to the Mask R-CNN framework in order to link the same instance over frames. The VIS task was formally proposed in [37], and most VIS methods follow the tacking-by-detection paradigm of Mask-Track R-CNN. SipMask-VIS [6] adopts a tracking branch similar to the one-stage FCOS [33] and YOLACT [4] framework. CompFeat [15] proposes a temporal attention module and a spatial attention module to extract contextual information in temporal and spatial dimensions. STMask [21] refines spatial features by aligning features between anchors and ground-truth bounding boxes, and designs a temporal fusion module to learn cross-frame information. CrossVIS [39] is built on CondInst [32] and exchanges dynamic filters in two different frames to learn a more robust video-based instance representation. Although these methods are online algorithms, they are not the optimal solution since detection and tracking are conducted in two independent steps.

Different from the tracking-by-detection paradigm, the state-of-the-art algorithm MaskProp [2] designs a mask propagation mechanism to perform detection and tracking simultaneously. MaskProp utilizes deformable convolution [13] and attention to propagate instance features across frames. VisTR [35] and IFC [18] take advantage of the superior sequence modeling of transformers, and extend the transformer-based detection model DETR [7] to solve VIS problem. Despite the promising performance of these methods, they are both offline methods. It remains a challenge to combine detection and tracking in an online paradigm.

3   Method

Given an input video $G \in R^{T \times 3 \times H \times W}$ comprising of T frames of spatial size $H \times W$, VIS task requires our method to detect, segment and track instances of a predefined category set $\omega = \{1, ..., P\}$ in video G. To be specific, our model predicts an instance mask track $M_G^i \in \{0, 1\}^{T \times H \times W}$ with a class label $c^i \in \{1, ..., P\}$ and a confidence score $s^i \in [0, 1]$ for each detected instance i in G.

In order to solve this challenging problem, we propose a new online paradigm named Instance as Identity (IAI). The detailed framework of IAI is illustrated in Fig. 2. The IAI paradigm designs two modules to extend the original image instance segmentation framework, *i.e.*, identification module and association module. In this section, we first offer a brief introduction to the basic image instance segmentation framework. Then we describe how the ID and association module are designed to combine detection and tracking in an online way.

3.1   Image Instance Segmentation

Commonly, there are two kinds of image instance segmentation frameworks: two-stage framework and one-stage framework. Since our IAI paradigm is not

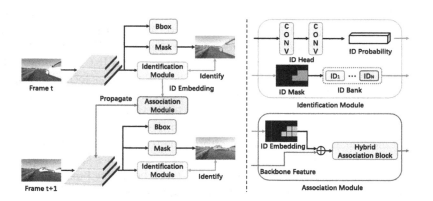

Fig. 2. Overview of the IAI paradigm. In IAI paradigm, we use identification module and association module for tracking and modeling temporal information. The identification consists of an ID head and an ID bank, the former is used to predict ID probability and the latter is used to encode ID mask into ID embedding. The association module is comprised of one hybrid association block and used to propagate information from previous frames to frame t+1.

designed on specific image segmentation framework, we take a simplified image segmentation framework for convenience. As it is presented in Fig. 2, the simplified image instance segmentation model contains backbone, bounding box head and mask head. For the image segmentation task, the model firstly uses the backbone to extract features from the image. Then the bounding box head utilizes the object features to classify and regress the bounding box. The mask head utilizes the object features to predict the mask for each instance.

3.2 Identification Module

As shown in Fig. 3, previous tracking-by-detection VIS algorithms always add a tracking branch to the image instance segmentation model to achieve instance association. In this way, temporal information is only utilized for tracking but not for detection. To overcome this disadvantage, MaskProp proposes a mask propagation paradigm to combine detection and tracking. Maskprop processes each instance independently for association across and aggregates all the single-object predictions into a multi-object prediction. However, this post ensemble paradigm is not efficient for multiple object association in video tasks.

Revisit ID Mechanism. As for multi-object learning and understanding, ID Mechanism [42] was recently proposed for associating and re-identifying multiple given objects in video. The ID mechanism consists of an ID embedding and an ID decoding. The ID embedding module utilizes an identity bank and a random permutation matrix to embed the mask of multiple different targets for propagation. The ID decoding module predicts the targets' probabilities using the aggregated feature. Although this ID mechanism provides a good idea for multiple object association in the video, it

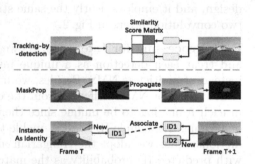

Fig. 3. Tracking patterns of different VIS paradigms.

is impractical to directly apply it to video instance segmentation. There are two main challenges for application in VIS: (1) No targets and ground truth will be given in the first frame, which means nothing to be encoded in the ID embedding; (2) Once a new instance appears in the intermediate frame, the ID decoding module is unable to recognize it and always treats it as background.

Improved ID Mechanism in VIS. To address these challenges, we propose an improved ID mechanism for VIS. In the improved ID mechanism, we will assign each instance a unique ID for the entire video. First, we will detect new instances and assign them unique IDs. Then we will use a similar ID embedding method to encode the mask of different objects. Finally, we will predict IDs for previously detected instances and recognize new instances in subsequent frames.

Through this improved ID mechanism, our IAI paradigm could achieve multiple object association more effectively.

Here, we take an example to illustrate our improved ID mechanism. In the first frame, we detect a new instances i and assign it a new ID d. Then we encode ID and mask information of instance i into ID embedding and save it to memory. In subsequent frames, the ID probability of detected instances will be predicted. With the predicted IDs, different objects are tracked across frames.

ID Head. In order to recognize new instances and match previous instances through ID, we design a new ID head to predict the ID probability for all object proposals. As seen in Fig. 2, the ID head is parallel with other heads, $i.e.$, classification head, bounding box head, and mask head, and shares the same features with them, which means no additional cues are required. The ID head predicts ID probability for all the object proposals. As the number of instances could be various in different frames, we set a number N which is large than the maximum amount of objects in a video of the benchmark (e.g. 20 for YouTube-VIS 2019) as the number of IDs in the ID head. Moreover, the ID head predicts a specific N-1-th ID for all the new instances and then assign specific IDs for them. We use IDs from 0 to N-2 to denote the detected instances in previous frames, and the N-th ID means the background class. The ID head does not need elaborate design, and it employs nearly the same structure as the classification head, $e.g.$ two convolution layers in Fig. 2.

Post Processing for Inference. As we directly predict the IDs for instances and treat each detection as a unique instance, we use a class-agnostic NMS instead of multiclass NMS. Besides, we use an average of ID score and classification score for NMS. Different from the category prediction, the ID prediction in each frame has to be unique since there could not be two same instances in one frame. The simple ID head is unable to guarantee the uniqueness of ID predictions. Thus we adopt the Hungarian algorithm [19] to assign the unique ID with predicted ID probability as the matching cost. Since there will be various new objects through the video, we set a previously detected object number S to assign ID for new objects. If the object ID is predicted as ID N-1, we will assign it a new ID S+1 and increase S by 1 accordingly. Once S equals $N-1$, we assume there could not be more new instances in the video, and discard the newly detected instances in following frames.

ID Embedding. Assume there are L detected objects in current frame, after the unique ID prediction $U \in \{0, 1, ..., N\}^L$ and mask prediction $M \in \{0, 1\}^{L \times HW}$ are obtained, we produce an ID embedding to propagate these information to following frames. In order to encode ID information and mask information of multiple instances together, we combine U and M to generate the one hot ID masks $Y \in \{0, 1\}^{(N+1) \times HW}$,

$$Y_{U_i} = M_i, \qquad 1 \le i \le L. \tag{1}$$

We employ a similar ID embedding method in AOT [42] to encode the ID masks Y. In AOT, an identity bank $D \in R^{(N+1) \times C}$ with C channel dimensions is used to project different instance features into the same feature space. The ID embedding $E \in R^{HW \times C}$ is generated by,

$$E = Y^T D. \tag{2}$$

3.3 Association Module

Revisit Previous VIS Methods. Previous tracking-by-detection methods do not propagate information of one frame to the next frame, and they store features of previous instances for tracking instead. To combine detection and tracking in a unified model, Maskprop utilizes an attention mechanism to propagate object information. However, this attention mechanism is not efficient since it requires propagating features of every instance in a frame independently, which generates many redundant computations. Moreover, MaskProp separates the video into densely overlapped clips and propagate features from the center frame to all other frames in the clip. Although this propagation manner could avoid information loss during long-term propagation, it takes tremendous computation and memory consumption to perform attention between numerous frames.

New Association Mechanism For VIS. With the compact ID embedding of previous objects, we propose a new association mechanism for VIS. In our new mechanism, we utilize a local memory to store object information of last frame, and maintains a global memory of initial frame to build long-term correlation. Based on these two memory, we use attention operation to get features that contains previous object information and current information. Moreover, we adopt a parallel classification branch

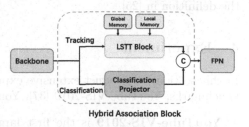

Fig. 4. Illustration of Hybrid Association Block.

to encode backbone features for further classification. This new association mechanism could propagate multiple object information from previous frames to current frame at once, which is much more efficient than MaskProp.

Hybrid Association Block (HAB). To implement this new association methcanism, as shown in Fig. 2, we propose an HAB, which is extended on the LSTT block in previous VOS method AOT [42]. As shown in Fig. 4, the new HAB contains a classification projector for additional classification in VIS task. In detail, the LSTT block conducts attention between backbone features and ID embeddings from global and local memory, which learns correlation between frames for object tracking. As for the classification projector, it is 1×1 convolution to encode backbone features for classification. For the output of HAB, we concat outputs from two branches to form the final output.

3.4 Training Details

As for training, we follow the sequential training strategy in [41], in which 5 frames in a video are randomly sampled to form a sequence. For each sequence, we first assign IDs for instances in the sequence since there are no ground truth IDs in the original YouTube-VIS dataset. We assign IDs for instances per frame from 0 to N-1, *e.g.* the first instance assigned 0, the second assigned 1 and so on. One important case should be mentioned is in the frame one instance first appears, the ground truth ID of it should be assigned N.

We train detection and tracking in an end-to-end way, and the loss is

$$L = L_{cls} + L_{bbox} + L_{mask} + L_{id}, \tag{3}$$

where L_{cls}, L_{bbox} and L_{mask} represent the classification loss, bounding box loss and mask loss in image instance segmentation model [16,32]. L_{id} denotes the ID loss, which is implemented with a similar function like L_{cls}. For example, we use focal loss [25] for ID loss when combining with CondInst,

$$L_{id} = -\alpha_t (1 - p_i(d))^\lambda log(p_i(d)), \tag{4}$$

where $p_i(d)$ is the probability of assigning ID d to instance i, α_t and λ follow the definition in [25].

4 Experiment

In this section, we conduct extensive experiments to evaluate IAI on three VIS benchmarks, YouTube-VIS-2019 [37], YouTube-VIS-2021 [36] and OVIS [31].

– **YouTube-VIS-2019** is the first large-scale benchmark to video instance segmentation, which consists of 2,883 high-resolution YouTube videos. The dataset is annotated with 4883 unique objects from 40 common categories and contains about 131k instance masks.
– **YouTube-VIS-2021** is extended on the basis of the YouTube-VIS-2019, with more videos and a modified category label set. The YouTube-VIS-2021 contains 3,859 high-resolution YouTube videos, 8,171 unique video instances and approximately 232k high-quality manual annotated masks.
– **OVIS** is a large-scale benchmark for video instance segmentation, which aims to perceive object occlusion in videos. The OVIS dataset consists of 901 videos with severe object occlusions. The dataset is annotated with 5,223 unique instances from 25 commonly seen categories.

We use the average precision (mAP) and average recall (AR) defined in [37] as the evaluation metric. Following previous works, we evaluate all results on validation set through official evaluation servers.

4.1   Implementation Details

Settings. The model is implemented with MMDetection-2.11 [10]. For training, we initialize our model with weights of corresponding instance segmentation model pre-trained on COCO train2017. The instance segmentation models are pretrained with 12 epochs. The VIS network is optimized with AdamW optimizer setting the initial learning rate to 10^{-4}, weight decay to 10^{-4}. The learning rate is reduced by a factor of 10 at 9 and 12 epochs. Specifically, the backbone's learning rate is set to 0.1 of the network's, and weight decay is set to 0.9 of the network's to avoid overfitting. The input size of each frame is resized to 360 × 640. The number of IDs in ID head is set to 20.

We randomly sample 5 frames from a video to form a sequence for training. We train our model on VIS datasets with 1× schedule, *i.e.*, 12 epochs. The models are trained on 4 T V100 GPUs with batch size of 16. During inference, the video is processed frame by frame without any test time augmentation. The FPS data of inference is measured on Tesla V100.

4.2   Main Results

YouTube-VIS-2019 Dataset. We apply our IAI paradigm on one-stage segmentation model CondInst, and compare it with state-of-the-art methods in Tab. 1. With simple multi-scale training augmentation, our method achieves 38.6 mAP with ResNet-50 backbone, which outperforms all the online methods in Tab. 1. Moreover, our method even outperforms STEm-seg, STMask and CrossVIS with a stronger Resnet-101 backbone. With the ResNet-101 backbone, our method surpasses the recently proposed online method STMask and CrossVIS by about 5 points in mAP with simple multi-scale augmentation. As for speed, our method achieves a nearly real-time speed at 27.4 FPS. Compared with other online algorithms, we argue that utilizing prior information for detection during inference partly slows down IAI paradigm.

As for the state-of-the-art offline method MaskProp, we argue the high performance of MaskProp partly comes from its combination with multiple strong networks, *e.g.* Spatiotemporal Sampling Network [3], Hybrid Task Cascade mask head [9], High-Resolution Mask Refinement post-process, and complex training augmentations, *e.g.* extra OpenImages [20] datasets and longer training schedule. Meanwhile, MaskProp requires huge computation and memory cost to achieve high performance, which impedes it from online scenarios. We aim to design an efficient online paradigm for VIS, and our method can be integrated with different image segmentation models to solve VIS in an online fashion. Overall, the experimental results prove the effectiveness of the new paradigm.

YouTube-VIS-2021 Dataset. YouTube-VIS-2021 dataset is an upgraded version of YouTube-VIS-2019 dataset, with more videos and an improved class set. We evaluate our method on this new dataset and compare it with some state-of-the-art approaches. As shown in Tab. 2, our algorithm outperforms SipMask-VIS and CrossVIS by a large margin without any training augmentations. The experiment results further demonstrate IAI's advantage over other paradigms.

Table 1. Comparisons with SOTA VIS methods on **YouTube-VIS-2019** val set. "√" under "Aug." means multi-scale augmentation for training, "√√" indicates stronger augmentation or additional data. "Temp." means modeling temporal information for detection. We follow [23] to define whether the VIS algorithm is online or offline.

Methods	Backbone	Aug.	Temp.	mAP	AP_{50}	AP_{75}	AR_1	AR_{10}	FPS
offline methods									
STEm-Seg [1]	ResNet-50	√√	√	30.6	50.7	33.5	31.6	37.1	4.4
	ResNet-101			34.6	55.8	37.9	34.4	41.6	2.1
MaskProp [2]	ResNet-50	√√	√	40.0	-	42.9	-	-	<6.2
	ResNet-101			42.5	-	-	45.6	-	<5.6
VisTR [35]	ResNet-50	√	√	36.2	59.8	36.9	37.2	42.4	30.0
	ResNet-101			40.1	64.0	45.0	38.3	44.9	27.7
Seq Mask R-CNN [22]	ResNet-50	√√	√	40.4	63.0	43.8	41.1	49.7	-
	ResNet-101			**43.8**	**65.5**	**47.4**	**43.0**	**53.2**	-
IFC [18]	ResNet-50	√	√	**41.2**	**65.1**	**44.6**	42.3	49.6	107.1
	ResNet-101			42.6	66.6	46.3	43.5	51.4	89.4
online methods									
MaskTrack R-CNN [37]	ResNet-50			30.3	51.1	32.6	31.0	35.5	32.8
	ResNet-101			31.9	53.7	32.3	32.5	37.7	28.6
SipMask-VIS [6]	ResNet-50	√		33.7	54.1	35.8	35.4	40.1	34.1
STMask [21]	ResNet-50			33.5	52.1	36.9	31.1	39.2	28.6
	ResNet-101	√		36.8	56.8	38.0	34.8	41.8	23.4
QueryInst-VIS [14,40]	ResNet-50	√		36.2	56.7	39.7	36.1	42.9	32.3
CompFeat [15]	ResNet-50	√√	√	35.3	56.0	38.6	33.1	40.3	-
SG-Net [27]	ResNet-50	√		34.8	56.1	36.8	35.8	40.8	23.0
	ResNet-101			36.3	57.1	39.6	35.9	43.0	19.8
CrossVIS [39]	ResNet-50	√		36.3	56.8	38.9	35.6	40.7	39.8
	ResNet-101			36.6	57.3	39.7	36.0	42.0	35.6
IAI+CondInst	ResNet-50		√	37.9	58.8	42.1	38.7	46.8	27.4
	ResNet-50	√		**38.6**	**60.1**	**41.9**	**38.4**	**45.6**	27.4
	ResNet-101		√	41.0	61.3	45.3	40.8	47.5	23.7
	ResNet-101	√		**41.9**	**63.7**	**47.5**	**41.1**	**49.6**	23.7

Table 2. Comparisons with SOTA VIS methods on **YouTube-VIS-2021** val set.

Methods	Backbone	Aug	mAP	AP_{50}	AP_{75}	AR_1	AR_{10}
MaskTrack R-CNN	ResNet-50		28.6	48.9	29.6	26.5	33.8
SipMask-VIS		√	31.7	52.5	34.0	30.8	37.8
CrossVIS			33.3	53.8	37.0	30.1	37.6
CrossVIS		√	34.2	54.4	37.9	30.4	38.2
IFC		√	35.2	57.2	37.5	-	-
IAI+CondInst			**37.7**	**58.0**	**42.3**	**34.6**	**45.6**

OVIS Dataset. To further prove the effectiveness and robustness of our method, we evaluate our method on the OVIS dataset. The OVIS dataset is

Table 3. Comparisons with SOTA VIS methods on **OVIS** val set.

Methods	Backbone	Aug	mAP	AP_{50}	AP_{75}	AR_1	AR_{10}
MaskTrack R-CNN	ResNet-50		10.8	25.3	8.5	7.9	14.9
SipMask-VIS		✓	10.2	24.7	7.8	7.9	15.8
STEm-Seg		✓	13.8	32.1	11.9	9.1	20.0
QueryInst-VIS		✓	14.7	34.7	11.6	9.0	21.2
STMask			15.4	33.8	12.5	8.9	21.3
CrossVIS		✓	14.9	32.7	12.1	10.3	19.8
CMaskTrack R-CNN		✓	15.4	33.9	13.1	9.3	20.0
IAI+CondInst			18.5	36.8	18.0	11.7	24.0
IAI+CondInst∗			**20.3**	**39.5**	**19.0**	**11.9**	**26.2**

∗ means the model is pretrained on YouTube-VIS 2019 dataset.

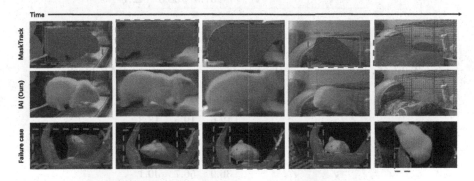

Fig. 5. Qualitative results. (top) Compared with MaskTrack R-CNN, IAI could make better use of temporal information, and performs well even in fast-moving scene. (bottom) Some errors are propagated once mistake happens in previous frames.

much harder than YouTube-VIS-2019 and YouTube-VIS-2021 dataset, which contains more instances and more occluded cases per video. As shown in Tab. 3, our methods outperforms SOTA VIS methods by a large margin (+4.9 mAP), which indicates strong ability of our methods on dealing with object occlusion.

Qualitative Results. Fig. 5 visualizes some qualitative results of IAI in comparison with MaskTrack R-CNN. As shown in Fig. 5, IAI segments and tracks object more accurately than MaskTrack R-CNN, especially in fast-moving scenes. IAI makes a better fuse of temporal information, which enables it to handle motion blur and fast-moving target tracking. However, IAI relies on the segmentation quality of the first frame, once a mistake happens in the first frame, IAI might propagate it to the next frames.

4.3   Ablation Study

In this section, we conduct extensive ablation study experiments to prove the general effectiveness of our method. All the experiments are conducted on the YouTube-VIS 2019 dataset. All models are with ResNet-50 FPN [24] as backbone, and trained in 1 × schedule without any augmentation.

Table 4. Experiments of the identification module and association module.

Identification	Association	mAP	AP$_{50}$	AP$_{75}$
		24.0	40.4	23.0
✓		12.6	19.5	13.3
	✓	24.5	40.9	25.0
✓	✓	37.9	58.8	42.1

Table 5. Experiments of three key components of the HAB block.

Local	Global	Class	mAP	AP$_{50}$	AP$_{75}$
			12.6	19.5	13.3
✓		✓	32.9	50.0	36.7
	✓	✓	34.0	53.1	38.2
✓	✓		36.9	56.7	40.1
✓	✓	✓	37.9	58.8	42.1

Identification module and association module. We conduct ablation study to prove the effectiveness of two key modules of our method. As shown in Tab. 4, the basic model without ID and association module performs poor, and with both two modules, our IAI achieves superior accuracy. Another important observation is that both two modules are necessary for our IAI paradigm. Without association module, the model could not model previous information and predict accurate IDs. Even worse, the identification module will lead to terrible performance because the ID supervision adds extra noise to model training. Without identification module, the model could not track the instances and gets similar performance to image model.

HAB block. As the HAB block is the basic component of our association module, we conduct some experiments to verify the effectiveness. We study three key components of the HAB block in Tab. 5: global memory, local memory and classification projector. From the results, we can find that both three components are effective in our IAI paradigm. The global, local memory and classification projector could bring an improvement of 5.0, 3.0 and 1.0 mAP separately.

Image Segmentation Model.
To prove the generality of
IAI paradigm, we experiment
with both one-stage and two-
stage models. In the experiment, we choose CondInst and
Mask R-CNN for the rep-
resentative of one-stage and
two-stage models separately.
In Tab. 6, we compare our
paradigm with the tracking-by-

Table 6. Comparisons with other paradigms on
different image instance segmentation frameworks.
"Track" means tracking-by-detection paradigm.

Image Model	Paradigm	mAP	AP_{50}	AP_{75}
Mask R-CNN	Track	30.3	51.1	32.6
	IAI	31.7	49.9	34.6
CondInst	Track	32.1	-	-
	IAI	37.9	58.8	42.1

detection paradigm on two image segmentation models. From the results, we
could see that IAI paradigm outperforms the tracking-by-detection paradigm
on both one-stage and two-stage segmentation models. As for why the IAI on
CondInst bring a larger improvement than IAI on MaskTrack, we argue that IAI
benefits more from a better image model because more accurate segmentation
of first frame (no previous information) can lead to better propagation for next
frames.

ID Loss Function. As we introduce a new ID loss in IAI paradigm, we study
the effect of the different ID loss functions in Tab. 7. From the results, we could
find that focal loss [25] brings a 1.5 mAP improvement over cross-entropy (CE)
loss. As the classification loss function is focal loss, this comparison proves that
keeping ID loss function consistent with the classification loss function is enough
for good performance, which indicates that no additional design is required for
the ID loss function.

ID Head Convolution Layer Number. As the ID head plays an important
role in the identification module, and we evaluate the effect of different ID head
convolution layer numbers on performance. As shown in Tab. 8, more convolution
layers do not bring obvious improvement, and using 4 convolution layers even
gets worse accuracy. A possible reason is that ID information is relatively simple
compared with appearance information. The appearance information might contain color, shape and other characteristics, while ID information only focuses on
similarity between instances. Since ID information is easy to capture, increasing
parameters is unable to boost the performance and might cause overfitting.

Table 7. Experiments of different ID loss functions.

L_{id}	mAP	AP_{50}	AP_{75}
CE loss	36.4	53.8	40.9
Focal loss	37.9	58.8	42.1

Table 8. Experiments of different ID head convolution layer numbers.

ID Head	mAP	AP_{50}	AP_{75}
2Conv	37.9	58.8	42.1
3Conv	38.1	61.3	40.9
4Conv	35.3	53.6	37.9

5   Conclusion

In this paper, we introduce IAI, a novel generic online paradigm for video instance segmentation. The new IAI paradigm successfully utilizes prior object information for both detection and tracking in an online way, and perform multiple object association at once. These advantages make IAI outperform all the online video instance segmentation methods in the challenging YouTube-VIS benchmarks. Notably, the IAI paradigm shows obvious advantages over previous tracking-by-detection paradigm on occluded scenes, outperforming these methods by a large margin on OVIS benchmark. We hope our IAI paradigm could perform as a strong baseline in the VIS and OVIS task, and contribute to future research on video understanding tasks.

Acknowledgment. This work was supported in part by the National NSF of China (No. 62120106009), the Fundamental Research Funds for the Central Universities (No. K22RC00010).

References

1. Athar, A., Mahadevan, S., Ošep, A., Leal-Taixé, L., Leibe, B.: STEm-Seg: Spatio-temporal embeddings for instance segmentation in videos. In: Vedaldi, A., Bischof, H., Brox, T., Frahm, J.-M. (eds.) ECCV 2020. LNCS, vol. 12356, pp. 158–177. Springer, Cham (2020). https://doi.org/10.1007/978-3-030-58621-8_10
2. Bertasius, G., Torresani, L.: Classifying, segmenting, and tracking object instances in video with mask propagation. In: CVPR (2020)
3. Bertasius, G., Torresani, L., Shi, J.: Object detection in video with spatiotemporal sampling networks. In: Ferrari, V., Hebert, M., Sminchisescu, C., Weiss, Y. (eds.) ECCV 2018. LNCS, vol. 11216, pp. 342–357. Springer, Cham (2018). https://doi.org/10.1007/978-3-030-01258-8_21
4. Bolya, D., Zhou, C., Xiao, F., Lee, Y.J.: YOLACT: real-time instance segmentation. In: ICCV (2019)
5. Bolya, D., Zhou, C., Xiao, F., Lee, Y.J.: YOLACT++: better real-time instance segmentation. TPAMI (2020)
6. Cao, J., Anwer, R.M., Cholakkal, H., Khan, F.S., Pang, Y., Shao, L.: SipMask: spatial information preservation for fast image and video instance segmentation. In: Vedaldi, A., Bischof, H., Brox, T., Frahm, J.-M. (eds.) ECCV 2020. LNCS, vol. 12359, pp. 1–18. Springer, Cham (2020). https://doi.org/10.1007/978-3-030-58568-6_1

7. Cao, J., Anwer, R.M., Cholakkal, H., Khan, F.S., Pang, Y., Shao, L.: SipMask: spatial information preservation for fast image and video instance segmentation. In: Vedaldi, A., Bischof, H., Brox, T., Frahm, J.-M. (eds.) ECCV 2020. LNCS, vol. 12359, pp. 1–18. Springer, Cham (2020). https://doi.org/10.1007/978-3-030-58568-6_1

8. Chen, H., Sun, K., Tian, Z., Shen, C., Huang, Y., Yan, Y.: BlendMask: top-down meets bottom-up for instance segmentation. In: CVPR (2020)

9. Chen, K., et al.: Hybrid task cascade for instance segmentation. In: CVPR (2019)

10. Chen, K., et al.: MMDetection: Open MMLAB detection toolbox and benchmark. arXiv preprint arXiv:1906.07155 (2019)

11. Chen, Y., Pont-Tuset, J., Montes, A., Van Gool, L.: Blazingly fast video object segmentation with pixel-wise metric learning. In: CVPR (2018)

12. Cheng, T., Wang, X., Huang, L., Liu, W.: Boundary-preserving mask R-CNN. In: Vedaldi, A., Bischof, H., Brox, T., Frahm, J.-M. (eds.) ECCV 2020. LNCS, vol. 12359, pp. 660–676. Springer, Cham (2020). https://doi.org/10.1007/978-3-030-58568-6_39

13. Dai, J., et al.: Deformable convolutional networks. In: ICCV (2017)

14. Fang, Y., et al.: Instances as queries. In: ICCV (2021)

15. Fu, Y., Yang, L., Liu, D., Huang, T.S., Shi, H.: CompFeat: comprehensive feature aggregation for video instance segmentation. In: AAAI (2021)

16. He, K., Gkioxari, G., Dollar, P., Girshick, R.: Mask R-CNN. In: ICCV (2017)

17. He, K., Zhang, X., Ren, S., Sun, J.: Deep residual learning for image recognition. In: CVPR (2016)

18. Hwang, S., Heo, M., Oh, S.W., Kim, S.J.: Video instance segmentation using inter-frame communication transformers. arXiv preprint arXiv:2106.03299 (2021)

19. Kuhn, H.W.: The Hungarian method for the assignment problem. Naval Res. Logist. Q. 2(1–2), 83–97 (1955)

20. Kuznetsova, A., et al.: The open images dataset v4: Unified image classification, object detection, and visual relationship detection at scale. IJCV. 128, 1956–1981 (2020)

21. Li, M., Li, S., Li, L., Zhang, L.: Spatial feature calibration and temporal fusion for effective one-stage video instance segmentation. In: CVPR (2021)

22. Lin, H., Wu, R., Liu, S., Lu, J., Jia, J.: Video instance segmentation with a propose-reduce paradigm (2021)

23. Lin, J., Gan, C., Han, S.: TSM: temporal shift module for efficient video under-standing. In: ICCV (2019)

24. Lin, T.Y., Dollar, P., Girshick, R., He, K., Hariharan, B., Belongie, S.: Feature pyramid networks for object detection. In: CVPR (2017)

25. Lin, T.Y., Goyal, P., Girshick, R., He, K., Dollar, P.: Focal loss for dense object detection. In: ICCV (2017)

26. Lin, T.-Y., et al.: Microsoft COCO: common objects in context. In: Fleet, D., Pajdla, T., Schiele, B., Tuytelaars, T. (eds.) ECCV 2014. LNCS, vol. 8693, pp. 740–755. Springer, Cham (2014). https://doi.org/10.1007/978-3-319-10602-1_48

27. Liu, D., Cui, Y., Tan, W., Chen, Y.: Sg-net: Spatial granularity network for one-stage video instance segmentation. In: CVPR (2021)

28. Oh, S.W., Lee, J.Y., Xu, N., Kim, S.J.: Video object segmentation using space-time memory networks. In: ICCV (2019)

29. Perazzi, F., Pont-Tuset, J., McWilliams, B., Van Gool, L., Gross, M., Sorkine-Hornung, A.: A benchmark dataset and evaluation methodology for video object segmentation. In: CVPR (2016)

30. Pont-Tuset, J., et al.: The 2017 Davis challenge on video object segmentation. arXiv:1704.00675 (2017)
31. Qi, J., et al.: Occluded video instance segmentation. arXiv preprint arXiv:2102.01558 (2021)
32. Tian, Z., Shen, C., Chen, H.: Conditional convolutions for instance segmentation. In: Vedaldi, A., Bischof, H., Brox, T., Frahm, J.-M. (eds.) ECCV 2020. LNCS, vol. 12346, pp. 282–298. Springer, Cham (2020). https://doi.org/10.1007/978-3-030-58452-8_17
33. Tian, Z., Shen, C., Chen, H., He, T.: FCOS: fully convolutional one-stage object detection. In: ICCV (2019)
34. Voigtlaender, P., Chai, Y., Schroff, F., Adam, H., Leibe, B., Chen, L.C.: FEELVOS: fast end-to-end embedding learning for video object segmentation. In: CVPR (2019)
35. Wang, Y., et al.: End-to-end video instance segmentation with transformers. In: CVPR (2021)
36. Xu, N., et al.: YouTube-vis dataset 2021 version (2021). https://youtube-vos.org/dataset/vis
37. Yang, L., Fan, Y., Xu, N.: Video instance segmentation. In: ICCV (2019)
38. Yang, L., Wang, Y., Xiong, X., Yang, J., Katsaggelos, A.K.: Efficient video object segmentation via network modulation. In: CVPR (2018)
39. Yang, S., et al.: Crossover learning for fast online video instance segmentation. In: ICCV (2021)
40. Yang, S., et al.: Tracking instances as queries. arXiv preprint arXiv:2106.11963 (2021)
41. Yang, Z., Wei, Y., Yang, Y.: Collaborative video object segmentation by foreground-background integration. In: Vedaldi, A., Bischof, H., Brox, T., Frahm, J.-M. (eds.) ECCV 2020. LNCS, vol. 12350, pp. 332–348. Springer, Cham (2020). https://doi.org/10.1007/978-3-030-58558-7_20
42. Yang, Z., Wei, Y., Yang, Y.: Associating objects with transformers for video object segmentation. In: NeurIPS (2021)
43. Yang, Z., Wei, Y., Yang, Y.: Collaborative video object segmentation by multi-scale foreground-background integration. TPAMI (2021)
44. Zhang, R., Tian, Z., Shen, C., You, M., Yan, Y.: Mask encoding for single shot instance segmentation. In: CVPR (2020)

Laplacian Mesh Transformer: Dual Attention and Topology Aware Network for 3D Mesh Classification and Segmentation

Xiao-Juan Li[1,2] , Jie Yang[1,2(✉)] , and Fang-Lue Zhang[3]

[1] Institute of Computing Technology, Chinese Academy of Sciences, Beijing, China
{lixiaojuan,yangjie01}@ict.ac.cn
[2] University of Chinese Academy of Sciences, Beijing, China
[3] Victoria University of Wellington, Wellington, New Zealand
fanglue.zhang@vuw.ac.nz

Abstract. Deep learning-based approaches for shape understanding and processing tasks have attracted considerable attention. Despite the great progress that has been made, the existing approaches fail to efficiently capture sophisticated structure information and critical part features simultaneously, limiting their capability of providing discriminative deep shape features. To address the above issue, we proposed a novel deep learning framework, Laplacian Mesh Transformer, to extract the critical structure and geometry features. We introduce a dual attention mechanism, where the 1st level self-attention mechanism is used to capture the critical partial/local structure and geometric information on the entire mesh, and the 2nd level is to fuse the geometrical and structural features together with the learned importance according to a specific downstream task. More particularly, Laplacian spectral decomposition is adopted as our basic structure representation given its ability to describe shape topology (connectivity of triangles). Our approach builds a hierarchical structure to process shape features from fine to coarse using the dual attention mechanism, which is stable under the isometric transformations. It enables an effective feature extraction that can tackle 3D meshes with complex structure and geometry efficiently in various shape analysis tasks, such as shape segmentation and classification. Extensive experiments on the standard benchmarks show that our method outperforms state-of-the-art methods.

Keywords: Laplacian EigenVector · Transformer · Attention mechanism · Topology aware · Shape segmentation & classification

Supplementary Information The online version contains supplementary material available at https://doi.org/10.1007/978-3-031-19818-2_31.

1 Introduction

3D data analysis has been an important topic in computer graphics and computer vision research. Numerous tasks in semantic understanding [3,28], 3D detection [10,90], shape abstraction [65,69] rely on the advanced 3D shape analysis and understanding technology, especially for the urgent requirements in autonomous driving, virtual/augmented reality, robotics, and model creation.

As an essential method to represent 3D shapes, polygonal meshes have been successfully used in the above applications for efficient modeling and rendering of 3D objects. To make it possible to learn the features of 3D meshes of neural networks, many large-scale datasets (*e.g.* ShapeNet [9], ModelNet [79]) are built and made available to the public. Considering that the polygonal meshes describe the detailed surfaces (including the geometry and structures) by a set of 2D polygons [6], some voxelized approaches [12,46] extend the 2D deep learning methods to the 3D domain. However, it suffers from massive computation and memory demands and thus has a limited capacity to cope with high-resolution mesh data. Other pioneering works focus on learning features from point clouds to perform 3D data analysis, such as PointNet [54] and PointNet++ [56]. They have achieved good performances on segmentation and classification by multi-layer perceptrons (MLPs) or dynamic graphs [75]. Although point cloud is lightweight and mitigates the computation cost issue, it lacks topological information compared to the polygonal meshes. Therefore, the prior deep learning-based methods fail to capture complex structural information and partial features for shape analysis

This work focuses on polygonal meshes and develops a novel deep architecture based on self-attention [70] to learn 3D shape features in a topology-aware manner. The design of our network is based on two key observations. Firstly, the eigenvectors obtained from a Laplacian spectral decomposition on meshes are used to shape analysis [41,44] and indicate some topological information(*e.g.* symmetry), which can be naturally used as a representation of the topology of 3D meshes. Secondly, the relationships among the elements of the structural and partial geometric features can well represent the 3D meshes and their parts in a discriminative way. The self-attention mechanism used in Transformer [13,14,70,83] has shown its capability of extracting the relationships among all the elements of input signals in natural language processing and image analysis [29,58,88], which can be adopted to analyze 3D shapes [18,86,89] more effectively, especially for 3D meshes [17].

This paper builds a dual attention architecture in a topology-aware fashion, Laplacian Mesh Transformer, to understand the complex structure and rich geometric information of 3D meshes. Our method takes raw mesh features as input and produces global features containing effective descriptions of topology and geometry information. There are two branches (see Fig. 1) that simultaneously learn critical geometric features and structural features from the Euclidean coordinates and Laplacian spectral decomposition, respectively. In both branches, we extract features of different scales by four self-attentions. Then we apply a final attention-based fusion module to learn the importance of the topology and geometry information and fuse them to form the final global features when applying to different downstream tasks. With the help of the dual attention mechanism and Laplacian spectral decomposition, we build a hierarchical structure from fine (partial) to coarse (global) to process shape features. Compared to alternative methods, our approach utilizes both spatial and spectral information by dual attention and is able to dynamically determine the contribution of topology and geometry features during inference for a better shape understanding.

Our network architecture is illustrated in Fig. 1. To demonstrate the aptitude of our approach to describe mesh features, we build two downstream networks to perform shape segmentation and classification on the ShapeNet [9] and COSEG [76] datasets, which are the fundamental shape analysis tasks in computer vision. Our extensive experiments demonstrate the robustness of our Laplacian Mesh Transformer to various vertex types and different triangulations of meshes. Our method achieves remarkable performance in both segmentation and classification with a lightweight network, and it can also be potentially applied to other tasks, such as shape retrieval.

The main contributions of our method are as follows:

- We design a dual attention mechanism for learning features on 3D polygonal meshes, which takes eigenvectors from Laplacian spectral decomposition as the raw topological description;
- We propose a deep architecture that focuses on sophisticated polygonal meshes and takes the partial geometric/structural features and their importance into consideration in a fine (partial) to coarse (global) manner;
- We conduct extensive experiments on multiple 3D shape analysis tasks to demonstrate our superior effectiveness compared to state-of-the-art methods.

2 Related Works

This section briefly reviews learning-based methods in the 3D domain and then summarizes the popular self-attention-based work, which is helpful for many applications.

Deep Learning on 3D Domain. With the increased availability of 3D models and the development of deep learning frameworks, there are various approaches to analysis and modeling 3D models, thanks to the mighty deep learning tools nowadays. For different representations of 3D data, recent works have been developed for voxels [12,46,55,78], multi-view images of 3D data [34,63,64], point cloud [1,19,39,54,56], meshes [20,23,32,66,73,82], and implicit functions [11,47,50,52]. Voxels represent values on a regular grid in three-dimensional space, which are similar to pixels inside of a 2D image. Some operators of deep learning could be extended and applied to the 3D voxels naturally. For the multi-view images of 3D data, a shape can be rendered into multiple images from different views. By applying the traditional 2D image CNNs to these 2D images from different views, the entire model is represented by aggregating the features of these images. The point cloud is a general representation of any 3D shape, which is easy to capture with portable devices. Many works solve the following challenges: noisy, sparse, and disordered. Compared to point clouds, meshes are considered a better representation of concrete geometric shapes and structures. Nevertheless, it is tough to learn on the meshes given their irregularity. [32] reconstructs a 3D mesh by the laplacian in an extrinsic/intrinsic manner, but ours uses the laplacian eigenvectors and attention to help the deep networks understand shapes. For the comprehensive and detailed review of deep learning on 3D data, we refer the readers to these surveys [2, 7, 33, 80].

Self-Attention Mechanism. The self-attention mechanism is widely used in many natural language processing (NLP) and computer vision tasks. The survey papers [25, 37, 42, 67] have comprehensive discussion on the attention mechanism. Bahdanau *et al.* [4] first adopts the attention mechanism (soft-search) into the neural machine translation, the attention map is predicted to summarize the contextual relationships by bidirectional RNN [60]. Lin *et al.* [43] introduces a model for learning an interpretable embedding by self-attention. Followed by this, [70] proposed the transformer and applied it to the machine translation, which does not depend on the convolution operator, and achieves promising results by utilizing the global context. Furthermore, the researchers made a great effort to develop and expand the transformer, such as XLNET [83], a two-way transformer – BERT [15]. However, in the NLP field, the sentences are sequential and semantic meaningful, while the vertex on 3D meshes are usually disordered and have no semantics.

At the same time, self-attention also makes great potential impacts and receives more and more attention in computer vision (*e.g.* object detection – DETR [8]). Wang *et al.* [72] proposed a residual attention module for image classification in a stacked manner. Zhang *et al.* [87] designed a generative model with self-attention, which enables attention-driven and long-range dependency modeling for image GAN [22]. Recently, Visual Transformer [77] and Vision Transformer (ViT) [16] interpret an image as semantic visual tokens and sequential patches, respectively, then apply the transformer to the above sequential data. They all exceed the performance of CNN-based methods on image processing tasks when the training data is sufficient.

Inspired by the local patch structure used in ViT and the basic semantic information in language words, we propose a dual attention module based on self-attention, which can take the geometry and topology into consideration, capturing the local partial criticism on the 3D meshes and obtaining crucial semantic information.

3   Methodology

3.1   Overview

Given a 3D polygonal mesh $\mathcal{M} = (\mathcal{V}, \mathcal{E})$ with $|\mathcal{V}|$ vertices and $|\mathcal{E}|$ edges, our goal is to let the network learn a function $f : V \to \mathbb{R}^{|S|}$ that maps vertex features to

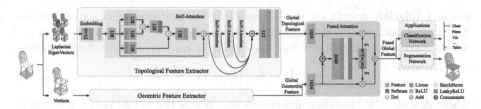

Fig. 1. Our framework for shape analysis. The Laplacian eigenvectors and vertex coordinates are fed the first level of our dual attention mechanism, where the two self-attention branches for topology and geometry features have the same architecture. The second level of attention in the fusing module merges the two sets of features with learned importance to generate the final global feature for the whole shape, which can be used to perform some downstream tasks.

vector space S. For the classification task, $|S|$ could be 1 since the whole shape has one attribute; For the segmentation task, $|S| = |\mathcal{V}|$ because each vertex of the shape has a attribute. The vertex feature typically contains coordinates, normal, curvatures, PCA, etc. We aim to design a network that can learn a general function f, describing the importance of geometric and structural contexts. So we proposed a dual attention mechanism to learn a reasonable fusion between geometry and topology to improve the performance according to the global context. Our network takes the coordinates, normal, Laplacian eigenvector of the vertices of a mesh model as input and predicts the probability matrix with size $|\mathcal{V}| \times l_s$ for each vertex for shape segmentation and probability score with size l_c for each category on the entire shape for classification. In the following sections, we briefly revisit the formulation of Laplacian spectral decomposition in Sect. 3.2. Then, we further present the dual attention mechanism (Sect. 3.3) on geometry and topology for 3D polygonal meshes, which aims to capture the partial critical features and the importance of geometry and topology. Lastly, we describe our entire network architecture (Sect. 3.4) that determines the importance of geometry and topology and feeds the fused feature to perform classification, segmentation, or other tasks. Our network considers the shape's geometry and topology simultaneously and adjusts their importance by the attention mechanism to achieve adequate shape understanding.

3.2 Laplacian Spectral Decomposition

The Laplacian Spectral Decomposition effectively describes the mesh topology and geometric properties, *i.e.* the connectivity of vertex or symmetry. Figure 2 visualizes Laplacian eigenvectors on some 3D meshes. An observation is that Laplacian eigenvectors are intuitive when visualized for the segmentation task. Hence, in most cases, if our networks can determine the instances' segments by laplacian features, our network can balance the weights by our dual attention to get more accurate segmentation. We have a discussion about that in Subsect. 4.4.

A mesh $\mathcal{M} = (\mathcal{V}, \mathcal{E})$ with arbitrary vertices and different connectivity can be regraded as a graph with $|\mathcal{V}|$ nodes and $|\mathcal{E}|$ relationships. We can adapt the graph Laplacian matrix on the 3D mesh to capture the topology of vertices. The Laplacian spectral decomposition depends on the number of vertices and different triangulation. In practice, the Laplacian matrix is formulated as follows:

$$\mathbf{L} = \mathbf{A}^{-1}(\mathbf{D} - \mathbf{W}) \tag{1}$$

where $\mathbf{A} \in \mathbb{R}^{|V| \times |V|}$ is a diagonal matrix that places the weights of each vertex on the diagonal of the matrix, the weight is defined as the Voronoi area of the one-ring triangles surrounding each vertex. $\mathbf{W} \in \mathbb{R}^{|V| \times |V|} = \{w_{i,j}\}$ is a cotangent weighted adjacent matrix that is sparse and a discretization of the continuous Laplacian on smooth surfaces [48]. $w_{i,j} \neq 0$ means that vertices v_i and v_j are connected by a edge on meshes, the value describes the cotangent weight [48,53] $w_{i,j} = \frac{1}{2}(cot\alpha_{ij} + cot\beta_{ij})$ of the edge, where α_{ij}, β_{ij} are the angles opposite of the mesh edge (v_i, v_j). $\mathbf{D} \in \mathbb{R}^{|V| \times |V|}$ is the degree matrix that is a diagonal matrix and each diagonal entry $d_{i,i} = \sum_{j=1}^{|V|} w_{i,j}$ is the sum of each row of the weighted adjacent matrix $\mathbf{A}$. After that, we calculate the eigenvector x of the Laplacian matrix $\mathbf{L}$: $det(\mathbf{L}x - \lambda\mathbf{I}) = 0$ according to [61]. We sort the absolute values of all the eigenvalues in an ascending order. In our paper, We use the eigenvectors corresponding to the first 12 eigenvalues as the descriptors of the topology of 3D meshes. In Sect. 4.4, we evaluate the performance when using different numbers of eigenvectors.

In Fig. 2, we illustrate some visualized Laplacian eigenvectors on different meshes. We visualize the Laplacian eigenvectors on the meshes with different vertex numbers for each example. From the results, we can see that the Laplacian eigenvectors are robust to different discretizations when the meshes can be discretized into reasonable sets of triangles, which is suitable for revealing the topology of meshes.

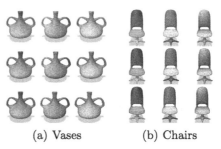

(a) Vases   (b) Chairs

Fig. 2. Laplacian eigenvector visualization. For the two examples from COSEG, we only visualize the first three eigenvectors of Laplacian spectral decomposition in different columns. From the top row to the bottom row, we show the eigeowvectors for the meshes with 2000, 3500, and 5000 vertices.

3.3 Dual Attention

Given a polygonal mesh $\mathcal{M}$, we use the above formulation to encode the topology $f_t \in \mathbb{R}^{|V| \times 12}$ of that mesh in the vertex feature. We also encode the shape geometry $f_g \in \mathbb{R}^{|V| \times 3}$ of 3D meshes represented by the vertex coordinates into the vertex feature. Now that each mesh can be represented as the vertex feature set $\{f_t, f_g\}$, which is defined on the vertex set $\mathcal{V}$ of mesh $\mathcal{M}$. Turning the 3D mesh graph into a vertex-wise feature set prevents using complex graph structures when training the network. Besides, the vertex-wise feature fits the self-attention operator [70], which is permutation-invariant and independent of the connection between vertices.

The structure of our dual attention is illustrated in Fig. 1. There are two attention modules: one is the self-attention (adopted from PCT [24]) with two branches, another is the fusion attention which learns the importance of geometry and topology adaptive according to the global context. The features are firstly fed to the self-attention encoder Enc_g for encoding the geometry feature f_g and the other self-attention encoder Enc_t for encoding the topology feature f_t. The two encoders Enc_g, Enc_t have a similar structure. For the encoder $Enc_t = \{enc^{emb}, enc^{sa}, enc^{cat}, enc^{slp}\}$ in the topology branch, it contains one embedding module enc^{emb}, four self-attention operators $\{enc_i^{sa}, i = 1, 2, 3, 4\}$, and one feature concatenation block enc^{cat}. At the end of the encoder, the feature map goes through a single layer of perceptrons enc^{slp} to generate the final feature f_t' for topology. We formulate the above process as:

$$f_t = enc^{emb}(f_t), f_t^1 = enc_1^{sa}(f_t)$$
$$f_t^i = enc_i^{sa}(f_t^{i-1}), i = 2, 3, 4 \tag{2}$$
$$f_t = enc^{cat}(f_t^1, f_t^2, f_t^3, f_t^4), \quad f_t' = enc^{slp}(f_t)$$

where enc^{emb} consists of two FC layers with batch-normalization and ReLU activation, which embeds the features into a 128-dimensional embedding space. enc^{sa} is a standard self-attention module, and its architecture is presented in Fig. 1. enc^{cat} performs the concatenation of multiple feature maps. Finally, for the topology branch, the network Enc^t maps the Laplacian eigenvector $f_t \in \mathbb{R}^{|V| \times 12}$ into the feature space $f_t' \in \mathbb{R}^{128}$. The geometry branch perform the same process, Enc_g takes the coordinates $f_g \in \mathbb{R}^{|V| \times 3}$ as input and generates the geometry feature $f_g' \in \mathbb{R}^{128}$. The two branches do not share weights.

Furthermore, we proposed an attention-based fusion module enc^{fus} to merge two features f_t', f_g' with an attention mask adaptively. The module learns to adjust the attention mask for achieving better performance for the given shape analysis tasks. It takes the generated features f_t', f_g' as inputs and predicts the attention weights w_t, w_g for the feature fusion. The weights w_g and w_t are predicted by three FC layers and batch-norm layers. For the final output, we use the exponential function $exp(\cdot)$ to ensure the weights are non-negative. Namely, the final output f is:

$$(w_t, w_g) = enc^{fus}(f_t', f_g')$$
$$f = w_t f_t' + w_g f_g' \tag{3}$$

where w_t, w_g are learnable weights and $w_t + w_g = 1$. w_t and w_g describe their contribution to the global feature f. In the end, the attention-based fusion module learns to determine the importance of geometry and topology according to the global context.

3.4   Network Architecture

We build deep network models based on our proposed dual attention module to perform shape analysis tasks. The overall pipeline is illustrated in Fig. 3.

For any mesh $\mathcal{M}$, we can obtain its geometry feature f_g and topology feature f_t by the coordinates and Laplacian spectral decomposition. Then, we can obtain the global feature f that fuses the geometry feature f_g and topology feature f_t according to their learned importance by our proposed dual attention. Since dual attention can understand more critical factors for shape understanding, it gives a reasonable criticism during inference, conforming to human intuition. We can then feed the global shape feature f to downstream network modules for shape analysis tasks. Particularly, we implement segmentation and classification networks to demonstrate the effectiveness of our approach.

Segmentation: Part segmentation predicts a vertex-wise function that can map the vertex features to semantic labels for each vertex, *i.e.* segment the whole shape into some meaningful semantic parts (*e.g.* arm, chair back, etc.). For learning the 3D shape in a cross-category manner, we simultaneously feed the category information as a one-hot vector and fused global feature to the segmentation network as shown in Fig. 3, which comprises five MLPs, each including a linear layer, a batch-norm layer, and an activation (ReLU/LeakyReLU). Nevertheless, for the final output, we use $sigmoid(\cdot)$ as activation to predict the probabilities ($\in \mathbb{R}^{|V| \times k}$) of all semantic labels for each vertex. Note that we add one dropout layer for avoiding overfitting, k is the number of part semantics. Finally, the probabilities are turned to a semantic label by an $argmax$ function.

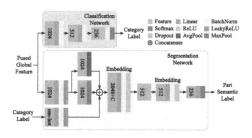

Fig. 3. Architecture of segmentation and classification networks. The segmentation network (bottom) first processes the fused global feature and the vector of labels. Then the processed global feature is duplicated, goes through average and max pooling, respectively, and is then concatenated with category features. Two embedding modules take the concatenated features to predict vertex-wise labels. In the classification network (top), the fused global features go through three MLPs, then a *softmax* operation is adopted to predict the probability.

Classification: This task aims to predict the probability of belonging to one semantic category for a given shape, which maps the input features to one semantic label for the whole mesh, such as chairs, tables, etc. Most parts of the classification network are the same as the segmentation network, but it predicts only one probability vector ($\in \mathbb{R}^k$) for the whole shape, where k is the number of categories.

4 Experiments and Evaluations

Laplacian Mesh Transformer is a general method for applying self-attention on triangular meshes to exploit shape features, enabling various applications, such as segmentation, classification, and shape retrieval. In this section, we present extensive quantitative and qualitative experiments on our shape classification and segmentation networks to evaluate the efficiency of the extracted features by Laplacian Mesh Transformer. We test our method and the existing deep models on three popular large scale datasets (i.e. ShapeNet [9], ModelNet [79], COSEG [84]). We also perform ablation studies to demonstrate the effectiveness of our key components. The experiments were conducted on a computer with an i9-9900K CPU and an RTX 2080Ti GPU.

4.1 Implementation Details

We primarily use the above large datasets for our experiments. ShapeNet provides 16 categories with semantic part labels, and ModelNet contains 40 categories of CAD models without any semantic part labels. COSEG dataset contains models segmented and labeled over 11 categories. For COSEG, we use the three largest and most commonly used categories, i.e. Vase, Chairs, and Tele-aliens. We follow the official splits of training and test set for all the above datasets. Due to the non-manifold nature of raw 3D data, we must ensure the shape is a manifold for the Laplacian spectral decomposition. Therefore, we follow the manifold algorithm [31] to preprocess the raw data and simplify [26] these watertight meshes to roughly the same number (2048) of vertices. Note that we have demonstrated that our network is independent of the number of vertices. The input features include coordinates (3) and Laplacian eigenvectors (12) for two branches. The 12-d Laplacian eigenvectors are the absolute Laplacian eigenvectors corresponding to the 12 lowest frequencies. For all the shapes, we scale them into a unit sphere. According to the evaluation (see Table 4), we achieve the best performance using four attention blocks and 12-d Laplacian eigenvectors.

We train the dual attention and downstream networks simultaneously. Our network is trained for 1000 epochs using the Adam solver [38] with a learning rate starting from 5e-4 and decaying every 100 epochs with a decay rate of 0.8. The trainable parameters are initialized randomly with Gaussian distribution. We implemented our network in PyTorch [51]. The backbone network of self-attention is borrowed from

Table 1. Comparison on shape classification of ModelNet10 (MN10), ModelNet40 (MN40), and ShapeNet (SN). All the alternative methods are classified into three clusters according to the input type. Note that '–' indicates the number is not reported.

Methods	Input type	MN10	MN40	SN
PointNet [54]	Point	–	89.2	–
PointNet++ [56]	Point	–	91.9	–
SO-Nett [40]	Point	95.7	93.4	–
PCT [24]	Point	–	93.2	–
3DShapeNets [79]	Volume	83.5	77.0	–
VoxNet [46]	Volume	91.0	84.5	–
ACNN [5]	Mesh	–	–	93.9
SyncSpecCNN [85]	Mesh	–	–	99.7
SPH [36]	Mesh	–	68.2	–
LaplacianNet [57]	Mesh	97.4	94.2	**99.8**
Ours	Mesh	**98.6**	**95.5**	99.4

PCT [24]. Most linear layers are composed of MLPs with ReLU activation. Empirically, our network converges in one day with a batch size of 32.

4.2   Shape Classification

We compare our shape classification network with state-of-the-art methods quantitatively. We evaluate all the methods on three datasets, ShapeNet (16 categories), ModelNet40 (40 categories), and ModelNet10 (10 categories), which are all widely used benchmarks for 3D shape classification. The output of our classification network is a probability score vector over all categories. We optimize the network by minimizing the cross-entropy loss between the ground truth one-hot vector and the probability logits. We observe that our method successfully beats all the other methods on the ModelNet Benchmark, including point-based methods [24,40,54,56], volume-based methods [46,79], and mesh-based methods [5,36,57,85]. Meanwhile, our method achievess comparable performance on the ShapeNet compared to SyncSpecCNN and LaplacianNet. The results are shown in Table 1, we report the overall accuracy across all categories. The mean overall accuracy on three large datasets is 97.9%, which outperforms the attention-based models such as PCT [24] and strong mesh-based models such as [57]. Note that our method only takes the 3-d coordinates and 12-d Laplacian eigenvectors as input, more inputs features (*e.g.* normal) could further improve the performance of our network. Please refer to our supplementary for more evaluations.

4.3   Shape Segmentation

Mesh segmentation is a critical and challenging task supporting methods for shape understanding and synthesis. Here, we evaluate our dual-attention mechanism on ShapeNet [9] and COSEG [84] datasets for part segmentation, which aims to divide a mesh into meaningful parts. Our network architecture for part segmentation is illustrated in Fig. 3. In ShapeNet, we train our network in the cross-categories setup where there are 16 categories and 50 different parts in total. Compared to ShapeNet, some of the categories in COSEG contain fewer data, bringing difficulties to deep learning methods. The three large categories of COSEG are: Vase, Chair, Tele-Alines, which contain 200,

Fig. 4.   Part segmentation results. We examples from different categories of ShapeNet. Note that the performance on Motor is lower than most other categories as in Table 2, due to its more complex topology and the larger number of mechanical parts.

300, 400 shapes, respectively. Moreover, its ground truth labels are annotated on point clouds sampled from the meshes. Since we need the Laplacian eigenvectors on the manifold meshes with graph structure in our input, we turn the raw meshes to manifold meshes [31] and transfer the labels on the point cloud to the nearest mesh vertices in the data preparation stage. More evaluations are presented in supplementary. The input features include three parts: coordinates (3), Laplacian eigenvectors (12), and the category label (one-hot vector). Our network generates vertex-level semantic probabilities on the input meshes. The cross-entropy loss is used to supervise the output of the network according to ground truth one-hot vectors. Following previous works, we evaluate the performance of each method by the widely used accuracy and IoU (Intersection-over-Union). We compare our method with the state-of-the-art shape segmentation methods [5,27,34,35,57,71,85].

Table 2. Comparison with different shape segmentation methods on ShapeNet. Based on the output of different methods, we compare our method with the others using accuracy and/or IoU. Ours outperforms SOTA algorithms in 13/16 categories on the accuracy metric.

	Method	Mean	Airplane	Bag	Cap	Car	Chair	Earphone	Guitar	Knife	Lamp	Laptop	Motorbike	Mug	Pistol	Rocket	Skateboard	Table
Accuracy	Shapeboost [35]	77.2	85.8	93.1	85.9	79.5	70.1	81.4	89.0	81.2	71.1	86.1	77.2	94.9	88.2	79.2	91.0	74.5
	Guo et al. [27]	77.6	87.4	91.0	85.7	80.1	66.8	79.8	89.9	77.1	71.6	82.7	80.1	95.1	84.1	76.9	89.6	77.8
	ShapePFCN [34]	85.7	90.3	94.6	94.5	90.2	82.9	84.9	91.8	82.8	78.0	95.3	87.0	96.0	91.5	81.6	91.9	84.8
	LaplacianNet [57]	91.5	89.6	90.2	88.2	88.2	83.2	82.3	95.6	88.7	87.4	96.3	70.6	97.0	92.7	82.2	94.7	92.6
	Ours	**92.6**	**90.7**	**96.5**	**95.0**	89.1	**92.7**	**93.2**	96.9	**93.5**	90.3	**97.1**	85.7	**98.6**	94.5	82.5	92.5	92.6
IoU	FeaStNet [71]	81.5	79.3	74.2	69.9	71.7	87.5	64.2	90.0	80.1	78.7	94.7	62.4	91.8	78.3	48.1	71.6	79.6
	ACNN [5]	79.6	76.4	72.9	70.8	72.7	86.1	71.1	87.8	82.0	77.4	95.5	45.7	89.5	77.4	49.2	82.1	76.7
	VoxelCNN [85]	79.4	75.1	72.8	73.3	70.0	87.2	63.5	88.4	79.6	74.4	93.5	58.7	91.8	76.4	51.2	65.3	77.1
	Yi et al. [85]	**84.7**	81.6	81.7	81.9	75.2	90.2	74.9	**93.0**	86.1	**84.7**	**95.6**	**66.7**	92.7	81.6	62.1	82.9	82.1
	LaplacianNet [57]	84.3	82.9	83.4	81.7	**80.0**	75.4	71.8	91.9	81.0	80.9	92.5	59.2	93.5	86.3	**74.3**	**90.3**	86.4
	Ours	83.7	**83.2**	**92.1**	**87.2**	71.0	**91.2**	**80.6**	91.3	**86.9**	81.9	93.4	60.6	**94.6**	**87.3**	63.9	85.6	**88.4**

In Table 2, we report the accuracy and IoU scores over ShapeNet of all the methods, which demonstrates that our method achieves the best performance on the average scores. For the accuracy and IoU on each category, our method outperforms the prior methods on 13 and 9 categories, respectively. Table 3 presents the accuracy score on COSEG dataset.

Table 3. Mesh segmentation accuracy on COSEG [76] of each method. Our method achieves the best performance.

Methods	Vases	Chairs	Tele-aliens	Mean
MeshCNN [26]	85.2	92.8	94.4	90.8
PD-MeshNet [49]	81.6	90.0	89.0	86.8
SubdivNet [30]	96.7	96.7	97.3	96.9
Xie et al. [81]	87.1	85.9	83.2	85.4
Wang et al. [74]	95.9	91.2	90.7	92.6
LaplacianNet [57]	94.2	92.2	93.9	93.4
Ours	**98.1**	**97.7**	**97.4**	**97.7**

We follow the SubdivNet for spliting training and testing sets for each category. In this table, we compare with some mesh-based segmentation methods [26,49,57,74,81]. We can observe that our method beats all the alternative methods on three datasets and achieves 97.7% on the average accuracy. For the Tele-alines dataset, we outperform SubdivNet [30] by a small margin. Figure 4 shows some examples of the shape segmentation task on ShapeNet (16 categories).

4.4 Ablation Studies

We perform five sets of ablation studies to demonstrate the necessity and effectiveness of our key designs. We first evaluate the dual attention and Laplacian features by checking the performance on the shape segmentation.Then, we validate the choice of the number of Laplacian eigenvectors and self-attention blocks. Finally, we demonstrate that our Laplacian mesh transformer is robust to various triangulation and different numbers of vertices. Table 4 shows all the ablation studies quantitatively on the COSEG dataset for the part segmentation.

With *v.s.* Without dual attention (DA). Our critical designs, dual attention, discriminates the importance of topology and geometry of the input shape for the specific task. Here, we aim to demonstrate the importance of using the two-stage attention for shape analysis quantitatively and qualitatively, including self-attention based topology/geometry feature extractor and attention-based fusion (FA) module. We built and trained two ablated networks: The first one (denoted as Ours-w/o DA) only has one attention-based feature extractor and no fusing module. We directly feed the concatenation of coordinates (3) and Laplacian eigenvectors (12) to the feature extractor and use the processed feature in the following segmentation network. The second ablated version (denoted as Ours-w/o FA) adopts the original two branches to process topological and geometric inputs separately and replaces the fusing module with a simple concatenation operation. Then, the concatenated features are fed to the following segmentation network. Table 4 reports their performance on part segmentation. The quantitative results demonstrate that the dual attention mechanism brings a large improvement.

Table 4. Ablation studies. We evaluate the architectures without Dual Attention (DA), Fused Attention (FA), and test different numbers of Laplacian Eigenvectors (LEV), Self-Attention (SA) blocks, and different triangulations.

Methods	Vases	Chairs	Tele aliens	Mean
Ours (#LEV 6)	95.3	96.2	90.0	93.8
Ours (#LEV 18)	95.9	96.6	96.7	96.4
Ours (#SA 2)	95.2	95.2	95.8	95.4
Ours (#SA 3)	97.6	97.0	96.1	96.9
Ours (#SA 5)	95.4	97.8	96.9	96.7
Ours (#V 2000)	97.9	97.5	97.3	97.5
Ours (#V 3500)	97.2	97.0	97.4	97.2
Ours (#V 5000)	97.9	97.1	97.2	97.4
Ours (Remesh)	98.0	97.7	97.1	97.6
Ours (w/o LEV)	95.5	95.7	88.0	93.1
Ours (w/o DA)	95.1	93.3	88.4	92.2
Ours (w/o FA)	95.4	96.2	89.3	93.6
Ours full-version (#SA 4, #LEV 12)	**98.1**	**97.7**	**97.4**	**97.7**

For our dual attention, we aim to learn the importance in a self-supervised fashion and use the important feature to determine more accurately. Figure 5 illustrates some segmentation results using our full network, where we also visualize the learned importance of the topology and geometry of different shapes, which contributes significantly to the better segmentation results of our method. The results and the visualization (Fig. 5) demonstrate that our model can simultaneously learn the structural information and critical features without any supervision. Some amount of supervision could be good guidance for training the network, but the data is hard to annotate and very time-consuming. Moreover, how to balance the importance of each task in multi-task learning is very difficult to supervise.

With _v.s._ Without Laplacian Eigenvectors (LEV). In this experiment, we tested a network where all the Laplcian EigenVectors are replaced by the vertex coordinates (3) in the upper branch. In Table 4, we see that removing the Laplacian eigenvectors gives worse performance than our full model, which shows the critical role of the Laplacian eigenvectors for representing the shape topology.

Number of LEVs and Self-Attention (SA) blocks. We tested different numbers of LEVs (4, 6, 18) and SA blocks (2, 3, 5) used in our network on part segmentation. The results in Table 4 show that the combination of 4 SAs and 12 LEVs achieves the best performance. Therefore, we set the default number of LEVs and SA blocks to 12 and 4, respectively. From the results, we find that more LEVs can result in more noises for the input features, and fewer LEVs are not sufficient to represent meaningful topological information. Besides, the network's performance reaches saturation as the number of SA increases.

Different Triangulation and Vertex Numbers. To demonstrate that our network is independent of triangulation and vertex numbers, we conducted experiments on processed meshes with different triangulation and various vertex numbers. We first subdivide [45] the meshes and simplify or sample [21,68] them to 2000, 3500, and 5000 vertices to train the models Ours-#V 2000, Ours-#V 3500, and Ours-# 5000 respectively. We use a mixture of two categories for training and the third for testing. The official splits of training and test data are applied to the three datasets. As shown in Table 4, the above models achieve similar performance as our original model.

Moreover, we re-mesh [59] the data of COSEG and simplify them to around 2048 vertices. Figure 2 shows that LEV is independent of the connectivity of triangles since it is induced from the geodesic dis-

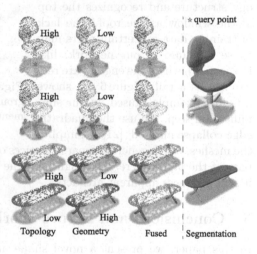

Fig. 5. Importance visualization. We visualize the attention maps from the self-attention (geometry & topology) and fused attention modules. For each shape, we show the three vertex-wise attention maps for two different query points in a row vertex-wise attention maps. We can observe that our dual attention is able to determine which is more important on the specific task, _e.g._ part segmentation. Blue to yellow means increasing weights. (Color figure online)

tance and invariant under the isometric transformation, so that our method can resist the instability of different discretizations. Table 4 reports the quantitative evaluation result. The performance on the re-meshed datasets is close to the original performance of our complete network.

4.5   Limitations and Failure Cases

Our approach is limited by the geometric properties of 3D meshes. Although there are many available mesh datasets, the meshes are non-manifold and have complex topological structures, which could lead to problematic/non-robust Laplacian spectral decomposition results. For example, most meshes from ShapeNet [9] are created by artists who do not consider geometric properties. Hence, all the meshes need a pre-process to be manifold to achieve successful decomposition and have to be simplified to a specific number of vertices before feeding into networks. We also show several representative failure cases as shown in Fig. 6: In Fig. 6(a), our method failed to cope with

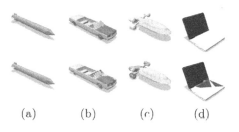

tiny parts and recognized them as the noises of the main body of the rocket. In Fig. 6(b), for a car with no roof, our network expects a complete topology structure and recognizes the top of the window as the roof. The lack of training data on certain parts may cause a failure of our network, like in Fig. 6 (c), where the engines are recognized as a tail. Figure 6 (d) shows a failure example caused by our simplification step. We use the quadratic edge collapse method [21] to simplify

(a)   (b)   (c)   (d)

Fig. 6. Failure Cases. The first row is groundtruth and the second row is segmented by our method.

the meshes, which generates sparse vertices on *flat* surfaces, such as the keyboard part of the laptop here. Although only the label of one vertex is mispredicted, it still produces apparent artifacts.

5   Conclusions and Future Works

In this paper, we present a novel shape analysis framework, *Laplacian Mesh Transformer*, which efficiently utilizes the shape topology and geometry information in deep feature extraction for polygonal meshes. More particularly, inspired by the recent advances of Transformer-based models in natural language processing and 2D image analysis, we propose the dual attention mechanism that achieves higher performance than prior works. In its two-stage process, we first explore the relationships between the elements of geometry features and topology features extracted by Laplacian spectral decomposition and then adopt a fusing attention module to merge the features effectively. Such a hierarchical structure to process features from fine to coarse can tackle 3D meshes with complex structure and geometry, benefiting shape analysis tasks. An avenue for future research is to apply the proposed learning framework to other potential tasks, such as shape retrieval or generative modeling. Another direction of future work is to design a transformer operator on irregular meshes, like the 2D CNN kernels in images. Furthermore, we hope to integrate the Laplacian spectral decomposition into our network architecture in an end-to-end manner, enabling the network to take raw non-manifold mesh data with arbitrary connectivities and vertex numbers as input, *e.g.* [62]. If the meshes have a large number of vertices, segmenting the meshes into patches would be a good solution to extend our scalability. For unstructured data such as point clouds, we can extend our work by constructing graphs based on proximity (as done by SyncSpecCN [85]) and using graph Laplacian, such as the raw LiDAR data for autonomous driving.

Acknowledgments. The work was supported by the National Natural Science Foundation of China (No. 61872440).

References

1. Achlioptas, P., Diamanti, O., Mitliagkas, I., Guibas, L.: Learning representations and generative models for 3D point clouds. In: ICML, pp. 40–49 (2018)
2. Ahmed, E., et al.: Deep learning advances on different 3D data representations: a survey, vol. 1. arXiv preprint arXiv:1808.01462 (2018)
3. Armeni, I., et al.: 3D semantic parsing of large-scale indoor spaces. In: Proceedings of the IEEE Conference on Computer Vision and Pattern Recognition, pp. 1534–1543 (2016)
4. Bahdanau, D., Cho, K., Bengio, Y.: Neural machine translation by jointly learning to align and translate. In: International Conference on Learning Representations (2015). https://arxiv.org/abs/1409.0473
5. Boscaini, D., Masci, J., Rodolà, E., Bronstein, M.: Learning shape correspondence with anisotropic convolutional neural networks. In: Advances in Neural Information Processing Systems, pp. 3189–3197 (2016)
6. Botsch, M., Kobbelt, L., Pauly, M., Alliez, P., Lévy, B.: Polygon Mesh Processing. CRC Press, Boca Raton (2010)
7. Bronstein, M.M., Bruna, J., LeCun, Y., Szlam, A., Vandergheynst, P.: Geometric deep learning: going beyond euclidean data. IEEE Signal Process. Mag. **34**(4), 18–42 (2017)
8. Carion, N., Massa, F., Synnaeve, G., Usunier, N., Kirillov, A., Zagoruyko, S.: End-to-End object detection with transformers. CoRR abs/2005.12872 (2020). https://arxiv.org/abs/2005.12872
9. Chang, A.X., et al.: ShapeNet: an information-rich 3D model repository. arXiv preprint arXiv:1512.03012 (2015)
10. Chen, X., Ma, H., Wan, J., Li, B., Xia, T.: Multi-view 3D object detection network for autonomous driving. In: Proceedings of the IEEE Conference on Computer Vision and Pattern Recognition, pp. 1907–1915 (2017)
11. Chen, Z., Zhang, H.: Learning implicit fields for generative shape modeling. In: Proceedings of the IEEE Conference on Computer Vision and Pattern Recognition, pp. 5939–5948 (2019)

12. Choy, C.B., Xu, D., Gwak, J.Y., Chen, K., Savarese, S.: 3D-R2N2: a unified approach for single and multi-view 3D object reconstruction. In: Leibe, B., Matas, J., Sebe, N., Welling, M. (eds.) ECCV 2016. LNCS, vol. 9912, pp. 628–644. Springer, Cham (2016). https://doi.org/10.1007/978-3-319-46484-8_38

13. Dai, Z., Yang, Z., Yang, Y., Carbonell, J., Le, Q.V., Salakhutdinov, R.: Transformer-xl: attentive language models beyond a fixed-length context. arXiv preprint arXiv:1901.02860 (2019)

14. Devlin, J., Chang, M.W., Lee, K., Toutanova, K.: Bert: pre-training of deep bidirectional transformers for language understanding. arXiv preprint arXiv:1810.04805 (2018)

15. Devlin, J., Chang, M., Lee, K., Toutanova, K.: BERT: pre-training of deep bidirectional transformers for language understanding. In: Burstein, J., Doran, C., Solorio, T. (eds.) North American Chapter of the Association for Computational Linguistics: Human Language Technologies, pp. 4171–4186. Association for Computational Linguistics (2019). https://doi.org/10.18653/v1/n19-1423

16. Dosovitskiy, A., et al.: An image is worth 16×16 words: transformers for image recognition at scale. CoRR abs/2010.11929 (2020). https://arxiv.org/abs/2010.11929

17. Dwivedi, V.P., Bresson, X.: A generalization of transformer networks to graphs. arXiv preprint arXiv:2012.09699 (2020)

18. Engel, N., Belagiannis, V., Dietmayer, K.: Point transformer. IEEE Access **9**, 134826–134840 (2021)

19. Fan, H., Su, H., Guibas, L.J.: A point set generation network for 3D object reconstruction from a single image. In: Proceedings of the IEEE Conference on Computer Vision and Pattern Recognition, pp. 605–613 (2017)

20. Gao, L., et al.: SDM-NET: deep generative network for structured deformable mesh. ACM Trans. Graph. (Proceedings of ACM SIGGRAPH Asia 2019) **38**(6), 243:1–243:15 (2019)

21. Garland, M., Heckbert, P.S.: Surface simplification using quadric error metrics. In: Proceedings of the 24th Annual Conference on Computer Graphics and Interactive Techniques, pp. 209–216 (1997)

22. Goodfellow, I.: Generative adversarial networks. Commun. ACM **63**(11), 139–144 (2020)

23. Groueix, T., Fisher, M., Kim, V.G., Russell, B.C., Aubry, M.: AtlasNet: a papier-mâché approach to learning 3D surface generation. In: Proceedings of the IEEE Conference on Computer Vision and Pattern Recognition (2018)

24. Guo, M.H., Cai, J.X., Liu, Z.N., Mu, T.J., Martin, R.R., Hu, S.M.: Pct: point cloud transformer. Comput. Visual Media **7**(2), 187–199 (2021)

25. Guo, M.H., et al.: Attention mechanisms in computer vision: a survey (2021)

26. Hanocka, R., Hertz, A., Fish, N., Giryes, R., Fleishman, S., Cohen-Or, D.: Meshcnn: a network with an edge. ACM Trans. Graph. (TOG) **38**(4), 1–12 (2019)

27. Henaff, M., Bruna, J., LeCun, Y.: Deep convolutional networks on graph-structured data. arXiv preprint arXiv:1506.05163 (2015)

28. Hou, J., Dai, A., Nießner, M.: 3d-sis: 3D semantic instance segmentation of rgb-d scans. In: Proceedings of the IEEE/CVF Conference on Computer Vision and Pattern Recognition, pp. 4421–4430 (2019)

29. Hu, H., Zhang, Z., Xie, Z., Lin, S.: Local relation networks for image recognition. In: Proceedings of the IEEE/CVF International Conference on Computer Vision, pp. 3464–3473 (2019)

30. Hu, S.M., et al.: Subdivision-based mesh convolution networks. arXiv preprint arXiv:2106.02285 (2021)

31. Huang, J., Su, H., Guibas, L.: Robust watertight manifold surface generation method for shapenet models. arXiv preprint arXiv:1802.01698 (2018)
32. Huang, R., Rakotosaona, M.J., Achlioptas, P., Guibas, L.J., Ovsjanikov, M.: Operatornet: recovering 3D shapes from difference operators. In: Proceedings of the IEEE/CVF International Conference on Computer Vision, pp. 8588–8597 (2019)
33. Ioannidou, A., Chatzilari, E., Nikolopoulos, S., Kompatsiaris, I.: Deep learning advances in computer vision with 3D data: a survey. ACM Comput. Surv. (CSUR) 50(2), 1–38 (2017)
34. Kalogerakis, E., Averkiou, M., Maji, S., Chaudhuri, S.: 3D shape segmentation with projective convolutional networks. In: Proceedings of the IEEE Conference on Computer Vision and Pattern Recognition, pp. 3779–3788 (2017)
35. Kalogerakis, E., Hertzmann, A., Singh, K.: Learning 3D mesh segmentation and labeling. ACM Trans. Graph. (TOG) 29(4), 102 (2010)
36. Kazhdan, M., Funkhouser, T., Rusinkiewicz, S.: Rotation invariant spherical harmonic representation of 3D shape descriptors. In: Symposium on Geometry Processing, vol. 6, pp. 156–164 (2003)
37. Khan, S., Naseer, M., Hayat, M., Zamir, S.W., Khan, F.S., Shah, M.: Transformers in vision: a survey. arXiv preprint arXiv:2101.01169 (2021)
38. Kingma, D.P., Ba, J.: Adam: a method for stochastic optimization. arXiv preprint arXiv:1412.6980 (2014)
39. Li, C.L., Zaheer, M., Zhang, Y., Poczos, B., Salakhutdinov, R.: Point cloud GAN. arXiv preprint arXiv:1810.05795 (2018)
40. Li, J., Chen, B.M., Lee, G.H.: So-net: self-organizing network for point cloud analysis. In: Proceedings of the IEEE Conference on Computer Vision and Pattern Recognition, pp. 9397–9406 (2018)
41. Lim, D., et al.: Sign and basis invariant networks for spectral graph representation learning. arXiv preprint arXiv:2202.13013 (2022)
42. Lin, T., Wang, Y., Liu, X., Qiu, X.: A survey of transformers. arXiv preprint arXiv:2106.04554 (2021)
43. Lin, Z., et al.: A structured self-attentive sentence embedding. In: International Conference on Learning Representations. OpenReview.net (2017). https://openreview.net/forum?id=BJC_jUqxe
44. Litany, O., Remez, T., Rodola, E., Bronstein, A., Bronstein, M.: Deep functional maps: Structured prediction for dense shape correspondence. In: Proceedings of the IEEE International Conference on Computer Vision, pp. 5659–5667 (2017)
45. Loop, C.: Smooth subdivision surfaces based on triangles. Master's thesis, University of Utah, Department of Mathematics (1987)
46. Maturana, D., Scherer, S.: VoxNet: a 3D convolutional neural network for realtime object recognition. In: 2015 IEEE/RSJ International Conference on Intelligent Robots and Systems (IROS), pp. 922–928. IEEE (2015)
47. Mescheder, L., Oechsle, M., Niemeyer, M., Nowozin, S., Geiger, A.: Occupancy networks: learning 3D reconstruction in function space. In: Proceedings of the IEEE Conference on Computer Vision and Pattern Recognition, pp. 4460–4470 (2019)
48. Meyer, M., Desbrun, M., Schröder, P., Barr, A.H.: Discrete differential-geometry operators for triangulated 2-manifolds. In: Visualization and Mathematics III, pp. 35–57. Springer, Heidelberg (2003). https://doi.org/10.1007/978-3-662-05105-4_2
49. Milano, F., Loquercio, A., Rosinol, A., Scaramuzza, D., Carlone, L.: Primal-dual mesh convolutional neural networks. arXiv preprint arXiv:2010.12455 (2020)

50. Park, J.J., Florence, P., Straub, J., Newcombe, R., Lovegrove, S.: DeepSDF: learning continuous signed distance functions for shape representation. In: Proceedings of the IEEE Conference on Computer Vision and Pattern Recognition, pp. 165–174 (2019)
51. Paszke, A., et al.: Automatic differentiation in pytorch (2017)
52. Peng, S., Niemeyer, M., Mescheder, L., Pollefeys, M., Geiger, A.: Convolutional occupancy networks (2020)
53. Pinkall, U., Polthier, K.: Computing discrete minimal surfaces and their conjugates. Exp. Math. **2**(1), 15–36 (1993)
54. Qi, C.R., Su, H., Mo, K., Guibas, L.J.: PointNet: deep learning on point sets for 3D classification and segmentation. In: Proceedings of the IEEE Conference on Computer Vision and Pattern Recognition, pp. 652–660 (2017)
55. Qi, C.R., Su, H., Nießner, M., Dai, A., Yan, M., Guibas, L.J.: Volumetric and multi-view cnns for object classification on 3D data. In: Proceedings of the IEEE Conference on Computer Vision and Pattern Recognition, pp. 5648–5656 (2016)
56. Qi, C.R., Yi, L., Su, H., Guibas, L.J.: PointNet++: deep hierarchical feature learning on point sets in a metric space. In: Advances in Neural Information Processing Systems, pp. 5099–5108 (2017)
57. Qiao, Y.L., Gao, L., Rosin, P., Lai, Y.K., Chen, X., et al.: Learning on 3D meshes with Laplacian encoding and pooling. IEEE Trans. Vis. Comput. Graph. **28**, 1317–1327 (2020)
58. Ramachandran, P., Parmar, N., Vaswani, A., Bello, I., Levskaya, A., Shlens, J.: Stand-alone self-attention in vision models. arXiv preprint arXiv:1906.05909 (2019)
59. Rineau, L., Yvinec, M.: A generic software design for delaunay refinement meshing. Comput. Geom. **38**(1–2), 100–110 (2007)
60. Schuster, M., Paliwal, K.K.: Bidirectional recurrent neural networks. IEEE Trans. Signal Process. **45**(11), 2673–2681 (1997)
61. Sharp, N., Crane, K.: A laplacian for nonmanifold triangle meshes. In: Computer Graphics Forum (SGP), vol. 39, no. 5, pp. 69–80 (2020)
62. Sharp, N., Crane, K.: A laplacian for nonmanifold triangle meshes. In: Computer Graphics Forum, vol. 39, pp. 69–80. Wiley Online Library (2020)
63. Su, H., et al.: SplatNet: sparse lattice networks for point cloud processing. In: Proceedings of the IEEE Conference on Computer Vision and Pattern Recognition, pp. 2530–2539 (2018)
64. Su, H., Maji, S., Kalogerakis, E., Learned-Miller, E.: Multi-view convolutional neural networks for 3D shape recognition. In: Proceedings of the IEEE International Conference on Computer Vision, pp. 945–953 (2015)
65. Sun, C.Y., Zou, Q.F., Tong, X., Liu, Y.: Learning adaptive hierarchical cuboid abstractions of 3d shape collections. ACM Trans. Graph. (TOG) **38**(6), 1–13 (2019)
66. Tan, Q., Gao, L., Lai, Y.K., Xia, S.: Variational autoencoders for deforming 3D mesh models. In: Proceedings of the IEEE Conference on Computer Vision and Pattern Recognition, pp. 5841–5850 (2018)
67. Tay, Y., Dehghani, M., Bahri, D., Metzler, D.: Efficient transformers: a survey. arXiv preprint arXiv:2009.06732 (2020)
68. Trappolini, G., Cosmo, L., Moschella, L., Marin, R., Melzi, S., Rodolà, E.: Shape registration in the time of transformers. Adv. Neural Inf. Process. Syst. **34**, 5731–5744 (2021)
69. Tulsiani, S., Su, H., Guibas, L.J., Efros, A.A., Malik, J.: Learning shape abstractions by assembling volumetric primitives. In: Proceedings of the IEEE Conference on Computer Vision and Pattern Recognition, pp. 2635–2643 (2017)

70. Vaswani, A., et al.: Attention is all you need. In: Advances in Neural Information Processing Systems, pp. 5998–6008 (2017)
71. Verma, N., Boyer, E., Verbeek, J.: Feastnet: feature-steered graph convolutions for 3D shape analysis. In: Proceedings of the IEEE Conference on Computer Vision and Pattern Recognition, pp. 2598–2606 (2018)
72. Wang, F., et al.: Residual attention network for image classification. In: IEEE Conference on Computer Vision and Pattern Recognition, pp. 6450–6458. IEEE Computer Society (2017). https://doi.org/10.1109/CVPR.2017.683
73. Wang, N., Zhang, Y., Li, Z., Fu, Y., Liu, W., Jiang, Y.G.: Pixel2mesh: generating 3D mesh models from single RGB images. In: Proceedings of the European Conference on Computer Vision (ECCV), pp. 52–67 (2018)
74. Wang, P., et al.: 3D shape segmentation via shape fully convolutional networks. Comput. Graph. **76**, 182–192 (2018)
75. Wang, Y., Sun, Y., Liu, Z., Sarma, S.E., Bronstein, M.M., Solomon, J.M.: Dynamic graph CNN for learning on point clouds. ACM Trans. Graph. (TOG) **38**(5), 1–12 (2019)
76. Wang, Y., Asafi, S., Van Kaick, O., Zhang, H., Cohen-Or, D., Chen, B.: Active co-analysis of a set of shapes. ACM Trans. Graph. (TOG) **31**(6), 1–10 (2012)
77. Wu, B., et al.: Visual transformers: token-based image representation and processing for computer vision. CoRR abs/2006.03677 (2020). https://arxiv.org/abs/2006.03677
78. Wu, J., Zhang, C., Xue, T., Freeman, W.T., Tenenbaum, J.B.: Learning a probabilistic latent space of object shapes via 3D generative-adversarial modeling. In: Proceedings of the 30th International Conference on Neural Information Processing Systems, pp. 82–90 (2016)
79. Wu, Z., et al.: 3D shapenets: a deep representation for volumetric shapes. In: Proceedings of the IEEE Conference on Computer Vision and Pattern Recognition, pp. 1912–1920 (2015)
80. Xiao, Y.P., Lai, Y.K., Zhang, F.L., Li, C., Gao, L.: A survey on deep geometry learning: from a representation perspective. Comput. Visual Media **6**(2), 113–133 (2020)
81. Xie, Z., Xu, K., Liu, L., Xiong, Y.: 3D shape segmentation and labeling via extreme learning machine. In: Computer Graphics Forum, vol. 33, pp. 85–95. Wiley Online Library (2014)
82. Yang, J., Mo, K., Lai, Y.K., Guibas, L.J., Gao, L.: Dsg-net: learning disentangled structure and geometry for 3D shape generation, vol. 3, p. 3. arXiv preprint arXiv:2008.05440 (2020)
83. Yang, Z., Dai, Z., Yang, Y., Carbonell, J., Salakhutdinov, R.R., Le, Q.V.: Xlnet: generalized autoregressive pretraining for language understanding. Adv. Neural Inf. Process. Syst. **32**, 1–11 (2019)
84. Yi, L., et al.: A scalable active framework for region annotation in 3D shape collections. ACM Trans. Graph. (ToG) **35**(6), 1–12 (2016)
85. Yi, L., Su, H., Guo, X., Guibas, L.J.: Syncspeccnn: synchronized spectral cnn for 3D shape segmentation. In: Proceedings of the IEEE Conference on Computer Vision and Pattern Recognition, pp. 2282–2290 (2017)
86. Yu, X., Rao, Y., Wang, Z., Liu, Z., Lu, J., Zhou, J.: Pointr: diverse point cloud completion with geometry-aware transformers. In: Proceedings of the IEEE/CVF International Conference on Computer Vision, pp. 12498–12507 (2021)

87. Zhang, H., Goodfellow, I.J., Metaxas, D.N., Odena, A.: Self-attention generative adversarial networks. In: Chaudhuri, K., Salakhutdinov, R. (eds.) International Conference on Machine Learning. Proceedings of Machine Learning Research, vol. 97, pp. 7354–7363. PMLR (2019). https://proceedings.mlr.press/v97/zhang19d.html

88. Zhao, H., Jia, J., Koltun, V.: Exploring self-attention for image recognition. In: Proceedings of the IEEE/CVF Conference on Computer Vision and Pattern Recognition, pp. 10076–10085 (2020)

89. Zhao, H., Jiang, L., Jia, J., Torr, P.H., Koltun, V.: Point transformer. In: Proceedings of the IEEE/CVF International Conference on Computer Vision, pp. 16259–16268 (2021)

90. Zhou, Y., Tuzel, O.: Voxelnet: end-to-end learning for point cloud based 3D object detection. In: Proceedings of the IEEE Conference on Computer Vision and Pattern Recognition, pp. 4490–4499 (2018)

Geodesic-Former: A Geodesic-Guided Few-Shot 3D Point Cloud Instance Segmenter

Tuan Ngo and Khoi Nguyen[✉]

VinAI Research, Hanoi, Vietnam
ducminhkhoi@gmail.com

Abstract. This paper introduces a new problem in 3D point cloud: few-shot instance segmentation. Given a few annotated point clouds exemplified a target class, our goal is to segment all instances of this target class in a query point cloud. This problem has a wide range of practical applications where point-wise instance segmentation annotation is prohibitively expensive to collect. To address this problem, we present Geodesic-Former – the first geodesic-guided transformer for 3D point cloud instance segmentation. The key idea is to leverage the geodesic distance to tackle the density imbalance of LiDAR 3D point clouds. The LiDAR 3D point clouds are dense near the object surface and sparse or empty elsewhere making the Euclidean distance less effective to distinguish different objects. The geodesic distance, on the other hand, is more suitable since it encodes the scene's geometry which can be used as a guiding signal for the attention mechanism in a transformer decoder to generate kernels representing distinct features of instances. These kernels are then used in a dynamic convolution to obtain the final instance masks. To evaluate Geodesic-Former on the new task, we propose new splits of the two common 3D point cloud instance segmentation datasets: ScannetV2 and S3DIS. Geodesic-Former consistently outperforms strong baselines adapted from state-of-the-art 3D point cloud instance segmentation approaches with a significant margin. The code is available at https://github.com/VinAIResearch/GeoFormer.

Keywords: Few-shot learning · 3D point cloud instance segmentation

1 Introduction

This paper introduces a new problem of few-shot 3D point cloud instance segmentation (3DFSIS). As Fig. 1 shows, given a few support point clouds (a.k.a. scenes) with their ground-truth masks to define a target class, we aim to segment all instances of the target class in a query scene. Compared to related vision tasks such as 3D point cloud instance segmentation (3DIS) and 3D point cloud few-shot semantic segmentation (3DF3S), 3DFSIS is fundamentally different. For 3DIS, the training and test classes are the same. One could reliably learn an instance segmenter with abundant annotated examples in training, then

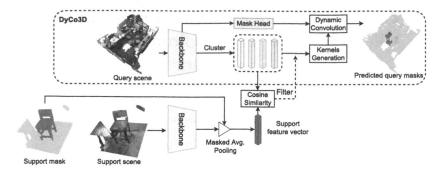

Fig. 1. Our baseline adapted on DyCo3D [16] for 3DFSIS. The query and support point clouds are first input to a shared backbone to extract their features. Then the query points are grouped into candidates based on their semantic and predicted object centroids while the support points are masked-average-pooled to obtain a support feature vector. The cosine similarity between the support feature vector and every candidate's average feature is used to filter out irrelevant candidates. The final candidates are used to generate kernels for dynamic convolution with the feature produced by the mask head in order to obtain the final instance masks of the query scene.

apply that segmenter to the test scenes. That is not the case in 3DFSIS where training and test classes are disjoint. For 3DF3S, we need to predict each point with a semantic label instead of an instance label as in 3DFSIS. That is, we do not need to distinguish different instances of the same class as in 3DF3S. Furthermore, unlike weakly/semi-supervised learning in 3DIS, where all classes are known in training, in the training of 3DFSIS, the new classes are not known in advance. Thus, the model needs to quickly learn from a few examples of new classes whenever they arrive.

3DFSIS is an important vision task and has a wide range of applications including autonomous driving, and augmented reality, especially in applications where training a reliable 3D instance segmenter is prohibitively impossible due to the expensive costs of collecting a sufficient amount of annotated point clouds. However, learning in 3D point clouds is very challenging due to: (1) 3D point clouds are unordered, imbalanced in density (dense near object surface but sparse elsewhere); and (2) the variance in appearance, size, and shape between the support and query scenes is significantly higher than that of 2D images. These two challenges are amplified in the few-shot setting where a very limited number of labeled examples of new classes are provided, e.g. 1 to 5 shots at most compared to 30 to 50 shots with ease in a 2D image. This is due to the reason that one has to label point-by-point in a 3D point cloud rather than labeling approximate polygons for instance masks as in a 2D image. Therefore, it is not trivial to adapt any 2DFSIS to 3DFSIS.

A simple but strong baseline for 3DFSIS can be adapted from a 3D point cloud instance segmenter, e.g. DyCo3D [16], to the few-shot setting. The baseline is depicted in Fig. 1. First, similar points are grouped into candidates based on their Euclidean centroids and semantic predictions. Then each candidate is

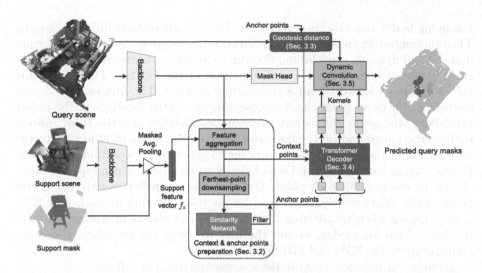

Fig. 2. Our proposed approach, Geodesic-Former, for 3DFSIS. Given support and query scenes, a shared backbone is used to extract their features. Support features are further masked average pooled with the support mask to obtain a support feature vector representing the target class. Then the query features and support feature vector are aggregated to obtain *context points* which are further sub-sampled in farthest-point-downsampling operation to obtain *anchor points* representing the initial prediction of the object location. Next, the geodesic distances between every anchor point to all the context points are computed taking into account the imbalance in point cloud density (distributed near object surface only). This geodesic distance is used as the positional encoding to guide the transformer decoder whose key/value and query are context and anchor points, respectively, so as to produce a kernel for each anchor point. Finally, each kernel dynamically convolves with the features produced by the mask head along with the geodesic distance embedding to obtain the final object instance mask.

passed to a subnetwork to generate a kernel for dynamic convolution [40] so as to obtain the final instance mask. To filter out the irrelevant candidates which do not belong to the target class, one can use cosine similarity between the support feature vector and the average feature vector of all points of each candidate. This framework has several limitations. First, as mentioned above, 3D point clouds are imbalanced in density and mostly distributed near the object surface so that the Euclidean distance for clustering is unreliable, i.e. points that are close together might not necessarily belong to the same object and vice versa. Second, the clustering in DyCo3D relies heavily on the performance of the offset centroid predictions, hence, it might be overfitting to some 3D shapes and sizes of the training object classes, resulting in poor generalization to the test classes.

To address these limitations, we propose a new geodesic-guided transformer decoder to generate the kernel for the dynamic convolution from a set of initial anchor points, giving the name of our approach, Geodesic-Former. The overview of Geodesic-Former is depicted in Fig. 2. First, geodesic distance embedding based on the geodesic distances between each of the anchor points to all context points is computed. In this way, the geodesic distance between two points belonging to different objects is very large, helping differentiate different objects. Then this embedding is used as positional encoding to guide the later transformer decoder and dynamic convolution. Second, to avoid overfitting to the shape and size of training classes, we use a combination of the Farthest Point Sampling [37], a similarity network, and a transformer decoder. The first samples initial seeds from the query point cloud representing the initial locations of the object candidates, the second filters out irrelevant candidates, and the third contextualizes the foreground (FG) candidates to precisely represent objects with the information of the context points in order to generate the convolution kernels. In this way, as long as an initial seed belongs to an object, it can represent that object. In contrast, for each point, DyCo3D has to predict exactly the center point of the object it belongs to in order for the clustering to work well. This is even harder when transferring to the new object classes in testing. Also, to the best of our knowledge, we are the first to adopt the transformer decoder architecture to the 3DIS and 3DFSIS.

In sum, our technical contributions are summarized as follows:

- We introduce a new 3D point cloud few-shot instance segmentation task.
- To evaluate the new task, we introduce new splits adapted from the ScannetV2 and S3DIS datasets.
- To address the new task, we propose a strong baseline (adapted from SOTA 3DIS methods) and our novel proposed approach, Geodesic-Former, combining the transformer decoder with dynamic convolution in respect of the geodesic distance encoding scene's geometry.

In the following, Sect. 2 reviews prior work; Sect. 3 specifies Geodesic-Former; and Sect. 4 presents our implementation details and experimental results. Sect. 5 concludes with some remarks and discussions.

2   Related Work

This section reviews closely related work in 2D and 3D instance segmentation.

3D point cloud instance segmentation (3DIS) approaches can be divided into two groups: proposal-based and proposal-free. The *proposal-based* approaches [17,51,52] first detect 3D bounding boxes, then segment the foreground region inside them. 3D-SIS [17] proposes a Mask R-CNN-based 3D instance segmentation architecture, jointly learns features from both RGB images and 3D point cloud. 3D-BoNet [51] simplifies the detection network by directly predicting a fixed number of unoriented 3D bounding boxes from a global feature vector, then segmenting foreground points inside each box. GSPN [52] generates proposals by reconstructing shapes from noisy observations and further refining these proposals with a Region-based PointNet [36]. On the other hand, the *proposal-free* approaches [3,8,16,20,44] learn embedding features then group points to instances. SGPN [44] adopts the double-hinge loss to learn discriminative features in order to compute the similarity matrix of paired points for grouping points. PointGroup [20] predicts the 3D offset from each point to its instance's centroid and generates clusters from two sets: original points and shifted points. HAIS [3] proposes a hierarchical clustering method where a small cluster can be either filtered out or absorbed by a larger cluster to become its part. DyCo3D [16] adopts the same clustering approach but leverages dynamic convolution [40,46] to generate 3D instance masks. SSTNet [25] constructs a super point tree based on the point cloud's semantic features and uses tree traversal to split nodes into instances. All the 3DIS approaches assume the training and test classes are the same, and there is a large number of annotated data for training. The setting of 3DFIS is fundamentally different: the training and test classes are disjoint and we only have a few annotated examples for each test class.

Few-shot 2D instance segmentation approaches [9,30,33,34,50] extend the Mask R-CNN [13] – a common 2D image instance segmenter – to the few-shot setting. The support features are extracted from a few labeled examples and incorporated into the query feature map to segment objects of the target class. [33] utilizes the anchor-free detector [41] to alleviate the overfitting problem of the anchor boxes to the training classes and assembles the predicted object's latent parts into an object mask. However, 2D images are structured, grid-based, and dense whereas 3D point clouds are unordered, irregular, and sparse. Therefore, these approaches cannot be applied directly to 3DFSIS.

Few-shot 3D Point Cloud Semantic Segmentation (3DF3S). Recently, [55] introduced the problem of few-shot 3D point cloud semantic segmentation. From the support scene, multiple prototypes are extracted and propagated to the query points based on their affinity matrix. However, this approach does not distinguish different instances of the same object class. 3DFSIS is arguably harder than 3DFSSS since we need to classify all points into instance labels instead of semantic labels only.

Vision transformer has been applied to 2D image classification [7,42,53], object detection [2,10,29,45,47,56], semantic segmentation [38,49], and instance segmentation [6,12,18,24]. Furthermore, the transformer architecture is naturally fit to process unordered 3D point clouds since its attention mechanism is permutation invariant. Some recent approaches [27,31,54] have shown the potential of transformers in some 3D tasks. [54] designs a self-attention network to process 3D point clouds and achieve good results on 3D semantic segmentation, object part segmentation, and object classification. [27,31,35] leverage transformer-based architecture for 3D object detection. We are the first to adopt the cross-attention transformer decoder with a special design for the 3DIS and 3DFSIS tasks.

3 Our Geodesic-Former

3.1 Problem Setting

In training, we are provided a sufficiently large training set of base classes C_{train}, i.e. $\{P^t, m^t\}_{t=1}^T$, where P^t, m^t are the 3D point cloud of the scene t and its ground-truth segmentation masks, respectively, and T is the number of training samples. In testing, given K support 3D point cloud scenes P_s and their ground-truth masks m_s to define a new target class c_{test}, we seek to segment all instances m_q of the target class in a query scene P_q. It is worth noting that the target class is different from the base classes, or $c_{test} \notin C_{train}$. In this paper, we explore two configurations: 1-shot and 5-shot instance segmentation.

To address this problem, we design our approach Geodesic-Former inspired by DyCo3D [16]. The overview of Geodesic-Former is illustrated in Fig. 2. To extract features F_s, F_q from the support and query point clouds P_s, P_q, respectively, we employ a 3D U-Net with sparse convolution [11] used in [16]. In addition, a support feature vector f_s is extracted from the support features F_s via a masked-average-pooling operation representing the target class.

In the following, Sect. 3.2 first describes how to prepare the context and anchor points for the transformer decoder. Sect. 3.3 specifies how to compute the geodesic distance between each anchor point to every context point to guide the transformer decoder and dynamic convolution. Sect. 3.4 discusses how the transformer decoder generates the convolution kernel for the dynamic convolution, which is presented in Sect. 3.5, in order to produce the final instance mask. Finally, Sect. 3.6 proposes the strategy to train our approach.

3.2   Context and Anchor Points Preparation

First, we aggregate support feature $f_s \in \mathbb{R}^{1 \times d}$ into the query features $F_q \in \mathbb{R}^{N_q \times d}$ inspired by [48], resulting in integrated features of the *context points* $F_c \in \mathbb{R}^{N_q \times d}$ as follows:

$$F_c = W_{proj} * [F_q \odot f_s; F_q - f_s; F_q], \tag{1}$$

where d is the number of feature channels, N_q is the number of query points, $W_{proj} \in \mathbb{R}^{3d \times d}$ is the linear layer weight; $*, [\cdot; \cdot], \odot, -$ are the convolution, concatenation, channel-wise multiplication, and subtraction operations, respectively. In this way, we preserve the original query point features along with the newly rectified and subtracted features from the support.

Next, from the context points, a smaller set of points is sampled by a farthest-point-down sampling to represent distinct object candidates. In our work, we sample a large enough number of candidates so that they can cover all objects in all cases. Then, a similarity network, which is a multi-layer perceptron (MLP), is used to filter relevant candidates as *anchor points* $F_a \in \mathbb{R}^{N_a \times d}$, N_a is the number of anchor points, having high appearance similarity with the support examples. After this step, we take the context points F_c and anchor points F_a as input to the transformer decoder to generate the kernel of each anchor point.

3.3   Geodesic Distance Embedding Computation

The 3D point clouds captured by LiDAR sensors have an important property that it is distributed unequally in the 3D space (dense near the object surface and sparse elsewhere). As a result, two points are close in 3D Euclidean distance but they might belong to two different objects. In this case, the geodesic distance

(a)   (b)

Fig. 3. (a) Comparison between Euclidean distance and geodesic distance. For each image, the green points are the top-2000 nearest neighbors of the red point in Euclidean distance (left) and geodesic distance (right). (b) An example of FG/BG flipping in training and testing making transformer classifier confused, i.e. sofa is labeled as BG in training (left) but FG in testing (right). (Color figure online)

[21] which encodes the scene's geometry would be a better choice as visualized in Fig. 3a. In other words, if two points are close in Euclidean distance but there is no path or their geodesic distance is too high, they clearly belong to two separate objects. Therefore, we propose to use the geodesic distance between the anchor points and every context point as geometry guidance to distinguish objects in subsequent modules.

To obtain the geodesic distance, we first employ the ball query algorithm [37] to get a directed sparse graph whose nodes are context points and each node only connects to at most κ other nodes. There exists a directed edge from node 1 to node 2 if node 2 is among the κ nearest neighbors of node 1 and within a radius τ, and the weight of the edge is always positive and equal to the local Euclidean distance between the two nodes. After that, we use the shortest path algorithm, i.e. Djikstra [5], to compute the length of the shortest path from each anchor point to every context point in the obtained sparse graph as its geodesic distance. Finally, the geodesic distance embedding $G^i \in \mathbb{R}^{N_q \times d}$ of an anchor point i is obtained by encoding its geodesic distance using the sine/cosine function in [43].

3.4 Transformer Decoder

The transformer decoder takes as input the anchor points $F_a \in \mathbb{R}^{N_a \times d}$ and context points $F_c \in \mathbb{R}^{N_q \times d}$ to produce the kernel $W^i \in \mathbb{R}^L$ for each anchor point i, where L is the number of parameters in dynamic convolution. The decoder follows the design of DETR [2] consisting of a multi-block of transformer layers with two kinds of attention: self-attention between anchor points and cross-attention between anchor and context points. Hence, each anchor point knows each other and captures a complete object structure to generate a kernel for the dynamic convolution. Notably, the attention mechanism in a transformer is inherently fitted to the 3D point cloud since they are both unordered.

Importantly, to address the 3DFSIS, we make substantial modifications to the positional encoding and output of the decoder. First, to guide the attention in the transformer with the geodesic geometry structure as discussed in Sect. 3.3, the geodesic distance embedding G is used as the positional encoding instead of the embedding of 3D point coordinates. Second, we do not predict the object class, i.e., the foreground (FG)/background (BG) classification, due to the FG and BG confusion of few-shot settings during the training and testing phase. In particular, a lot of new classes presenting in training scenes but are labeled as BG causing the trained classification head to predict them as BG (false negative) in testing as depicted in Fig. 3b. Instead, the similarity network filters the FG anchor points as described in Sect. 3.2.

3.5   Dynamic Convolution

To prepare features for dynamic convolution whose weights are predicted by the transformer decoder, a **mask head** takes as input the query point features $F_q \in \mathbb{R}^{N_q \times d}$ to produce the mask features $F_{mask} \in \mathbb{R}^{N_q \times d}$ by applying two blocks of MLP with Batch Norm [19], and ReLU [32] in between. Also, the geodesic distance is critical geometric cue to distinguish instances, we directly append the geodesic distance embedding $G^i \in \mathbb{R}^{N_q \times d}$ of each anchor point i to the mask features in order to obtain the final instance mask $\widehat{m}^i \in [0, 1]^{N_q \times 1}$ in a **dynamic convolution** as follows :

$$\widehat{m}^i = \text{Conv}\left([F_{mask}; G^i]; W^i\right), \tag{2}$$

where $[\cdot; \cdot]$ is the concatenation operation, and Conv is implemented with several convolutional layers as in DyCo3D [16].

3.6   Training Strategy

Pretraining: First, we pretrain the U-Net backbone, mask head, and the transformer decoder with the standard 3D point cloud instance segmentation task on the base classes. In this stage, the feature aggregation in Fig. 2 is not used since we do not have support feature, instead, we copy the features of query points to the context points directly. Also, we add a classification head on top of the output of the transformer decoder to predict the semantic category $\widehat{\gamma}^i$ along with the kernel generation to predict the mask $\widehat{m}^i$ for each anchor point i. The number of classes is $\Gamma + 1$ where Γ is the total number of base classes of C_{train} and one additional background class. The matching cost $C_{match}^{pretrain} \in \mathbb{R}_+^{N_a \times N_{gt}}$ between the prediction $(\widehat{\gamma}^i, \widehat{m}^i)$ and the ground truth (γ^j, m^j) is computed as:

$$C_{match}^{pretrain}(i, j) = L_{seg}(\widehat{m}^i, m^j) + L_{cls}(\widehat{\gamma}^i, \gamma^j), \tag{3}$$

where L_{seg} is the dice loss [39], and L_{cls} is the sigmoid focal loss [26]. Based on the matching cost $C_{match}^{pretrain}$, the Hungarian algorithm [23] is leveraged to find the optimal 1-to-1 matching π^*, then the following loss is used for training:

$$L_{Hungarian}^{pretrain} = \sum_{i=1}^{N_{GT}} L_{seg}(\widehat{m}^i, m^{\pi^*(i)}) + \sum_{i=1}^{N_a} L_{cls}(\widehat{\gamma}^i, \gamma^{\pi^*(i)}). \tag{4}$$

If a class prediction $\widehat{\gamma}^i$ has no ground truths matched, it will be matched with the background class.

Table 1. Class splits of the ScannetV2 and S3DIS datasets. Fold 0 is used for training while fold 1 is used for testing.

ScannetV2		S3DIS	
Fold 0	Fold 1	Fold 0	Fold 1
Cabinet	Sofa	Beam	Door
Bed	Table	Board	Floor
Chair	Window	Bookcase	Sofa
Door	Picture	Ceiling	Table
Bookshelf	Shower curtain	Chair	Wall
Counter	Refrigerator	Column	Window
Desk	Toilet		
Curtain	Sink		
Bathtub	Other furniture		

Episodic Training: We leverage the episodic training strategy – a common approach for few-shot image classification – to mimic the test scenario in training. That is, for each episode, we randomly sample a pair of support and query point clouds P_s, P_q and their masks m_s, m_q from training examples of the base classes. In this stage, the classification head is removed and we add feature aggregation and similarity network to train with the transformer decoder while freezing the backbone and mask head. This is the final architecture of Geodesic-Former as depicted in Fig. 2. The following matching cost and loss are used to train and Approach in this stage:

$$C_{match}^{episodic}(i,j) = L_{seg}(\widehat{m}^i, m^j), \quad L_{Hungarian}^{episodic} = \sum_{j=1}^{N_{GT}} L_{seg}(\widehat{m}^i, m^{\pi^*(i)}). \quad (5)$$

For $K > 1$ shots, we additionally apply the episodic training on a set of balanced support-query pairs of the base and new classes to further fine-tune the Geodesic-Former. In testing, the final support feature vector f_s is the average vector of all feature vectors f_s^k of K support scenes.

4   Experiments

Datasets: To evaluate Geodesic-Former on the new 3DFSIS task, we introduce two new datasets derived from ScannetV2 [4] and S3DIS [1] used for 3D point cloud instance segmentation. ScannetV2 consists of 1613 point clouds of scans from 707 unique indoor scenes with 20 semantic classes in total and 18 classes for instance segmentation. We follow the common split of 1201, 312, and 100 for training, evaluating, and testing, respectively [16]. Inspired by [55] for 3D few-shot semantic segmentation, we split the 18 foreground classes into two non-overlapping folds based on the alphabetical order with nine classes each, one for training classes (fold 0) and the other for test classes (fold 1). S3DIS is another benchmark for 3D indoor scenes which contains 272 point clouds collected from 6 large-scale areas with 13 semantic categories. We only keep 12 main categories and remove the "clutter" class. We also split it into two folds with six classes each. Area 5 containing 68 point clouds is used for testing while the rest is used for training. Table 1 summarizes the class splits of ScannetV2 and S3DIS.

We report the results for the test classes in the following procedure: (1) we randomly sample $K = \{1, 5\}$ support examples, with their binary masks for every class in the training set (with the purpose of saving the whole test set for the query scenes only) and apply them to the whole test set, a.k.a the fixed support set; (2) for each query scene in the test set, if a test class does not present in the scene, we skip the evaluation of that class for that scene. To improve the reliability of the measured metrics, we sample and evaluate all the approaches on ten disjoint fixed support sets, and report the average with standard deviation. In this setting, we consider the unlabeled points of new classes in the training set as unseen points commonly used in 2DFSIS.

Evaluation Metrics: For ScannetV2, we adopt the mean average precision (mAP) and AP50 used in the instance segmentation task. For S3DIS, we apply the metrics that are used in [14–16, 20] to test classes: mCov, mPrec, and mRec. They are the mean instance-wise IoU, mean precision, and mean recall.

Implementation Details: We adopt the sparse convolution [11] to implement the backbone network. The voxel size is set to 0.02 m for ScannetV2 and 0.05 m for S3DIS, and the output channel of the backbone network is set to 16. To calculate the geodesic distance, we employ the FAISS[1] library for ball-query search, then we re-implement by vectorizing the shortest path algorithm, i.e., Dijkstra's algorithm, in Pytorch to speed up the processing time. The transformer decoder is the same as [31] consisting of four layers, each uses multi-head attention with four heads, and the output dimension and the hidden dimension are set to 64. We train our model using the Adam optimizer [22] with a cosine learning rate scheduler [28]. During the pretraining phase, the initial learning rate is set to 10^{-2}, and the number of training epochs is 500. After that, we train for another 200 epochs in episodic training with the learning rate of 5×10^{-3}. Our data augmentation is the same as [20]'s.

4.1   Ablation Study

We conduct several experiments on the validation set of ScannetV2 to study the contribution of various components of our method with one shot, $K = 1$.

Similarity Network, Transformer Decoder, Geodesic Distance. In Table 2, the first and second rows show the performance of our baseline in Fig. 1,

---

[1] https://github.com/facebookresearch/faiss.

Table 2. Ablation study on each component's contribution to the final results. "SN", "TD", and "GDE" denote similarity network, transformer decoder, and geodesic distance embedding, respectively. (*) denotes the baseline of per-point classification.

	Combination			Metric	
	SN	TD	GDE	mAP	AP50
Baseline (DyCo3D [16])				6.2	11.7
Baseline (*)				4.9	9.7
	✓			6.6	12.5
		✓		6.7	13.1
			✓	7.8	14.2
	✓	✓		7.6	14.3
	✓		✓	8.7	14.9
		✓	✓	9.4	17.1
Geodesic-Former	✓	✓	✓	**10.6**	**19.8**
Geodesic-Former w/cls.		✓	✓	4.5	10.2

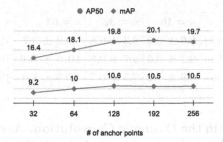

Fig. 4. Study on the number of anchor points N_a.

and a per-point classification variant where we use cosine similarity to filter out irrelevant points before clustering by predicted objects' centers. This variant performs poorly as each point is classified independently without geometric cues of objects, and the classified points are so cluttered to form a complete shape. When replacing the cosine similarity in the baseline with a similarity network, the performance slightly increases, +0.4 in row 3. When the clustering algorithm in the baseline is replaced by the transformer decoder, the performance also slightly improves, +0.5 in row 4. Especially, when adding the geodesic distance embedding to the dynamic convolution of the baseline, the performance is significantly boosted, +1.6 in row 5. This justifies the importance of geodesic distance to the segmentation. When combining each pair of the three components, the performance improves substantially over each component alone. Finally, our full approach, Geodesic-Former achieves the best performance, 10.6 in mAP and 19.8 in AP50. These results show that when combining these components together, the performance gain is much larger than using them separately. We also have an ablation when turning off the similarity network and using the classification head in the pretraining phase, the performance drops significantly, −6.1 in row 9. This justifies our claims that using the classification head in our 3DFSIS is sub-optimal due to the FG/BG confusion as described in Sect. 3.4.

Number of Anchor Points. The results are summarized in Fig. 4. Using the number of anchor points of 128 gives the best results. This is because using too few anchor points cannot capture the diversity of objects in the scene, whereas using too many does not boost the performance significantly.

Table 3. Study on the number of dynamic convolution layers.

# of layers	1	2	3	4
mAP on training set	22.6 ± 1.4	28.1 ± 1.7	28.0 ± 1.3	28.3 ± 1.5
mAP on testing set	3.4 ± 0.2	10.6 ± 0.6	9.3 ± 1.3	6.4 ± 1.9

Table 4. Study on ball query settings in Sect. 3.3 to form sparse directed graph.

mAP	$\kappa = 16$	$\kappa = 32$	$\kappa = 64$	$\kappa = 128$
$\tau = 0.03$ m	9.0 ± 0.9	9.3 ± 0.8	9.9 ± 0.7	10.1 ± 0.8
$\tau = 0.05$ m	9.2 ± 0.6	9.7 ± 0.8	$\mathbf{10.6 \pm 0.6}$	10.6 ± 0.7
$\tau = 0.1$ m	8.9 ± 1.2	9.3 ± 0.7	10.5 ± 1.0	10.3 ± 0.9

Number of Layers in the Dynamic Convolution. As can be seen in Table 3, using only a single layer of dynamic convolution leads to a significant drop in performance (−7.2 in mAP). On the other hand, using too many layers may be prone to overfitting the training data and harder to optimize due to a large number of generated parameters. Using two layers gives the best results.

Ball Query Configuration. Table 4 reports the results with different nearest neighbors κ and radii τ to form the directed sparse graph (as described in Sect. 3.3) in order to compute the geodesic distance. From this table, $\kappa = 64$ and $\tau = 0.05$ m give the best results.

4.2   Comparison with Prior Work

Since there is no prior work on 3DFSIS, we adapt three state-of-the-art (SOTA) approaches on 3DIS: DyCo3D [16], PointGroup [20], and HAIS [3] to the few-shot setting for comparing with our approach. The adapted version of DyCo3D is exactly our baseline as depicted in Fig. 1. We apply the cosine similarity filter to all methods to remove irrelevant proposals after the clustering stage and the other modules are kept exactly the same as in their original papers. The similarity thresholds for these methods are carefully fine-tuned to achieve the best performance for a fair comparison, i.e. 0.95, 0.9, and 0.8 for DyCo3D, PointGroup, and HAIS, respectively. Notably, the set aggregation module in HAIS requires another statistical class-specific radius to aggregate fragments into larger components. We calculate this radius based on the support scenes and then apply it to the query scene.

Table 5 and Tab. 6 show the comparison results on the S3DIS and ScannetV2 datasets, respectively. For ScannetV2, HAIS performs worst among the four, probably due to the sensitive class-specific radius in its set aggregation module. Geodesic-Former consistently outperforms all of them by a large margin in all metrics, i.e., +4.4 for one shot and +6.8 for five shots in the mAP. Moreover, our method is more robust across different runs where the standard variations of mAP and AP50 are only 0.7 and 1.4, respectively, compared with 2.0 and 3.1 of the second-best DyCo3D's. For S3DIS, Geodesic-Former outperforms others with a significant margin, i.e. in mCov and mRec, about +4 for one shot and +7 for five shots. HAIS's results are slightly better than ours in mPre due to its strict threshold to get high precision but low recall rate.

Table 5. Comparison of Geodesic-Former and the strong baselines on ScannetV2.

	$K = 1$		$K = 5$	
	mAP	AP50	mAP	AP50
DyCo3D [16]	6.2 ± 2.0	11.7 ± 3.1	6.4 ± 1.2	11.9 ± 2.2
PointGroup [20]	5.3 ± 1.2	10.3 ± 2.5	5.3 ± 0.5	11.7 ± 0.8
HAIS [3]	1.6 ± 0.6	3.5 ± 0.8	1.0 ± 0.2	2.3 ± 0.4
Geodesic-Former	$\mathbf{10.6 \pm 0.7}$	$\mathbf{19.8 \pm 1.4}$	$\mathbf{13.2 \pm 0.9}$	$\mathbf{24.8 \pm 1.3}$

Table 6. Comparison of Geodesic-Former and the strong baselines on S3DIS.

	$K = 1$			$K = 5$		
	mCov	mPre	mRec	mCov	mPre	mRec
DyCo3D [16]	13.5 ± 2.1	2.9 ± 1.0	4.1 ± 1.4	14.5 ± 1.3	3.1 ± 0.5	4.1 ± 1.4
PointGroup [20]	12.9 ± 2.8	4.6 ± 1.4	3.8 ± 1.3	13.7 ± 0.8	4.6 ± 0.6	3.8 ± 0.8
HAIS [3]	4.6 ± 1.2	$\mathbf{8.1 \pm 0.9}$	3.9 ± 1.3	5.0 ± 1.9	$\mathbf{11.8 \pm 2.0}$	4.1 ± 0.4
Ours	$\mathbf{17.8 \pm 1.5}$	7.0 ± 0.4	$\mathbf{8.5 \pm 1.7}$	$\mathbf{20.2 \pm 2.1}$	10.8 ± 1.3	$\mathbf{12.2 \pm 1.8}$

4.3 Qualitative Results

Figure 5 shows the qualitative results of our approach and others on ScannetV2. For the training class "chair" shown on row 1, all approaches perform well. For the test classes (rows 2–5), there are differences in the segmentation results. Geodesic-Former outperforms others in the hard cases such as in the thin object ("show curtain" - row 2), in the big object ("table" - row 3), and in the incomplete object ("window" - row 4). These examples demonstrate the strong capability of our approach when handling objects to various extent thanks to the transformer decoder and the geodesic distance embedding. However, Geodesic-Former mis-segments the sofa-stool as sofa due to their similar appearance (row 5).

Also, Fig. 6 illustrates the quality of the computed geodesic distance. For each red point, we visualize the top reachable geodesic-distance nearest neighbors (green points) and unreachable points (gray points) which have *infinite* geodesic distance. It justifies that the geodesic distance helps distinguish objects much better than Euclidean distance.

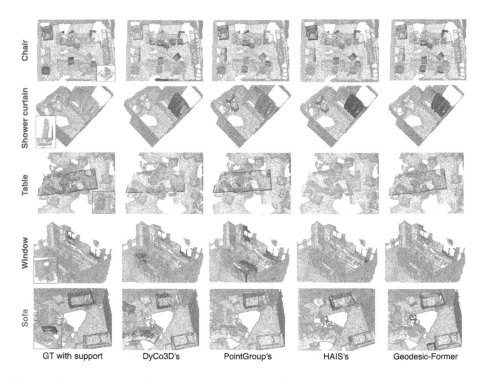

Fig. 5. Qualitative results of Geodesic-Former and the strong baselines on the Scan-netV2 dataset. Each row shows an example of the query scene with its GT mask and the support scene with its GT mask (the smaller red-border box) on the first column. The name of the support class is on the left next to GT. (Color figure online)

Fig. 6. Representative examples of computed geodesic distance. For each image, the green points is the top reachable geodesic-distance nearest neighbors of the red point. (Color figure online)

5   Discussion and Conclusion

Discussion. We have succeeded in applying our approach on a lower number of shots only, i.e., 1 and 5 shots. For a higher number of shots $(K > 5)$, the improvement is insignificant due to the simple averaging operation. The study on how to aggregate features from multiple supports in a 3D point cloud to leverage their geometric structure would be an interesting research topic.

Conclusion. In this work, we have introduced the new few-shot 3D point cloud instance segmentation task and have proposed the Geodesic-Former – a new geodesic-guided transformer with dynamic convolution to address it. Extensive experiments have been conducted on the newly introduced splits of ScannetV2 and S3DIS datasets showing that our approach achieves robust and significant performance gain on both datasets from the very strong baselines adapted from the state-of-the-art approaches in 3D instance segmentation, i.e., +4.4 for one shot and +6.8 for five shots in mAP on ScannetV2; +4.3 for one shot and +5.7 for five shots in mCov on S3DIS. We hope that our proposed problem, datasets, and approach could facilitate future work in this direction.

References

1. Armeni, I., et al.: 3D semantic parsing of large-scale indoor spaces. In: Proceedings of the IEEE/CVF Conference on Computer Vision and Pattern Recognition (2016)
2. Carion, N., Massa, F., Synnaeve, G., Usunier, N., Kirillov, A., Zagoruyko, S.: End-to-end object detection with transformers. In: Vedaldi, A., Bischof, H., Brox, T., Frahm, J.-M. (eds.) ECCV 2020. LNCS, vol. 12346, pp. 213–229. Springer, Cham (2020). https://doi.org/10.1007/978-3-030-58452-8_13
3. Chen, S., Fang, J., Zhang, Q., Liu, W., Wang, X.: Hierarchical aggregation for 3D instance segmentation. In: Proceedings of the IEEE/CVF International Conference on Computer Vision (2021)
4. Dai, A., Chang, A.X., Savva, M., Halber, M., Funkhouser, T., Nießner, M.: Scannet: richly-annotated 3D reconstructions of indoor scenes. In: Proceedings of the IEEE/CVF Conference on Computer Vision and Pattern Recognition (2017)
5. Dijkstra, E.W., et al.: A note on two problems in connexion with graphs. Numerische mathematik 1(1), 269–271 (1959)
6. Dong, B., Zeng, F., Wang, T., Zhang, X., Wei, Y.: Solq: segmenting objects by learning queries. arXiv preprint arXiv:2106.02351 (2021)
7. Dosovitskiy, A., et al.: An image is worth 16×16 words: transformers for image recognition at scale. arXiv preprint arXiv:2010.11929 (2020)
8. Engelmann, F., Bokeloh, M., Fathi, A., Leibe, B., Nießner, M.: 3D-mpa: multi-proposal aggregation for 3D semantic instance segmentation. In: Proceedings of the IEEE/CVF Conference on Computer Vision and Pattern Recognition (2020)
9. Fan, Z., et al.: FGN: fully guided network for few-shot instance segmentation. In: Proceedings of the IEEE/CVF Conference on Computer Vision and Pattern Recognition (2020)
10. Fang, Y., et al.: You only look at one sequence: rethinking transformer in vision through object detection. arXiv preprint arXiv:2106.00666 (2021)

11. Graham, B., Engelcke, M., Van Der Maaten, L.: 3D semantic segmentation with submanifold sparse convolutional networks. In: Proceedings of the IEEE/CVF Conference on Computer Vision and Pattern Recognition (2018)

12. Guo, R., Niu, D., Qu, L., Li, Z.: Sotr: segmenting objects with transformers. In: Proceedings of the IEEE/CVF International Conference on Computer Vision (2021)

13. He, K., Gkioxari, G., Dollár, P., Girshick, R.: Mask r-cnn. In: Proceedings of the IEEE/CVF International Conference on Computer Vision (2017)

14. He, T., Gong, D., Tian, Z., Shen, C.: Learning and memorizing representative prototypes for 3D point cloud semantic and instance segmentation. In: Vedaldi, A., Bischof, H., Brox, T., Frahm, J.-M. (eds.) ECCV 2020. LNCS, vol. 12363, pp. 564–580. Springer, Cham (2020). https://doi.org/10.1007/978-3-030-58523-5_33

15. He, T., Liu, Y., Shen, C., Wang, X., Sun, C.: Instance-aware embedding for point cloud instance segmentation. In: Vedaldi, A., Bischof, H., Brox, T., Frahm, J.-M. (eds.) ECCV 2020. LNCS, vol. 12375, pp. 255–270. Springer, Cham (2020). https://doi.org/10.1007/978-3-030-58577-8_16

16. He, T., Shen, C., van den Hengel, A.: Dyco3d: robust instance segmentation of 3D point clouds through dynamic convolution. In: Proceedings of the IEEE/CVF Conference on Computer Vision and Pattern Recognition (2021)

17. Hou, J., Dai, A., Nießner, M.: 3d-sis: 3D semantic instance segmentation of rgb-d scans. In: Proceedings of the IEEE/CVF Conference on Computer Vision and Pattern Recognition, pp. 4421–4430 (2019)

18. Hu, J., et al.: Istr: end-to-end instance segmentation with transformers. arXiv preprint arXiv:2105.00637 (2021)

19. Ioffe, S., Szegedy, C.: Batch normalization: accelerating deep network training by reducing internal covariate shift. In: International Conference on Machine Learning. PMLR (2015)

20. Jiang, L., Zhao, H., Shi, S., Liu, S., Fu, C.W., Jia, J.: Pointgroup: dual-set point grouping for 3D instance segmentation. In: Proceedings of the IEEE/CVF Conference on Computer Vision and Pattern Recognition (2020)

21. Kimmel, R., Sethian, J.A.: Computing geodesic paths on manifolds. Proc. Natl. Acad. Sci. **95**(15), 8431–8435 (1998)

22. Kingma, D.P., Ba, J.: Adam: a method for stochastic optimization. arXiv preprint arXiv:1412.6980 (2014)

23. Kuhn, H.W.: The Hungarian method for the assignment problem. Naval Res. Logist. Q. **2**(1–2), 83–97 (1955)

24. Li, Z., et al.: Panoptic segformer. arXiv preprint arXiv:2109.03814 (2021)

25. Liang, Z., Li, Z., Xu, S., Tan, M., Jia, K.: Instance segmentation in 3D scenes using semantic superpoint tree networks. In: Proceedings of the IEEE/CVF International Conference on Computer Vision (2021)

26. Lin, T.Y., Goyal, P., Girshick, R., He, K., Dollár, P.: Focal loss for dense object detection. In: Proceedings of the IEEE/CVF International Conference on Computer Vision (2017)

27. Liu, Z., Zhao, X., Huang, T., Hu, R., Zhou, Y., Bai, X.: Tanet: robust 3D object detection from point clouds with triple attention. In: Proceedings of the AAAI Conference on Artificial Intelligence, vol. 34 (2020)

28. Loshchilov, I., Hutter, F.: Sgdr: stochastic gradient descent with warm restarts. arXiv preprint arXiv:1608.03983 (2016)

29. Meng, D., et al.: Conditional detr for fast training convergence. In: Proceedings of the IEEE/CVF International Conference on Computer Vision (2021)

30. Michaelis, C., Ustyuzhaninov, I., Bethge, M., Ecker, A.S.: One-shot instance segmentation. arXiv preprint arXiv:1811.11507 (2018)
31. Misra, I., Girdhar, R., Joulin, A.: An end-to-end transformer model for 3D object detection. In: Proceedings of the IEEE/CVF International Conference on Computer Vision (2021)
32. Nair, V., Hinton, G.E.: Rectified linear units improve restricted boltzmann machines. In: International Conference on Machine Learning (2010)
33. Nguyen, K., Todorovic, S.: Fapis: A few-shot anchor-free part-based instance segmenter. In: Proceedings of the IEEE/CVF Conference on Computer Vision and Pattern Recognition, pp. 11099–11108 (2021)
34. Nguyen, K., Todorovic, S.: ifs-rcnn: an incremental few-shot instance segmenter. In: Proceedings of the IEEE/CVF Conference on Computer Vision and Pattern Recognition, pp. 7010–7019 (2022)
35. Pan, X., Xia, Z., Song, S., Li, L.E., Huang, G.: 3D object detection with pointformer. In: Proceedings of the IEEE/CVF Conference on Computer Vision and Pattern Recognition (2021)
36. Qi, C.R., Su, H., Mo, K., Guibas, L.J.: Pointnet: deep learning on point sets for 3D classification and segmentation. In: Proceedings of the IEEE/CVF Conference on Computer Vision and Pattern Recognition (2017)
37. Qi, C.R., Yi, L., Su, H., Guibas, L.J.: Pointnet++: deep hierarchical feature learning on point sets in a metric space. In: Advances in Neural Information Processing Systems (2017)
38. Strudel, R., Garcia, R., Laptev, I., Schmid, C.: Segmenter: transformer for semantic segmentation. arXiv preprint arXiv:2105.05633 (2021)
39. Sudre, C.H., Li, W., Vercauteren, T., Ourselin, S., Jorge Cardoso, M.: Generalised dice overlap as a deep learning loss function for highly unbalanced segmentations. In: Cardoso, M.J., et al. (eds.) DLMIA/ML-CDS -2017. LNCS, vol. 10553, pp. 240–248. Springer, Cham (2017). https://doi.org/10.1007/978-3-319-67558-9_28
40. Tian, Z., Shen, C., Chen, H.: Conditional convolutions for instance segmentation. In: Vedaldi, A., Bischof, H., Brox, T., Frahm, J.-M. (eds.) ECCV 2020. LNCS, vol. 12346, pp. 282–298. Springer, Cham (2020). https://doi.org/10.1007/978-3-030-58452-8_17
41. Tian, Z., Shen, C., Chen, H., He, T.: Fcos: fully convolutional one-stage object detection. In: Proceedings of the IEEE/CVF International Conference on Computer Vision (2019)
42. Touvron, H., Cord, M., Douze, M., Massa, F., Sablayrolles, A., Jégou, H.: Training data-efficient image transformers & distillation through attention. In: International Conference on Machine Learning. PMLR (2021)
43. Vaswani, A., et al.: Attention is all you need. In: Advances in Neural Information Processing Systems (2017)
44. Wang, W., Yu, R., Huang, Q., Neumann, U.: Sgpn: similarity group proposal network for 3D point cloud instance segmentation. In: Proceedings of the IEEE/CVF Conference on Computer Vision and Pattern Recognition, pp. 2569–2578 (2018)
45. Wang, W., et al.: Pyramid vision transformer: a versatile backbone for dense prediction without convolutions. arXiv preprint arXiv:2102.12122 (2021)
46. Wang, X., Zhang, R., Kong, T., Li, L., Shen, C.: Solov2: dynamic and fast instance segmentation. In: Advances in Neural Information Processing Systems, vol. 33 (2020)
47. Wang, Y., Zhang, X., Yang, T., Sun, J.: Anchor detr: query design for transformer-based detector. arXiv preprint arXiv:2109.07107 (2021)

48. Xiao, Y., Marlet, R.: Few-shot object detection and viewpoint estimation for objects in the wild. In: Vedaldi, A., Bischof, H., Brox, T., Frahm, J.-M. (eds.) ECCV 2020. LNCS, vol. 12362, pp. 192–210. Springer, Cham (2020). https://doi.org/10.1007/978-3-030-58520-4_12

49. Xie, E., Wang, W., Yu, Z., Anandkumar, A., Alvarez, J.M., Luo, P.: Segformer: simple and efficient design for semantic segmentation with transformers. arXiv preprint arXiv:2105.15203 (2021)

50. Yan, X., Chen, Z., Xu, A., Wang, X., Liang, X., Lin, L.: Meta r-cnn: towards general solver for instance-level low-shot learning. In: Proceedings of the IEEE International Conference on Computer Vision (2019)

51. Yang, B., et al.: Learning object bounding boxes for 3D instance segmentation on point clouds. In: Advances in Neural Information Processing Systems (2019)

52. Yi, L., Zhao, W., Wang, H., Sung, M., Guibas, L.J.: Gspn: generative shape proposal network for 3D instance segmentation in point cloud. In: Proceedings of the IEEE/CVF Conference on Computer Vision and Pattern Recognition (2019)

53. Yuan, L., et al.: Tokens-to-token vit: training vision transformers from scratch on imagenet. arXiv preprint arXiv:2101.11986 (2021)

54. Zhao, H., Jiang, L., Jia, J., Torr, P.H., Koltun, V.: Point transformer. In: Proceedings of the IEEE/CVF International Conference on Computer Vision (2021)

55. Zhao, N., Chua, T.S., Lee, G.H.: Few-shot 3D point cloud semantic segmentation. In: Proceedings of the IEEE/CVF Conference on Computer Vision and Pattern Recognition (2021)

56. Zhu, X., Su, W., Lu, L., Li, B., Wang, X., Dai, J.: Deformable detr: deformable transformers for end-to-end object detection. arXiv preprint arXiv:2010.04159 (2020)

Union-Set Multi-source Model Adaptation for Semantic Segmentation

Zongyao Li📵, Ren Togo📵, Takahiro Ogawa📵, and Miki Haseyama(✉)📵

Hokkaido University, Sapporo, Japan
{li,togo,ogawa,mhaseyama}@lmd.ist.hokudai.ac.jp

Abstract. This paper solves a generalized version of the problem of multi-source model adaptation for semantic segmentation. Model adaptation is proposed as a new domain adaptation problem which requires access to a pre-trained model instead of data for the source domain. A general multi-source setting of model adaptation assumes strictly that each source domain shares a common label space with the target domain. As a relaxation, we allow the label space of each source domain to be a subset of that of the target domain and require the union of the source-domain label spaces to be equal to the target-domain label space. For the new setting named union-set multi-source model adaptation, we propose a method with a novel learning strategy named model-invariant feature learning, which takes full advantage of the diverse characteristics of the source-domain models, thereby improving the generalization in the target domain. We conduct extensive experiments in various adaptation settings to show the superiority of our method.

Keywords: Model adaptation · Domain adaptation · Semantic segmentation

1 Introduction

Learning with unlabeled data is a long-term problem in the field of machine learning, and it has shown great significance more than ever since the rise of deep learning which heavily relies on well-labeled data. Due to the difficulty of unsupervised learning, some studies import labeled data from a different domain, and the problem accordingly becomes unsupervised domain adaptation (UDA) [9,19]. Typically, a method of UDA borrows the knowledge from a labeled source domain for learning with an unlabeled target domain by using data of both the domains and reducing the domain gap [14]. However, considering the fact that regulations are increasingly and strictly constituted for protecting private data, sharing the source-domain data may be impractical in some applications. For such situations, an alternative way for borrowing the source-domain knowledge without direct use of the source-domain data is necessary.

This study was partly supported by JSPS KAKENHI Grant Number JP21H03456 and conducted on the Data Science Computing System of Education and Research Center for Mathematical and Data Science, Hokkaido University.

Model adaptation [15] (also named source-free domain adaptation) has been proposed for solving the above problem. As a derivative of UDA, model adaptation replaces the source-domain data with a source-domain pre-trained model and is thus no longer limited by the restriction of data sharing. Since private information can be hardly recovered from the pre-trained models, model adaptation faces less limitations and is practical in a wider range of applications than traditional UDA. Furthermore, in addition to the less difficulty of getting the access permission, pre-trained models also require considerably less storage size than training data, and therefore using multiple pre-trained models of different source domain, i.e., multi-source model adaptation (MSMA), seems to be a cost-efficient option when multiple appropriate source domains exist. However, unlike multi-source UDA which has been widely studied [11,35,36], the study on MSMA is insufficient despite its promising prospect. To the best of our knowledge, only one work studied on MSMA for image classification [1].

This paper focuses on the problem of MSMA for semantic segmentation which still remains to be solved. Similar to multi-source UDA, in a general MSMA setting, a common label space is expected to be shared by all the source domains and the target domain, which is a too stringent assumption to be practical in some real-world scenarios. Therefore, in this paper, we relax the requirement for the source-domain label spaces and propose a generalized version of MSMA named union-set multi-source model adaptation (US-MSMA). Specifically, in our US-MSMA setting, the union of all the source-domain label spaces instead of the label space of each source domain is required to be equal to the target domain label space. In other words, the label space of each source domain is just expected to be a subset but not necessarily the same as that of the target domain. Such a relaxation considerably extends the applicability of MSMA. Moreover, it also allows selection of source domains from a larger candidate set which may improve the adaptation performance. For example, a high-performance model that is trained in a source domain with high-quality labels of part of the target domain classes, can be used in the training of US-MSMA and contributes to improving the adaptation performance for certain classes. The generalized multi-source setting is especially compatible with model adaptation due to the low cost for introducing pre-trained models.

For handling the problem of US-MSMA for semantic segmentation, we propose a two-stage method which consists of a model adaptation stage and a model integration stage. For the model adaptation stage, we propose a novel learning strategy, model-invariant feature learning. Specifically, to take full advantage of diverse characteristics of the source-domain pre-trained models, we train the source-domain models to produce target-domain features with similar distributions which are referred to as model-invariant features. The conception of the model-invariant feature learning is to reduce the domain biases and improve the generalization ability of the source-domain models by harmonizing the model characteristics derived from different source domains. Moreover, to obtain predictions in the target-domain label space, we introduce a classifier ensemble strategy which combines predictions of all the classifiers of the source-domain models as the complete prediction. To integrate the adapted source-domain models, we further introduce the model integration stage which distills knowledge of the adapted models to train a final model. We validate the effectiveness of our method in various situations of the union-set multi-source setting by conducting extensive experiments.

This paper's contributions are summarized as follows.

- We propose the problem setting of US-MSMA, which relaxes the requirement for the source-domain label spaces in the general multi-source setting and is thus applicable to a wider range of practical scenarios.
- We propose a two-stage method to handle the problem of US-MSMA for semantic segmentation. In the first stage, we propose the novel model-invariant feature learning for better generalization in the target domain. And as the second stage, we introduce the model integration to train a final model which absorbs knowledge from the adapted source-domain models.
- We conduct experiments in extensive adaptation settings that use several source-domain sets with different label space settings. Experimental results demonstrate the superiority of our method to previous adaptation methods.

2 Related Works

2.1 UDA for Semantic Segmentation

UDA for semantic segmentation has been widely studied in recent years. Typical technologies used in the methods mainly include image-to-image translation [4,14,16,29,32], adversarial learning [7,20,25,26,34], and semi-supervised learning [3,5,27,37,39]. Image-to-image translation is used for reducing the visual domain gap by modifying some image characteristics and is mainly performed with a GAN [10]-based model [38]. Adversarial learning introduces a domain discriminator for recognizing the domain of intermediate features [34] or final outputs [25]. Training the segmentation network and the discriminator against each other can align the feature distributions of the domains. Semi-supervised learning technologies can be readily applied to UDA due to the similar problem setting, including pseudo-label learning [37,39], self-ensembling [5], and entropy minimization [3,27]. In this paper, we also use the pseudo-label learning as the baseline of our method and introduce the entropy minimization for further improving the adaptation performance.

2.2 Multi-source UDA

Multi-source UDA methods have been developed for image classification [13, 21,35], semantic segmentation [11,36], and object detection [33]. The technologies used in single-source UDA play a dominant role also in multi-source UDA. Zhao et al. [35] introduce multiple domain discriminators on the top of the feature extractor and conduct the adversarial learning between the feature extractor and the discriminators via a gradient reversal layer. Peng et al. [21] align moments of feature distributions of the source domains and the target domain to transfer the source-domain knowledge. Zhao et al. [36] align the domains at both the pixel level and the feature level with the image-to-image translation and the adversarial learning. He et al. [11] also perform the pixel-level adaptation and further train with target-domain pseudo labels. Yao et al. [33] train a domain-adaptive object detector with multiple source subnets and obtain a target subnet by weighting and combining the parameters of the source subnets. The above methods are all developed for the general multi-source setting which is a particular case of our union-set multi-source setting.

2.3   Model Adaptation

Previous studies on model adaptation mainly focus on image classification, using some technologies similar to those used in UDA, such as generative models [15], adversarial learning [30], information maximization [17], and class prototypes [31]. As to semantic segmentation, Liu et al. [18] generate fake source-domain samples with real source-domain distribution to transfer the source-domain knowledge and introduce a patch-level self-supervision module for target-domain pseudo labels. Fleuret et al. [8] reduce uncertainty of predictions from multiple classifiers suffering from random noise to enhance robustness of the learned feature representation. Stan et al. [24] learn a prototypical distribution to encode source-domain knowledge and align the distributions across domains with the learned distribution. Model adaptation in the multi-source setting is studied poorly, with only one work [1] for image classification to the best our knowledge. Ahmed et al. [1] learn an optimal combination of multiple source-domain models with trainable weights, whereas the inference time is increased by several times due to the model combination.

3   Proposed Method

3.1   Problem Setting of US-MSMA

First, we detail the US-MSMA problem setting of our method. Let $\{D_i^S\}_{i=1}^k$ denote k labeled source domains with class sets $\{\Phi_i^S\}_{i=1}^k$ and D^T the unlabeled target domain with a class set Φ^T. Different from the general multi-source setting in which each of $\{\Phi_i^S\}_{i=1}^k$ must be absolutely the same as Φ^T, in our US-MSMA setting, the union of $\{\Phi_i^S\}_{i=1}^k$ is assumed to be equal to Φ^T, i.e., $\Phi_1^S \bigcup \Phi_2^S \cdots \bigcup \Phi_k^S = \Phi^T$. Given access to unlabeled data $\{x_i^T\}_{i=1}^n$ of D^T and k models $\{M_i\}_{i=1}^k$ pre-trained with $\{D_i^S\}_{i=1}^k$ respectively, we aim to obtain a model that learns knowledge transferred from $\{D_i^S\}_{i=1}^k$ to D^T and consequently achieves reasonable performance in D^T.

3.2   Overview of the Proposed Method

Two-Stage Architecture. Figure 1 shows an overview of the proposed method which consists of the following two stages: a model adaptation stage and a model

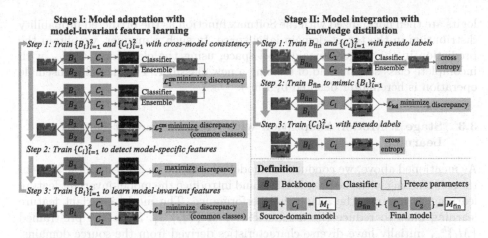

Fig. 1. An overview of the proposed method. For the ease of understanding, we show the case of using two source-domain models. The self-training with pseudo labels in Stage I is omitted in the figure.

integration stage. In Stage I, we conduct the model adaptation by retraining the pre-trained source-domain models $\{M_i\}_{i=1}^k$ with the target domain D^T. The source-domain knowledge is transferred to the target domain by training $\{M_i\}_{i=1}^k$ with pseudo labels of D^T in the manner of self-training (omitted in Fig. 1), and we improve the adaptation with the model-invariant feature learning. Then, in Stage II, we train a final model M_{fin} by distilling and integrating knowledge from $\{M_i\}_{i=1}^k$ trained in Stage I. As defined in Fig. 1, $\{M_i\}_{i=1}^k$ are individual models composed of a backbone B_i and a classifier C_i, and M_{fin} is an ensemble model composed of an integration backbone B_{fin} and all the classifiers $\{C_i\}_{i=1}^k$. The classifiers of the ensemble model are combined with a classifier ensemble strategy described below.

Classifier Ensemble. In both Stages I and II, we predict the probability distribution over all the target-domain classes Φ^T with the classifiers $\{C_i\}_{i=1}^k$. However, since the source-domain class sets $\{\Phi_i^S\}_{i=1}^k$ are not necessarily equal to Φ^T, the prediction over Φ^T may not be available from any individual C_i. And due to the possibly different class sets, the predictions of $\{C_i\}_{i=1}^k$ cannot be simply averaged. Therefore, to obtain the complete prediction, we use a classifier ensemble strategy which simultaneously combines and averages the predictions of $\{C_i\}_{i=1}^k$. Specifically, we calculate the unnormalized logits of the complete prediction by averaging the output logits of each class over $\{C_i\}_{i=1}^k$ as the following equation:

$$l_c(\cdot) = \frac{1}{\sum_{i=1}^k \mathbb{1}(c \in \Phi_i^S)} \sum_{i=1}^k C_{i,c}(\cdot), \ \forall c \in \Phi^T, \tag{1}$$

where $\mathbb{1}(\cdot)$ denotes the indicator function, and $C_{i,c}(\cdot)$ denotes the logits of class c predicted by C_i if $c \in \Phi_i^S$ and is regarded as zero otherwise. The calculated logits are then normalized with the Softmax function as the predicted probability distribution in the target-domain label space. In such a manner, we obtain the prediction in the target-domain label space, using the classifiers $\{C_i\}_{i=1}^k$ with incomplete class sets instead of training a new classifier. The classifier ensemble operation is hereinafter denoted by $\text{En}(\cdot)$.

3.3   Stage I: Model Adaptation with Model-invariant Feature Learning

As mentioned above, we conduct the model adaptation in Stage I on the basis of the self-training with the pseudo labels and introduce the model-invariant feature learning to improve the adaptation performance. The model-invariant feature learning aims to reduce the domain biases of $\{M_i\}_{i=1}^k$. Since the pre-trained $\{M_i\}_{i=1}^k$ initially have diverse characteristics derived from the source domains, the domain bias of each M_i can be reduced by harmonizing the characteristics of $\{M_i\}_{i=1}^k$. We realize it by training the backbones $\{B_i\}_{i=1}^k$ to produce features with similar distributions which we refer to as model-invariant features. The model-invariant features are learned by a iterative process composed of three steps: the first step for cross-model consistency and the subsequent two steps for adversarial learning between the backbones $\{B_i\}_{i=1}^k$ and the classifiers $\{C_i\}_{i=1}^k$, as shown in the left side of Fig. 1. We detail each component of Stage I as follows.

Self-training with Pseudo Labels. To transfer the source-domain knowledge to the target domain, we perform the self-training for each of the individual models $\{M_i\}_{i=1}^k$ and also ensemble models that are composed of a backbone B_i $(i = 1, \ldots, k)$ and all the classifiers $\{C_i\}_{i=1}^k$. To this end, we use the pre-trained $\{M_i\}_{i=1}^k$ to generate pseudo labels $\{y_i\}_{i=1}^k$ in the source-domain label spaces and pseudo labels y^T in the target-domain label space for each image of D^T. To generate the pseudo labels in the target-domain label space, we combine and average the predictions of $\{M_i\}_{i=1}^k$ but not like the classifier ensemble described above since $\{M_i\}_{i=1}^k$ are pre-trained independently. Specifically, we first cast the probability distributions predicted by $\{M_i\}_{i=1}^k$ over the target-domain label space as $p_{i,c}(\cdot) = M_{i,c}(\cdot)$ if $c \in \Phi_i^S$ and $p_{i,c}(\cdot) = \frac{M_{i,0}(\cdot)}{\sum_{c' \in \Phi^T} \mathbb{1}(c' \notin \Phi_i^S)}$ otherwise, where $M_{i,c}(\cdot)$ is the probability of class c predicted by M_i and $M_{i,0}(\cdot)$ is the probability of the other classes not in Φ_i^S. Then we average the probability distributions $\{p_i\}_{i=1}^k$ and assign pseudo labels according to the average prediction. In the self-training, we use the cross-entropy loss function $\text{CE}(logits, target)$ to train $\{M_i\}_{i=1}^k$ as follows:

$$\mathcal{L}_{\text{pl}} = \mathbb{E}_{x^T \in D^T} \sum_{i=1}^k [\text{CE}(C_i(B_i(x^T)), y_i) + \text{CE}(\text{En}(\{C_j(B_i(x^T))\}_{j=1}^k), y_T)], \quad (2)$$

where $\text{En}(\cdot)$ is the classifier ensemble operation described in Sect. 3.2.

Cross-model Consistency. On the basis of the assumption that a backbone that produces model-invariant features should be compatible with any classifier, we randomly recombine $\{B_i\}_{i=1}^k$ and $\{C_i\}_{i=1}^k$ as k new models $\{C_{\mathrm{ma}(i)}(B_i(\cdot))\}_{i=1}^k$ where $\mathrm{ma}(i)$ is the index of the classifier matched with B_i. The recombined models are trained in terms of cross-model consistency which includes overall consistency of the ensemble predictions and per-class consistency of the logits output by individual classifiers. For the overall consistency, we perform the classifier ensemble operation for predictions of the original models $\{C_i(B_i(\cdot))\}_{i=1}^k$ and the recombined models $\{C_{\mathrm{ma}(i)}(B_i(\cdot))\}_{i=1}^k$ and minimize the discrepancy between the ensemble predictions with the following loss function:

$$\mathcal{L}_1^{\mathrm{cm}} = \mathbb{E}_{x^T \in D^T} ||\sigma(\mathrm{En}(\{C_i(B_i(x^T))\}_{i=1}^k)) - \sigma(\mathrm{En}(\{C_{\mathrm{ma}(i)}(B_i(x^T))\}_{i=1}^k))||_1, \quad (3)$$

where $\sigma(\cdot)$ is the Softmax function. For the per-class consistency, we train the recombined models to output consistent logits of each class. We calculate the average logits of each class and minimize the discrepancy between the output logits and the average logits as follows:

$$\mathcal{L}_2^{\mathrm{cm}} = \mathbb{E}_{x^T \in D^T} \sum_{i=1}^k \sum_c^{\Phi_i^S} ||C_{\mathrm{ma}(i),c}(B_i(x^T)) - \delta_c(x^T)||_1, \quad (4)$$

where $\delta_c(\cdot)$ is the average logits of class c and calculated with the following equation:

$$\delta_c(\cdot) = \frac{1}{\sum_{i=1}^k \mathbb{1}(c \in \Phi_i^S)} \sum_{i=1}^k C_{\mathrm{ma}(i),c}(B_i(\cdot)), \quad (5)$$

where $C_{\mathrm{ma}(i),c}(\cdot)$ is the logits of class c output by $C_{\mathrm{ma}(i)}$ if $c \in \Phi_{\mathrm{ma}(i)}^S$ and zero otherwise. By recombining and training the models with $\mathcal{L}_1^{\mathrm{cm}}$ and $\mathcal{L}_2^{\mathrm{cm}}$ for the overall and the per-class consistency respectively, the features produced by the backbones are constrained to have similar distributions, i.e., to be model-invariant.

Adversarial Learning. In addition to the cross-model consistency, we further introduce adversarial learning between the backbones $\{B_i\}_{i=1}^k$ and the classifiers $\{C_i\}_{i=1}^k$ to enhance the model-invariant feature learning. Specifically, the adversarial learning consists of two steps: training $\{C_i\}_{i=1}^k$ to detect model-specific features and training $\{B_i\}_{i=1}^k$ to produce model-invariant features. The two steps are performed iteratively, and the parameters of $\{B_i\}_{i=1}^k$ ($\{C_i\}_{i=1}^k$) are frozen while updating those of $\{C_i\}_{i=1}^k$ ($\{B_i\}_{i=1}^k$).

For the training of C_i ($i = 1, \ldots, k$), we feed the features from B_i and B_j ($j \neq i$) into C_i and maximize the discrepancy between the predictions by minimizing the following loss function:

$$\mathcal{L}_C = \mathbb{E}_{x^T \in D^T} \sum_{i=1}^k [\sum_{j=1}^k -||C_i(B_i(x^T)) - C_i(B_j(x^T))||_1 + \mathrm{CE}(C_i(B_i(x^T)), y_i)], \quad (6)$$

where we add a cross-entropy term to prevent the recognition ability of $\{C_i\}_{i=1}^k$ from degradation while maximizing the discrepancy. The features from B_j that induce inconsistent predictions with those using the features from B_i, are considered domain-specific and detected by updating C_i with $\mathcal{L}_C$.

We train $\{B_i\}_{i=1}^k$ with a loss that calculates the discrepancy between the output logits and the average logits with features from each of the backbones respectively, as defined in the following equation:

$$\mathcal{L}_B = \mathbb{E}_{x^T \in D^T} \sum_{i=1}^k \sum_{j=1}^k [\sum_c^{\Phi_j^S} ||C_{j,c}(B_i(x^T)) - \delta_{i,c}(x^T)||_1 + \mathrm{CE}(C_j(B_i(x^T)), y_j)], \quad (7)$$

where $\delta_{i,c}(\cdot)$ is the average logits of class c using the features from B_i and calculated as follows:

$$\delta_{i,c}(\cdot) = \frac{1}{\sum_{j=1}^k \mathbb{1}(c \in \Phi_j^S)} \sum_{j=1}^k C_{j,c}(B_i(\cdot)). \quad (8)$$

By minimizing the L1-norm term of $\mathcal{L}_B$, each of the backbones is trained to produce features that induce per-class consistent logits from different classifiers and are thus considered domain-invariant. However, since the L1-norm term involves only the classes shared by multiple classifiers, we additionally add a cross-entropy term to train each backbone to be compatible with all the classifiers.

Unlike the typical adversarial learning, we conduct the adversarial learning between two groups of the backbones $\{B_i\}_{i=1}^k$ and the classifiers $\{C_i\}_{i=1}^k$ rather than two specific opponents. Moreover, we train $\{B_i\}_{i=1}^k$ with a loss dissimilar to that for $\{C_i\}_{i=1}^k$, instead of minimizing the term $||C_i(B_i(x^T)) - C_i(B_j(x^T))||_1$ of $\mathcal{L}_C$ which may lead to excessive similarity among $\{B_i\}_{i=1}^k$. The adversarial learning is compatible with the cross-model consistency and can further enhance the model-invariance of the learned features.

With the components described above, the training of Stage I is performed by repeating a process of three steps. Step 1 simultaneously updates $\{B_i\}_{i=1}^k$ and $\{C_i\}_{i=1}^k$ with the loss $\mathcal{L}_{pl}$ of the self-training and the losses $\mathcal{L}_1^{cm}$ and $\mathcal{L}_2^{cm}$ of the cross-model consistency. Then Steps 2 and 3 updates $\{C_i\}_{i=1}^k$ and $\{B_i\}_{i=1}^k$ with $\mathcal{L}_C$ and $\mathcal{L}_B$, respectively.

3.4   Stage II: Model Integration with Knowledge Distillation

The models adapted in Stage I have slightly different performances, and it is generally unknown which one would perform best in the test. To achieve the best performance not inferring with all the models, we introduce the model integration stage that distills the knowledge of $\{M_i\}_{i=1}^k$ and trains a final model M_{fin}. M_{fin} is composed of an integration backbone B_{fin} and all the classifiers $\{C_i\}_{i=1}^k$, and the ensemble prediction $\mathrm{En}(\{C_i(B_{\mathrm{fin}}(\cdot))\}_{i=1}^k)$ is regarded as the final prediction. The parameters of the backbones $\{B_i\}_{i=1}^k$ are frozen in this stage.

Similar to Stage I, M_{fin} is trained with a loss that updates B_{fin} and $\{C_i\}_{i=1}^k$ together and losses that update B_{fin} and $\{C_i\}_{i=1}^k$ respectively. Specifically, B_{fin} and $\{C_i\}_{i=1}^k$ are trained together by minimizing $\text{CE}(\text{En}(\{C_i(B_{\text{fin}}(\cdot))\}_{i=1}^k), y_T)$, the cross-entropy loss for the ensemble prediction using pseudo labels that are generated with $\{M_i\}_{i=1}^k$ trained in Stage I. Then, B_{fin} is trained to mimic $\{B_i\}_{i=1}^k$ with a knowledge distillation loss defined as follows:

$$\mathcal{L}_{\text{kd}} = \mathbb{E}_{x^T \in D^T} \sum_{i=1}^k \text{KLD}(C_i(B_{\text{fin}}(x^T)), C_i(B_i(x^T))), \tag{9}$$

where $\text{KLD}(input, target)$ is the Kullback-Leibler divergence function. Moreover, to maintain the compatibility of $\{B_i, C_i\}_{i=1}^k$, we train $\{C_i\}_{i=1}^k$ by minimizing $\sum_{i=1}^k \text{CE}(C_i(B_i(x^T)), y_i)$ for the predictions of individual classifiers. By training with the above losses, M_{fin} absorbs the knowledge of $\{M_i\}_{i=1}^k$ trained in Stage I and consequently outperforms all of $\{M_i\}_{i=1}^k$. Moreover, since $\{C_i\}_{i=1}^k$ are very light networks, the inference speed of M_{fin} is almost the same as that of a single model of $\{M_i\}_{i=1}^k$.

4 Experiments

4.1 Implementation Details

As the segmentation network in all the experiments, we used Deeplab V2 [2] with ResNet101 [12] of which the ResNet101 backbone and the atrous spatial pyramid pooling (ASPP) classifier were used as the backbones $\{B_i\}_{i=1}^k$ and the classifiers $\{C_i\}_{i=1}^k$, respectively. We trained the networks with a stochastic gradient descent (SGD) optimizer with an initial learning rate of 2.5×10^{-4} for 80,000 iterations. During the training, the learning rate was decreased with the poly policy with a power of 0.9, and the mini-batch size was set to 1. We used an equal weight of 1.0 for all the losses. Moreover, we also introduced a maximum squares loss [3] which minimizes the uncertainty of predictions, as a supplement to the losses $\mathcal{L}_{\text{pl}}$ and $\mathcal{L}_B$ in Stage I and also the loss $\text{CE}(\text{En}(\{C_i(B_{\text{fin}}(\cdot))\}_{i=1}^k), y_T)$ in Stage II to improve the adaptation performance. The code will be released.[1]

4.2 Datasets and Adaptation Settings

We used Synscapes dataset [28], GTA5 dataset [22], and Synthia dataset [23] as the source domains and Cityscapes dataset [6] as the target domain. All the source domains are synthetic datasets that are composed of photo-realistic images of street scenes, and Cityscapes is a real-world dataset of street-scene images. Synscapes, GTA5, and Cityscapes share a common label space containing 19 classes, while Synthia shares a subset containing 16 classes with the others. Synscapes, GTA5, and Synthia are hereinafter referred to as S, G, and T, respectively.

---

[1] https://github.com/lzy7976/union-set-model-adaptation.

Table 1. Class distributions of the non-overlapping setting and the partly-overlapping setting. The target domain contains 19 classes: road, sidewalk, building, wall, fence, pole, traffic light, traffic sign, vegetation, terrain, sky, person, rider, car, truck, bus, train, motorcycle, bicycle.

Setting		Source domain	Background classes											Foreground classes							
			road	side.	buil.	wall	fence	pole	t-lig.	t-sign	vege.	terr.	sky	pers.	rider	car	truck	bus	train	moto.	bicy.
Non-overlapping	S+G	S												✓	✓	✓	✓	✓	✓	✓	✓
		G	✓	✓	✓	✓	✓	✓	✓	✓	✓	✓	✓								
	S(G)+T	S(G)										✓		✓	✓	✓	✓	✓	✓	✓	✓
		T	✓	✓	✓	✓	✓	✓	✓	✓	✓		✓								
	S+G+T	S														✓	✓	✓	✓		
		G	✓	✓	✓	✓	✓	✓	✓	✓	✓	✓	✓								
		T												✓	✓					✓	✓
Partly-overlapping	S+G	S	✓	✓	✓	✓	✓	✓	✓	✓	✓	✓	✓			✓	✓	✓	✓		
		G	✓	✓	✓	✓	✓	✓	✓	✓	✓	✓	✓	✓	✓					✓	✓
	S(G)+T	S(G)	✓	✓	✓	✓	✓	✓	✓	✓	✓	✓	✓			✓	✓	✓	✓		
		T	✓	✓	✓	✓	✓	✓	✓	✓	✓		✓	✓	✓					✓	✓

We conducted experiments with all the possible combinations of the source domains including $S+G$, $S+T$, $G+T$, and $S+G+T$. Furthermore, since our union-set multi-source setting is flexible in terms of the source-domain label spaces, we evaluated our method in various settings of the source-domain label spaces to demonstrate its effectiveness in a wide range of scenarios. Specifically, we conducted experiments in three label space settings: non-overlapping, partly-overlapping, and fully-overlapping. In the non-overlapping setting, the target-domain classes are divided into subsets with no common classes and assigned to the source domains. In the partly-overlapping setting, background classes are shared as common classes, while foreground classes are divided for the source domains. We detail the class distributions of the non-overlapping setting and the partly-overlapping setting in Table 1. In the fully-overlapping setting, all classes are shared expect three classes (terrain, truck, train) that are absent in Synthia, but we still involved these classes in the experiments of $S+T$ and $G+T$.

4.3   Results in Non-overlapping Setting

Since no existing methods are proposed for such a problem setting, we slightly modified three methods proposed for related problem settings for comparisons: a UDA method [25] for semantic segmentation, a single-source model adaptation (SSMA) method [8] for semantic segmentation, and an MSMA method [1] for image classification. For the UDA method based on adversarial learning, we used domain-specific discriminators and classifiers for the source domains and trained one shared backbone. For both the UDA method and the SSMA method, the complete predictions were obtained by first casting the predictions in the source-domain label spaces over the target-domain label space and then averaging the predictions, in the same manner as that for generating the pseudo labels. For the MSMA method which learns a set of weights for combining the models, we still cast the prediction of each model over the target-domain label space in the same manner and calculated the weighted average prediction with the learned weights. Moreover, for explicit comparisons with the SSMA method and

Table 2. Results in the non-overlapping setting. ST: self-training. CMC: cross-model consistency. ADV: adversarial learning. MSL: maximum squares loss. MI: model integration. PM: proposed method.

Method	$S+G$	$S+T$	$G+T$	$S+G+T$
ST	42.3	38.8	35.8	40.4
ST+CMC	43.2	39.1	36.1	40.6
ST+CMC+ADV	44.0	39.8	36.4	41.5
ST+CMC+ADV+MSL (=**Stage I of PM**)	45.8	41.4	37.2	42.1
ST+CMC+ADV+MSL+MI (=**PM**)	**46.6**	**42.3**	**37.9**	**44.2**
SSMA [8]	43.5	40.6	37.0	41.9
MSMA [1]	30.2	26.0	20.7	22.7
UDA [25]	45.7	39.2	35.9	41.1

Target-domain image	Ground truth	Prediction (our method)	Prediction (non-adaptation)

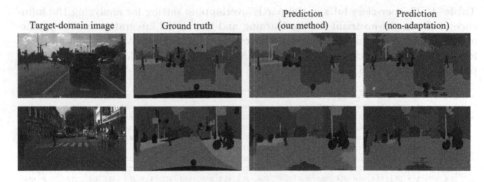

Fig. 2. Examples of the qualitative results of our method and the source-domain models without adaptation in the non-overlapping setting of $S+G$.

the MSMA method, we performed the training of the methods using the same maximum squares loss and the cross-entropy loss with the same pseudo labels as those used in our method.

Table 2 shows the results in the non-overlapping setting. The mean Intersection over Union (IoU) over all the target-domain classes were used as the evaluation metric. For the incomplete versions of our method which trained multiple models but no final model, we independently evaluated the ensemble models composed of one backbone and multiple classifiers and reported the average performance. As shown in Table 2, the cross-model consistency and the adversarial learning, which are the two components of the model-invariant feature learning, successively improved the performance. By introducing the maximum squares loss, we obtained the results of Stage I of our method with further improvements. Finally, we performed the model integration of Stage II and achieved the best performance with significant improvements compared to the baseline trained with only the self-training. The above results for the ablation study validated the effectiveness of each component of our method.

Table 3. Results in the partly-overlapping setting. ST: self-training. CMC: cross-model consistency. ADV: adversarial learning. MSL: maximum squares loss. MI: model integration. PM: proposed method.

Method	$S+G$	$S+T$	$G+T$
ST	44.2	44.0	39.1
ST+CMC	46.0	44.8	39.6
ST+CMC+ADV	46.6	45.2	40.6
ST+CMC+ADV+MSL (= **Stage I of PM**)	47.4	45.9	42.2
ST+CMC+ADV+MSL+MI (= **PM**)	**48.3**	**47.2**	**43.5**
SSMA [8]	47.2	46.3	42.7
MSMA [1]	46.5	44.1	41.9
UDA [25]	47.9	46.9	41.6

Table 4. The per-class IoUs in the partly-overlapping setting for analyzing the influence of the model-invariant feature learning and the model integration on the performance over common and uncommon classes. ST: self-training. MIF: model-invariant feature learning, which is equal to ST+CMC+ADV in Table 3. S-I: Stage I of the proposed method. PM: proposed method.

Setting/Method		IoUs of background classes (common)												IoUs of foreground classes (uncommon)								
		road	side.	buil.	wall	fence	pole	t-lig.	t-sign	vege.	terr.	sky	mean	pers.	rider	car	truck	bus	train	moto.	bicy.	mean
$S+G$	ST	81.1	43.7	79.4	28.0	19.7	38.0	42.3	45.6	84.2	32.1	80.3	52.2	58.3	25.5	80.0	18.9	16.3	1.7	20.3	45.2	33.3
	MIF	86.4	46.2	82.1	34.6	27.4	37.7	42.5	43.4	85.0	40.2	84.2	55.4	58.8	26.8	78.4	23.6	20.9	2.2	21.0	44.0	34.5
	E I	85.1	46.0	82.6	34.6	27.3	37.5	41.6	42.5	84.9	39.5	84.0	55.1	58.4	25.2	80.3	29.9	31.2	2.7	24.5	42.6	36.8
	PM	87.3	47.5	83.4	34.1	28.1	37.5	42.1	43.8	85.2	41.8	84.3	55.9	58.2	25.0	81.0	34.3	33.2	2.2	24.4	47.4	37.9
$S+T$	ST	73.8	36.2	82.1	26.3	19.4	36.8	37.2	38.7	84.4	27.6	82.7	49.6	60.8	22.3	82.1	22.2	18.3	1.9	27.5	55.9	36.4
	MIF	74.1	39.5	83.1	27.3	18.3	37.1	36.6	37.4	84.5	24.9	85.0	49.8	62.0	26.6	82.1	27.2	24.1	1.6	31.6	55.0	38.8
	S-I	76.8	39.4	83.5	26.6	20.8	37.3	38.0	37.3	85.0	21.3	83.7	50.0	62.7	31.3	81.7	27.3	27.0	1.4	27.0	55.9	40.3
	PM	80.6	42.5	84.1	27.2	23.4	37.6	37.0	39.6	85.2	23.2	84.6	51.4	62.4	31.9	82.2	31.8	37.1	1.9	30.4	54.0	41.5
$G+T$	ST	67.9	29.1	83.4	28.1	23.5	32.6	29.0	32.2	79.2	14.2	84.3	45.8	31.7	19.5	72.9	33.2	12.8	0.0	21.8	47.5	29.9
	MIF	73.9	32.3	83.5	28.8	22.3	33.2	26.8	26.5	78.6	15.9	86.4	46.2	35.1	20.6	82.0	36.1	16.3	0.0	24.2	48.8	32.9
	S-I	82.2	37.3	83.3	27.7	17.2	33.7	25.8	23.8	80.1	20.3	85.6	47.0	35.5	20.2	84.4	33.2	30.2	0.0	32.1	49.5	35.6
	PM	84.4	39.3	83.5	26.5	22.5	33.9	26.2	26.3	80.1	21.5	84.6	48.1	38.5	21.5	84.6	35.1	36.7	0.0	30.9	51.0	37.3

Compared to other adaptation methods, our method outperformed all the others even with only Stage I in the non-overlapping setting. Moreover, both the SSMA and the MSMA methods need to infer with multiple models to obtain the complete predictions while the final model of our method has only one backbone and thus spends much less time for inference than the two methods. The MSMA method failed in the non-overlapping setting because the weighted model ensemble of the method is meaningless without any common classes.

Figure 2 shows two examples of the qualitative results of our method and the source-domain models without adaptation using source domains of $S+G$. It can be seen in the figure that the segmentation of both background and foreground classes was improved by our method. In the upper example, the predictions for sky, sidewalk, and vegetation became clearly better after the adaptation. And in the lower example, the classes including sidewalk, traffic sign, vegetation, and rider were segmented more precisely in the result of our method.

Table 5. Results in the fully-overlapping setting. ST: self-training. CMC: cross-model consistency. ADV: adversarial learning. MSL: maximum squares loss. MI: model integration. PM: proposed method.

Method	$S+G$	$S+T$	$G+T$
ST	47.6	45.7	44.0
ST+CMC	49.4	46.6	44.7
ST+CMC+ADV	50.9	47.7	45.7
ST+CMC+ADV+MSL (=**Stage I of PM**)	51.0	47.8	45.7
ST+CMC+ADV+MSL+MI (=**PM**)	**51.7**	48.7	**46.2**
SSMA [8]	49.8	47.6	45.5
MSMA [1]	**51.7**	47.2	44.2
UDA [25]	51.6	**49.0**	45.8

4.4 Results in Partly-Overlapping Setting

For the partly-overlapping setting, we conducted the same experiments as those of the non-overlapping setting except $S+G+T$. The results are shown in Table 3. Similar to the results of the ablation study in the non-overlapping setting, each component contributed to the improvement consistently. Since both common and uncommon classes of the source domains exist in the partly-overlapping setting, we also show the per-class IoUs in Table 4 to analyze the influence of the model-invariant feature learning and the model integration on the performance over common and uncommon classes. By comparing "MIF" with "ST" in Table 4, the effectiveness of the model-invariant feature learning for both common and uncommon classes was indicated by the clear improvements except for the common classes of $S+T$. We think the slight improvement for the common classes of $S+T$ was because S has a much less domain gap with the target domain for the background classes than that of T, and consequently harmonizing the model characteristics derived from S and T is not helpful to the generalization in the target domain. Moreover, the comparison between "S-I" and "PM" showed that the model integration of Stage II also improved the performance over both common and uncommon classes.

As to the comparisons to the other adaptation methods, our method achieved the best performance again as shown in Table 3. However, unlike the non-overlapping setting, the version with only Stage I of our method failed to outperform the SSMA and the UDA methods. Due to the presence of the common classes, the SSMA method benefited considerably from the model ensemble for obtaining complete predictions, which, however, decreased the inference speed by several times. Similarly, with the presence of the common classes, the MSMA method achieved reasonable performance with the weighted model ensemble. The UDA method achieved close performance to ours in $S+G$ and $S+T$ but required the access to the source-domain data.

4.5   Results in Fully-Overlapping Setting

We conducted experiments in also the fully-overlapping setting which is exactly the general multi-source model adaptation. Due to the identical label space of the source domains and the target domain, we no longer performed model ensemble for the SSMA method and averaged the independent performances of all the trained models as the final performance. The MSMA method was evaluated with no changes since the weighted model ensemble acts as the core of the method. And for the UDA method, we used domain-specific discriminators but only one classifier. We did not conduct experiments in the $S+G+T$ setting since compared to the performance in $S+G$, no improvements can be gained by introducing T due to the much larger domain gap between T and the target domain.

Table 5 shows the results in the fully-overlapping setting. Similarly to the other two settings, the results of the ablation study validated the effectiveness of each component of our method, with the only difference that the maximum squares loss made almost no contributions in the fully-overlapping setting. It can be explained by the fact that the pseudo labels generated in the fully-overlapping setting were more accurate than those in the other settings, which diminished the significance of the maximum squares loss.

In the comparisons to the other adaptation methods, our method still outperformed the SSMA method, while the UDA method achieved slightly better performance than ours in $S+T$. Moreover, the MSMA method achieved the same performance as ours in $S+G$ using the two source domains with closer domain gaps with the target domain, which indicated that the efficiency of the weighted model ensemble is maximized with source domains that have the same label space and similar domain gaps with the target domain. Overall, our method has the best cost-performance ratio considering the inference speed and the requirement for the access to source-domain data.

5   Conclusion

This paper has presented a novel problem named union-set multi-source model adaptation, which requires the union of the source-domain label spaces to be equal to the target-domain label space and is thus applicable to a wider range of practical scenarios than that with the general multi-source setting. To tackle the problem of union-set multi-source model adaptation for semantic segmentation, we proposed a method with a novel learning strategy, model-invariant feature learning, to improve the generalization in the target domain by harmonizing the diverse characteristics of the source-domain models. Moreover, we further performed the model integration which distills the knowledge from the adapted models and trains a final model with improved performance. The effectiveness of each component of our method was validated by the results of the elaborate ablation studies, and the superiority of our method compared to previous adaptation methods was demonstrated by the results of the extensive experiments in various settings.

References

1. Ahmed, S.M., Raychaudhuri, D.S., Paul, S., Oymak, S., Roy-Chowdhury, A.K.: Unsupervised multi-source domain adaptation without access to source data. In: Proceedings of the IEEE/CVF Conference on Computer Vision and Pattern Recognition, pp. 10103–10112 (2021)
2. Chen, L.C., Papandreou, G., Kokkinos, I., Murphy, K., Yuille, A.L.: DeepLab: semantic image segmentation with deep convolutional nets, atrous convolution, and fully connected CRFs. IEEE Trans. Pattern Anal. Mach. Intell. **40**(4), 834–848 (2017)
3. Chen, M., Xue, H., Cai, D.: Domain adaptation for semantic segmentation with maximum squares loss. In: Proceedings of the IEEE/CVF International Conference on Computer Vision, pp. 2090–2099 (2019)
4. Chen, Y.C., Lin, Y.Y., Yang, M.H., Huang, J.B.: CrDoCo: pixel-level domain transfer with cross-domain consistency. In: Proceedings of the IEEE/CVF conference on computer vision and pattern recognition, pp. 1791–1800 (2019)
5. Choi, J., Kim, T., Kim, C.: Self-ensembling with GAN-based data augmentation for domain adaptation in semantic segmentation. In: Proceedings of the IEEE/CVF International Conference on Computer Vision, pp. 6830–6840 (2019)
6. Cordts, M., et al.: The cityscapes dataset for semantic urban scene understanding. In: Proceedings of the IEEE Conference on Computer Vision and Pattern Recognition, pp. 3213–3223 (2016)
7. Du, L., Tan, J., Yang, H., Feng, J., Xue, X., Zheng, Q., Ye, X., Zhang, X.: Ssf-dan: Separated semantic feature based domain adaptation network for semantic segmentation. In: Proceedings of the IEEE/CVF International Conference on Computer Vision. pp. 982–991 (2019)
8. Fleuret, F., et al.: Uncertainty reduction for model adaptation in semantic segmentation. In: Proceedings of the IEEE/CVF Conference on Computer Vision and Pattern Recognition, pp. 9613–9623 (2021)
9. Ganin, Y., Lempitsky, V.: Unsupervised domain adaptation by backpropagation. In: Proceedings of the International Conference on Machine Learning, pp. 1180–1189 (2015)
10. Goodfellow, I., et al.: Generative adversarial nets. In: Advances in Neural Information Processing Systems 27 (2014)
11. He, J., Jia, X., Chen, S., Liu, J.: Multi-source domain adaptation with collaborative learning for semantic segmentation. In: Proceedings of the IEEE/CVF Conference on Computer Vision and Pattern Recognition, pp. 11008–11017 (2021)
12. He, K., Zhang, X., Ren, S., Sun, J.: Deep residual learning for image recognition. In: Proceedings of the IEEE Conference on Computer Vision and Pattern Recognition, pp. 770–778 (2016)
13. Hoffman, J., Mohri, M., Zhang, N.: Algorithms and theory for multiple-source adaptation. In: Advances in Neural Information Processing Systems 31 (2018)
14. Hoffman, J., et al.: CyCADA: cycle-consistent adversarial domain adaptation. In: Proceedings of the International Conference on Machine Learning, pp. 1989–1998 (2018)
15. Li, R., Jiao, Q., Cao, W., Wong, H.S., Wu, S.: Model adaptation: unsupervised domain adaptation without source data. In: Proceedings of the IEEE/CVF Conference on Computer Vision and Pattern Recognition, pp. 9641–9650 (2020)
16. Li, Y., Yuan, L., Vasconcelos, N.: Bidirectional learning for domain adaptation of semantic segmentation. In: Proceedings of the IEEE/CVF Conference on Computer Vision and Pattern Recognition, pp. 6936–6945 (2019)

17. Liang, J., Hu, D., Feng, J.: Do we really need to access the source data? source hypothesis transfer for unsupervised domain adaptation. In: Proceedings of the International Conference on Machine Learning, pp. 6028–6039 (2020)
18. Liu, Y., Zhang, W., Wang, J.: Source-free domain adaptation for semantic segmentation. In: Proceedings of the IEEE/CVF Conference on Computer Vision and Pattern Recognition, pp. 1215–1224 (2021)
19. Long, M., Cao, Y., Wang, J., Jordan, M.: Learning transferable features with deep adaptation networks. In: Proceedings of the International Conference on Machine Learning, pp. 97–105 (2015)
20. Luo, Y., Zheng, L., Guan, T., Yu, J., Yang, Y.: Taking a closer look at domain shift: category-level adversaries for semantics consistent domain adaptation. In: Proceedings of the IEEE/CVF Conference on Computer Vision and Pattern Recognition, pp. 2507–2516 (2019)
21. Peng, X., Bai, Q., Xia, X., Huang, Z., Saenko, K., Wang, B.: Moment matching for multi-source domain adaptation. In: Proceedings of the IEEE/CVF International Conference on Computer Vision, pp. 1406–1415 (2019)
22. Richter, S.R., Vineet, V., Roth, S., Koltun, V.: Playing for data: ground truth from computer games. In: Proceedings of the European Conference on Computer Vision, pp. 102–118 (2016)
23. Ros, G., Sellart, L., Materzynska, J., Vazquez, D., Lopez, A.M.: The synthia dataset: a large collection of synthetic images for semantic segmentation of urban scenes. In: Proceedings of the IEEE Conference on Computer Vision and Pattern Recognition, pp. 3234–3243 (2016)
24. Stan, S., Rostami, M.: Unsupervised model adaptation for continual semantic segmentation. In: Proceedings of the AAAI Conference on Artificial Intelligence, vol. 35, pp. 2593–2601 (2021)
25. Tsai, Y.H., Hung, W.C., Schulter, S., Sohn, K., Yang, M.H., Chandraker, M.: Learning to adapt structured output space for semantic segmentation. In: Proceedings of the IEEE Conference on Computer Vision and Pattern Recognition, pp. 7472–7481 (2018)
26. Tsai, Y.H., Sohn, K., Schulter, S., Chandraker, M.: Domain adaptation for structured output via discriminative patch representations. In: Proceedings of the IEEE/CVF International Conference on Computer Vision, pp. 1456–1465 (2019)
27. Vu, T.H., Jain, H., Bucher, M., Cord, M., Pérez, P.: Advent: adversarial entropy minimization for domain adaptation in semantic segmentation. In: Proceedings of the IEEE/CVF Conference on Computer Vision and Pattern Recognition, pp. 2517–2526 (2019)
28. Wrenninge, M., Unger, J.: Synscapes: a photorealistic synthetic dataset for street scene parsing. arXiv preprint arXiv:1810.08705 (2018)
29. Wu, Z., et al.: DCAN: dual channel-wise alignment networks for unsupervised scene adaptation. In: Ferrari, V., Hebert, M., Sminchisescu, C., Weiss, Y. (eds.) ECCV 2018. LNCS, vol. 11209, pp. 535–552. Springer, Cham (2018). https://doi.org/10.1007/978-3-030-01228-1_32
30. Xia, H., Zhao, H., Ding, Z.: Adaptive adversarial network for source-free domain adaptation. In: Proceedings of the IEEE/CVF International Conference on Computer Vision, pp. 9010–9019 (2021)
31. Yang, S., Wang, Y., van de Weijer, J., Herranz, L., Jui, S.: Unsupervised domain adaptation without source data by casting a bait. arXiv preprint arXiv:2010.12427 (2020)

32. Yang, Y., Soatto, S.: FDA: fourier domain adaptation for semantic segmentation. In: Proceedings of the IEEE/CVF Conference on Computer Vision and Pattern Recognition, pp. 4085–4095 (2020)
33. Yao, X., Zhao, S., Xu, P., Yang, J.: Multi-source domain adaptation for object detection. In: Proceedings of the IEEE/CVF International Conference on Computer Vision, pp. 3273–3282 (2021)
34. Zhang, Y., Qiu, Z., Yao, T., Liu, D., Mei, T.: Fully convolutional adaptation networks for semantic segmentation. In: Proceedings of the IEEE Conference on Computer Vision and Pattern Recognition, pp. 6810–6818 (2018)
35. Zhao, H., Zhang, S., Wu, G., Moura, J.M., Costeira, J.P., Gordon, G.J.: Adversarial multiple source domain adaptation. Adv. Neural. Inf. Process. Syst. **31**, 8559–8570 (2018)
36. Zhao, S., et al.: Multi-source domain adaptation for semantic segmentation. Adv. Neural. Inf. Process. Syst. **32**, 7287–7300 (2019)
37. Zheng, Z., Yang, Y.: Rectifying pseudo label learning via uncertainty estimation for domain adaptive semantic segmentation. Int. J. Comput. Vision **129**(4), 1106–1120 (2021)
38. Zhu, J.Y., Park, T., Isola, P., Efros, A.A.: Unpaired image-to-image translation using cycle-consistent adversarial networks. In: Proceedings of the IEEE International Conference on Computer Vision, pp. 2223–2232 (2017)
39. Zou, Y., Yu, Z., Kumar, B., Wang, J.: Unsupervised domain adaptation for semantic segmentation via class-balanced self-training. In: Proceedings of the European Conference on Computer Vision, pp. 289–305 (2018)

Point MixSwap: Attentional Point Cloud Mixing via Swapping Matched Structural Divisions

Ardian Umam[1]([envelope]) [iD], Cheng-Kun Yang[2][iD], Yung-Yu Chuang[2][iD],
Jen-Hui Chuang[1][iD], and Yen-Yu Lin[1,3][iD]

[1] National Yang Ming Chiao Tung University, Hsinchu, Taiwan
`ardianumam.ee09@nycu.edu.tw`, `jchuang@cs.nctu.edu.tw`, `lin@cs.nycu.edu.tw`
[2] National Taiwan University, Taipei City, Taiwan
`d08922002@csie.ntu.edu.tw`, `cyy@csie.ntu.edu.tw`
[3] Academia Sinica, Taipei City, Taiwan

Abstract. Data augmentation is developed for increasing the amount and diversity of training data to enhance model learning. Compared to 2D images, point clouds, with the 3D geometric nature as well as the high collection and annotation costs, pose great challenges and potentials for augmentation. This paper presents a 3D augmentation method that explores the structural variance across multiple point clouds, and generates more diverse point clouds to enrich the training set. Specifically, we propose an attention module that decomposes a point cloud into several disjoint point subsets, called divisions, in a way where each division has a corresponding division in another point cloud. The augmented point clouds are synthesized by swapping matched divisions. They exhibit high diversity since both intra- and inter-cloud variations are explored, hence useful for downstream tasks. The proposed method for augmentation can act as a module and be integrated into a point-based network. The resultant framework is end-to-end trainable. The experiments show that it achieves state-of-the-art performance on the ModelNet40 and ModelNet10 benchmarks. The code for this work is publicly available (The source code is available at: https://github.com/ardianumam/PointMixSwap).

1 Introduction

Recent advance in deep neural networks (DNN) has been made for 3D point cloud analysis ranging from classification [15,16,27], segmentation [25,33] to detection [14,17]. However, the issue of data hungry in DNN becomes even

---

A. Umam and C.-K. Yang—The authors have equal contribution to this work.

Supplementary Information The online version contains supplementary material available at https://doi.org/10.1007/978-3-031-19818-2_34.

S. Avidan et al. (Eds.): ECCV 2022, LNCS 13689, pp. 596–611, 2022.
https://doi.org/10.1007/978-3-031-19818-2_34

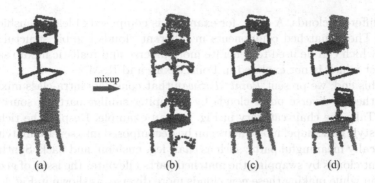

Fig. 1. Given (a) two source point clouds, the augmented samples are synthesized by (b) PointMixup [3], (c) RSMix [11], and (d) our method. The augmented point clouds by our method are diverse in the sense that the structural variance across different point clouds is utilized for synthesis, which is achieved by swapping the matched structural divisions. Colors show the identities of source points. Note that points generated by PointMixup do not have corresponding points in the source clouds, thereby drawn in another color. (Color figure online)

worse for point clouds due to the high collection and annotation costs [28,29]. Existing point cloud datasets are typically limited in both object quantity and category diversity. For example, ModelNet40 [26] (12K objects of 40 categories) and ScanObjectNN [21] (15K objects of 15 categories), two benchmarks for point cloud classification, are much smaller than image classification benchmarks, such as the ImageNet [10] dataset (1.4M images of 1K categories). Limited training data often make 3D point cloud networks suffer from overfitting and poor generalization to unseen data.

Data augmentation aims to increase the size and diversity of training data and can alleviate the unfavorable effects caused by the lack of annotated data. Compared to 2D images, 3D point clouds with rich geometric shapes and deformations offer great potentials for developing structural data augmentation techniques, which have been relatively underutilized. As an effective data augmentation technique, *mixup* [32] has made significant progress on 2D image augmentation. It targets at expanding the data distribution based on the assumption that a linearly interpolated data sample also leads to linearly interpolated label. However, the literature about point cloud mixup is rare. The permutation invariant property of 3D point clouds results in no point-to-point correspondences across clouds. It follows that linear interpolation commonly used in mixup method is not applicable.

To address this issue, PointMixup [3] computes the shortest paths to match points across clouds, and then applies linear interpolation to the coordinates of matched points. Meanwhile, RSMix [11] mixes two point cloud samples by replacing a specific part of one sample with a shape-preserved part from another sample. The synthesized point clouds by PointMixup suffer from the geometric shape distortion problem, while RSMix generates discontinuous and less realistic samples, especially in the areas with points from different clouds. We observe that point clouds of the same class are usually composed of matchable components across different clouds. A chair, for example, is composed of legs, a cushion, and a back. These matched components in different clouds exhibit structural variability, which can be used to generate more diverse and realistic mixup samples, an aspect which is not explored in PointMixup and RSMix.

To this end, we present *Point MixSwap* that considers intra-class mixup and can synthesize diverse point clouds by swapping similar parts of source point clouds. Take the chair category in Fig. 1 as an example. Despite the rich variations in style and shape, most chairs can be decomposed into several matched and semantically meaningful parts, such as chair leg, cushion, and back. Synthesizing new point clouds by swapping the matched parts alleviates the issue of geometric distortion while making these new clouds more diverse, as shown in Fig. 1.

Specifically, our goal is to divide a point cloud, a set of points, into a few disjoint and meaningful subsets, called *divisions* in this work, in a way where each division has the corresponding division in another point cloud of the same class. To this end, we introduce an encoder-decoder module. The encoder is applied to each cloud with its points as tokens, and captures both short- and long-range dependency. Inspired by [1], the decoder takes as input both *division queries* and the point-specific outputs of the encoder. Suppose the predefined number of divisions is R. There will be R division queries, one for each division. Via proper designs, the R division queries in the decoder can divide each point cloud into R divisions, with each covering similar points that are attended by the same division query. In addition, divisions which are from different point clouds but are associated with the same query are considered matched. In this way, not only intra-cloud division decomposition but also inter-cloud division variance are utilized for mixup.

This work makes the following contributions. First, we introduce an effective technique that explores structural variance for point clouds of the same class for synthesizing diverse point clouds by swapping matched divisions. Second, a novel encoder-decoder architecture is introduced to decompose a point cloud into semantically meaningful divisions with cross-cloud correspondences. Third, the synthesized point clouds lead to significant improvement for classification, reaching the state-of-the-art performance, and shape retrieval.

2   Related Work

Data Augmentation on 2D and 3D Data. Various methods have been proposed for data augmentation on 2D images, ranging from conventional approaches, such as random crop and color jittering [10,18,20] to advanced ones, such as AutoAugment [5,6] and generative adversarial networks (GAN) based methods [19,35,36]. In contrast, literature on 3D point cloud augmentation is relatively scarce [4,9,12]. Li *et al.* [12] propose the augmentor network to derive a rotation matrix and a point-wise translation to transform the point clouds in the batch. Choi *et al.* [4] come up with part-aware data augmentation for point cloud object detection. Given 3D object bounding boxes, they set the number of divisions and apply separate operations, such as random drop and random jittering. As their divisions are predefined, such an approach cannot ensure a consistent division meaning and its correspondence across point clouds within a class, which is a key factor in motivating our MixSwap.

Data Augmentation via Mixup. Existing methods [8,24,30–32] make significant progress on mixup for generating 2D images. For example, Kim *et al.* [8] consider saliency maps in the process of mixup, ensuring augmented data with sufficient information. Yun *et al.* [31] perform random cut in an image and replace the cut region with a patch from another image. Nonetheless, these methods are designed for 2D images and are inapplicable to data in geometric domains, including point clouds.

PointMixup [3] generalizes the idea of mixup to 3D data by seeking the optimal *interpolant* defined by the shortest path interpolation. Nonetheless, the interpolants, being locally generated virtual samples, suffer from the structural or shape distortion. Although this issue has been partly addressed in RSMix [11], where a subset of a point cloud is replaced by a subset of another cloud, the resultant augmented clouds preserve geometric structure within individual subsets, but with less realistic global appearance, as shown in Fig. 1(c), especially in the boundary of different subsets. In addition, none of these two methods have explored the structural variance within the point clouds of the same class. On the other hand, our method focuses on synthesizing point clouds by developing an encoder-decoder module, which carries out intra-cloud division decomposition and leverages inter-cloud division variance to enrich mixup samples.

Point Cloud Structure Division. Parsing point clouds into semantic parts reveals crucial information for point cloud analysis. Chen *et al.* [2] encode the shape structure intrinsically for 3D points in an unsupervised manner. Zhu *et al.* [34] develop an adaptive learning module for shape co-segmentation using the group consistency loss and an additional shape part dataset. However, these methods usually derive one model only for each category. In contrast, our method can decompose a pair of point clouds of the same class into geometrically consistent and matched divisions. Furthermore, our method does not require part-wise annotations and, more importantly, is applicable to point clouds of an arbitrary class by using a single model. In light of the differences between shape co-segmentation and our technique, our aim is to create consistent divisions within samples in order to improve augmentation. As such, perfect decomposition is not a requirement for the proposed method.

3 Proposed Method

3.1 Overview

We are given a training set of point clouds of C categories, $\mathcal{D} = \{(P_n, \mathbf{y}_n)\}$. Without loss of generality, we assume that the number of points in each cloud is M, *i.e.*, $P_n = \{\mathbf{p}_n^m\}_{m=1}^M$, where point $\mathbf{p}_n^m \in \mathbb{R}^3$ is represented by its 3D coordinate

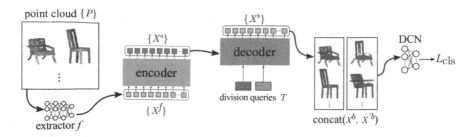

Fig. 2. Overview of the proposed network architecture. Point MixSwap leverages the attention mechanism to identify structural divisions for point clouds of the same category via an encoder-decoder module. This module is integrated into a point cloud framework. It receives the per-point features compiled by a feature extraction backbone. It generates augmented point clouds, which serve as the input to the downstream classifier network (DCN). The whole network can be end-to-end trained via the objective function of DCN.

and $\mathbf{y}_n \in \{0,1\}^C$ is a C-dimensional binary vector indicating the category of P_n. The downstream task in this work is to train a function that is capable of mapping a point cloud to its class label, *i.e.*, point cloud classification.

This work proposes a data augmentation that utilizes structural variance posed within point clouds of the same class to synthesize training data by decomposing each point cloud into structural divisions and enriching the source dataset via cross-cloud combinations of these structural divisions. The proposed method is depicted in Fig. 2, which considers *intra-class mixup*. In mini-batch optimization, each batch consists of N point clouds of an arbitrary class, *i.e.*, $\{P_n\}_{n=1}^N$. After applying a feature extractor f, the features $\{X_n^f\}_{n=1}^N$ are obtained, where $X_n^f = \{\mathbf{x}_n^{m,f}\}_{m=1}^M$, with $\mathbf{x}_n^{m,f} \in \mathbb{R}^E$ represents the per-point feature vector of embedding size E.

An encoder-decoder architecture is introduced to discover the divisions that are geometrical parts shared across point clouds. The per-point features of point cloud P, *i.e.*, X^f, are fed into the encoder to produce its self-attention features X^a. Let R denote the predefined number of divisions that R *division queries*, $T \in \mathbb{R}^{R \times E}$, are created. The decoder takes as input both division queries T and the output of the encoder for the point cloud X^a, before the division-point attention features X^b is generated for the downstream classification task. In the process, the R division queries jointly decompose point cloud P_n into R disjoint subsets. After division swapping, the augmented point clouds are generated to facilitate the downstream classifier network.

Figure 2 shows an example with $R = 2$, where coloring is used to illustrate the mapping between divisions and points. The details about the encoder-decoder module and division swapping are provided in the following.

3.2 Encoder-Decoder Architecture

The encoder-decoder architecture, *i.e.*, transformer [23], can offer an effective way to capture the correlation across samples. The encoder is composed of several self-attention layers to capture long-range dependency and improve the features given the extractor f. As for the decoder, we are inspired by DETR [1], where the learned positional embedding can be utilized as anchor boxes for object detection, and extend the idea to 3D point clouds to capture the similar geometrical parts shared across point clouds. Specifically, we create R *division queries* initialized using the Xavier method [7], where R is the number of divisions for decomposition. The decoder takes the self-attention features X^a from encoder and division queries as input, and produces the division-point attention features X^b. After optimization, these R division queries represent the R divisions shared across point clouds.

In the original decoder layer [23], the number of output feature vectors is the same as the number of the queries. Since we pass the division queries $T \in \mathbb{R}^{R \times E}$ as tokens into the decoder, the generated feature vectors are query-specific, instead of point-specific, which is ineffectual for classification. Hence, we introduce a designed decoder, which is composed of two coupled cross-attention layers, as illustrated in Fig. 3. In the first layer, the division queries T serve as the queries while the point-specific features X^a act as the key-value pairs. Their roles are switched in the second layer and jointly produce point-specific features X^b, where $X^b \in \mathbb{R}^{M \times E}$.

Following by the common practice in [23], the intermediate attention features X^o in Fig. 3 are computed as follows

$$Attention(Q, K, V) = \text{softmax}(\frac{QK^T}{\sqrt{E}})V = SV = X^o, \tag{1}$$

where $Q = TQ_w$, $K = X^a K_w$, and $V = X^a V_w$, while Q_w, K_w, and V_w are three matrices for linear projection. The softmax operation is applied along the last dimension. The generated features X^o softly attend to all points and are then passed into the second attention layer.

In Eq. 1, the point-division attention matrix $S \in \mathbb{R}^{R \times M}$, which we call it the *division map*, is obtained and will be utilized for point cloud decomposition. In contrast to the first cross-attention layer, we use self-attention features X^a as queries and X^o from the first layer as key-value pairs. Through the similar cross-attention operation used in the first layer, we obtain the features X^b, which encode the correlation between the division queries T and the per-point features X^a, with residual learning adopted in the two coupled cross-attention layers.

The division-point attention features X^b can be fed into the downstream classifier network (DCN) for training. The whole network is end-to-end trainable in accordance with the task of classification. In this way, these division queries and the division map S are learned to minimize the cross-entropy loss,

$$L_{cls} = -\sum_{n=1}^{N}\sum_{c=1}^{C} \mathbf{y}_{n,c}\log(\hat{\mathbf{y}}_{n,c}), \tag{2}$$

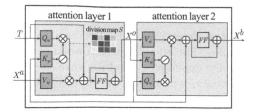
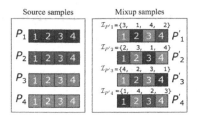

Fig. 3. Architecture of the coupled decoder layers in Point MixSwap. Symbols $\otimes$, $\oplus$, and $\oslash$ denote matrix multiplication, element-wise sum, and matrix transposition, respectively. The pink boxes (FF) represent multi-layer perceptron.

Fig. 4. Division mixswap is depicted to synthesize R new mixup point clouds with complete, non-repeating divisions from R source point clouds. An example of $R = 4$ is given in the figure.

where $\mathbf{y}_n$ is one-hot encoded label vector of point cloud P_n, $\hat{\mathbf{y}}_n$ is predicted probability distribution, and C is the number of classes in the training set.

The structural division can be inferred from division map S in Eq. 1. Specifically, the m-th point is assigned to division $d(m)$ if

$$d(m) = \underset{r}{\mathrm{argmax}}(S(r, m)). \tag{3}$$

In this way, the division map S can be considered as the division segmentation map and is used to retrieve the R structural divisions for each point cloud. Furthermore, since the division queries are shared for all samples, every r-th division query attends similar subsets of points across point clouds. As a result, a division in a point cloud has its corresponding division in each of other point clouds.

3.3   Division Mixswap

New mixup point clouds are synthesized by swapping matched divisions, and each of them contains non-repeating divisions. Specifically, to synthesize R new point clouds $\{P'_r\}_{r=1}^R$, we randomly pick R source point clouds in a batch, $\{P_r\}_{r=1}^R$. Figure 4 illustrates how our mixswap synthesizes R new point clouds from R source point clouds with $R = 4$. The first mixing index array $\mathcal{I}_{P'_1}$ is defined as a random permutation vector of integer numbers ranging from 1 to R, and the following mixing index arrays are specified as one time cyclic rotation of their previous one. Using these mixing index arrays, new mixup point clouds are then synthesized, where the r-th element in the mixing index array gives the source of the r-th division. Since the divison S may decompose point clouds into division of diverse sizes, therefore, we further sample each of the synthesized point clouds into a fixed number of points.

$$P' = \Gamma^M \left(\mathrm{concat}(\{\mathbb{S}_r \odot P_{\mathcal{I}(r)}\}_{r=1}^R) \right). \tag{4}$$

where $\mathbb{S}_r$ denotes a binary mask used to select all points of the r-th division, Γ^M denotes a sampling operation which returns M points, and $\odot$ represents element-wise multiplication. Mixup point clouds acquired via Eq. 4 keep the orderless property, thus consistent with point cloud data.

Mixup operation can be carried out in both the input level, *i.e.*, performed among point cloud samples P_n, and in the feature level, *i.e.*, performed among point features X^b. The former requires aligned training sets whose point clouds have the same pose in order to achieve its optimal improvement gain. Meanwhile, considering point features are computed from the original point clouds, the latter is more robust to unaligned training sets, which can be performed by replacing P with X^b in Eq. 4. In the following, we discuss the cases where our method is applied to point clouds with unaligned poses.

3.4 Alignment Mechanism

As shown in Fig. 2, our method adopts an existing point cloud feature extractor. Most extractors such as [15, 25] are designed to work with unaligned point clouds and can implicitly address pose variations with some specified mechanism, such as T-Net in PointNet [15]. As a result, the resultant features are somewhat robust to variations. To further improve the performance on unaligned cases, we present a mechanism, called principal axis alignment (PAA), to pre-process the given point clouds. We compute the largest principal axes of each point cloud. In a batch of point clouds, one is randomly chosen as the reference, while the rest are aligned to the reference according to the principal axes.

3.5 Implementation Details

We train the network with 500 epochs, where the first warm-up 20 epochs are run without executing Point MixSwap, to stabilize the learning of division queries. For DGCNN [25], the SGD solver is adopted with a momentum of 0.9 and a learning rate of 0.001 scheduled using the cosine annealing strategy [13]. For PointNet [15], the Adam optimizer is employed with an initial learning rate of 0.001 and is gradually reduced with a decay rate of 0.5 every 20 epochs. Unless further specified, we set the number of divisions to three, $R = 3$, and use feature-level augmentation in the experiments.

Limitations. The proposed Point MixSwap works for point clouds of the same category. It is not applicable to point clouds of different categories.

4 Experimental Results

4.1 Datasets

We evaluate the proposed Point MixSwap on the ModelNet40 (**M40**) [26], ModelNet10 (**M10**) [26], and ScanObjectNN (**SON**) [22] datasets, which are widely

Table 1. Accuracy scores of the proposed Point MixSwap on 20%, 50%, and 100% of the ModelNet40 (M40) and ModelNet10 (M10) datasets.

Method	Rate 20%		Rate 50%		Rate 100%	
	M40	M10	M40	M10	M40	M10
PointNet	82.1	89.4	85.9	92.7	88.6	93.2
PointNet + Ours	**86.3** (4.2↑)	**91.3** (1.9↑)	**88.7** (2.8↑)	**93.6** (0.9↑)	**90.2** (1.6↑)	**93.9** (0.7↑)
DGCNN	87.5	93.2	91.5	94.3	92.7	94.8
DGCNN + Ours	**91.3** (3.8↑)	**94.6** (1.4↑)	**92.8** (1.3↑)	**94.9** (0.6↑)	**93.5** (0.8↑)	**96.0** (1.2↑)

Table 2. Accuracy scores of the proposed Point MixSwap on 20%, 50%, and 100% of the rotated ModelNet40 (RM40) and ScanObjectNN (SON) datasets.

Method	Rate 20%		Rate 50%		Rate 100%	
	RM40	SON	RM40	SON	RM40	SON
PointNet	82.0	62.5	85.5	71.3	88.5	76.2
PointNet + Ours	85.2 (3.2↑)	66.1 (3.6↑)	87.7 (2.2↑)	74.0 (2.7↑)	89.5 (1.0↑)	78.8 (2.6↑)
PointNet + Ours + PAA	**86.2** (4.2↑)	**67.0** (4.5↑)	**87.9** (2.4↑)	**74.3** (3.0↑)	**89.7** (1.2↑)	**78.9** (2.7↑)
DGCNN	87.0	73.7	90.3	81.6	91.5	86.2
DGCNN + Ours	89.3 (2.3↑)	76.3 (2.6↑)	91.1 (0.8↑)	84.1 (2.5↑)	**92.3** (0.8↑)	88.6 (2.4↑)
DGCNN + Ours + PAA	**90.1** (3.1↑)	**76.8** (3.1↑)	**91.3** (1.0↑)	**84.8** (3.2↑)	**92.3** (0.8↑)	**89.0** (2.8↑)

used for point cloud recognition. The OBJ_ONLY version is adopted for SON. M40 and M10 are synthetic datasets, while SON is a real-world dataset. Following previous works [15,16,25], we uniformly sample 1,024 points on the mesh faces according to the face areas and then normalize them into a unit sphere. We discard the normals of these samples and only use their 3D point coordinates.

We evaluate the proposed method on the reduced datasets, to investigate the effectiveness of our method when less training data are available. The dataset size is reduced to 20% and 50% with stratified sampling.

4.2   Shape Classification

To evaluate our method, we consider PointNet [15] and DGCNN [25] as the backbones for feature extraction, and report the performance of the models trained with (ours) and without (baseline) the proposed Point MixSwap. We first evaluate the proposed method on M40 and M10, where most of the samples are well aligned. As demonstrated in Table 1, Point MixSwap consistently boosts the accuracy regardless of the backbone networks and training data sizes. With only 50% of the training set, it achieves slightly better performance than the baseline model trained on the full dataset, in all backbones and datasets. While at the reduction rate of 20%, the accuracy is also comparable to the baseline with 50% of training data. The results reveal the effectiveness of Point MixSwap to work with different point cloud classification network architectures. More experiments on different backbones can be found in the supplementary material.

before alignment   after alignment   2 divs.   3 divs.   4 divs.   5 divs.

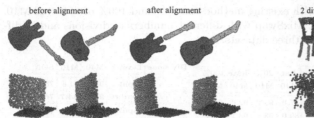 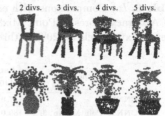

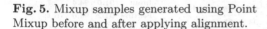

Fig. 5. Mixup samples generated using Point Mixup before and after applying alignment.

Fig. 6. Mixswap samples with different numbers of divisions.

To further demonstrate the generality of the proposed method to unaligned and real-world datasets, we evaluate the proposed method on the rotated Model-Net40 (RM40), where random rotation is applied to each point cloud of the training and testing sets, and the unaligned dataset, SON. The proposed method is evaluated with and without using the proposed principal axis alignment (PAA). Table 2 summarizes the results. In all settings, the proposed method without alignment produces notable improvement compared to the baseline although the source samples are unaligned. This is because the adopted backbones are developed to address pose variations. Also, each derived division query attends to point tokens described by per-point local features, and tolerates a certain degree of unalignment. With the alignment mechanism PAA, the proposed method yields further and consistent improvement in all settings. Visualization of some mixup samples before and after applying alignment is given in Fig. 5.

4.3 Competing Methods and Comparisons

We compare the proposed method with the state-of-the-art point cloud augmentation methods on the reduced and full training datasets. The competing methods include PointMixup [3], PointAugment [12], RSMix [11] and Point-WOLF [9]. For the accuracy scores already reported in the original papers, we take the numbers directly from the papers. For those that are not given in the papers, particularly for those by PointAugment and RSMix on the reduced training sets, we run their official released codes for obtaining the accuracy scores. In addition, we note that PointAugment's performance is unstable on ModelNet10; Thus, we run the official codes several times and report the average accuracy scores instead of the ones from the paper.

Table 3(a) reports the accuracy scores of all compared methods on both M40 and M10, with 20% of the dataset and the full dataset. The proposed Point MixSwap outperforms the state-of-the-art methods in all settings. In addition, our method shows a good performance gain when the training data is insufficient, 20% of the dataset in this case. It shows that the proposed method is effective for data augmentation.

Table 3. (a) Comparisons with existing methods on 20% and 100% of M40 and M10. (b) Accuracy scores of Point MixSwap with different numbers of divisions and in different mixup levels on 20% of three datasets.

Method	Rate 20%		Rate 100%	
	M40	M10	M40	M10
DGCNN	87.5	93.2	92.6	94.8
DGCNN + PointMixup [3]	89.0	93.8	93.1	95.1
DGCNN + PointAugment [12]	88.6	92.8	93.4	95.2
DGCNN + RSMix [11]	90.1	93.7	**93.5**	95.9
DGCNN + PointWOLF [9]	89.3	93.5	93.2	95.1
DGCNN + Ours	**91.3**	**94.6**	**93.5**	**96.0**

(a)

Divisions	Level	M40	M10	SON
2	Input	91.0	94.6	75.9
	Feature	91.1	**94.7**	76.2
3	Input	91.2	94.5	76.1
	Feature	**91.3**	94.6	**76.3**
4	Input	91.0	94.4	75.7
	Feature	91.2	94.6	76.1
5	Input	91.0	94.3	75.5
	Feature	91.2	94.6	76.0

(b)

Table 4. Accuracy scores by using the baselines, different variants of our method, and three trivial division methods.

CDA	Point MixSwap			Trivial division			Accuracy		
	Enc-dec	Input-level	Feature-level	Hor.	Ver.	Random	M40	M10	SON
							87.5	93.2	73.0
✓							88.7 (1.2 ↑)	93.5 (0.3)	73.7 (0.7)
✓	✓						88.8 (1.3 ↑)	93.6 (0.4 ↑)	73.6 (0.6 ↑)
	✓	✓					89.5 (2.0 ↑)	94.0 (0.8 ↑)	74.7 (1.7 ↑)
	✓		✓				89.7 (2.2 ↑)	94.2 (1.0 ↑)	75.0 (2.0 ↑)
	✓	✓	✓				89.5 (2.0 ↑)	94.1 (0.9 ↑)	74.9 (1.9 ↑)
✓	✓	✓					91.1 (3.6 ↑)	94.5 (1.3 ↑)	76.1 (3.1 ↑)
✓	✓		✓				**91.3 (3.8 ↑)**	**94.6 (1.4 ↑)**	**76.3 (3.3 ↑)**
✓	✓	✓	✓				91.2 (3.7 ↑)	94.5 (1.3 ↑)	76.1 (3.1 ↑)
✓				✓			89.2 (1.7 ↑)	93.7 (0.5 ↑)	73.9 (0.9 ↑)
✓					✓		89.0 (1.5 ↑)	93.6 (0.4 ↑)	73.8 (0.8 ↑)
✓						✓	88.9 (1.4 ↑)	93.4 (0.2 ↑)	73.8 (0.8 ↑)

4.4 Ablation Study and Analysis

We perform ablation studies to evaluate the impacts of the proposed components and present performance analysis. Here, the experiments are conducted on 20% of the training sets.

Contributions of Components. To evaluate the effectiveness of the proposed method, we first report the performance of the baseline by training without using any data augmentations. Here, DGCNN is adopted as the baseline. Then we evaluate the contribution of the conventional data augmentation (CDA) and the proposed Point MixSwap. The adopted CDA comprises random scaling, random translation and random drop, following [11]. Moreover, we perform the mixup operation at the input level, the feature level, or both, to see the performance with different component combinations. In addition, to check if the accuracy improvement comes from Point MixSwap rather than trivial data decomposition and reconstruction, three simple division approaches are investigated. In the first two approaches, we uniformly divide a point cloud horizontally and vertically, respectively. The third approach uses random division.

Table 4 reports the results of the ablation studies. First, we compare the performance of data augmentation by using CDA, input-level and feature-level Point MixSwap. Both input-level and feature-level Point MixSwap achieve notably higher accuracy than CDA. To further investigate the source of the performance gain, we combine Point MixSwap with CDA, but neither input-level nor feature-level mixup is enabled. In this case, the difference from the CDA-only configuration lies in the attention mechanism enabled by the encoder-decoder blocks (Enc-dec) to process the features. Table 4 shows that, on M40, the Point MixSwap+CDA configuration (88.8) yields slightly better performance than the CDA configuration (88.7). Similar trend is also encountered on the M10 and SON datasets. It indicates that the major source of performance gain is not the attention mechanism, but the effective divisions derived by the proposed encoder-decoder block for mixup augmentation.

Second, we investigate the impacts of performing mixup at the input level and the feature level. In Table 4, the feature-level mixup achieves relatively higher accuracy than the input-level mixup, i.e., 89.7 versus 89.5 on M40, 94.2 versus 94.0 on M10, and 75.0 versus 74.7 on SON. Combining both of them yields slightly lower accuracy than using the feature-level mixup alone. Third, we combine Point MixSwap with CDA. The result demonstrates that Point MixSwap can be complementary to other augmentation methods, and can work together with other types of data augmentation for further performance enhancement. Specifically, the combination of feature-level Point MixSwap and CDA achieves significant performance gains compared to the baseline.

Finally, we consider the performance by using the three trivial division approaches. In Table 4, notably inferior improvements are obtained by using the three division approaches compared to the proposed Point MixSwap. The results indicate that accuracy improvements are not due to trivial data decomposition and reconstruction for augmentation, but rather to the effective divisions derived by using the proposed method.

Analysis on the Number of Divisions. We analyze the impact of division numbers on Point MixSwap. Table 3(b) reports the results by setting the division number to 2, 3, 4, and 5, respectively. For each number, we measure the accuracy with mixup at the input level and feature level. It can be observed that feature-level mixup yields better performance for all division numbers. This is reasonably well grounded because point features are computed in the original samples, i.e., before mixup is performed. Meanwhile, for the input-level mixup, the performance could degrade with a higher division number. Figure 6 visualizes some mixswap results with different numbers of divisions.

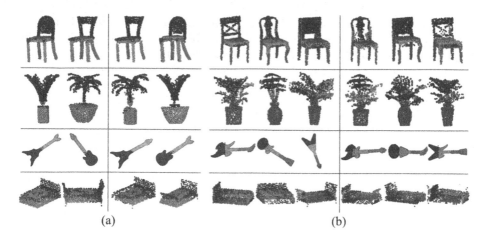

(a)   (b)

Fig. 7. Mixup examples generated by Point MixSwap. (a) Four examples with two divisions. For a pair of source point clouds on the left, we show the generated mixup samples. (b) Four examples with three divisions. For an input triplet on the left, the three generated mixup samples are displayed. Points are colored according to their divisions. (Color figure online)

4.5   Qualitative Results of Mixup Samples

Figure 7 shows the synthesized examples via the proposed Point MixSwap. We set the number of division queries to 2 and 3, to generate new samples shown in Fig. 7(a) and (b), respectively. For each setting, the second column depicts the generated mixswap samples from the source sample pair/triplet given in the first column. Note that for the guitar and bed categories, different poses of source samples are provided, and the generated mixup samples after applying alignment mechanism are shown, in which the first source sample is set as the reference. The generated mixup samples before alignment mechanism is applied, can be found in the supplementary material.

In Fig. 7(a), Point MixSwap successfully identifies chair leg and back as the two major structural divisions in the chair category, and poses a consistent correspondence across samples, where points of the same division are colored with the same color. For the plant, guitar and bed categories, the two major structural divisions are also consistently identified across samples. Hence, the generated mixup samples accomplished via Point MixSwap show not only diverse geometric shapes but also structure-preserved characteristics.

When the number of divisions is set to 3, the chair is segmented by its leg, cushion and back, as the three structural divisions. As for the plant, guitar and bed categories, they have meaningful divisions as shown in the figure. A higher division number enables the mixup process to possibly generate more diverse samples in the sense that each new sample can be synthesized with more structural divisions from more different samples in the input set. We further discuss the case where the given division number exceeds the number of

Table 5. Shape retrieval performance in mAP (%) of different data augmentation methods on the M40 dataset.

Backbone	CDA	PointAugment	Ours
PointNet	70.5	75.8 (5.3 ↑)	**78.4** (7.9 ↑)
DGCNN	85.3	89.0 (3.7 ↑)	**90.6** (5.3 ↑)

structural divisions posed by certain categories. Take the plant category as an example. According to the geometrical structure, each sample naturally poses two structural divisions, the pot and plant. Given three division queries, two of these tokens attend to similar structural divisions, the plant part in this case, as depicted in Fig. 7(b) with green and brown colors. Nonetheless, our method can still generate diverse and structure-preserved mixup samples by utilizing the division cross-correspondence. More visualization examples of other categories can be found in the supplementary material.

4.6 Shape Retrieval

To demonstrate the advantage of the proposed method to another downstream task, following PointAugment [12], we also examine the proposed method for shape retrieval which retrieves the most similar shape based on cosine similarity of the global features. We regard every sample in the testing set as a query shape, and the retrieval performance in mean average precision (mAP) is reported on the M40 dataset, as shown in Table 5. The proposed method produces significant improvement margin compared to CDA in both PointNet and DGCNN, while a notable margin is also observed compared to PointAugment.

5 Conclusion

This paper proposes Point MixSwap, a novel data augmentation technique for 3D point clouds. It is developed to exploit structural variations among point clouds of the same class to synthesize diverse and structure-preserved augmented samples. Point MixSwap introduces an intuitive idea of data augmentation by decomposing a point cloud into several disjoint divisions. Each division has a consistently corresponding division in other point clouds. Thus, augmented mixup data can be synthesized by swapping one or more matched divisions among the source point clouds. As a mixup augmentation technique, Point MixSwap is guided by an attention mechanism, and to the best of our knowledge, it is the first augmentation technique that utilizes an attention mechanism to explore matchable divisions across source data. Point MixSwap is end-to-end trainable and can be employed by any point-based networks. Comprehensive experiments demonstrate the effectiveness of Point MixSwap on boosting the model accuracy, especially when only limited data are available.

Acknowledgments. This work was supported in part by the Ministry of Science and Technology (MOST) under grants 109–2221-E-009–113-MY3, 111–2628-E-A49-025-MY3, 111–2634-F-007–002, 110–2634-F-002–050, 110–2634-F-002–051, 110–2634-F-006–022 and 110–2634-F-A49-006. This work was funded in part by Qualcomm through a Taiwan University Research Collaboration Project and by MediaTek. We thank the National Center for High-performance Computing (NCHC) of National Applied Research Laboratories (NARLabs) in Taiwan for providing computational and storage resources.

References

1. Carion, N., Massa, F., Synnaeve, G., Usunier, N., Kirillov, A., Zagoruyko, S.: End-to-end object detection with transformers. In: Vedaldi, A., Bischof, H., Brox, T., Frahm, J.-M. (eds.) ECCV 2020. LNCS, vol. 12346, pp. 213–229. Springer, Cham (2020). https://doi.org/10.1007/978-3-030-58452-8_13
2. Chen, N., et al.: Unsupervised learning of intrinsic structural representation points. In: CVPR (2020)
3. Chen, Y., et al.: PointMixup: augmentation for point clouds. In: Vedaldi, A., Bischof, H., Brox, T., Frahm, J.-M. (eds.) ECCV 2020. LNCS, vol. 12348, pp. 330–345. Springer, Cham (2020). https://doi.org/10.1007/978-3-030-58580-8_20
4. Choi, J., Song, Y., Kwak, N.: Part-aware data augmentation for 3D object detection in point cloud (2021)
5. Cubuk, E.D., Zoph, B., Mane, D., Vasudevan, V., Le, Q.V.: Autoaugment: learning augmentation strategies from data. In: CVPR (2019)
6. Cubuk, E.D., Zoph, B., Shlens, J., Le, Q.: Randaugment: practical automated data augmentation with a reduced search space. In: NIPS
7. Glorot, X., Bengio, Y.: Understanding the difficulty of training deep feedforward neural networks. In: Proceedings of the thirteenth international conference on artificial intelligence and statistics, pp. 249–256. JMLR Workshop and Conference Proceedings (2010)
8. Kim, J.H., Choo, W., Song, H.O.: Puzzle mix: exploiting saliency and local statistics for optimal mixup. In: ICLR (2020)
9. Kim, S., Lee, S., Hwang, D., Lee, J., Hwang, S.J., Kim, H.J.: Point cloud augmentation with weighted local transformations. In: ICCV (2021)
10. Krizhevsky, A., Sutskever, I., Hinton, G.E.: ImageNet classification with deep convolutional neural networks
11. Lee, D., et al.: Regularization strategy for point cloud via rigidly mixed sample. In: CVPR (2021)
12. Li, R., Li, X., Heng, P.A., Fu, C.W.: Pointaugment: an auto-augmentation framework for point cloud classification. In: CVPR (2020)
13. Loshchilov, I., Hutter, F.: SGDR: stochastic gradient descent with warm restarts. arXiv preprint arXiv:1608.03983 (2016)
14. Qi, C.R., Litany, O., He, K., Guibas, L.J.: Deep hough voting for 3D object detection in point clouds. In: ICCV (2019)
15. Qi, C.R., Su, H., Mo, K., Guibas, L.J.: Pointnet: deep learning on point sets for 3D classification and segmentation. In: CVPR (2017)
16. Qi, C.R., Yi, L., Su, H., Guibas, L.J.: Pointnet++: deep hierarchical feature learning on point sets in a metric space. NIPS (2017)
17. Shi, S., et al.: PV-RCNN: point-voxel feature set abstraction for 3D object detection. In: CVPR (2020)

18. Simonyan, K., Zisserman, A.: Very deep convolutional networks for large-scale image recognition. arXiv preprint arXiv:1409.1556 (2014)
19. Sixt, L., Wild, B., Landgraf, T.: RenderGAN: generating realistic labeled data. Frontiers in Robotics and AI (2018)
20. Szegedy, C., et al.: Going deeper with convolutions. In: ICCV (2015)
21. Uy, M.A., Pham, Q.H., Hua, B.S., Nguyen, D.T., Yeung, S.K.: Revisiting point cloud classification: a new benchmark dataset and classification model on real-world data. In: ICCV (2019)
22. Uy, M.A., Pham, Q.H., Hua, B.S., Nguyen, T., Yeung, S.K.: Revisiting point cloud classification: A new benchmark dataset and classification model on real-world data. In: Proceedings of the IEEE/CVF international conference on computer vision, pp. 1588–1597 (2019)
23. Vaswani, A., et al.: Attention is all you need. In: NIPS (2017)
24. Verma, V., et al.: Manifold Mixup: better representations by interpolating hidden states. In: ICML (2019)
25. Wang, Y., Sun, Y., Liu, Z., Sarma, S.E., Bronstein, M.M., Solomon, J.M.: Dynamic graph CNN for learning on point clouds. TOG (2019)
26. Wu, Z., et al.: 3D shapeNets: a deep representation for volumetric shapes. In: CVPR (2015)
27. Xiang, T., Zhang, C., Song, Y., Yu, J., Cai, W.: Walk in the cloud: learning curves for point clouds shape analysis. In: ICCV (2021)
28. Yang, C.K., Chuang, Y.Y., Lin, Y.Y.: Unsupervised point cloud object co-segmentation by co-contrastive learning and mutual attention sampling. In: ICCV (2021)
29. Yang, C.K., Wu, J.J., Chen, K.S., Chuang, Y.Y., Lin, Y.Y.: An mil-derived transformer for weakly supervised point cloud segmentation. In: CVPR (2022)
30. Yoo, J., Ahn, N., Sohn, K.A.: Rethinking data augmentation for image super-resolution: a comprehensive analysis and a new strategy. In: CVPR (2020)
31. Yun, S., Han, D., Oh, S.J., Chun, S., Choe, J., Yoo, Y.: CutMix: regularization strategy to train strong classifiers with localizable features. In: ICCV (2019)
32. Zhang, H., Cisse, M., Dauphin, Y.N., Lopez-Paz, D.: Mixup: beyond empirical risk minimization. ICLR (2018)
33. Zhao, H., Jiang, L., Jia, J., Torr, P.H., Koltun, V.: Point transformer. In: ICCV (2021)
34. Zhu, C., Xu, K., Chaudhuri, S., Yi, L., Guibas, L.J., Zhang, H.: AdaCoSeg: adaptive shape co-segmentation with group consistency loss. In: CVPR (2020)
35. Zhu, X., Liu, Y., Li, J., Wan, T., Qin, Z.: Emotion classification with data augmentation using generative adversarial networks. In: KDD (2018)
36. Zhu, Y., Aoun, M., Krijn, M., Vanschoren, J., Campus, H.T.: Data augmentation using conditional generative adversarial networks for leaf counting in arabidopsis plants. In: BMVC (2018)

BATMAN: Bilateral Attention Transformer in Motion-Appearance Neighboring Space for Video Object Segmentation

Ye Yu[1]([✉])[ID], Jialin Yuan[2][ID], Gaurav Mittal[1][ID], Li Fuxin[2][ID], and Mei Chen[1][ID]

[1] Microsoft, Bellevue, USA
{yu.ye,gaurav.mittal,mei.chen}@microsoft.com
[2] Oregon State University, Corvallis, USA
{yuanjial,lif}@oregonstate.edu

Abstract. Video Object Segmentation (VOS) is fundamental to video understanding. Transformer-based methods show significant performance improvement on semi-supervised VOS. However, existing work faces challenges segmenting visually similar objects in close proximity of each other. In this paper, we propose a novel Bilateral Attention Transformer in Motion-Appearance Neighboring space (BATMAN) for semi-supervised VOS. It captures object motion in the video via a novel optical flow calibration module that fuses the segmentation mask with optical flow estimation to improve within-object optical flow smoothness and reduce noise at object boundaries. This calibrated optical flow is then employed in our novel bilateral attention, which computes the correspondence between the query and reference frames in the neighboring bilateral space considering both motion and appearance. Extensive experiments validate the effectiveness of BATMAN architecture by outperforming all existing state-of-the-art on all four popular VOS benchmarks: Youtube-VOS 2019 (85.0%), Youtube-VOS 2018 (85.3%), DAVIS 2017Val/Test-dev (86.2%/82.2%), and DAVIS 2016 (92.5%).

Keywords: Bilateral attention · Motion-appearance space · Optical flow calibration · Video object segmentation · Vision transformer

At Oregon State University, Jialin Yuan and Li Fuxin are supported in part by NSF grant 1911232.

Supplementary Information The online version contains supplementary material available at https://doi.org/10.1007/978-3-031-19818-2_35.

1   Introduction

Video Object Segmentation (VOS) is fundamental to video understanding with broad applications in content creation, content moderation, and autonomous driving. In this paper, we focus on the semi-supervised VOS task, where we segment target objects in each frame of the entire video sequence (query frames) given their segmentation masks in the first frame (reference frame) only. Moreover, the task is class-agnostic in that we do not have any class annotation for any object to be segmented in either training or testing phases. The key challenge in semi-supervised VOS is how to propagate the mask from the reference frame to all the query frames in the rest of the sequence without any class annotation.

Due to the absence of class-specific features, VOS models need to match features of the reference frame to that of the query frames both spatially and temporally to capture the class-agnostic correspondence and propagate the segmentation masks. Previous methods attempt to store features from preceding frames in memory networks and match the query frame through a non-local attention mechanism [7,27], or compute a global-to-global attention through an encoder-decoder transformer [25], or propagate and calibrate features from the reference frame to the query frames using a propagation-correction scheme [47]. These methods employ a global attention mechanism to establish correspondence between the full reference frame and the full query frame. This can lead to failure in distinguishing the target object(s) from the background particularly when there are multiple objects with a similar visual appearance. A spatial local attention is proposed in [50] to mitigate this problem, where the attention is only computed between each query token and its surrounding key tokens within a spatial local window. However, it still suffers from incorrectly segmenting visually similar objects in close proximity of each other.

In addition to spatial correspondence, it is essential to match features temporally for optimal object segmentation across video frames. To this end, some VOS methods [8,45] leverage optical flow to capture object motion. [45] warps the memory frame mask using optical flow before performing local matching between memory and query features based on the warped mask, while [8] simultaneously trains the model for object segmentation and optical flow estimation by bidirectionally fusing feature maps from the two branches. However, these methods are not able to perform optimally as optical flow is usually noisy and warping features/masks to match objects across frames accumulates errors in both optical flow and segmentation mask along the video sequence.

To overcome the above challenges, we propose Bilateral Attention Transformer in Motion-Appearance Neighboring space (BATMAN). BATMAN introduces a novel bilateral attention module that computes the local attention map between the query frame and memory frames with both motion and appearance in consideration. Unlike the conventional spatial local attention mechanism (Fig. 1(a)) that computes the attention within a predefined fixed local window, our bilateral attention adaptively computes the local attention based on the tokens' spatial distance, appearance similarity, and optical flow smoothness, as shown in Fig. 1(b). Observing that optical flow may be especially noisy for fast-

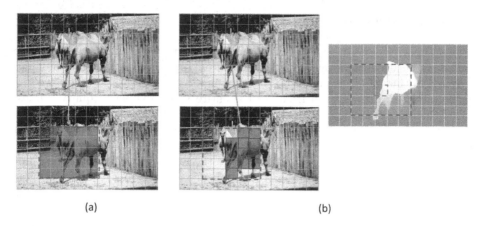

(a)   (b)

Fig. 1. Spatial local attention vs. bilateral attention. (a) Conventional spatial local attention. For any given token in the query frame (top), compute the attention with the neighboring tokens within a predefined fixed local window from the memory frame (bottom). (b) Our proposed bilateral attention. Given a token in the query frame (top), adaptively select the most relevant tokens (bottom), based on the distance in the bilateral space of appearance and motion (right), for cross attention computation

moving object(s), BATMAN introduces a novel optical flow calibration module that leverages the mask information from the memory frame to smooth the optical flow within the same object while reducing noise at the object boundary.

We conduct extensive experiments on four popular VOS benchmarks: Youtube-VOS 2019 [46], Youtube-VOS 2018 [46], DAVIS 2017 [32], and DAVIS 2016 [30] to validate the BATMAN architecture. We show that BATMAN achieves superior performance on all benchmarks and outperforms all previous state-of-the-art methods. We summarize the main contributions of our work below,

- A novel bilateral attention module that computes attention between query and memory features in the bilateral space of motion and appearance, which improves the correspondence matching by adaptively focusing on relevant object features while reducing the noise from the background.
- A novel optical flow calibration module that fuses the object segmentation mask and the initial optical flow estimation to smooth the within-object optical flow and reduce noise at the object boundary.
- Incorporating the optical flow calibration and bilateral attention mechanisms, we design a novel BATMAN architecture. BATMAN establishes new state-of-the-art performance on Youtube-VOS 2019 / 2018 and DAVIS 2017 / 2016 benchmarks. To the best of our knowledge, BATMAN is the first work to compute attention in the bilateral space of motion and appearance for VOS.

2   Related Work

Semi-supervised VOS. The task aims to segment the particular object instances throughout the entire video sequence given one or more annotated frames (the first frame in general). Early DNN works [3, 29, 44] fine-tune the pre-trained networks on the first frame using multiple data augmentations on the given mask at test time to adapt to specific instances. Therefore, these methods are extremely slow during inference due to excessive fine-tuning. Later tracking-based works [5, 17, 40] adopt object tracking technologies to indicate the target location of objects for segmentation to improve inference time. However, these approaches are not robust to occlusion and drifting with error accumulated during the propagation. "Tracking-by-detection" paradigm is introduced into VOS in [16] to take object segmentation as a subtask of tracking, in which the accuracy of tracking often limits the performance. To handle occlusion and drifting, matching-based methods [6, 39] perform feature matching to find objects that are similar to the target objects in the reference frames. STM [27] and its following works [34, 45] leverage an external memory to store past frames' features and then distinguish objects with a similar appearance by pixel-level attention-based matching from the memory.

Vision Transformer. Initially proposed for machine translation, Transformers [38] replace the recurrence and convolutions entirely with hierarchical attention-based mechanisms and achieve outstanding performance. Later, transformer networks became dominant models used in natural language processing (NLP) tasks [42, 52]. Recently, with the observance of its strength in parallel modeling global correlation or attention, transformer blocks were introduced to computer vision tasks, such as image recognition [10], saliency prediction [53], object detection [4, 54], and object segmentation [41], where vision transformers have achieved excellent performance compared to the CNN-based counterparts. Researchers then employed transformer architecture into the VOS task [11, 23, 25, 50]. SST [11] adopts the transformer's encoder to compute attention based on the spatial-temporal information among multiple history frames. In [23], a transductive branch is used to capture the spatial-temporal information, which is integrated with an online inductive branch within a unified framework. TransVOS [25] introduces a transformer-based VOS framework with intuitive structure from the transformer networks in NLP. AOT [50] proposes an Identification Embedding to construct multi-object matching and computes attention for multiple objects simultaneously. In this paper, we introduce a novel bilateral attention transformer framework, where it computes the attention with both the encoded appearance features and the motion features in consideration. Therefore, it is robust to occlusion, drift, and ambiguity between objects with a similar appearance.

Optical Flow. Applying optical flow to VOS can encourage motion consistency through the entire video sequence. Early approaches [8, 37, 48] consider VOS and

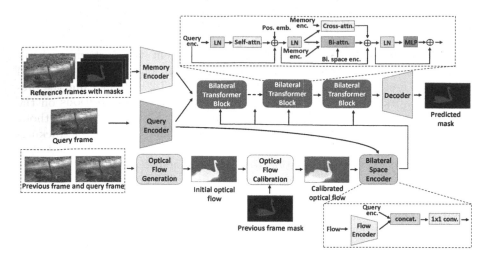

Fig. 2. Overview of the BATMAN architecture. Frame-level features of the reference frames and the query frame are extracted through the memory and query encoders, respectively. A pre-trained FlowNet is used to generate an initial optical flow estimation between the previous frame and the query frame, which is then improved by the optical flow calibration module. A bilateral space encoder is used to encode the query features and the calibrated optical flow into a bilateral space encoding, which is used by the bilateral attention. Multiple layers of bilateral transformer blocks are stacked for matching the correspondence between the reference and query features. Lastly, a decoder is used to predict the query frame segmentation mask

optical flow estimation simultaneously with the assumption that the two tasks are complementary. Recently, RMNet [45] introduces using optical flow generated with an offline model to warp object mask from the previous frame to the query frame and then performing regional matching. It avoids unnecessary matching in regions without target objects or mismatching of objects with a similar appearance. Instead of simply warping the object's mask to indicate the target area, our BATMAN computes the correlation of each pair of tokens considering their optical flow estimation, appearance similarity, and spatial distance simultaneously. Thus, it is more effective in removing irrelevant matching tokens compared to [45]. Meanwhile, our method is more robust to the accumulated error in warping from the optical flow estimation.

3   Method

In this section, we first introduce the proposed BATMAN architecture, and then discuss in depth its core modules: bilateral attention and optical flow calibration.

3.1   Bilateral Attention Transformer in Motion-Appearance Neighboring Space (BATMAN)

Figure 2 provides an overview of the proposed BATMAN architecture. We first extract frame-level features through the memory and query encoders (details in Sect. 4.1) to capture the target object features for establishing correspondence in the later transformer layers. Meanwhile, we compute the initial optical flow between the query frame and its previous frame through a frozen pre-trained FlowNet [36]. Then, we feed the object mask from the previous frame, together with the initial optical flow estimation, into our optical flow calibration module to improve the optical flow (Sect. 3.3). We then encode the calibrated optical flow and the query frame features into tokens in the bilateral space of motion and appearance. Following this, we stack multiple bilateral transformer blocks to model the spatial-temporal relationships among the reference and query frames at pixel-level, based on the bilateral space encoding tokens (Sect. 3.2). After aggregating the spatial-temporal information, the decoder predicts an object mask for the query frame.

3.2   Bilateral Transformer and Bilateral Attention

As shown in Fig. 2, in each bilateral transformer block, the query frame features first go through a self-attention [38] to aggregate the information within the query frame followed by adding a sinusoidal position embedding [38] encoding the tokens' relative positions. Then we apply cross-attention and bilateral attention (described below) to it with the reference frame features and add the results. Following the common practice in vision transformers [25,50], we insert layer normalization [1] before and after each attention module. Finally, we employ a two-layer feed-forward MLP block before feeding the output to the next layer.

Bilateral Space Encoding (E) is used to index each position (token) of the query frame features in the bilateral space. As shown in Fig. 2, we first encode the calibrated optical flow using a flow encoder (details in Sect. 4.1). Then we concatenate the optical flow encoding and the query image encoding (from query encoder) in channel dimension. Finally, we use a 1×1 convolutional layer to project the concatenation to a 1-dimensional space (in channel) where each position (token) has a single scalar coordinate for the bilateral space of motion and appearance. Bilateral space encoding is employed in bilateral attention below.

Bilateral Attention is used to aggregate spatial-temporal information between the query tokens and neighboring key tokens from the reference frames in the bilateral space of motion and appearance. Unlike global cross-attention where each query token computes attention with all key tokens from the reference frames, our bilateral attention adaptively selects the most relevant key tokens for each query token based on the bilateral space encoding. To formulate, we define query tokens $Q \in \mathbb{R}^{HW \times C}$, key tokens $K \in \mathbb{R}^{HW \times C}$, and value embedding tokens $V \in \mathbb{R}^{HW \times C}$, where Q is from the query frame and K and V are aggregated from multiple reference frames. H, W, and C represent the height, width, and channel dimensions of the tokens, respectively. Mathematically, we define bilateral attention as,

$$BiAttn(Q, K, V) = softmax(\frac{QK^T M}{\sqrt{C}})V \tag{1}$$

where $M \in [0, 1]^{HW \times HW}$ is the bilateral space binary mask that defines the attention scope for each query token. For each query token $Q_{h,w}$ at (h, w) position, we define the corresponding bilateral space binary mask $M_{h,w}$ as,

$$M_{h,w}(i, j, E) = \begin{cases} 1 & \text{if } |i - h| \leqslant W_d \text{ and } |j - w| \leqslant W_d \\ & \text{and } |argsort_{W_d}(E_{h,w}) - argsort_{W_d}(E_{i,j})| \leqslant W_b \\ 0 & \text{otherwise} \end{cases} \tag{2}$$

where (i, j) is the position for each key token, $E \in \mathbb{R}^{HW \times 1}$ is the bilateral space encoding of the queries discussed above, W_d and W_b are predefined local windows in spatial and bilateral domains, respectively. $argsort_{W_d}(E_{i,j})$ denotes sorting all bilateral space encoding E within the spatial local window W_d and finding the corresponding index at position (i, j). To train the bilateral space encoding E by stochastic gradient descent directly, in practice, instead of computing $QK^T M$ as shown in Eq. 1, we compute $QK^T + E$ if $M = 1$, while computing $QK^T - L$ if $M = 0$, where $L \in \mathbb{R}$ is a large positive number. This approximates to $QK^T M$ in Eq. 1 after using $softmax$. Equation 2 shows that for each query token, it computes the attention with another key token only if they are close to each other spatially and share similar bilateral space encoding (similar motion and appearance). We further analyze the bilateral space binary mask with visualization in Sect. 4.3.

We implement the bilateral attention modules via a multi-headed formulation [38] where we linearly project queries, keys, and values multiple times with different learnable projections, and we feedforward the multiple heads of bilateral attention in parallel followed by concatenation and a linear projection. Mathematically, we define the multi-head bilateral attention as,

$$MultiHead(Q, K, V) = Concat(head_1, ..., head_h)W^O$$
$$where \; head_i = BiAttn(QW_i^Q, KW_i^K, VW_i^V) \tag{3}$$

where projection matrices are $W_i^Q \in \mathbb{R}^{C \times d_{hidden}}$, $W_i^K \in \mathbb{R}^{C \times d_{hidden}}$, $W_i^V \in \mathbb{R}^{C \times d_{hidden}}$, and $W^O \subset \mathbb{R}^{C \times C}$. In this work, we set the number of heads ($h = C/d_{hidden}$) to 8 [38], where d_{hidden} is the hidden dimension of each head.

3.3   Optical Flow Calibration

As mentioned in the introduction, optical flow estimation can be noisy for objects with large motion and in texture-less areas. We introduce an optical flow calibration module to improve flow estimation by leveraging the segmentation mask

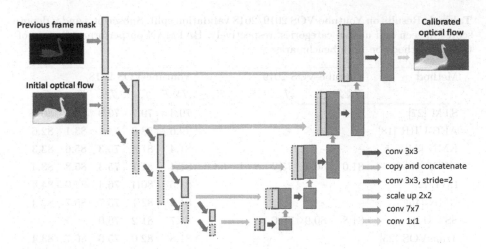

Fig. 3. The optical flow calibration module. A CNN in the U-Net architecture [33] is used to fuse the segmentation mask into the optical flow

from the previous frame. As shown in Fig. 3, the module employs an architecture similar to U-Net [33] with 11-layers total. To train this module to improve optical flow, we compute the Mean Square Error (MSE) between the initial optical flow and the output optical flow in training. Without the MSE loss, mask information can dominate the calibration module and thereby generate an embedding feature for the mask instead.

4   Experiments

We validate BATMAN on popular benchmark datasets YouTube-VOS 2019/2018 and DAVIS 2017/2016. We first provide implementation details, followed by the experimental results. We then present the ablation study on our design.

4.1   Implementation Details

We use ResNet50 [14] as the feature extractor for memory/query/flow encoder. We follow the identification embedding in [50] to encode multiple object masks in the memory encoding simultaneously. We use a RAFT [36] model pre-trained on FlyingThings3D [24] for optical flow generation. We use FPN [19] with Group Normalization [43] as the decoder. We employ 12 bilateral transformer blocks with W_d and W_b set to 7 [50] and 84 (details in supplementary), respectively.

We implement our model in PyTorch [28] and train with a batch size of 16 distributed on 8 V100 GPUs. Following previous works [22,25,45,50], we first pre-train our model on synthetic video sequences generated from static image datasets (COCO [20], ECSSD [35], MSRA10K [9], SBD [13], PASCALVOC2012

Table 1. Results on Youtube-VOS 2019/2018 validation split. Subscript s and u denote scores in seen and unseen categories, respectively. BATMAN outperforms all state-of-the-art methods on both benchmarks

Method	Youtube-VOS 2019					Youtube-VOS 2018				
	$\mathcal{J}\&\mathcal{F}$	$\mathcal{J}_s$	$\mathcal{J}_u$	$\mathcal{F}_s$	$\mathcal{F}_u$	$\mathcal{J}\&\mathcal{F}$	$\mathcal{J}_s$	$\mathcal{J}_u$	$\mathcal{F}_s$	$\mathcal{F}_u$
STM [27]	–	–	–	–	–	79.4	79.7	72.8	84.2	80.9
AFB-URR [18]	–	–	–	–	–	79.6	78.8	74.1	83.1	82.6
KMN [34]	–	–	–	–	–	81.4	81.4	75.3	85.6	83.3
CFBI [49]	81.0	80.6	75.2	85.1	83.0	81.4	81.1	75.3	85.8	83.4
LWL [2]	–	–	–	–	–	81.5	80.4	76.4	84.9	84.4
RMN [45]	–	–	–	–	–	81.5	82.1	75.7	85.7	82.4
SST [11]	81.8	80.9	76.6	–	–	81.7	81.2	76.0	–	–
TransVOS [25]	–	–	–	–	–	81.8	82.0	75.0	86.7	83.4
LCM [15]	–	–	–	–	–	82.0	82.2	75.7	86.7	83.4
CFBI+ [51]	82.6	81.7	77.1	86.2	85.2	82.8	81.8	77.1	86.6	85.6
STCN [7]	82.7	81.1	78.2	85.4	85.9	83.0	81.9	77.9	86.5	85.7
RPCMVOS [47]	83.9	82.6	**79.1**	86.9	87.1	84.0	83.1	78.5	87.7	86.7
AOT [50]	84.1	83.5	78.4	88.1	86.3	84.1	83.7	78.1	88.5	86.1
BATMAN	**85.0**	**84.5**	79.0	**89.3**	**87.2**	**85.3**	**84.7**	**79.2**	**89.8**	**87.4**

[12]) by applying random augmentations. We then train the model on the VOS benchmarks. The loss function is a combination of bootstrapped cross-entropy loss, soft Jaccard loss [26], and mean squared error loss. The training is optimized using AdamW [21] optimizer and Exponential Moving Average (EMA) [31]. The learning rate for training is set to 2×10^{-4} with a weight decay of 0.07. We train the model for 100,000 iterations.

4.2   Experimental Results

We present validation results on the popular Youtube 2019/2018 and DAVIS 2017/2016 benchmarks compared to existing state-of-the-art methods.

Metrics. The region similarity ($\mathcal{J}$) and the boundary accuracy ($\mathcal{F}$) are computed following the standard evaluation setting proposed in [30]. On DAVIS, we report the two metrics and their mean value ($\mathcal{J}\&\mathcal{F}$). On YouTube-VOS, we report all the metrics on seen categories and unseen categories separately as generated by the evaluation server at CodaLab.

Youtube-VOS   [46] is a large-scale dataset for multi-object video segmentation with objects in multiple categories. In YouTube-VOS 2018, the *training* set

Table 2. Comparisons to the state-of-the-art methods on DAVIS benchmarks. (**Y**) indicates including Youtube-VOS dataset in training. BATMAN outperforms all state-of-the-art methods on all three DAVIS benchmarks

Method	DAVIS 2017 val			DAVIS 2017 test-dev			DAVIS 2016 val		
	$\mathcal{J}\&\mathcal{F}$	$\mathcal{J}$	$\mathcal{F}$	$\mathcal{J}\&\mathcal{F}$	$\mathcal{J}$	$\mathcal{F}$	$\mathcal{J}\&\mathcal{F}$	$\mathcal{J}$	$\mathcal{F}$
AFB-URR [18]	74.6	73.0	76.1	–	–	–	–	–	–
LWL [2]	81.6	79.1	84.1	–	–	–	–	–	–
STM [27](**Y**)	–	79.2	84.3	–	–	–	–	88.7	89.9
CFBI [49](**Y**)	81.9	79.3	84.5	75.0	71.4	78.7	89.4	88.3	90.5
SST [11](**Y**)	82.5	79.9	85.1	–	–	–	–	–	–
KMN [34](**Y**)	82.8	80.0	85.6	77.2	74.1	80.3	90.5	89.5	91.5
CFBI+ [51](**Y**)	82.9	80.1	85.7	75.6	71.6	79.6	89.9	88.7	91.1
RMN [45](**Y**)	83.5	81.0	86.0	75.0	71.9	78.1	88.8	88.9	88.7
LCM [15](**Y**)	83.5	80.5	86.5	78.1	74.4	81.8	90.7	89.9	91.4
RPCMVOS [47](**Y**)	83.7	81.3	86.0	79.2	75.8	82.6	90.6	87.1	94.0
TransVOS [25](**Y**)	83.9	81.4	86.4	76.9	73.0	80.9	90.5	89.8	91.2
AOT [50](**Y**)	84.9	82.3	87.5	79.6	75.9	83.3	91.1	90.1	92.1
STCN [7](**Y**)	85.4	82.2	88.6	76.1	72.7	79.6	91.6	**90.8**	92.5
BATMAN(Y)	**86.2**	**83.2**	**89.3**	**82.2**	**78.4**	**86.1**	**92.5**	90.7	**94.2**

contains 3,471 videos with 5,945 unique objects in 65 categories and the *validation* set has 474 videos containing of 894 unique objects in 65 seen categories and additional 26 unseen categories. YouTube-VOS 2019 expands the YouTube-VOS 2018 dataset with more videos and object annotations. Its *training* set contains the same 3,471 videos but has 6,459 objects. Its *validation* set has 507 videos containing of 1,063 objects. With the existence of the unseen object categories, the YouTube-VOS is useful to evaluate the generalization capability of the VOS model on unseen object categories. We evaluate all the results on the official YouTube-VOS evaluation servers on CodaLab.

Table 1 shows that BATMAN outperforms all state-of-the-art on Youtube-VOS 2019 and 2018 benchmarks. The higher region similarity ($\mathcal{J}$) and better boundary accuracy ($\mathcal{F}$) validate that bilateral attention is able to learn the most informative features from the reference frames and match the query frames.

DAVIS is one of the most popular benchmarks for video object segmentation with high-quality masks for salient objects. As part of DAVIS, DAVIS 2016 [30] is a single-object benchmark and DAVIS 2017 [32] is a multi-object extension of DAVIS 2016. In DAVIS 2016, the *training* and *validation* sets contain 30 and 20 videos, respectively. In DAVIS 2017, the *training* set consists of 60 videos, and the *validation* set consists of 30 videos, and the *test-dev* set consists of 30 videos with only the first frame annotated.

Fig. 4. Qualitative results. Compared to spatial local attention, bilateral attention segments objects better especially when background shares similar appearance with the target object (Color figure online)

Table 2 compares BATMAN with existing state-of-the-art methods on DAVIS 2017 validation set, test-dev set, and DAVIS 2016 validation set. Note that KMN [34] only reports the results of DAVIS 2017 test-dev split with images resized to 600p. We follow the standard practice of most previous works and keep the images in the original 480p resolution in evaluation. On both multi-object datasets (DAVIS2017 val/test) and single-object dataset (DAVIS 2016), BATMAN outperforms all existing state-of-the-art methods. Moreover, BATMAN achieves the largest absolute accuracy improvement (2.6%) on the hardest DAVIS 2017 test-dev split, which validates the robustness of our model for VOS.

4.3   Ablation Study

In this section, we analyze the effectiveness of the bilateral attention and compare it to the conventional spatial local attention, as well as the efficacy of the calibrated optical flow. For qualitative analysis, we visualize the bilateral space binary mask generated by the bilateral attention, and the optical flow output from our calibration module.

Bilateral Attention. Table 3 compares the accuracy ($\mathcal{J}\&\mathcal{F}$) between our proposed bilateral attention and the conventional spatial local attention, and validates that the bilateral attention achieves superior performance on all benchmarks. We also visualize the segmentation masks from the two attention mechanisms in Fig. 4 for both DAVIS 2017 and Youtube-VOS 2019. We can see that with spatial local attention, the model tends to fail to segment objects with

Table 3. Ablation on bilateral attention. The model with bilateral attention outperforms that with spatial local attention on all benchmarks

Attention type	DAVIS 2017 val	DAVIS 2017 test-dev	DAVIS 2016 val	Youtube-VOS 2019	Youtube-VOS 2018
Spatial local	84.9	77.5	91.6	84.1	83.8
Bilateral	**86.2**	**82.2**	**92.5**	**85.0**	**85.3**

Fig. 5. Visualization of bilateral space binary masks from the bilateral attention. The bilateral attention adaptively generates binary masks for on and off object query tokens. Better view in color version (Color figure online)

similar appearances (e.g., the second camel is included in the mask of the first camel (Fig. 4a); part of the red pig is segmented as the green pig (Fig. 4c); the right hand of the man in green is mistakenly segmented as part of the man in red (Fig. 4e); the tail of the zebra in the green mask is mistakenly segmented as that of the zebra in yellow (Fig. 4f)). Besides, when the appearance features (especially at the object boundary) are fuzzy (e.g., the shade of the goat (Fig. 4b), the reflection on the TV box (Fig. 4d), and the reflection of the bird's legs in the water (Fig. 4g)), the model with spatial local attention finds it difficult to segment the object properly. In contrast, the bilateral attention and the resultant adaptive bilateral space binary masks enables our model to segment target objects correctly, especially when the target object exhibits salient motion (e.g., the Frisbee (Fig. 4h) and the skydiving men (Fig. 4i)). We provide additional visualizations for segmentation in the supplementary.

Figure 5 shows some examples of the binary masks generated from the bilateral attention. The first row shows the optical flow of the query frames. One off-object (background) query token is highlighted (in red) in the second row for

(a) Comparison of the initial optical flow (middle) and the calibrated optical flow (bottom). The calibrated optical flow is smoother within the same object, and sharper at object boundary

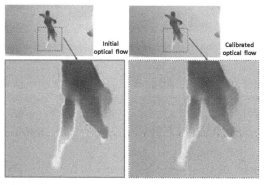

(b) Blocky artifact on the initial optical flow is decreased on the calibrated optical flow

Fig. 6. Visualization of optical flow on Davis 2017 *val.* set. The calibrated optical flow is smoother within the object and sharper at the boundary

each scene. The corresponding bilateral space binary mask is highlighted in the third row. In comparison, we also show an on-object query token in the fourth row, and the corresponding binary mask is given in the last row. We can see that for an off-object query token, the bilateral attention module tends to focus on the background locations (e.g., the water around the swan neck (Fig. 5a) or the sky around the woman with dogs (Fig. 5f)). On the other hand, when the query token is on the object, it tends to select the neighboring on-object tokens (e.g., the leg of the dancing man (Fig. 5c) or the camel hump (Fig. 5d)) for the attention computation. This qualitatively validates that adaptive attention computation enables propagating segmentation masks from the reference frames to the query frame more accurately.

Table 4. Comparisons of bilateral attention w/ and w/o optical flow calibration. Calibrating the optical flow leads to higher accuracy on all benchmarks

Optical flow type	DAVIS 2017 val	DAVIS 2017 test-dev	DAVIS 2016 val	Youtube-VOS 2019	Youtube-VOS 2018
w/o calibration	86.0	81.7	92.4	84.6	84.8
w/ calibration	**86.2**	**82.2**	**92.5**	**85.0**	**85.3**

Optical flow   On-object query token   Bilateral mask

Fig. 7. Failure cases of the bilateral binary mask generation. The bilateral attention may lose focus when a background object exhibits dominant motion and/or the target object does not exhibit salient motion

Optical Flow Calibration. The optical flow calibration module leverages the predicted previous frame mask to improve the optical flow estimation for the current frame. Table 4 compares the bilateral attention w/ and w/o calibrated optical flow. With the calibrated optical flow, BATMAN achieves higher accuracy on all benchmarks, validating that optical flow is improved with the help of the previous frame segmentation mask. As shown in Fig. 6, the calibrated optical flow is smoother, both within the same object and within the background. Meanwhile, the object boundary is sharper. Specifically, the blocky artifacts along the object boundary, which exists in the initial optical flow, are reduced effectively without affecting the object boundary sharpness.

Limitations. The bilateral space binary mask generation is influenced by the motion in the scene. Therefore, if the target objects do not exhibit salient motion, or some background object(s) exhibit salient motion and/or share(s) a similar appearance to the target object(s), the bilateral mask can be noisy and the bilateral attention may lose focus. Figure 7 shows two failure cases: in the upper row, a man on a motorcycle moves quickly across the scene, which overwhelms the motion of the target woman. Hence, the bilateral attention fails to focus on the target object. Similarly, in the bottom row, the motion of the target woman is not salient (especially on the boundary) so the bilateral mask scatters. We plan to extend our method to better handle such scenarios.

5   Conclusions

This paper proposes a novel architecture, BATMAN, for semi-supervised VOS by adaptively computing attention between the query frame and reference frames based on the bilateral encoding of motion and appearance. Compared to conventional spatial local attention, bilateral attention adaptively selects the most relevant tokens to compute the correlation attention which helps to match the object correspondence spatially and temporally with the help of calibrated optical flow. Extensive experiments validate that BATMAN outperforms all existing state-of-the-art on all popular Youtube-VOS and DAVIS benchmarks.

References

1. Ba, J.L., Kiros, J.R., Hinton, G.E.: Layer normalization. arXiv preprint arXiv:1607.06450 (2016)
2. Bhat, G., et al.: Learning what to learn for video object segmentation. In: Vedaldi, A., Bischof, H., Brox, T., Frahm, J.-M. (eds.) ECCV 2020. LNCS, vol. 12347, pp. 777–794. Springer, Cham (2020). https://doi.org/10.1007/978-3-030-58536-5_46
3. Caelles, S., Maninis, K.K., Pont-Tuset, J., Leal-Taixé, L., Cremers, D., Van Gool, L.: One-shot video object segmentation. In: Proceedings of the IEEE Conference on Computer Vision and Pattern Recognition, pp. 221–230 (2017)
4. Carion, N., Massa, F., Synnaeve, G., Usunier, N., Kirillov, A., Zagoruyko, S.: End-to-end object detection with transformers. In: Vedaldi, A., Bischof, H., Brox, T., Frahm, J.-M. (eds.) ECCV 2020. LNCS, vol. 12346, pp. 213–229. Springer, Cham (2020). https://doi.org/10.1007/978-3-030-58452-8_13
5. Chen, X., Li, Z., Yuan, Y., Yu, G., Shen, J., Qi, D.: State-aware tracker for real-time video object segmentation. In: Proceedings of the IEEE/CVF Conference on Computer Vision and Pattern Recognition, pp. 9384–9393 (2020)
6. Chen, Y., Pont-Tuset, J., Montes, A., Van Gool, L.: Blazingly fast video object segmentation with pixel-wise metric learning. In: Proceedings of the IEEE Conference on Computer Vision and Pattern Recognition, pp. 1189–1198 (2018)
7. Cheng, H.K., Tai, Y.W., Tang, C.K.: Rethinking space-time networks with improved memory coverage for efficient video object segmentation. Adv. Neural Inf. Process. Syst. **34**, 1–14 (2021)
8. Cheng, J., Tsai, Y.H., Wang, S., Yang, M.H.: Segflow: joint learning for video object segmentation and optical flow. In: 2017 IEEE International Conference on Computer Vision (ICCV), pp. 686–695 (2017)
9. Cheng, M.M., Mitra, N.J., Huang, X., Torr, P.H., Hu, S.M.: Global contrast based salient region detection. IEEE Trans. Pattern Anal. Mach. Intell. **37**(3), 569–582 (2014)
10. Dosovitskiy, A., et al.: An image is worth 16×16 words: transformers for image recognition at scale. arXiv preprint arXiv:2010.11929 (2020)
11. Duke, B., Ahmed, A., Wolf, C., Aarabi, P., Taylor, G.W.: Sstvos: sparse spatiotemporal transformers for video object segmentation. In: Proceedings of the IEEE/CVF Conference on Computer Vision and Pattern Recognition, pp. 5912–5921 (2021)

12. Everingham, M., Van Gool, L., Williams, C.K., Winn, J., Zisserman, A.: The pascal visual object classes (voc) challenge. Int. J. Comput. Vision **88**(2), 303–338 (2010)
13. Hariharan, B., Arbeláez, P., Bourdev, L., Maji, S., Malik, J.: Semantic contours from inverse detectors. In: 2011 International Conference on Computer Vision, pp. 991–998. IEEE (2011)
14. He, K., Zhang, X., Ren, S., Sun, J.: Deep residual learning for image recognition. In: Proceedings of the IEEE Conference on Computer Vision and Pattern Recognition, pp. 770–778 (2016)
15. Hu, L., Zhang, P., Zhang, B., Pan, P., Xu, Y., Jin, R.: Learning position and target consistency for memory-based video object segmentation. In: Proceedings of the IEEE/CVF Conference on Computer Vision and Pattern Recognition, pp. 4144–4154 (2021)
16. Huang, X., Xu, J., Tai, Y.W., Tang, C.K.: Fast video object segmentation with temporal aggregation network and dynamic template matching. In: Proceedings of the IEEE/CVF Conference on Computer Vision and Pattern Recognition, pp. 8879–8889 (2020)
17. Khoreva, A., Benenson, R., Ilg, E., Brox, T., Schiele, B.: Lucid data dreaming for video object segmentation. Int. J. Comput. Vision **127**(9), 1175–1197 (2019)
18. Liang, Y., Li, X., Jafari, N., Chen, J.: Video object segmentation with adaptive feature bank and uncertain-region refinement. Adv. Neural Inf. Process. Syst. **33**, 3430–3441 (2020)
19. Lin, T.Y., Dollár, P., Girshick, R., He, K., Hariharan, B., Belongie, S.: Feature pyramid networks for object detection. In: Proceedings of the IEEE Conference on Computer Vision and Pattern Recognition, pp. 2117–2125 (2017)
20. Lin, T.-Y., et al.: Microsoft COCO: common objects in context. In: Fleet, D., Pajdla, T., Schiele, B., Tuytelaars, T. (eds.) ECCV 2014. LNCS, vol. 8693, pp. 740–755. Springer, Cham (2014). https://doi.org/10.1007/978-3-319-10602-1_48
21. Loshchilov, I., Hutter, F.: Decoupled weight decay regularization. arXiv preprint arXiv:1711.05101 (2017)
22. Lu, X., Wang, W., Danelljan, M., Zhou, T., Shen, J., Van Gool, L.: Video object segmentation with episodic graph memory networks. In: Vedaldi, A., Bischof, H., Brox, T., Frahm, J.-M. (eds.) ECCV 2020. LNCS, vol. 12348, pp. 661–679. Springer, Cham (2020). https://doi.org/10.1007/978-3-030-58580-8_39
23. Mao, Y., Wang, N., Zhou, W., Li, H.: Joint inductive and transductive learning for video object segmentation. In: Proceedings of the IEEE/CVF International Conference on Computer Vision, pp. 9670–9679 (2021)
24. Mayer, N., et al.: A large dataset to train convolutional networks for disparity, optical flow, and scene flow estimation. In: IEEE International Conference on Computer Vision and Pattern Recognition (CVPR) (2016)
25. Mei, J., Wang, M., Lin, Y., Yuan, Y., Liu, Y.: Transvos: video object segmentation with transformers. arXiv preprint arXiv:2106.00588 (2021)
26. Nowozin, S.: Optimal decisions from probabilistic models: the intersection-over-union case. In: IEEE Conference on Computer Vision and Pattern Recognition, pp. 548–555 (2014)
27. Oh, S.W., Lee, J.Y., Xu, N., Kim, S.J.: Video object segmentation using space-time memory networks. In: 2019 IEEE/CVF International Conference on Computer Vision (ICCV), pp. 9225–9234 (2019)
28. Paszke, A., et al.: Automatic differentiation in pytorch (2017)
29. Perazzi, F., Khoreva, A., Benenson, R., Schiele, B., Sorkine-Hornung, A.: Learning video object segmentation from static images. In: Proceedings of the IEEE Conference on Computer Vision and Pattern Recognition, pp. 2663–2672 (2017)

30. Perazzi, F., Pont-Tuset, J., McWilliams, B., Van Gool, L., Gross, M., Sorkine-Hornung, A.: A benchmark dataset and evaluation methodology for video object segmentation. In: Proceedings of the IEEE Conference on Computer Vision and Pattern Recognition, pp. 724–732 (2016)

31. Polyak, B.T., Juditsky, A.B.: Acceleration of stochastic approximation by averaging. SIAM J. Control Optim. **30**(4), 838–855 (1992)

32. Pont-Tuset, J., Perazzi, F., Caelles, S., Arbeláez, P., Sorkine-Hornung, A., Van Gool, L.: The 2017 davis challenge on video object segmentation. arXiv preprint arXiv:1704.00675 (2017)

33. Ronneberger, O., Fischer, P., Brox, T.: U-Net: convolutional networks for biomedical image segmentation. In: Navab, N., Hornegger, J., Wells, W.M., Frangi, A.F. (eds.) MICCAI 2015. LNCS, vol. 9351, pp. 234–241. Springer, Cham (2015). https://doi.org/10.1007/978-3-319-24574-4_28

34. Seong, H., Hyun, J., Kim, E.: Kernelized memory network for video object segmentation. In: Vedaldi, A., Bischof, H., Brox, T., Frahm, J.-M. (eds.) ECCV 2020. LNCS, vol. 12367, pp. 629–645. Springer, Cham (2020). https://doi.org/10.1007/978-3-030-58542-6_38

35. Shi, J., Yan, Q., Xu, L., Jia, J.: Hierarchical image saliency detection on extended CSSD. IEEE Trans. Pattern Anal. Mach. Intell. **38**(4), 717–729 (2015)

36. Teed, Z., Deng, J.: RAFT: recurrent all-pairs field transforms for optical flow. In: Vedaldi, A., Bischof, H., Brox, T., Frahm, J.-M. (eds.) ECCV 2020. LNCS, vol. 12347, pp. 402–419. Springer, Cham (2020). https://doi.org/10.1007/978-3-030-58536-5_24

37. Tsai, Y.H., Yang, M.H., Black, M.J.: Video segmentation via object flow. In: Proceedings of the IEEE Conference on Computer Vision and Pattern Recognition, pp. 3899–3908 (2016)

38. Vaswani, A., et al.: Attention is all you need. Adv. Neural Inf. Process. Syst. **30**, 1–11 (2017)

39. Voigtlaender, P., Chai, Y., Schroff, F., Adam, H., Leibe, B., Chen, L.C.: Feelvos: fast end-to-end embedding learning for video object segmentation. In: Proceedings of the IEEE/CVF Conference on Computer Vision and Pattern Recognition, pp. 9481–9490 (2019)

40. Wang, Q., Zhang, L., Bertinetto, L., Hu, W., Torr, P.H.: Fast online object tracking and segmentation: a unifying approach. In: Proceedings of the IEEE/CVF Conference on Computer Vision and Pattern Recognition, pp. 1328–1338 (2019)

41. Wang, Y., et al.: End-to-end video instance segmentation with transformers. In: Proceedings of the IEEE/CVF Conference on Computer Vision and Pattern Recognition, pp. 8741–8750 (2021)

42. Wolf, T., et al.: Transformers: State-of-the-art natural language processing. In: Proceedings of the 2020 Conference on Empirical Methods in Natural Language Processing: System Demonstrations, pp. 38–45 (2020)

43. Wu, Y., He, K.: Group normalization. In: Ferrari, V., Hebert, M., Sminchisescu, C., Weiss, Y. (eds.) ECCV 2018. LNCS, vol. 11217, pp. 3–19. Springer, Cham (2018). https://doi.org/10.1007/978-3-030-01261-8_1

44. Xiao, H., Feng, J., Lin, G., Liu, Y., Zhang, M.: Monet: deep motion exploitation for video object segmentation. In: Proceedings of the IEEE Conference on Computer Vision and Pattern Recognition, pp. 1140–1148 (2018)

45. Xie, H., Yao, H., Zhou, S., Zhang, S., Sun, W.: Efficient regional memory network for video object segmentation. In: 2021 IEEE/CVF Conference on Computer Vision and Pattern Recognition (CVPR), pp. 1286–1295 (2021)

46. Xu, N., et al.: Youtube-vos: a large-scale video object segmentation benchmark. arXiv preprint arXiv:1809.03327 (2018)

47. Xu, X., Wang, J., Li, X., Lu, Y.: Reliable propagation-correction modulation for video object segmentation. arXiv preprint arXiv:2112.02853 (2021)

48. Xu, Y.S., Fu, T.J., Yang, H.K., Lee, C.Y.: Dynamic video segmentation network. In: Proceedings of the IEEE Conference on Computer Vision and Pattern Recognition, pp. 6556–6565 (2018)

49. Yang, Z., Wei, Y., Yang, Y.: Collaborative video object segmentation by foreground-background integration. In: Vedaldi, A., Bischof, H., Brox, T., Frahm, J.-M. (eds.) ECCV 2020. LNCS, vol. 12350, pp. 332–348. Springer, Cham (2020). https://doi.org/10.1007/978-3-030-58558-7_20

50. Yang, Z., Wei, Y., Yang, Y.: Associating objects with transformers for video object segmentation. Adv. Neural Inf. Process. Syst. **34**, 1–12 (2021)

51. Yang, Z., Wei, Y., Yang, Y.: Collaborative video object segmentation by multi-scale foreground-background integration. IEEE Trans. Pattern Anal. Mach. Intell. (2021)

52. Zaheer, M., et al.: Big bird: transformers for longer sequences. Adv. Neural Inf. Process. Syst. **33**, 17283–17297 (2020)

53. Zhang, J., Xie, J., Barnes, N., Li, P.: Learning generative vision transformer with energy-based latent space for saliency prediction. Adv. Neural Inf. Process. Syst. **34**, 1–16 (2021)

54. Zhu, X., Su, W., Lu, L., Li, B., Wang, X., Dai, J.: Deformable detr: deformable transformers for end-to-end object detection. arXiv preprint arXiv:2010.04159 (2020)

SPSN: Superpixel Prototype Sampling Network for RGB-D Salient Object Detection

Minhyeok Lee, Chaewon Park, Suhwan Cho, and Sangyoun Lee[✉]

Yonsei University, Seoul, Korea
{hydragon516,chaewon28,chosuhwan,syleee}@yonsei.ac.kr

Abstract. RGB-D salient object detection (SOD) has been in the spotlight recently because it is an important preprocessing operation for various vision tasks. However, despite advances in deep learning-based methods, RGB-D SOD is still challenging due to the large domain gap between an RGB image and the depth map and low-quality depth maps. To solve this problem, we propose a novel superpixel prototype sampling network (SPSN) architecture. The proposed model splits the input RGB image and depth map into component superpixels to generate component prototypes. We design a prototype sampling network so that the network only samples prototypes corresponding to salient objects. In addition, we propose a reliance selection module to recognize the quality of each RGB and depth feature map and adaptively weight them in proportion to their reliability. The proposed method makes the model robust to inconsistencies between RGB images and depth maps and eliminates the influence of non-salient objects. Our method is evaluated on five popular datasets, achieving state-of-the-art performance. We prove the effectiveness of the proposed method through comparative experiments. Code and models are available at https://github.com/Hydragon516/SPSN.

Keywords: RGB-D salient object detection · Superpixel · Prototype learning · Reliance selection

1 Introduction

The salient object detection (SOD) task detects and segments objects that visually attract the most human interest from a single image or video. The SOD task is a useful preprocessing operation for various computer vision tasks such as few-shot learning, weakly-supervised semantic segmentation, object recognition, tracking, and image parsing. However, despite recent advances in deep learning,

---

M. Lee and C. Park—These authors contributed equally.

Supplementary Information The online version contains supplementary material available at https://doi.org/10.1007/978-3-031-19818-2_36.

it is still challenging due to camouflaged objects, extreme lighting conditions, and scenes containing multiple objects with complex shapes. To potentially improve performance for such difficult scenes, RGB-D SOD, using an additional depth map, has recently been in the spotlight.

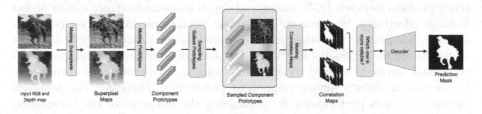

Fig. 1. The overall flow of the proposed model. Our model generates and samples component prototypes from superpixel maps. It also compares the reliability of correlation maps created from component prototypes to generate the predicted mask.

Recent deep learning-based studies [5–7,21,51] achieve significant RGB-D SOD performance by fusing RGB information and additional depth information. However, due to the large domain gap between an RGB image containing rich detail information and a depth image containing geometric information, previous works [7,8,19,36,39,41] focus on the process of effectively fusing these two pieces of information. These methods show that they can effectively extract feature information about salient objects from RGB images and depth maps, but they have two major limitations.

First, they perform inconsistently due to mismatches between the RGB image and the depth map. For example, in the case of a picture hung on the wall, the depth map lacks saliency information compared to the RGB image due to the pictures thinness. Furthermore, the RGB image contains complex texture information about the background scene despite the particularly monotonous depth map background. This unnecessary additional information acts as noise in the network and makes it difficult to generate an accurate saliency mask. This often causes conventional methods to fail in challenging scenes involving complex background structures and multiple foreground objects.

Second, the quality of depth maps is inconsistent due to the limitations of the depth sensor. Some studies [23,42] suggest additional processes for depth map refinement to solve this problem. Although these methods can improve the consistency of low-quality depth maps, they are inefficient due to the additional network or computational costs.

To solve the problems described above, we propose a novel superpixel prototype sampling network (SPSN) architecture. Figure 1 shows the overall SPSN process. First, we note that RGB images and depth maps provide different kinds of information and can complement each other. RGB images have various detail and texture information in the foreground and background, which provides rich context information to the network as it passes through the encoder. In comparison, the depth map lacks detailed information, but it is more robust than an RGB image in extreme lighting conditions. For a preprocessing operation to effectively fuse and complement the advantages of an RGB image and depth map, we use the simple linear iterative clustering (SLIC) algorithm [2] to segment the RGB image and depth map into superpixel components. Moreover, we propose a prototype sampling network module (PSNM) to solve the inconsistency problem between RGB images and depth maps and extract salient object features effectively. We generate component prototypes from superpixel components, inspired by prototype learning, which is used extensively in few-shot segmentation tasks [28,30,43]. PSNM, composed of transformers and graph convolutional layers, is trained to selectively sample only prototypes corresponding to salient objects among component prototypes. Therefore, the proposed method improves performance by minimizing the influence of the background and extracting consistent salient features from RGB images and depth maps. Furthermore, for the network to be flexible enough to handle low-quality depth maps, we propose a reliance selection module (RSM). The RSM is trained to evaluate the quality of the features generated from the RGB component prototypes and depth component prototypes. As a result, the RSM adaptively changes the RGB image and depth map dependence of the network. In other words, the proposed model minimizes performance degradation in situations such as low-quality depth maps and low-light RGB images and effectively creates a saliency mask.

The experimental results over five benchmark datasets show that our model significantly outperforms previous state-of-the-art approaches. Finally, we demonstrate the validity of our method through various ablation studies.

2   Related Work

2.1   RGB-D SOD

Recent RGB-based SOD methods [16,34,44,52,53] have demonstrated outstanding performance. However, they are still challenged by insufficient information to express the complex characteristics of scenes with multiple objects, transparent objects, ambiguous borders between the foreground and background, and extreme light conditions. Meanwhile, owing to the development of various consumer-grade depth cameras, additional depth cues of abundant structural and geometrical information have been enabled for SOD studies. Therefore, RGB-D SOD has gained significant attention and has been widely studied to supplement the limits of RGB-based methods on the scenarios mentioned above.

2.2   Traditional RGB-D Methods

Traditional RGB-D SOD algorithms [11–13,15,22,25,27] focused on utilizing various hand-crafted features, such as contrast, center or boundary prior, and center-surround difference. Lang et al. [27] introduced the depth prior by modeling the relationship between depth and saliency with a mixture of Gaussians.

Additionally, Cheng et al. [11] grouped the pixels in the input image with k-means clustering and obtained three saliency cues-color contrast, depth contrast and spatial bias-from each cluster to generate saliency maps. Moreover, Ju et al. [25] proposed an anisotropic center-surround difference based on the assumption that salient objects tend to stand out from the surroundings. Because these methods rely heavily on hand-crafted features of relatively limited information, their performance deteriorates in complex scenes.

2.3 Deep Learning-Based RGB-D Methods

Existing deep learning-based RGB-D SOD methods focus more on fusing the complementary features extracted from the RGB and depth channels because the domain gap represented by each channel is significant. The merging strategies can be grouped into three categories based on when the fusion takes place: early fusion [39,41], middle fusion [7,8], and late fusion [19,36]. Early fusion methods concatenate the RGB image and depth image at the earliest stage and regard the integrated four-channel matrix as a single input. For example, Qu et al. [39] introduced this method by generating hand-crafted feature vectors from each RGB-D pair, which were fed as input to a CNN-based model. Middle fusion methods fuse the two different feature maps extracted from individual networks. For example, Chen et al. [7] suggested a two-stream complementary-aware network in which the features from the same stages of each modality are fused with the help of a complementary-aware fusion block. Finally, late fusion methods produce individual saliency prediction maps from both the RGB and depth channels, and the two predicted maps are merged by a post-processing operation such as pixel-wise summation and multiplication. For example, Piao et al. [36] proposed a depth-induced multiscale recurrent attention network to extract the features from an RGB image and depth image individually and designed depth refinement blocks for integration.

However, these methods neglect the problem of mismatches between the two modalities. For example, in some scenarios such as a thin calendar hung on the wall, the RGB image more accurately discriminates the salient object and the background, whereas all the pixel values are similar to each other in the depth image. To deal with this problem, several studies have proposed methods to enhance such unreliable input data by utilizing hand-crafted techniques to improve the accuracy. Zhao et al. [51] suggested a contrast prior loss to increase the color difference between the foreground and background of the depth input. Ji et al. [23] proposed an effective depth calibration strategy that corrects the latent bias of the raw depth maps. Furthermore, Zhang et al. [48] presented a depth correction network to decrease the noise in unreliable depth data, assuming the object boundaries in the depth map align with those in the RGB map.

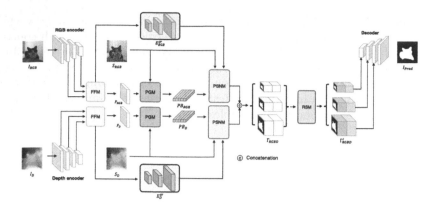

Fig. 2. Overall architecture of the superpixel prototype sampling network (SPSN). The proposed network has one RGB encoder and one depth encoder. Our model consists of a feature fusion model (FFM) for effective fusion of encoder features, prototype generating module (PGM) for prototype extraction, prototype sampling network module (PSNM) for prototype sampling, and reliance selection module (RSM) for reliability selection of RGB and depth features.

3   Proposed Method

3.1   Overview

Figure 2 shows the overall architecture of the proposed SPSN. The proposed model uses an RGB image $\mathbf{I_{RGB}}$, depth map $\mathbf{I_D}$, and their superpixel maps $\mathbf{S_{RGB}}$, $\mathbf{S_D}$ as inputs. Our model is composed primarily of four parts: the feature fusion module (FFM), prototype generating module (PGM), prototype sampling network module (PSNM), and reliance selection module (RSM). The SPSN also has two encoders for RGB images and depth maps and one decoder.

3.2   Feature Fusion Module

As shown in Fig. 3(a), FFM fuses multiscale features from the encoder. We extract three features $\mathbf{E_1} \in \mathbb{R}^{C_{(1/8)} \times \frac{H}{8} \times \frac{W}{8}}$, $\mathbf{E_2} \in \mathbb{R}^{C_{(1/16)} \times \frac{H}{16} \times \frac{W}{16}}$, and $\mathbf{E_3} \in \mathbb{R}^{C_{(1/32)} \times \frac{H}{32} \times \frac{W}{32}}$ from the encoder, where H and W are the height and width of the input image, respectively, and $C_{(1/8)}$, $C_{(1/16)}$, and $C_{(1/32)}$ are the number of channels of the multiscale encoder feature. Because the architectures of the RGB encoder and depth encoder are identical, the size of the extracted features is the same. The FFM consists of a 1×1 convolution layer and upsampling layers, integrating the multiscale features of the encoder and extracting the global contextual information through atrous spatial pyramid pooling (ASPP) [9] layer. As a result, The FFM generates RGB fusion feature $\mathbf{F_{RGB}} \in \mathbb{R}^{128 \times \frac{H}{8} \times \frac{W}{8}}$ and depth fusion feature $\mathbf{F_D} \in \mathbb{R}^{128 \times \frac{H}{8} \times \frac{W}{8}}$, as shown in Fig. 2. In addition, the channel-reduced encoder feature $\mathbf{E^{cr}}$ after the 1×1 convolution layer is used as the input for PSNM.

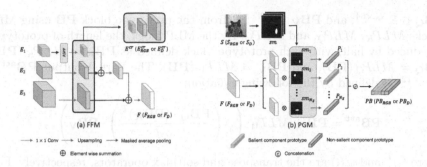

Fig. 3. Structure of (a) FFM and (b) PGM. The FFM fuses the multiscale features of the encoder. The PGM generates a prototype block from the superpixel mask **sm** and the fusion feature F generated from the FFM.

3.3 Prototype Generating Module

The PGM aims to generate component prototypes from fusion features $\mathbf{F_{RGB}}$ and $\mathbf{F_D}$, obtained from the FFM. As shown in Fig. 3(b), we first create a superpixel map $\mathbf{S}$ from each RGB image $\mathbf{I_{RGB}}$ and depth map $\mathbf{I_D}$ using the SLIC algorithm [2]. Next, we create a superpixel mask group **sm** where each channel is a binary mask for each superpixel. Therefore, if the number of superpixels is N_S, the size of **sm** is $N_S \times H \times W$. **sm** is down-sampled to the same size as the fusion feature $\mathbf{F}$ (*i.e.* $\mathbf{F_{RGB}}$ or $\mathbf{F_D}$) generated by the FFM, so the size of the superpixel mask $\mathbf{sm_i}$ constituting each channel of **sm** is $1 \times \frac{H}{8} \times \frac{W}{8}$, where $i = 1, 2, ..., N_S$. Like the prototype learning of few-shot segmentation tasks [28,30,43], the PGM creates a component prototype from each superpixel mask. Thus, the prototype $\mathbf{P_i}$ generated by $\mathbf{sm_i}$ is defined as $\mathbf{P_i} = MAP(\mathbf{F}, \mathbf{sm_i})$, where $MAP(.)$ is the masked average pooling operator. Finally, we define a prototype block $\mathbf{PB} \in \mathbb{R}^{N_S \times 128}$ as a concatenation of the component prototypes generated. As shown in Fig. 3 (b), we define prototypes created from superpixel masks on salient objects as salient component prototypes, and prototypes created from superpixel masks at other locations as non-salient component prototypes.

3.4 Prototype Sampling Network Module

The PSNM aims to sample only the salient component prototypes from all the component prototypes $\mathbf{P_i}$ created from the RGB images and depth maps. Therefore, the PSNM should focus on correlations between $\mathbf{P_i}$ that contain consistent characteristics for salient objects, and it must be able to distinguish them from inconsistent background components. Figure 4 shows the structure of the proposed PSNM, which consists of Parts A, B, C, and D.

Part A. Part A is a transformer module with multi-head attention to enhance the correlation between $\mathbf{P_i}$. Inspired by the previous key, query, and value-based multi-head attention method [20,45,47], we first generate $\mathbf{PB_K} \in \mathbb{R}^{N_S \times 64}$,

$\mathbf{PB_Q} \in \mathbb{R}^{N_S \times 64}$, and $\mathbf{PB_V} \in \mathbb{R}^{N_S \times 64}$ from the prototype block $\mathbf{PB}$ using MLP blocks MLP_K, MLP_Q, and MLP_V. By the MLP block, the length of prototypes is reduced by half, with each prototype block defined as $\mathbf{PB_K} = MLP_K (\mathbf{PB})$, $\mathbf{PB_Q} = MLP_Q (\mathbf{PB})$, and $\mathbf{PB_V} = MLP_V (\mathbf{PB})$. The Part A output $\mathbf{PB^{att}} \in \mathbb{R}^{N_S \times 128}$ is defined by the following equation:

$$\mathbf{PB^{att}} = \mathbf{PB} + MLP_W \left(\psi \left(\frac{\mathbf{PB_Q} \cdot (\mathbf{PB_K})^T}{\sqrt{d}} \right) \cdot \mathbf{PB_V} \right), \tag{1}$$

where $(.)^T$ and $\psi (.)$ are the transpose and softmax operators, respectively. Furthermore, $d = 64$, the length of $\mathbf{PB_K}$, $\mathbf{PB_Q}$, and $\mathbf{PB_V}$. In addition, MLP_W is an MLP block that increases the length of the reduced prototype to the length of the original.

Part B. Part B is a network for sampling salient component prototypes from $\mathbf{PB^{att}}$ with enhanced correlation between component prototypes. Since the component prototype is the result of masked average pooling of the encoder features, it is a one-dimensional vector representing each component feature. These feature shapes are similar to the embedded point features of 3D point cloud networks [37,38,46]. Therefore, we propose a graph convolution network based on the feature distance between the prototypes, inspired by EdgeConv [46], used in 3D point cloud networks. As shown in Part B of Fig. 4, the proposed module consists of three EdgeConv layers and one MLP block. First, we define the input component prototype block of EdgeConv as $\mathbf{PB^{in}} \in \mathbb{R}^{N_S \times 128}$ containing prototypes $\mathbf{P_1^{in}}, \mathbf{P_2^{in}}, ..., \mathbf{P_{N_S}^{in}}$. Then, EdgeConv uses the k-nearest-neighbor (k-NN) algorithm to create a graph between a target prototype $\mathbf{P_i^{in}}$ and a_k prototypes $\mathbf{P_{j1}^{in}}, \mathbf{P_{j2}^{in}}, ..., \mathbf{P_{ja_k}^{in}}$ that are most close to each other in the feature space. Next, as shown in Fig. 4, EdgeConv extracts edge features χ_{ijx} between each node generated in the graph, where $x = 1, 2, ..., a_k$. The edge features χ_{ijx} are defined as follows:

$$\chi_{ijx} = h_\theta \left(\mathbf{P_i^{in}}, \mathbf{P_{jx}^{in}} - \mathbf{P_i^{in}} \right), \tag{2}$$

where $h_\theta : \mathbb{R}^{c_e} \times \mathbb{R}^{c_e} \to \mathbb{R}^{c_e}$ is a nonlinear function with a set of learnable parameters θ, and $c_e = 128$. Therefore, as shown in Fig. 4, a total of a_k χ_{ijx} are generated, so the size of χ_{ij} is $a_k \times N_S \times 128$. This process is equivalent to generating dynamic graphs proposed by [46]. The final output $\mathbf{PB^{out}} \in \mathbb{R}^{N_S \times 128}$ of the EdgeConv layer is defined as $\mathbf{PB^{out}} = MAX (\chi_{ij})$, where $MAX (.)$ is the channel-wise symmetric aggregation operator max pooling, according to [46]. The symmetric aggregation operator makes the network independent of the prototype order. As a result, Part B generates a prototype sampler vector $\mathbf{S_{pred}} \in \mathbb{R}^{N_S}$, defined as follows:

$$\mathbf{S_{pred}} = Sigmoid \left(MLP_F \left(\mathbf{PB^{out}} \right) \right), \tag{3}$$

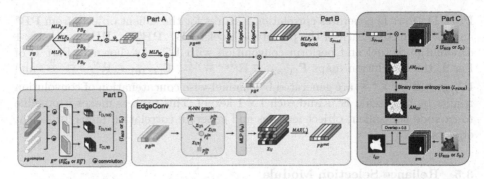

Fig. 4. Structure of the PSNM, composed primarily of four subparts. The PSNM selectively samples only prototypes corresponding to salient objects.

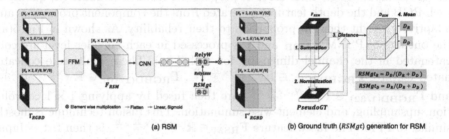

Fig. 5. (a) Structure of the RSM and (b) ground truth generating process for the RSM. The RSM aims to discriminate the reliability of each RGB feature and depth feature to adaptively balance the contribution of the two when generating the final saliency map.

where $Sigmoid(.)$ is the sigmoid operator and MLP_F is an MLP block that reduces the length of the prototype block. Therefore, as shown in Fig. 4, $\mathbf{S_{pred}}$ has values between 0 and 1, and we multiply $\mathbf{S_{pred}}$ by $\mathbf{PB^{att}}$ to create a $\mathbf{PB^s}$, where only the salient object is sampled.

Part C. Part C is an auxiliary module for training the network of Part B. It is used only in the training phase and is removed in the testing phase. As shown in Part C in Fig. 4, we compute the channel-wise sum of the multiplication of $\mathbf{S_{Pred}}$ and $\mathbf{m_S}$ to generate the auxiliary prediction superpixel map $\mathbf{AM_{Pred}} \in \mathbb{R}^{H \times W}$. In other words, $\mathbf{AM_{Pred}}$ is the set of superpixel masks sampled by $\mathbf{S_{Pred}}$. We also generate an auxiliary ground truth superpixel map $\mathbf{AM_{GT}} \in \mathbb{R}^{H \times W}$ from the $\mathbf{sm}$ and the ground truth salient object mask $\mathbf{I_{GT}}$. $\mathbf{AM_{GT}}$ is the channel-wise sum of $\mathbf{sm}$ satisfying $\frac{\sum(\mathbf{m_{s_k}} \times \mathbf{I_{GT}})}{\sum(\mathbf{m_{s_k}})} > 0.5$, where $\sum(.)$ is the sum of all pixel values. Therefore, as shown in Part C of Fig. 4, $\mathbf{AM_{GT}}$ is similar to $\mathbf{I_{GT}}$. We use the binary cross-entropy loss between $\mathbf{AM_{Pred}}$ and $\mathbf{AM_{GT}}$ as an objective function so that Part B can learn to sample only the salient object prototypes.

Part D. Part D generates correlation features for the salient objects from $\mathbf{PB^s}$ and $\mathbf{E^{cr}}$. We treat each of the prototype blocks $\mathbf{PB_1^s}, \mathbf{PB_2^s}, ..., \mathbf{PB_{N_s}^s}$ as a 1×1 convolution kernel and perform convolution with $\mathbf{E^{cr}}$. As shown in Part D of Fig. 4, the correlation maps $\mathbf{\Gamma_{(1/32)}} \in \mathbb{R}^{N_S \times \frac{H}{32} \times \frac{W}{32}}$, $\mathbf{\Gamma_{(1/16)}} \in \mathbb{R}^{N_S \times \frac{H}{16} \times \frac{W}{16}}$, and $\mathbf{\Gamma_{(1/8)}} \in \mathbb{R}^{N_S \times \frac{H}{8} \times \frac{W}{8}}$ are generated by channel-wise concatenation of convolution results by multiscale $\mathbf{E^{cr}}$s and each 1×1 kernel. This process makes it possible to exclude non-salient object features and generate correlation maps for salient objects.

3.5   Reliance Selection Module

As previously mentioned, reliable modality varies depending on the characteristics of the input image. Therefore, we propose the RSM to evaluate the quality of each RGB and the depth features generated from the component prototypes and adaptively weight them in proportion to their reliability. As shown in Fig. 5(a), the outputs of PSNM $\mathbf{\Gamma_{RGB}}$ and $\mathbf{\Gamma_D}$ processed in each encoder level are concatenated in the channel dimension for each level. These multiscale concatenated features $\mathbf{\Gamma_{RGBD(1/8)}} \in \mathbb{R}^{(2 \times N_s) \times \frac{H}{8} \times \frac{W}{8}}$, $\mathbf{\Gamma_{RGBD(1/16)}} \in \mathbb{R}^{(2 \times N_s) \times \frac{H}{16} \times \frac{W}{16}}$, and $\mathbf{\Gamma_{RGBD(1/32)}} \in \mathbb{R}^{(2 \times N_s) \times \frac{H}{32} \times \frac{W}{32}}$ are then fused by applying 1×1 convolution, upsampling, and element-wise summation. This fusion technique is mostly similar to FFM. The fused feature $\mathbf{F_{RSM}} \in \mathbb{R}^{(2 \times N_s) \times \frac{H}{8} \times \frac{W}{8}}$ is then fed as input to RSM. The RSM network consists of three convolutional layers. Each layer is composed of convolution, batch normalization, and ReLU [3] activation. After extracting the features, we flatten the output of the last layer and apply linear function and sigmoid function. In this way, we obtain a vector $\mathbf{RelyW} \in \mathbb{R}^2$ of two reliance values $RelyW_R$ and $RelyW_D$, lying between 0 and 1, which represent the reliability of each $\mathbf{\Gamma_{RGB}}$ and $\mathbf{\Gamma_D}$ respectively. The more reliable the feature is, the higher the reliance value. Finally, we obtain a reliance-weighted RGB-D feature matrix $\mathbf{\Gamma'_{RGBD}}$ by multiplying $RelyW_R$ and $RelyW_D$ with $\mathbf{\Gamma_{RGB}}$. The equations are as follows:

$$\mathbf{\Gamma'_{RGBD}} = \begin{cases} RelyW_R \times \mathbf{\Gamma_{RGBD}^k} & ,0 \leq k < N_s \\ RelyW_D \times \mathbf{\Gamma_{RGBD}^k} & ,N_s \leq k < 2 \times N_s \end{cases}, \tag{4}$$

where k indicates the channel dimension.

Ground Truth for RSM. To optimize RSM, we generate a ground truth vector $\mathbf{RSM_{gt}}$. The process is demonstrated in Fig. 5(b). First, we process the channel-wise summation of $\mathbf{F_{RSM}}$ and apply min-max normalization. Thereby, we obtain a one-channel matrix $\mathbf{PseudoGT} \in \mathbb{R}^{1 \times \frac{H}{8} \times \frac{W}{8}}$ which contains the channel-wise statistics for width $\times$ height dimensions in $\mathbf{F_{RSM}}$. Because each channel of $\mathbf{F_{RSM}}$ represents a candidate for the correlation map compressed to a small size, the bigger the pixel value is, the more likely that pixel belongs to the salient object. Next, we calculate the L_1 distance between $\mathbf{PseudoGT}$ and each channel of $\mathbf{F_{RSM}}$ to obtain a distance matrix $\mathbf{D_{RSM}} \in \mathbb{R}^{(2 \times N_s) \times \frac{H}{8} \times \frac{W}{8}}$. From

$\mathbf{D_{RSM}}$, we obtain the mean distance values D_R and D_D by averaging the values where the channel index k is $0 \le k < N_s$ and $N_s \le k < 2 \times N_s$, respectively. Finally, we acquire the two elements of $\mathbf{RSM_{gt}}$ by the following equations:

$$RSM_{gt_R} = D_D/(D_R + D_D) \tag{5}$$

$$RSM_{gt_D} = D_R/(D_R + D_D) \tag{6}$$

Therefore, RSM_{gt_R} and RSM_{gt_D} represent the similarity between the RGB correlation maps and $\mathbf{PseudoGT}$, and the similarity between the depth correlation maps and $\mathbf{PseudoGT}$, respectively, which are in other words, the reliability. With the generated $\mathbf{RSM_{gt}}$, RSM is optimized by minimizing the L_1 distance between $\mathbf{RelyW}$ and $\mathbf{RSM_{gt}}$.

3.6 Model Optimization

We optimize the model with three object functions L_{mask}, L_{PSNM} and L_{RSM}. First, L_{mask} is the intersection over union (IOU) loss between the predicted saliency map I_{pred} and the ground truth mask I_{GT}, expressed as:

$$L_{mask}\left(I_{pred}, I_{GT}\right) = 1 - \frac{\sum_{x,y} min\left(I_{pred(x,y)}, I_{GT(x,y)}\right)}{\sum_{x,y} max\left(I_{pred(x,y)}, I_{GT(x,y)}\right)}, \tag{7}$$

where (x, y) are the pixel coordinates. Next, as described in Sect. 3.4, L_{PSNM} is the binary cross entropy loss between $\mathbf{AM_{Pred}}$ and $\mathbf{AM_{GT}}$. Finally, the loss function L_{RSM} for RSM is defined by measuring the L_1 distance between $\mathbf{RelyW}$ and $\mathbf{RSM_{gt}}$. L_{RSM} is expressed as:

$$L_{RSM}\left(\mathbf{RelyW}, \mathbf{RSM_{gt}}\right) = |\mathbf{RelyW}, \mathbf{RSM_{gt}}| \tag{8}$$

As a result, we combine all these constraints regarding PSNM, RSM, and I_{pred}, and obtain the following objective function L:

$$L_{total} = \lambda_m L_{mask} + \lambda_p L_{PSNM} + \lambda_r L_{RSM}, \tag{9}$$

where λ_m, λ_p, and λ_r denote the weights controlling the contribution of each multiplied loss function.

4 Experiments

4.1 Datasets

We perform our experiments on the following five popular RGB-D SOD benchmarks to validate the effectiveness of our proposed method: NJU2K [25], NLPR [35], STERE [33], DES [11] and SIP [19]. NJU2K [25] and NLPR [35] consists of 1985 and 1000 paired stereoscopic images, respectively. STERE [33] consists of 1000 stereo images collected from the Internet. DES [11], which is also called RGBD135 in some other papers captures seven indoor scenes and contains 135 indoor images acquired by Microsoft Kinect. SIP [19] is a high-quality dataset with 929 images. To make a fair comparison with previous works, we conduct experiments with two different training setups. First, we use 1485 samples from NJU2K [25] and 700 samples from NLPR [35] following the same setup as [29,32,51]. Second, we follow the same training settings as existing works [10,23,24,36,42,49,50], using 800 samples from DUT-RGBD [36], 1485 samples from NJU2K [25] and 700 samples from NLPR [35].

4.2 Evaluation Metrics

We evaluate the performance of our method and other methods using five widely used evaluation metrics: the mean F-measure (F_β) [1], mean absolute error (MAE, M) [4], S-measure (S_α) [17], E-measure (E_ξ) [18], and precision-recall (PR) curve.

Table 1. Quantitative comparison on five representative large-scale benchmark datasets. ↑ indicates that higher is better and ↓ indicates that lower is better. $*$ denotes the models are trained on NJU2K [25] and NLPR [35]; the rest are trained on DUT-RGBD [36], NJU2K [25], and NLPR [35]. The best and second best are highlighted in red and blue, respectively.

	Metric	DMRA [36]	CPFP [51] $*$	CIM [50]	CoN [24]	CMWN [29] $*$	PGAR [10]	CasG [49]	ATS [32] $*$	D2F [42]	DCF [23]	Ours $*$	Ours
	Venue	ICCV 2019	CVPR 2019	CVPR 2020	ECCV 2020	ECCV 2020	ECCV 2020	ECCV 2020	ECCV 2020	CVPR 2021	CVPR 2021		
NJU2K [25]	E_ξ ↑	.908	.895	–	.912	.936	.916	.877	.921	.923	.922	.943	.950
	S_α ↑	.886	.878	.899	.894	.903	.909	.849	.901	.903	–	.912	.918
	F_β ↑	.872	.837	.886	.872	.902	.893	.864	.893	.901	.897	.912	.920
	M ↓	.051	.053	.043	.047	.046	.042	.073	.040	.039	.038	.033	.032
NLPR [35]	E_ξ ↑	.941	.924	–	.936	.951	.955	.952	.945	.950	.956	.962	.958
	S_α ↑	.899	.888	.914	.907	.917	.930	.919	.907	.918	–	.926	.923
	F_β ↑	.854	.822	.875	.848	.903	.885	.904	.876	.897	.893	.914	.910
	M ↓	.031	.036	.026	.031	.029	.024	.025	.028	.024	.023	.022	.023
STERE [33]	E_ξ ↑	.920	.903	–	.923	.941	.919	.930	.921	.933	.931	.942	.943
	S_α ↑	.886	.879	.893	.908	.905	.913	.899	.897	.904	–	.906	.907
	F_β ↑	.867	.830	.880	.885	.901	.880	.901	.884	.898	.890	.898	.900
	M ↓	.047	.051	.044	.041	.043	.041	.039	.039	.036	.037	.035	.035
DES [11]	E_ξ ↑	.944	.927	–	.945	.969	.939	.947	.952	.962	–	.976	.974
	S_α ↑	.900	.872	.905	.910	.934	.913	.905	.907	.920	–	.938	.937
	F_β ↑	.866	.829	.876	.861	.930	.880	.906	.885	.896	–	.943	.936
	M ↓	.030	.038	.025	.027	.022	.026	.028	.024	.021	–	.016	.016
SIP [19]	E_ξ ↑	.863	.899	–	.909	.913	.908	–	–	–	.920	.936	.934
	S_α ↑	.806	.850	–	.858	.867	.876	–	–	–	–	.890	.892
	F_β ↑	.819	.819	–	.842	.874	.854	–	–	–	.877	.896	.890
	M ↓	.085	.064	–	.063	.062	.055	–	–	–	.051	.042	.042

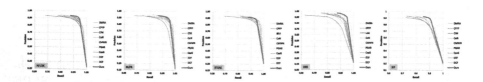

Fig. 6. Precision-recall curve comparison on five datasets.

4.3 Implementation Details

We set the number of superpixels N_S to 100 and a_k of EdgeConv to 10. We also set λ_m, λ_p, and λ_r in Eq. 9 to 1, 1, and 10, respectively, for balanced training. We implement the proposed method using the open deep learning framework PyTorch. The backbone network is equipped with VGG-16 [40], with initial parameters pre-trained in ImageNet [14]. All images are uniformly resized to 352×352 pixels for training and inferring. For network training, we used the Adam optimizer [26] with $\beta_1 = 0.9$, $\beta_2 = 0.999$, and $\epsilon = 10^{-8}$. The learning rate decayed from 8×10^{-5} to 8×10^{-6} with the cosine annealing scheduler [31]. The total number of epochs was set to 200 with batch size 16 with two NVIDIA RTX 3090 GPUs for all experiments in this study.

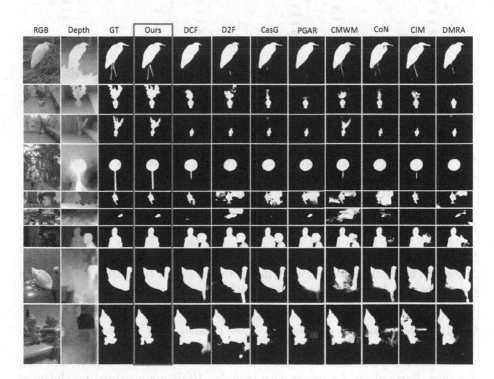

Fig. 7. Qualitative comparison with eight state-of-the-art methods.

4.4 Comparison with State-of-the-Art Methods

Quantitative Comparison. Table 1 shows our quantitative performance compared with 10 recently published state-of-the-art RGB-D SOD methods, DMRA [36], CPFP [51], CIM [50], CoN [24], CMWN [29], GAR [10], CasG [32], ATS [49], D2F [42], DCF [23], on five popular benchmark datasets. Because the training data in these comparative studies differ slightly, as some used NJU2K [25] and

Table 2. Performance with different combinations of our contributions. RE and DE represent the encoders for the RGB and depth, respectively. RS and DS are the set of FFM, PGM, and PSNM in the RGB and depth streams, respectively.

Index	Method RE	DE	RS	DS	RSM	NJU2K [25] $E_\xi\uparrow$	$S_\alpha\uparrow$	$F_\beta\uparrow$	$M\downarrow$	NLPR [35] $E_\xi\uparrow$	$S_\alpha\uparrow$	$F_\beta\uparrow$	$M\downarrow$	STERE [33] $E_\xi\uparrow$	$S_\alpha\uparrow$	$F_\beta\uparrow$	$M\downarrow$	DES [11] $E_\xi\uparrow$	$S_\alpha\uparrow$	$F_\beta\uparrow$	$M\downarrow$	SIP [19] $E_\xi\uparrow$	$S_\alpha\uparrow$	$F_\beta\uparrow$	$M\downarrow$
(a)	✓					.872	.836	.848	.058	.871	.852	.793	.044	.883	.852	.843	.061	.843	.775	.764	.048	.825	.738	.734	.069
(b)	✓	✓				.904	.863	.869	.051	.912	.877	.842	.037	.908	.870	.864	.052	.888	.831	.820	.038	.868	.796	.793	.062
(c)	✓	✓			✓	.934	.888	.887	.044	.952	.905	.888	.028	.934	.889	.883	.042	.932	.887	.877	.031	.913	.855	.853	.057
(d)	✓	✓	✓			.937	.903	.901	.039	.950	.915	.896	.026	.938	.902	.894	.037	.950	.905	.900	.027	.918	.870	.878	.051
(e)	✓	✓	✓	✓		.944	.912	.910	.035	.954	.919	.902	.025	.940	.905	.898	.036	.963	.922	.917	.022	.926	.881	.887	.047
(f)	✓	✓	✓	✓	✓	.950	.918	.920	.032	.958	.923	.910	.023	.943	.907	.900	.035	.974	.937	.936	.016	.934	.892	.899	.042

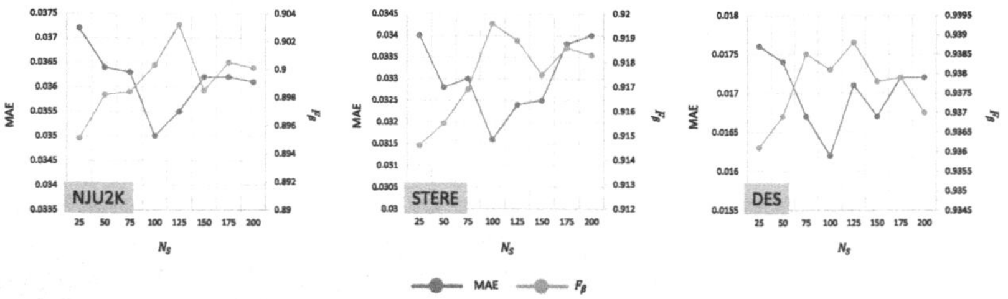

Fig. 8. Comparison of performance characteristics with respect to N_S for the NJU2K [25], STERE [33], and DES [11] datasets.

NLPR [35] whereas others also used DUT-RGBD [36], we show the performance on both settings for a fair comparison. It is observed that our model notably outperforms the other methods. In particular, our model exceeds the counterpart methods by a dramatic margin in terms of all four evaluation metrics on NJU2K [25] and DES [11], which are considered more challenging than to the others due to the low contrast and objects cluttering the background. This result further indicates that our network can perform well on various complex scenes. Moreover, we plotted the PR curves in Fig. 6 for a better comparison. The results show that ours lies above most of the methods compared.

Qualitative Comparison. In Fig. 7, we compare our qualitative results to those of eight top-ranking RGB-D SOD approaches on several challenging scenarios, including low contrast, reflection, thin objects, multiple objects, and long distance. Particularly for scenes with complex RGB maps caused by cluttered objects and patterns in the background (e.g., the second, third, and fourth row), our model utilized more information from the reliable depth maps to generate an accurate saliency map. Furthermore, the accuracy of such scenes is boosted by the PSNM, which effectively discriminates the foreground from the background. Similarly, our model can handle samples with depth maps that are ambiguous because of light reflection and long distance (e.g., the sixth and eighth rows) because our model adaptively decides to rely on the more accurate RGB maps. Furthermore, it is observed that our model is robust to scenes with multiple objects (e.g., the fifth and ninth rows).

4.5   Ablation Analysis

We verify the performance of our model through various ablation studies. Table 2 shows the effects of the proposed modules in various combinations. RE and DE in Table 2 represent the VGG-16 [40] encoders for the RGB images and depth maps, respectively. In addition, RS and DS are the set of FFM, PGM, and PSNM in the RGB and depth streams, respectively. The proposed RSM only applies when both RE and DE are used. Figure 8 also shows the performance of our model according to the number of superpixels.

Impact of Prototype Sampling. As shown in Table 2, (d) and (e), to which the prototype sampling method is applied, our method achieves better performance than (a) and (b) on all datasets. This is because the encoder-decoder-based network delivers not only the features for the salient object but also the background and non-salient object feature information extracted from the encoder to the decoder, preventing accurate mask generation. In contrast, the proposed SPSN model performs well because the network can selectively extract only the important salient object feature information by PSNM. Furthermore, Fig. 9 shows the salient prototype sampling results of the proposed method.

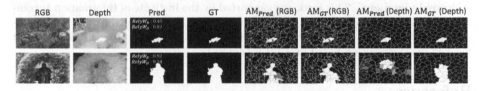

Fig. 9. Visualization of our results in several challenging situations. AM_{pred} and AM_{GT} are described in Sect. 3.4, and $RelyW_R$ and $RelyW_D$ are described in Sect. 3.5

Impact of RSM. When the proposed RSM module is applied, as shown in Table 2 (c) and (f), it shows significant performance improvement when the RGB image and depth map are used together. The performance improves because RSM selects feature maps generated from RGB and depth streams based on their reliability. Therefore, as shown in Fig. 9, $RelyW_R$ is small for RGB images with camouflaged objects, and $RelyW_D$ is small for low-quality depth maps. This structure shows that the model reduces the biased dependence and makes it robust to low-quality depth maps.

Number of Superpixels. We conduct ablation studies to observe how the MAE and F_β values change according to the number of superpixels N_S. Figure 8 shows the changes in performance using the NJU2K [25], STERE [33], and DES [11] datasets according to the number of superpixels. As shown in Fig. 8, the proposed model performs best near $N_S = 100$. Additionally, if N_S is too small or too large, the performance will decrease. This degraded performance results because if N_S is too small, the superpixel masks cannot effectively separate salient and non-salient objects and cannot provide a sufficient number of component prototypes. Conversely, if N_S is too large, it is difficult to create coherent features for the salient object by creating too small superpixel masks that are too small.

5   Conclusion

In this paper, we aim to segment salient objects by designing an SPSN, which suppresses the effects of background objects and effectively takes advantage of RGB and depth maps. Specifically, our network is composed of four novel modules-the FFM, which fuses the multiscale features extracted from the encoder; the PGM, which renders the fused feature maps to component prototypes; the PSNM, which discriminates the prototype that belongs to the salient object; and the RSM, which adaptively selects the contribution of RGB and depth features. The results demonstrate the outstanding improvement of our method over the previous studies, indicating that our model can capture salient objects in various challenging scenes. Furthermore, extensive ablation studies show the contribution and effectiveness of each of the proposed modules.

Acknowledgement. This work was supported by the Institute of Information & communications Technology Planning & Evaluation(IITP) grant funded by the Korea government(MSIT) (No. 2021-0-00172, The development of human Re-identification and masked face recognition based on CCTV camera).

References

1. Achanta, R., Hemami, S., Estrada, F., Susstrunk, S.: Frequency-tuned salient region detection. In: 2009 IEEE Conference on Computer Vision and Pattern Recognition, pp. 1597–1604. IEEE (2009)
2. Achanta, R., Shaji, A., Smith, K., Lucchi, A., Fua, P., Süsstrunk, S.: Slic superpixels compared to state-of-the-art superpixel methods. IEEE Trans. Pattern Anal. Mach. Intell. **34**(11), 2274–2282 (2012)
3. Agarap, A.F.: Deep learning using rectified linear units (relu). arXiv preprint arXiv:1803.08375 (2018)
4. Borji, A., Cheng, M.M., Jiang, H., Li, J.: Salient object detection: a benchmark. IEEE Trans. Image Process. **24**(12), 5706–5722 (2015)
5. Chen, H., Li, Y.F., Su, D.: M 3 net: multi-scale multi-path multi-modal fusion network and example application to rgb-d salient object detection. In: 2017 IEEE/RSJ International Conference on Intelligent Robots and Systems (IROS), pp. 4911–4916. IEEE (2017)
6. Chen, H., Li, Y.F., Su, D.: Attention-aware cross-modal cross-level fusion network for rgb-d salient object detection. In: 2018 IEEE/RSJ International Conference on Intelligent Robots and Systems (IROS), pp. 6821–6826. IEEE (2018)
7. Chen, H., Li, Y.: Progressively complementarity-aware fusion network for rgb-d salient object detection. In: Proceedings of the IEEE Conference on Computer Vision and Pattern Recognition, pp. 3051–3060 (2018)

8. Chen, H., Li, Y.: Three-stream attention-aware network for rgb-d salient object detection. IEEE Trans. Image Process. **28**(6), 2825–2835 (2019)

9. Chen, L.C., Papandreou, G., Kokkinos, I., Murphy, K., Yuille, A.L.: Deeplab: semantic image segmentation with deep convolutional nets, atrous convolution, and fully connected crfs. IEEE Trans. Pattern Anal. Mach. Intell. **40**(4), 834–848 (2017)

10. Chen, S., Fu, Y.: Progressively guided alternate refinement network for RGB-D salient object detection. In: Vedaldi, A., Bischof, H., Brox, T., Frahm, J.-M. (eds.) ECCV 2020. LNCS, vol. 12353, pp. 520–538. Springer, Cham (2020). https://doi.org/10.1007/978-3-030-58598-3_31

11. Cheng, Y., Fu, H., Wei, X., Xiao, J., Cao, X.: Depth enhanced saliency detection method. In: Proceedings of International Conference on Internet Multimedia Computing and Service, pp. 23–27 (2014)

12. Cong, R., Lei, J., Fu, H., Hou, J., Huang, Q., Kwong, S.: Going from RGB to RGBD saliency: a depth-guided transformation model. IEEE Trans. Cybern. **50**(8), 3627–3639 (2019)

13. Cong, R., Lei, J., Zhang, C., Huang, Q., Cao, X., Hou, C.: Saliency detection for stereoscopic images based on depth confidence analysis and multiple cues fusion. IEEE Signal Process. Lett. **23**(6), 819–823 (2016)

14. Deng, J., Dong, W., Socher, R., Li, L.J., Li, K., Fei-Fei, L.: Imagenet: a large-scale hierarchical image database. In: 2009 IEEE Conference on Computer Vision and Pattern Recognition, pp. 248–255. Ieee (2009)

15. Desingh, K., Krishna, K.M., Rajan, D., Jawahar, C.: Depth really matters: improving visual salient region detection with depth. In: BMVC, pp. 1–11 (2013)

16. Fan, D.-P., Cheng, M.-M., Liu, J.-J., Gao, S.-H., Hou, Q., Borji, A.: Salient objects in clutter: bringing salient object detection to the foreground. In: Ferrari, V., Hebert, M., Sminchisescu, C., Weiss, Y. (eds.) ECCV 2018. LNCS, vol. 11219, pp. 196–212. Springer, Cham (2018). https://doi.org/10.1007/978-3-030-01267-0_12

17. Fan, D.P., Cheng, M.M., Liu, Y., Li, T., Borji, A.: Structure-measure: a new way to evaluate foreground maps. In: Proceedings of the IEEE International Conference on Computer Vision, pp. 4548–4557 (2017)

18. Fan, D.P., Gong, C., Cao, Y., Ren, B., Cheng, M.M., Borji, A.: Enhanced-alignment measure for binary foreground map evaluation. arXiv preprint arXiv:1805.10421 (2018)

19. Fan, D.P., Lin, Z., Zhang, Z., Zhu, M., Cheng, M.M.: Rethinking rgb-d salient object detection: models, data sets, and large-scale benchmarks. IEEE Trans. Neural Netw. Learn. Syst. **32**(5), 2075–2089 (2020)

20. Fu, J., et al.: Dual attention network for scene segmentation. In: Proceedings of the IEEE/CVF Conference on Computer Vision and Pattern Recognition, pp. 3146–3154 (2019)

21. Fu, K., Fan, D.P., Ji, G.P., Zhao, Q.: Jl-dcf: joint learning and densely-cooperative fusion framework for rgb-d salient object detection. In: Proceedings of the IEEE/CVF Conference on Computer Vision and Pattern Recognition, pp. 3052–3062 (2020)

22. Guo, J., Ren, T., Bei, J.: Salient object detection for rgb-d image via saliency evolution. In: 2016 IEEE International Conference on Multimedia and Expo (ICME), pp. 1–6. IEEE (2016)

23. Ji, W., et al.: Calibrated rgb-d salient object detection. In: Proceedings of the IEEE/CVF Conference on Computer Vision and Pattern Recognition, pp. 9471–9481 (2021)

24. Ji, W., Li, J., Zhang, M., Piao, Y., Lu, H.: Accurate RGB-D salient object detection via collaborative learning. In: Vedaldi, A., Bischof, H., Brox, T., Frahm, J.-M. (eds.) ECCV 2020. LNCS, vol. 12363, pp. 52–69. Springer, Cham (2020). https://doi.org/10.1007/978-3-030-58523-5_4

25. Ju, R., Ge, L., Geng, W., Ren, T., Wu, G.: Depth saliency based on anisotropic center-surround difference. In: 2014 IEEE International Conference on Image Processing (ICIP), pp. 1115–1119. IEEE (2014)

26. Kingma, D.P., Ba, J.: Adam: a method for stochastic optimization. arXiv preprint arXiv:1412.6980 (2014)

27. Lang, C., Nguyen, T.V., Katti, H., Yadati, K., Kankanhalli, M., Yan, S.: Depth matters: influence of depth cues on visual saliency. In: Fitzgibbon, A., Lazebnik, S., Perona, P., Sato, Y., Schmid, C. (eds.) ECCV 2012. LNCS, vol. 7573, pp. 101–115. Springer, Heidelberg (2012). https://doi.org/10.1007/978-3-642-33709-3_8

28. Li, G., Jampani, V., Sevilla-Lara, L., Sun, D., Kim, J., Kim, J.: Adaptive prototype learning and allocation for few-shot segmentation. In: Proceedings of the IEEE/CVF Conference on Computer Vision and Pattern Recognition, pp. 8334–8343 (2021)

29. Li, G., Liu, Z., Ye, L., Wang, Y., Ling, H.: Cross-modal weighting network for RGB-D salient object detection. In: Vedaldi, A., Bischof, H., Brox, T., Frahm, J.-M. (eds.) ECCV 2020. LNCS, vol. 12362, pp. 665–681. Springer, Cham (2020). https://doi.org/10.1007/978-3-030-58520-4_39

30. Liu, Y., Zhang, X., Zhang, S., He, X.: Part-aware prototype network for few-shot semantic segmentation. In: Vedaldi, A., Bischof, H., Brox, T., Frahm, J.-M. (eds.) ECCV 2020. LNCS, vol. 12354, pp. 142–158. Springer, Cham (2020). https://doi.org/10.1007/978-3-030-58545-7_9

31. Loshchilov, I., Hutter, F.: Sgdr: stochastic gradient descent with warm restarts. arXiv preprint arXiv:1608.03983 (2016)

32. Luo, A., Li, X., Yang, F., Jiao, Z., Cheng, H., Lyu, S.: Cascade graph neural networks for RGB-D salient object detection. In: Vedaldi, A., Bischof, H., Brox, T., Frahm, J.-M. (eds.) ECCV 2020. LNCS, vol. 12357, pp. 346–364. Springer, Cham (2020). https://doi.org/10.1007/978-3-030-58610-2_21

33. Niu, Y., Geng, Y., Li, X., Liu, F.: Leveraging stereopsis for saliency analysis. In: 2012 IEEE Conference on Computer Vision and Pattern Recognition, pp. 454–461. IEEE (2012)

34. Park, C., Lee, M., Cho, M., Lee, S.: Saliency detection via global context enhanced feature fusion and edge weighted loss. arXiv preprint arXiv:2110.06550 (2021)

35. Peng, H., Li, B., Xiong, W., Hu, W., Ji, R.: RGBD salient object detection: a benchmark and algorithms. In: Fleet, D., Pajdla, T., Schiele, B., Tuytelaars, T. (eds.) ECCV 2014. LNCS, vol. 8691, pp. 92–109. Springer, Cham (2014). https://doi.org/10.1007/978-3-319-10578-9_7

36. Piao, Y., Ji, W., Li, J., Zhang, M., Lu, H.: Depth-induced multi-scale recurrent attention network for saliency detection. In: Proceedings of the IEEE/CVF International Conference on Computer Vision, pp. 7254–7263 (2019)

37. Qi, C.R., Su, H., Mo, K., Guibas, L.J.: Pointnet: deep learning on point sets for 3D classification and segmentation. In: Proceedings of the IEEE Conference on Computer Vision and Pattern Recognition, pp. 652–660 (2017)

38. Qi, C.R., Yi, L., Su, H., Guibas, L.J.: Pointnet++: deep hierarchical feature learning on point sets in a metric space. Adv. Neural Inf. Process. Syst. 30, 1–10 (2017)

39. Qu, L., He, S., Zhang, J., Tian, J., Tang, Y., Yang, Q.: Rgbd salient object detection via deep fusion. IEEE Trans. Image Process. 26(5), 2274–2285 (2017)

40. Simonyan, K., Zisserman, A.: Very deep convolutional networks for large-scale image recognition. arXiv preprint arXiv:1409.1556 (2014)
41. Song, H., Liu, Z., Du, H., Sun, G., Le Meur, O., Ren, T.: Depth-aware salient object detection and segmentation via multiscale discriminative saliency fusion and bootstrap learning. IEEE Trans. Image Process. **26**(9), 4204–4216 (2017)
42. Sun, P., Zhang, W., Wang, H., Li, S., Li, X.: Deep rgb-d saliency detection with depth-sensitive attention and automatic multi-modal fusion. In: Proceedings of the IEEE/CVF Conference on Computer Vision and Pattern Recognition, pp. 1407–1417 (2021)
43. Wang, K., Liew, J.H., Zou, Y., Zhou, D., Feng, J.: Panet: few-shot image semantic segmentation with prototype alignment. In: Proceedings of the IEEE/CVF International Conference on Computer Vision, pp. 9197–9206 (2019)
44. Wang, W., Zhao, S., Shen, J., Hoi, S.C., Borji, A.: Salient object detection with pyramid attention and salient edges. In: Proceedings of the IEEE/CVF Conference on Computer Vision and Pattern Recognition, pp. 1448–1457 (2019)
45. Wang, X., Girshick, R., Gupta, A., He, K.: Non-local neural networks. In: Proceedings of the IEEE Conference on Computer Vision and Pattern Recognition, pp. 7794–7803 (2018)
46. Wang, Y., Sun, Y., Liu, Z., Sarma, S.E., Bronstein, M.M., Solomon, J.M.: Dynamic graph cnn for learning on point clouds. ACM Trans. Graph. (TOG) **38**(5), 1–12 (2019)
47. Zhang, H., Goodfellow, I., Metaxas, D., Odena, A.: Self-attention generative adversarial networks. In: International conference on machine learning, pp. 7354–7363. PMLR (2019)
48. Zhang, J., et al.: Uc-net: Uncertainty inspired rgb-d saliency detection via conditional variational autoencoders. In: Proceedings of the IEEE/CVF Conference on Computer Vision and Pattern Recognition, pp. 8582–8591 (2020)
49. Zhang, M., Fei, S.X., Liu, J., Xu, S., Piao, Y., Lu, H.: Asymmetric two-stream architecture for accurate RGB-D saliency detection. In: Vedaldi, A., Bischof, H., Brox, T., Frahm, J.-M. (eds.) ECCV 2020. LNCS, vol. 12373, pp. 374–390. Springer, Cham (2020). https://doi.org/10.1007/978-3-030-58604-1_23
50. Zhang, M., Ren, W., Piao, Y., Rong, Z., Lu, H.: Select, supplement and focus for rgb-d saliency detection. In: Proceedings of the IEEE/CVF Conference on Computer Vision and Pattern Recognition, pp. 3472–3481 (2020)
51. Zhao, J.X., Cao, Y., Fan, D.P., Cheng, M.M., Li, X.Y., Zhang, L.: Contrast prior and fluid pyramid integration for rgbd salient object detection. In: Proceedings of the IEEE/CVF Conference on Computer Vision and Pattern Recognition, pp. 3927–3936 (2019)
52. Zhao, J.X., Liu, J.J., Fan, D.P., Cao, Y., Yang, J., Cheng, M.M.: Egnet: edge guidance network for salient object detection. In: Proceedings of the IEEE/CVF International Conference on Computer Vision, pp. 8779–8788 (2019)
53. Zhao, R., Ouyang, W., Li, H., Wang, X.: Saliency detection by multi-context deep learning. In: Proceedings of the IEEE Conference on Computer Vision and Pattern Recognition, pp. 1265–1274 (2015)

Global Spectral Filter Memory Network for Video Object Segmentation

Yong Liu[1], Ran Yu[1], Jiahao Wang[1], Xinyuan Zhao[3], Yitong Wang[2],
Yansong Tang[1], and Yujiu Yang[1(✉)]

[1] Tsinghua Shenzhen International Graduate School, Tsinghua University,
Beijing, China
{liu-yong20,yu-r19}@mails.tsinghua.edu.cn,
{tang.yansong,yang.yujiu}@sz.tsinghua.edu.cn
[2] ByteDance Inc., Beijing, China
[3] Northwestern University, Evanston, USA

Abstract. This paper studies semi-supervised video object segmentation through boosting intra-frame interaction. Recent memory network-based methods focus on exploiting inter-frame temporal reference while paying little attention to intra-frame spatial dependency. Specifically, these segmentation model tends to be susceptible to interference from unrelated nontarget objects in a certain frame. To this end, we propose Global Spectral Filter Memory network (GSFM), which improves intra-frame interaction through learning long-term spatial dependencies in the spectral domain. The key components of GSFM is 2D (inverse) discrete Fourier transform for spatial information mixing. Besides, we empirically find low frequency feature should be enhanced in encoder (backbone) while high frequency for decoder (segmentation head). We attribute this to semantic information extracting role for encoder and fine-grained details highlighting role for decoder. Thus, Low (High) Frequency Module is proposed to fit this circumstance. Extensive experiments on the popular DAVIS and YouTube-VOS benchmarks demonstrate that GSFM noticeably outperforms the baseline method and achieves state-of-the-art performance. Besides, extensive analysis shows that the proposed modules are reasonable and of great generalization ability. Our source code is available at https://github.com/workforai/GSFM.

Keywords: Video object segmentation · Spectral domain

1 Introduction

Video Object Segmentation (VOS) [35,36,58,63] aims at identifying and segmenting objects in videos. It is one of the most challenging tasks in computer vision with many potential applications, including interactive video editing, augmented reality [31], and autonomous driving [71]. In this paper, we focus on the semi-supervised setting where target objects are defined by the given masks of

---

Y. Liu—This work was done during an internship at ByteDance Inc.

S. Avidan et al. (Eds.): ECCV 2022, LNCS 13689, pp. 648–665, 2022.
https://doi.org/10.1007/978-3-031-19818-2_37

(a) Query pixel in current frame   (b) Matched pixels (STCN)   (c) Matched pixels (Ours)

Fig. 1. Illustration of the disadvantages of lacking semantic global information. The highlight red pixels in the first column are target pixels. The second column shows that previous method [6] would incorrectly match similar pixels of other objects. In the third column, our model relieves the confusion problem by enhancing low-frequency components and updating features from spectral domain. (Color figure online)

the first frame. It is crucial for semi-supervised VOS to fully utilize the available reference information to distinguish targets from background objects.

Since the critical problem of this task lies in how to make full use of the spatial-temporal dependency to recognize the targets, matching-based methods, which perform pixel-level matching with historical reference frames, have received tremendous attention. The Space-Time Memory Network [33] memorizes intermediate frames with segmentation masks as references and performs pixel-level matching between them with the current frame to segment target objects in a bottom-up manner, which has been proved effective and has served as the current mainstream framework. Some works [5,6,15,22,24,38,39,44,48, 56,56,59] further develop STM and have achieved excellent performance.

Although these methods have made great progress in the field of VOS, they pay little attention to excavating intra-frame dependency and only utilize local representation for matching and prediction due to the inductive bias of convolution. Lacking global dependency would cause low efficacy in distinguishing similar pixels, *e.g.*, pixels of similar color or objects of the same category. We take the typical method STCN [6] for illustration. In Fig. 1 (b), some pixels belonging to background objects are mismatched with the target pixel due to their similar local features. Ignoring long-range dependency for matching would lead to a high risk of interference from other objects. Since the matching-based approaches rely on the matching process to identify the targets, incorrectly matched pixels would negatively affect the final segmentation and even lead to error accumulation. Therefore, it is necessary to excavate the intra-frame spatial dependency to enhance the representation of features.

According to the Fourier theory [19], FFT function generates outputs based on pixels from all spatial locations when processing input feature. Thus, the spectral domain representation contains rich global information. Inspired by this, we introduce a Global Spectral Filter Memory network (GSFM), which fuses global dependency from spectral domain and distinguishes the high-frequency and low-frequency components for targeted enhancement. In GSFM, we propose the Low Frequency Module (LFM) and High Frequency Module (HFM) to enhance different representation according to the characteristics of the encoder-decoder network structure.

The role of encoder is to extract deep features for subsequent modules, and the encoded features need to contain rich semantic information. Intuitively, low-frequency components correspond to high-level semantic information while ignoring details. Some theoretical researches on CNN from spectral domain [49,61,67] also point out similar observations. Inspired by the above analysis, we propose a Low-Frequency Module (LFM) for the encoding process to update the features in the spectral domain and emphasize their low-frequency components. Figure 1 (c) illustrates that with LFM enhancing global semantic information, the distinguishability of similar pixels is greatly improved. Extensive experiments also demonstrate the rationality of emphasizing low-frequency in the encoder.

Different from encoding, features in the decoding process need to contain more fine-grained information for accurate prediction. And high-frequency components correspond to the image parts that change drastically, *e.g.*, object boundaries and texture details. Combined with the above analysis, we believe that focusing on high-frequency components would help to rich the fine-grained representation of features and make more accurate predictions of boundaries or ambiguous regions. Therefore, we introduce a High Frequency Module (HFM) in the decoding process, which enhances the high-frequency components of features to better capture detailed information. Besides, to take full advantage of HFM, we combine it with an additional boundary prediction branch to provide better localization and shape guidance.

Experiments show that the proposed model noticeably outperforms the baseline method and achieves state-of-the-art performance on DAVIS [35,36] and YouTube-VOS [58] datasets. The contribution of this paper can be summarized as follows. Firstly, we propose to leverage the spectral domain to enhance the global spatial dependency of features for semi-supervised VOS. Secondly, considering the differences between the process of encoding and decoding, we propose LFM and HFM to perform targeted enhancement, respectively. Thirdly, we combine object boundaries and high-frequency to provide better localization and shape information while keeping the decoding features are fine-grained.

2   Related Work

Semi-supervised Video Object Segmentation. Since the masks for the first frame are given, early methods [3,28,47,47,54] take the strategy that online fine-tune the network according to the object mask of the first frame, which suffers from slow inference speed. Propagation-based methods [1,7,8,13,16,18,21,45,60] forward propagate the segmentation masks as a reference to the next frame, and they are difficult to handle complicated scenarios. Some other researchers have decoupled VOS into three independent subtasks of detection, tracking, and segmentation [17,21,27,43]. Although this approach balances running time and accuracy, it is extremely dependent on the performance of the detectors and makes the entire pipeline complex.

In recent years matching-based methods have received great attention for excellent performance and robustness. FEELVOS [46], CFBI [64] and CFBI+ [66] perform global and local matching with the first frame and the previous adjacent frame, respectively. AOT [65] associates multiple target objects into the same embedding space by employing an identification mechanism. STM [33] leverages the memory network to memorize intermediate frames as references, which has been proved effective and has served as the current mainstream framework. Based on STM, KMN [38] and RMNet [56] perform local-to-local matching by using the Gaussian kernel and hard crop strategy. SwiftNet [48] and AFB-URR [24] reduce memory redundancy by calculating the similarity between query and memory. LCM [15] and SCM [70] proposes spatial constraint to enhance spatial location information. EGMN [26] employs an episodic memory network to memorize frames as nodes and capture cross-frame correlations by edges. MiVOS [5] further developed KMN [38] by utilizing the top-k strategy to reduce noise information in the memory read block. STCN [6] improves the feature extraction and performs more reasonable matching by decoupling the image and masks.

Despite the great performance achieved by these methods, they ignore the importance of fully excavating the intra-frame global information, which may lead to a high risk of interference by pixels with similar local features.

Spectral Domain Learning. Recent years have witnessed increasing research enthusiasm on combining spectral domain and deep learning [10,37,40,49,55,67]. Among them, some researches [49,61,67] attempt to explain the behavior of convolution neural network from the perspective of spectral domain. They point out that the features of different frequency bands represent different types of information and observe some properties of deep neural networks related to it. With the guidance of these works and rethinking about the characteristics of the encoder-decoder structure, we propose separating the high-frequency and low-frequency components for reasonably utilizing them. In this paper, we introduce a low-frequency module (LFM) and a high-frequency module (HFM). LFM enhances the low-frequency components during encoding to fuse global semantic features, while HFM enhances the high-frequency components in the decoder to make features contain more fine-grained details.

Some previous methods [24,62,73] applying spatial prior filter or introducing boundary to features can also be explained from the perspective of spectral domain. Applying filter kernels or highlighting boundaries in the spatial domain is essentially a special way to distinguish between high and low frequencies. While this approach can also serve the purpose of targeted enhancement, it loses the advantage of global perception in the spectral domain. Therefore, our approach that updates features in the spectral domain is more generalized and effective.

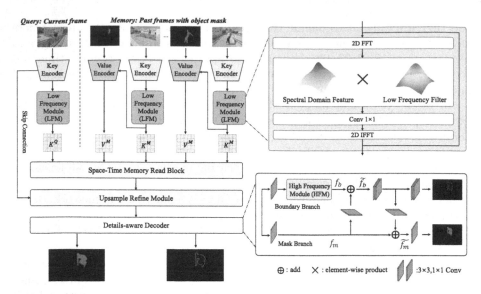

Fig. 2. Overview of GSFM. The network takes both query (current frame) and memory (past frames with masks) as input. LFM enhances low-frequency components of features and fuses global information in the spectral domain. Having K^M, K^Q and V^M extracted from the encoder, the memory read block calculates similarity between query and memory. The refine module upsamples the features and outputs to the decoder. With HFM enhancing high-frequency components, the decoder jointly predicts object masks and boundaries.

3   Method

3.1   Overview

The overall architecture of our GSFM is shown in Fig. 2. Given a video sequence and the annotation of the first frame, we process it frame by frame. During processing, the current frame is considered a query, and the past reference frames with segmentation masks are memory. Following the baseline method STCN [6], a Key Encoder extracts key features for each frame, and a Value Encoder extracts value features only for memory frames. By performing matching between query and memory in Space-Time Memory Read Block, the decoder identifies and segments the target object in a bottom-up manner. In the encoder, for exploiting the intra-frame semantic information to improve the representative capacity of features and promote the effectiveness of matching, a low-frequency module (LFM) enhances the low-frequency components of the features and performs global information updating from the spectral domain. In the decoder, the high-frequency module (HFM) enhances high-frequency components to highlight fine-grained information for accurate prediction. Besides, we take the strategy that jointly learning object boundaries and masks in an end-to-end manner [9,25,42,72]. With the interaction between mask branch and boundary branch features, the network can better perceive the localization and shape information, which also helps identify the target objects.

3.2 Frequency Modules

According to the spectral convolution theorem [2] in Fourier theory, updating a single value in the spectral domain affects globally all original data, which sheds light on design operations with the non-local receptive field. Intuitively, the high-frequency components correspond to the pixels varying drastically, such as object boundaries and textures, while the low-frequency components correspond to the general semantic information. Some previous theoretical studies [49,61,67] on spectral-domain and deep learning also point similar observation. Besides, to show the information represented by different frequency components more vividly, we take Fig. 3 as an example (for convenience, we use the grayscale image). In Fig. 3, the remaining informa-

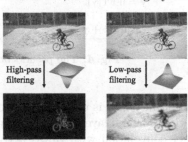

tion after high-pass filtering is the edges and details of objects. After low-pass filtering, the result is an image that retains the general semantic information (some details and noise are blurred). Considering that the role of the encoder is to extract high-level semantic information, while the decoder pursues focus on detailed features, we believe that this difference is similar to the difference between the high and low-frequency components. Thus, we propose a low-frequency module (LFM) for the encoder and a high-frequency module (HFM) for the decoder.

Fig. 3. Illustration of different frequency components. The top line is the original image.

The architecture of LFM and HFM is the same, and their difference is the frequency domain filter (LFM is a low-pass filter, and HFM is a high-pass filter). In our experiments, the filter is set in the form of Gaussian. Here we take LFM as an example to introduce the process. As shown in Fig. 4, having the image feature tensor x, LFM first transfers it to the spectral domain by FFT. Then the spectral features y will be passed through a low-pass filter to enhance low-frequency components, which helps to make features rich in global semantic information. Specifically, we generate a coefficient map g with the same spatial size of the feature y and perform element-wise multiplication between them with the help of broadcast mechanism. For LFM, the center of the coefficient map has the value of 0 and increases around in the form of Gaussian (without spectrum centralization, the center of the spectrum after FFT is high frequency, and the surrounding is low frequency). Before updating the spectral domain, note that the spectral features are complex numbers for the FFT operation. To make the complex number features compatible with the neural layers, we split the complex number into a real part y_r and an imaginary part y_i. For ease of computation, we append the imaginary part to the real part by concatenating them along the channel dimension, forming a new tensor with double channels. Essentially, the

---

Algorithm: Pseudocode of LFM

```
# x: input feature
# B: batchsize, C: dimension of channel, H, W: spatial size of input feature
# y_r, y_i: the real and imaginary part of the spectral features, respectively

# generate the Gaussian frequency filter
g = Make_gaussian_filter(H, W)   # g: (B, 1, H, W)
y = FFT(x)                              # y: (B, C, H, W)
y = y * g
# convert complex number features to real number
y_r, y_i = y.real, y.imag               # y r, y i: (B, C, H, W)
y = Concatenate([y_r, y_i], dim=1)      # y: (B, 2*C, H, W)
y = ReLU(Conv(y))
# convert back to complex number
y_r, y_i = Split(y, dim=1)              # y r, y i: (B, C, H, W)
y = Complex(y_r, y_i)
y = iFFT(y)                             # y: (B, C, H, W)

return x + y
```

Fig. 4. The pseudocode of low frequency module (LFM)

resultant tensor is treated as a vanilla real number tensor, and we can perform a series of neural layers on it. To update features in the spectral domain, we utilize 1×1 convolution with ReLU activation function. According to the convolution theorem [19], convolution in one domain equals point-wise multiplication in the other domain, which implies that the 1×1 convolution in spectral-domain incurs a global update in the spatial domain. After that, the results are converted back to complex numbers by splitting them into real and imaginary parts along the channel dimension. Inverse 2-D FFT operation transfers the spectral features back to the spatial domain. Finally, LFM outputs the enhanced features by adding the updated features y with initial tensor x.

3.3   Details-Aware Decoder

Only utilizing local information for pixel-level mask prediction may lead to a lack of overall perception of the objects and an over-reliance on pixel appearance information such as pixel color. Intuitively, object boundaries and object masks have a close relation. It would be helpful to locate and identify target objects from the background if the model has some sense of the shape or boundary of the objects, especially with HFM highlighting detailed information. Besides, since semi-supervised VOS is a pixel-level tracking task, accurate boundary segmentation is significant. Otherwise, it is easy to cause error accumulation. Therefore, we propose to combine HFM and object boundaries to provide localization and more detailed guidance.

The architecture of the decoder is shown in Fig. 2. Compared to the vanilla mask decoder of other memory network-based approaches, we add a branch dedicated to predicting object boundaries so that the model gives more attention to the object boundaries and shapes. The input feature of the boundary branch is first processed by HFM to enhance its high-frequency components, which helps to better perceive fine-grained information for accurate prediction. In addition, due to the special relationship between object boundaries and object masks, there is a lot of mutually exploitable information between their features. Specifically, features from the mask branch can provide basic information for localizing boundaries. After making sense of object boundaries, the shape and location information in boundary features is also conducive to guiding more precise mask predictions. To take full advantage of the special relationship between them, we take a fusion module [9] for the interaction between the mask branch and the boundary branch. Take the Mask → Boundary (M2B) Fusion as example, the fusion process can be formulated as follows:

$$\widetilde{f_b} = \mathcal{F}(f_m) + f_b, \tag{1}$$

where $\widetilde{f_b}$ denotes the fused boundary features, f_m is the mask branch feature, and f_b is the boundary branch feature. $\mathcal{F}$ is a 1×1 convolution with ReLU function. The fusion block is the same for the boundary → Mask (M2B) Fusion.

Boundary Ground Truth. Following previous works [57,68], we take the boundary prediction as a pixel-level classification problem. Since only the ground truth of the mask is available in the video object segmentation dataset, we use the Laplacian operator to generate the boundary ground truth. The Laplacian operator is a second-order gradient operator. As it is regarded as a classification problem, the resultant boundaries need to be converted into binary maps, and we binarize them with a threshold of 0.1.

Boundary Loss. Following previous work [9], we use dice loss [30] and binary cross-entropy to optimize the boundary predictions. Dice loss measures the overlap between predicted boundaries and ground truth. More importantly, dice loss can better handle category imbalance and focus on foreground pixels, which is compatible with boundary prediction (the number of boundary points is much less than points of non-boundary). The boundary loss $\mathcal{L}_b$ can be formulated as follows:

$$\mathcal{L}_b = \mathcal{L}_{Dice} + \mathcal{L}_{BCE}. \tag{2}$$

The dice loss is given as follows:

$$\mathcal{L}_{Dice} = 1 - \frac{2\sum_i p^i q^i}{\sum_i (p^i)^2 + \sum_i (q^i)^2 + \epsilon}, \tag{3}$$

where p and q denote the predictions and ground truth, respectively. i denotes the i-th pixel and ϵ is a smooth term to avoid zero division.

3.4 Other Modules

Encoder. Following STCN [6], we construct a Key Encoder and a Value Encoder. For each frame, the key features are extracted only once. In other words, we would reuse the "query key" as the "memory key" if one frame is memorized into the memory during video sequences. For memory frames, since both memory keys and memory values are extracted from the same image, it is natural to reuse existing key features as the input of value encoder. Specifically, a backbone first extracts memory features from images with segmentation masks and the resultant features are concatenated with the last layer features from key encoder. Then two ResBlocks [14] and a CBAM block [53] process them and output the final memory value features V^M.

Space-Time Memory Read Block. The query frame and T memory frames are encoded into the followings: memory key $K^M \in \mathbb{R}^{C^k \times T \times H/16 \times W/16}$, memory value $V^M \in \mathbb{R}^{C^v \times T \times H/16 \times W/16}$, query key $K^Q \in \mathbb{R}^{C^k \times H/16 \times W/16}$

In the Space-Time Memory Read block, activation weights are computed by measuring the similarities between K^Q and K^M. Then the V^M is retrieved by a weighted summation with the weights to get the output M. This operation can be summarized as:

$$M_i = \frac{1}{Z} \sum_j \mathcal{D}(K_i^Q, K_j^M) V_j^M, \tag{4}$$

where i and j are the index of the query and the memory location, $Z = \sum_j \mathcal{D}(K_i^Q, K_j^M)$ is the normalizing factor. $\mathcal{D}$ denotes similarity measure (following [6], in our experiments we take the L2 distance as measurement).

Refine Module. We use the same refinement module as previous works [5,6,32]. The role of the refinement modules is to process the matched value features and merge the detail information from the shallow layer of the encoder.

4 Implementation Details

Following the training strategy in previous works [5,6,65], we first pretrain our model on static image datasets [4,20,41,50,69] and then perform main training on YouTube-VOS and DAVIS datasets. During pretraining, each image is expanded into a pseudo video of three frames by data augmentation. For main training, we randomly pick three frames in chronological order (with a ground-truth mask for the first frame) from a video to form a training sample. The range of random sampling varies with the training process. In the intermediate period of training, the sampling range is set larger to improve the robustness of the model, while at the end of training, it is set smaller to narrow the gap between training and inference. We use randomly cropped 384×384 patches for training.

Our models are trained with eight 32 GB Tesla V100 GPUs with the Adam optimizer using PyTorch. The batch size is set to 16 for each GPU during pretraining and 8 during main training. It takes about 18 h to perform pretraining and 6 h for main training. We adopt ResNet50 [14] as backbone for key encoders

Table 1. The quantitative evaluation on DAVIS dataset. '*' indicates our re-implementation version. The results of baseline method are underlined

Method	DAVIS2016			DAVIS2017 val			DAVIS2017 test-dev			FPS
	$\mathcal{J}$	$\mathcal{F}$	$\mathcal{J}\&\mathcal{F}$	$\mathcal{J}$	$\mathcal{F}$	$\mathcal{J}\&\mathcal{F}$	$\mathcal{J}$	$\mathcal{F}$	$\mathcal{J}\&\mathcal{F}$	
RANet [51]	86.6	87.6	87.1	63.2	68.2	65.7	53.4	56.2	55.3	–
FEELVOS [46]	81.1	82.2	81.7	69.1	74.0	71.5	55.2	60.5	57.8	–
RGMP [32]	81.5	82.0	81.8	64.8	68.6	66.7	51.3	54.4	52.8	–
DMVOS [52]	88.0	87.5	87.8	–	–	–	–	–	–	–
STM [33]	88.7	89.9	89.3	79.2	84.3	81.8	69.3	75.2	72.2	7.9
KMN [38]	89.5	91.5	90.5	80.0	85.6	82.8	74.1	80.3	77.2	7.1
CFBI [64]	88.3	90.5	89.4	79.1	84.6	81.9	71.1	78.5	74.8	3.4
GCM [22]	87.6	85.7	86.6	69.3	73.5	71.4	–	–	–	–
G-FRTM [34]	–	–	84.3	–	–	76.4	–	–	–	–
GIEL [12]	–	–	–	80.2	85.3	82.7	72.0	78.3	75.2	–
SwiftNet [48]	90.5	90.3	90.4	78.3	83.9	81.1	–	–	–	20.6
RMNet [56]	88.9	88.7	88.8	81.0	86.0	83.5	71.9	78.1	75.0	< 11.9
SSTVOS [11]	–	–	–	79.9	85.1	82.5	–	–	–	–
LCM [15]	89.9	91.4	90.7	80.5	86.5	83.5	74.4	81.8	78.1	< 9.5
MiVOS [5]	87.8	90.0	88.9	80.5	85.8	83.1	72.6	79.3	76.0	6.5
JOINT [29]	–	–	–	80.8	86.2	83.5	–	–	–	3.8
RPCMVOS [59]	87.1	94.0	90.6	81.3	86.0	83.7	75.8	82.6	79.2	–
DMN-AOA [23]	–	–	–	81.0	87.0	84.0	74.8	81.7	78.3	< 6.2
HMMN [39]	89.6	92.0	90.8	81.9	87.5	84.7	74.7	82.5	78.6	6.8
AOT-L [65]	89.7	92.3	91.0	80.3	85.7	83.0	75.3	82.3	78.8	15.2
STCN* [6]	<u>90.1</u>	<u>92.2</u>	<u>91.1</u>	<u>81.5</u>	<u>87.9</u>	<u>84.7</u>	<u>72.7</u>	<u>79.6</u>	<u>76.1</u>	11.7
GSFM (Ours)	**90.1**	**92.7**	**91.4**	**83.1**	**89.3**	**86.2**	**74.0**	**80.9**	**77.5**	8.9

and ResNet18 for value encoder. Bootstrapped cross-entropy loss (hard example mining) is used for mask segmentation. Binary cross-entropy loss and Dice loss are used for boundary prediction. The weight of boundary prediction loss is 0.05. For inference, we adopt top-k filtering [5,6] in our experiment with $k = 50$ in default. We memorize every 3 frame, and no previous temporary frame is used. Unless otherwise specified, we utilize the DAVIS2017 val set for experiment analysis.

5   Experiments

5.1   Comparisons with State-of-the-Art Methods

DAVIS 2016 [35] is a densely annotated video object segmentation benchmark which contains 20 high-quality annotated video sequences. We compare GSFM with state-of-the-art methods in Table 1. Since the scenarios in this dataset are relatively simple and only focus on a single target object, the segmentation results of most of the methods are excellent. Based on the STCN [6], our method achieves the performance of 91.4 $\mathcal{J}\&\mathcal{F}$.

Table 2. Evaluation on YouTube-VOS 2018 validation set. Seen and Unseen denote the presence or absence of these categories in the training set, respectively. $\mathcal{G}$ is the averaged score of all $\mathcal{J}$ and $\mathcal{F}$.

Methods	Seen		Unseen		$\mathcal{G}$
	$\mathcal{J}$	$\mathcal{F}$	$\mathcal{J}$	$\mathcal{F}$	
OnAVOS [47]	60.1	62.7	46.6	51.4	55.2
PReMVOS [27]	71.4	75.9	56.5	63.7	66.9
STM [33]	79.7	84.2	72.8	80.9	79.4
AFB-URR [24]	78.8	83.1	74.1	82.6	79.6
GCM [22]	72.6	75.6	68.9	75.7	73.2
KMN [38]	81.4	85.6	75.3	83.3	81.4
G-FRTM [34]	68.6	71.3	58.4	64.5	65.7
SwiftNet [48]	77.8	81.8	72.3	79.5	77.8
SSTVOS [11]	80.9	–	76.6	–	81.8
RMNet [56]	82.1	85.7	75.7	82.4	81.5
LCM [15]	82.2	86.7	75.7	83.4	82.0
MiVOS [5]	80.0	84.6	74.8	82.4	80.4
JOINT [29]	81.5	85.9	78.7	86.5	83.1
HMMN [39]	82.1	87.0	76.8	84.6	82.6
RPCMVOS [59]	83.1	87.7	78.5	86.7	84.0
DMN-AOA [23]	82.5	86.9	76.2	84.2	82.5
AOT-L [65]	82.5	87.5	77.9	86.7	83.7
STCN* [6]	<u>81.8</u>	<u>86.4</u>	<u>77.8</u>	<u>85.6</u>	<u>82.9</u>
GSFM (Ours)	**82.8**	**87.5**	**78.3**	**86.5**	**83.8**

DAVIS 2017 [36] is a multiple objects benchmark. The validation set contains 59 objects in 30 videos. In the Table 1, GSFM achieves an average score of 86.2, which outperforms baseline methods by 1.5 $\mathcal{J}\&\mathcal{F}$. What's more, we also test our model on the more challenging DAVIS 2017 test-dev split set. It also significantly surpasses the baseline method (1.4 $\mathcal{J}\&\mathcal{F}$).

YouTube-VOS [58] is the largest benchmark available for video object segmentation. It contains 3471 videos in the training set (65 categories), 507 videos in the valid set (additional 26 categories not in the training set), and 541 videos in the test set. As shown in Table 2, our method achieves competitive results (83.8) on YouTube-VOS and outperforms the baseline methods by 0.9 $\mathcal{J}\&\mathcal{F}$.

Qualitative Results. Figure 5 shows some comparison examples between ours GSFM and STCN [6]. In the first example, similar pixels of the dogs are easily mis-segmented by STCN because only local information is used for matching. While with LFM enhancing global semantic information, GSFM can identify

Fig. 5. Visualization results of our proposed method. (a) denotes the segmentation results of our baseline method [6]. (b) is the results of our GSFM. The first example shows that our model can better perceive the overall semantic information of the target and thus identify similar objects. The second example shows that our approach makes a better determination of the ambiguous areas

targets more robustly. This is also illustrated in Fig. 1. The second example shows that with HFM enhancing fine-grained information, the proposed model has a better judgment for details and ambiguous areas.

5.2 Ablation Study

Table 3. Enhancing different frequency. $freq_L$, $freq_H$, and $freq_F$ denotes enhancing low, high, and full-frequency, respectively

LFM	HFM	$\mathcal{J}$	$\mathcal{F}$	$\mathcal{J}\&\mathcal{F}$
$freq_F$	$freq_H$	81.6	88.0	$84.8^{\downarrow 0.5}$
$freq_H$	$freq_H$	80.8	87.6	$84.2^{\downarrow 1.1}$
Attn	$freq_H$	81.6	88.2	$84.9^{\downarrow 0.4}$
$freq_L$	$freq_F$	81.4	87.9	$84.7^{\downarrow 0.6}$
$freq_L$	$freq_L$	81.2	87.7	$84.5^{\downarrow 0.8}$
$freq_L$	$freq_H$	**81.9**	**88.7**	**85.3**

Table 4. The quantitative results of generalization effect. FM denotes the proposed LFM and HFM and ✓ indicates deployed

Method	FM	$\mathcal{J}$	$\mathcal{F}$	$\mathcal{J}\&\mathcal{F}$
STM [33]		78.8	84.2	81.5
	✓	80.8	86.2	$\mathbf{83.5}^{\uparrow}$
KMN [38]		79.7	85.5	82.6
	✓	81.6	87.8	$\mathbf{84.7}^{\uparrow}$
MiVOS [5]		79.8	85.6	82.7
	✓	81.7	87.4	$\mathbf{84.6}^{\uparrow}$

Analysis on LFM and HFM. In addition to the observation in some theoretical works [49], we conduct experiments to verify the rationality of enhancing low-frequency in encoder and high-frequency in decoder. The results are shown in Table 3. Note that enhancing full frequency is different from removing the module since it still updates the features in the spectral domain. From the

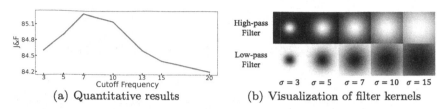

(a) Quantitative results   (b) Visualization of filter kernels

Fig. 6. Analysis on the selection of frequency filters

table we can see that, when the high-frequency components are enhanced in the encoder, there is a significant decrease on performance (1.1 $\mathcal{J}\&\mathcal{F}$), which illustrates the encoded features need contain enough high-level semantic information. Conversely, decoder features need have fine-grained detail information. Besides, we have also tried other strategy that fusing global information, *e.g.*, attention, and LFM works better.

Generalizability Analysis. To demonstrate the generalization ability of our frequency modules and prove that the lack of intra-frame global dependency is a common problem of memory network-based methods, we conduct experiments by applying our modules on some other methods as well. As shown in Table 4, the effectiveness of these methods is significantly improved by adding the frequency modules, which further shows the rationality of enhancing different frequency components separately in different parts of the network.

Selection of Frequency Filters. When performing frequency enhancement, we need to choose the cutoff frequency σ that distinguishes high and low frequency (the value of the cutoff frequency affects the frequency filter). After visualizing and experimenting with Gaussian filter kernels of different cutoff frequencies, finally, we choose $\sigma = 7$ as the cutoff frequency in default. Figure 6(a) shows that too large or too small cutoff frequency will have a bad effect. From Fig. 6(b) we can see that if σ is set too large, the high-pass filter will pass almost all frequencies while the low-pass filter will filter out all frequencies, which losses the function of selective enhancement. Same thing if σ is set too small.

Component Analysis. We experiment with the effectiveness of the proposed LFM, HFM, and Boundary Decoder. As shown in Table 5, all of them bring performance improvement and their combination works better (upgraded 1.2).

Effect of Small Objects. Although the LFM takes a residual structure to enhance low-frequency components during encoding, it does not result in information loss. To prove that, we analyze the segmentation effect of small objects on YouTubeVOS dataset. Figure 7 and Table 6 show the qualitative results and quantitative results respectively. In Table 6, we count the results for objects with

Table 5. Ablation study of the proposed modules

LFM	HFM	Boundary Branch	$\mathcal{J}$	$\mathcal{F}$	$\mathcal{J}\&\mathcal{F}$	FPS
			80.8	87.4	84.1	11.7
✓			81.5	87.9	$\mathbf{84.7}^{\uparrow 0.6}$	11.3
	✓		81.2	87.8	$\mathbf{84.5}^{\uparrow 0.4}$	10.9
✓	✓		81.8	88.2	$\mathbf{85.0}^{\uparrow 0.9}$	9.5
	✓	✓	81.4	87.7	$\mathbf{84.6}^{\uparrow 0.5}$	9.1
✓	✓	✓	81.9	88.7	$\mathbf{85.3}^{\uparrow 1.2}$	8.9

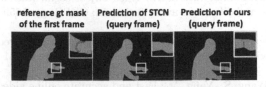

Fig. 7. Qualitative results on small objects.

Table 6. Quantitative results on small objects.

Area	5%	1%	0.5%
STCN	80.6	76.3	73.5
Ours	81.4	78.1	75.0

area less than 5%, 1% and 0.5% of the image. It can be seen that the segmentation results of small objects are not worse, but better due to the enhanced discrimination of features.

6   Conclusions

To fully utilize the intra-frame spatial dependency, we propose a Global Spectral Filter Memory network (GSFM) for semi-supervised video object segmentation in this paper. According to the different characteristics of encoding and decoding, GSFM separately enhances corresponding frequency components. With LFM integrating high-level semantic information and HFM highlighting fine-grained details, GSFM shows excellent performance on the popular DAVIS [35,36] and YouTube-VOS [58]. Besides, extensive experiments also demonstrate the rationality and generalization ability of our frequency modules. We hope that the strategy enhancing low-frequency for encoding and high-frequency for decoding would inspire some research in related fields.

Acknowledgement. This work was partially supported by the National Natural Science Foundation of China under Grant No. U1903213 and the Shenzhen Key Laboratory of Marine IntelliSense and Computation (NO. ZDSYS20200811142605016.)

References

1. Bao, L., Wu, B., Liu, W.: CNN in MRF: video object segmentation via inference in a cnn-based higher-order spatio-temporal MRF. In: CVPR, pp. 5977–5986 (2018)
2. Bergland, G.D.: A guided tour of the fast fourier transform. IEEE Spect. **6**(7), 41–52 (1969)
3. Caelles, S., Maninis, K., Pont-Tuset, J., Leal-Taixé, L., Cremers, D., Gool, L.V.: One-shot video object segmentation. In: CVPR, pp. 5320–5329 (2017)
4. Cheng, H.K., Chung, J., Tai, Y., Tang, C.: Cascadepsp: toward class-agnostic and very high-resolution segmentation via global and local refinement. In: CVPR, pp. 8887–8896 (2020)
5. Cheng, H.K., Tai, Y., Tang, C.: Modular interactive video object segmentation: interaction-to-mask, propagation and difference-aware fusion. In: CVPR, pp. 5559–5568 (2021)
6. Cheng, H.K., Tai, Y., Tang, C.: Rethinking space-time networks with improved memory coverage for efficient video object segmentation. In: NIPS, pp. 11781–11794 (2021)
7. Cheng, J., Tsai, Y., Hung, W., Wang, S., Yang, M.: Fast and accurate online video object segmentation via tracking parts. In: CVPR, pp. 7415–7424 (2018)
8. Cheng, J., Tsai, Y., Wang, S., Yang, M.: Segflow: joint learning for video object segmentation and optical flow. In: ICCV, pp. 686–695 (2017)
9. Cheng, T., Wang, X., Huang, L., Liu, W.: Boundary-preserving mask R-CNN. In: Vedaldi, A., Bischof, H., Brox, T., Frahm, J.-M. (eds.) ECCV 2020. LNCS, vol. 12359, pp. 660–676. Springer, Cham (2020). https://doi.org/10.1007/978-3-030-58568-6_39
10. Chi, L., Jiang, B., Mu, Y.: Fast fourier convolution. In: NIPS (2020)
11. Duke, B., Ahmed, A., Wolf, C., Aarabi, P., Taylor, G.W.: Sstvos: sparse spatiotemporal transformers for video object segmentation. In: CVPR, pp. 5912–5921 (2021)
12. Ge, W., Lu, X., Shen, J.: Video object segmentation using global and instance embedding learning. In: CVPR, pp. 16836–16845 (2021)
13. Han, J., Yang, L., Zhang, D., Chang, X., Liang, X.: Reinforcement cutting-agent learning for video object segmentation. In: CVPR, pp. 9080–9089 (2018)
14. He, K., Zhang, X., Ren, S., Sun, J.: Deep residual learning for image recognition. In: CVPR, pp. 770–778 (2016)
15. Hu, L., Zhang, P., Zhang, B., Pan, P., Xu, Y., Jin, R.: Learning position and target consistency for memory-based video object segmentation. arXiv preprint arXiv:2104.04329 (2021)
16. Hu, Y., Huang, J., Schwing, A.G.: Maskrnn: instance level video object segmentation. In: NIPS, pp. 325–334 (2017)
17. Huang, X., Xu, J., Tai, Y., Tang, C.: Fast video object segmentation with temporal aggregation network and dynamic template matching. In: CVPR, pp. 8876–8886 (2020)
18. Jang, W., Kim, C.: Online video object segmentation via convolutional trident network. In: CVPR, pp. 7474–7483 (2017)
19. Katznelson, Y.: An Introduction to Harmonic Analysis. Cambridge University Press, Cambridge (2004)
20. Li, X., Wei, T., Chen, Y.P., Tai, Y., Tang, C.: FSS-1000: a 1000-class dataset for few-shot segmentation. In: CVPR, pp. 2866–2875 (2020)

21. Li, X., Loy, C.C.: Video object segmentation with joint re-identification and attention-aware mask propagation. In: Ferrari, V., Hebert, M., Sminchisescu, C., Weiss, Y. (eds.) ECCV 2018. LNCS, vol. 11207, pp. 93–110. Springer, Cham (2018). https://doi.org/10.1007/978-3-030-01219-9_6

22. Li, Yu., Shen, Z., Shan, Y.: Fast video object segmentation using the global context module. In: Vedaldi, A., Bischof, H., Brox, T., Frahm, J.-M. (eds.) ECCV 2020. LNCS, vol. 12355, pp. 735–750. Springer, Cham (2020). https://doi.org/10.1007/978-3-030-58607-2_43

23. Liang, S., Shen, X., Huang, J., Hua, X.S.: Video object segmentation with dynamic memory networks and adaptive object alignment. In: ICCV, pp. 8065–8074 (2021)

24. Liang, Y., Li, X., Jafari, N.H., Chen, J.: Video object segmentation with adaptive feature bank and uncertain-region refinement. In: NIPS (2020)

25. Lin, S., Yang, L., Saleemi, I., Sengupta, S.: Robust high-resolution video matting with temporal guidance. In: WACV (2022)

26. Lu, X., Wang, W., Danelljan, M., Zhou, T., Shen, J., Van Gool, L.: Video object segmentation with episodic graph memory networks. In: Vedaldi, A., Bischof, H., Brox, T., Frahm, J.-M. (eds.) ECCV 2020. LNCS, vol. 12348, pp. 661–679. Springer, Cham (2020). https://doi.org/10.1007/978-3-030-58580-8_39

27. Luiten, J., Voigtlaender, P., Leibe, B.: Premvos: proposal-generation, refinement and merging for video object segmentation. In: ACCV, pp. 565–580 (2018)

28. Maninis, K.: Video object segmentation without temporal information. TPAMI 41(6), 1515–1530 (2019)

29. Mao, Y., Wang, N., Zhou, W., Li, H.: Joint inductive and transductive learning for video object segmentation. arXiv preprint arXiv:2108.03679 (2021)

30. Milletari, F., Navab, N., Ahmadi, S.: V-net: Fully convolutional neural networks for volumetric medical image segmentation. In: Fourth International Conference on 3D Vision, 3DV 2016, Stanford, CA, USA, 25–28 October 2016 (2016)

31. Ngan, K.N., Li, H.: Video Segmentation and its Applications. Springer, Heidelberg (2011). https://doi.org/10.1007/978-1-4419-9482-0

32. Oh, S.W., Lee, J., Sunkavalli, K., Kim, S.J.: Fast video object segmentation by reference-guided mask propagation. In: CVPR, pp. 7376–7385 (2018)

33. Oh, S.W., Lee, J., Xu, N., Kim, S.J.: Video object segmentation using space-time memory networks. In: ICCV, pp. 9225–9234 (2019)

34. Park, H., Yoo, J., Jeong, S., Venkatesh, G., Kwak, N.: Learning dynamic network using a reuse gate function in semi-supervised video object segmentation. In: CVPR. pp. 8405–8414 (2021)

35. Perazzi, F., Pont-Tuset, J., McWilliams, B., Gool, L.V., Gross, M.H., Sorkine-Hornung, A.: A benchmark dataset and evaluation methodology for video object segmentation. In: CVPR, pp. 724–732 (2016)

36. Pont-Tuset, J., Perazzi, F., Caelles, S., Arbelaez, P., Sorkine-Hornung, A., Gool, L.V.: The 2017 DAVIS challenge on video object segmentation. arXiv preprint arXiv:1704.00675 (2017)

37. Qin, Z., Zhang, P., Wu, F., Li, X.: Fcanet: frequency channel attention networks. arXiv preprint arXiv:2012.11879 (2020)

38. Seong, H., Hyun, J., Kim, E.: Kernelized memory network for video object segmentation. In: Vedaldi, A., Bischof, H., Brox, T., Frahm, J.-M. (eds.) ECCV 2020. LNCS, vol. 12367, pp. 629–645. Springer, Cham (2020). https://doi.org/10.1007/978-3-030-58542-6_38

39. Seong, H., Oh, S.W., Lee, J., Lee, S., Lee, S., Kim, E.: Hierarchical memory matching network for video object segmentation. arXiv preprint arXiv:2109.11404 (2021)

40. Shen, X., et al.: Dct-mask: discrete cosine transform mask representation for instance segmentation. In: CVPR (2021)
41. Shi, J., Yan, Q., Xu, L., Jia, J.: Hierarchical image saliency detection on extended CSSD. TPAMI **38**, 717–729 (2016)
42. Suh, S., et al.: Weighted mask R-CNN for improving adjacent boundary segmentation. J. Sensors **2021**, 8872947:1–8872947:8 (2021)
43. Sun, M., Xiao, J., Lim, E.G., Zhang, B., Zhao, Y.: Fast template matching and update for video object tracking and segmentation. In: CVPR, pp. 10788–10796 (2020)
44. Tang, Y., et al.: Breaking shortcut: exploring fully convolutional cycle-consistency for video correspondence learning. arXiv preprint arXiv:2105.05838 (2021)
45. Tsai, Y., Yang, M., Black, M.J.: Video segmentation via object flow. In: CVPR, pp. 3899–3908 (2016)
46. Voigtlaender, P., Chai, Y., Schroff, F., Adam, H., Leibe, B., Chen, L.: FEELVOS: fast end-to-end embedding learning for video object segmentation. In: CVPR, pp. 9481–9490 (2019)
47. Voigtlaender, P., Leibe, B.: Online adaptation of convolutional neural networks for video object segmentation. In: BMVC (2017)
48. Wang, H., Jiang, X., Ren, H., Hu, Y., Bai, S.: Swiftnet: real-time video object segmentation. In: CVPR, pp. 1296–1305 (2021)
49. Wang, H., Wu, X., Huang, Z., Xing, E.P.: High-frequency component helps explain the generalization of convolutional neural networks. In: CVPR (2020)
50. Wang, L., et al.: Learning to detect salient objects with image-level supervision. In: CVPR, pp. 3796–3805 (2017)
51. Wang, Z., Xu, J., Liu, L., Zhu, F., Shao, L.: Ranet: ranking attention network for fast video object segmentation. In: ICCV, pp. 3977–3986 (2019)
52. Wen, P., et al.: DMVOS: discriminative matching for real-time video object segmentation. In: ACMMM, pp. 2048–2056 (2020)
53. Woo, S., Park, J., Lee, J.-Y., Kweon, I.S.: CBAM: convolutional block attention module. In: Ferrari, V., Hebert, M., Sminchisescu, C., Weiss, Y. (eds.) ECCV 2018. LNCS, vol. 11211, pp. 3–19. Springer, Cham (2018). https://doi.org/10.1007/978-3-030-01234-2_1
54. Xiao, H., Feng, J., Lin, G., Liu, Y., Zhang, M.: Monet: deep motion exploitation for video object segmentation. In: CVPR, pp. 1140–1148 (2018)
55. Xiao, M., et al.: Invertible image rescaling. In: Vedaldi, A., Bischof, H., Brox, T., Frahm, J.-M. (eds.) ECCV 2020. LNCS, vol. 12346, pp. 126–144. Springer, Cham (2020). https://doi.org/10.1007/978-3-030-58452-8_8
56. Xie, H., Yao, H., Zhou, S., Zhang, S., Sun, W.: Efficient regional memory network for video object segmentation. arXiv preprint arXiv:2103.12934 (2021)
57. Xie, S., Tu, Z.: Holistically-nested edge detection. Int. J. Comput. Vis. **125**(1–3), 3–18 (2017)
58. Xu, N., et al.: Youtube-vos: a large-scale video object segmentation benchmark. arXiv preprint arXiv:1809.03327 (2018)
59. Xu, X., Wang, J., Li, X., Lu, Y.: Reliable propagation-correction modulation for video object segmentation. In: AAAI, pp. 2946–2954 (2022)
60. Xu, Y., Fu, T., Yang, H., Lee, C.: Dynamic video segmentation network. In: CVPR, pp. 6556–6565 (2018)
61. Xu, Z.J., Zhang, Y., Xiao, Y.: Training behavior of deep neural network in frequency domain. In: ICONIP (2019)
62. Yang, L., Wang, Y., Xiong, X., Yang, J., Katsaggelos, A.K.: Efficient video object segmentation via network modulation. In: CVPR, pp. 6499–6507 (2018)

63. Yang, Z., Tang, Y., Bertinetto, L., Zhao, H., Torr, P.H.S.: Hierarchical interaction network for video object segmentation from referring expressions. In: BMVC, p. 254 (2021)
64. Yang, Z., Wei, Y., Yang, Y.: Collaborative video object segmentation by foreground-background integration. In: Vedaldi, A., Bischof, H., Brox, T., Frahm, J.-M. (eds.) ECCV 2020. LNCS, vol. 12350, pp. 332–348. Springer, Cham (2020). https://doi.org/10.1007/978-3-030-58558-7_20
65. Yang, Z., Wei, Y., Yang, Y.: Associating objects with transformers for video object segmentation. arXiv preprint arXiv:2106.02638 (2021)
66. Yang, Z., Wei, Y., Yang, Y.: Collaborative video object segmentation by multi-scale foreground-background integration. IEEE TPAMI (2021)
67. Yin, D., Lopes, R.G., Shlens, J., Cubuk, E.D., Gilmer, J.: A fourier perspective on model robustness in computer vision. In: NIPS (2019)
68. Yu, Z., Feng, C., Liu, M., Ramalingam, S.: Casenet: deep category-aware semantic edge detection. In: CVPR (2017)
69. Zeng, Y., Zhang, P., Lin, Z.L., Zhang, J., Lu, H.: Towards high-resolution salient object detection. In: ICCV, pp. 7233–7242 (2019)
70. Zhang, P., Hu, L., Zhang, B., Pan, P.: Spatial constrained memory network for semi-supervised video object segmentation. In: CVPR Workshops (2020)
71. Zhang, Z., Fidler, S., Urtasun, R.: Instance-level segmentation for autonomous driving with deep densely connected mrfs. In: CVPR (2016)
72. Zhao, K., Kang, J., Jung, J., Sohn, G.: Building extraction from satellite images using mask r-cnn with building boundary regularization. In: CVPR Workshops (2018)
73. Zhou, T., Li, J., Wang, S., Tao, R., Shen, J.: Matnet: motion-attentive transition network for zero-shot video object segmentation. IEEE Trans. Image Process. **29**, 8326–8338 (2020)

Video Instance Segmentation via Multi-Scale Spatio-Temporal Split Attention Transformer

Omkar Thawakar[1]([✉]), Sanath Narayan[2], Jiale Cao[3], Hisham Cholakkal[1],
Rao Muhammad Anwer[1], Muhammad Haris Khan[1], Salman Khan[1],
Michael Felsberg[4], and Fahad Shahbaz Khan[1,4]

[1] MBZUAI, Abu Dhabi, UAE
omkar.thawakar@mbzuai.ac.ae
[2] IIAI, Abu Dhabi, UAE
[3] Tianjin University, Tianjin, China
[4] Linköping University, Linköping, Sweden

Abstract. State-of-the-art transformer-based video instance segmentation (VIS) approaches typically utilize either single-scale spatio-temporal features or per-frame multi-scale features during the attention computations. We argue that such an attention computation ignores the multi-scale spatio-temporal feature relationships that are crucial to tackle target appearance deformations in videos. To address this issue, we propose a transformer-based VIS framework, named MS-STS VIS, that comprises a novel multi-scale spatio-temporal split (MS-STS) attention module in the encoder. The proposed MS-STS module effectively captures spatio-temporal feature relationships at multiple scales across frames in a video. We further introduce an attention block in the decoder to enhance the temporal consistency of the detected instances in different frames of a video. Moreover, an auxiliary discriminator is introduced during training to ensure better foreground-background separability within the multi-scale spatio-temporal feature space. We conduct extensive experiments on two benchmarks: Youtube-VIS (2019 and 2021). Our MS-STS VIS achieves state-of-the-art performance on both benchmarks. When using the ResNet50 backbone, our MS-STS achieves a mask AP of 50.1%, outperforming the best reported results in literature by 2.7% and by 4.8% at higher overlap threshold of AP_{75}, while being comparable in model size and speed on Youtube-VIS 2019 val. set. When using the Swin Transformer backbone, MS-STS VIS achieves mask AP of 61.0% on Youtube-VIS 2019 val. set. Source code is available at https://github.com/OmkarThawakar/MSSTS-VIS.

1 Introduction

Video instance segmentation (VIS) is a challenging computer vision problem with numerous real-world applications, including intelligent video analysis and

Supplementary Information The online version contains supplementary material available at https://doi.org/10.1007/978-3-031-19818-2_38.

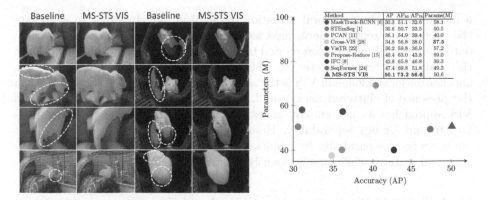

Baseline   MS-STS VIS   Baseline   MS-STS VIS

Method	AP	AP₅₀	AP₇₅	Params(M)
MaskTrack-RCNN [6]	30.3	51.1	32.6	58.1
STEmSeg [1]	30.6	50.7	33.5	50.5
PCAN [11]	36.1	54.9	39.4	40.0
Cross-VIS [28]	34.8	56.8	38.0	37.5
VisTR [22]	36.2	59.8	36.9	57.2
Propose-Reduce [15]	40.4	63.0	43.8	69.0
IFC [8]	42.8	65.8	46.8	39.3
SeqFormer [24]	47.4	69.8	51.8	49.3
MS-STS VIS	50.1	73.2	56.6	50.6

Fig. 1. On the left: Qualitative comparison of our MS-STS VIS with the baseline in the case of target appearance deformations on two example videos. For each method, four sample video frames are shown along the columns. The baseline struggles to accurately predict the *rabbit* class instance undergoing appearance deformations due to scale variation, aspect-ratio change and fast motion. As a result, the predicted mask quality is hampered (marked by white dotted region). Similarly, the mask quality is also deteriorated for the *mouse* class. Our MS-STS VIS addresses these issues by capturing multi-scale spatio-temporal feature relationships, leading to improved mask quality. Best viewed zoomed in. **On the right:** Accuracy (AP) *vs.* model size (params) comparison with existing methods using a single model and single-scale inference on YouTube-VIS 2019 val. set. We also report performance at AP₅₀ and AP₇₅. All methods here utilize the same ResNet50 backbone. Compared to the best reported results in literature [24], our MS-STS VIS achieves absolute gains of 2.7% and 4.8% in terms of overall mask AP and at higher overlap threshold of AP₇₅, respectively, while being comparable in model size and speed ([24]: 11 FPS *vs.* **Ours:** 10 FPS).

autonomous driving. Given a video sequence, the task is to simultaneously segment and track all object instances from a set of semantic categories. The problem is particularly challenging since the target object needs to be accurately segmented and tracked despite appearance deformations due to several real-world issues such as, target size variation, aspect-ratio change and fast motion.

Recently, transformers [20] have shown promising results on several vision tasks, including VIS [22,24]. The recent transformer-based VIS approaches [22, 24] are built on DETR [4] and Deformable DETR [29] frameworks, utilizing an encoder-decoder architecture along with instance sequence matching and segmentation mechanisms to generate final video mask predictions. These approaches typically employ either single-scale spatio-temporal features [22] or per-frame multi-scale features [24] during attention computations at the encoder and decoder. However, such an attention computation ignores the *multi-scale spatio-temporal feature relationships*, which are crucial towards handling target appearance deformations due to real-world challenges such as, scale variation, change in aspect-ratio and fast motion in videos.

In this work, we investigate the problem of designing an attention mechanism, within the transformer-based VIS framework, to effectively capture multi-scale spatio-temporal feature relationships in a video. With this aim, we introduce a multi-scale spatio-temporal attention mechanism, which learns to aggregate the necessary attentions performed along the spatial and temporal axes without losing crucial information related to target appearance deformations in both the spatial and temporal axes. In addition to target appearance deformations, another major challenge in VIS is the accurate delineation of the target object in the presence of cluttered background. Surprisingly, existing transformer-based VIS approaches do not employ an explicit mechanism to enforce foreground-background (fg-bg) separability. Here, we introduce a loss formulation that improves fg-bg separability by emphasizing the fg regions in multi-scale spatio-temporal features while simultaneously suppressing the bg regions.

Contributions: We propose a transformer-based video instance segmentation framework, MS-STS VIS, with the following contributions.

- We propose a novel multi-scale spatio-temporal split (MS-STS) attention module in the transformer encoder for effectively capturing spatio-temporal feature relationships. Our MS-STS module first attends to features across frames at a given spatial scale via an intra-scale temporal attention block, and then progressively attends to neighboring spatial scales across frames via an inter-scale temporal attention block to obtain enriched feature representations. To further improve the video mask prediction, we introduce an attention block in the decoder that enhances the temporal consistency of the detected instances in different frames of a video.
- We introduce an auxiliary discriminator network during training to enhance the fg-bg separability within the multi-scale spatio-temporal feature space. Here, the discriminator network is trained to distinguish between ground-truth and predicted mask, while the encoder is learned to fool the discriminator by generating features that result in better mask predictions.
- Comprehensive experiments are performed on Youtube-VIS 2019 and 2021 datasets. Our proposed MS-STS VIS sets a new state-of-the-art on both datasets. When using the ResNet50 backbone, our MS-STS VIS outperforms all existing methods with an overall mask AP of 50.1% on Youtube-VIS 2019 val. set, while being on par in terms of model size and speed compared to the state-of-the-art method (see Fig. 1 (right)). Specifically, the proposed MS-STS VIS achieves a significant performance improvement over the baseline, in case of target appearance deformations due to scale variation, aspect-ratio change and fast motion (see Fig. 1 (left)).

2   Baseline Framework

We base our approach on the recently introduced SeqFormer [24]. SeqFormer is built on the Deformable DETR [29] framework, which comprises a CNN backbone followed by a transformer encoder-decoder with deformable attention. We choose SeqFormer as our base framework to demonstrate the impact of our proposed contributions on a *strong baseline*. Here, a video clip $\mathbf{x}$ consisting T frames of spatial size $H^0 \times W^0$ with a set of object instances is input to the backbone. Latent feature maps for each frame are obtained from the backbone at L multiple scales 2^{-l-2} with $1 \leq l \leq L$ and passed through separate convolution filters for each scale. The output feature dimension for each of these convolution filters is set to C. The resulting multi-scale feature maps of each frame are then input to a transformer encoder comprising multi-scale deformable attention blocks. For each frame, the transformer encoder outputs multi-scale feature maps with the same size as its input. These encoder output features maps from each frame along with n learnable instance query embeddings $\mathbf{I}^Q \in \mathbb{R}^C$ are then input to the transformer decoder comprising a series of self- and cross-attention blocks. The n instance queries are further decomposed into n box queries $\mathbf{B}^Q$ per-frame and are used to query the box features from the encoder feature maps of the corresponding frame. The learned box queries across T frames are then aggregated temporally to obtain n instance features $\mathbf{I}^O \in \mathbb{R}^C$. These instance features output by the decoder are then used for video instance mask prediction. We refer to [24] for additional details.

Limitation: As discussed above, the aforementioned SeqFormer framework independently utilizes *per-frame* multi-scale features during attention computations. As a result, it ignores the spatio-temporal feature relationships during attention computation that is crucial for the VIS problem. Different from Seq-Former that utilizes *per-frame* spatial features at multiple scales, our approach performs *multi-scale spatio-temporal attention* computation. Such a multi-scale spatio-temporal attention is especially desired in cases when the target object undergoes appearance deformations due to real-world challenges such as, scale variation, aspect-ratio change and fast motion in videos (see Figs. 1 and 3). Furthermore, distinct from the baseline SeqFormer, our approach employs an explicit mechanism to ensure accurate delineation of foreground objects from the cluttered background by enhancing fg-bg separability.

3   Method

3.1   Overall Architecture

Figure 2(a) shows the overall architecture of the proposed MS-STS VIS approach built on the baseline framework described above. Our MS-STS VIS comprises a backbone network, a transformer encoder-decoder and a sequence matching and segmentation block. The focus of our design is the introduction of a novel multi-scale spatio-temporal split (MS-STS) attention module (Fig. 2(b)) in the transformer encoder to effectively capture spatio-temporal feature relationships at multiple scales across frames in a video. The MS-STS attention module comprises intra- and inter-scale temporal attention blocks (Fig. 2(c) and (d)). The intra-scale block takes C-dimensional backbone features $\mathbf{F}$ as input and enriches features across frames in a video within a given scale, whereas the inter-scale block then progressively attends to multiple spatial scales across frames to generate C-dimensional spatio-temporally enriched features $\hat{\mathbf{Z}}$. These features $\hat{\mathbf{Z}}$

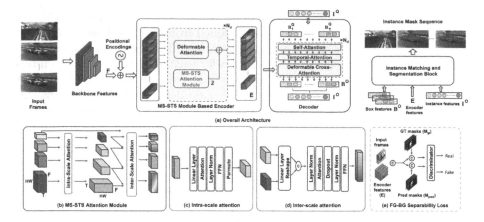

Fig. 2. (a) Our MS-STS VIS architecture comprises a backbone, a transformer encoder-decoder and a instance matching and segmentation block. Here, our key contributions are: (i) a novel MS-STS attention module in the encoder to capture spatio-temporal feature relationships at multiple scales across frames, (ii) a temporal attention block in the decoder for enhancing the temporal consistency of the box queries and (iii) an adversarial loss for enhancing foreground-background (fg-bg) separability. Specifically, the MS-STS attention module (b) attends to the input backbone features by employing intra- and inter-scale temporal attentions to generate enriched features which are then fused with standard baseline features. While the intra-scale temporal attention block (c) enriches features across frames within a given scale, its inter-scale counterpart (d) progressively attends to multiple spatial scales across frames to obtain spatio-temporally enriched features. These enriched features are further improved using an adversarial loss (e) to enhance fg-bg separability. The resulting encoder features along with the temporally consistent instance features from the decoder are used within the matching and segmentation block for the video instance mask prediction.

are fused in each encoder layer with the base features output by the standard deformable attention. While backbone features $\mathbf{F}$ are used as input to the first encoder layer, the subsequent layers utilize the outputs from the preceding layer as input. As a result, the encoder outputs C-dimensional multi-scale spatio-temporally enriched features $\mathbf{E}$ that are further input to the transformer decoder. To achieve temporal consistency among box queries from different frames, we introduce a temporal attention block within the transformer decoder. Next, the encoder features $\mathbf{E}$ along with the instance features $\mathbf{I}^O$ (aggregated temporally attended box queries) from the decoder are utilized within the instance matching and segmentation block to obtain the video instance mask prediction. To further improve the predicted video instance mask quality, we introduce an adversarial loss during training to enhance foreground-background (fg-bg) separability. The adversarial loss (Fig. 2(e)) strives to enhance the encoder features by discriminating between the predicted and ground-truth masks, utilizing the encoder features $\mathbf{E}$, the input frames $\mathbf{x}$ and the binary object mask $\mathbf{M}$. Next, we describe our MS-STS attention module-based encoder.

3.2   MS-STS Attention Module Based Encoder

Within the proposed MS-STS VIS framework, we introduce a novel multi-scale spatio-temporal split (MS-STS) attention module in the transformer encoder to effectively capture spatio-temporal feature relationships at multiple scales across frames. To this end, the MS-STS module takes the backbone features as input and produces multi-scale spatio-temporally enriched features, which are then fused with the standard features within the base framework. The MS-STS module (see Fig. 2(b)) comprises an intra-scale and an inter-scale attention block (see Fig. 2(c) and (d)) described next.

Intra-scale Temporal Attention Block: Given the backbone features as input, our intra-scale block independently attends to each scale (spatial resolution) temporally across frames (see Fig. 2(c)). Let $\mathbf{z}_{s,t}^l$ be the feature at spatial scale l, position s and frame t, where $s \in [0, S^l-1]$, $t \in [0, T-1]$, and $S^l = H^l \cdot W^l$. The intra-scale temporal attention block processes the features $\mathbf{Z}^l \in \mathbb{R}^{S^l \times T \times C}$ using intra-scale self-attention ($\mathrm{SA_{intra}}$), layer normalization (LN), and MLP layers as:

$$\mathbf{Y}^l = \mathrm{SA_{intra}}(\mathrm{LN}(\mathbf{Z}^l)) + \mathbf{Z}^l, \qquad \tilde{\mathbf{Z}}^l = \mathrm{MLP}(\mathrm{LN}(\mathbf{Y}^l)) + \mathbf{Y}^l, \qquad (1)$$

$$\text{where} \quad \mathrm{SA_{intra}}(\mathrm{LN}(\mathbf{z}_{s,t}^l)) = \sum_{t=0}^{T-1} \mathrm{Softmax}\left(\mathbf{q}_{s,t}^l \cdot \mathbf{k}_{s,t}^l / \sqrt{C}\right) \mathbf{v}_{s,t}^l, \qquad (2)$$

where $\mathbf{q}_{s,t}^l, \mathbf{k}_{s,t}^l, \mathbf{v}_{s,t}^l \in \mathbb{R}^D$ are the query, key, and value vectors obtained from $\mathbf{z}_{s,t}^l$ (after LN) using the embedding matrices $\mathbf{W}_q, \mathbf{W}_k, \mathbf{W}_v \in \mathbb{R}^{C \times C}$. The intra-scale temporal attention operates on each spatial scale l across frames and produces temporally relevant intermediate features $\{\tilde{\mathbf{Z}}^l\}_{l=1}^L$.

Inter-scale Temporal Attention Block: The inter-scale temporal attention block takes the intermediate features output by the intra-scale block and aims to learn similarities between the spatio-temporal features across two neighbouring spatial scales (see Fig. 2(e)). Let $\tilde{\mathbf{Z}}^l \in \mathbb{R}^{S^l \times T \times C}$ and $\tilde{\mathbf{Z}}^{l+1}$ be the intra-scale attended features at spatial scales l and $l + 1$. To compute inter-scale temporal attention between the two neighbouring scale features, we first upsample the lower resolution features by $\times 2$ using bilinear interpolation, concatenate along the feature dimensions and project it into C dimensions using $\mathbf{W}_p \in \mathbb{R}^{2C \times C}$ as $\mathbf{H}^l = \mathbf{W}_p(\mathrm{CONCAT}(\tilde{\mathbf{Z}}^l, \mathrm{UPSAMPLE}(\tilde{\mathbf{Z}}^{l+1}))$. We reshape $\mathbf{H}^l \in \mathbb{R}^{S^l \times T \times C}$ to $\mathbb{R}^{(S^l \cdot T) \times C}$ and then compute the joint spatio-temporal attention to obtain enriched features $\hat{\mathbf{Z}}^l$ given by

$$\mathbf{Y}^l = \mathrm{SA_{inter}}(\mathrm{LN}(\mathbf{H}^l)) + \mathbf{H}^l, \qquad \hat{\mathbf{Z}}^l = \mathrm{MLP}(\mathrm{LN}(\mathbf{Y}^l)) + \mathbf{Y}^l, \qquad (3)$$

where the inter-scale self-attention ($\mathrm{SA_{inter}}$) computation is given by

$$\mathrm{SA_{inter}} = \sum_{\hat{s}=0}^{S-1} \sum_{t=0}^{T-1} \mathrm{Softmax}\left(\mathbf{q}_{s,t}^l \cdot \mathbf{k}_{\hat{s},t}^l / \sqrt{C}\right) \mathbf{v}_{\hat{s},t}^l. \qquad (4)$$

Input Frame   Baseline   Ours   Input Frame   Baseline   Ours   Input Frame   Baseline   Ours

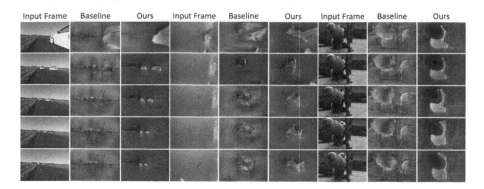

Fig. 3. Example attention map visualizations obtained at the output of the baseline encoder and our MS-STS attention module-based encoder. Here, we show the comparison on example videos from Youtube-VIS 2019 val. set. The baseline struggles to accurately focus on the *car* instance undergoing significant scale variation, where the size becomes extremely small in the later frames. Similarly, it fails to accurately focus on the *shark* instance undergoing aspect-ratio change and the *person* instance partially visible only in the first two frames of the middle video. In the last video, the baseline inaccurately highlights the irrelevant object (in orange) occluding the target *panda* instance. Our MS-STS attention module-based encoder, which strives to capture spatio-temporal feature relationships at multiple scales, successfully focuses on these challenging targets despite scale variations (*car*), aspect-ratio change (*shark*), partial visibility (*person*) and appearance deformations due to occlusion (*panda*). Best viewed zoomed in.

To summarize, our MS-STS attention module utilizes an intra-scale temporal attention block to first attend to features across frames at a given spatial scale. It then employs an inter-scale temporal attention block to progressively attend to neighboring spatial scales across frames for obtaining enriched feature representations $\hat{\mathbf{Z}}^l$. The resulting features $\hat{\mathbf{Z}}^l$ are fused with the standard baseline features through a convolution operation. Finally, the encoder outputs enriched multi-scale spatio-temporal features $\mathbf{E}$ after processing the input features through N_d attention layers. Figure 3 shows example attention maps obtained at the output of our MS-STS attention module-based encoder. The attention maps are computed as the average activation strength across the C features at a particular spatial position for the largest scale considered, *i.e.*, $H/8 \times W/8$. The attention maps are shown for example videos from Youtube-VIS 2019 val. set depicting target appearance deformations. We also compare our encoder output to that of the baseline encoder. Compared to the baseline, our MS-STS module-based encoder better focuses on the *car* instance (left video) undergoing scale variations, the *shark* instance (middle video) undergoing aspect-ratio change, the *person* instance (middle video) partially visible only in the first two frames and the *panda* instance (right video) exhibiting target deformations due to the irrelevant object (in orange) occluding the target *panda* instance.

3.3 Enhancing Temporal Consistency in Decoder

As discussed earlier, the transformer decoder in the base framework comprises a series (layers) of alternating self- and cross-attention blocks, operating on the box queries $\mathbf{B}^Q$ of individual frames. Although the video-level instance features $\mathbf{I}^O$ are obtained through an aggregation of the corresponding frame-level box features, the temporal consistency of the detected instances is hampered likely due to the per-frame attention computation of the box queries. To alleviate this issue, we introduce a temporal attention block in between the self- and cross-attention blocks of a decoder layer. Our temporal attention block attends to the sequence of box queries $\{\mathbf{B}_i^Q\}_{i=1}^T$ from T frames and produces temporally consistent queries at its output. Such a temporal attention between the per-frame box queries of an instance enables information flow between the probable box locations across frames, thereby improving the temporal consistency of the video mask predictions. The resulting temporally-attended box queries are then utilized to query the multi-scale spatio-temporally enriched features $\mathbf{E}$, output by the encoder, for computing the box features $\mathbf{B}^O$ in a per-frame manner. Afterwards, these box features are aggregated to obtain video-level instance features $\mathbf{I}^O$. Finally, the resulting instance features along with the box features and multi-scale spatio-temporally enriched features $\mathbf{E}$ are input to an instance matching and segmentation block for generating the video mask predictions.

3.4 Enhancing Foreground-Background Separability

Both our MS-STS attention in encoder and temporal attention in decoder promote multi-scale spatio-temporal feature enrichment and temporal consistency, respectively, across frames. This helps to address the critical issue of appearance deformations in the target object due to challenges such as, scale variation, aspect-ratio change and fast motion. In addition to addressing target appearance deformations across frames, another common challenge in the VIS problem is the accurate delineation of foreground objects from the cluttered background. To this end, we introduce an adversarial loss during training of our MS-STS VIS framework for enhancing foreground-background (fg-bg) separability. To the best of our knowledge, we are the first to explore an adversarial loss within a transformer-based VIS framework for enhancing fg-bg separability.

In our adversarial formulation, the objective is to discriminate between the ground-truth mask and the predicted mask output by our MS-STS VIS. With this objective, we introduce an auxiliary discriminator network during training. The discriminator takes the input frames along with the corresponding encoder features and binary masks as its input. Here, the binary mask $\mathbf{M}$ is obtained either from the ground-truth or predictions, such that all object instances (category-agnostic) within a frame are marked as foreground. While the discriminator D attempts to distinguish between ground-truth and predicted binary masks ($\mathbf{M}_{gt}$ and $\mathbf{M}_{pred}$, respectively), the encoder learns to output enhanced features $\mathbf{E}$ such that the predicted masks $\mathbf{M}_{pred}$ are close to ground-truth $\mathbf{M}_{gt}$. Let $\mathbf{F}_{gt} = \text{CONCAT}(\mathbf{x}, \mathbf{E}, \mathbf{M}_{gt})$ and $\mathbf{F}_{pr} = \text{CONCAT}(\mathbf{x}, \mathbf{E}, \mathbf{M}_{pred})$

denote the real and fake input, respectively, to the discriminator D. Similar to [9], the adversarial loss is then given by

$$\min_{Enc} \max_{D} \mathbb{E}[\log D(\mathbf{F}_{gt})] + \mathbb{E}[\log(1 - D(\mathbf{F}_{pr}))] + \lambda_1 \mathbb{E}[||D(\mathbf{F}_{gt}) - D(\mathbf{F}_{pr})||_1]. \quad (5)$$

Since the mask prediction depends on the quality of the encoder features that are decoded by the queries, we treat our encoder Enc as a generator in our adversarial formulation above. As a result, the encoder learns to better delineate foreground and background regions leading to improved video instance mask prediction. Note that the discriminator network is utilized only during training.

4   Experiments

Datasets: The *YouTube-VIS 2019* [26] dataset contains 2883 high-quality videos with $131K$ annotated object instances belonging to 40 different categories. The *YouTube-VIS 2021* [25] dataset contains 3,859 high-quality videos with $232K$ annotated object instances belonging to 40 different improved categories. YouTube-VIS 2021 [25] has a 40-category label set by merging *eagle* and *owl* into *bird*, *ape* into *monkey*, deleting *hands*, and adding *flying disc*, *squirrel* and *whale*, maintaining the same number of categories as YouTube-VIS 2019 [26] set.

Evaluation Metrics: We follow the standard protocol, where the evaluation metrics, Average Precision (AP) and Average Recall (AR), are adapted from image instance segmentation with the video Intersection over Union (IoU) of the mask sequences as the threshold.

Implementation Details: We choose ResNet-50 [7] as the default backbone, unless otherwise specified, for our MS-STS VIS framework. Outputs from conv$_3$, conv$_4$ and conv$_5$ of the backbone are utilized to obtain multi-scale feature inputs to our transformer encoder-decoder, as in [29]. Both encoder and decoder layers are set to $N_d = 6$. The feature dimension C is set to 256, while the number of instance queries is set to 300 and length of video clip $T = 5$, as in [24]. We use the AdamW optimizer with a base learning rate (LR) of 2×10^{-4}, $(\beta_1, \beta_2) = (0.9, 0.999)$ and a weight decay of 10^{-4}. LR of linear projections of deformable attention modules and the backbone are scaled by 0.1. The model is first pretrained on COCO [16] for 24 epochs with a batch size of 2. Similar to [24], the pretrained weights are then used to train the model on Youtube-VIS and COCO dataset for 12 epochs with $T = 5$ and batch size set to 2. The LR is scaled by a factor of 0.1 at 4th and 10th epochs. The framework is trained on 8 Nvidia V100 GPUs using PyTorch-1.9 [18].

4.1   State-of-the-Art Comparison

Table 1 presents the state-of-the-art comparison on the YouTube-VIS 2019 val. set. When using the ResNet-50 backbone, the recent one-stage PCAN [11] and

Table 1. State-of-the-art comparison on **YouTube-VIS 2019 val** set. Our MR-STS VIS consistently outperforms the state-of-the-art results reported in literature. When using the ResNet-50 backbone, MS-STS VIS achieves overall mask AP score of 50.1% with an absolute gain of 2.7% over the best existing SeqFormer, while being comparable in terms of model size and speed (SeqFormer: 11 FPS *vs.* Ours: 10 FPS). Similarly, when using the ResNet-101 backbone, our MS-STS VIS achieves overall mask AP of 51.1%. Further, MS-STS VIS achieves the best accuracy reported on this dataset with a mask AP of 61.0% and outperforms SeqFormer with an absolute gain of 1.7%, using the same Swin-L backbone.

Method	Venue	Backbone	Type	AP	AP_{50}	AP_{75}	AR_1	AR_{10}
IoUTracker+ [26]	ICCV 2019	ResNet-50	–	23.6	39.2	25.5	26.2	30.9
OSMN [27]	CVPR 2018	ResNet-50	Two-Stage	27.5	45.1	29.1	28.6	33.1
DeepSORT [23]	ICIP 2017	ResNet-50	Two-stage	26.1	42.9	26.1	27.8	31.3
FEELVOS [21]	CVPR 2019	ResNet-50	Two-stage	26.9	42.0	29.7	29.9	33.4
SeqTracker [26]	ICCV 2019	ResNet-50	–	27.5	45.7	28.7	29.7	32.5
MaskTrack R-CNN [26]	ICCV 2019	ResNet-50	Two-stage	30.3	51.1	32.6	31.0	35.5
MaskProp [2]	CVPR 2020	ResNet-50	–	40.0	–	42.9	–	–
SipMask-VIS [3]	ECCV 2020	ResNet-50	One-stage	32.5	53.0	33.3	33.5	38.9
SipMask-VIS [3]	ECCV 2020	ResNet-50	One-stage	33.7	54.1	35.8	35.4	40.1
STEm-Seg [1]	ECCV 2020	ResNet-50	–	30.6	50.7	33.5	31.6	37.1
Johnander *et al.* [10]	GCPR 2021	ResNet-50	–	35.3	–	–	–	–
CompFeat [5]	AAAI 2021	ResNet-50	–	35.3	56.0	38.6	33.1	40.3
CrossVIS [28]	ICCV 2021	ResNet-50	One-stage	36.3	56.8	38.9	35.6	40.7
PCAN [11]	NeurIPS 2021	ResNet-50	One-stage	36.1	54.9	39.4	36.3	41.6
VisTR [22]	CVPR 2021	ResNet-50	Transformer	35.6	56.8	37.0	35.2	40.2
SeqFormer [24]	Arxiv 2021	ResNet-50	Transformer	47.4	69.8	51.8	45.5	54.8
MS-STS VIS (Ours)		ResNet-50	Transformer	**50.1**	**73.2**	**56.6**	**46.1**	**57.7**
MaskTrack R-CNN [26]	ICCV 2019	ResNet-101	Two-stage	31.9	53.7	32.3	32.5	37.7
MaskProp [2]	CVPR 2020	ResNet-101	–	42.5	–	45.6	–	–
STEm-Seg [1]	ECCV, 2020	ResNet-101	–	34.6	55.8	37.9	34.4	41.6
CrossVIS [28]	ICCV 2021	ResNet-101	One-stage	36.6	57.3	39.7	36.0	42.0
PCAN [11]	NeurIPS 2021	ResNet-101	One-stage	37.6	57.2	41.3	37.2	43.9
VisTR [22]	CVPR 2021	ResNet-101	Transformer	38.6	61.3	42.3	37.6	44.2
SeqFormer [24]	Arxiv 2021	ResNet-101	Transformer	49.0	71.1	55.7	46.8	56.9
MS-STS VIS (Ours)		ResNet-101	Transformer	**51.1**	**73.2**	**59.0**	**48.3**	**58.7**
SeqFormer [24]	Arxiv 2021	Swin-L	Transformer	59.3	82.1	66.6	51.7	64.4
MS-STS VIS (Ours)		Swin-L	Transformer	**61.0**	**85.2**	**68.6**	**54.7**	**66.4**

CrossVIS [28] approaches achieve an overall mask accuracy (AP) of 36.1% and 36.3%, respectively. With the same ResNet-50 backbone, the first transformer-based VIS approach, VisTR [22], built on DETR framework achieves an overall mask AP of 35.6%. Among existing methods, the recently introduced Seq-Former [24] based on Deformable DETR framework achieves the best overall accuracy with a mask AP of 47.4%. Our proposed MS-STS VIS approach outperforms SeqFormer [24] by achieving an overall mask AP of 50.1%, using the same ResNet-50 backbone. Specifically, our MS-STS VIS provides an absolute gain of 4.8% at a higher overlap threshold of AP_{75} over SeqFormer. Similarly, our MS-STS VIS consistently outperforms SeqFormer with an overall mask AP of 51.1%, when using the ResNet-101 backbone. Finally, when using the recent Swin Transformer backbone, the proposed MS-STS VIS achieves the best accuracy reported in literature with an overall mask AP of 61.0%.

Table 2. State-of-the-art comparison on **YouTube-VIS 2021** val set. All results are reported using the same ResNet-50 backbone. Our MS-STS VIS achieves state-of-the-art results with an overall mask AP of 42.2% and an absolute gain of 2.8% over the best existing SeqFormer at a higher overlap threshold of AP_{75}.

Method	AP	AP_{50}	AP_{75}	AR_1	AR_{10}
MaskTrack R-CNN [25]	28.6	48.9	29.6	26.5	33.8
SipMask-VIS [3]	31.7	52.5	34.0	30.8	37.8
VisTR [22]	31.8	51.7	34.5	29.7	36.9
CrossVIS [28]	34.2	54.4	37.9	30.4	38.2
IFC [8]	36.6	57.9	39.9	–	–
SeqFormer [24]	40.5	62.4	43.7	36.1	48.1
MS-STS VIS (Ours)	**42.2**	**63.7**	**46.5**	**41.7**	**51.1**

Table 3. On the left: Impact of our contributions when progressively integrating them into the baseline on the Youtube-VIS 2019 val. set. We observe a consistent performance improvement due to the integration of our proposed contributions. Our final MS-STS VIS (row 4) achieves an absolute gain of 3.7% over the baseline. **On the right:** Attribute-based performance comparison between the baseline and our MS-STS VIS on the custom set comprising 706 videos in total. Here, we present the comparison on fast motion, target size change and aspect-ratio change attributes. Our MS-STS VIS achieves consistent improvement in performance over the baseline on all attributes. See Sect. 4.2 for details.

Model	AP	Attribute Type	#Videos	Baseline	MS-STS VIS
Baseline	46.4	Custom Set	706	49.5	**53.7**
Baseline + MS-STS	48.4	Fast Motion	605	48.9	**54.2**
Baseline + MS-STS + T-Dec	49.1	Target Size Change	706	49.5	**53.7**
Baseline + MS-STS + T-Dec + FG-BG Loss	50.1	Aspect-Ratio Change	687	49.2	**53.9**

Table 2 reports the state-of-the-art comparison on the YouTube-VIS 2021 val. set. Among existing methods, CrossVIS [28] and IFC [8] achieve overall mask AP of 34.2% and 36.6%, respectively. SeqFormer [24] obtains a mask AP of 40.5%. Our MS-STS VIS sets a new state-of-the-art with an overall mask AP of 42.2%. Specifically, MS-STS VIS provides an absolute gain of 2.8% over SeqFormer at higher overlap threshold of AP_{75}, when using the same ResNet-50 backbone.

4.2 Ablation Study

Here, we first evaluate the merits of our three proposed contributions: MS-STS attention module-based encoder (Sect. 3.2), temporal attention in the decoder (Sect. 3.3) and the adversarial loss for enhancing fg-bag separability (Sect. 3.4). Table 3 (left) shows the baseline comparison on the YouTube-VIS 2019 val. set. All results reported in Table 3 (left) are obtained using the same ResNet50 backbone. As discussed earlier, our MS-STS VIS employs SeqFormer as its base framework. We train the baseline SeqFormer (denoted here as Baseline) using the official implementation and achieve an overall mask AP score of 46.4%. The introduction of our MS-STS attention module-based encoder within the baseline (referred as Baseline + MS-STS) significantly improves the overall performance to 48.4% with an absolute gain of 2.0%. The overall performance is further improved to 49.1% with the integration of the temporal attention in the decoder (denoted as Baseline + MS-STS + T-Dec). Finally, the introduction of the adversarial loss during the training for enhancing the fg-bg separability provides an absolute gain of 1.0% (denoted as Baseline + MS-STS + T-Dec + FG-BG Loss). Our final MS-STS VIS achieves an absolute gain of 3.7% over the baseline.

We further analyze the performance of our method (MS-STS VIS) under three specific challenging scenarios: fast motion, target size change (scale variation) and aspect-ratio change. To this end, we classified the videos into three categories: (i) fast motion of the object, (ii) object size changes, and (iii) aspect-ratio changes. In particular, we follow the well-established VOT-2015 benchmark [12] to label a particular video from above categories as follows: (i) **fast motion:** if object center in current frame moves by at least 30% of its size in previous frame. (ii) **change in object size:** if the ratio of the maximum size to the minimum size of an object in the video is greater than 1.5. (iii) **change in aspect-ratio:** if the ratio of the maximum to the minimum aspect (width/height) of the bounding box enclosing an object in the video is greater than 1.5.

Based on the aforementioned criteria, we first select the newly added videos in Youtube-VIS 2021 training set while retaining the same set of classes as in Youtube-VIS 2019 dataset. We refer these selected videos as custom set and classify them into the aforementioned attributes. To evaluate the performance of our MS-STS VIS and the baseline, we use the models trained on Youtube-VIS 2019 training set with the same ResNet-50 backbone. Note that we ensure that there is no overlap between the videos in Youtube-VIS 2019 training set and our custom set (only comprising the newly added videos from the Youtube-VIS 2021 training set). Table 3 (right) shows the comparison between the baseline SeqFormer and our MS-STS VIS. On the entire custom set, our MS-STS VIS obtains significantly improved performance over the baseline. Specifically, MS-STS VIS achieves absolute gains of 5.3%, 4.2% and 4.7% over the baseline on fast motion, target size change and aspect-ratio change attributes, respectively.

4.3 Qualitative Analysis

Figures 4 and 5 show qualitative results obtained by our MS-STS VIS framework on example videos from the Youtube-VIS 2019 val. and 2021 val. sets, respectively. We observe our MS-STS VIS framework to obtain promising video mask prediction in various challenging scenarios involving target appearance deformations due to fast motion, aspect-ratio change and scale variation. *E.g.*, in Fig. 4, video masks are predicted accurately for *hand* in row 1 (scale change), *eagle* in row 4 (fast motion, aspect-ratio change, scale variation), *panda* in row 2 (aspect-ratio change), *etc.*. Similarly, we observe promising video masks predictions for *leopard* in row 1 (fast motion), *person* in row 3 (scale variation), *dog* in row 2 (aspect-ratio change, scale variation) in Fig. 5. These results show the efficacy

Fig. 4. Qualitative results obtained by our MS-STS VIS framework on six example videos in the Youtube-VIS 2019 val set. Our MS-STS VIS achieves promising video mask prediction in various challenging scenarios including, fast motion (*eagle* in row 4), scale change (*hand* in row 1), aspect-ratio change (*panda* in row 2, *person* in row 3), multiple instances of same class (*tiger* in row 6).

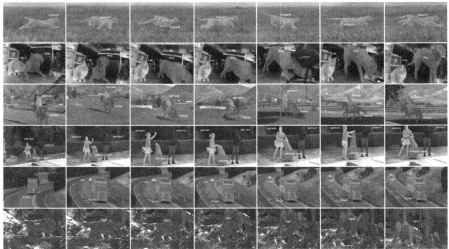

Fig. 5. Qualitative results on six example videos in the Youtube-VIS 2021 val set. Our MS-STS VIS achieves favorable video mask prediction in various scenarios involving target appearance deformations: fast motion (*leopard* in row 1), scale variation (*person* and *seal* in row 4), aspect-ratio change (*dog* in row 2, *tiger* in row 6). Also see the supplementary for additional results.

of our MS-STS VIS framework under different challenges for the task of video instance segmentation. Additional results are presented in the supplementary.

5   Related Work

Two-stage and Single-stage VIS: Several VIS methods [2,14,26] adapt the two-stage pipeline, such as Mask R-CNN [6] by introducing an additional tracking branch for target association. The work of [26] introduces MaskTrack R-CNN that jointly performs detection, segmentation and tracking. Bertasius *et al.* [2] utilize a branch in Mask R-CNN to propagate frame-level instance masks. The work of [14] introduces a modified variational autoencoder (VAE) on top of Mask R-CNN for VIS. On the other hand, several works [1,3,11,13,17] adapt the one-stage pipeline, such as FCOS [19], where a linear combination of mask bases are directly predicted as final segmentation. SipMask [3] introduces a spatial information preservation module for real-time VIS. The work of [1] introduces an approach where a short 3D convolutional spatio-temporal volume is adopted to learn pixel-level embedding. The work of [11] proposes to refine a space-time memory into a set of instance and frame-level prototypes.

Transformers-Based VIS: Wang *et al.* [22] introduce a transformer-based encoder-decoder architecture, named VisTR, that formulates VIS as a direct end-to-end parallel sequence prediction task. In the encoder, VisTR utilizes a single-scale attention that computes similarities between all pairs of features from multiple spatial locations of a low-resolution feature map, across frames. The work of [8] introduces inter-frame communication transformers, where memory tokens are used to communicate between frames. The recent SeqFormer [24], built on Deformable DETR framework [29], utilizes per-frame multi-scale features during attention computations. While demonstrating promising results, SeqFormer struggles in case of target deformations likely due to not explicitly capturing the spatio-temporal feature relationships during attention computation. To address these issues, we proposed a framework comprising an encoder that captures multi-scale spatio-temporal feature relationships. We also introduced an attention block in decoder to enhance temporal consistency of detected instance in different frames and an adversarial loss during training that ensures better fg-bg separability within multi-scale spatio-temporal feature space.

6   Conclusions

We proposed a transformer-based VIS framework, MS-STS VIS, comprising a novel multi-scale spatio-temporal split attention (MS-STS) module to effectively capture spatio-temporal feature relationships at multiple scales across frames in a video. We further introduced an auxiliary discriminator during training that strives to enhance fg-bg separability within the multi-scale spatio-temporal feature space. Our MS-STS VIS specifically tackles target appearance deformations due to real-world challenges. Extensive experiments reveal the benefits of the proposed contributions, achieving state-of-the-art performance on two benchmarks.

Acknowledgements. The work was partially supported by VR grants 2016–05543 and 2018–04673, WASP, ELLIIT, and SNIC funded via grant 2018–05973.

References

1. Athar, A., Mahadevan, S., Ošep, A., Leal-Taixé, L., Leibe, B.: STEm-Seg: spatio-temporal embeddings for instance segmentation in videos. In: Vedaldi, A., Bischof, H., Brox, T., Frahm, J.-M. (eds.) ECCV 2020. LNCS, vol. 12356, pp. 158–177. Springer, Cham (2020). https://doi.org/10.1007/978-3-030-58621-8_10
2. Bertasius, G., Torresani, L.: Classifying, segmenting, and tracking object instances in video with mask propagation. In: CVPR (2020)
3. Cao, J., Anwer, R.M., Cholakkal, H., Khan, F.S., Pang, Y., Shao, L.: SipMask: spatial information preservation for fast image and video instance segmentation. In: Vedaldi, A., Bischof, H., Brox, T., Frahm, J.-M. (eds.) ECCV 2020. LNCS, vol. 12359, pp. 1–18. Springer, Cham (2020). https://doi.org/10.1007/978-3-030-58568-6_1
4. Carion, N., Massa, F., Synnaeve, G., Usunier, N., Kirillov, A., Zagoruyko, S.: End-to-end object detection with transformers. In: Vedaldi, A., Bischof, H., Brox, T., Frahm, J.-M. (eds.) ECCV 2020. LNCS, vol. 12346, pp. 213–229. Springer, Cham (2020). https://doi.org/10.1007/978-3-030-58452-8_13
5. Fu, Y., Yang, L., Liu, D., Huang, T.S., Shi, H.: CompFeat: comprehensive feature aggregation for video instance segmentation. In: AAAI (2021)
6. He, K., Gkioxari, G., Dollár, P., Girshick, R.: Mask R-CNN. In: ICCV (2017)
7. He, K., Zhang, X., Ren, S., Sun, J.: Deep residual learning for image recognition. In: CVPR (2016)
8. Hwang, S., Heo, M., Oh, S.W., Kim, S.J.: Video instance segmentation using inter-frame communication transformers. In: NeurIPS (2021)
9. Isola, P., Zhu, J.Y., Zhou, T., Efros, A.A.: Image-to-image translation with conditional adversarial networks. In: CVPR (2017)
10. Johnander, J., Brissman, E., Danelljan, M., Felsberg, M.: Learning video instance segmentation with recurrent graph neural networks. In: GCPI (2021)
11. Ke, L., Li, X., Danelljan, M., Tai, Y.W., Tang, C.K., Yu, F.: Prototypical cross-attention networks for multiple object tracking and segmentation. In: NeurIPS (2021)
12. Kristan, M., et al.: The visual object tracking vot2015 challenge results. In: ICCV workshops (2015)
13. Li, M., Li, S., Li, L., Zhang, L.: Spatial feature calibration and temporal fusion for effective one-stage video instance segmentation. In: CVPR (2021)
14. Lin, C., Hung, Y., Feris, R., He, L.: Video instance segmentation tracking with a modified VAE architecture. In: CVPR (2020)
15. Lin, H., Wu, R., Liu, S., Lu, J., Jia, J.: Video instance segmentation with a propose-reduce paradigm. In: arXiv preprint arXiv:2103.13746 (2021)
16. Lin, T.-Y., et al.: Microsoft COCO: common objects in context. In: Fleet, D., Pajdla, T., Schiele, B., Tuytelaars, T. (eds.) ECCV 2014. LNCS, vol. 8693, pp. 740–755. Springer, Cham (2014). https://doi.org/10.1007/978-3-319-10602-1_48
17. Liu, D., Cui, Y., Tan, W., Chen, Y.: SG-Net: spatial granularity network for one-stage video instance segmentation. In: CVPR (2021)
18. Paszke, A., et al.: An imperative style, high-performance deep learning library. In: NeurIPS (2019). https://papers.neurips.cc/paper/9015-pytorch-an-imperative-style-high-performance-deep-learning-library.pdf
19. Tian, Z., Shen, C., Chen, H., He, T.: FCOS: fully convolutional one-stage object detection. In: ICCV (2019)
20. Vaswani, A., et al.: Pattention is all you need. In: NeurIPS (2017)

21. Voigtlaender, P., Chai, Y., Schroff, F., Adam, H., Leibe, B., Chen, L.: FEELVOS: fast end-to-end embedding learning for video object segmentation. In: CVPR (2019)
22. Wang, Y., et al.: End-to-end video instance segmentation with transformers. In: CVPR (2021)
23. Wojke, N., Bewley, A., Paulus, D.: Simple online and realtime tracking with a deep association metric. In: ICIP (2017)
24. Wu, J., Jiang, Y., Zhang, W., Bai, X., Bai, S.: SeqFormer: a frustratingly simple model for video instance segmentation. In: arXiv preprint arXiv:2112.08275 (2021)
25. Xu, N., et al.: Youtube-vis dataset 2021 version. https://youtube-vos.org/dataset/vis (2021)
26. Yang, L., Fan, Y., Xu, N.: Video instance segmentation. In: ICCV (2019)
27. Yang, L., Wang, Y., Xiong, X., Yang, J., Katsaggelos, A.K.: Efficient video object segmentation via network modulation. In: CVPR (2018)
28. Yang, S., et al.: Crossover learning for fast online video instance segmentation. In: ICCV (2021)
29. Zhu, X., Su, W., Lu, L., Li, B., Wang, X., Dai, J.: Deformable detr: Deformable transformers for end-to-end object detection. In: ICLR (2021)

RANKSEG: Adaptive Pixel Classification with Image Category Ranking for Segmentation

Haodi He[1], Yuhui Yuan[3(✉)], Xiangyu Yue[2], and Han Hu[3(✉)]

[1] University of Science and Technology of China, Hefei, China
[2] UC Berkeley, Berkeley, USA
[3] Microsoft Research Asia, Beijing, China
{yuhui.yuan,hanhu}@microsoft.com

Abstract. The segmentation task has traditionally been formulated as a complete-label (We use the term "complete label" to represent the set of all predefined categories in the dataset.) pixel classification task to predict a class for each pixel from a fixed number of predefined semantic categories shared by all images or videos. Yet, following this formulation, standard architectures will inevitably encounter various challenges under more realistic settings where the scope of categories scales up (e.g., beyond the level of 1k). On the other hand, in a typical image or video, only a few categories, i.e., a small subset of the complete label are present. Motivated by this intuition, in this paper, we propose to decompose segmentation into two sub-problems: (i) image-level or video-level multi-label classification and (ii) pixel-level rank-adaptive selected-label classification. Given an input image or video, our framework first conducts multi-label classification over the complete label, then sorts the complete label and selects a small subset according to their class confidence scores. We then use a rank-adaptive pixel classifier to perform the pixel-wise classification over only the selected labels, which uses a set of rank-oriented learnable temperature parameters to adjust the pixel classifications scores. Our approach is conceptually general and can be used to improve various existing segmentation frameworks by simply using a lightweight multi-label classification head and rank-adaptive pixel classifier. We demonstrate the effectiveness of our framework with competitive experimental results across four tasks, including image semantic segmentation, image panoptic segmentation, video instance segmentation, and video semantic segmentation. Especially, with our RankSeg, Mask2Former gains +0.8%/+0.7%/+0.7% on ADE20K panoptic segmentation/YouTubeVIS 2019 video instance segmentation/VSPW video semantic segmentation benchmarks respectively. Code is available at: https://github.com/openseg-group/RankSeg.

H. He and Y. Yuan—Equal contribution.

Supplementary Information The online version contains supplementary material available at https://doi.org/10.1007/978-3-031-19818-2_39.

Keywords: Rank-adaptive · Selected-label · Image semantic segmentation · Image panoptic segmentation · Video instance segmentation · Video semantic segmentation

1   Introduction

Image and video segmentation, i.e., partitioning images or video frames into multiple meaningful segments, is a fundamental computer vision research topic that has wide applications including autonomous driving, surveillance system, and augmented reality. Most recent efforts have followed the path of fully convolutional networks [54] and proposed various advanced improvements, e.g., high-resolution representation learning [60,70], contextual representation aggregation [12,23,35,81,87], boundary refinement [43,66,84], and vision transformer architecture designs [53,61,82,88].

Most of the existing studies formulate the image and video segmentation problem as a complete-label pixel classification task. In the following discussion, we'll take image semantic segmentation as an example for convenience. For example, image semantic segmentation needs to select the label of each pixel from the complete label[1] set that is predefined in advance. However, it is unnecessary to consider the complete label set for every pixel in each image as most standard images only consist of objects belonging to a few categories. Figure 2 plots the statistics on the percentage of images that contain no more than the given class number in the entire dataset vs. the number of classes that appear within each image. Accordingly, we can see that 100.00%, 99.99%, 99.14%, and 99.85% of images contain less than 25 categories on PASCAL-Context [57], COCO-Stuff [8], ADE20K-Full [19,89], and COCO+LVIS [29,39] while each of them contains 60, 171, 847, and 1,284 predefined semantic categories respectively. Besides, Fig. 1 shows an example image that only contains 7 classes while the complete label set consists of 171 predefined categories.

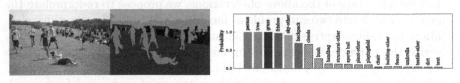

(a) Image   (b) Ground-truth   (c) Multi-label image classification prediction

Fig. 1. Illustrating the motivation of exploiting the multi-label image classification: (a) An example image selected from COCO-Stuff. (b) The ground-truth segmentation map consisting of 7 classes. (c) The histogram of existence probability over the selected top 20 categories sorted by their confidence scores, which are predicted with our method. Our method only needs to identify the label of each pixel from these selected 20 categories instead of all 171 categories. The names of true/false positive categories are marked with black/red color, respectively. The bars associated with true positive categories share the same color as the ones adopted in the ground-truth segmentation map. (Color figure online)

[1] We use "label", "category", and "class" interchangeably.

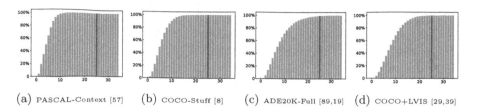

(a) PASCAL-Context [57]   (b) COCO-Stuff [8]   (c) ADE20K-Full [89,19]   (d) COCO+LVIS [29,39]

Fig. 2. Illustrating the cumulative distribution of the number of images that contain no more than the given class number: The x-axis represents the class number presented in the image and the y-axis represents the percentage of images that contain no more than the given class number in the entire dataset. We plot the cumulative distribution on four benchmarks. We can see that more than 99% of all images only contain less than 25 categories on four benchmarks, which are marked with dark purple bars. The above four benchmarks contain 60, 171, 847, and 1,284 predefined semantic categories respectively. (Color figure online)

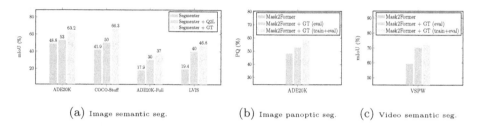

(a) Image semantic seg.   (b) Image panoptic seg.   (c) Video semantic seg.

Fig. 3. Illustrating the effectiveness of multi-label classification: The segmentation results of exploiting the ground-truth multi-label during only evaluation or both training and evaluation on (a) PASCAL-Context, COCO-Stuff, ADE20K-Full, and COCO+LVIS (image semantic segmentation)/(b) ADE20K (image panoptic segmentation)/(c) VSPW (video semantic segmentation). Refer to Sect. 3.4 for more details. YouTubeVIS results are not included as we can not access the ground-truth.

To take advantage of the above observations, we propose to re-formulate the segmentation task into two sub-problems including multi-label image/video classification and rank-adaptive pixel classification over a subset of selected labels. To verify the potential benefits of our method, we investigate the gains via exploiting the ground-truth multi-label of each image or video. In other words, the pixel classifier only needs to select the category of each pixel from a collection of categories presented in the current image or video, therefore, we can filter out all other categories that do not appear. Figure 3 summarizes the comparison results based on Segmenter [65] and Mask2Former [18]. We can see that the segmentation performance is significantly improved given the ground-truth multi-label prediction. In summary, multi-label classification is an important but long-neglected sub-problem on the path toward more accurate segmentation.

Motivated by the significant gains obtained with the ground-truth multi-label, we propose two different schemes to exploit the benefit of multi-label image predictions including the *independent single-task scheme* and *joint multi-task scheme*. For the independent single-task scheme, we train one model for multi-label image/video classification and another model for segmentation. Specifically, we first train a model to predict multi-label classification probabilities for each image/video, then we estimate the existing label subset for each image/video based on the multi-label predictions, last we use the predicted label subset to train the segmentation model for rank-adaptive selected-label pixel classification during both training and testing. For the joint multi-task scheme, we train one model to support both multi-label image/video classification and segmentation based on a shared backbone. Specifically, we apply a multi-label prediction head and a segmentation head, equipped with a rank-adaptive adjustment scheme, over the shared backbone and train them jointly. For both schemes, we need to send the multi-label predictions into the segmentation head for rank-adaptive selected-label pixel classification, which enables selecting a collection of categories that appear and adjusting the pixel-level classification scores according to the image/video content adaptively.

We demonstrate the effectiveness of our approach on various strong baseline methods including DeepLabv3 [12], Segmenter [65], Swin-Transformer [53], BEiT [4], MaskFormer [19], Mask2Former [18], and ViT-Adapter [15] across multiple segmentation benchmarks including PASCAL-Context [57], ADE20K [89], COCO-Stuff [8], ADE20K-Full [19,89], COCO+LVIS [29,39], YouTubeVIS [78], and VSPW [56].

2   Related Work

Image Segmentation. We can roughly categorize the existing studies on image semantic segmentation into two main paths: (i) region-wise classification methods [1,7,7,26,27,58,67,74], which first organize the pixels into a set of regions (usually super-pixels), and then classify each region to get the image segmentation result. Several very recent methods [19,69,86] exploit the DETR framework [10] to conduct region-wise classification more effectively; (ii) pixel-wise classification methods, which predict the label of each pixel directly and dominate most previous studies since the pioneering FCN [54]. There exist extensive follow-up studies that improve the pixel classification performance via constructing better contextual representations [12,81,83,87] or designing more effective decoder architectures [3,13,64]. Image panoptic segmentation [41,42] aims to unify image semantic segmentation and image instance segmentation tasks. Some recent efforts have introduced various advanced architectures such as Panoptic FPN [41], Panoptic DeepLab [17], Panoptic Segformer [48], K-Net [86], and Mask2Former [18]. Our RankSeg is complementary with various paradigms and consistently improves several representative state-of-the-art methods across both image semantic segmentation and image panoptic segmentation tasks.

Video Segmentation. Most of the previous works address the video segmentation task by extending the existing image segmentation models with temporal consistency constraint [71]. Video semantic segmentation aims to predict the semantic category of all pixels in each frame of a video sequence, where the main efforts focus on two paths including exploiting cross-frame relations to improve the prediction accuracy [11,25,36,40,44,59] and leveraging the information of neighboring frames to accelerate computation [34,38,47,52,55,77]. Video instance segmentation [78] requires simultaneous detection, segmentation and tracking of instances in videos and there exist four mainstream frameworks including tracking-by-detection [9,24,32,49,68], clip-and-match [2,6], propose-and-reduce [50], and segment-as-a-whole [16,37,72,75]. We show the effectiveness of our method on both video semantic segmentation and video instance segmentation tasks via improving the very recent state-of-the-art method Mask2Former [16].

Multi-label Classification. The goal of multi-label classification is to identity all the categories presented in a given image or video over the complete label set. The conventional multi-label image classification literature partitions the existing methods into three main directions: (i) improving the multi-label classification loss functions to handle the imbalance issue [5,76], (ii) exploiting the label co-occurrence (or correlations) to model the semantic relationships between different categories [14,33,46,79], and (iii) localizing the diverse image regions associated with different categories [28,45,51,73,80]. In our independent single-task scheme, we choose the very recent state-of-the-art method Query2Label [51] to perform multi-label classification on various semantic segmentation benchmarks as it is a very simple and effective method that exploits the benefits of both label co-occurrence and localizing category-dependent regions. There exist few efforts that apply multi-label image classification to address segmentation task. To the best of our knowledge, the most related study EncNet [85] simply adds a multi-label image classification loss w/o changing the original semantic segmentation head that still needs to select the label of each pixel from all predefined categories. We empirically show the advantage of our method over EncNet in the ablation experiments. Besides, our proposed method is naturally suitable to solve large-scale semantic segmentation problem as we only perform rank-adaptive pixel classification over a small subset of the complete label set based on the multi-label image prediction. We also empirically verify the advantage of our method over the very recent ESSNet [39] in the ablation experiments.

3   Our Approach

We first introduce the overall framework of our approach in Sect. 3.1, which is also illustrated in Fig. 4. Second, we introduce the details of the independent single-task scheme in Sect. 3.2 and those of joint multi-task scheme in Sect. 3.3. Last, we conduct analysis experiments to investigate the detailed improvements of our method across multiple segmentation tasks in Sect. 3.4.

3.1   Framework

The overall framework of our method is illustrated in Fig. 4, which consists of one path for multi-label image classification and one path for semantic segmentation. The multi-label prediction is used to sort and select the top κ category embeddings with the highest confidence scores which are then sent into the rank-adaptive

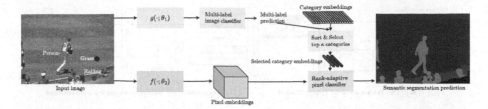

Fig. 4. Illustrating the framework of our approach: Given an input image that contains person, grass, and railing, which are marked with ●, ●, and ●, respectively. First, we use the multi-label image classification model $g(\cdot; \theta_1)$ and multi-label image classifier to predict the presence probabilities of all categories. Second, we sort the labels according to the predicted presence probabilities and select the top κ most probable category embeddings. Last, we send the selected subset of category embeddings into the rank-adaptive selected-label pixel classifier to identify the label of each pixel based on the pixel embeddings output by the semantic segmentation model $f(\cdot; \theta_2)$. (Color figure online)

selected-label pixel classifier to generate the semantic segmentation prediction. We explain the mathematical formulations of both multi-label image classification and rank-adaptive selected-label pixel classification as follows:

Multi-label Image Classification. The goal of multi-label image classification is to predict the set of existing labels in a given image $\mathbf{x} \in \mathbb{R}^{H \times W \times 3}$, where H and W represent the input height and width. We generate the image-level multi-label ground truth $\mathbf{y}^{gt}$ from the ground truth segmentation map and we represent $\mathbf{y}^{gt}$ with a vector of K binary values $[y_1^{gt}, y_2^{gt}, ..., y_K^{gt}]^{\mathsf{T}}, y_i^{gt} \in \{0, 1\}$, where K represents the total number of predefined categories and $y_i^{gt} = 1$ represents the existence of pixels belonging to i-th category and $y_i^{gt} = 0$ otherwise.

The prediction $\mathbf{y} \in \mathbb{R}^K$ is a vector that records the existence confidence score of each category in the given image $\mathbf{x}$. We use $g(\cdot; \theta_1)$ to represent the backbone for the multi-label image classification model. We estimate the multi-label predictions of input $\mathbf{x}$ with the following sigmoid function:

$$y_k = \frac{e^{\xi(g(\mathbf{x}; \theta_1), \mathbf{h}_k)}}{e^{\xi(g(\mathbf{x}; \theta_1), \mathbf{h}_k)} + 1}, \tag{1}$$

where y_k is the k-th element of $\mathbf{y}$, $g(\mathbf{x}; \theta_1)$ represents the output feature map, $\mathbf{h}_k$ represents the multi-label image classification weight associated with the k-th category, and $\xi(\cdot)$ represents a transformation function that estimates the similarity between the output feature map and the multi-label image classification weights. We supervise the multi-label predictions with the asymmetric loss that operates differently on positive and negative samples by following [5,51].

Rank-Adaptive Selected-Label Pixel Classification. The goal of semantic segmentation is to predict the semantic label of each pixel and the label is selected from all predefined categories. We use $\mathbf{z} \in \mathbb{R}^{H \times W \times K}$ to represent the predicted pixel classification probability map for the input image $\mathbf{x}$. We use $f(\cdot; \theta_2)$ to represent the semantic segmentation backbone and $\mathbf{z}^{gt} \in \mathbb{R}^{H \times W}$ to represent the ground-truth segmentation map. Instead of choosing the label of

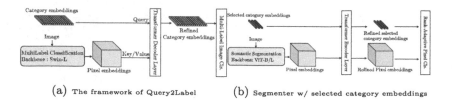

(a) The framework of Query2Label   (b) Segmenter w/ selected category embeddings

Fig. 5. (a) **Illustrating the framework of Query2Label**: The multi-label image classification backbone $g(\cdot; \theta_1)$ is set as Swin-L by default. The transformation $\xi(\cdot)$ is implemented as two transformer decoder layers followed by a linear layer that prepares the refined category embeddings for the multi-label image classifier. (b) **Illustrating the framework of Segmenter w/ selected category embeddings**: The semantic segmentation backbone $f(\cdot; \theta_2)$ is set as ViT-B/16 or ViT-L/16. The transformation $\psi(\cdot)$ is implemented as two transformer encoder layers followed by ℓ_2-normalization before estimating the segmentation map.

each pixel from all K predefined categories, based on the previous multi-label prediction $\mathbf{y}$ for image $\mathbf{x}$, we introduce a more effective rank-adaptive selected-label pixel classification scheme:

- Sort and select the top κ elements of the classifier weights according to the descending order of multi-label predictions $\mathbf{y} = [y_1, y_2, \cdots, y_K]$:

$$[\overline{\mathbf{w}}_1, \overline{\mathbf{w}}_2, \cdots, \overline{\mathbf{w}}_\kappa] = \text{Top-}\kappa([\mathbf{w}_1, \mathbf{w}_2, \cdots, \mathbf{w}_K], \mathbf{y}), \qquad (2)$$

- Rank-adaptive classification of pixel (i, j) over the top κ selected categories:

$$\mathbf{z}_{i,j,k} = \frac{e^{\psi(f(\mathbf{x};\theta_2)_{i,j}, \overline{\mathbf{w}}_k)/\tau_k}}{\sum_{l=1}^{\kappa} e^{\psi(f(\mathbf{x};\theta_2)_{i,j}, \overline{\mathbf{w}}_l)/\tau_l}}, \qquad (3)$$

where $[\mathbf{w}_1, \mathbf{w}_2, \cdots, \mathbf{w}_K]$ represents the pixel classification weights for all K predefined categories and $[\overline{\mathbf{w}}_1, \overline{\mathbf{w}}_2, \cdots, \overline{\mathbf{w}}_\kappa]$ represents the top κ selected pixel classification weights associated with the largest multi-label classification scores. $f(\mathbf{x}; \theta_2)$ represents the output feature map for semantic segmentation. $\psi(\cdot)$ represents a transformation function that estimates the similarity between the pixel features and the pixel classification weights. κ represents the number of selected category embeddings and κ is chosen as a much smaller value than K. We apply a set of rank-adaptive learnable temperature parameters $[\tau_1, \tau_2, \cdots, \tau_\kappa]$ to adjust the classification scores over the selected top κ categories. The temperature parameters across different selected classes are shared in all of the baseline experiments by default[2]. We analyze the influence of κ choices and the benefits of such a rank-oriented adjustment scheme in the following discussions and experiments.

---

[2] We set $\tau_1 = \tau_2 = \cdots = \tau_\kappa$ for all baseline segmentation experiments.

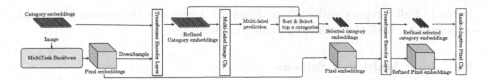

Fig. 6. Illustrating the framework of joint multi-task scheme: $g(\cdot; \theta_1)$ and $f(\cdot; \theta_2)$ are set as the shared multi-task backbone and $\theta_1 = \theta_2$. $\xi(\cdot)$ is implemented as one transformer encoder layer or two transformer decoder layers or global average pooling + linear projection. $\psi(\cdot)$ is implemented as two transformer encoder layers followed by L2-normalization before estimating the segmentation map.

3.2 Independent Single-Task Scheme

Under the independent single-task setting, the multi-label image classification model $g(\cdot; \theta_1)$ and the semantic segmentation model $f(\cdot; \theta_2)$ are trained separately and their model parameters are not shared, i.e., $\theta_1 \neq \theta_2$. Specifically, we first train the multi-label image classification model $g(\cdot; \theta_1)$ to identify the top κ most likely categories for each image. Then we train the rank-adaptive selected-label pixel classification model, i.e., semantic segmentation model, $f(\cdot; \theta_2)$ to predict the label of each pixel over the selected top κ classes.

Multi-label Image Classification Model. We choose the very recent SOTA multi-label classification method Query2Label [51] as it performs best on multiple multi-label classification benchmarks by the time of our submission according to paper-with-code.[3] The key idea of Query2Label is to use the category embeddings as the query to gather the desired pixel embeddings as the key/value, which is output by an ImageNet-22K pre-trained backbone such as Swin-L, adaptively with one or two transformer decoder layers. Then Query2Label scheme applies a multi-label image classifier over the refined category embeddings to predict the existence of each category. Figure 5a illustrates the framework of Query2Label framework. Refer to [51] and the official implementation for more details. The trained weights of Query2Label model are fixed during both training and inference of the following semantic segmentation model.

Rank-Adaptive Selected-Label Pixel Classification Model. We choose a simple yet effective baseline Segmenter [65] as it achieves even better performance than Swin-L[4] when equipped with ViT-L. Segmenter first concatenates the category embedding with the pixel embeddings output by a ViT model together and then sends them into two transformer encoder layers. Last, based on the refined pixel embeddings and category embeddings, Segmenter computes their ℓ_2-normalized scalar product as the segmentation predictions. We select the top κ most likely categories for each image according to the predictions of the Query2Label model and only use the selected top κ category embeddings

---

[3] https://paperswithcode.com/task/multi-label-classification.
[4] Segmenter w/ ViT-L: 53.63% vs. Swin-L: 53.5% on ADE20K.

instead of all category embeddings. Figure 5b illustrates the overall framework of Segmenter with the selected category embeddings.

3.3   Joint Multi-task Scheme

Considering that the independent single-task scheme suffers from extra heavy computation overhead as the Query2Label method relies on a large backbone, e.g., Swin-L, we introduce a joint multi-task scheme that shares the backbone for both sub-tasks, in other words, $\theta_1 = \theta_2$ and the computations of $g(\cdot; \theta_1)$ and $f(\cdot; \theta_2)$ are also shared.

Figure 6 shows the overall framework of the joint multi-task scheme. First, we apply a shared multi-task backbone to process the input image and output the pixel embeddings. Second, we concatenate the category embeddings with the down-sampled pixel embeddings[5], send them into one transformer encoder layer, and apply the multi-label image classifier on the refined category embeddings to estimate the multi-label predictions. Last, we sort and select the top κ category embeddings, concatenate the selected category embeddings with the pixel embeddings, send them into two transformer encoder layers, and compute the semantic segmentation predictions based on ℓ_2-normalized scalar product between the refined selected category embeddings and the refined pixel embeddings. We empirically verify the advantage of the joint multi-task scheme over the independent single-task scheme in the ablation experiments.

3.4   Analysis Experiments

Oracle Experiments. We first conduct several groups of oracle experiments based on Segmenter w/ ViT-B/16 on four challenging image semantic segmentation benchmarks (PASCAL-Context/COCO-Stuff/ADE20K-Full/COCO+LVIS), Mask2Former w/ Swin-L on both ADE20K panoptic segmentation benchmark[6] and VSPW video semantic segmentation benchmark.

- *Segmenter/Mask2Former + GT (train + eval):* the upper-bound segmentation performance of Segmenter/Mask2Former when training & evaluating equipped with the ground-truth multi-label of each image or video, in other words, we only need to select the category of each pixel over the ground-truth existing categories in a given image or video.
- *Segmenter/Mask2Former + GT (eval):* the upper-bound segmentation performance of Segmenter/Mask2Former when only using the ground-truth multi-label of each image or video during evaluation.

Figure 3 illustrates the detailed comparison results. We can see that only applying the ground-truth multi-label during evaluation already brings considerable improvements and further applying the ground-truth multi-label during

---

[5] Different from the semantic segmentation task, the multi-label image classification task does not require high-resolution representations.

[6] We choose Swin-L by following the MODEL_ZOO of the official Mask2Former implementation: https://github.com/facebookresearch/Mask2Former.

Table 1. Ablation of the improvements with our method. MT: multi-task learning with auxiliary multi-label image classification scheme. LS: label sort and selection, in other words, sort and select the top κ classes. RA: rank-adaptive-τ, which applies independent τ for pixel classification scores associated with the different ranking positions.

Method.	Image semantic seg.				Image panoptic seg.	Video semantic seg.
	PASCAL-Context	COCO-Stuff	ADE20K-Full	COCO+LVIS	ADE20K	VSPW
Baseline	53.85	41.85	17.93	19.41	48.1	59.4
+ MT	54.05	42.38	17.81	20.26	48.2	59.5
+ MT + LS	54.27	44.31	18.26	21.13	48.8	59.6
+ MT + LS + RA	**54.76**	**44.98**	**18.78**	**21.26**	**48.9**	**60.1**

training significantly improves the segmentation performance across all benchmarks. For example, when compared to the baseline Segmenter or Mask2Former, Segmenter + GT (train + eval) gains +17%/+24%/+19%/+27% absolute mIoU scores across PASCAL-Context/COCO-Stuff/ADE20K-Full/COCO+LVIS and Mask2Former + GT (train + eval) gains +11%/+13% on ADE20K/VSPW.

Improvement Analysis of RankSeg. Table 1 reports the results with different combinations of the proposed components within our joint multi-task scheme. We can see that: (i) multi-task learning (MT) introduces the auxiliary multi-label image classification task and brings relatively minor gains on most benchmarks, (ii) combining MT with label sorting & selection (LS) achieves considerable gains, and (iii) applying the rank-adaptive-τ manner (shown in Eq. 3) instead of shared-τ achieves better performance. We investigate the possible reasons by analyzing the value distribution of learned $1/\tau$ with the rank-adaptive-τ manner in Fig. 7, which shows that the learned $1/\tau$ is capable of adjusting the pixel classification scores based on the order of multi-label classification scores. In summary, we choose "MT + LS + RA" scheme by default, which gains +0.91%/+3.13%/+0.85%/+1.85%/+0.8%/+0.7% over the baseline methods across these six challenging segmentation benchmarks respectively.

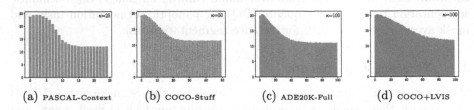

(a) PASCAL-Context   (b) COCO-Stuff   (c) ADE20K-Full   (d) COCO+LVIS

Fig. 7. Illustrating the values of learned $1/\tau$ with rank-adaptive-τ manner on four semantic segmentation benchmarks. We can see that the values of $1/\tau$ are almost monotonically decreasing, thus meaning that the pixel classification scores of the category associated with larger multi-label classification scores are explicitly increased.

4   Experiment

4.1   Implementation Details

We illustrate the details of the datasets, including ADE20K [89], ADE20K-Full [19,89], PASCAL-Context [57], COCO-Stuff [8], COCO+LVIS [29,39], VSPW [56], and YouTubeVIS [78], in the supplementary material.

Multi-label Image Classification. For the independent single-task scheme, following the official implementation[7] of Query2Label, we train multi-label image classification models, e.g., ResNet-101 [31], TResNetL [62], and Swin-L [53], for 80 epochs using Adam solver with early stopping. Various advanced tricks such as cutout [22], RandAug [21] and EMA [30] are also used. For the joint multi-task scheme, we simply train the multi-label image classification models following the same settings as the segmentation models w/o using the above-advanced tricks that might influence the segmentation performance. We illustrate more details of the joint multi-task scheme in the supplementary material.

Segmentation. We adopt the same settings for both independent single-task scheme and joint multi-task scheme. For the segmentation experiments based on Segmenter [65], DeepLabv3 [12], Swin-Transformer [53], BEiT [4], and ViT-Adapter [15]. We follow the default training & testing settings of their reproduced version based on `mmsegmentation` [20]. For the segmentation experiments based on MaskFormer or Mask2Former, we follow their official implementation[8].

Hyper-parameters. We set κ as 25, 50, 50, 100, and 100 on PASCAL-Context, ADE20K, COCO-Stuff, ADE20K-Full, and COCO+LVIS respectively as they consist of a different number of semantic categories. We set their multi-label image classification loss weights as 5, 10, 10, 100, and 300. The segmentation loss weight is set as 1. We illustrate the hyper-parameter settings of experiments on MaskFormer [19] or Mask2Former [18] in the supplementary material.

Metrics. We report mean average precision (mAP) for multi-label image classification task, mean intersection over union (mIoU) for image/video semantic segmentation task, panoptic quality (PQ) for panoptic segmentation task, and mask average precision (AP) for instance segmentation task.

4.2   Ablation Experiments

We conduct all ablation experiments based on the Segmenter w/ ViT-B and report their single-scale evaluation results on COCO-Stuff `test` and COCO+LVIS `test` if not specified. The baseline with Segmenter w/ ViT-B achieves 41.85% and 19.41% mIoU on COCO-Stuff and COCO+LVIS, respectively.

Influence of the Multi-label Classification Accuracy. We investigate the influence of multi-label classification accuracy on semantic segmentation tasks

---

[7] https://github.com/SlongLiu/query2labels.
[8] https://github.com/facebookresearch/Mask2Former.

Table 2. Influence of the multi-label image classification accuracy (mAP) on semantic segmentation accuracy (mIoU) based on independent single-task manner.

Backbone	COCO-Stuff			COCO+LVIS		
	ResNet101 [63]	TResNetL [63]	Swin-L [53]	ResNet101 [63]	TResNetL [63]	Swin-L [53]
mAP (%)	55.17	60.10	**64.79**	26.54	31.01	**34.93**
mIoU (%)	39.34	42.87	**44.42**	16.30	19.88	**21.19**
△	−2.51	+1.02	**+2.57**	−3.11	+0.47	**+1.78**

Table 3. Independent single-task scheme vs. Joint multi-task scheme: we adopt Swin-L as the backbone for the multi-label predictions in the independent single-task scheme.

Method	COCO-Stuff				COCO+LVIS			
	#params.	FLOPs	mIoU (%)	△	#params.	FLOPs	mIoU (%)	△
Indep. single-task	343.21M	182.8G	44.42	+2.52	347.33M	233.62G	21.19	+1.78
Joint multi-task	**109.73M**	**78.71G**	**44.98**	**+3.08**	**111.45M**	**99.59G**	**21.26**	**+1.85**

Table 4. Influence of the size of the selected label set, i.e., κ.

κ	COCO-Stuff							COCO+LVIS			
	25	50	75	100	125	150	171	50	100	200	1,284
mIoU (%)	44.65	**44.98**	44.58	44.67	44.41	44.37	44.20	20.36	**21.26**	20.99	20.93
△	+2.80	**+3.13**	+2.73	+2.82	+2.56	+2.52	+2.35	+0.95	**+1.85**	+1.58	+1.52

Table 5. Influence of the multi-label classification loss weight.

Multi-label cls. loss weight	1	5	10	20
mAP (%)	59.14	62.38	**62.52**	62.18
mIoU (%)	44.11	44.90	**44.98**	44.04
△	+2.26	+3.05	**+3.13**	+2.19

Table 6. Influence of multi-label prediction head architecture.

Method	#params.	FLOPs	mAP	mIoU (%)	△
GAP+Linear	103.24M	76.94G	60.62	43.19	+1.34
2× TranDec	114.27M	78.83G	61.15	44.07	+2.22
1× TranEnc	109.73M	78.71G	**62.52**	44.98	**+3.13**

based on the independent single-task scheme. We train multiple Query2Label models based on different backbones, e.g., ResNet101, TResNetL, and Swin-L. Then we train three Segmenter w/ ViT-B segmentation models based on their multi-label predictions independently. According to the results in Table 2, we can see that more accurate multi-label classification prediction brings more semantic segmentation performance gains and less accurate multi-label classification prediction even results in worse results than baseline.

Independent Single-Task Scheme vs. Joint Multi-task Scheme. We compare the performance and model complexity of the independent single-task scheme (w/ Swin-L) and joint multi-task scheme in Table 3. To ensure fairness, we choose the # of labels in the selected label set, i.e., κ, as 50 for both schemes, and the segmentation models are trained & tested under the same settings. According to the comparison results, we can see that joint multi-task scheme achieves better performance with fewer parameters and FLOPs. Thus, we choose the joint multi-task scheme in the following experiments for efficiency if not specified.

Influence of Different Top κ**.** We study the influence of the size of selected label set, i.e., κ, as shown in Table 4. According to the results, our method achieves the best performance on COCO-Stuff/COCO+LVIS when $\kappa = 50/\kappa = 100$, which achieves a better trade-off between precision and recall for multi-label predictions. Besides, we attempt to fix $\kappa = 50$ during training and report the results when changing κ during evaluation on COCO-Stuff: $\kappa = 25 : 44.89\%/\kappa = 50 : 44.98\%/\kappa = 75 : 45.00\%/\kappa = 100 : 45.01\%$. Notably, we also report the results with $\kappa = K$, in other words, we only sort the classifier weights, which also achieve considerable gains. Therefore, we can see that sorting the classes and rank-adaptive adjustment according to multi-label predictions are the key to the gains. In summary, our method consistently outperforms baseline with different κ values. We also attempt to use dynamic κ for different images during evaluation but observe no significant gains. More details are provided in the supplementary material.

Influence of the Multi-label Classification Loss Weight. We study the influence of the multi-label image classification loss weights with the joint multi-task scheme on COCO-Stuff and report the results in Table 5. We can see that setting the multi-label classification loss weight as 10 achieves the best performance.

Influence of $\xi(\cdot)$ **Choice.** Table 6 compares the results based on different multi-label prediction head architecture choices including "GAP+Linear" (applying global average pooling followed by linear projection), "2× TranDec" (using two transformer decoder layers), and "1× TranEnc" (using one transformer encoder layer) on COCO-Stuff. Both "2× TranDec" and "1× TranEnc" operate on feature maps with $\frac{1}{32}$ resolution of the input image. According to the results, we can see that "1× TranEnc" achieves the best performance and we implement $\xi(\cdot)$ as one transformer encoder layer if not specified. We also attempt generating multi-label prediction (mAP $= 58.55\%$) from the semantic segmentation prediction directly but observe no performance gains, thus verifying the importance of relatively more accurate multi-label classification predictions.

More comparison results with the previous EncNet [85] and ESSNet [39] are summarized in the supplementary material.

4.3   State-of-the-Art Experiments

Image Semantic Segmentation. We apply our method to various state-of-the-art image semantic segmentation methods including DeepLabv3, Seg-Mask-L/16, Swin-Transformer, BEiT, and ViT-Adapter-L. Table 7 summarizes the detailed comparison results across three semantic segmentation benchmarks including ADE20K, COCO-Stuff, and COCO+LVIS, where we evaluate the multi-scale segmentation results on ADE20K/COCO-Stuff and single-scale segmentation results on COCO+LVIS (due to limited GPU memory) respectively. More details of how to apply our joint multi-task scheme to these methods are provided in the supplementary material. According to the results in Table 7, we can see that our RankSeg consistently improves DeepLabv3, Seg-Mask-L/16,

Table 7. Combination with DeepLabv3, Seg-Mask-L, Swin-B, BEⅈT, and ViT-Adapter.

Method	ADE20K			COCO-Stuff			COCO+LVIS		
	#params.	FLOPs	mIoU (%)	#params.	FLOPs	mIoU (%)	#params.	FLOPs	mIoU (%)
DeepLabv3	87.21M	347.64G	45.19	87.22M	347.68G	38.42	88.08M	350.02G	11.04
+ RankSeg	91.87M	349.17G	46.61	91.75M	349.09G	39.86	93.21M	359.47G	12.76
Seg-Mask-L/16	333.23M	377.83G	53.63	333.26M	378.57G	47.12	334.4M	420.2G	23.71
+ RankSeg	345.99M	377.18G	54.47	346.03M	377.46G	47.93	348.31M	422.6G	24.60
Swin-B	121.42M	299.81G	52.4	121.34M	299.98G	47.16	122.29M	309.34G	20.33
+ RankSeg	125.43M	300.48G	53.01	125.46M	300.56G	47.85	126.89M	306.53G	20.81
BEⅈT	441.27M	1745.99G	57.0	441.30M	1746.54G	49.9	OOM		
+ RankSeg	456.28M	1751.04G	57.8	456.35M	1751.34G	50.3	OOM		
ViT-Adapter-L	570.74M	2743.20G	60.5	N/A			N/A		
+ RankSeg	584.71M	2747.81G	60.7	N/A			N/A		

OOM means out of memory error on 8 × 32G V100 GPUs.
N/A means the results are available due to limited GPU resources.

Table 8. Combination with Mask2Former based on Swin-L.

Method	Image semantic seg.			Image panoptic seg.			Video semantic seg.			Video instance seg.		
	ADE20K			ADE20K			VSPW (T=2)			YouTubeVIS 2019 (T=2)		
	#params.	FLOPs	mIoU (%)	#params.	FLOPs	PQ (%)	#params.	FLOPs	mIoU (%)	#params.	FLOPs	AP (%)
Mask2Former [18]	205.51M	369.02G	57.3	205.55M	377.98G	48.1	205.50M	737.38G	59.4	205.52M	753.79G	60.4
+ RankSeg	208.79M	369.83G	58.0	208.83M	379.11G	48.9	208.75M	738.14G	60.1	208.70M	754.59G	61.1

$T = 2$ means each video clip is composed of 2 frames during training/evaluation and we report the GFLOPs over 2 frames.

Swin-Transformer, BEⅈT, and ViT-Adapter-L across three evaluated benchmarks. For example, with our RankSeg, BEⅈT gains 0.8% on ADE20K with slightly more parameters and GFLOPs.

Image Panoptic Segmentation & Video Semantic Segmentation & Video Instance Segmentation. To verify the generalization ability of our method, we extend RankSeg to "rank-adaptive selected-label region classification" and apply it to the very recent Mask2Former [18]. According to Table 8, our RankSeg improves the image semantic segmentation/image panoptic segmentation/video semantic segmentation/video instance segmentation performance by +0.7%/+0.8%/+0.7%/+0.7% respectively with slightly more parameters and GFLOPs. More details about the Mask2Former experiments and the results of combining MaskFormer with RankSeg are provided in the supplementary material.

5   Conclusion

This paper introduces a general and effective rank-oriented scheme that formulates the segmentation task into two sub-problems including multi-label classification and rank-adaptive selected-label pixel classification. We first verify the potential benefits of exploiting multi-label image/video classification to improve pixel classification. We then propose a simple joint multi-task scheme that is capable of improving various state-of-the-art segmentation methods across multiple benchmarks. We hope our initial attempt can inspire more efforts towards using a rank-oriented manner to solve the challenging segmentation problem with a large number of categories. Last, we want to point out that designing & exploiting more accurate multi-label image/video classification methods is a long-neglected but very important sub-problem towards more general and accurate segmentation.

References

1. Arbeláez, P., Hariharan, B., Gu, C., Gupta, S., Bourdev, L., Malik, J.: Semantic segmentation using regions and parts. In: CVPR (2012)
2. Athar, A., Mahadevan, S., Ošep, A., Leal-Taixé, L., Leibe, B.: STEm-Seg: spatio-temporal embeddings for instance segmentation in videos. In: Vedaldi, A., Bischof, H., Brox, T., Frahm, J.-M. (eds.) ECCV 2020. LNCS, vol. 12356, pp. 158–177. Springer, Cham (2020). https://doi.org/10.1007/978-3-030-58621-8_10
3. Badrinarayanan, V., Kendall, A., Cipolla, R.: SegNet: a deep convolutional encoder-decoder architecture for image segmentation. PAMI **39**, 2481–2495 (2017)
4. Bao, H., Dong, L., Wei, F.: BEiT: BERT pre-training of image transformers. arXiv preprint arXiv:2106.08254 (2021)
5. Ben-Baruch, E., et al.: Asymmetric loss for multi-label classification. arXiv preprint arXiv:2009.14119 (2020)
6. Bertasius, G., Torresani, L.: Classifying, segmenting, and tracking object instances in video with mask propagation. In: CVPR, pp. 9739–9748 (2020)
7. Caesar, H., Uijlings, J., Ferrari, V.: Region-based semantic segmentation with end-to-end training. In: Leibe, B., Matas, J., Sebe, N., Welling, M. (eds.) ECCV 2016. LNCS, vol. 9905, pp. 381–397. Springer, Cham (2016). https://doi.org/10.1007/978-3-319-46448-0_23
8. Caesar, H., Uijlings, J., Ferrari, V.: COCO-Stuff: thing and stuff classes in context. In: CVPR (2018)
9. Cao, J., Anwer, R.M., Cholakkal, H., Khan, F.S., Pang, Y., Shao, L.: SipMask: spatial information preservation for fast image and video instance segmentation. In: Vedaldi, A., Bischof, H., Brox, T., Frahm, J.-M. (eds.) ECCV 2020. LNCS, vol. 12359, pp. 1–18. Springer, Cham (2020). https://doi.org/10.1007/978-3-030-58568-6_1
10. Carion, N., Massa, F., Synnaeve, G., Usunier, N., Kirillov, A., Zagoruyko, S.: End-to-end object detection with transformers. In: Vedaldi, A., Bischof, H., Brox, T., Frahm, J.-M. (eds.) ECCV 2020. LNCS, vol. 12346, pp. 213–229. Springer, Cham (2020). https://doi.org/10.1007/978-3-030-58452-8_13
11. Chandra, S., Couprie, C., Kokkinos, I.: Deep spatio-temporal random fields for efficient video segmentation. In: CVPR, pp. 8915–8924 (2018)
12. Chen, L.C., Papandreou, G., Schroff, F., Adam, H.: Rethinking atrous convolution for semantic image segmentation. arXiv:1706.05587 (2017)
13. Chen, L.-C., Zhu, Y., Papandreou, G., Schroff, F., Adam, H.: Encoder-decoder with atrous separable convolution for semantic image segmentation. In: Ferrari, V., Hebert, M., Sminchisescu, C., Weiss, Y. (eds.) ECCV 2018. LNCS, vol. 11211, pp. 833–851. Springer, Cham (2018). https://doi.org/10.1007/978-3-030-01234-2_49
14. Chen, T., Xu, M., Hui, X., Wu, H., Lin, L.: Learning semantic-specific graph representation for multi-label image recognition. In: ICCV (2019)
15. Chen, Z., et al.: Vision transformer adapter for dense predictions. arXiv preprint arXiv:2205.08534 (2022)

16. Cheng, B., Choudhuri, A., Misra, I., Kirillov, A., Girdhar, R., Schwing, A.G.: Mask2Former for video instance segmentation. arXiv preprint arXiv:2112.10764 (2021)
17. Cheng, B., et al.: Panoptic-DeepLab. arXiv:1910.04751 (2019)
18. Cheng, B., Misra, I., Schwing, A.G., Kirillov, A., Girdhar, R.: Masked-attention mask transformer for universal image segmentation. arXiv preprint arXiv:2112.01527 (2021)
19. Cheng, B., Schwing, A.G., Kirillov, A.: Per-pixel classification is not all you need for semantic segmentation. arXiv preprint arXiv:2107.06278 (2021)
20. Contributors, M.: MMSegmentation: OpenMMLab semantic segmentation toolbox and benchmark. https://github.com/open-mmlab/mmsegmentation (2020)
21. Cubuk, E.D., Zoph, B., Shlens, J., Le, Q.V.: RandAugment: practical automated data augmentation with a reduced search space. In: CVPRW, pp. 702–703 (2020)
22. DeVries, T., Taylor, G.W.: Improved regularization of convolutional neural networks with cutout. arXiv preprint arXiv:1708.04552 (2017)
23. Fu, J., et al.: Dual attention network for scene segmentation. In: CVPR, pp. 3146–3154 (2019)
24. Fu, Y., Yang, L., Liu, D., Huang, T.S., Shi, H.: CompFeat: comprehensive feature aggregation for video instance segmentation. arXiv preprint arXiv:2012.03400, 6 (2020)
25. Gadde, R., Jampani, V., Gehler, P.V.: Semantic video CNNs through representation warping. In: ICCV, pp. 4453–4462 (2017)
26. Gould, S., Fulton, R., Koller, D.: Decomposing a scene into geometric and semantically consistent regions. In: ICCV (2009)
27. Gu, C., Lim, J.J., Arbelaez, P., Malik, J.: Recognition using regions. In: CVPR (2009)
28. Guo, H., Zheng, K., Fan, X., Yu, H., Wang, S.: Visual attention consistency under image transforms for multi-label image classification. In: CVPR (2019)
29. Gupta, A., Dollar, P., Girshick, R.: LVIS: a dataset for large vocabulary instance segmentation. In: CVPR, pp. 5356–5364 (2019)
30. He, K., Fan, H., Wu, Y., Xie, S., Girshick, R.: Momentum contrast for unsupervised visual representation learning. In: CVPR, pp. 9729–9738 (2020)
31. He, K., Zhang, X., Ren, S., Sun, J.: Deep residual learning for image recognition. In: CVPR (2016)
32. Hu, A., Kendall, A., Cipolla, R.: Learning a spatio-temporal embedding for video instance segmentation. arXiv preprint arXiv:1912.08969 (2019)
33. Hu, H., Zhou, G.T., Deng, Z., Liao, Z., Mori, G.: Learning structured inference neural networks with label relations. In: CVPR (2016)
34. Hu, P., Caba, F., Wang, O., Lin, Z., Sclaroff, S., Perazzi, F.: Temporally distributed networks for fast video semantic segmentation. In: CVPR, pp. 8818–8827 (2020)
35. Huang, Z., Wang, X., Huang, L., Huang, C., Wei, Y., Liu, W.: CCNet: criss-cross attention for semantic segmentation. In: CVPR, pp. 603–612 (2019)
36. Hur, J., Roth, S.: Joint optical flow and temporally consistent semantic segmentation. In: Hua, G., Jégou, H. (eds.) ECCV 2016. LNCS, vol. 9913, pp. 163–177. Springer, Cham (2016). https://doi.org/10.1007/978-3-319-46604-0_12
37. Hwang, S., Heo, M., Oh, S.W., Kim, S.J.: Video instance segmentation using inter-frame communication transformers. In: NIPS 34 (2021)
38. Jain, S., Wang, X., Gonzalez, J.E.: Accel: a corrective fusion network for efficient semantic segmentation on video. In: CVPR, pp. 8866–8875 (2019)
39. Jain, S., Paudel, D.P., Danelljan, M., Van Gool, L.: Scaling semantic segmentation beyond 1k classes on a single GPU. In: ICCV, pp. 7426–7436 (2021)

40. Jin, X., et al.: Video scene parsing with predictive feature learning. In: ICCV, pp. 5580–5588 (2017)
41. Kirillov, A., Girshick, R., He, K., Dollár, P.: Panoptic feature pyramid networks. In: CVPR, pp. 6399–6408 (2019)
42. Kirillov, A., He, K., Girshick, R., Rother, C., Dollár, P.: Panoptic segmentation. In: CVPR, pp. 9404–9413 (2019)
43. Kirillov, A., Wu, Y., He, K., Girshick, R.: PointRend: image segmentation as rendering. In: CVPR, pp. 9799–9808 (2020)
44. Kundu, A., Vineet, V., Koltun, V.: Feature space optimization for semantic video segmentation. In: CVPR, pp. 3168–3175 (2016)
45. Lanchantin, J., Wang, T., Ordonez, V., Qi, Y.: General multi-label image classification with transformers. In: CVPR (2021)
46. Li, Q., Qiao, M., Bian, W., Tao, D.: Conditional graphical lasso for multi-label image classification. In: CVPR (2016)
47. Li, Y., Shi, J., Lin, D.: Low-latency video semantic segmentation. In: CVPR, pp. 5997–6005 (2018)
48. Li, Z., et al.: arXiv preprint arXiv:2109.03814 (2021)
49. Lin, C.C., Hung, Y., Feris, R., He, L.: Video instance segmentation tracking with a modified VAE architecture. In: CVPR, pp. 13147–13157 (2020)
50. Lin, H., Wu, R., Liu, S., Lu, J., Jia, J.: Video instance segmentation with a propose-reduce paradigm. In: CVPR, pp. 1739–1748 (2021)
51. Liu, S., Zhang, L., Yang, X., Su, H., Zhu, J.: Query2Label: a simple transformer way to multi-label classification. arXiv preprint arXiv:2107.10834 (2021)
52. Liu, Y., Shen, C., Yu, C., Wang, J.: Efficient semantic video segmentation with per-frame inference. In: Vedaldi, A., Bischof, H., Brox, T., Frahm, J.-M. (eds.) ECCV 2020. LNCS, vol. 12355, pp. 352–368. Springer, Cham (2020). https://doi.org/10.1007/978-3-030-58607-2_21
53. Liu, Z., et al.: Swin transformer: hierarchical vision transformer using shifted windows. arXiv preprint arXiv:2103.14030 (2021)
54. Long, J., Shelhamer, E., Darrell, T.: Fully convolutional networks for semantic segmentation. In: CVPR (2015)
55. Mahasseni, B., Todorovic, S., Fern, A.: Budget-aware deep semantic video segmentation. In: CVPR, pp. 1029–1038 (2017)
56. Miao, J., Wei, Y., Wu, Y., Liang, C., Li, G., Yang, Y.: VSPW: a large-scale dataset for video scene parsing in the wild. In: Proceedings of the IEEE/CVF Conference on Computer Vision and Pattern Recognition, pp. 4133–4143 (2021)
57. Mottaghi, R., et al.: The role of context for object detection and semantic segmentation in the wild. In: CVPR (2014)
58. Neuhold, G., Ollmann, T., Rota Bulo, S., Kontschieder, P.: The mapillary vistas dataset for semantic understanding of street scenes. In: CVPR (2017)
59. Nilsson, D., Sminchisescu, C.: Semantic video segmentation by gated recurrent flow propagation. In: CVPR, pp. 6819–6828 (2018)
60. Pohlen, T., Hermans, A., Mathias, M., Leibe, B.: Full-resolution residual networks for semantic segmentation in street scenes. In: CVPR, pp. 4151–4160 (2017)
61. Ranftl, R., Bochkovskiy, A., Koltun, V.: Vision transformers for dense prediction. In: ICCV, pp. 12179–12188 (2021)
62. Ridnik, T., Ben-Baruch, E., Noy, A., Zelnik-Manor, L.: ImageNet-21k pretraining for the masses (2021)
63. Ridnik, T., Lawen, H., Noy, A., Ben Baruch, E., Sharir, G., Friedman, I.: TResNet: high performance GPU-dedicated architecture. In: WACV, pp. 1400–1409 (2021)

64. Ronneberger, O., Fischer, P., Brox, T.: U-Net: convolutional networks for biomedical image segmentation. In: Navab, N., Hornegger, J., Wells, W.M., Frangi, A.F. (eds.) MICCAI 2015. LNCS, vol. 9351, pp. 234–241. Springer, Cham (2015). https://doi.org/10.1007/978-3-319-24574-4_28

65. Strudel, R., Garcia, R., Laptev, I., Schmid, C.: Segmenter: transformer for semantic segmentation. arXiv preprint arXiv:2105.05633 (2021)

66. Takikawa, T., Acuna, D., Jampani, V., Fidler, S.: Gated-SCNN: gated shape CNNs for semantic segmentation. In: ICCV, pp. 5229–5238 (2019)

67. Uijlings, J.R., Van De Sande, K.E., Gevers, T., Smeulders, A.W.: Selective search for object recognition. IJCV **104**, 154–171 (2013). https://doi.org/10.1007/s11263-013-0620-5

68. Voigtlaender, P., et al.: MOTS: multi-object tracking and segmentation. In: CVPR, pp. 7942–7951 (2019)

69. Wang, H., Zhu, Y., Adam, H., Yuille, A., Chen, L.C.: Max-DeepLab: end-to-end panoptic segmentation with mask transformers. In: CVPR, pp. 5463–5474 (2021)

70. Wang, J., et al.: Deep high-resolution representation learning for visual recognition. TPAMI **43**, 3349–3364 (2019)

71. Wang, W., Zhou, T., Porikli, F., Crandall, D., Van Gool, L.: A survey on deep learning technique for video segmentation. arXiv preprint arXiv:2107.01153 (2021)

72. Wang, Y., et al.: End-to-end video instance segmentation with transformers. In: CVPR, pp. 8741–8750 (2021)

73. Wang, Z., Chen, T., Li, G., Xu, R., Lin, L.: Multi-label image recognition by recurrently discovering attentional regions. In: ICCV (2017)

74. Wei, Y., Feng, J., Liang, X., Cheng, M.M., Zhao, Y., Yan, S.: Object region mining with adversarial erasing: a simple classification to semantic segmentation approach. In: CVPR (2017)

75. Wu, J., Jiang, Y., Zhang, W., Bai, X., Bai, S.: SeqFormer: a frustratingly simple model for video instance segmentation. arXiv preprint arXiv:2112.08275 (2021)

76. Wu, T., Huang, Q., Liu, Z., Wang, Yu., Lin, D.: Distribution-balanced loss for multi-label classification in long-tailed datasets. In: Vedaldi, A., Bischof, H., Brox, T., Frahm, J.-M. (eds.) ECCV 2020. LNCS, vol. 12349, pp. 162–178. Springer, Cham (2020). https://doi.org/10.1007/978-3-030-58548-8_10

77. Xu, Y.S., Fu, T.J., Yang, H.K., Lee, C.Y.: Dynamic video segmentation network. In: CVPR, pp. 6556–6565 (2018)

78. Yang, L., Fan, Y., Xu, N.: Video instance segmentation. In: ICCV, pp. 5188–5197 (2019)

79. Ye, J., He, J., Peng, X., Wu, W., Qiao, Yu.: Attention-driven dynamic graph convolutional network for multi-label image recognition. In: Vedaldi, A., Bischof, H., Brox, T., Frahm, J.-M. (eds.) ECCV 2020. LNCS, vol. 12366, pp. 649–665. Springer, Cham (2020). https://doi.org/10.1007/978-3-030-58589-1_39

80. You, R., Guo, Z., Cui, L., Long, X., Bao, Y., Wen, S.: Cross-modality attention with semantic graph embedding for multi-label classification. In: AAAI (2020)

81. Yuan, Y., Chen, X., Wang, J.: Object-contextual representations for semantic segmentation. In: Vedaldi, A., Bischof, H., Brox, T., Frahm, J.-M. (eds.) ECCV 2020. LNCS, vol. 12351, pp. 173–190. Springer, Cham (2020). https://doi.org/10.1007/978-3-030-58539-6_11

82. Yuan, Y., et al.: HRFormer: high-resolution transformer for dense prediction. arXiv preprint arXiv:2110.09408 (2021)

83. Yuan, Y., Huang, L., Guo, J., Zhang, C., Chen, X., Wang, J.: OCNet: object context network for scene parsing. arXiv preprint arXiv:1809.00916 (2018)

84. Yuan, Y., Xie, J., Chen, X., Wang, J.: SegFix: model-agnostic boundary refinement for segmentation. In: Vedaldi, A., Bischof, H., Brox, T., Frahm, J.-M. (eds.) ECCV 2020. LNCS, vol. 12357, pp. 489–506. Springer, Cham (2020). https://doi.org/10.1007/978-3-030-58610-2_29

85. Zhang, H., et al.: Context encoding for semantic segmentation. In: CVPR, pp. 7151–7160 (2018)

86. Zhang, W., Pang, J., Chen, K., Loy, C.C.: K-Net: towards unified image segmentation. arXiv preprint arXiv:2106.14855 (2021)

87. Zhao, H., Shi, J., Qi, X., Wang, X., Jia, J.: Pyramid scene parsing network. In: CVPR (2017)

88. Zheng, S., et al.: Rethinking semantic segmentation from a sequence-to-sequence perspective with transformers. In: CVPR, pp. 6881–6890 (2021)

89. Zhou, B., Zhao, H., Puig, X., Fidler, S., Barriuso, A., Torralba, A.: Scene parsing through ADE20K dataset. In: CVPR (2017)

Learning Topological Interactions for Multi-Class Medical Image Segmentation

Saumya Gupta(✉) ⓘ, Xiaoling Hu, James Kaan, Michael Jin, Mutshipay Mpoy,
Katherine Chung, Gagandeep Singh, Mary Saltz, Tahsin Kurc, Joel Saltz,
Apostolos Tassiopoulos, Prateek Prasanna ⓘ, and Chao Chen

Stony Brook University, Stony Brook, NY, USA
{saumya.gupta,xiaoling.hu,chao.chen.1}@stonybrook.edu

Abstract. Deep learning methods have achieved impressive performance for multi-class medical image segmentation. However, they are limited in their ability to encode topological interactions among different classes (e.g., containment and exclusion). These constraints naturally arise in biomedical images and can be crucial in improving segmentation quality. In this paper, we introduce a novel *topological interaction module* to encode the topological interactions into a deep neural network. The implementation is completely convolution-based and thus can be very efficient. This empowers us to incorporate the constraints into end-to-end training and enrich the feature representation of neural networks. The efficacy of the proposed method is validated on different types of interactions. We also demonstrate the generalizability of the method on both proprietary and public challenge datasets, in both 2D and 3D settings, as well as across different modalities such as CT and Ultrasound. Code is available at: https://github.com/TopoXLab/TopoInteraction.

Keywords: Medical imaging · Segmentation · Topological interaction

1 Introduction

Instead of using hand-crafted features, state-of-the-art deep segmentation methods [4–6,16,29] learn powerful feature representations automatically and achieve satisfactory performances. However, standard deep neural networks cannot learn global structural constraints regarding semantic labels, which can often be critical in biomedical domains. While existing works mostly focus on encoding the topology of a single label [8,18,19,35], limited progress has been made addressing the constraints regarding interactions between different labels. Even strong

---

S. Gupta and X. Hu—Equal contribution.

Supplementary Information The online version contains supplementary material available at https://doi.org/10.1007/978-3-031-19818-2_40.

methods (e.g., nnUNet [21]) may fail to preserve the constraints as they only optimize per-pixel accuracy. For example, in the segmentation of abdominal aorta, we know a priori that the aorta wall always encloses the lumen. Exploiting this constraint can help us segment the wall correctly, providing accurate geometric measures (e.g., wall thickness and aorta volume) for the prediction of aortic aneurysm eruption risk [11]. See Fig. 1 for an illustration. Another kind of global constraint is mutual exclusion of different labels. For example, in multi-organ segmentation, ensuring different organs to not touch each other can help improve the segmentation quality.

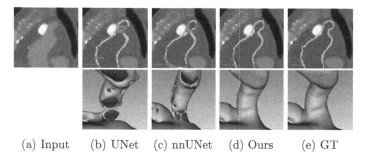

(a) Input   (b) UNet   (c) nnUNet   (d) Ours   (e) GT

Fig. 1. Motivating examples for aorta segmentation. Red and yellow represent aortic lumen and wall, respectively. Anatomically, the lumen is always enclosed by the wall, separated from the background (illustrated in (e) ground truth *GT*). Even strong baselines, e.g., (b) *UNet* and (c) *nnUNet*, fail to respect this anatomically important topological constraint because often the intensity of the wall in the input is similar to that of the background. Our proposed method explicitly encodes the constraint, thereby improving the segmentation quality. (Color figure online)

In this paper, we investigate how to help deep neural networks learn these global structural constraints, which we call *topological interactions*, between different semantic labels. To encode such interaction constraints into convolutional neural networks is challenging; it is hard to directly encode hard constraints into kernels while keeping them learnable. Traditional methods [3,10,22,26,31,38] solve the segmentation problem as a combinatorial optimization problem (e.g., graph-cut or multicut) and encode these topological interactions as constraints of the solution. However, these approaches do not apply to deep neural networks, which do not rely on a global optimization for the inference. Even if one can encode the constrained optimization as a post-processing step, it will be very inefficient. More importantly, the optimization is not differentiable and thus cannot be incorporated into training.

We propose a novel method to learn the topological interactions for multi-class segmentation tasks. A desirable solution should be efficient. Furthermore, it should be incorporated into training to help the network learn. Our key observation is that a broad class of topological interactions, namely, enclosing and exclusion, boils down to certain impermissible label combinations of adjacent pixels/voxels. Inspired by such observation, we propose a *topological interaction module* that encodes the constraints into a neural network through a series of convolutional operations. Instead of directly encoding the constraint into the convolutional kernels, the proposed module directly identifies locations where the constraints are violated. Our module is extremely efficient due to the convolution-based design. Furthermore, it can naturally be incorporated into the training of neural networks, e.g., through an extra loss penalizing the constraint-violating pixels/voxels. As shown in Fig. 1, incorporated with our module, the network can learn to segment aortic walls correctly even when strong baselines, such as nnUNet, fail.

We evaluate the proposed method by performing experiments on both proprietary and public challenge datasets, in both 2D and 3D settings, and across different modalities. The results show that our method is generalizable and can be employed in various scenarios where topological interactions apply. It not only enforces the constraints, but also improves the segmentation quality significantly in standard metrics such as DICE, Hausdorff distance, etc. This is as expected; a network that encodes the constraints also learns a better representation for segmentation. In summary, our contributions are as follows:

- We propose an efficient convolution-based module to encode the topological interactions in a multi-class segmentation setting.
- The proposed module is very efficient and generic. It can be incorporated into any backbone to encode the constraints in an end-to-end training pipeline.
- Through extensive experiments on multiple medical imaging datasets, we show our method effectively improves the segmentation quality without increasing computational cost.

2   Related Work

Multi-Class Image Segmentation. Numerous graph or energy based methods have been proposed to deal with multi-class image segmentation in the pre-deep learning era. Some of these methods integrate fuzzy spatial relations [9] or encode spatial interactions via inter-object distances [28]. Others encode spatial relationships for hierarchical segmentation [12,36]. For example, Strekalovskiy et al. [36] enforce geometric constraints by introducing a label ordering constraint. Li et al. [27] propose to segment nested objects with graph-based approaches. Delong et al. [10] propose to encode geometric constraints between different regions into a graph cut framework for multi-class image segmentation.

Geometric and Topological Constraints. Early works, using classic frameworks such as level set or Markov random field, enforce topological or geometric constraints while solving the energy minimization problem [3,10,14,22,25,26, 31,38]. However, these methods cannot be easily incorporated into the training of deep neural networks. In recent years, new methods have been proposed to incorporate geometric/topological constraints into the training of deep neural networks (DNNs) [8,18,19,35,39]. These methods enable the DNNs to learn

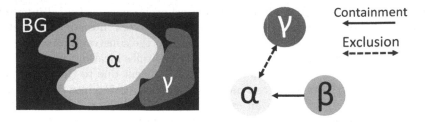

Fig. 2. Schematic illustration of the topological interactions: containment and exclusion. *BG* denotes the background class. **Containment:** β contains α. **Exclusion:** α and γ are mutually exclusive.

geometry-/topology-aware representations and to deliver better segmentation results. However, all these methods are focusing on the topology, e.g., connections, loops and branches, of a single foreground class. They cannot enforce topological interactions between different classes. For example, in aorta segmentation, forcing the aortic wall to be a tube in 3D cannot guarantee that the wall contains the lumen and separates it from the background. This gap motivates our investigation on encoding the inter-class topological interactions in DNN training.

The method closest to ours is [2], which we refer to as TopoCRF. It encodes the mutual exclusion constraint as a constraint on the posterior probability (softmax layer output) at each pixel/voxel, without taking neighborhoods into account. Therefore, this approach cannot really exclude the case when adjacent pixels have a forbidden label combination. The explicit construction of 2^c constraint-encoding priors for a c-class problem is also very expensive and does not scale. Additional methods similar to TopoCRF are [32] which we refer to as MIDL, and [13] which we refer to as NonAdj. MIDL is a direct application of TopoCRF by simply adding a DICE loss term. NonAdj extends TopoCRF by taking the adjacent pixels into consideration, however, it requires a strong pre-trained model to perform well. Both MIDL and NonAdj focus on modeling joint distributions, and thus suffer from similar issues as TopoCRF.

3 Methodology

Broadly speaking, topological interactions between different foreground classes include two types, containment and exclusion. In Fig. 2, we illustrate these constraints using three class labels, α, β and γ.

- **Containment:** Class β contains α if β completely surrounds α. We use solid arrow from β to α to denote the containment relationship. In real applications, e.g., aorta segmentation, the aortic wall contains the lumen. See Fig. 3(a) for an illustration.
- **Exclusion:** Classes α and γ are mutually exclusive if the pixels/voxels of class α and class γ cannot be adjacent to each other. We use dashed double-arrow

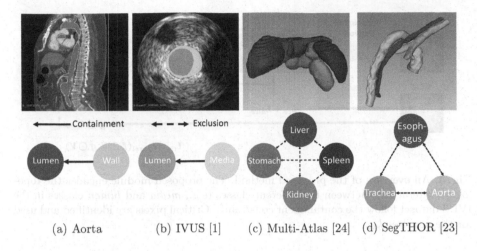

(a) Aorta   (b) IVUS [1]   (c) Multi-Atlas [24]   (d) SegTHOR [23]

Fig. 3. Multi-Class topological interactions for each dataset.

to denote the exclusion relationship. In multi-organ segmentation, there is clear separation between stomach and liver. They are mutually exclusive. See Fig. 3(c) for an illustration.

These constraints are quite general and can be observed in different medical imaging applications. See Fig. 3 for more examples. We can also enforce stronger constraints. For containment, we may require the surrounding class (β in Fig. 2) to be at least d-pixel thick. For exclusion, we may require the gap between two mutually exclusive classes to be at least d-pixel wide. We call these generalized constraints *d-containment* and *d-exclusion*.

Overview of the Proposed Method. Though the aforementioned topological interactions are global constraints, we observe that they can be encoded in a localized manner. Specifically, both containment and exclusion constraints can be rewritten as forbidding certain label combinations for adjacent pixels/voxels. In the example in Fig. 2, β contains α equals to the constraint that a pixel/voxel of label α cannot be adjacent to a pixel/voxel of any label other than β and itself. Exclusion is more straightforward, α and γ are mutually exclusive if any two adjacent pixels/voxels do not have the label pair (α, γ) or (γ, α).

We enforce these constraints into DNN training by proposing a novel topological interaction module. The idea is to go through all pairs of adjacent pixels/voxels and identify the pairs that violate the desired constraints. Pixels belonging to these pairs are the ones inducing errors into the topological interaction. We will refer to them as *critical pixels*. Our topological interaction module will output these critical pixels. Then, we can incorporate the module into training by designing a loss paying extra penalty to these critical pixels.

An efficient implementation of the module, however, is not trivial. Simply looping through all pixels is too expensive to serve as a frequent operation during training. To this end, we propose an efficient implementation of the constraints

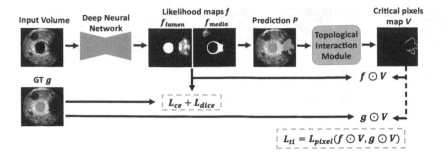

Fig. 4. An overview of the proposed method. The proposed module encodes the topological interactions between the different classes (e.g., *media* and *lumen* classes in the IVUS dataset follow the containment constraint). Critical pixels are identified and used for the new loss L_{ti}.

purely based on convolutional operations (Sect. 3.1). The method is much more efficient and can easily generalize to more challenging d-containment and d-exclusion without much extra computational expense. Finally, in Sect. 3.2, we incorporate the proposed module into training by formulating a loss function penalizing the critical pixels. This ensures the DNNs learn better feature representation while respecting the imposed constraints, as we will demonstrate empirically. Figure 4 provides an overview of the proposed method.

3.1 Topological Interaction Module

The *topological interaction module* encodes the topological interactions defined above. Recall the key is to forbid certain label combinations appearing in any pair of adjacent pixels. Our module identifies the pairs that violate the constraints.

Next, we explain how to map the constraints into the local constraints regarding two labels that should not appear in adjacent pixels. For exclusion constraint, the forbidding label pair is obvious. In Fig. 2, labels α and γ are mutually exclusive. We create new labels $A = \alpha$ and $C = \gamma$, and forbid them to appear in adjacent pixels. For containment constraints, say label β contains label α (as in Fig. 2), we create a new label $A = \alpha$ and a new label C being the union of all other labels except for α and β. Then β containing α is equivalent to $A = \alpha$ not touching C.

For the rest of this section, we focus on how to create a module identifying adjacent pixel pair having the label pair (A, C) or (C, A). For ease of exposition, we assume a 2D 4-connectivity neighborhood (i.e., each pixel is only adjacent to 4 neighboring pixels), and so $d = 1$. The approach can be naturally generalized to other connectivities as formalized in the classic digital topology [34].

Naive Solution. Given a discretized segmentation map predicted by the network, the naive solution is simply looping over all pixels and for each pixel, scan all its neighbors. For every pair of adjacent pixels with the label pair (A, C) or (C, A), we flag both of the pixels as critical. The obvious issue with this naive

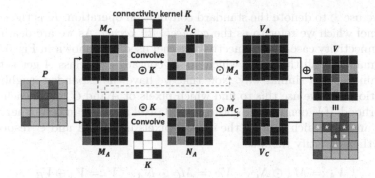

Fig. 5. 2D illustration of the **proposed** strategy to detect the set V of topological critical pixels. We use 4-connectivity kernel. The entire critical pixel map V is highlighted with *'s.

solution is that it is very expensive. Furthermore, such computation can only run on a CPU, and so is rather slow; this is detailed in the Supplementary Material.[1]

Convolution-Based Solution. Let $P \subseteq \mathbf{R}^d$ denote the d-dimensional discrete segmentation map predicted by the network. We want to generate a critical pixel map in which only those label-A pixels with a label-C neighbor are activated and vice-versa. We achieve this goal through manipulations of different semantic masks. First, to determine the critical pixels in A, we expand the C mask by d pixels, and then find out the intersection of the expanded mask with the A mask ($d = 1$ for 2D 4-connectivity). In this way, we obtain the set of all the critical A pixels: they fall within the expanded C mask, and thus must be a neighbor of some C pixels. In top row of Fig. 5, second to fourth columns, we show the C mask (denoted by M_C), its expansion, and the intersection with the A mask (denoted by M_A), resulting in the critical A pixels. In a similar manner, we can obtain the set of critical C pixels by expanding the A mask and finding its intersection with the C mask. This is illustrated in the bottom row, second to fourth columns of Fig. 5.

In practice, expanding a mask can be done efficiently using the *dilation* morphological operation [15]. In dilation, we convolve a given binary mask with a kernel K. The kernel defines the neighbors of a given voxel. Formally, let M_A and M_C be the class masks for A and C respectively. We then obtain neighborhood information N_A and N_C via dilation/convolution as follows:

$$N_A := M_A \circledast K, \quad N_C := M_C \circledast K \qquad (1)$$

---

[1] There is an alternate way to better implement this naive solution by creating extra maps representing neighboring pixels. The issue of such a method is it does not scale well with larger neighborhood (which is necessary for more general constraints assuming a gap of width d between forbidden label pairs). See the Supplementary Material for more details.

where we use ⊛ to denote the standard convolution operation. K is the convolution kernel which we refer to as the *connectivity kernel*. As we are dealing with 2D 4-connectivity case, the connectivity kernel used is as shown in Fig. 5. Notice that in map N_A, all the pixels which are in contact with class A get activated. We obtain N_C in a similar way. Now that we have the expanded neighborhood information, and we use this to find which pixels of A and C fall in each other's neighborhood. If V denotes the entire critical pixel map, it can be further divided into V_A and V_C which contain the critical pixels in class A and C respectively. We can then quantify them as:

$$V_A := M_A \odot N_C, \quad V_C := M_C \odot N_A, \quad V := V_A \oplus V_B \tag{2}$$

where $\oplus$ denotes the union operation, and $\odot$ denotes the Hadamard product.

Figure 5 gives an overview of our method to compute topological critical pixels in the form of a binary mask V. Thus through the manipulation of maps obtained via standard convolution, we are able to augment existing information by deriving information relevant to topological interactions.

Remark on the Connectivity Kernel K. We remark that the connectivity kernel K depends on the definition of neighborhood. Our current choice of K corresponds to the 4-connectivity neighborhood (illustrated in Fig. 5). In general, we can choose different neighborhood definitions corresponding to different kernels. Following the classic digital topology [34], in 2D, we can have 4- and 8-connectivities. In 3D, we can have 6- and 26-connectivities. We can also specify different connectivity kernels for classes A and C. See Supplementary Material for illustrations.[2]

We also note it is natural to generalize the neighborhood definition and modify the kernel accordingly to enforce the more general/stronger constraints: d-containment and d-exclusion. These constraints essentially boil down to the constraint that labels A and C cannot appear on two pixels within distance d. To encode such constraints, we simply define the neighborhood of a pixel p to be all pixels within a $(2d+1) \times (2d+1)$ local patch centered at p. The connectivity kernel is then an all-one kernel of the same size.

Computational Efficiency. We analyze the computational efficiency of the proposed method by determining its complexity as a function of the input and neighborhood size. Let the image size be $N \times N$. Suppose we enforce a separation of d pixels, then the neighborhoods to be inspected for each pixel will be $k \times k$, where $k = 2d + 1$. In the naive solution, we require scanning the neighborhood of each pixel via loops and so the time complexity is in the order of $O(N^2 k^2)$, not really scalable. This is apart from the fact that such a solution can only run on a CPU. On the contrary, the convolution-based solution has a time complexity

---

[2] In digital topology, to ensure the Jordan curve theorem is correct, one needs to have either 4-conn. for foreground and 8-conn. for background, or the opposite. This is not in conflict with our method. A and C are both considered foreground labels. In 2D, they can use either 4-conn. or 8-conn. as long as they are the same. Similar rules apply to 3D.

$O(N^2 \log N)$. Here $\log N$ is due to the FFT (Fast Fourier Transform) implementation of convolution. While the naive solution's running time is quadratic to k, our proposed is independent of k due to FFT. In practice, deep learning frameworks are highly optimized for convolution operations, and so they are several orders of magnitude cheaper than the naive solution. The memory requirement for both methods is similar in the order of $O(N^2)$ to store the map V.

3.2 Incorporating into End-to-End Training

To incorporate the proposed topological interaction module into end-to-end training, we propose a topological interaction loss to correct the violations by penalizing the critical pixels.

Let $f \in \mathbb{R}^{c \times H \times W}$ be the multi-class likelihood map predicted by the network, where c, H and W denote the number of classes, height and width of the image, respectively. $g \in \mathbb{R}^{H \times W}$ is the ground truth segmentation map with discrete labels, $0, 1, ..., c - 1$. We use L_{pixel} to denote the pixel-wise loss function, such as, cross-entropy, mean-squared-error, or dice losses. We use the binary mask V obtained from Sect. 3.1, to define L_{ti}, denoting the additional topological interaction loss, as:

$$L_{ti} = L_{pixel}(f \odot V, g \odot V) \tag{3}$$

L_{ti} can essentially encode the topological interactions, correct the topological interaction errors, and eventually produce a segmentation that is topologically correct. The final loss of our method, L_{total}, is given by:

$$L_{total} = L_{ce} + \lambda_{dice} L_{dice} + \lambda_{ti} L_{ti} \tag{4}$$

where L_{ce} and L_{dice} denote the cross-entropy and dice loss. The loss is controlled by the weights λ_{dice} and λ_{ti}.

4 Experiments

Datasets. We validate our method on four datasets: The proprietary **Aorta** dataset contains 3D CT scans of 28 patients from an institutional database of patients with thoracic and/or abdominal aortic aneurysm. The **IVUS** (IntraVascular Ultrasound) [1] is a 2D dataset of human coronary arteries and contains lumen and media-adventitia labels. The **Multi-Atlas BTCV** [24] is a multi-organ segmentation challenge, containing 3D CT scans of the cervix and abdomen. We use the abdomen dataset and segment four classes, namely, spleen, left kidney, liver, and stomach which appear in close proximity. We have clinically verified that the exclusion constraint holds among these four classes. The **SegTHOR** [23] 2019 challenge contains 3D CT scans of thoracic organs at risk (OAR). In this dataset, the OARs are the heart, trachea, aorta and esophagus. The exclusion constraint holds among three classes, that is, the trachea, the aorta, and the esophagus do not touch each other. We do not take the heart class into consideration.

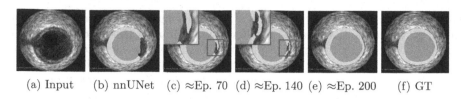

(a) Input   (b) nnUNet   (c) ≈Ep. 70 (d) ≈Ep. 140 (e) ≈Ep. 200   (f) GT

Fig. 6. Epoch (*Ep.*) progression of the proposed method. Critical pixel map identified by the module is marked in red. (Color figure online)

The containment constraint holds for the Aorta and IVUS datasets, while the exclusion constraint holds for the remaining two. Figure 3 gives an overview of the classes in each dataset and the topological interactions among them.

Baselines and Implementation Details. We use the PyTorch framework, a single NVIDIA Tesla V100-SXM2 GPU (32G Memory) and a Dual Intel Xeon Silver 4216 CPU@2.1 GHz (16 cores) for all the experiments. The comparison baselines consist of the UNet [7,33], FCN [30], nnUNet [21], TopoCRF [2], MIDL [32], and NonAdj [13]. We use the publicly available codes for UNet, FCN, nnUNet, and NonAdj. For TopoCRF and MIDL, we implemented it by ourselves in PyTorch. Specifically, for TopoCRF, MIDL, NonAdj and our proposed method, we fine-tune the models pre-trained by nnUNet. To support our claim that our method can be incorporated into any backbone, we train our module on FCN and UNet backbones as well. More details and additional results are included in the Supplementary Material.

The connectivity kernel K, in 2D, is a 3×3 kernel filled with 1's to enforce 8-connectivity. Similarly in 3D, K is a $3 \times 3 \times 3$ kernel filled with 1's to enforce 26-connectivity. We also perform an ablation study on the connectivity kernel in the Supplementary Material.

Evaluation Metrics. Dice score [40], Hausdorff distance (HD) [20], and average symmetric surface distance (ASSD) [17] are used as the performance metrics. We introduce a new metric called the *% violations*. The % violations is calculated by the number of pixels violating the constraint as a fraction of the total number of foreground class pixels/voxels. We report the % violations for all the pixels/voxels together instead of separately per class. For all metrics, we report the means and standard deviations. We also perform the unpaired t-test [37] to determine the statistical significance of the improvement. The statistically significant better performances are highlighted with bold in all the tables. The t-test [37] used to determine the statistical significance of the improvement has a confidence interval of 95%. The best, while not statistically significant, performances are highlighted with italics.

4.1   Results

Table 1 shows the quantitative results for the containment constraint on the Aorta and IVUS datasets, while Table 2 shows the quantitative results for the

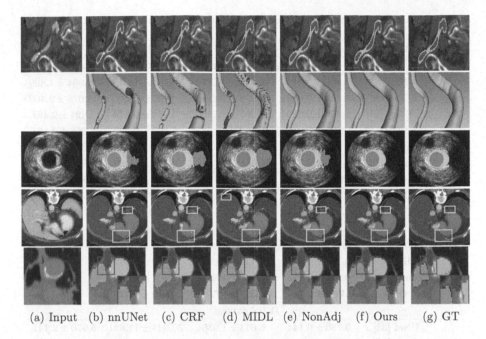

(a) Input   (b) nnUNet   (c) CRF   (d) MIDL   (e) NonAdj   (f) Ours   (g) GT

Fig. 7. Qualitative results compared with the baselines. Top three rows deal with the containment constraint, while bottom two rows deal with the exclusion constraint. Aorta: rows 1–2, IVUS: row 3, Multi-Atlas: row 4, SegTHOR: row 5. The second row is the 3D view of the first row. It is hard to visualize the input 3D volumetric image and so we leave it blank in the second row. Colors for the classes correspond to the ones used in Fig. 3. (Color figure online)

exclusion constraint on the Multi-Atlas (Abdominal) and SegTHOR datasets. In Fig. 7, we show the qualitative comparison of different methods. The comprehensive quantitative and qualitative results of our method on the UNet and FCN backbones, along with different connectivity kernels can be found in the Supplementary Material. In general, we observe that learning the topological constraint leads to better feature representation and thus better segmentations both qualitatively and quantitatively. We discuss the results for both interactions below.

Quantitative and Qualitative Results for Containment Constraint. From Table 1, we observe that the proposed method improves the quality of segmentations by improving all the metrics significantly. In Fig. 7, we see that the networks trained with the proposed method have considerably fewer topological violations compared to the other baseline networks. In the top two rows of the figure, we see that the proposed method fixes the topological interaction errors by enforcing the lumen always be enclosed by the wall. By enforcing this constraint, our method is able to reconstruct the broken lumen and wall structures, thereby significantly improving the segmentation quality. In the third row,

Table 1. Quantitative comparison for containment constraint.

Class	Model	Dice↑	HD↓	ASSD↓	% Violations↓
			Aorta dataset		
Lumen	UNet [7]	0.900 ± 0.016	64.392 ± 16.874	9.315 ± 1.749	13.994 ± 1.809
	FCN [30]	0.894 ± 0.013	57.974 ± 19.756	9.77 ± 1.421	15.675 ± 2.409
	nnUNet [21]	0.906 ± 0.020	36.368 ± 12.559	4.563 ± 0.675	5.424 ± 2.461
	Topo-CRF [2]	0.897 ± 0.057	40.162 ± 18.687	5.952 ± 0.999	8.358 ± 2.151
	MIDL [32]	0.912 ± 0.008	32.157 ± 16.270	6.405 ± 0.524	6.377 ± 1.661
	NonAdj [13]	0.916 ± 0.030	32.465 ± 18.848	4.771 ± 1.129	4.932 ± 1.479
	Ours	**0.922 ± 0.009**	**25.959 ± 13.574**	**3.920 ± 0.765**	**3.526 ± 1.244**
Wall	UNet [7]	0.677 ± 0.015	71.109 ± 24.653	12.497 ± 1.372	/
	FCN [30]	0.651 ± 0.015	66.059 ± 17.188	12.339 ± 0.959	/
	nnUNet [21]	0.741 ± 0.026	42.486 ± 15.139	8.005 ± 0.811	/
	Topo-CRF [2]	0.739 ± 0.010	46.873 ± 17.636	7.914 ± 0.877	/
	MIDL [32]	0.742 ± 0.028	43.132 ± 15.624	6.420 ± 1.242	/
	NonAdj [13]	0.748 ± 0.017	38.197 ± 19.598	4.887 ± 0.702	/
	Ours	**0.758 ± 0.017**	**31.137 ± 17.772**	**5.799 ± 0.737**	/
			IVUS dataset		
Lumen	UNet [33]	0.786 ± 0.144	6.643 ± 1.936	30.944 ± 11.631	5.970 ± 2.141
	FCN [30]	0.824 ± 0.071	5.319 ± 1.519	22.551 ± 7.882	3.766 ± 1.444
	nnUNet [21]	0.893 ± 0.066	3.464 ± 0.917	11.152 ± 3.954	2.708 ± 1.032
	Topo-CRF [2]	0.887 ± 0.096	4.138 ± 1.454	10.497 ± 2.487	2.371 ± 0.960
	MIDL [32]	0.891 ± 0.073	4.226 ± 1.390	10.641 ± 2.322	2.394 ± 0.918
	NonAdj [13]	0.897 ± 0.081	3.140 ± 1.154	9.628 ± 3.221	2.173 ± 0.994
	Ours	**0.949 ± 0.070**	**2.046 ± 1.079**	**6.057 ± 2.746**	**0.157 ± 0.808**
Media	UNet [33]	0.651 ± 0.130	7.391 ± 1.072	21.984 ± 6.634	/
	FCN [30]	0.782 ± 0.144	6.806 ± 1.147	13.863 ± 4.511	/
	nnUNet [21]	0.856 ± 0.090	5.646 ± 1.228	6.491 ± 2.314	/
	Topo-CRF [2]	0.843 ± 0.106	5.409 ± 1.166	5.929 ± 1.785	/
	MIDL [32]	0.841 ± 0.121	5.461 ± 1.214	6.071 ± 1.837	/
	NonAdj [13]	0.848 ± 0.117	5.983 ± 1.342	6.615 ± 1.937	/
	Ours	**0.910 ± 0.089**	**3.873 ± 0.933**	**3.171 ± 1.871**	/

we show results on the IVUS dataset. Due to artifacts in the input (like shadow), nnUNet erroneously classifies extraneous lumen regions beyond the media. Due to the smoothness loss component in TopoCRF, the boundaries of its segmentations are a lot smoother compared to nnUNet, however, it also fails to correct the lumen prediction. MIDL performs similarly as TopoCRF, and while NonAdj performs better than both of them, it still fails in several places. By enforcing the containment constraint, our method is able to learn better features and gets rid of such extraneous lumen regions.

For both the Aorta and IVUS datasets, by identifying the critical pixels, our method improves the learning capability of the network through the epochs. In

Table 2. Quantitative comparison for exclusion constraint.

Class	Model	Dice↑	HD↓	ASSD↓	% Violations↓
		Multi-Atlas dataset			
Spleen	UNet [7]	0.919 ± 0.041	47.037 ± 17.365	4.323 ± 0.367	1.857 ± 0.123
	FCN [30]	0.909 ± 0.037	134.915 ± 65.623	17.646 ± 10.604	3.041 ± 0.181
	nnUNet [21]	0.950 ± 0.041	6.084 ± 1.078	0.573 ± 0.131	0.819 ± 0.064
	Topo-CRF [2]	0.947 ± 0.028	6.403 ± 1.039	1.844 ± 0.517	0.934 ± 0.032
	MIDL [32]	0.944 ± 0.015	5.597 ± 1.374	0.565 ± 0.124	0.725 ± 0.151
	NonAdj [13]	0.952 ± 0.058	5.621 ± 1.065	0.513 ± 0.175	0.521 ± 0.082
	Ours	*0.960 ± 0.009*	**5.340 ± 1.049**	**0.484 ± 0.109**	**0.464 ± 0.043**
Kidney	UNet [7]	0.908 ± 0.079	61.602 ± 13.168	9.992 ± 2.461	/
	FCN [30]	0.892 ± 0.018	187.472 ± 36.096	11.583 ± 2.396	/
	nnUNet [21]	0.931 ± 0.018	27.252 ± 5.406	5.352 ± 0.199	/
	Topo-CRF [2]	0.928 ± 0.059	30.209 ± 5.317	6.308 ± 0.905	/
	MIDL [32]	0.935 ± 0.071	25.208 ± 5.440	4.885 ± 0.421	/
	NonAdj [13]	0.934 ± 0.012	24.182 ± 5.561	4.692 ± 0.657	/
	Ours	*0.936 ± 0.026*	**20.013 ± 2.785**	**4.298 ± 0.798**	/
Liver	UNet [7]	0.912 ± 0.016	64.556 ± 13.894	2.324 ± 0.513	/
	FCN [30]	0.885 ± 0.034	183.870 ± 49.796	29.061 ± 13.484	/
	nnUNet [21]	0.951 ± 0.008	38.931 ± 12.161	1.922 ± 0.506	/
	Topo-CRF [2]	0.949 ± 0.006	46.449 ± 14.188	2.072 ± 0.313	/
	MIDL [32]	0.955 ± 0.005	34.276 ± 11.253	1.344 ± 0.431	/
	NonAdj [13]	0.957 ± 0.003	33.671 ± 13.543	1.185 ± 0.372	/
	Ours	**0.962 ± 0.005**	**30.341 ± 9.111**	**0.985 ± 0.386**	/
Stomach	UNet [7]	0.846 ± 0.084	76.000 ± 24.352	5.023 ± 1.508	/
	FCN [30]	0.708 ± 0.156	172.855 ± 43.735	11.328 ± 3.178	/
	nnUNet [21]	0.895 ± 0.015	45.767 ± 7.960	2.720 ± 0.430	/
	Topo-CRF [2]	0.888 ± 0.015	46.877 ± 9.861	3.675 ± 0.358	/
	MIDL [32]	0.899 ± 0.012	40.282 ± 6.437	2.567 ± 0.431	/
	NonAdj [13]	0.907 ± 0.028	41.749 ± 8.630	2.184 ± 0.325	/
	Ours	**0.910 ± 0.018**	**35.514 ± 10.295**	**1.644 ± 0.311**	/
		SegTHOR dataset			
Esophagus	UNet [7]	0.827 ± 0.038	11.357 ± 2.709	1.186 ± 0.113	3.212 ± 0.720
	FCN [30]	0.800 ± 0.031	10.770 ± 2.085	1.303 ± 0.128	3.616 ± 0.709
	nnUNet [21]	0.841 ± 0.014	8.018 ± 2.085	0.950 ± 0.070	1.947 ± 0.525
	Topo-CRF [2]	0.839 ± 0.029	8.602 ± 2.363	0.991 ± 0.081	2.070 ± 0.687
	MIDL [32]	0.840 ± 0.020	7.266 ± 2.132	0.921 ± 0.136	1.271 ± 0.912
	NonAdj [13]	0.843 ± 0.020	6.293 ± 2.703	0.897 ± 0.078	1.215 ± 0.211
	Ours	**0.858 ± 0.019**	**5.582 ± 2.250**	**0.798 ± 0.042**	**0.749 ± 0.428**
Trachea	UNet [7]	0.897 ± 0.027	10.656 ± 4.047	0.728 ± 0.146	/
	FCN [30]	0.891 ± 0.031	11.789 ± 5.291	0.953 ± 0.221	/
	nnUNet [21]	0.910 ± 0.018	9.423 ± 2.393	0.478 ± 0.152	/
	Topo-CRF [2]	0.909 ± 0.022	10.435 ± 2.334	0.473 ± 0.167	/
	MIDL [32]	0.914 ± 0.027	7.929 ± 2.305	0.456 ± 0.143	/
	NonAdj [13]	0.913 ± 0.028	7.866 ± 2.343	0.440 ± 0.113	/
	Ours	**0.929 ± 0.020**	**7.280 ± 2.109**	**0.316 ± 0.186**	/

Fig. 6, we show how our method improves the network predictions through the epochs on an IVUS data sample. Our results demonstrate that our proposed method is able to significantly improve the segmentation quality without the need for any additional post-processing.

Quantitative and Qualitative Results for Exclusion Constraint. For the Multi-Atlas dataset, our method brings in the greatest improvement for the stomach and liver classes. As can be seen in fourth row of Fig. 7, it is correctly able to separate these two classes while the other methods fail to do so. This correlates with the quantitative metrics as well. In the case of the spleen and kidney classes, nnUNet itself predicts separation between these two classes. Our method improves the dice score slightly, but significantly improves other metrics like HD and ASSD. For the SegTHOR dataset, our method brings in the greatest improvement for the esophagus and trachea classes which tend to come in contact at several points across their lengths. In the final row of Fig. 7, we show that our proposed method is able to impose the exclusion constraint among the three classes. For the aorta class, nnUNet is largely able to separate it from the other classes, and so our method's performance on this class is comparable to nnUNet. We include results for the aorta and heart classes in the Supplementary Material.

4.2   Ablation Studies

To further demonstrate the efficacy of the proposed method, we conduct several ablation studies. The following ablation studies have been performed on the IVUS dataset (containment constraint). We perform identical ablation studies on the Multi-Atlas dataset (exclusion constraint) in the Supplementary Material.

Ablation Study for Loss Functions. Our additional topological interaction loss L_{ti} is a general term, and can adopt any existing pixel-wise loss function. We conduct an ablation study using three different loss functions for L_{pixel}, the cross-entropy loss (CE), the mean-squared-error loss (MSE), and the dice loss. The results are tabulated in the top half of Table 3, where the *None* entry denotes nnUNet trained without L_{ti}. Using CE for L_{pixel} gives the best performance. However, using any of the choices for L_{pixel} results in improvement across all metrics compared to the vanilla nnUNet. Thus L_{ti} is a generic term which works towards its intended purpose of correcting topological errors irrespective of the choice of L_{pixel}.

Ablation Study for Loss Weights. Since the topological loss is the main contribution of this paper, we conduct another ablation study in terms of its weight λ_{ti}. We run the experiments with different weights for the additional topological interaction loss and report the results in the bottom half of Table 3. When $\lambda_{ti} = 1e-4$, the proposed method achieves the best performance. However, a reasonable range of λ_{ti} always results in improvement. This demonstrates the efficacy and robustness of the proposed method.

Table 3. Ablation study for L_{pixel} and λ_{ti} (IVUS).

Class	L_{pixel}	Dice↑	HD↓	ASSD↓	% Violations↓
Lumen	None	0.893 ± 0.066	3.464 ± 0.917	11.152 ± 3.954	2.708 ± 1.032
	MSE	0.915 ± 0.073	3.162 ± 0.937	9.963 ± 3.086	0.835 ± 0.907
	DICE	0.937 ± 0.067	2.385 ± 1.065	6.520 ± 2.845	0.320 ± 0.811
	CE	$\mathbf{0.949 \pm 0.070}$	$\mathbf{2.046 \pm 1.079}$	$\mathbf{6.057 \pm 2.746}$	$\mathbf{0.157 \pm 0.808}$
Media	None	0.856 ± 0.090	5.646 ± 1.228	6.491 ± 2.314	/
	MSE	0.893 ± 0.087	4.042 ± 0.986	3.874 ± 1.912	/
	DICE	0.896 ± 0.088	3.964 ± 1.112	3.445 ± 1.681	/
	CE	$\mathbf{0.910 \pm 0.089}$	$\mathbf{3.873 \pm 0.933}$	$\mathbf{3.171 \pm 1.871}$	/

Class	λ_{ti}	Dice↑	HD↓	ASSD↓	% Violations↓
Lumen	0	0.893 ± 0.066	3.464 ± 0.917	11.152 ± 3.954	2.708 ± 1.032
	5.0e−5	0.913 ± 0.071	3.249 ± 0.998	9.338 ± 3.649	0.964 ± 0.893
	1.0e−4	$\mathbf{0.949 \pm 0.070}$	$\mathbf{2.046 \pm 1.079}$	$\mathbf{6.057 \pm 2.746}$	$\mathbf{0.157 \pm 0.808}$
	1.5e−4	0.941 ± 0.069	2.124 ± 1.062	6.426 ± 2.976	0.187 ± 0.814
	2.0e−4	0.938 ± 0.070	2.428 ± 1.041	6.558 ± 2.780	0.252 ± 0.830
Media	0	0.856 ± 0.090	5.646 ± 1.228	6.491 ± 2.314	/
	5.0e−5	0.877 ± 0.088	5.099 ± 0.997	5.024 ± 2.100	/
	1.0e−4	$\mathbf{0.910 \pm 0.089}$	$\mathbf{3.873 \pm 0.933}$	$\mathbf{3.171 \pm 1.871}$	/
	1.5e−4	0.905 ± 0.088	3.889 ± 0.919	3.257 ± 1.877	/
	2.0e−4	0.885 ± 0.089	4.319 ± 1.059	4.364 ± 1.943	/

5   Conclusion

We introduce a new convolution-based module for multi-class image segmentation that focuses on topological interactions. The module consists of an efficient algorithm to identify critical pixels which induce topological errors. We also introduce an additional topologically constrained loss function. By incorporating the module as well as the loss function into the training of deep neural networks, we enforce the network to learn better feature representations, resulting in improved segmentation quality. Results suggest that the method is generalizable to both 2D and 3D settings, and across modalities such as US and CT.

Acknowledgements. We thank the anonymous reviewers for their constructive feedback. The reported research was partly supported by grants NSF IIS-1909038 and NIH 1R21CA258493-01A1.

References

1. Balocco, S., et al.: Standardized evaluation methodology and reference database for evaluating IVUS image segmentation. Comput. Med. Imaging Graph. **38**(2), 70–90 (2014)

2. BenTaieb, A., Hamarneh, G.: Topology aware fully convolutional networks for histology gland segmentation. In: Ourselin, S., Joskowicz, L., Sabuncu, M.R., Unal, G., Wells, W. (eds.) MICCAI 2016. LNCS, vol. 9901, pp. 460–468. Springer, Cham (2016). https://doi.org/10.1007/978-3-319-46723-8_53
3. Chen, C., Freedman, D., Lampert, C.H.: Enforcing topological constraints in random field image segmentation. In: CVPR (2011)
4. Chen, L.C., Papandreou, G., Kokkinos, I., Murphy, K., Yuille, A.L.: Semantic image segmentation with deep convolutional nets and fully connected CRFs. arXiv preprint arXiv:1412.7062 (2014)
5. Chen, L.C., Papandreou, G., Kokkinos, I., Murphy, K., Yuille, A.L.: DeepLab: semantic image segmentation with deep convolutional nets, atrous convolution, and fully connected CRFs. IEEE Trans. Pattern Anal. Mach. Intell. **40**(4), 834–848 (2017)
6. Chen, L.C., Papandreou, G., Schroff, F., Adam, H.: Rethinking atrous convolution for semantic image segmentation. arXiv preprint arXiv:1706.05587 (2017)
7. Çiçek, Ö., Abdulkadir, A., Lienkamp, S.S., Brox, T., Ronneberger, O.: 3D U-Net: learning dense volumetric segmentation from sparse annotation. In: Ourselin, S., Joskowicz, L., Sabuncu, M.R., Unal, G., Wells, W. (eds.) MICCAI 2016. LNCS, vol. 9901, pp. 424–432. Springer, Cham (2016). https://doi.org/10.1007/978-3-319-46723-8_49
8. Clough, J., Byrne, N., Oksuz, I., Zimmer, V., Schnabel, J., King, A.: A topological loss function for deep-learning based image segmentation using persistent homology. TPAMI (2020)
9. Colliot, O., Camara, O., Bloch, I.: Integration of fuzzy spatial relations in deformable models-application to brain MRI segmentation. Pattern Recogn. **39**(8), 1401–1414 (2006)
10. Delong, A., Boykov, Y.: Globally optimal segmentation of multi-region objects. In: 2009 IEEE 12th International Conference on Computer Vision, pp. 285–292. IEEE (2009)
11. Doweidar, M.H.: Advances in Biomechanics and Tissue Regeneration. Academic Press, Cambridge (2019)
12. Felzenszwalb, P.F., Veksler, O.: Tiered scene labeling with dynamic programming. In: 2010 IEEE Computer Society Conference on Computer Vision and Pattern Recognition, pp. 3097–3104. IEEE (2010)
13. Ganaye, P.A., Sdika, M., Triggs, B., Benoit-Cattin, H.: Removing segmentation inconsistencies with semi-supervised non-adjacency constraint. Med. Image Anal. **58**, 101551 (2019)
14. Han, X., Xu, C., Prince, J.L.: A topology preserving level set method for geometric deformable models. TPAMI (2003)
15. Haralick, R.M., Sternberg, S.R., Zhuang, X.: Image analysis using mathematical morphology. TPAMI (1987)
16. He, K., Gkioxari, G., Dollár, P., Girshick, R.: Mask R-CNN. In: Proceedings of the IEEE International Conference on Computer Vision, pp. 2961–2969 (2017)
17. Heimann, T., et al.: Comparison and evaluation of methods for liver segmentation from CT datasets. IEEE Trans. Med. Imaging **28**(8), 1251–1265 (2009)
18. Hu, X., Li, F., Samaras, D., Chen, C.: Topology-preserving deep image segmentation. In: NeurIPS (2019)
19. Hu, X., Wang, Y., Fuxin, L., Samaras, D., Chen, C.: Topology-aware segmentation using discrete Morse theory. In: ICLR (2021)

20. Huttenlocher, D.P., Klanderman, G.A., Rucklidge, W.J.: Comparing images using the Hausdorff distance. IEEE Trans. Pattern Anal. Mach. Intell. **15**(9), 850–863 (1993)

21. Isensee, F., Jaeger, P.F., Kohl, S.A., Petersen, J., Maier-Hein, K.H.: nnU-Net: a self-configuring method for deep learning-based biomedical image segmentation. Nat. Methods **18**, 203–211 (2021)

22. Kappes, J.H., Speth, M., Reinelt, G., Schnörr, C.: Higher-order segmentation via multicuts. Comput. Vis. Image Underst. **143**, 104–119 (2016)

23. Lambert, Z., Petitjean, C., Dubray, B., Ruan, S.: SegTHOR: segmentation of thoracic organs at risk in CT images (2019)

24. Landman, B., Xu, Z., Igelsias, J., Styner, M., Langerak, T., Klein, A.: MICCAI multi-atlas labeling beyond the cranial vault-workshop and challenge. In: Proceedings of MICCAI Multi-Atlas Labeling Beyond Cranial Vault-Workshop Challenge, vol. 5, p. 12 (2015)

25. Le Guyader, C., Vese, L.A.: Self-repelling snakes for topology-preserving segmentation models. TIP **17**, 767–779 (2008)

26. Leon, L.M.C., De Miranda, P.A.V.: Multi-object segmentation by hierarchical layered oriented image foresting transform. In: 2017 30th SIBGRAPI Conference on Graphics, Patterns and Images (SIBGRAPI), pp. 79–86. IEEE (2017)

27. Li, K., Wu, X., Chen, D.Z., Sonka, M.: Optimal surface segmentation in volumetric images-a graph-theoretic approach. IEEE Trans. Pattern Anal. Mach. Intell. **28**(1), 119–134 (2005)

28. Litvin, A., Karl, W.C.: Coupled shape distribution-based segmentation of multiple objects. In: Christensen, G.E., Sonka, M. (eds.) IPMI 2005. LNCS, vol. 3565, pp. 345–356. Springer, Heidelberg (2005). https://doi.org/10.1007/11505730_29

29. Long, J., Shelhamer, E., Darrell, T.: Fully convolutional networks for semantic segmentation. In: Proceedings of the IEEE Conference on Computer Vision and Pattern Recognition, pp. 3431–3440 (2015)

30. Long, J., Shelhamer, E., Darrell, T.: Fully convolutional networks for semantic segmentation. In: CVPR (2015)

31. Nosrati, M.S., Hamarneh, G.: Local optimization based segmentation of spatially-recurring, multi-region objects with part configuration constraints. IEEE Trans. Med. Imaging **33**(9), 1845–1859 (2014)

32. Reddy, C., Gopinath, K., Lombaert, H.: Brain tumor segmentation using topological loss in convolutional networks. In: International Conference on Medical Imaging with Deep Learning-Extended Abstract Track (2019)

33. Ronneberger, O., Fischer, P., Brox, T.: U-Net: convolutional networks for biomedical image segmentation. In: Navab, N., Hornegger, J., Wells, W.M., Frangi, A.F. (eds.) MICCAI 2015. LNCS, vol. 9351, pp. 234–241. Springer, Cham (2015). https://doi.org/10.1007/978-3-319-24574-4_28

34. Rosenfeld, A.: Digital topology. Am. Math. Mon. **86**(8), 621–630 (1979)

35. Shit, S., et al.: clDice-a novel topology-preserving loss function for tubular structure segmentation. In: CVPR (2021)

36. Strekalovskiy, E., Cremers, D.: Generalized ordering constraints for multilabel optimization. In: 2011 International Conference on Computer Vision, pp. 2619–2626. IEEE (2011)

37. Student: The probable error of a mean. Biometrika pp. 1–25 (1908)

38. Ulén, J., Strandmark, P., Kahl, F.: An efficient optimization framework for multi-region segmentation based on Lagrangian duality. IEEE Trans. Med. Imaging **32**(2), 178–188 (2012)

39. Yang, J., Hu, X., Chen, C., Tsai, C.: A topological-attention ConvLSTM network and its application to EM images. In: de Bruijne, M., et al. (eds.) MICCAI 2021. LNCS, vol. 12901, pp. 217–228. Springer, Cham (2021). https://doi.org/10.1007/978-3-030-87193-2_21
40. Zou, K.H., et al.: Statistical validation of image segmentation quality based on a spatial overlap index1: scientific reports. Acad. Radiol. **11**(2), 178–189 (2004)

Unsupervised Segmentation in Real-World Images via Spelke Object Inference

Honglin Chen[1]($\boxtimes$), Rahul Venkatesh[1], Yoni Friedman[4], Jiajun Wu[1],
Joshua B. Tenenbaum[4], Daniel L. K. Yamins[1,2,3], and Daniel M. Bear[2,3]

[1] Department of Computer Science, Stanford, USA
honglinc@stanford.edu
[2] Department of Psychology, Stanford, USA
dbear@stanford.edu
[3] Wu Tsai Neurosciences Institute, Stanford, USA
[4] Department of Brain and Cognitive Sciences, CBMM, MIT, Cambridge, USA

Abstract. Self-supervised, category-agnostic segmentation of real-world images is a challenging open problem in computer vision. Here, we show how to learn static grouping priors from motion self-supervision by building on the cognitive science concept of a Spelke Object: a set of physical stuff that moves together. We introduce the Excitatory-Inhibitory Segment Extraction Network (EISEN), which learns to extract pairwise affinity graphs for static scenes from motion-based training signals. EISEN then produces segments from affinities using a novel graph propagation and competition network. During training, objects that undergo correlated motion (such as robot arms and the objects they move) are decoupled by a bootstrapping process: EISEN explains away the motion of objects it has already learned to segment. We show that EISEN achieves a substantial improvement in the state of the art for self-supervised image segmentation on challenging synthetic and real-world robotics datasets.

1 Introduction

Most approaches to image segmentation rely heavily on supervised data that is challenging to obtain and are largely trained in a category-specific way [6,14,17, 31]. Thus, even state of the art segmentation networks struggle with recognizing untrained object categories and complex configurations [10]. A self-supervised, category-agnostic segmentation algorithm would be of great value.

But how can a learning signal for such an algorithm be obtained? The cognitive science of perception in babies provides a clue, via the concept of a *Spelke*

D. L. K. Yamins and D. M. Bear—Equal contribution.

Supplementary Information The online version contains supplementary material available at https://doi.org/10.1007/978-3-031-19818-2_41.

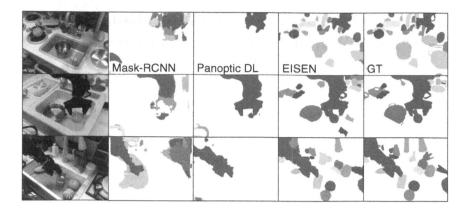

Fig. 1. Unsupervised Segmentation of Spelke Objects. Two standard object segmentation architectures, Mask-RCNN and Panoptic DeepLab, largely fail to learn to detect Spelke objects in the **Bridge** dataset without dense, categorical supervision. In contrast, our approach (EISEN) can detect these objects, without any supervision, via motion-based bootstrapping: learning to predict what moves together, then using top-down inference to segregate arm from object motion.

object [33]: a collection of physical stuff that moves together under the application of everyday physical actions.[1] Perception of Spelke objects is category-agnostic and acquired by infants without supervision [33]. In this work we build a neural network that learns from motion signals to segment Spelke objects in still images (Fig. 1). To achieve this goal, we make two basic innovations.

First, we design a pairwise affinity-based grouping architecture that is optimized for learning from motion signals. Most modern segmentation networks are based on pixelwise background-foreground categorization [6,17]. However, Spelke objects are fundamentally relational, in that they represent whether *pairs* of scene elements are likely to move together. Moreover, this physical connectivity must be learned from real-world video data in which motion is comparatively sparse, as only one or a few Spelke objects is typically moving at a time (Fig. 2, top). Standard pixelwise classification problems that attempt to approximate these pairwise statistics (such as the "Spelke-object-or-not" task) induce large numbers of false negatives for temporarily non-moving objects. Directly learning pairwise affinities avoids these problems.

To convert affinities into actual segmentations, we implement a fully differentiable grouping network inspired by the neuroscience concepts of recurrent

---

[1] More formally, two pieces of stuff are considered to be in the same Spelke object if and only if, under the application of any sequence of actions that causes sustained motion of one of the pieces of stuff, the magnitude of the motion that the other piece of stuff experiences relative to the first piece is approximately zero compared to the magnitude of overall motion. Natural action groups arise from the set of all force applications exertable by specific physical actuator, such as (e.g.) a pair of human hands or a robotic gripper.

Fig. 2. Two Challenges of Learning Spelke Objects. (Top Row) Motion in real-world video is sparse. Thus, pairwise inferences about whether two points are moving together (e.g. the yellow points on the robot arm) or are not moving together (e.g. any yellow-cyan pairing) are valid. However, pointwise motion-based inferences of whether a point is in a Spelke object or not will have many false negatives (e.g. points in the cyan object). **(Bottom row)** Inanimate objects (e.g. magenta lid) only move when moved by something else (e.g. the robotic arm), requiring explaining away of the apparent motion correlation (e.g. yellow-magenta pairs in the bottom row). (Color figure online)

label propagation and border ownership cells [28,41]. This network consists of (i) a quasi-local, recurrent affinity Propagation step that creates (soft) segment identities across pixels in an image and (ii) a winner-take-all Competition step that assigns a unique group label to each segment. We find through ablation studies that this specific grouping mechanism yields high-quality segments from affinities.

A second innovation is an iterative scheme for network training. In real-world video, most objects are inanimate, and thus only seen in motion when caused to move by some other animate object, such as a human hand or robotic gripper (Fig. 2, bottom). This correlated motion must therefore be dissociated to learn to segment the mover from the moved object. Cognitive science again gives a clue to how this may be done: babies first learn to localize hands and arms, then later come to understand external objects [38]. We implement this concept as a confidence-thresholded bootstrapping procedure: motion signals that are already well-segmented by one iteration of network training are explained away, leaving unexplained motions to be treated as independent sources for supervising the next network iteration. For example, in natural video datasets with robotic grippers, the gripper arm will naturally arise as a high-confidence segment first, allowing for the object in the gripper to be recognized as a separate object via explaining-away. The outputs of this explaining away train the next network iteration to recognize inanimate-but-occasionally-moved objects in still images, even when they not themselves being moved.

We train this architecture on optical flow from unlabeled real-world video datasets, producing a network that estimates high-quality Spelke-object segmentations on still images drawn from such videos. We call the resulting network the Excitatory-Inhibitory Segment Extraction Network (EISEN). We show EISEN to be robust even when the objects and configurations in the training videos and test images are distinct. In what follows, we review the literature on related works, describe the EISEN architecture and training methods in detail, show results on both complex synthetic datasets and real-world videos, and analyze algorithmic properties and ablations.

2   Related Work

Segmentation as Bottom-Up Perceptual Grouping. The Gestalt psychologists discovered principles according to which humans group together elements of a scene, such as feature similarity, boundary closure, and correlated motion ("common fate") [36]. This inspired classical computer vision efforts to solve segmentation as a bottom-up graph clustering problem [26,31]. Although these approaches achieved partial success, they have proved difficult to adapt to the enormous variety of objects encountered in real-world scenes like robotics environments. Thus, today's most successful algorithms instead aim to segment objects by learning category-specific cues on large, labeled datasets [6,17,42].

Unsupervised and Category-Agnostic Segmentation. Several recent approaches have tried to dispense with supervision by drawing on advances in self-supervised object *categorization*. DINO, LOST, and Token-Cut perform "object discovery" by manipulating the attention maps of self-supervised Vision Transformers, which can be considered as maps of affinity between an image patch and the rest of the scene [5,32,39]. PiCIE learns to group pixels in an unsupervised way by encouraging particular invariances and equivariances across image transformations. While these early results are encouraging, they apply more naturally to *semantic* segmentation than to grouping individual Spelke objects (instance segmentation): to date, they are mostly limited either to detecting a single object per image or to grouping together all the objects of each category. The GLOM proposal [19] sketches out an unsupervised approach for constructing "islands" of features to represent object parts or wholes, which is similar to our grouping mechanism; but it does not provide a specific algorithmic implementation. We find the architectural particulars of EISEN are essential for successful object segmentation in real-world images (see **Ablations**).

Object Discovery from Motion. A number of unsupervised object discovery methods can segment relatively simple synthetic objects but struggle on realistic scenes [8,15,20,24]. When applied to the task of *video object segmentation*, Slot Attention-based architectures can segment realistic moving objects [21,40], but none of these methods uses motion to learn to segment the majority of objects that are static at any given time. Several approaches discover objects via motion signals, making a similar argument to ours for motion revealing physical

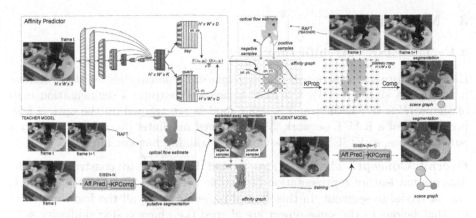

Fig. 3. The EISEN architecture and training process. (Top: Architecture)
The EISEN architecture consists of (i) an Affinity Predictor module which extracts a
pairwise affinity graph for each scene, and (ii) the KProp-Competition module, which
converts the affinity graph into an actual segmentation. The affinity predictor is trained
to predict thresholded optical flow estimates computed via the RAFT algorithm, with
positive samples corresponding to pairs of moving points (green affinity graph edges),
and negative samples corresponding to moving-nonmoving point pairs (red edges).
Edges are computed for all pairs of close-by points and a sampling of further-separated
point pairs. Segments are extracted from the affinity graph via a two-stage mechanism consisting of Kaleidoscopic Propagation and inter-node Competition (see text and
Fig. 4 for more details). **(Bottom: Iterative Training)** Differences between RAFT
optical flow estimates and high-confidence segments from static stage-N EISEN outputs are "explained away" by positing the existence of new Spelke objects, which are
then used to supervised the stage-$(N+1)$ EISEN model. (Color figure online)

structure [1,7,8,29,30,34,37]. However, they have been limited to segmenting a
narrow range of objects or scenes.

We hypothesize that generalization to realistic, complex scenes benefits
greatly from affinity-based grouping and learning. In this respect, our work is
heavily inspired by PSGNet, an unsupervised affinity-based network that learns
to segment scenes from both object motion and other grouping principles [2].
We make two critical advances on that work: (1) replacing its problematic (and
non-differentiable) Label Propagation algorithm with a neural network; and (2)
introducing a bootstrapping procedure that uses top-down inference to explain
raw motion observations in terms of confidently grouped objects. In combination, these novel contributions allow EISEN to accurately perform a challenging
task: the static segmentation of real-world objects without supervision.

3   Methods

3.1   The EISEN Architecture

EISEN performs *unsupervised, category-agnostic segmentation of static scenes*: it takes in a single $H \times W \times 3$ RGB image and outputs a segmentation map of shape $H' \times W'$. EISEN and baseline models are trained on the optical flow predictions of a RAFT network [35] pretrained on Sintel [3]. RAFT takes in a pair of frames, so EISEN requires videos for training but not inference.

Overall Concept. The basic idea behind EISEN is to construct a high-dimensional feature representation of a scene (of shape $H' \times W' \times Q$) that is almost trivial to segment. In this desired representation, all the feature vectors $\mathbf{q}_{ij}$ that belong to the same object are aligned (i.e., have cosine similarity ≈ 1) and all feature vectors that belong to distinct objects are nearly orthogonal (cosine similarity ≈ 0). A spatial slice of this feature map looks like a set of flat object segment "plateaus," so we call it the *plateau map* representation. Object segments can be extracted from a plateau map by finding clusters of vectors pointing in similar directions.

The plateau map is inherently relational: both building and extracting segments from it are straightforward given accurate pairwise affinities between scene elements. EISEN therefore consists of three modules applied sequentially to a convolutional feature extractor backbone (Fig. 3):

1. *Affinity Prediction*, which computes pairwise affinities between features;
2. *Kaleidoscopic Propagation* (KProp), a graph RNN that aligns the vectors of a plateau map by passing messages on the extracted affinity graph;
3. *Competition*, an RNN that imposes winner-take-all dynamics on the plateau map to extract object segments and suppress redundant activity.

All three modules are differentiable, but only Affinity Prediction has trainable parameters. We use the ResNet50-DeepLab backbone in Panoptic-DeepLab [6], which produces output features of shape $H/4 \times W/4 \times 128$.

Affinity Prediction. This module computes affinities $A(i, j, i', j')$ between pairs of extracted feature vectors $\mathbf{f}_{ij}, \mathbf{f}_{i'j'}$. Each feature vector is embedded in $\mathbb{R}^D$ with linear key and query functions, and the affinities are given by standard softmax self-attention plus row-wise normalization:

$$\tilde{A}_{i'j'}^{ij} = \text{Softmax}\left(\frac{1}{\sqrt{D}}(W_k\mathbf{f}_{ij})(W_q\mathbf{f}_{i'j'})^T\right), \quad A_{i'j'}^{ij} = \tilde{A}_{i'j'}^{ij} \Big/ \max_{\{i',j'\}} \tilde{A}_{i'j'}^{ij}. \quad (1)$$

To save memory, we typically compute affinities only within a 25×25 grid around each feature vector plus a random sample of long-range "global" affinities.

Kaleidoscopic Propagation. The KProp graph RNN (Fig. 4) is a smooth relaxation of the discrete Label Propagation (LProp) algorithm [16]. Besides being nondifferentiable, LProp suffers from a "label clashing" problem: once a cluster forms, the discreteness of labels makes it hard for another cluster to merge

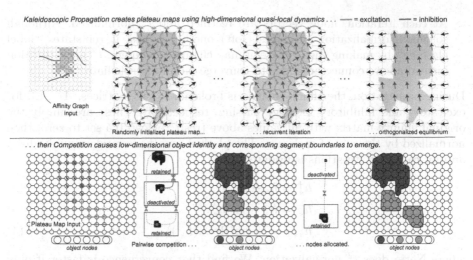

Fig. 4. Kaleidoscopic Propagation and Competition. (Top Row: KProp) For each node in an input affinity graph, a random normalized Q-dimensional vector is allocated (blue and black arrows). The KProp module is a graph RNN that implements high-dimensional quasi-local dynamics, with affinities A corresponding to excitatory connections and inverted affinities $1 - A$ corresponding to inhibitory connections. These dynamics are repeated a fixed number iterations, quickly coming to equilibrium at a "plateau map" in which candidate segments correspond to nearly-orthogonal domains in the Q-dimensional vector field each of which is nearly-flat. **(Bottom Row: Competition)** Plateau maps are converted into segmentations by having "object nodes" compete for ownership of points within the plateau map. A set of putative object nodes are initialized with randomly located basepoints (red highlighted nodes). Each object node corresponds to an object mask consisting of plateau map locations with high Q-vector correlation to the vector at the basepoint. Pairs of object nodes with overlapping masks compete, with the overall-more-aligned node winning and suppressing alternates. Reinitialization then occurs only over non-covered territory. After a small number of iterations, the process equilibrates with the masks containing segment estimates, and the object nodes describing the scene graph. (Color figure online)

with it. This is pernicious when applied to image graphs, as the equilibrium clusters are more like superpixels than object masks [2]. KProp is adapted to the specific demands of image segmentation through the following changes:

- Instead of integers, each node is labeled with a continuous vector $\mathbf{q}_{ij} \in \mathbb{R}^Q$; the full hidden state at iteration s is $h_s \in \mathbb{R}^{N \times Q}$.
- The nondifferentiable message passing in LProp is replaced with two smooth operations: each node sends (1) *excitatory* messages to its high-affinity neighbors, which encourages groups of connected nodes to align; and (2) *inhibitory* messages to its low-affinity neighbors, which orthogonalizes disconnected node pairs. These messages cause clusters of nodes to merge, split, and shift in a pattern reminiscent of a kaleidoscope, giving the algorithm its name.

– At each iteration, node vectors are rectified and ℓ^2 normalized. Although softmax normalization produces (soft) one-hot labels, it reinstates "label clashing" by making the Q plateau map channels compete. ℓ^2 normalization instead allows connected nodes to converge on an intermediate value.

During propagation, the affinity matrix is broken into two matrices, A^+, A^-, for excitatory and inhibitory message passing, respectively. These are simply the original affinity matrix with all values above (resp., below) 0.5 set to zero, then normalized by the sum of each row. The plateau map h_0 is randomly initialized and for each of S iterations is updated by

$$h_s^+ = h_s + A^+ h_s, \tag{2}$$
$$h_s^- = h_s^+ - A^- h_s^+, \tag{3}$$
$$h_{s+1} = \mathrm{Norm}(\mathrm{ReLu}(h_s^-)), \tag{4}$$

where Norm does ℓ^2 normalization. We find that convergence is faster if only one random node passes messages at the first iteration.

Competition. Vector clusters in the final plateau map are generic points on the $(Q-1)$-sphere, not the (soft) one-hot labels desired of a segmentation map. The Competition module resolves this by identifying well-formed clusters, converting them to discrete *object nodes*, and suppressing redundant activity (Fig. 4 bottom.) First, K *object pointers* $\{p^k\} \in \mathbb{R}^2$ are randomly placed at (h, w) locations on the plateau map and assigned *object vectors* $\mathbf{p}^k \in \mathbb{R}^Q$ according to their positions; an *object segment* $m^k \in \mathbb{R}^{H \times W}$ for each vector is then given by its cosine similarity with the full plateau map:

$$p^k = (p_h^k, p_w^k), \qquad \mathbf{p}^k = h_S(p_h^k, p_w^k), \qquad m^k = h_S \cdot \mathbf{p}^k. \tag{5}$$

Some of the masks may overlap, and some regions of the plateau map may not be covered by any mask. We use recurrent winner-take-all dynamics to select a minimal set of object nodes that fully explains the map. Let $\mathcal{J}(\cdot, \cdot)$ denote the Jaccard index and let θ be a threshold hyperparameter (set at 0.2 in all our experiments). Competition occurs between each pair of object vectors with masks satisfying $\mathcal{J}(m^k, m^{k'}) > \theta$; the winner is the vector with greater total mask weight $\sum_{i,j} m^k$. An object that wins every pairwise competition is *retained*, while all others are *deactivated* by setting their masks to zero (Fig. 4 bottom.) This process is repeated for a total of R iterations by re-initializing each deactivated object $(p^l, \mathbf{p}^l, m^l = 0)$ on parts of the plateau map that remain uncovered, $U = 1 - \sum_k m^k$. Thus the Competition module retains a set of $M <= K$ nonzero (soft) masks, which are then softmax-normalized along the M dimension to convert them into a one-hot pixelwise segmentation of the scene.

3.2   Training EISEN via Spelke Object Inference

Because KProp and Competition have no trainable parameters, training EISEN is tantamount to training the affinity matrix A. This is done with a single loss function: the row-wise KL divergence between A and a *target connectivity matrix*, $\mathcal{C}$, restricted to the node pairs determined by *loss mask*, $\mathcal{D}$:

$$\mathcal{L}_{\text{EISEN}} = \sum_{i,j} \text{KLDiv}(\mathcal{D}_{i'j'}^{ij} \odot A_{i'j'}^{ij}, \mathcal{D}_{i'j}^{ij} \odot \mathcal{C}_{i'j'}^{ij}). \tag{6}$$

To compute $\mathcal{C}$ and $\mathcal{D}$ we consider pairs of scene elements (a, b) that project into image coordinates (i, j) and (i', j'), respectively. If only one element of the pair is moving (over long enough time scales), it is likely the two elements do not belong to the same Spelke object; when neither is moving, there is no information about their connectivity, so no loss should be computed on this pair. This is the core physical logic – "Spelke object inference" – by which we train EISEN.

Computing Connectivity Targets from Motion. Let $\mathcal{I}(\cdot)$ be a motion indicator function, here $\mathcal{I}(a) = (|\textbf{flow}_{ij}| > 0)$, where **flow** is a map of optical flow. The logic above dictates setting

$$C_{i'j'}^{ij} \leftarrow 0 \text{ if } (\mathcal{I}(a) \textbf{ xor } \mathcal{I}(b)), \tag{7}$$

$$\mathcal{D}_{i'j'}^{ij} \leftarrow 1 \text{ if } (\mathcal{I}(a) \textbf{ or } \mathcal{I}(b)) \textbf{ else } 0. \tag{8}$$

To learn accurate affinities there must also pairs with $C_{i'j'}^{ij} = 1$ that indicate when two scene elements belong to the same object.[2] When a scene contains multiple uncorrelated motion sources, the optical flow map has an appearance similar to a plateau map (e.g. Figure S2, second column.) This allows the flow map to be segmented into multiple motion sources as if it *were* a plateau map using the Competition algorithm (see Supplement for details.) The positive pairs in the connectivity target can then be set according to

$$\tilde{C}_{i'j'}^{ij} \leftarrow 1 \text{ if } (\mathcal{S}_M(a) == \mathcal{S}_M(b)) \textbf{ and}(\mathcal{I}(a) == \mathcal{I}(b) == 1) \textbf{ else } 0, \tag{9}$$

where $\mathcal{S}_M$ is the estimated map of motion segments. Any elements of the background are assumed to be static with $\mathcal{S}_M(a) == \mathcal{I}(a) == 0$ (see Supplement.)

Segmenting Correlated Motion Sources by Top-Down Inference. Naïve application of Eq. (9) cannot handle the case of an agent moving a Spelke object (as in Fig. 2) because agent and object will be moving in concert and thus will appear as a single "flow plateau." However, a *non-naïve observer* might have already seen the agent alone moving and have learned to segment it via static cues. If this were so, the agent's pixels could be "explained away" from the raw motion signal, isolating the Spelke object as its own target segment (Fig. 3, lower panels.) Concretely, let $\mathcal{S}_T$ be a map of *confidently segmented objects* output by a teacher model, $\mathcal{T}$ (see Supplement for how EISEN computes confident segments.) Any scene elements that do not project to confident segments have $\mathcal{S}_T(a) = 0$.

---

[2] If scenes are assumed to have at most one independent motion source, these are simply the pairs with $\mathcal{I}(a) == \mathcal{T}.(b) == 1$. This often holds in robotics scenes (and is perhaps the norm in a baby's early visual experience) but not in many standard datasets (e.g. busy street scenes.) We therefore handle the more general case.

Then the final loss mask is modified to include all pairs with at least one moving or confidently segmented scene element,

$$\hat{\mathcal{D}}_{i'j'}^{ij} \leftarrow 1 \text{ if } ((\mathcal{S}_M(a) + \mathcal{S}_T(a) > 0) \text{ or } (\mathcal{S}_M(b) + \mathcal{S}_T(b) > 0)) \text{ else } 0. \quad (10)$$

Explaining away is performed by overwriting pairs in Eq. (9) according to whether two scene elements belong to the same or different confident segments, *regardless of whether they belong to the same motion segment*:

$$\hat{\mathcal{C}}_{i'j'}^{ij} \leftarrow (\mathcal{S}_T(a) == \mathcal{S}_T(b)) \text{ if } (\mathcal{S}_T(a) + \mathcal{S}_T(b) > 0) \text{ else } \tilde{\mathcal{C}}_{i'j'}^{ij}. \quad (11)$$

Thus the final connectivity target, $\hat{\mathcal{C}}$, combines Spelke object inference with the confident teacher predictions, defaulting to the latter in case of conflict.

Bootstrapping. Since objects that appear moving more often should be confidently segmented earlier in training, it is natural to *bootstrap*, using one (frozen) EISEN model as teacher for another student EISEN (Fig. 3.) After some amount of training, the student is frozen and becomes the teacher for the next round, as a new student is initialized with the final weights of the prior round. Although bootstrapping could be continued indefinitely, we find that EISEN confidently segments the majority of Spelke objects after three rounds.

4   Results

4.1   Datasets, Training, and Evaluation

Full details of datasets, training, and evaluation are in the Supplement. Briefly, we train EISEN and baseline models on motion signals from three datasets: **Playroom**, a ThreeDWorld [12] dataset of realistically simulated and rendered objects (2000 total) that are invisibly pushed; the **DAVIS2016** [27] video object segmentation dataset, which we repurpose to test *static* segmentation learning in the presence of background motion; and **Bridge** [9], a robotics dataset in which human-controlled robot arms move a variety of objects.

We compare EISEN to the (non-differentiable) affinity-based SSAP [13], the Transformer-based DETR [4], the centroid prediction-based Panoptic DeepLab (PDL) [6], and the region proposal-based Mask-RCNN [17]. All baselines require pixelwise segmentation supervision, for which we use the same motion signals as EISEN except for the conversion to pairwise connectivity. Because they were not designed to handle sparse supervision, we tune baseline object proposal hyperparameters to maximize recall. All models are evaluated on mIoU between ground truth and best-matched predicted segments; DETR and Mask-RCNN are not penalized for low precision.

Table 1. Performance (mIoU) of instance segmentation models on the TDW-Playroom dataset. Models with **full** supervision receive masks for all movable objects in the scene at training time; models with **motion** supervision receive the optical flow predicted by RAFT.

Model	Full supervision		Motion supervision	
	Val	Test	Val	Test
SSAP	0.802	0.575	0.295	0.235
DETR	0.860	0.647	0.297	0.258
Panoptic DeepLab	**0.870**	0.608	0.620	0.373
Mask-RCNN	0.713	0.387	0.629	0.467
EISEN	0.788	**0.675**	**0.730**	**0.638**

4.2 Learning to Segment from Sparse Object Motion

EISEN Outperforms Standard Architectures at Motion-Based Learning. Baseline segmentation architectures easily segment the **Playroom**-*val* set when given full supervision of all objects (Table 1, Full supervision.) When supervised only on RAFT-predicted optical flow, however, these models perform substantially worse (Table 1, Motion supervision) and exhibit characteristic qualitative failures (Fig. 5), such as missing or lumping together objects.

EISEN, which treats object motion as an exclusively *relational* learning signal, performs well whether given full or motion-only supervision (Table 1.) Moreover, in contrast to the baselines, EISEN also accurately segments most objects in *test* scenes that differ from its training distribution in background, object number, and multi-object occlusion patterns (Fig. 5; see Supplement.) These results suggest that only EISEN learns to detect the class of *Spelke objects* – the category-agnostic concept of "physical stuff that moves around together." Interestingly, the strongest motion-supervised baseline is Mask-RCNN, which may *implicitly* use relational cues in its region proposal and non-maximal suppression modules to partly exclude false negative static regions of the scene.

4.3 Self-supervised Segmentation of Real-World Scenes

Learning to Segment in the Presence of Background Motion. The **Playroom** dataset has realistically complex Spelke objects but unrealistically simple motion. In particular, its scenes lack background motion and do not show the agentic mover of an object. Most video frames in the **DAVIS2016** dataset [27] have both object and (camera-induced) background motion, so we use it to test whether a useful segmentation learning signal can be extracted and used to train EISEN in this setting. Applying Competition to flow plateau maps often exposes a large background segment, which can be suppressed to yield a target object motion segment (Fig. 6A; also see Supplement.) When this motion signal is used to train EISEN, the *static* segmentation performance on *held-out scenes* is 0.52,

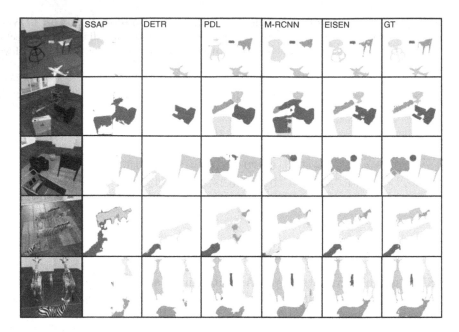

Fig. 5. EISEN outperforms baselines at learning to segment from motion. Segmentation predictions of EISEN and each baseline are shown for examples from the **Playroom** *val* set (top three rows) and *test* set (bottom two rows.) Baselines frequently lump, miss, and distort the shapes of objects. EISEN is able to capture fine details (e.g. the chair and giraffe legs) and segment closely spaced objects of similar appearance, (e.g. the zebras.)

demonstrating that motion-based self-supervision supports learning of complex Spelke objects real scenes (Fig. 6B.)

Unsupervised Segmentation of the Bridge Robotics Dataset. We train EISEN for three rounds of bootstrapping to segment Spelke objects in **Bridge** (see Methods). EISEN's segmentation of **Bridge** scenes dramatically improves with each round (Table 2 and Fig. 7). In the first round, the model mainly learns to segment the robot arm, which is expected because this object is seen moving more than any other and the untrained EISEN teacher outputs few confident segments that could overwrite the raw motion training signal. In the subsequent rounds, top-down inference from the pretrained EISEN teacher modifies the raw motion signal; the improvement during these rounds suggests that top-down inference about physical scene structure can extract better learning signals than what is available from the raw image or motion alone. In contrast to EISEN, neither Mask-RCNN nor Panoptic DeepLab segment most of the objects either after applying the same bootstrapping procedure or when pretrained on COCO with categorical supervision (Table 2 and Fig. 1.) EISEN's combination of bottom-up grouping with top-down inference thus enables unsupervised segmentation of Spelke objects in real scenes.

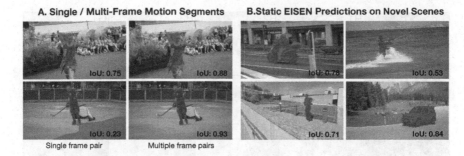

Fig. 6. EISEN learns to segment objects in *static, held-out scenes* on DAVIS2016. (A) Confident *teacher* segments computed from multiple frame pairs are better than those from a single frame pair. (B) Without any motion information, EISEN segments objects in *single RGB images* from held-out scenes.

Table 2. Performance on Bridge after each round of bootstrapping. EISEN improves at segmentation across three rounds by using its own inference pass to create better supervision signals. Neither Mask-RCNN nor Panoptic DeepLab perform well whether bootstrapped or pretrained on COCO.

Model	Round 1	Round 2	Round 3	Pretrained
MaskRCNN	0.053	0.081	0.102	0.070
Panoptic DeepLab	0.051	0.056	0.057	0.175
EISEN	0.336	0.453	0.551	–

4.4 Ablations of the EISEN Architecture

Ablating KProp and Competition. EISEN performance on **Playroom** is nearly equal when using all affinity pairs versus using local and a small sample of long-range pairs (<7% of total), though it drops slightly if long-range pairs are omitted (Table 3). This suggests that plateau map alignment is mainly a local phenomenon and that grouping with EISEN relies heavily on local cues.

In contrast, the architectural components of KProp and Competition are essential for EISEN's function. When either excitatory or inhibitory messages are ablated, or when using Softmax rather than ℓ^2-normalization, performance drops nearly to zero (Table 3.) Moreover, the Competition module is better at extracting segments from the final plateau map than simply taking the Argmax over the channel dimension; this is expected, since the ℓ^2-normalization in KProp does not encourage plateau map clusters to be one-hot vectors.

KProp and Competition are both RNNs, so their function may change with the number of iterations. Performance saturates only with >30 KProp iterations and drops to near zero with a single iteration, implying that sustained message passing is essential: EISEN cannot operate as a feedforward model. In contrast, Competition requires only a single iteration on **Playroom** data (Table 3).

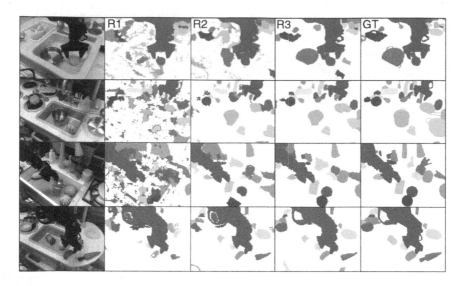

Fig. 7. EISEN improves at segmenting Spelke objects with each round of bootstrapping. After the first round of bootstrapping (R1), EISEN can segment the arm but few other objects well. Subsequent rounds (R2 and R3) substantially improve both the number of objects detected and their segmentation quality.

Table 3. Ablations of EISEN. Altering the architectural components of KProp or Competition drastically degrades performance, but only a small sample of long-range affinities are necessary (Left). Lowering the number of RNN iterations for KProp or Comp gradually degrades performance (Right).

messages	affinity	norm	readout	mIoU	KProp iters	Comp iters	mIoU
Ex+Inb	Loc+Glob	ℓ^2	Comp	0.730	40	3	0.730
Ex+Inb	Loc	ℓ^2	Comp	0.700	30	3	0.720
Ex+Inb	Full	ℓ^2	Comp	0.732	20	3	0.697
Ex+Inb	Loc+Glob	ℓ^2	Argmax	0.676	10	3	0.389
Ex	Loc+Glob	ℓ^2	Comp	0.036	1	3	0.052
Inb	Loc+Glob	ℓ^2	Comp	0.036	40	2	0.730
Ex+Inb	Loc+Glob	softmax	Comp	0.036	40	1	0.729

Table 4. Comparison of DINO and EISEN affinities with different graph clustering algorithms. EISEN affinities are downsampled to the same size as DINO affinities for a fair comparison

Model	Spectral clustering	LabelProp	AffinityProp	KProp+comp
DINO	0.354	0.135	0.255	0.545
EISEN	0.062	0.084	0.319	0.684

Ablating Affinity Prediction. Finally, we compare EISEN's affinities to other affinity-like model representations. Object segments can be extracted from the attention maps of a self-supervised Vision Transformer (DINO [5]) using KProp and Competition, but their accuracy is well below EISEN's; prior graph-based segmentation methods [11,16,31] do not detect **Playroom** objects as well as KProp-Competition (Table 4; see Supplement.) These experiments imply that EISEN is a better source of affinities than the (statically trained) DINO attention maps and that EISEN's grouping network best makes use of both sources.

5 Conclusion

We have proposed EISEN, a fully differentiable, graph-based grouping architecture for learning to segment Spelke objects. While our algorithm performs on par with prior segmentation models when fully supervised (Table 1), its main strength is an ability to learn *without supervision*: by applying top-down inference with its own segmentation predictions, it progressively improves motion-based training signals. These key architecture and learning innovations are critical for dealing with the challenges of unsupervised, category-agnostic object segmentation in real-world scenes (Fig. 2.) Since EISEN is based on the principle of grouping things that move together, it cannot necessarily address higher-level notions of "objectness" that include things rarely seen moving (e.g., houses and street signs.) It will therefore be important in future work to explore the relationship between motion-based and motion-independent object learning and identify deeper principles of grouping that extend to both.

Acknowledgements. J.B.T is supported by NSF Science Technology Center Award CCF-1231216. D.L.K.Y is supported by the NSF (RI 1703161 and CAREER Award 1844724) and hardware donations from the NVIDIA Corporation. J.B.T. and D.L.K.Y. are supported by the DARPA Machine Common Sense program. J.W. is in part supported by Stanford HAI, Samsung, ADI, Salesforce, Bosch, and Meta. D.M.B. is supported by a Wu Tsai Interdisciplinary Scholarship and is a Biogen Fellow of the Life Sciences Research Foundation. We thank Chaofei Fan and Drew Linsley for early discussions about EISEN.

References

1. Arora, T., Li, L.E., Cai, M.B.: Learning to perceive objects by prediction. In: SVRHM 2021 Workshop@ NeurIPS (2021)
2. Bear, D., et al.: Learning physical graph representations from visual scenes. In: Advances in Neural Information Processing Systems 33, pp. 6027–6039 (2020)
3. Butler, D.J., Wulff, J., Stanley, G.B., Black, M.J.: A naturalistic open source movie for optical flow evaluation. In: Fitzgibbon, A., Lazebnik, S., Perona, P., Sato, Y., Schmid, C. (eds.) ECCV 2012. LNCS, vol. 7577, pp. 611–625. Springer, Heidelberg (2012). https://doi.org/10.1007/978-3-642-33783-3_44

4. Carion, N., Massa, F., Synnaeve, G., Usunier, N., Kirillov, A., Zagoruyko, S.: End-to-end object detection with transformers. In: Vedaldi, A., Bischof, H., Brox, T., Frahm, J.-M. (eds.) ECCV 2020. LNCS, vol. 12346, pp. 213–229. Springer, Cham (2020). https://doi.org/10.1007/978-3-030-58452-8_13

5. Caron, M., et al.: Emerging properties in self-supervised vision transformers. In: Proceedings of the IEEE/CVF International Conference on Computer Vision, pp. 9650–9660 (2021)

6. Cheng, B., et al.: Panoptic-DeepLab: a simple, strong, and fast baseline for bottom-up panoptic segmentation. In: Proceedings of the IEEE/CVF Conference on Computer Vision and Pattern Recognition, pp. 12475–12485 (2020)

7. Dorfman, N., Harari, D., Ullman, S.: Learning to perceive coherent objects. In: Proceedings of the Annual Meeting of the Cognitive Science Society, vol. 35 (2013)

8. Du, Y., Smith, K., Ulman, T., Tenenbaum, J., Wu, J.: Unsupervised discovery of 3D physical objects from video. arXiv preprint arXiv:2007.12348 (2020)

9. Ebert, F., et al.: Bridge data: boosting generalization of robotic skills with cross-domain datasets. arXiv preprint arXiv:2109.13396 (2021)

10. Follmann, P., Böttger, T., Härtinger, P., König, R., Ulrich, M.: MVTec D2S: densely segmented supermarket dataset. In: Ferrari, V., Hebert, M., Sminchisescu, C., Weiss, Y. (eds.) ECCV 2018. LNCS, vol. 11214, pp. 581–597. Springer, Cham (2018). https://doi.org/10.1007/978-3-030-01249-6_35

11. Frey, B.J., Dueck, D.: Clustering by passing messages between data points. Science **315**(5814), 972–976 (2007)

12. Gan, C., et al.: ThreeDWorld: a platform for interactive multi-modal physical simulation. arXiv preprint arXiv:2007.04954 (2020)

13. Gao, N., et al.: SSAP: single-shot instance segmentation with affinity pyramid. In: Proceedings of the IEEE/CVF International Conference on Computer Vision, pp. 642–651 (2019)

14. Girshick, R.: Fast R-CNN. In: Proceedings of the IEEE International Conference on Computer Vision, pp. 1440–1448 (2015)

15. Greff, K., et al.: Multi-object representation learning with iterative variational inference. In: International Conference on Machine Learning, pp. 2424–2433. PMLR (2019)

16. Gregory, S.: Finding overlapping communities in networks by label propagation. New J. Phys. **12**(10), 103018 (2010)

17. He, K., Gkioxari, G., Dollár, P., Girshick, R.: Mask R-CNN. In: Proceedings of the IEEE International Conference on Computer Vision, pp. 2961–2969 (2017)

18. He, K., Zhang, X., Ren, S., Sun, J.: Delving deep into rectifiers: surpassing human-level performance on ImageNet classification. In: Proceedings of the IEEE International Conference on Computer Vision, pp. 1026–1034 (2015)

19. Hinton, G.: How to represent part-whole hierarchies in a neural network. arXiv preprint arXiv:2102.12627 (2021)

20. Kabra, R., et al.: SIMONe: view-invariant, temporally-abstracted object representations via unsupervised video decomposition. In: Advances in Neural Information Processing Systems 34 (2021)

21. Kipf, T., et al.: Conditional object-centric learning from video. arXiv preprint arXiv:2111.12594 (2021)

22. Lin, T.Y., Dollár, P., Girshick, R., He, K., Hariharan, B., Belongie, S.: Feature pyramid networks for object detection. In: Proceedings of the IEEE Conference on Computer Vision and Pattern Recognition, pp. 2117–2125 (2017)

23. Liu, W., Rabinovich, A., Berg, A.C.: ParseNet: looking wider to see better. arXiv preprint arXiv:1506.04579 (2015)

24. Locatello, F., et al.: Object-centric learning with slot attention. In: Advances in Neural Information Processing Systems 33, pp. 11525–11538 (2020)
25. Luo, L., Xiong, Y., Liu, Y., Sun, X.: Adaptive gradient methods with dynamic bound of learning rate. arXiv preprint arXiv:1902.09843 (2019)
26. Peng, B., Zhang, L., Zhang, D.: A survey of graph theoretical approaches to image segmentation. Pattern Recogn. **46**(3), 1020–1038 (2013)
27. Perazzi, F., et al.: A benchmark dataset and evaluation methodology for video object segmentation. In: Proceedings of the IEEE Conference on Computer Vision and Pattern Recognition, pp. 724–732 (2016)
28. Roelfsema, P.R., et al.: Cortical algorithms for perceptual grouping. Ann. Rev. Neurosci. **29**(1), 203–227 (2006)
29. Ross, M.G., Kaelbling, L.P.: Segmentation according to natural examples: learning static segmentation from motion segmentation. IEEE Trans. Pattern Anal. Mach. Intell. **31**(4), 661–676 (2008)
30. Sabour, S., Tagliasacchi, A., Yazdani, S., Hinton, G., Fleet, D.J.: Unsupervised part representation by flow capsules. In: International Conference on Machine Learning, pp. 9213–9223. PMLR (2021)
31. Shi, J., Malik, J.: Normalized cuts and image segmentation. IEEE Trans. Pattern Anal. Mach. Intell. **22**(8), 888–905 (2000)
32. Siméoni, O., et al.: Localizing objects with self-supervised transformers and no labels. arXiv preprint arXiv:2109.14279 (2021)
33. Spelke, E.S.: Principles of object perception. Cogn. Sci. **14**(1), 29–56 (1990)
34. Tangemann, M., et al.: Unsupervised object learning via common fate. arXiv preprint arXiv:2110.06562 (2021)
35. Teed, Z., Deng, J.: RAFT: recurrent all-pairs field transforms for optical flow. In: Vedaldi, A., Bischof, H., Brox, T., Frahm, J.-M. (eds.) ECCV 2020. LNCS, vol. 12347, pp. 402–419. Springer, Cham (2020). https://doi.org/10.1007/978-3-030-58536-5_24
36. Todorovic, D.: Gestalt principles. Scholarpedia **3**(12), 5345 (2008)
37. Tsao, T., Tsao, D.Y.: A topological solution to object segmentation and tracking. arXiv preprint arXiv:2107.02036 (2021)
38. Ullman, S., Harari, D., Dorfman, N.: From simple innate biases to complex visual concepts. Proc. Natl. Acad. Sci. **109**(44), 18215–18220 (2012)
39. Wang, Y., Shen, X., Hu, S., Yuan, Y., Crowley, J., Vaufreydaz, D.: Self-supervised transformers for unsupervised object discovery using normalized cut. arXiv preprint arXiv:2202.11539 (2022)
40. Yang, C., Lamdouar, H., Lu, E., Zisserman, A., Xie, W.: Self-supervised video object segmentation by motion grouping. In: Proceedings of the IEEE/CVF International Conference on Computer Vision, pp. 7177–7188 (2021)
41. Zhou, H., Friedman, H.S., Von Der Heydt, R.: Coding of border ownership in monkey visual cortex. J. Neurosci. **20**(17), 6594–6611 (2000)
42. Zhu, X., Su, W., Lu, L., Li, B., Wang, X., Dai, J.: Deformable DETR: deformable transformers for end-to-end object detection. arXiv preprint arXiv:2010.04159 (2020)

A Simple Baseline for Open-Vocabulary Semantic Segmentation with Pre-trained Vision-Language Model

Mengde Xu[1,3], Zheng Zhang[1,3], Fangyun Wei[3], Yutong Lin[2,3], Yue Cao[3], Han Hu[3], and Xiang Bai[1(✉)]

[1] Huazhong University of Science and Technology, Wuhan, China
{mdxu,mdxu}@hust.edu.cn
[2] Xi'an Jiaotong University, Xi'an, China
yutonglin@stu.xjtu.edu.cn
[3] Microsoft Research Asia, Beijing, China
{zhez,fawe,hanhu}@microsoft.com

Abstract. Recently, open-vocabulary image classification by vision language pre-training has demonstrated incredible achievements, that the model can classify arbitrary categories without seeing additional annotated images of that category. However, it is still unclear how to make the open-vocabulary recognition work well on broader vision problems. This paper targets open-vocabulary semantic segmentation by building it on an off-the-shelf pre-trained vision-language model, i.e., CLIP. However, semantic segmentation and the CLIP model perform on different visual granularity, that semantic segmentation processes on pixels while CLIP performs on images. To remedy the discrepancy in processing granularity, we refuse the use of the prevalent one-stage FCN based framework, and advocate a two-stage semantic segmentation framework, with the first stage extracting generalizable mask proposals and the second stage leveraging an image based CLIP model to perform open-vocabulary classification on the masked image crops which are generated in the first stage. Our experimental results show that this two-stage framework can achieve superior performance than FCN when trained only on COCO Stuff dataset and evaluated on other datasets without fine-tuning. Moreover, this simple framework also surpasses previous state-of-the-arts of zero-shot semantic segmentation by a large margin: +29.5 hIoU on the Pascal VOC 2012 dataset, and +8.9 hIoU on the COCO Stuff dataset. With its simplicity and strong performance, we hope this framework to serve as a baseline to facilitate future research. The code are made publicly available at https://github.com/MendelXu/zsseg.baseline.

M. Xu, Z. Zhang and F. Wei—Equal contribution.

Supplementary Information The online version contains supplementary material available at https://doi.org/10.1007/978-3-031-19818-2_42.

1 Introduction

Semantic segmentation is a fundamental computer vision task that assigns every pixel of an image with category labels. Accompanied by the development of deep learning [13,23,27,35,42], the semantic segmentation has also evolved tremendously under the supervised learning paradigm [3,7,36]. However, unlike common image-level datasets such as ImageNet-1K/ImageNet-22K image classification which are easily scaled up to tens of thousands of categories, existing semantic segmentation tasks involve usually up to tens or hundreds of categories due to the significantly higher annotation cost, and thus limit the segmentors' capability in handling rich semantics.

Zero-shot semantic segmentation [5] is an attempt to break the bottleneck of limited categories. However, the narrowly defined zero-shot semantic segmentation usually only takes a small amount of labeled segmentation data and refuses to make use of any other data/information, consequently resulting in poor performance. In this work, we focus on another more practical setting: *open-vocabulary* semantic segmentation, as a generalized zero-shot semantic segmentation, concentrates more on establishing a feasible method to segment arbitrary classes and allows the use of additional data/information except the segmentation data. Specifically, we propose to leverage a recent advance of image-level vision-language learning model, i.e., CLIP [41].

While the vision-language learning model has learnt a strong vision-category alignment model using rich image-caption data, how to effectively transfer its image-level recognition capability to pixel-level is unclear. An natural idea is to integrate the vision-language model with a fully convolutional networks (FCN) [36], an architecture widely used for fully supervised semantic segmentation. A main difficulty of the integration is that the CLIP model is learnt at image-level, which differs from the granularity of FCN that models semantic segmentation as a pixel classification problem, where a linear classifier is applied on each pixel feature to produce the classification results, with each column of the linear classifier weight matrix representing each category. Empirically, we found the granularity inconsistency lead unsatisfactory performance.

To better leverage the strong vision-category correspondence capability involved in the image-level CLIP model, we pursue mask proposal based semantic segmentation approaches such as MaskFormer [9], which first extracts a set of class-agnostic mask proposals and then classifies each mask proposal into a different category. This two-stage approach decouples the semantic segmentation task into two sub-tasks of class-agnostic mask generation and mask category classification. Both sub-tasks prove well adaptation to handle unseen classes: firstly, the class-agnostic mask proposal generation trained using *seen* classes is observed well generalizable to *unseen* classes; secondly, the second mask proposal classification stage is at a same recognition granularity than that used in a CLIP model. To further bridge the gap with a CLIP model, the masked image crop of each proposal is used as input to the CLIP model for *unseen* classes classification. In addition, we employ a prompt-learning approach [34] to further improve the *unseen* classes classification accuracy given a pre-trained CLIP model.

We evaluate the proposed approach under two different settings: 1) *Cross-dataset* setting where the model is trained on one dataset and evaluated on other datasets without fine-tuning. Under this setting, our two-stage framework demonstrate well generalization capability. It outperforms FCN approach by **+13.1** mIoU on Cityscapes, **+19.6** mIoU on Pascal Context, **+5.6** mIoU on ADE20k with 150 classes and **+2.9** mIoU on ADE20k with 847 classes. 2) *Zero-shot* setting where the model is trained on a part of *seen* class of a dataset and evaluated on all classes (including *seen* and *unseen* classes). We use this setting for comparing with other zero-shot semantic segmentation methods. We show that the proposed approach, though simple and straightforward, can surpass previous state-of-the-arts zero-shot segmentation approaches [5,20,40,48] by a large margin. On Pascal VOC 2012 [14], this approach outperforms previous best methods that w/o self-training by **+37.8** hIoU, and by **+29.5** hIoU when an additional self-training process is involved. On COCO Stuff [6], the approach outperforms previous best methods that w/o self-training by **+19.6** hIoU and by **+8.9** hIoU when an additional self-training process is involved. We hope our simple but effective approach can encourage more study in this direction.

2   Related Works

Vision-Language Pre-training. Vision-language pre-training focuses on how to connect visual concepts and language concepts. Early approaches [8,31,33, 37,44] were performed on some cleaned datasets with relatively small data scale. Therefore, those models usually need to be fine-tuned on some specific downstream tasks. Some recent works [25,41] have explored the benefits of large-scale noisy data obtained from web pages for vision-language pre-training. CLIP [41], as a representative work, employs a contrastive learning approach to distinguish the correct image-text pair in each training batch. Because many vision/language concepts are covered in large-scale data, the CLIP illustrates surprisingly strong capability on zero-shot/open-vocabulary image classification and image-text retrieval. This work introduces the CLIP model as a strong vision-category correspondent for open-vocabulary semantic segmentation.

Semantic Segmentation. Semantic segmentation is a fundamental task in computer vision that aims to assign a category to each pixel. Fully convolutional network [36] and its variants [3,7,49], as a practical and straightforward approach to model the semantic segmentation as a pixel-wise classification problem, have dominated this field in the past few years. Recently, MaskFormer [9] explored to model the semantic segmentation as two sub-tasks: segment generation and segment classification and has shown competitive performance compared to FCN based approaches.

Zero-Shot Learning and Open-Vocabulary Learning. Zero-shot learning has been widely studied in recent years. A narrowly defined zero-shot learning focuses on learning transferable representations from the annotated data of seen classes to represent unseen classes. For example, [1,47] proposed to learn a joint embedding space between the images and the name/description of the category for image classification, and [28] explored taking the advantages of mid-level semantic representation. Recently, the open-vocabulary learning has attracted more attentions. As a generalized zero-shot learning, the open-vocabulary learning is more concerned with establishing a feasible method for arbitrary class recognition and allows the use of any additional information. For example, Visual N-Grams [29] and CLIP [41] explored the use of web-crawled data for image classification and [19] introduced the vision-language pre-training model for the open-vocabulary object detection and showed it could significantly improve the long-tile object detection [22].

Zero-Shot Semantic Segmentation. Some pioneer works to study the zero-shot learning for semantic segmentation. ZS3Net [5] uses generative models to synthesize pixel-level features by word embeddings of *unseen* classes. CSRL [32] further incorporating the structural relation in feature synthesize. CaGNet [20, 21] introduce a contextual module for better feature generation. Different from [5, 20,32], SPNet [48] attempt to mapping vision feature to the semantic space via word embedding. JoEm [4] a joint embedding strategy between the vision encoder and semantic encoder. In [26], variational mapping is used to learn semantic features. In [24], the uncertainty-aware losses are proposed to eliminate noisy samples. Other works explored other directions or aspects of zero-shot semantic segmentation. In [11], the super-pixel pooling is utilized to improve the region grouping generalization. In [40], the self-training for zero-shot semantic segmentation are carefully studied. In [38,43], the transductive learning setting are explored. In [45], they explore the utilization of image caption. However, all those methods have not explored the utilization of the vision-language pre-training model in zero-shot semantic segmentation. There are two concurrent work [16,30] try to utilize the vision-language pre-training model in semantic segmentation. However, LSeg [30] is an FCN-based approach focus on few shot setting. Openseg [16], which is a similar work to ours, utilizes external grounding dataset while we don't. In addition, Openseg is based on ALIGN [25] while we adopt CLIP [41].

3 Preliminary

In this section, we first introduce the setting of open-vocabulary semantic segmentation and revisit CLIP as preliminary.

3.1 Open-Vocabulary Semantic Segmentation

Zero-Shot Setting. Open-vocabulary is an generalized zero-shot task, so the zero-shot semantic segmentation protocol can also evaluate open-vocabulary semantic segmentation. In this setting, model predicts masks for *unseen* classes

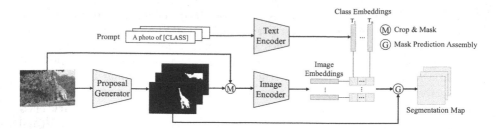

Fig. 1. Overview of our two-stage open-vocabulary semantic segmentation framework. We reformulate and break down the open-vocabulary semantic segmentation into two steps: 1) training a mask proposal generator to generate a set of binary masks; 2) leveraging the pre-trained CLIP to classify each mask proposal.

$\mathcal{C}^{\text{unseen}}$ by learning from some labeled data of *seen* classes $\mathcal{C}^{\text{seen}}$, and the *seen* classes and *unseen* classes are disjoint, i.e., $\mathcal{C}^{\text{unseen}} \cap \mathcal{C}^{\text{seen}} = \varnothing$. Usually, $\mathcal{C}^{\text{seen}}$ and $\mathcal{C}^{\text{unseen}}$ are often represented with semantic words like *dog, cat, apple*, and sometimes the description of the classes are also provided.

During training, a training set $\mathcal{X}_{\text{train}} = \{(\mathcal{I}_k, \mathcal{M}_k)\}$ with input images $\mathcal{I}_k$ and the ground-truth semantic segmentation annotations $\mathcal{M}_k$ is provided, and the training annotations $\{\mathcal{M}_k\}$ contains only the *seen* classes. The trained model is evaluated on a testing set $\mathcal{X}_{\text{test}}$, both *seen* classes and *unseen* classes need to be predicted in testing set $\mathcal{X}_{\text{test}}$.

Cross-Dataset Setting. In this setting, the model is trained on one dataset and evaluated on another dataset without fine-tuning. This is a more challenging setting than the *zero-shot setting*, where the model not only deals with the *unseen* classes, but also has to address the domain gap among different datasets.

3.2   Revisiting CLIP

CLIP [41] is a powerful pre-trained vision-language model, which shows surprisingly strong performance in associating the visual and textual concepts. CLIP is a two-stream method: it contains an image encoder $\mathcal{E}_{\text{image}}$ and a text encoder $\mathcal{E}_{\text{text}}$. For any given image-text paired data $\{\mathcal{I}, \mathcal{T}\}$, their semantic similarity can be estimated by computing the cosine distance between $\mathcal{E}_{\text{image}}(\mathcal{I})$ and $\mathcal{E}_{\text{text}}(\mathcal{T})$.

The pre-trained CLIP model can be used to classify images by a given set of classes without fine-tuning, which is also known as zero-shot/open-vocabulary image classification. Specifically, the class names are injected into the pre-defined prompt template and fed into CLIP's text encoder to generate the class embeddings, e.g., a typical prompt template is 'a photo of [CLASS]', where [CLASS] is replaced by the specific class name such as 'person' and 'cat'. The generated class embeddings are used as the classifier and the similarity with image embedding is computed for classification.

In this work, we extend the compatibility of CLIP from *image-level* zero-shot/open-vocabulary classification to *pixel-level* open-vocabulary semantic segmentation, by exploring the use of a pre-trained CLIP model as a strong vision-category correspondent.

4 Two-Stage Open-Vocabulary Semantic Segmentation

Figure 1 shows an overview of our two-stage framework. Given an image, a set of mask proposals are first generated, and then each proposals is fed into an image encoder and compared with the class weights obtained by applying text encoder on the prompt class description to perform the classification. Finally, the mask prediction are assembled together to produce the final segmentation results. We will describe each component of our framework in the following.

4.1 Mask Proposal Generation

We first introduce the mask proposal generation. In our work, we try three different methods to generate the mask proposals $\{\mathcal{M}_k^p\}$:

GPB-UCM [2]. This is a classical method to generate hierarchical segments by considering multiple low-level cues, e.g., brightness, color, texture, and local gradients. The generated segments of this approach are usually well aligned with the contour of objects.

Selective Search [46]. This method can also generate hierarchical segments. Since this method can effectively localize objects, it is widely used in object detection systems [17,18].

MaskFormer [9]. This is a recently proposed method for supervised semantic segmentation. Unlike a fully convolution network that models the semantic segmentation as the pixel-wise classification problem, MaskFormer disentangles the semantic segmentation into two sub-tasks: predicting the segments at first and then classifying the category of each segment. We observe that the predicted segments by MaskFormer can be used as the mask proposals, and we empirically demonstrate (see Table 6) that the MaskFormer trained on *seen* classes can produce high-quality mask proposals on the *unseen* classes. Therefore, we take this advantage of MaskFormer as our default mask proposal generator.

4.2 Region Classification via CLIP

Two Strategies for Using CLIP. There are two strategies to perform the region classification by utilizing the pre-trained CLIP:

– The first strategy is to directly apply the CLIP image encoder on each mask proposals for classification. Specifically, given an image $\mathcal{I}$ and a mask proposal $\mathcal{M}^p$, the mask proposals are first binarized with a threshold of 0.5, and then apply the binarized $\mathcal{M}^p$ to image $\mathcal{I}$, erase the unused background and only crop foreground area. The masked image crop is resized to 224^2 and then fed into CLIP for classification. However, since there is no extra training process, the training data of *seen* classes cannot be utilized, resulting in inferior performance on *seen* classes in the inference (see Table 7).

- To utilize the training data of *seen* classes, another approach is to retrain an image encoder. However, if we simply learn a set of new classifiers on the training data of *seen* classes, the retrained image encoder has no generalization ability on *unseen* classes since these classes have no corresponding classifiers. Therefore, we propose to use the features generated from the text encoder of the pre-trained CLIP model as the fixed classifier weights for the retrained image encoder. In this approach, the image encoder has a certain generalization ability to the *unseen* classes since the image encoder is encouraged to embed the vision features into the same embedding space of the text encoder through the *seen* classes. Notably, this approach can be easily integrated into the training process of the MaskFormer, by simply using the CLIP generated text features as the classifier weights of the MaskFormer, thus avoiding the need of training an additional image encoder.

The two strategies complement each other (see Table 7), therefore we ensemble the results of these two strategies by default. Given a mask proposal $\mathcal{M}^p$, we crop the foreground area $A_{fg} = \text{crop}(\mathcal{M}^p, \mathcal{I})$ (See Appendix for details), and compute its classification probability via CLIP vision encoder E_{vision} and text encoder E_{text}:

$$C_i(A_{fg}) = \frac{\exp(\text{cosine}(E_{\text{vision}}(A_{fg})), E_{\text{text}}(\mathcal{C}_i)/\tau)}{\sum_i^{\#\text{class}} \exp(\text{cosine}(E_{\text{vision}}(A_{fg}), E_{\text{text}}(\mathcal{C}_i))/\tau)}, \quad (1)$$

where $\mathcal{C}_i$ is name of i-th class and temperature $\tau=100$. The classification probability of CLIP can be ensembled with supervised model trained on seen classes and then generate final mask results according to Sect. 4.3.

Prompt Design. The original CLIP is not designed for open-vocabulary semantic segmentation. How to design feasible text prompts need to be explored.

Hand-Crafted Prompt. A simple approach is to re-use the hand-crafted prompts provided by CLIP which is originally designed for image classification on ImageNet-1K [12]. There are 80 different prompts, each consisting of a natural sentence with a blank position for injecting the category names. Since these prompts are not originally designed for semantic segmentation, some of them may have a adverse effect. So we evaluate each of these prompts on training data to select one most helpful prompt for open-vocabulary semantic segmentation.

Learning-Based Prompt. Prompt learning [34,51] recently showed great potential for adapting the pre-trained language/vision-language models on specific downstream tasks. We also explore this technique. Specifically, a prompt is a sequence of tokens. Each token belongs to one of the two types: $[P]$ indicates the prompt token and $[CLS]$ indicates the class token. A generalized prompt can be formulated as $[P]_0...[P]_m[CLS]$, where m is the number of prompt token. In prompt learning, the prompt tokens $[P]_0...[P]_m$ are set as learnable parameters that can be trained on the *seen* classes and generalized to the *unseen* classes.

4.3 Mask Prediction Assembly

Since the mask proposals may overlap each other, resulting in the possibility of some pixels being covered by several different mask proposals. Therefore, we employ a simple aggregation mechanism to generate semantic segmentation results from the mask predictions. Specifically, for a given pixel q, its predicted probability of being i-th category is defined as:

$$C_i(q) = \frac{\mathcal{M}_k^p(q)C_k^p(i)}{\sum \mathcal{M}_k^p(q)}, \tag{2}$$

where $\mathcal{M}_k^p(q)$ denotes the predicted probability of pixel q in k-th mask proposal $\mathcal{M}_k^p$, and $C_k^p(i)$ is the predicted probability of mask proposals $\mathcal{M}_k^p$ belonging to i-th category. Note that the sum of $C_i(q)$ over all categories is not guaranteed to be 1, and pixel q is classified to the category with highest predicted value.

5   Fully Convolution Network Approach

In addition to our proposed two-stage framework, a more conventional approach is to use the widely-used fully convolution network (FCN). As a dominant method in supervised semantic segmentation, FCN formulates the semantic segmentation as a pixel-wise classification problem. Specifically, given an image, FCN generates a high-resolution feature map, and a set of learned classifiers is applied on each pixel to produce segmentation predictions. Similar to our proposed two-stage framework, there are also two strategies to apply the CLIP on FCN framework:

- Directly using the feature map generated by the CLIP vision encoder to perform pixel-wise classification. Note that in the original CLIP model, the feature of an image are represented by the feature of [CLS] token, not the feature map, and this difference may lead to performance degradation. In addition, the original CLIP model uses the image size of 224×224 during pre-training, while semantic segmentation usually requires a higher image resolution (e.g., shorter size is 640). Therefore, the direct use of high-resolution image during inference may lead to inferior performance due to inconsistency in image size. To alleviate this problem, we try to use the sliding window technique, which is widely used in previous works [7] for performing multi-scale inference. We empirically found that it can improve performance and thus use it by default.
- The training data of the *seen* classes cannot be utilized in the first strategy. Instead, we retrain an FCN-based vision encoder on *seen* classes via the similar method introduced in Sect. 4.2. Specifically, we use the CLIP text encoder to generate a fixed classifier weight. Therefore, the retrained model can obtain a certain generalization ability to the *unseen* classes.

As the same as the two-stage framework, we also ensemble the prediction of these two strategies by default if not specified.

6   Experiments

6.1   Dataset and Evaluation Protocol

Dataset. We conduct extensive experiments on five challenging datasets to evaluate our method: COCO Stuff [6], Pascal VOC 2012 [14], Pascal Context [39], Cityscapes [10], and ADE20K [50].

COCO Stuff is a large-scale dataset that contains 117k training images and 5k validation images. It contains annotations of 171 classes, 80 thing classes and 91 sutff classes respectively.

Pascal VOC 2012 contains 11,185 training images and 1,449 validation images from 20 classes. The provided augmented annotations are used.

Cityscapes is a scene parsing dataset collected on urban streets, containing 5,000 finely annotated images and 20,000 coarsely annotated images. According to the common practices [10], we use 1,525 images of 19 classes in the finely annotated set for validation.

Pascal Context is an extensive dataset of Pascal VOC 2010, containing 4,998 training images and 5,005 validation images. We use the frequent 59 classes for validation.

ADE20K contains 20k training images, 2k validation images, and 3k testing images. There are two settings of 150 classes and 857 classes.

Data Split. For *Cross-dataset setting*, we train our model on the COCO Stuff dataset and test on the validation set of the others. For *Zero-shot setting*, we evaluate our method on COCO Stuff and Pascal VOC 2012. Following [48], we divide the COCO Stuff dataset into 156 *seen* classes and 15 *unseen* classes and the Pascal VOC 2012 dataset into 15 *seen* classes and 5 *unseen* classes.

Evaluation Protocol. For *cross-dataset setting*, we use the mean of class-wise intersection over union (mIoU) as major metric. For *zero-shot setting*, we use harmonic mean IoU (hIoU) among the *seen* classes and *unseen* classes as major metric by following previous works [40,48] (see Appendix for detail definition). We also report the pixel-wise classification accuracy (pAcc) as a reference.

6.2   Implementation Details

We conduct all experiments on 8×Nvidia V100 GPUs. We train a MaskFormer [9] model on the COCO Stuff dataset with ResNet-101 as the default backbone. An AdamW optimizer with the initial learning rate of 1e-4, weight decay of 1e-4 and a backbone multiplier of 0.1, and a poly learning rate policy with a power of 0.9 are used. The batch size is set to 32 for each GPU, and the total training iteration is 60K/120K for zero-shot setting and cross-dataset setting, respectively.

Table 1. We train our model on COCO Stuff dataset and evaluate on other datasets (cross-dataset setting). The number in the parentheses after the dataset name represents class number. Both methods are tuned through the same prompt engineering [19].

Method	Dataset			
	Cityscapes (19)	Pascal Context (59)	ADE20K (150)	ADE20K (847)
FCN	21.4	28.2	14.9	4.1
Ours	34.5	47.7	20.5	7.0

Table 2. Comparison with other methods on COCO Stuff in the zero-shot setting.

Table 3. Comparison with other methods on Pascal VOC in the zero-shot setting.

Method	hIoU	mIoU	
		Seen	Unseen
SPNet [48]	16.8	20.5	14.3
ZS3 [5]	15.0	34.7	9.5
CaGNet [20]	18.2	35.5	12.2
FCN	20.9	30.1	16.0
Ours	37.8	39.3	36.3
SPNet+ST [48]	30.3	34.6	26.9
ZS5 [5]	16.2	34.9	10.6
CaGNet+ST [20]	19.5	35.6	13.4
STRICT [40]	32.6	35.3	30.3
Ours+ST	**41.5**	**39.6**	**43.6**

Method	hIoU	mIoU	
		Seen	Unseen
SPNet [48]	25.1	73.3	15.0
ZS3 [5]	28.7	77.3	17.7
CaGNet [20]	39.7	78.4	25.6
FCN	50.7	85.5	36.0
Ours	77.5	**83.5**	72.5
SPNet+ST [48]	38.8	77.80	25.8
ZS5 [5]	33.3	78	21.2
CaGNet+ST [20]	43.7	78.6	30.3
STRICT [40]	49.8	82.7	35.6
Ours+ST	**79.3**	79.2	**78.1**

If not specified, the MaskFormer model is only trained on *seen* classes, and we use 100 mask proposals for both training and testing. For all other settings and hyper-parameters, we keep the original setting of MaskFormer without changes. CLIP with ViT-B/16 backbone is used by default if not specified. In text prompt tuning, the prompts are randomly initialized, and a SGD optimizer is used to train the learnable prompts. The learning rate is set to 0.02 and decayed according to the cosine learning rate policy, and the batch size is set to 32. We train 50 and 100 epochs for Pascal VOC and COCO Stuff, respectively. For Pascal VOC 2012 dataset, we use a batch size of 16 and a total training iteration of 20K, and keep all other setting as the same as the COCO Stuff dataset.

6.3 Comparison in Cross-Dataset Setting

We first evaluate our method on the cross-dataset setting. The model is trained on the COCO Stuff dataset and then evaluated on other datasets without fine-tuning. Table 1 clearly shows that our two-stage approach outperforms the FCN approach by a noticeable margin, demonstrating that our two-stage approach can better leverage the pre-trained CLIP model than the FCN approach. We do

Cow
Giraffe
Suitcase
Frisbee
Skateboard
Carrot
Scissors
Cardboard
Clouds
Grass
Playing field
River
Road
Tree
Wall-concrete
Seen

Fig. 2. Qualitative results on COCO Stuff dataset. Only results of unseen classes are visualized. Predictions misclassified to seen classes are labeled with *Seen* color.

not list the result on Pascal VOC as its categories overlap much with the COCO Stuff dataset, and our method can achieve 88.4 mIoU.

6.4   Comparison in Zero-Shot Setting

We then compare our method with previous state-of-the-arts on Pascal VOC 2012 dataset and COCO Stuff dataset. Since some works reported the performance by applying the self-training techniques (denoted as "ST"), we follow this practice and report the performance with or without self-training.

COCO Stuff. Table 2 shows the results. Compared with Pascal VOC 2012 dataset, COCO Stuff is more challenging. However, our approach still outperforms state-of-the-arts by a large margin. Specifically, without using the self-training, our method achieves 37.8 hIoU and 36.3 mIoU-unseen, outperforming the previous best method CaGNet [20] by +19.5 hIoU and +24.1 mIoU-unseen. By further employing the self-training, our method achieves 41.5 hIoU and 43.6 mIoU-unseen, outperforming the previous best method STRICT [40] by +8.9 hIoU and +13.3 mIoU-unseen. The qualitative results are shown in Fig. 2.

Pascal VOC 2012. The results are shown in Table 3. Without using the self-training, our method achieves 77.5 hIoU and 72.5 mIoU-unseen, outperforming the previous best method CaGNet [20] by a huge margin of +37.7 hIoU and +46.8 mIoU-unseen. By further employing the self-training, our method achieves 79.3 hIoU and 78.1 mIoU-unseen, outperforming the previous best method STRICT [40] by +29.5 hIoU and +42.5 mIoU-unseen.

While our method outperforms other state-of-the-art zero-shot semantic segmentation methods, **how the larger pre-trained data and image encoder affects the performance is still unclear**. To study these impacts, we design a new implementation that enables our approach to only leverage ImageNet-1K

Table 4. Study on how image encoder and pre-trained data affects the performance on Pascal VOC 2012 in the zero-shot setting.

Method	Image encoder	Pre-train data	hIoU	mIoU	
				Seen	Unseen
ZS3 [5]	ResNet-101	ImageNet	28.7	77.3	17.7
CaGNet [20]	ResNet-101	ImageNet	39.7	78.4	25.6
SPNet [48]	ResNet-101	ImageNet	25.1	73.3	15.0
	ResNet-101	CLIP-VL	33.4	74.1	21.5
Ours	ResNet-101	ImageNet	49.5	71.1	38.0
	ResNet-101	CLIP-VL	74.2	**84.6**	66.1
	VIT/B-16	CLIP-VL	**77.5**	83.5	**72.5**

Table 5. Study of different proposal generation methods on COCO Stuff dataset.

Method	hIoU	pAcc	mIoU-unseen
GPB-UCM [2]	10.9	9.5	11.6
Sel. Search [46]	11.0	23.5	13.3
MaskFormer [9]	**28.2**	**48.4**	**29.7**

Table 6. Evaluate the generalization ability of mask proposal generator.

Training set	Test set	mIoU	pAcc
COCO Stuff	COCO Stuff	69.4	87.7
ADE20K	COCO Stuff	62.5	84.6
ADE20K	ADE20K	71.6	90.2
COCO Stuff	ADE20K	64.4	87.7

classification data. Specifically, we train a vision-language model by only using ImageNet-1K: the class names of ImageNet-1K are treated as language inputs, and are encoded through a pure text encoder[1] to generate the classification weights. As shown in Table 4, our method achieves 49.5 hIoU with ResNet-101 as backbone, which is much higher than other approaches. On the other hand, we also try to integrate the CLIP with SPNet, and we find it only achieves 33.4 hIoU, which is far from our method by using the same ResNet-101 backbone. Those experiments indicate that the surpassing performance of our method does not only come from larger pre-training data, but also our two-stage framework.

6.5 Ablation Studies

In this section, we validate the key designs of our method. If not specified, we report the performance on the COCO Stuff dataset with the MaskFormer model of ResNet-101 and CLIP of ViT-B/16 by using the *zero-shot setting*.

Different Mask Proposal Generation Methods. We evaluate the performance of the mask proposal generation methods by plugging them into our pipeline. To avoid the impact of the learnable classifier trained on *seen* classes,

---

[1] We use SimCSE [15] as the text encoder trained on text data only.

Table 7. Study of different region classification methods on COCO Stuff dataset.

Method	hIoU	mIoU	
		Seen	Unseen
Retrained Vision Enc	8.7	38.7	4.9
CLIP Vision Enc	28.2	26.8	29.7
Ensemble	**37.8**	**39.3**	**36.3**

Table 8. Evaluate the performance of different CLIP variants in our framework on COCO Stuff dataset with manual prompt.

Backbone	hIoU	mIoU-unseen
ResNet-50	15.2	16.0
ResNet-101	13.8	12.5
ViT-B/32	15.3	15.7
ViT-B/16	**18.3**	**19.5**

Table 9. Comparison with supervised method on COCO Stuff validation dataset. Sup: MaskFormer trained on both *seen* classes and *unseen* classes.

Method	hIoU	mIoU	mIoU-unseen
Sup	**49.4**	**42.6**	**62.6**
Ours	37.8	39.2	36.3
Ours+ST	41.5	39.9	43.6

Table 10. Comparison with supervised method on the *unseen* classes of COCO Stuff validation set. Δ is the difference in mIoU between things and stuff of *unseen* classes.

Method	mIoU on *unseen* classes			
	All	Thing	Stuff	Δ
Sup	**62.6**	**67.8**	**58.3**	9.5
Ours	36.3	44.3	29.5	14.9
Ours+ST	43.6	48.4	39.5	8.9

we perform the comparison by directly classifying the masked regions with the CLIP model. The results are shown in Table 5, and the MaskFormer achieves better performance than the Selective Search and GPB-UCM. Note that even the other two methods are worse than the MaskFormer, they still achieve comparable performance compared with state-of-the-arts on mIoU-unseen.

Generalization of Mask Proposal Generator. Although using MaskFormer to generate the mask proposals achieves excellent performance on zero-shot setting, it is still unknown whether Maskformer can produce good performance on the cross-dataset setting, i.e., training on one and testing on another dataset. Therefore, we evaluate the generalization ability of using MaskFormer to generate mask proposals between the COCO Stuff dataset and the ADE20K dataset.

In this experiment, we want to evaluate only the quality of the proposal without the effects of the region classifier. However, it is difficult to design a simple "recall" metric for mask proposals in semantic segmentation similar to object detection. Because a segment can consist of multiple mask proposals, this may lead low recall while the final semantic segmentation result is still correct. Therefore, we designed an *"oracle"* experiment to evaluate how these proposals affect the final performance of semantic segmentation. Specifically, for each mask proposal, its category is specified as the same as the ground-truth segment in which it has the largest overlap. In this case, the segmentation performance can fully reflect the proposal quality.

The results are shown in Table 6. We directly report the mIoU in this experiment because the *seen* class cannot be defined between different datasets. We note that the MaskFormer model trained on COCO Stuff can produce good performance on ADE20K compared to the MaskFormer model directly trained on ADE20K with acceptable performance degradation, and vice versa. That demonstrates the generalization ability to use MaskFormer as the proposal generator.

Different Strategies of Using CLIP. We study the two different strategies of using CLIP discussed in Sect. 4.2: retrained vision encoder or directly using CLIP vision encoder without tuning. The results are shown in Table 7. The retrained vision encoder shows excellent performance on *seen* classes, while its performance on *unseen* classes is relatively low. In contrast, the CLIP vision encoder shows strong performance on *unseen* classes while worse than retrained vision encoder on *seen* classes by a large margin. By ensembling the two strategies, the performance on both *seen* and *unseen* classes is significantly improved, indicating the two strategies are complementary.

Different CLIP Variants. CLIP provides several variants with different network architectures and model sizes. We study how these models affect the performance of our method when using them as the region classifiers. We report the results without using the learnable prompt due to the high experimental overhead. The results are shown in Table 8. We find that all models perform well and CLIP with ViT-B/16 achieves the best performance.

Comparison with Supervised Baseline. We also compare our method with the supervised baseline on COCO Stuff. The supervised model is MaskFormer with ResNet-101 backbone which is trained on all classes, including *seen* and *unseen* classes. We report the results in Table 9. It is remarkable that while our method is worse than the supervised baseline by a large margin on mIoU-unseen and hIoU, the gap in mIoU is much close. That is because there are only 15 *unseen* classes in the current dataset partition. For reference, there are 156 *seen* classes.

To further explore the performance gap between our method and the supervised baseline, we split the *unseen* classes into *things* and *stuff*. The results are reported in Table 10. We find the performance gap of our method between *things* classes and the *stuff* classes is significantly large than the supervised baseline, and self-training can significantly reduce the gap. This observation suggests that the classification ability of CLIP models is different in *things* and *stuff*, which may be due to the bias of the pre-trained dataset used by CLIP.

7   Conclusion

In this work, we propose a simple and effective two-stage framework for open-vocabulary semantic segmentation with the advanced pre-trained vision-language model. We reformulate and break down the open-vocabulary semantic segmentation into two steps: 1) training a mask proposal generator to generate a set of binary masks and 2) leveraging the pre-trained CLIP to classify each mask proposal. We conduct extensive experiments to verify our approach. Notably, the proposed framework outperforms previous state-of-the-arts of zero-shot semantic segmentation on Pascal VOC 2012 and COCO Stuff by large margins. Our work reveals the potential for using pre-trained vision-language models on open-vocabulary/zero-shot semantic segmentation and provides a strong baseline for this community to facilitate future research.

References

1. Akata, Z., Perronnin, F., Harchaoui, Z., Schmid, C.: Label-embedding for image classification. IEEE Trans. Pattern Anal. Mach. Intell. **38**(7), 1425–1438 (2015)
2. Arbelaez, P., Maire, M., Fowlkes, C., Malik, J.: Contour detection and hierarchical image segmentation. IEEE Trans. Pattern Anal. Mach. Intell. **33**(5), 898–916 (2010)
3. Badrinarayanan, V., Kendall, A., Cipolla, R.: SegNet: a deep convolutional encoder-decoder architecture for image segmentation. IEEE Trans. Pattern Anal. Mach. Intell. **39**(12), 2481–2495 (2017)
4. Baek, D., Oh, Y., Ham, B.: Exploiting a joint embedding space for generalized zero-shot semantic segmentation. In: Proceedings of the IEEE/CVF International Conference on Computer Vision, pp. 9536–9545 (2021)
5. Bucher, M., Vu, T.H., Cord, M., Pérez, P.: Zero-shot semantic segmentation. In: Advances in Neural Information Processing Systems 32, pp. 468–479 (2019)
6. Caesar, H., Uijlings, J., Ferrari, V.: COCO-stuff: thing and stuff classes in context. In: Proceedings of the IEEE Conference on Computer Vision and Pattern Recognition, pp. 1209–1218 (2018)
7. Chen, L.C., Papandreou, G., Kokkinos, I., Murphy, K., Yuille, A.L.: DeepLab: semantic image segmentation with deep convolutional nets, atrous convolution, and fully connected CRFs. IEEE Trans. Pattern Anal. Mach. Intell. **40**(4), 834–848 (2017)
8. Chen, Y.-C., et al.: UNITER: UNiversal Image-TExt Representation learning. In: Vedaldi, A., Bischof, H., Brox, T., Frahm, J.-M. (eds.) ECCV 2020. LNCS, vol. 12375, pp. 104–120. Springer, Cham (2020). https://doi.org/10.1007/978-3-030-58577-8_7
9. Cheng, B., Schwing, A.G., Kirillov, A.: Per-pixel classification is not all you need for semantic segmentation. arXiv preprint arXiv:2107.06278 (2021)
10. Cordts, M., et al.: The cityscapes dataset for semantic urban scene understanding. In: Proceedings of the IEEE Conference on Computer Vision and Pattern Recognition, pp. 3213–3223 (2016)
11. Das, A., Xian, Y., He, Y., Schiele, B., Akata, Z.: (SP)^{2}Net for generalized zero-label semantic segmentation. In: Bauckhage, C., Gall, J., Schwing, A. (eds.) DAGM GCPR 2021. LNCS, vol. 13024, pp. 235–249. Springer, Cham (2021). https://doi.org/10.1007/978-3-030-92659-5_15

12. Deng, J., Dong, W., Socher, R., Li, L.J., Li, K., Fei-Fei, L.: ImageNet: a large-scale hierarchical image database. In: 2009 IEEE Conference on Computer Vision and Pattern Recognition, pp. 248–255. IEEE (2009)

13. Dosovitskiy, A., et al.: An image is worth 16×16 words: transformers for image recognition at scale. arXiv preprint arXiv:2010.11929 (2020)

14. Everingham, M., Winn, J.: The PASCAL visual object classes challenge 2012 (VOC2012) development kit. Pattern Analysis, Statistical Modelling and Computational Learning, Technical report 8, 5 (2011)

15. Gao, T., Yao, X., Chen, D.: SimCSE: simple contrastive learning of sentence embeddings. arXiv preprint arXiv:2104.08821 (2021)

16. Ghiasi, G., Gu, X., Cui, Y., Lin, T.Y.: Open-vocabulary image segmentation. arXiv preprint arXiv:2112.12143 (2021)

17. Girshick, R.: Fast R-CNN. In: Proceedings of the IEEE International Conference on Computer Vision, pp. 1440–1448 (2015)

18. Girshick, R., Donahue, J., Darrell, T., Malik, J.: Rich feature hierarchies for accurate object detection and semantic segmentation. In: Proceedings of the IEEE Conference on Computer Vision and Pattern Recognition, pp. 580–587 (2014)

19. Gu, X., Lin, T.Y., Kuo, W., Cui, Y.: Zero-shot detection via vision and language knowledge distillation. arXiv preprint arXiv:2104.13921 (2021)

20. Gu, Z., Zhou, S., Niu, L., Zhao, Z., Zhang, L.: Context-aware feature generation for zero-shot semantic segmentation. In: Proceedings of the 28th ACM International Conference on Multimedia, pp. 1921–1929 (2020)

21. Gu, Z., Zhou, S., Niu, L., Zhao, Z., Zhang, L.: From pixel to patch: synthesize context-aware features for zero-shot semantic segmentation. arXiv preprint arXiv:2009.12232 (2020)

22. Gupta, A., Dollar, P., Girshick, R.: LVIS: a dataset for large vocabulary instance segmentation. In: Proceedings of the IEEE/CVF Conference on Computer Vision and Pattern Recognition, pp. 5356–5364 (2019)

23. He, K., Zhang, X., Ren, S., Sun, J.: Deep residual learning for image recognition. In: Proceedings of the IEEE Conference on Computer Vision and Pattern Recognition, pp. 770–778 (2016)

24. Hu, P., Sclaroff, S., Saenko, K.: Uncertainty-aware learning for zero-shot semantic segmentation. In: Advances in Neural Information Processing Systems 33 (2020)

25. Jia, C., et al.: Scaling up visual and vision-language representation learning with noisy text supervision. arXiv preprint arXiv:2102.05918 (2021)

26. Kato, N., Yamasaki, T., Aizawa, K.: Zero-shot semantic segmentation via variational mapping. In: Proceedings of the IEEE/CVF International Conference on Computer Vision Workshops (2019)

27. Krizhevsky, A., Sutskever, I., Hinton, G.E.: ImageNet classification with deep convolutional neural networks. In: Pereira, F., Burges, C.J.C., Bottou, L., Weinberger, K.Q. (eds.) Advances in Neural Information Processing Systems, vol. 25. Curran Associates, Inc. (2012)

28. Lampert, C.H., Nickisch, H., Harmeling, S.: Attribute-based classification for zero-shot visual object categorization. IEEE Trans. Pattern Anal. Mach. Intell. **36**(3), 453–465 (2013)

29. Li, A., Jabri, A., Joulin, A., Van Der Maaten, L.: Learning visual n-grams from web data. In: Proceedings of the IEEE International Conference on Computer Vision, pp. 4183–4192 (2017)

30. Li, B., Weinberger, K.Q., Belongie, S., Koltun, V., Ranftl, R.: Language-driven semantic segmentation. In: International Conference on Learning Representations (2022)

31. Li, G., Duan, N., Fang, Y., Gong, M., Jiang, D.: Unicoder-VL: a universal encoder for vision and language by cross-modal pre-training. In: Proceedings of the AAAI Conference on Artificial Intelligence, vol. 34, pp. 11336–11344 (2020)

32. Li, P., Wei, Y., Yang, Y.: Consistent structural relation learning for zero-shot segmentation. In: Advances in Neural Information Processing Systems 33 (2020)

33. Li, X., et al.: OSCAR: object-semantics aligned pre-training for vision-language tasks. In: Vedaldi, A., Bischof, H., Brox, T., Frahm, J.-M. (eds.) ECCV 2020. LNCS, vol. 12375, pp. 121–137. Springer, Cham (2020). https://doi.org/10.1007/978-3-030-58577-8_8

34. Liu, P., Yuan, W., Fu, J., Jiang, Z., Hayashi, H., Neubig, G.: Pre-train, prompt, and predict: a systematic survey of prompting methods in natural language processing. arXiv preprint arXiv:2107.13586 (2021)

35. Liu, Z., et al.: Swin transformer: hierarchical vision transformer using shifted windows. arXiv preprint arXiv:2103.14030 (2021)

36. Long, J., Shelhamer, E., Darrell, T.: Fully convolutional networks for semantic segmentation. In: Proceedings of the IEEE Conference on Computer Vision and Pattern Recognition, pp. 3431–3440 (2015)

37. Lu, J., Batra, D., Parikh, D., Lee, S.: ViLBERT: pretraining task-agnostic visiolinguistic representations for vision-and-language tasks. arXiv preprint arXiv:1908.02265 (2019)

38. Lv, F., Liu, H., Wang, Y., Zhao, J., Yang, G.: Learning unbiased zero-shot semantic segmentation networks via transductive transfer. IEEE Signal Process. Lett. **27**, 1640–1644 (2020)

39. Mottaghi, R., et al.: The role of context for object detection and semantic segmentation in the wild. In: Proceedings of the IEEE Conference on Computer Vision and Pattern Recognition, pp. 891–898 (2014)

40. Pastore, G., Cermelli, F., Xian, Y., Mancini, M., Akata, Z., Caputo, B.: A closer look at self-training for zero-label semantic segmentation (2021)

41. Radford, A., et al.: Learning transferable visual models from natural language supervision. arXiv preprint arXiv:2103.00020 (2021)

42. Simonyan, K., Zisserman, A.: Very deep convolutional networks for large-scale image recognition. arXiv preprint arXiv:1409.1556 (2014)

43. Song, J., Shen, C., Yang, Y., Liu, Y., Song, M.: Transductive unbiased embedding for zero-shot learning. In: Proceedings of the IEEE Conference on Computer Vision and Pattern Recognition, pp. 1024–1033 (2018)

44. Su, W., et al.: VL-BERT: pre-training of generic visual-linguistic representations. arXiv preprint arXiv:1908.08530 (2019)

45. Tian, G., Wang, S., Feng, J., Zhou, L., Mu, Y.: Cap2Seg: inferring semantic and spatial context from captions for zero-shot image segmentation. In: Proceedings of the 28th ACM International Conference on Multimedia, pp. 4125–4134 (2020)

46. Uijlings, J.R., Van De Sande, K.E., Gevers, T., Smeulders, A.W.: Selective search for object recognition. Int. J. Comput. Vis. **104**(2), 154–171 (2013)

47. Xian, Y., Akata, Z., Sharma, G., Nguyen, Q., Hein, M., Schiele, B.: Latent embeddings for zero-shot classification. In: Proceedings of the IEEE Conference on Computer Vision and Pattern Recognition, pp. 69–77 (2016)

48. Xian, Y., Choudhury, S., He, Y., Schiele, B., Akata, Z.: Semantic projection network for zero-and few-label semantic segmentation. In: Proceedings of the IEEE/CVF Conference on Computer Vision and Pattern Recognition, pp. 8256–8265 (2019)

49. Rahman, S., Wang, L., Sun, C., Zhou, L.: ReDro: efficiently learning large-sized SPD visual representation. In: Vedaldi, A., Bischof, H., Brox, T., Frahm, J.-M. (eds.) ECCV 2020. LNCS, vol. 12360, pp. 1–17. Springer, Cham (2020). https:// doi.org/10.1007/978-3-030-58555-6_1
50. Zhou, B., Zhao, H., Puig, X., Fidler, S., Barriuso, A., Torralba, A.: Scene parsing through ADE20K dataset. In: Proceedings of the IEEE Conference on Computer Vision and Pattern Recognition, pp. 633–641 (2017)
51. Zhou, K., Yang, J., Loy, C.C., Liu, Z.: Learning to prompt for vision-language models. arXiv preprint arXiv:2109.01134 (2021)

Author Index

Printed in the United States
by Baker & Taylor Publisher Services